Computer Methods in Image Analysis

OTHER IEEE PRESS BOOKS

Microprocessors: Fundamentals and Applications, *Edited by W. C. Lin*
Machine Recognition of Patterns, *Edited by A. K. Agrawala*
Turning Points in American Electrical History, *Edited by J. E. Brittain*
Charge-Coupled Devices: Technology and Applications, *Edited by R. Melen and D. Buss*
Spread Spectrum Techniques, *Edited by R. C. Dixon*
Electronic Switching: Central Office Systems of the World, *Edited by A. E. Joel, Jr.*
Electromagnetic Horn Antennas, *Edited by A. W. Love*
Waveform Quantization and Coding, *Edited by N. S. Jayant*
Communication Satellite Systems: An Overview of the Technology, *Edited by R. G. Gould and Y. F. Lum*
Literature Survey of Communication Satellite Systems and Technology, *Edited by J. H. W. Unger*
Solar Cells, *Edited by C. E. Backus*
Computer Networking, *Edited by R. P. Blanc and I. W. Cotton*
Communications Channels: Characterization and Behavior, *Edited by B. Goldberg*
Large-Scale Networks: Theory and Design, *Edited by F. T. Boesch*
Optical Fiber Technology, *Edited by D. Gloge*
Selected Papers in Digital Signal Processing, II, *Edited by the Digital Signal Processing Committee*
A Guide for Better Technical Presentations, *Edited by R. M. Woelfle*
Career Management: A Guide to Combating Obsolescence, *Edited by H. G. Kaufman*
Energy and Man: Technical and Social Aspects of Energy, *Edited by M. G. Morgan*
Magnetic Bubble Technology: Integrated-Circuit Magnetics for Digital Storage and Processing, *Edited by H. Chang*
Frequency Synthesis: Techniques and Applications, *Edited by J. Gorski-Popiel*
Literature in Digital Processing: Author and Permuted Title Index (Revised and Expanded Edition), *Edited by H. D. Helms, J. F. Kaiser, and L. R. Rabiner*
Data Communications via Fading Channels, *Edited by K. Brayer*
Nonlinear Networks: Theory and Analysis, *Edited by A. N. Willson, Jr.*
Computer Communications, *Edited by P. E. Green, Jr. and R. W. Lucky*
Stability of Large Electric Power Systems, *Edited by R. T. Byerly and E. W. Kimbark*
Automatic Test Equipment: Hardware, Software, and Management, *Edited by F. Liguori*
Key Papers in the Development of Coding Theory, *Edited by E. R. Berlekamp*
Technology and Social Institutions, *Edited by K. Chen*
Key Papers in the Development of Information Theory, *Edited by D. Slepian*
Computer-Aided Filter Design, *Edited by G. Szentirmai*
Laser Devices and Applications, *Edited by I. P. Kaminow and A. E. Siegman*
Integrated Optics, *Edited by D. Marcuse*
Laser Theory, *Edited by F. S. Barnes*
Digital Signal Processing, *Edited by L. R. Rabiner and C. M. Rader*
Minicomputers: Hardware, Software, and Applications, *Edited by J. D. Schoeffler and R. H. Temple*
Semiconductor Memories, *Edited by D. A. Hodges*
Power Semiconductor Applications, Volume II: Equipment and Systems, *Edited by J. D. Harnden, Jr. and F. B. Golden*
Power Semiconductor Applications, Volume I: General Considerations, *Edited by J. D. Harnden and F. B. Golden*
A Practical Guide to Minicomputer Applications, *Edited by F. F. Coury*
Active Inductorless Filters, *Edited by S. K. Mitra*
Clearing the Air: The Impact of the Clean Air Act on Technology, *Edited by J. C. Redmond, J. C. Cook, and A. A. J. Hoffman*

Computer Methods in Image Analysis

Edited by
J. K. Aggarwal
Professor of Electrical Engineering and Computer Science
The University of Texas at Austin

Richard O. Duda
Staff Scientist
Stanford Research Institute

Azriel Rosenfeld
Research Professor, Computer Science Center
University of Maryland

A volume in the IEEE PRESS Selected Reprint Series, prepared under the sponsorship of the IEEE Computer Society.

IEEE PRESS

The Institute of Electrical and Electronics Engineers, Inc. New York

PRINTED IN THE UNITED STATES OF AMERICA

IEEE International Standard Book Numbers: Clothbound: 0-87942-089-8
Paperbound: 0-87942-090-1

Library of Congress Catalog Card Number 76-50335

Sole Worldwide Distributor (Exclusive of the IEEE):

JOHN WILEY & SONS, INC.
605 Third Ave.
New York, NY 10016

Wiley Order Numbers: Clothbound: 0-471-03112-7
Paperbound: 0-471-03113-5

Contents

Introduction . . . 1

Part I—Correlation and Matching . . . 3

Signal Detection by Complex Spatial Filtering, *A. Vander Lugt* (*IEEE Transactions on Information Theory*, April 1964) . . 5

Pattern Recognition Using Autocorrelation, *L. P. Horwitz and G. L. Shelton, Jr.* (*Proceedings of the IRE*, January 1961) . . 12

A Class of Algorithms for Fast Digital Image Registration, *D. I. Barnea and H. F. Silverman* (*IEEE Transactions on Computers*, February 1972) . . . 23

The Representation and Matching of Pictorial Structures, *M. A. Fischler and R. A. Elschlager* (*IEEE Transactions on Computers*, January 1973) . . . 31

Part II—Segmentation: Regions . . . 57

The IBM 1975 Optical Page Reader, Part II: Video Thresholding System, *M. R. Bartz* (*IBM Journal of Research and Development*, September 1968) . . . 59

Adjacency in Digital Pictures, *A. Rosenfeld* (*Information and Control*, September 1974) . . . 69

Scene Analysis Using Regions, *C. R. Brice and C. L. Fennema* (*Artificial Intelligence*, Fall 1970) . . . 79

Picture Segmentation by a Directed Split-and-Merge Procedure, *S. L. Horowitz and T. Pavlidis* (*Proceedings of the Second International Joint Conference on Pattern Recognition*, August 1974) . . . 101

Part III—Region Representation and Description . . . 111

Visual Pattern Recognition by Moment Invariants, *M. K. Hu* (*IRE Transactions on Information Theory*, February 1962) . . 113

Diffraction-Pattern Sampling for Automatic Pattern Recognition, *G. G. Lendaris and G. L. Stanley* (*Proceedings of the IEEE*, February 1970) . . . 122

Textural Features for Image Classification, *R. M. Haralick, K. Shanmugam, and I. Dinstein* (*IEEE Transactions on Systems, Man, and Cybernetics*, November 1973) . . . 141

A Transformation for Extracting New Descriptors of Shape, *H. Blum* (*Models for the Perception of Speech and Visual Form, W. Wathen-Dunn, Ed.*, 1967) . . . 153

Hexagonal Parallel Pattern Transformations, *M. J. E. Golay* (*IEEE Transactions on Computers*, August 1969) . . . 172

Part IV—Segmentation: Boundaries . . . 181

Edge and Curve Detection for Visual Scene Analysis, *A. Rosenfeld and M. Thurston* (*IEEE Transactions on Computers*, May 1971) . . . 183

An Operator Which Locates Edges in Digitized Pictures, *M. H. Hueckel* (*Journal of the ACM*, January 1971) . . . 191

Use of the Hough Transformation to Detect Lines and Curves in Pictures, *R. O. Duda and P. E. Hart* (*Communications of the ACM*, January 1972) . . . 204

The Recognition of Handwritten Numerals by Contour Analysis, *E. C. Greanias, P. F. Meagher, R. J. Norman, and P. Essinger* (*IBM Journal of Research and Development*, January 1963) . . . 209

An Application of Heuristic Search Methods to Edge and Contour Detection, *A. Martelli* (*Communications of the ACM*, February 1976) 217
Edge Detection in Pictures by Computer Using Planning, *M. D. Kelly* (*Machine Intelligence, B. Meltzer and D. Michie, Eds.*, 1971) 228

Part V—Boundary Representation and Description 245
On the Encoding of Arbitrary Geometric Configurations, *H. Freeman* (*IRE Transactions on Electronic Computers*, June 1961) 247
Fourier Preprocessing for Hand Print Character Recognition, *G. H. Granlund* (*IEEE Transactions on Computers,* February 1972) 256
A Formal Description for Two-Dimensional Patterns, *C. T. Zahn* (*Proceedings of the International Joint Conference on Artificial Intelligence,* May 1969) 263
Computer Analysis of Moving Polygonal Images, *J. K. Aggarwal and R. O. Duda* (*IEEE Transactions on Computers*, October 1975) 271

Part VI—Scene Analysis: The Blocks World 283
Machine Perception of Three-Dimensional Solids, *L. G. Roberts* (*Optical and Electro-Optical Information Processing, J. T. Tippett et al., Eds.*, 1965) 285
Decomposition of a Visual Scene into Three-Dimensional Bodies, *A. Guzmán* (*Proceedings of the AFIPS Fall Joint Computer Conference*, December 1968) 324
Impossible Objects as Nonsense Sentences, *D. A. Huffman* (*Machine Intelligence, B. Meltzer and D. Michie, Eds.*, 1971) . . 338

Part VII—Scene Analysis: The Natural World 367
On Syntactic Pattern Recognition, *K. S. Fu and P. H. Swain* (*Software Engineering, Vol. 2, J. T. Tou, Ed.*, 1969) 369
Some Techniques for Recognising Structures in Pictures, *H. G. Barrow, A. P. Ambler, and R. M. Burstall* (*Frontiers of Pattern Recognition, S. Watanabe, Ed.*, 1972) 397
A Semantics-Based Decision Theory Region Analyzer, *Y. Yakimovsky and J. A. Feldman* (*Proceedings of the Third International Joint Conference on Artificial Intelligence*, August 1973) 426
IGS: A Paradigm for Integrating Image Segmentation and Interpretation, *J. M. Tenenbaum and H. G. Barrow* (*Proceedings of the Third International Joint Conference on Pattern Recognition,* November 1976) 435
Relaxation Labelling and the Reduction of Local Ambiguities, *S. W. Zucker* (*Proceedings of the Third International Joint Conference on Pattern Recognition*, November 1976) 445

Bibliography 455

Author Index 460

Subject Index 462

Editors' Biographies 465

Introduction

The emergence of general-purpose digital computer systems in the late 1940's and early 1950's led to a keen appreciation of both their power and their revolutionary potential. It was soon recognized that the phrase "data processing" was not limited to operations on numerical data, but could include operations on symbolic, pictorial, and other kinds of data as well. An immediate practical motivation for processing pictorial data was the need for finding better ways to input the large volume of data encountered in business applications. This led directly to the development of optical character recognition systems, and with it to the research areas of pattern recognition and digital image analysis.

While optical character recognition is still the most important commercial application, many other applications either have reached or are on the threshold of reaching commercial reality. Examples of such applications include computerized tomography (the reconstruction of images from X-ray or ultrasonic projections), differential counting of blood cells, chest X-ray analysis, fingerprint identification, metallurgical image analysis, processing of particle chamber photographs, detection of targets and/or changes in aerial photographs, processing of remotely sensed earth resources imagery, visual inspection for manufacturing, and visual feedback and control for industrial robots.

The development of these applications has been paralleled by theoretical developments that generalize these particular results and provide the basis for future applications. Several textbooks are now available that provide systematic presentations of theoretical foundations. The purpose of this book is to provide engineers with ready access to some of the landmark papers that both supply further information and help bring the topic to life.

The major problem facing the Editors of a reprint collection is one of selecting a relatively few papers from a large number of relevant candidates. To some extent our problem has been eased by the fact that another IEEE Press reprint book, entitled *Machine Recognition of Patterns*, edited by A. K. Agrawala, has been published recently and covers statistically oriented methods. We have restricted our scope by focusing on computer methods for analyzing images. We have excluded such important topics as the physiology and psychology of visual perception, image digitization and coding, image filtering and enhancement, and special-purpose image processing hardware. Instead, we have concentrated on basic techniques for extracting descriptive information from pictorial data.

Even with this specialization, we often had to make difficult choices, and we are keenly aware of the many excellent papers that had to be excluded to produce a bounded and balanced collection. Hopefully, the Bibliography and the additional citations given in the introductions to each part will alleviate this problem, and will provide the interested reader with additional helpful pointers into the literature.

The collection itself is divided into seven major parts, providing the following progression from fundamental to advanced analysis techniques:

1) Correlation and Matching
2) Segmentation: Regions
3) Region Representation and Description
4) Segmentation: Boundaries
5) Boundary Representation and Description
6) Scene Analysis: The Blocks World
7) Scene Analysis: The Natural World

Since there are several other ways in which this collection could have been organized, a few comments about this grouping are in order.

Historically, template matching was the first technique used for pattern recognition, and even the most advanced techniques can often be interpreted as generalizations of template matching. Thus, it is natural to begin with papers devoted to correlation and matching. In many applications, however, the objects of interest cannot be recognized by searching the image for a direct match with a prototype. Even when a simple classification into one of a number of predefined categories is adequate, the objects may be too variable to allow the use of whole-object templates. For images in which one needs a description rather than a classification, a more sophisticated approach to scene analysis is certainly mandatory.

Most of the more advanced scene analysis techniques require that the picture be subdivided or segmented in some way. One general approach is to segment the image into regions which correspond to objects or parts of objects, or to background. Once formed, these regions can be represented and described in various ways. These descriptions provide properties or features that are useful either directly for object classification or indirectly as a step toward scene description. Thus, we have devoted Part II to methods for region formation, and Part III to methods for region representation and description.

Another widely used approach to segmentation is based on detecting edges or curves in the picture. Such linear features are often the boundaries or parts of the boundaries of regions of interest, or they may represent features of interest in themselves. Once found, they too can be represented and described in various ways. Methods for boundary detection are given in Part IV, and methods for boundary representation and description are given in Part V.

The final two parts are concerned with approaches to synthesizing complete scene descriptions from more primitive region and boundary information. Part VI concerns the analysis of idealized scenes composed of polyhedral objects. This so-called "blocks world" has received considerable attention, primarily because it clearly exhibits the problems that arise in analyzing scenes of three-dimensional objects in a setting that is sufficiently limited to allow general solutions. The real world of irregularly shaped objects is both much richer and much more complex than the blocks world, and its many unsolved problems remain as active areas for current research. However, particular solutions to particular problems have been obtained, and some general approaches to scene analysis are given in the concluding part.

While the foregoing summary gives a generally correct overview of the present collection, it must be said that the nature of both the field and the individual papers prevents such a clean separation of topics. Relatively few technical papers in this field are limited to just one topic, and some important papers do not fit comfortably in any of our seven categories. Indeed, it is this very interaction of concerns that gives the best papers a sense of liveliness that is lost in linearly organized, methodical expositions of general principles. It is our hope that this collection will provide the reader not only with valuable conceptual and technical information, but also with a comprehensive view of the general field of image analysis and its problems, accomplishments, and future potential.

J. K. Aggarwal
R. O. Duda
A. Rosenfeld
Editors

Part I
Correlation and Matching

Matching is one of the basic techniques for pictorial pattern recognition. In its most elementary form, matching consists of a pixel-by-pixel comparison of two pictures, exemplified by template matching for binary images and by cross correlation or matched filtering for gray-scale images. Common applications for such procedures include single-font machine printed character recognition, multispectral image registration, tracking of moving objects, map matching for navigation and guidance, automatic stereogrammetry, and the detection of changes in images.

Because correlation techniques are both fundamental and widely applicable, considerable attention has been devoted to their implementation. The first paper, by Vander Lugt, takes advantage of the fact that cross correlation is a linear, shift-invariant operation, and thus reduces to multiplication in the Fourier domain. This in turn leads to an optical implementation which has the advantage of great speed, but the disadvantage of requiring the image to be a transparency. [A recent survey of coherent optical processing is given by Vander Lugt (1974)[1]; for a digital fast-Fourier-transform approach, see Anuta (1970).]

Vander Lugt's examples illustrate clearly the properties of cross correlation. In particular, the immediate result of cross correlation is the production of another "image," the correlogram, which must be processed further to find the peaks corresponding to matches. In general, cross correlation is most appropriate for detecting objects that are essentially completely known except for an arbitrary translation, and determining that translation requires a two-dimensional search.

Searching the two-dimensional correlation surface for peaks is a nontrivial task, as can be appreciated from the graphical examples given by Smith and Phillips (1972). The common practice in optical character recognition is to use a shift register to displace the character through all possible discrete positions and to record matches exceeding some threshold. An alternative is to measure features or properties of the character that are invariant to translation. The second paper, by Horwitz and Shelton, treats both techniques, with emphasis on the latter. In particular, they exploit the fact that the autocorrelation function of a character produces an "image" which contains information about the character that is invariant to translation.

The Fourier transform of the autocorrelation function is the power spectrum (or wave number spectrum), which can be obtained optically as the Fraunhofer diffraction pattern. Both Horwitz and Shelton and Lendaris and Stanley (in Part II) discuss this equivalent alternative translation invariant transformation. In particular, Lendaris and Stanley suggest sampling the power spectrum to obtain translation invariant features for recognition. [An early example of this basic idea is given by Kain (1961).]

The idea of forming translation invariant transformations has the appeal of elegance. While the most common transformations lose information (for example, mirror-image figures have the same autocorrelation function), McLaughlin and Raviv (1968) showed that higher-order autocorrelation functions preserve all information except for translation. In principle, one can also find transformations that are invariant under other kinds of geometrical operations, such as rotation and scaling [see Doyle (1962)]. The main reasons that such ideas are not more widely used are that they are computationally expensive, require the object to be isolated, and transform a familiar domain into an unfamiliar one. However, for special applications, such techniques can be extremely effective.

The third paper, by Barnea and Silverman, is concerned with the problem of finding the position of best match. They point out that when the search is done on a conventional computer, very significant time savings can be obtained by quickly abandoning areas in which a few calculations show that a good match is very unlikely. This approach is analogous to techniques used in sequential decision theory and operations research; for a complementary approach, see Nagel and Rosenfeld (1972).

From a theoretical viewpoint, one can derive the optimal procedure for matching two images if one has a description of the ways in which the given image can differ from the reference image. The conventional matched filter is the optimal solution when the images differ by translation plus statistically independent, additive Gaussian noise, or when an appropriate criterion is the minimization of the mean-squared error between the two images. Unfortunately, the variability encountered in many problems is not well modeled by independent Gaussian noise, and mean-squared error is frequently not an appropriate matching criterion. For example, hand-printed characters are not so much noisy versions as they are distorted versions of ideal prototypes, and it is meaningless to compare a profile view with a frontal view of a face in terms of mean-squared error.

The remainder of the papers in this collection are concerned with problems for which elementary template matching is not adequate. The notion of matching may remain, but it is in terms of comparing properties or descriptions of the images rather than comparing individual brightness values. Perhaps one of the conceptually simplest generalizations concerns the matching of moderately deformed images. The outlines of a

[1] Dated references are given in the Bibliography at the end of this collection.

general theory for such problems have been developed by Grenander (1970). An intuitively appealing solution is to use "elastic" templates, where the degree of mismatch is measured by the energy needed to deform the template to match the image. This is a different kind of mean-squared-error criterion, one in which one measures spatial rather than gray-level error. The last paper in this part, by Fischler and Elschlager, expresses this error criterion more precisely in terms of an embedding metric that involves decomposing the object into components. [An alternative rubber-template approach is given by Widrow (1973).] Unfortunately, minimization of the error criterion leads to a computationally very expensive dynamic programming problem. Fischler and Elschlager give a much more efficient approximate procedure, and provide examples that demonstrate the versatility of their approach.

In all of the papers on matching, the topics of computational cost and special hardware implementations are never far away. In fact, the literature on optical character recognition is dominated by implementation considerations; see, for example, Fischer *et al.* (1962), Holt (1968), Kovalevsky (1968), and Ullmann (1973). In addition, a large literature exists on special-purpose electronic and optical image processing systems; see, for example, Preston (1972). While the first two papers in this part and the paper by Golay in Part III have clear hardware implications, this collection of papers is much more concerned with techniques than with their implementation. Some representative references on special hardware are Unger (1958), McCormick (1963), Gray (1972), and Duff *et al.* (1973).

Signal Detection By Complex Spatial Filtering

A. VANDER LUGT

Summary—In the past, spatial filtering in coherent optical systems has been limited by the inability to realize practically a general complex filter. This paper describes a technique for realizing such a filter, and gives an application of spatial filtering to the problem of detecting isolated signals in a variety of noise backgrounds. The experimental results obtained to date indicate that this technique provides an excellent two-dimensional filtering capability that will play a key role in such problems as shape recognition and signal detection.

Introduction

THE FUNDAMENTAL theory of optical spatial filtering has been formulated by several writers.[1,2] The close analogy of spatial filtering in coherent optical systems to optimum linear filtering in communication theory promised to open up new techniques for realizing those frequency domain filters which could not be synthesized in the time domain because of the realizability constraint; but spatial filtering has not advanced as far as one might have expected. Perhaps the main reason is that, though the formulation of the theory tacitly assumed that complex filters could be realized, the practical realization has presented a difficult problem. A major part of this problem is that photographic film, the most prominent candidate for use in the realization of spatial filters, can record only non-negative functions, so that auxiliary phase filters must be used to realize the complex filters.

The modulus of the transfer function for spatial filters is restricted to values between zero and one.[3] To demonstrate the theory, early experiments in spatial filtering used occluding filters with transfer functions having the values (0, 1) as band-pass filters.[2,3] Later, continuous amplitude control (*e.g.*, Gaussian weighting) was used to show how equalization could be accomplished in a coherent optical system. The next step was to add binary phase control to extend the range of filter values to the real axis between plus and minus one.[4,5] The binary phase functions were constructed by using ruled phase plates, evaporative techniques, or film reliefing techniques. These techniques are rather awkward to apply. Clearly, if more general filtering operations are to be performed, a fairly easy method for constructing a general complex filter must be found. This paper describes such a technique and gives some experimental results of major importance. This technique is especially attractive because it realizes the complex filter by a special process for recording on film without the use of an auxiliary phase filter.

Mathematical Nature of the Problem

The mathematical model shown in Fig. 1 describes the problem of sensing, recording and processing imagery. The scene will be denoted by $g(x, y)$, an intensity function of two space coordinates. The primary sensor makes a two-dimensional transformation of $g(x, y)$, and the recording process causes a further transformation. Because the recording medium is usually film, this transformation introduces film grain as a secondary source of noise. It has become common practice to refer to $f(x, y) = T_2\{T_1[g(x, y)]\}$ as the "scene." Though this is not strictly correct, transformation T_2 is usually a nonlinear process (and a poorly controlled one at that). Therefore, it becomes necessary to operate on the available data $f(x, y)$ rather than to attempt to recover $g(x, y)$ before the processing operation.

Processing of the imagery could involve several distinctly different operations, depending on the type of information desired. One such operation might be to differentiate the signal to emphasize large gradients in the image. In other cases, band-pass filtering will improve the imagery. The particular problem treated in this paper is that of maximizing the ratio of peak signal energy to mean square noise energy in the output of the processing system. The signal is any object of interest in the scene, and the noise is the rest of the imagery. This problem is familiar in signal detection theory, and the mathematical solution is well known. Suppose $n(x, y)$ is a homogeneous, isotropic random process with spectral density $N(p, q)$, and $s(x, y)$ is a known Fourier transformable function of space coordinates. If $h(x, y)$ denotes the impulse response of a linear space invariant filter with frequency response $H(p, q)$, we wish to find the optimum linear filter which operates on $f(x, y) = s(x, y) + n(x, y)$ to maximize the ratio of peak signal to rms noise.

Several writers have given the solution to this problem, but usually for the case where $N(p, q)$ is uniform. The general solution[6] is that the required filter response is

Manuscript received May 31, 1963. This work was jointly supported by Project MICHIGAN, under U. S. Army Electronics Command, Contract DA-36-039 SC-78801, and by Aeronautical Systems Division, Air Force Contract AF 33(616)-8433. A more detailed discussion of this subject is given in Report 2900-394-T/4594-22-T.

The author is with the Institute of Science and Technology, The University of Michigan, Ann Arbor, Mich.

[1] T. P. Cheatham, Jr. and A. Kohlenberg, "Optical filters: their equivalence to and differences from electrical networks, 1954 IRE Convention Record, pt. 4, pp. 6–12.

[2] E. O'Neill, "Spatial filtering in optics," IRE Trans. on Information Theory, vol. IT-2, pp. 56–65; June, 1956.

[3] L. J. Cutrona *et al.*, "Optical data processing and filtering systems," IRE Trans. on Information Theory, vol. IT-6, pp. 386–400; June, 1960.

[4] A. Marechal, "Filtering of optical images," in "*Communication and Information Theory Aspects of Modern Optics*," General Electric Co., Electronics Laboratory, Syracuse, N. Y.; August, 1962.

[5] J. Tsujiuchi, "Correction of optical images by compensation of aberrations and by spatial frequency filtering," in "Progress in Optics," E. Wolf, Ed., North-Holland Publishing Co., Amsterdam, The Netherlands; vol. 2, pp. 133–180; 1963.

[6] W. M. Brown, "Analysis of Linear Time-Invariant Systems," McGraw-Hill Book Co., Inc., New York, N. Y.; 1963.

Reprinted from *IEEE Trans. Inform. Theory*, vol. IT-10, pp. 139–145, Apr. 1964.

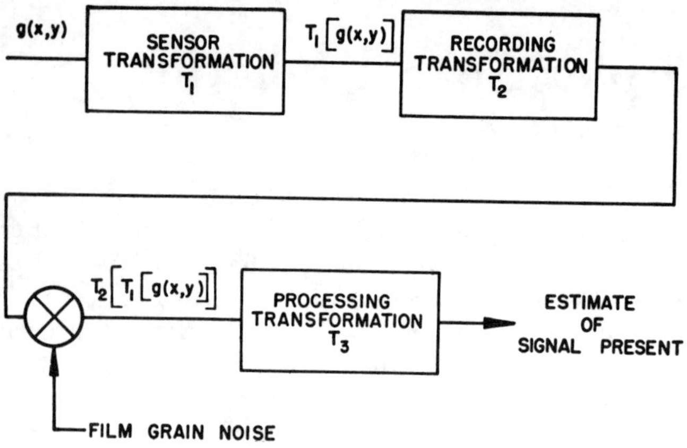

Fig. 1—Mathematical model of image processing.

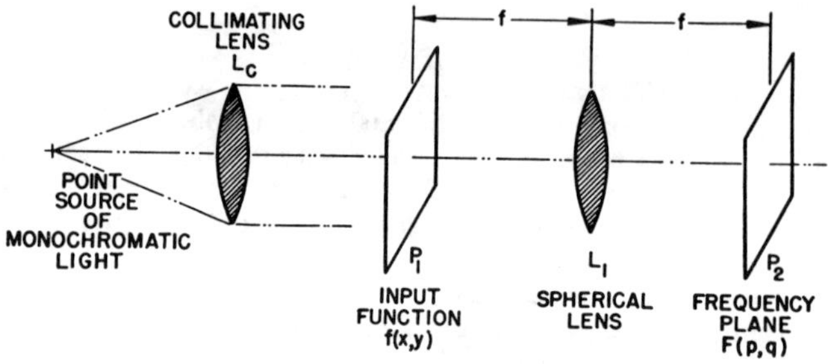

Fig. 2—Coherent Fourier analyzer.

$$H(p, q) = \frac{kS^*(p, q)}{N(p, q)} \tag{1}$$

where

$S^*(p, q)$ is the complex conjugate of the signal spectrum,
$N(p, q)$ is the spectral density of the background noise and
k is a constant chosen to make $|H(p, q)| < 1$ for passive spatial filters.

Eq. (1) reduces to the so-called matched filter case when $N(p, q)$ is uniform for all (p, q), as is usual in treatments of this problem for electronic systems. Almost always, the spectral density of the noise background in photographs is far from uniform. This fact might impose a limitation on the realizability of electronic filters, but, as we shall see later, it poses no problem in the realization of the optimum filter in coherent optical systems.

The output from the linear filter represented by (1), given here for future reference, is

$$r(x, y) = \frac{1}{4\pi^2} \iint_{-\infty}^{\infty} F(p, q)H(p, q)e^{j(px+qy)}\, dp\, dq. \tag{2}$$

Coherent Optical Processing Systems

Rhodes has shown that the optical system sketched in Fig. 2 functions as a two-dimensional Fourier analyzer.[7] If $f(x, y)$ denotes the specular amplitude transmission of the transparency in plane P_1, and $F(p, q)$ denotes the complex distribution of light in plane P_2, then

$$F(p, q) = \iint_{-\infty}^{\infty} f(x, y) \exp\,[j(px + qy)]\, dx\, dy. \tag{3}$$

In (3), p and q represent spatial frequency variables having the dimensions of radians/unit distance. The variables in plane P_2, which are in units of distance, are related to the frequency variables by

$$\xi = \frac{\lambda f}{2\pi} p \tag{4a}$$

$$\eta = \frac{\lambda f}{2\pi} q \tag{4b}$$

where

ξ is the direction parallel to x,
η is the direction parallel to y,
λ is the wavelength of the illumination and
f is the focal length of the spherical lens.

When a distribution of light is a function of frequency, the variables (p, q) will be used to emphasize that fact, as well as to simplify the notation associated with Fourier transform theory.

Since a spherical lens can take the Fourier transform of a complex distribution of light, one can construct an optical system by arranging a sequence of lenses which forms a succession of Fourier transform planes. An image of the input plane can be effected by placing a lens behind plane P_2 (see Fig. 3) which takes the transform of $F(p, q)$. A positive spherical lens always introduces a positive kernel in the transform relationship, hence the distribution in plane P_3 is

$$r(x, y) = \frac{1}{4\pi^2} \iint_{-\infty}^{\infty} F(p, q)e^{j(px+qy)}\, dp\, dq$$

$$= f(-x, -y). \tag{5}$$

We have assumed that the system has unity magnification and sufficient bandwidth to pass the highest spatial frequency of the input function. Note that the output is an inverted image of the input function, which is what one expects from an imaging system operating under any type of illumination.

It is convenient to realize separately the numerator and denominator of the optimum filter. If two transparencies are placed in contact, their complex transmissions

[7] J. E. Rhodes, Jr., "Analysis and synthesis of optical images," *Am. J. of Phys.*, vol. 21, pp. 337–343; May, 1953.

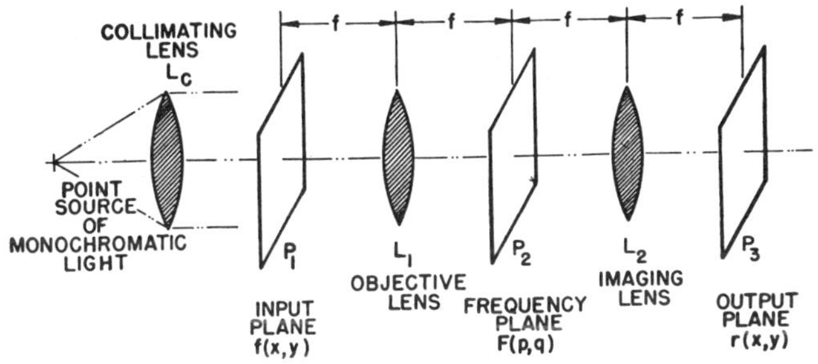

Fig. 3—Coherent optical processing system.

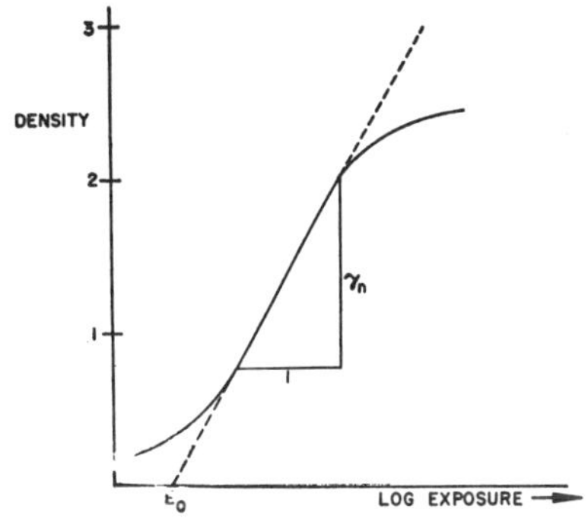

Fig. 4—Typical transfer curve for photographic film.

are multiplicative.[8] Consequently, to realize the filter described by (1), we need to insert in plane P_2 a transparency whose transmission is $1/N(p, q)$, as well as a transparency representing $S^*(p, q)$. When this filter is used, the output of the system essentially represents the probability that a signal has occurred. A bright spot in the output indicates a high probability, and low-light levels indicate a low one. (The output may not resemble the signal to any extent because the phase content of $S(p, q)$ is modified.) This process simultaneously detects signals with similar orientations regardless of their position in the input, as a result of the fact that the spectrum of a translated signal is modified by a linear phase factor. The phase factor contains precisely the information required to image the signal at the proper position in the output. In contrast, the system is sensitive to the orientation of the signals, but a rotation of the filter relative to the input provides a sequential search through all possible orientations.

Realization of the Optimum Filter

Having determined that the optimum filter which maximizes the ratio of peak signal energy to mean square energy is given by (1), we must find some method to realize $H(p, q)$. The denominator of $H(p, q)$ is always a non-negative function which can be realized on photographic film. The first step in this realization is a brief review of the transfer characteristics of films. Fig. 4 shows a typical curve of density vs log exposure, characterized in its linear region by

$$D_n = \gamma_n[\log E_n - \log E_0] \tag{6}$$

where

γ_n is the slope of the straight line,
E_n is the exposure,
E_0 is the intercept of the straight line,
D_n is the density and
n means that a negative transparency is used.

The coherent system operates on the specular amplitude transmission of the film, but using the relationship that $T = \exp[-D/2]$ and substituting (6), we have

$$T = CE_n^{-\gamma_n/2}. \tag{7}$$

If we make E_n proportional to $N(p, q)$ and require that $\gamma_n = 2$, we have realized the denominator of $H(p, q)$.

Since the numerator of (1) may represent a complex quantity, a different technique must be used to realize $S^*(p, q)$. One problem in realizing the numerator of the optimum filter is to determine both the amplitude and the phase of

$$S(p, q) = \iint_{-\infty}^{\infty} s(x, y) \exp[j(px + qy)]\, dx\, dy. \tag{8}$$

One cannot merely use a spherical lens, as previously described, to take the Fourier transform of $s(x, y)$, because any physical detector measures only the intensity of $S(p, q)$, not its phase. A Mach-Zehnder interferometer, however, can be used to determine the phase in a distribution of light by combining that distribution with a reference wave whose amplitude and phase distributions are known. This interferometer is modified for our purposes as shown in Fig. 5.

The signal $s(x, y)$ whose Fourier transform is to be found, and a spherical lens, are inserted in one beam of the interferometer. The lens displays the Fourier transform of $s(x, y)$ at its back focal plane outside the interferometer. If we neglect aberrations in the interferometer, the observed output in the back focal plane of the lens is

$$G(p, q) = |R(p, q) + S(p, q)|^2 = |R(p, q)|^2 + |S(p, q)|^2 + R^*(p, q)S(p, q) + R(p, q)S^*(p, q) \tag{9}$$

where $R(p, q) = |R(p, q)| \exp[j\phi(p, q)]$ represents the light coming from the reference beam, and $S(p, q) = |S(p, q)| \exp[j\theta(p, q)]$ is the signal spectrum. We can rewrite (9) as

$$G(p, q) = |R(p, q)|^2 + |S(p, q)|^2 + 2Re[R(p, q)S^*(p, q)] = |R(p, q)|^2 + |S(p, q)|^2 + 2\,|R(p, q)|\,|S(p, q)| \cos[\phi(p, q) - \theta(p, q)]. \tag{10}$$

[8] Unless otherwise noted, "transmission" will herein signify the specular amplitude transmission of the transparency.

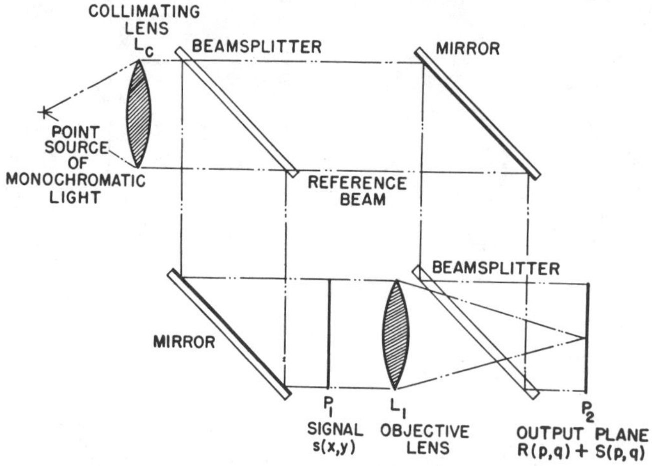

Fig. 5—Modified mach-zehnder interferometer.

Both the amplitude and phase of $R(p, q)$ could be adjusted to determine $\theta(p, q)$, but since the phase information of $S^*(p, q)$ is contained in $G(p, q)$, a non-negative function, $G(p, q)$ can be recorded on film to realize the optimum filter.

The film is exposed so that its transmission is proportional to $G(p, q)$. If we combine this film with the film on which $1/N(p, q)$ is realized, the transmission of the combination is

$$\frac{G(p, q)}{N(p, q)} = \frac{|R(p, q)|^2 + |S(p, q)|^2}{N(p, q)} + \frac{R^*(p, q)S(p, q)}{N(p, q)} + \frac{R(p, q)S^*(p, q)}{N(p, q)}. \tag{11}$$

Suppose we require $|R(p, q)|$ to be a constant, k, and $\phi(p, q)$ to be linear in (p, q), *i.e.*, $\phi(p, q) = bp + cq$. Then

$$\frac{G(p, q)}{N(p, q)} = A(p, q) + H^*(p, q)e^{-i(bp+cq)} + H(p, q)e^{i(bp+cq)} \tag{12}$$

where

$$A(p, q) = \frac{|R(p, q)|^2 + |S(p, q)|^2}{N(p, q)}$$

$$H(p, q) = \frac{kS^*(p, q)}{N(p, q)},$$

b, c are constants.

Thus, the third term of (12) represents the desired filter function, multiplied by a linear phase factor. The problem is to separate this term from the other two terms of (12). This can be accomplished by inserting the filter represented by (12) into the optical system (Fig. 3) at plane P_2. Lens L_2 separates the three terms in the output by taking the inverse transform of the light distribution in P_2; *i.e.*,

$$\text{output} = \frac{1}{4\pi^2} \iint_{-\infty}^{\infty} F(p, q) \frac{G(p, q)}{N(p, q)} e^{i(px+qy)} dp\, dq. \tag{13}$$

Substituting (12) in (13), we have

$$\text{output} = \frac{1}{4\pi^2} \iint_{-\infty}^{\infty} F(p, q)A(p, q)e^{i(px+qy)}\, dp\, dq$$
$$+ \frac{1}{4\pi^2} \iint_{-\infty}^{\infty} F(p, q)H^*(p, q)e^{i[(x-b)p+(y-c)q]}\, dp\, dq$$
$$+ \frac{1}{4\pi^2} \iint_{-\infty}^{\infty} F(p, q)H(p, q)e^{i[(x+b)p+(y+c)q]}\, dp\, dq. \tag{14}$$

The first term of (14), which appears on the optical axis, is is of no particular interest in this discussion. Nor is the second term, which appears at $x = b$, $y = c$. The third term, $r(x + b, y + c)$, where $r(x, y)$ is defined by (2), is exactly the output expected from an optimum filter. This term appears with its center off the optical axis by an amount $x = -b$, $y = -c$. At this point it will be convenient to set $c = 0$ since it is arbitrary. To prevent overlap of the three outputs, the value of b must be such that $b \geq B$, where B is the length of the signal in the x direction. The fact that this output occurs off axis by a distance $x = -b$ is not important, since the output sensor can be located properly to detect it.[9]

In passing, a comment will be made on the significance of the second term of (14). If $N(p, q) = N$ for all (p, q),

[9] This method is the equivalent of recording a complex valued function as a real valued function on a carrier frequency.[6]

we can write the third term of (14) as a cross-correlation integral (for $c = 0$);

$$r(x + b, y) = \iint_{-\infty}^{\infty} f(u, v)s^*(x + b + u, y + v)\, du\, dv. \quad (15)$$

The second term of (14) can then be written as a convolution integral;

$$r(x - b, y) = \iint_{-\infty}^{\infty} f(u, y)s(x - b - u, y - v)\, du\, dv. \quad (16)$$

Thus, when $N(p, q)$ is uniform, the cross correlation and convolution of the signal with the input function are both displayed in the output plane.

Experimental Results

A few experimental results will serve to illustrate the theory and indicate the potential of spatial filtering when complex filters can be realized.

Detection of Simple Geometrical Shapes

The first example presented is the detection of one of the elementary geometrical shapes shown in Fig. 6(a). Any one of the shapes could be selected; we chose the small rectangle. We realized the filter for this signal (which is real because of symmetry), by the method described. The output of interest is shown in Fig. 6(b). Note the simultaneous detection of all signals with the proper shape and orientation.

The second signal selected was the "L" shape, which has a complex spectrum. The Fourier transform of the complex filter, shown in Fig. 7, illustrates the fidelity with which the filter was realized. The light distribution in the center of the output is the transform of the first two terms of (9). The "L" on the left is the transform of the third term of (9). The other "L" is the transform of the last term. These two images are inverted and reversed relative to each other, which graphically demonstrates that $\mathfrak{F}[S(p, q)] = s(x, y)$ and $\mathfrak{F}[S^*(p, q)] = s^*(-x, -y)$. Of course, since $s(x, y)$ is real, $s^*(-x, -y) = s(-x, -y)$.

Fig. 8(a) shows the output, which is the cross correlation of the "L" with the input function. Note the symmetry in the output which is a necessary feature of cross correlation. (The other "L" 's in the input, having different orientations, do not give as large an output, but a rotation of the filter relative to the input would sequentially detect them.) Fig. 8(b) shows the output which is the convolution of the signal with the input function. It is asymmetrical, as a consequence of convolution.

Detection of Alphanumerics

The second example is the detection of alphanumerics. An interesting variation of the first example is to record the alphabet shown in Fig. 9(a) via its complex spectrum as the filter function. It is a simple matter to select any one of the alphanumerics as the signal to be detected and use it as the input signal. The output, when the letter "g" is the input, is shown in Fig. 9(b). Since the filter (which is the normal input function in disguise) does not have to be changed while the search is being carried out, this technique suggests a method for scanning a printed page for the presence of any particular alphanumeric.

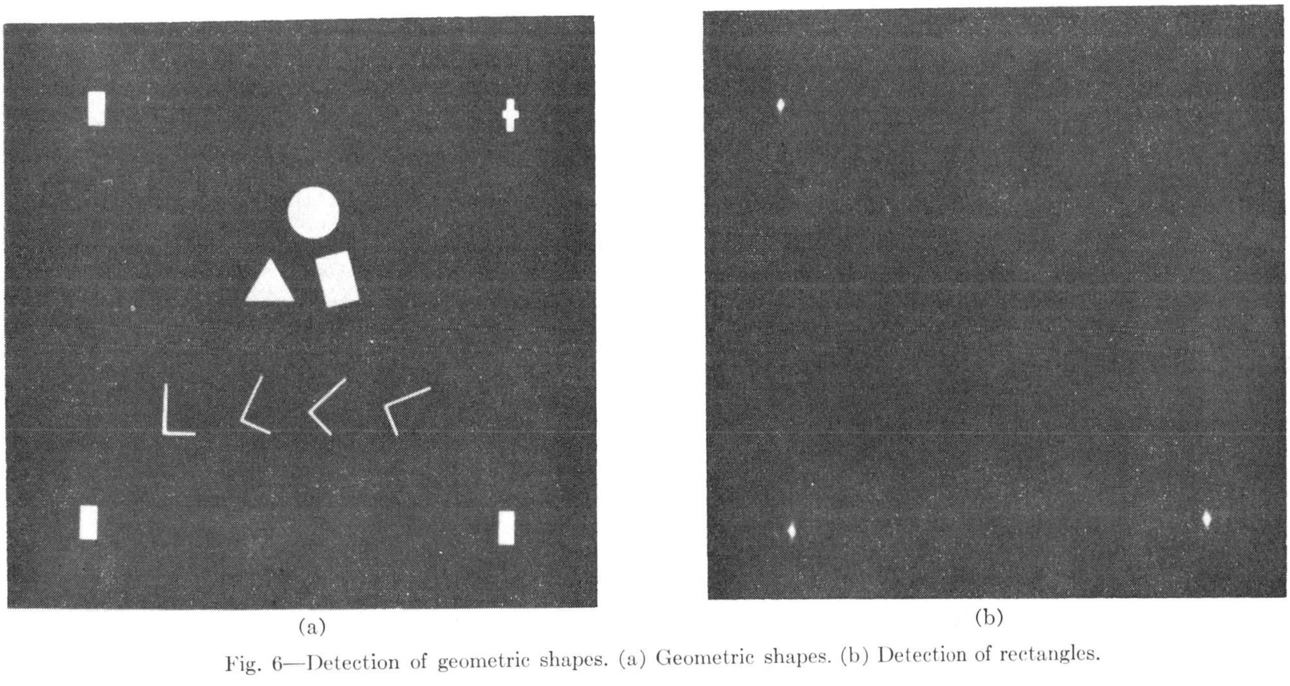

(a) (b)

Fig. 6—Detection of geometric shapes. (a) Geometric shapes. (b) Detection of rectangles.

Fig. 7—Fourier transform of complex spatial filter.

(a)

(b)

Fig. 8—Detection of signal having complex spectrum. (a) Cross-correlation of "L" with geometric shapes. (b) Convolution of "L" with geometric shapes.

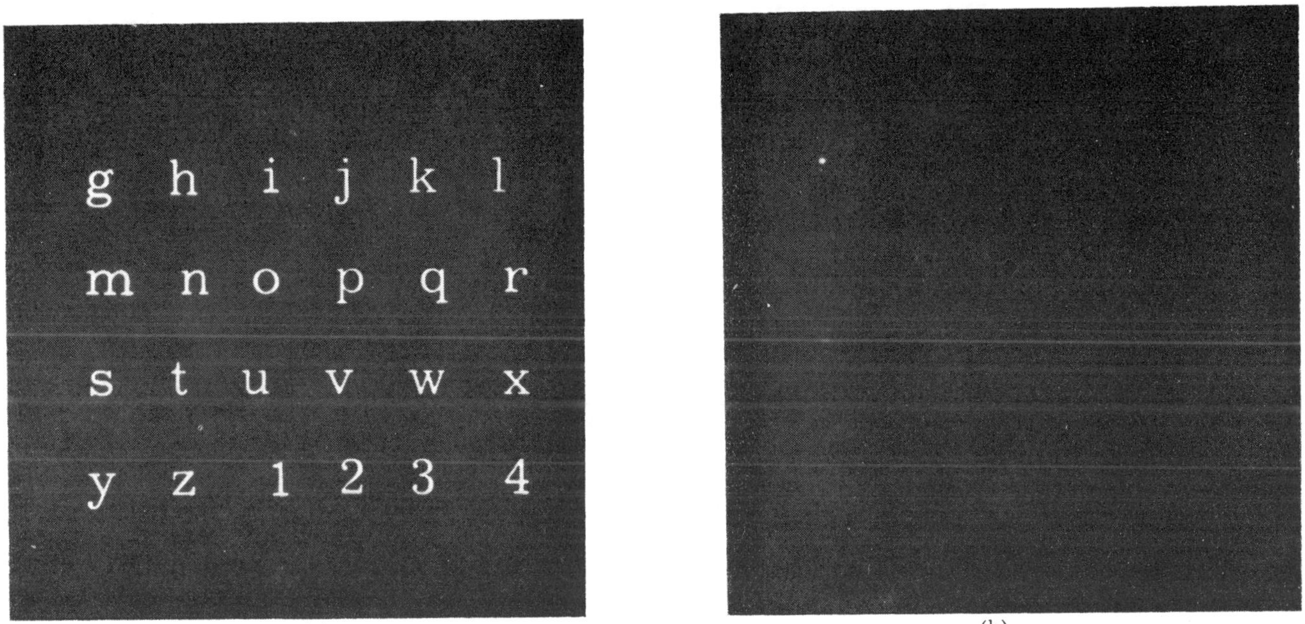

(a) (b)

Fig. 9—Detection of letter. (a) Alphanumerics. (b) Detection of the letter "*g*".

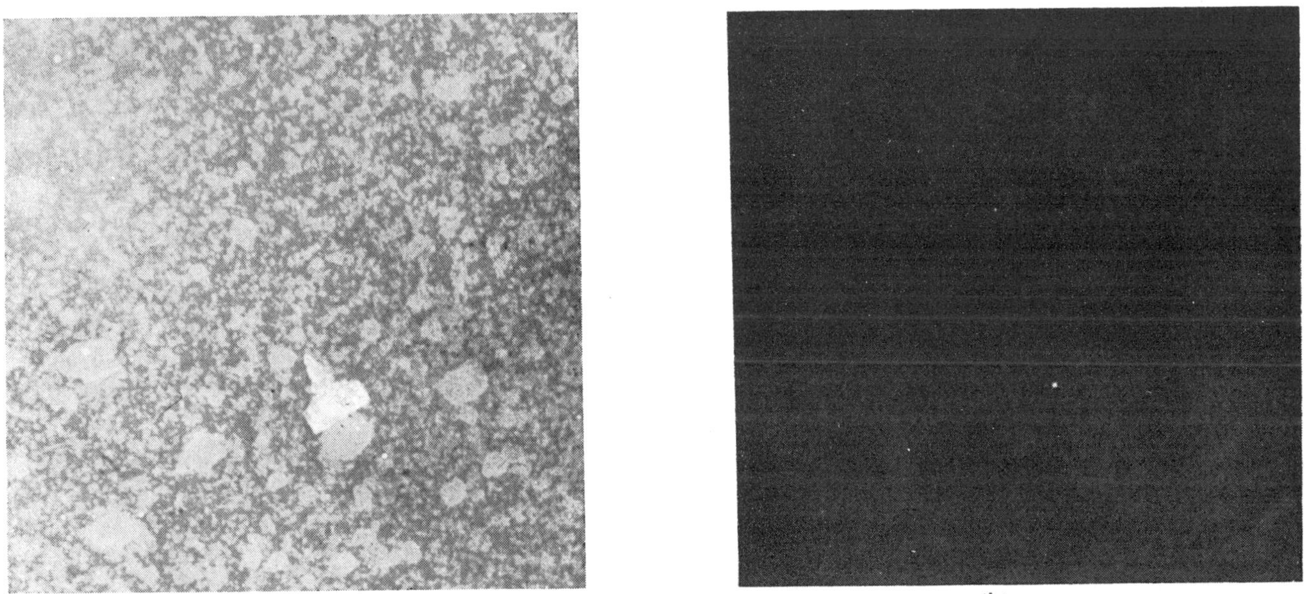

(a) (b)

Fig. 10—Detection of isolated signal in random noise. (a) Input scene. (b) Detection of signal.

Detection of Isolated Signal in Random Noise

In the preceding two examples the "noise" spectral density was uniform enough to be considered white. This final example shows the detection of a signal which is immersed in a noise background with nonuniform spectral density and demonstrates the power of using a coherent system for signal detection. Fig. 10(a) shows a signal in background noise. Since the noise spectral density is nonuniform, the denominator of the filter must be realized. Fig. 10(b) shows that the background noise has been completely suppressed and the signal has been detected.

Acknowledgment

The author wishes to thank A. Klooster, Jr., who carried out most of the experimental work. Helpful conversations with staff members, particularly E. N. Leith, are also acknowledged.

Pattern Recognition Using Autocorrelation*

L. P. HORWITZ†, MEMBER, IRE, AND G. L. SHELTON, JR.†

Summary—**A class of techniques for character recognition is described. These techniques are characterized by the property that the only parameters of the input which are used are those which are independent of the position of the character; that is, these techniques are registration invariant.**

In Section I of this paper, we describe the registration invariant mathematical formalism which underlies these techniques, and in Section II we describe in more detail the physical realization of several recognition schemes based on these concepts. Some results from our computer simulation of these procedures are given.

Introduction

DURING the past year, considerable progress has been made in optical data processing and character recognition. Some theoretical techniques have been developed which seem appropriate for handling at least a few aspects of the problem, and several machines have been proposed as physical realizations of the prescribed mathematical operations.

Since the theoretical techniques that have been worked out have played an important role in the experimental program, they will be described in some detail in Section I of this paper. The basic idea underlying these techniques is that of registration invariance; *i.e.*, all patterns are described in such a way that shifting translationally in any direction does not affect any observed parameter. This mode of description has strong implications for the engineering side of the situation; a machine designed to deal with information in this form does not have to search in a character field for the precise location of the character. There are, of course, other ways of dealing with character misregistration, but the essential point to be made here is that this particular aspect of the problem is dispensed with from the start, and an appropriate mathematical formalism is pointed out by requiring this invariance property. The starting point for the development of these ideas is the construction of a discrete binary representation of the patterns to be recognized. This is done, in principle, by placing a discrete mesh over the character and setting each mesh point to unity if it covers a black part of the character and to zero if it covers white. The patterns are then described by counting specified pairs of ones regardless of their positions in the matrix for all possible kinds of pairs.

In Section II of this paper, several recognition systems will be proposed in some detail as physical realizations of the mathematical operations prescribed in the first part. In particular, if the limiting case of a large number of mesh points is examined (that is, the continuous case), it becomes apparent that one procedure for making decisions within this mathematical framework is equivalent to an optical scheme utilizing diffraction effects proposed by one of the authors (GLS) before the theoretical work was started. This parallel registration invariant optical scheme will be described, and some of the modifications for increasing discrimination which follow from the theoretical analysis will also be discussed. An alternative optical scheme will then be presented which is based on a configuration of Kovásznay and Arman[1] at Johns Hopkins. This device uses geometrical optic effects entirely, and has properties that differ markedly from the diffraction scheme with regard to the modifications that can easily be imposed to aid discrimination. It is possible, for example, to change the nature of the transformation performed by the optical system in such a way that the autocorrelation (two-point combination count) operation is carried out in the vertical direction (for registration invariance in this direction) and an autoconvolution operation is carried out horizontally.

One of the chief disadvantages of such optical schemes for some application areas, however, is the necessity for having available transparent images of the patterns to be detected. Photographic negatives work quite well, but it is desirable in some cases, particularly when batch processing is not feasible, to work directly from the printed document. In addition, the size of the alphabet that can be handled in a parallel optical scheme is limited by the total light energy available from the source; the energy in each channel goes down with the reciprocal of the number of channels. We have therefore considered the design of electronic machines which operate on an electronic input from an optical scanner of some sort.

I. Theory of Pattern Recognition

A. *Registration Invariance and Autocorrelation*

One of the outstanding engineering problems encountered in the construction of character recognition devices has been the accommodation of character misregistration. The problem has not, of course, been insoluble; prescanning techniques and shifting search modes have been used effectively, but always result in a loss of time or increase in storage. For example, the simplest and most straightforward character recogni-

* Received by the IRE, August 1, 1960; revised manuscript received, October 27, 1960.

† IBM Mohansic Lab. Res. Ctr., Yorktown Heights, N. Y.

[1] L. S. Kovásznay and A. Arman, "Optical autocorrelation measurement of two-dimensional random patterns," *Rev. Sci. Instr.*, vol. 28, pp. 793–797; October, 1957.

Reprinted from *Proc. IRE*, vol. 49, pp. 175–185, Jan. 1961.

tion device consists of placing a mask over the pattern in question which consists of, say, a photographic negative of a known character. The amount of light passing through the mask is an indication of the similarity of the known and unknown characters, if the two are properly lined up. In general, it is impossible to prescribe the exact location of printed characters in advance, so that a search through various positions becomes necessary. Some of the distinct advantages of masking (stability against noise is identical to the use of matched filters or correlation in communication theory) are thereby minimized because of the additional uncertainties introduced.

It has, therefore, been our purpose to develop a description of patterns which is, by construction, independent of the absolute position of a character in a field of observation within translations or shifts. Since any two-dimensional pattern of black and white can be represented to any degree of approximation by an $N \times N$ matrix of zeros and ones for sufficiently large N, it will be adequate for our purposes to consider such binary matrices rather than the source patterns initially. The ones correspond to matrix points falling on black portions of the pattern, and the zeros to matrix points falling on white (see, for example, Fig. 1).

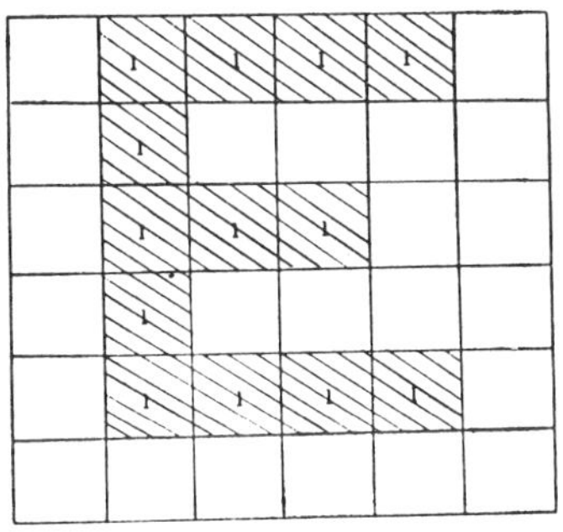

Fig. 1—"Ones" correspond to matrix points falling on the black portion of the pattern, and "zeros" to points off the pattern.

It is convenient to define the function $f(\boldsymbol{r})$, where $\boldsymbol{r}$ is the (x, y) coordinate of a matrix point, as unity on black points, zero on white. The number of pairs of ones with relative separation $\boldsymbol{R}=(\Delta x, \Delta y)$, if known for each $\boldsymbol{R}$, provides a registration invariant description (within a 180° rotation) of the pattern.

This representation has some intuitive appeal in that important geometrical characteristics are not obscured as in the crossing-type schemes or in the single-slit scan bank systems. If the number of pairs of each kind, $D(\boldsymbol{R})$, is plotted as a function of $\boldsymbol{R}=(\Delta x, \Delta y)$, then the resulting surface has information concerning local structure close to $|\boldsymbol{R}|=0$ and gross long-range structure for larger $|\boldsymbol{R}|$.

In order to examine the properties of the function $D(\boldsymbol{R})$ in greater detail, it is necessary to consider its formal expression in terms of $f(\boldsymbol{r})$. The existence of a pair of ones with separation $\boldsymbol{R}$ terminating at the matrix point $\boldsymbol{r}$ is established by the nonvanishing of the product $f(\boldsymbol{r})f(\boldsymbol{r}-\boldsymbol{R})$. Hence, the total number of pairs with separation $\boldsymbol{R}$ is given by

$$D(\boldsymbol{R}) = \sum_{\boldsymbol{r}} f(\boldsymbol{r})f(\boldsymbol{r}-\boldsymbol{R}). \tag{1}$$

Since the vector $\boldsymbol{r}-\boldsymbol{R}$ does not necessarily lie in the $N \times N$ matrix, the computation prescribed by (1) can be carried out only if $f(\boldsymbol{r})$ is defined over a more extensive domain. We therefore think of the pattern $f(\boldsymbol{r})$ defined initially in the first $N \times N$ quadrant of a $2N-1 \times 2N-1$ matrix, all other quadrants containing zeros. The structure of the larger matrix is to be repeated periodically over the plane, and the sum in (1) is taken over $-(N-1) \leq x,\ y \leq (N-1)$. It is clear from (1) that the description $D(\boldsymbol{R})$ is independent of the position of the pattern $f(\boldsymbol{r})$ so long as it is contained in some $N \times N$ matrix within the larger $2N-1 \times 2N-1$ matrix; it corresponds, in fact, to the discrete autocorrelation function of $f(\boldsymbol{r})$.

It is interesting to consider the harmonic analysis of $D(\boldsymbol{R})$. To do this, we define the discrete "Fourier transform" of $f(\boldsymbol{r})$:

$$f(\boldsymbol{r}) = \frac{1}{(2N-1)^2} \sum_{\boldsymbol{Q}} \bar{f}(\boldsymbol{Q}) \exp\left\{\frac{2\pi i}{2N-1}(\boldsymbol{r}\cdot\boldsymbol{Q})\right\}, \tag{2}$$

where

$$\bar{f}(\boldsymbol{Q}) = \sum_{\boldsymbol{r}} f(\boldsymbol{r}) \exp\left\{\frac{-2\pi i}{2N-1}(\boldsymbol{r}\cdot\boldsymbol{Q})\right\}. \tag{3}$$

The proof of the consistency of these relationships is a simple exercise resting on the identity

$$\sum_{\boldsymbol{r}} \exp\left\{\frac{2\pi i}{2N-1}(\boldsymbol{Q}-\boldsymbol{Q}')\cdot\boldsymbol{r}\right\} = \begin{cases} 0, & \boldsymbol{Q} \neq \boldsymbol{Q}' \text{ mod periodicity} \\ (2N-1)^2, & \boldsymbol{Q} = \boldsymbol{Q}' \text{ mod periodicity.} \end{cases} \tag{4}$$

All summations are taken over the $2N-1 \times 2N-1$ box. Substituting (2) into (1), one finds that

$$D(\boldsymbol{R}) = \frac{1}{(2N-1)^2} \sum_{\boldsymbol{Q}} |\bar{f}(\boldsymbol{Q})|^2 \cdot \exp\left\{\frac{2\pi i}{2N-1}(\boldsymbol{R}\cdot\boldsymbol{Q})\right\}, \tag{5}$$

and comparing this result with (2), we learn that the transform of $D(\boldsymbol{R})$ is given by

$$\bar{D}(\boldsymbol{Q}) = |\bar{f}(\boldsymbol{Q})|^2. \tag{6}$$

B. Optics and a Procedure for Decision

Although $D(\boldsymbol{R})$ can be calculated electronically by techniques such as those which we describe in Section II, the harmonic representation is especially convenient for describing a parallel optical technique suitable for certain areas of application. To see this most simply, we shall pass to the limit of the continuous distribution.

Let

$$N\Delta l = L, \tag{7}$$

where Δl is the spacing between mesh points and L the size of the character field. Let also

$$\begin{aligned} \mathbf{s} &= \mathbf{r}\Delta l \\ \mathbf{S} &= \mathbf{R}\Delta l \end{aligned} \tag{8}$$

and

$$\begin{aligned} F(\mathbf{s}) &= f(\mathbf{r}), \\ \theta(\mathbf{S}) &= (\Delta l)^2 D(\mathbf{R}). \end{aligned} \tag{9}$$

It then follows that, in the limit,

$$\begin{aligned} \theta(\mathbf{S}) &= \int d\tau F(\mathbf{s})F(\mathbf{s}-\mathbf{S}) \\ F(\mathbf{s}) &= \frac{1}{(2L)^2}\sum_{\mathbf{Q}} \overline{F}(\mathbf{Q}) \exp\left\{\frac{\pi i}{L}(\mathbf{s}\cdot\mathbf{Q})\right\} \\ \overline{F}(\mathbf{Q}) &= \int d\tau F(\mathbf{s}) \exp\left\{-\frac{\pi i}{L}(\mathbf{s}\cdot\mathbf{Q})\right\}, \end{aligned} \tag{10}$$

where $\overline{F}(\mathbf{Q}) = \lim\ (\Delta l)^2 \bar{f}(Q)$, and $d\tau$ is the differential of area. The integrals are taken over a $2L \times 2L$ square centered at $\mathbf{s}=0$, and $\mathbf{Q}$ is discretely summed over the entire plane. In this limiting process, the pattern remains stationary, but the number of defining points increases indefinitely. The point function $F(\mathbf{s})$ is piecewise constant at values 0 or 1. The theorem expressed by (6) is now of the form

$$\bar{\theta}(\mathbf{Q}) = |\overline{F}(\mathbf{Q})|^2, \tag{11}$$

a result that can be obtained by multiplying both sides of (6) by $(\Delta l)^4$ and taking the limit. The variable $\mathbf{Q}$ in (11) is still discrete; when $L \to \infty$, the relations applying to the usual Fourier transform are obtained, but it will suffice for our purposes to restrict the discussion to a periodic input pattern. Within such a final trivial limiting process, the function $\bar{\theta}(\mathbf{Q})$ is precisely the intensity pattern spatially displayed by the Fraunhofer diffraction pattern of an optical filter with transmission factor $F(\mathbf{s})$. The physical arrangement for producing this pattern will be discussed in Section II; we wish to concern ourselves here only with some of the mathematical properties of this representation of the input data.

The utility of this optical procedure for obtaining the Fourier transform of the autocorrelation function is most apparent from the point of view of a particular method of making decisions, *i.e.*, of identifying unknown patterns. One procedure for identifying patterns, analogous to the selected-point or region schemes such as that used in the IBM 1210 Sorter-Reader, would be to require the presence of important pairs of points or regions defined in a relative sense (as given in the autocorrelation function). Since there is no uncertainty in the registration of the pattern, this procedure could easily be developed for any given font in a small alphabet application with a careful computer analysis. Since each particular font would require individual analysis, and no particular design of type seems to have an intrinsic and universal interest at present, this technique will not be discussed further here. Instead, we shall consider a procedure which is analogous to the matched filter detection techniques of communication theory or the well-known masking technique used in the early work in character recognition. This method weights all configurations equally, and is in fact a calculation of the cosine of the angle between two multidimensional vectors.[2] For the discrete autocorrelation functions, we define the measure of similarity as

$$S_{AB} = \frac{\sum_{\mathbf{R}} D_A(\mathbf{R}) D_B(\mathbf{R})}{\left(\sum_{\mathbf{R}} D_A^2(\mathbf{R})\right)^{1/2}\left(\sum_{\mathbf{R}} D_B^2(\mathbf{R})\right)^{1/2}}, \tag{12}$$

where A and B refer to, say, an unknown and a known character respectively. According to the Schwartz inequality, $0 \leq S_{AB} \leq 1$, and $S_{AB}=1$ only if $D_A(\mathbf{R})$ and $D_B(\mathbf{R})$ are identical. This implies that A and B are either identical or differ only by a 180° rotation or a shift.

With the help of (4) and (5), we find that

$$\sum_{\overline{\mathbf{R}}} D_A(\mathbf{R}) D_B(\mathbf{R}) = \frac{1}{(2N-1)^2} \sum_{\mathbf{Q}} |\bar{f}_A(\mathbf{Q})|^2 |\bar{f}_B(\mathbf{Q})|^2, \tag{13}$$

so that an alternative expression for S_{AB} (computation in the "frequency" domain) is

$$S_{AB} = \frac{\sum_{\overline{\mathbf{R}}} |\bar{f}_A(\mathbf{Q})|^2 |\bar{f}_B(\mathbf{Q})|^2}{\left(\sum_{\mathbf{R}} |\bar{f}_A(\mathbf{Q})|^4\right)^{1/2}\left(\sum_{\mathbf{R}} |\bar{f}_B(\mathbf{Q})|^4\right)^{1/2}}. \tag{14}$$

If the numerator and denominator are multiplied by $(\Delta l)^8$, the limit $N \to \infty$ can be carried out without changing the value of S_{AB} except for the change due to the approximation of representing the continuous pattern by a discrete finite set of points initially. In the optical realization of this calculation, $|\overline{F}_B(\mathbf{Q})|^2$ corresponds to a filter in the Fraunhofer plane with the transmission factor given by the value of the function at the spatial point corresponding to the wave number integer pair $\mathbf{Q}$. The factor $|\overline{F}_A(\mathbf{Q})|^2$ is obtained from the intensity distribution that is the Fraunhofer diffraction pattern of the unknown character, and the sum is performed by collecting the total amount of light energy transmitted through the filter in a photovoltaic device.

In actual practice, a number of filters $|\overline{F}_B(\mathbf{Q})|^2$ would

[2] We are grateful to Dr. H. H. Goldstine for suggesting this very useful criterion for application here.

be used in parallel or series, each B corresponding to a separate recognition channel. Since the normalization factor

$$\left(\sum_{Q} | \bar{F}_A(Q) |^4 \right)^{-1/2}$$

is the same for all channels, recognition may be accomplished equally well if this factor is deleted from (14) and the maximum voltage over all channels is detected. It therefore remains for us to define the factor

$$\left(\sum_{Q} | \bar{F}_B(Q) |^4 \right)^{-1/2}$$

operationally. This quantity can be represented by an adjustable uniform attenuating filter T_B for each channel. If a uniform collimated light beam impinges upon the filters $|F_B(Q)|^2$ and T_B in series, the transmitted energy density is proportional to

$$| F_B(Q) |^2 T_B.$$

This distribution may then be reflected back through the pair of filters resulting in a total transmitted energy proportional to

$$\left(\sum_{Q} | F_B(Q) |^4 \right) T_B^2. \tag{15}$$

T_B may then be adjusted so that the quantity (15) is the same for each channel, resulting in the appropriate set of normalizing filters.

In order to get a feeling for the kind of discrimination that we would get with a real alphabet, some characters were processed digitally on the basis of (12). The relatively high resolution used insures a reasonable approximation to what the analogous optical results would be, and the results indicate a minimum discrimination of a few per cent (Fig. 2).

We have obtained comparable results in our studies of direct masking (no autocorrelation transform first) with the same alphabet in perfect registration. We have also simulated the effect of noise (Fig. 3) and the procedures of registered masking of the character and filtered autocorrelation functions compare very well in a rough way; the amount of noise that is added can amount to roughly as much as the area of the character before discrimination is lost in either case. We have, therefore, concluded that with respect to noise sensitivity and discrimination, nothing is lost by going to the autocorrelation description, and a great deal is gained in the registration problem. By use of appropriate nonlinear detection techniques, it should be possible to do better against noise, but this is still a subject for investigation.

It has been our desire to improve the discrimination by making distinct autocorrelation functions more nearly orthogonal with the same linear transformation on every pattern. If this linear transformation has the form

$$D'(R) = \sum_{R'} M(R - R')D(R'), \tag{16}$$

then an optical realization in a simple form is assured. Consider, for example, the application of the criterion (12) to functions of the form (16). The numerator of (12) is given by

$$\sum_{RR'R''} D_A(R'')M(R - R'')M(R - R')D_B(R') = \sum_{RR'} D_A(R')L(R' - R)D_B(R), \tag{17}$$

where

$$L(R) = \sum_{R'} M(R')M(R' - R). \tag{18}$$

In the wave number representation, (17) can be written as

$$\frac{1}{(2N - 1)^2} \sum_{Q} \bar{D}_A(Q)\bar{D}_B(Q)\bar{L}(Q) = \frac{1}{(2N-1)^2} \sum_{Q} | \bar{f}_A(Q) |^2 | \bar{f}_B(Q) |^2 | \bar{M}(Q) |^2. \tag{19}$$

The normalization can be carried out in the same way as described in connection with (15). It follows from (19) that whatever the form of $M(R)$ (it may contain negative terms), an optical decision procedure can be fabricated with positive transmission filters.

In particular, the use of a transformation of the form

$$M(R) = \begin{pmatrix} & 1 & \\ 1 & 1 & 1 \\ & 1 & \end{pmatrix} \tag{20}$$

A	1.000	.941	.940
B	.941	1.000	.987
R	.940	.987	1.000

Fig. 2—Comparison of autocorrelation functions.

	IDEALS		DELETION NOISE		ADDITIVE NOIS:.	

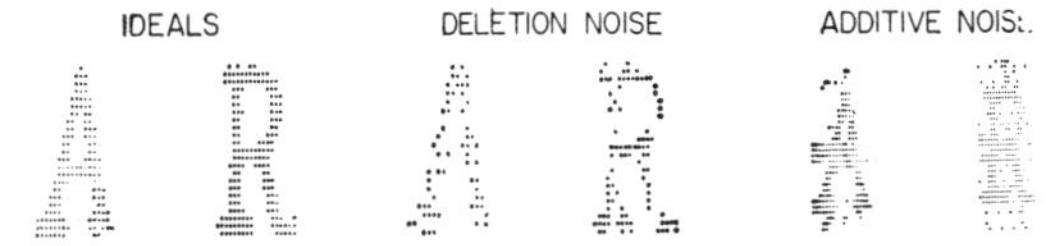

A	1.000	0.931	0.941	0.888	0.972	0.921
B	0.930	0.982	0.856	0.928	0.954	0.953
R	0.931	1.000	0.852	0.945	0.958	0.967
5	0.891	0.902	0.824	0.890	0.896	0.880
S	0.904	0.951	0.826	0.924	0.932	0.930
Q	0.867	0.910	0.800	0.863	0.891	0.906
O	0.890	0.895	0.835	0.857	0.890	0.883

Fig. 3—Comparison of autocorrelation functions with noisy input.

produced good results going across a few printed fonts, but decreased the discrimination within a given fixed font (Fig. 4). On the other hand, the second-difference operation[3] (the opposite of averaging, so to speak)

$$M(R) = \begin{bmatrix} & -1 & \\ -1 & 4 & -1 \\ & -1 & \end{bmatrix}, \tag{21}$$

produced poor results across fonts but increased the discrimination within a given font considerably. Computer simulation showed a discrimination of the order of 20 per cent to 30 per cent in bad cases as compared to 10 per cent with a comparison of autocorrelation functions (Fig. 5). In a test for noise sensitivity, we found the surprising result that the autocorrelation recognition technique with a second-differencing was just as insensitive with respect to the introduction of errors as without second-differencing, but discrimination was considerably improved, even in very noisy situations (Fig. 6).

IDEALS	A	B	R	INPUT	A	B	R
A	1.000	.941	.940	A	.953	.897	.907
B	.941	1.000	.987	B	.935	.975	.970
R	.940	.987	1.000	R	.936	.962	.977

Fig. 4—Comparison of smoothed autocorrelation functions.

IDEALS	A	B	R	INPUT	A	B	R
A	1.000	.493	.525	A	.571	.437	.458
B	.493	1.000	.700	B	.532	.696	.642
R	.525	.700	1.000	R	.545	.519	.696

Fig. 5—Comparison of second-differenced autocorrelation functions.

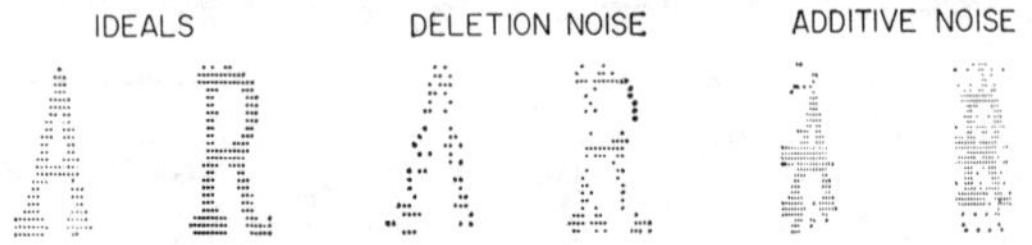

	IDEALS A	IDEALS R	DELETION NOISE A	DELETION NOISE R	ADDITIVE NOISE A	ADDITIVE NOISE R
A	1.000	0.526	0.420	0.392	0.676	0.283
B	0.494	0.695	0.300	0.450	0.478	0.320
R	0.526	1.000	0.338	0.539	0.507	0.364
5	0.522	0.576	0.305	0.434	0.506	0.258
S	0.515	0.654	0.327	0.451	0.508	0.285
Q	0.528	0.556	0.322	0.395	0.510	0.280
O	0.581	0.587	0.326	0.411	0.523	0.259

Fig. 6—Noise sensitivity for second-differenced autocorrelation functions.

[3] $|\overline{M}(Q)|^2 \propto |Q|^4$ in the optical analog.

II. Physical Realization

In the first part of this paper, registration invariant counting functions were described based on a binary matrix description of the characters, and it was shown that as the matrix mesh becomes finer and finer the continuous representation of the pair counting function is

$$\theta(S) = \int_{-\infty}^{+\infty} F(s) \cdot F(s - S) d\tau;$$

and since $\theta(S)$ is also given by

$$\theta(S) = \frac{1}{(2\pi)^2} \int_{-\infty}^{+\infty} \bar{\theta}(W) \cos(2\pi W \cdot S) dW \quad \text{where } W = \frac{Q}{2L}$$

one can either work with the wave number spectrum $\bar{\theta}(W)$ or with the autocorrelation function $\theta(S)$.

A. Wave Number Spectrum Model

The wave number spectrum of a pattern $F(x, y)$ is given by $\overline{F}(W_1W_2) \cdot \overline{F}(W_1W_2)^*$, where $\overline{F}(W_1W_2)$ is the Fourier transform of $F(x, y)$. It is known that the Fraunhofer diffraction pattern of a transmitting aperture is given by the square of the Fourier transform of the aperture;[4] consequently, the wave number spectrum of an input character can be obtained in parallel by a rather simple optical arrangement. Fig. 7 is a schematic of the optical arrangement required to produce the wave number spectrum, and Fig. 8 shows the wave number spectra for a set of numerals. The coordinates of the Fraunhofer plane are wave numbers, and the origin of the coordinate system is always located at the intersection of the optic axis and the Fraunhofer plane. If Cartesian coordinates (μ) and (ν) are set up in the Fraunhofer plane, the transform coordinates W_1 and W_2 are related as follows:

$$W_1 = \frac{2\pi}{\lambda} \frac{\mu}{f_2} \qquad W_2 = \frac{2\pi}{\lambda} \frac{\nu}{f_2}.$$

The size of the diffraction pattern is therefore proportional to the wavelength of the source and the focal length of L_2, and inversely proportional to the size of the aperture. The latter fact can be seen by computing the transform of a simple rectangular slit of height h and width w, which results in

$$I = A_0^2 \frac{\sin^2 W_1 \dfrac{w}{2}}{W_1^2 \left(\dfrac{w}{2}\right)^2} \cdot \frac{\sin^2 W_2 \dfrac{h}{2}}{W_2^2 \left(\dfrac{h}{2}\right)^2}$$

so that the distance between the origin and the first zero along the W_1 direction is

$$\mu_0 = \frac{\lambda f_2}{w}$$

[4] G. Joos, "Theoretical Physics," Blackie & Son, Ltd., London, Eng., p. 369; 1949.

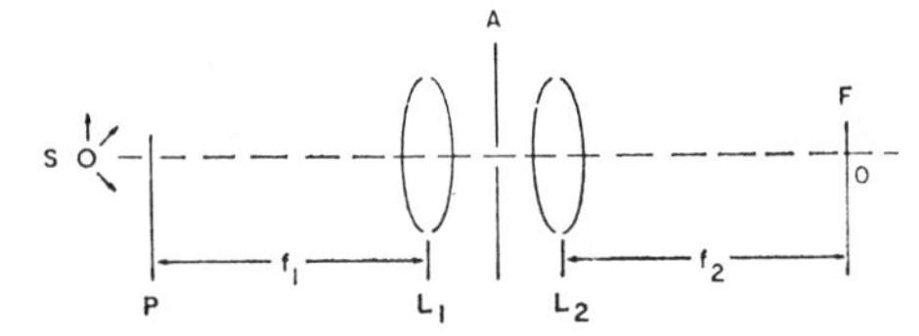

Fig. 7—Schematic of the optical arrangement. s = monochromatic point source; P = pin hole aperture; L_1 = lens, f_1 = focal length of L_1; O = optic axis; A = transmitting aperture of input; L_2 = lens; f_2 = focal length of L_2; F = Fraunhofer plane.

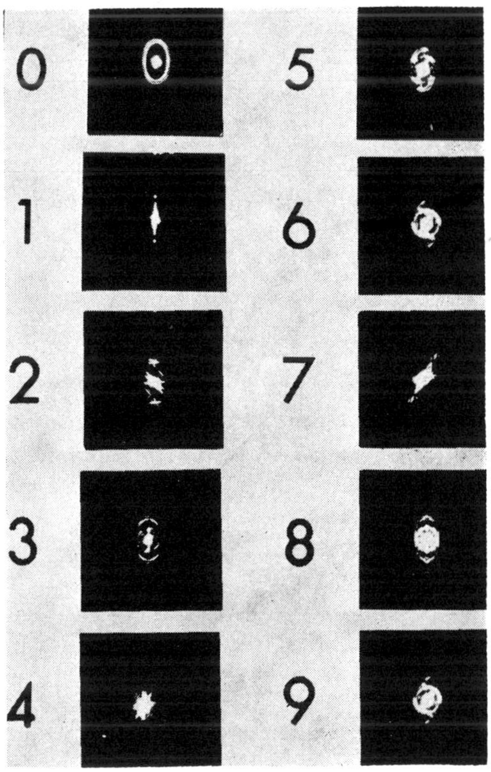

Fig. 8—Wave number spectrum of a set of numerals.

and along the W_2 direction is

$$\nu_0 = \frac{\lambda f_2}{h} \cdot$$

It can be appreciated from the diagram that translations of the aperture A do not alter the Fraunhofer pattern or its location. The Fraunhofer pattern can therefore be considered as a representation of the input character in an exactly known location. Fig. 9 shows schematically a recognition system which compares the Fraunhofer pattern of the input with normalized comparison masks for ideal characters. The comparison masks are obtained by photographing the Fraunhofer patterns for ideal characters, being careful to maintain unity gamma for the photographic procedure. The channel normalization masks are obtained by using a pair of crossed polaroids in each channel to adjust the channel transmission by the method described in Section I. The total light being transmitted through each channel is collected and converted to an electrical signal by a suitable photo-electric detector, and identification is made by determining the channel having the maximum signal.

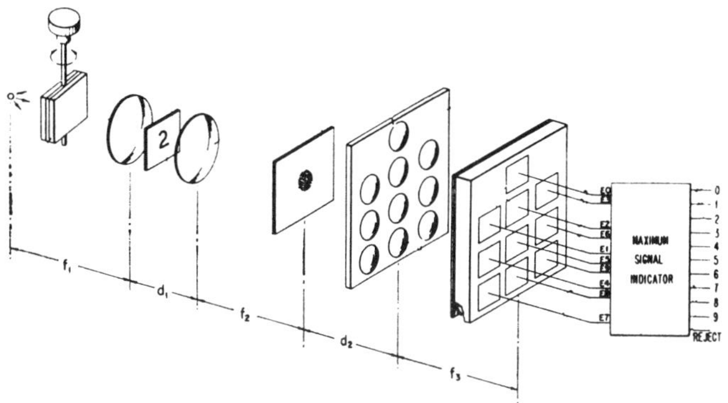

Fig. 9—Schematic of optical recognition procedure using wave number spectrum.

Other linear transformations of the form described in Section I can be carried out in the wave number space either to improve the discrimination or to improve the stability. In particular, the equivalent of the second-difference operation can be realized by inserting a transmission plate in the Fraunhofer plane which has transmission characteristics that vary as the fourth power of the wave number. In this case, the discrimination between characters in a fixed alphabet is improved and, contrary to one's intuitive reasoning, the sensitivity to random noise is essentially unchanged; however, the tolerance to systematic noise is decreased. Systematic noise in this context means the variations between two or more type styles which are very similar.

It was shown earlier that the size of the diffraction pattern was proportional to the wavelength of the source and the focal length of the lens. Accordingly, a convenient search mode can be employed to accommodate similar patterns which differ from each other only by a scale factor. In practice, it is somewhat easier to vary the wavelength by using a polychromatic source and a variable narrow band-pass filter. The band pass of a conventional interference filter can be shifted by inclining the plane of the filter to the direction of a collimated light beam. Other variable wavelength filters are discussed in most standard optic texts. A practical range for scale factor variation achieved in this way is approximately two to one.

A parallel comparison system has been described, but it is clear that the comparisons could be made sequentially if desired. Also, since the Fraunhofer pattern is inversely proportional to the character size, a practical system requires that normal-size characters be reduced to microfilm sizes in order to maintain a reasonable focal distance for lens L_2.

B. Autocorrelation Function Model

Of course, one could construct an autocorrelation function by taking the diffraction pattern of the wave number spectrum of the input, but some method of generating a transparency of the spectrum in real time would be required. No practical method for doing this is currently known; nonetheless, an experiment of this type was carried out using photographic means to generate the transparencies. Fig. 10 shows the correlation functions for the set of numerals in Fig. 8 that were obtained in this experiment. A spurious multiple line structure is present in these functions which was shown by analysis to be caused by the limited range of the photographic emulsion used to generate the transparencies. The range of the film has the effect of passing the wave number spectrum through a low-pass filter. This situation is quite analogous to the signal obtained by passing a square pulse through a low-pass filter in a communication network. A survey of the available photographic emulsions revealed that no emulsions were available with the required range.

The above experiment was instructive in that it made apparent that the photographic construction of filters, such as the second-difference filters previously discussed, cannot be obtained witn sufficient fidelity by a continuously varying photographic transparency. Construction of this type of filter will probably require a mosaic pattern in which the transmission per square is controlled by varying the size and spacing of the mosaic elements.

An optical arrangement illustrated in Fig. 11 will directly produce the autocorrelation function of the input pattern. This arrangement was described by Kovásznay and Arman.[1] It is shown in this article that the intensity distribution in the focal plane of the lens L_1 is given by

$$I(\mu_1'\nu') = \frac{I_0}{A}\iint f(x, y)\cdot f\left(x + \frac{2d}{F}\mu'_1 y + \frac{2d}{F}\nu'\right)dxdy,$$

where (μ') and (ν') are the coordinates in the focal plane, and (x) and (y) are the coordinates in the character aperture plane. If the following substitutions are introduced,

$$\mu = \frac{2d}{F}\mu', \qquad \nu = \frac{2d}{F}\nu'$$

so that

$$\mu' = \frac{F}{2d}\mu, \qquad \nu' = \frac{F}{2d}\nu,$$

the intensity distribution becomes

$$I\left(\frac{F}{2d}\mu, \frac{F}{2d}\nu\right) = \frac{I_0}{A}\iint f(x, y)\cdot f(x + \mu_1 y + \nu)dxdy,$$

where the size of autocorrelation function has a magnification factor $(F/2d)$.

In the above method, the peak of the autocorrelation function always appears at the intersection of the optic axis and the focal plane, independent of the location of the character aperture plate. For comparison, a well-known optical arrangement for generating cross correlations is shown in Fig. 12. The intensity distribution in the output plane is the cross correlation of the input character $f(x, y)$ and the reference character $g(x', y')$. If the input character is the same as the reference character, the intensity distribution is the same as the autocorrelation; however, in this arrangement one can see that a movement of the input causes a movement of the output. This point has been brought out to eliminate the confusion that has occurred with regard to these two methods.

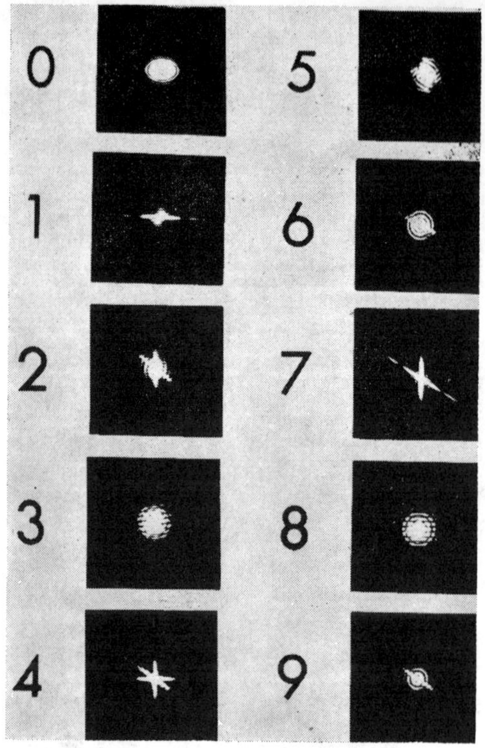

Fig. 10—The autocorrelation functions of a set of numerals generated by taking the Fraunhofer diffraction pattern of the wave number spectrum of the original input patterns.

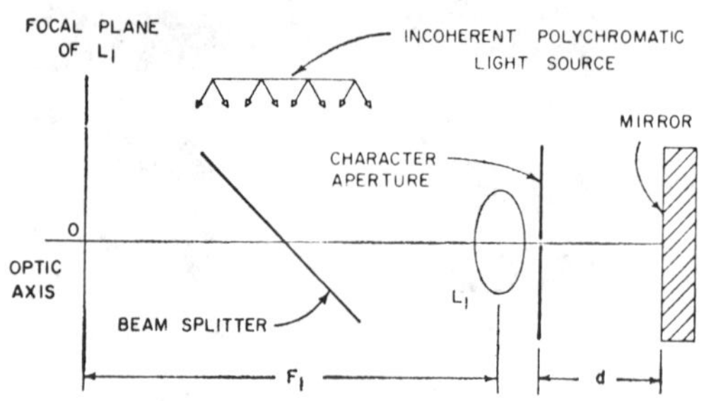

Fig. 11—Optical arrangement to produce directly the autocorrelation function of the input pattern.

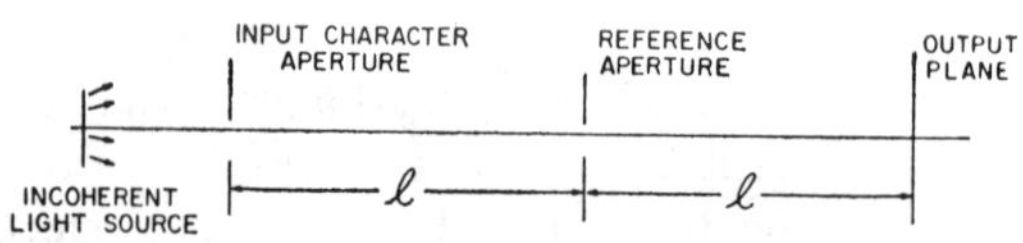

Fig. 12—A well-known optical arrangement for generating cross correlations.

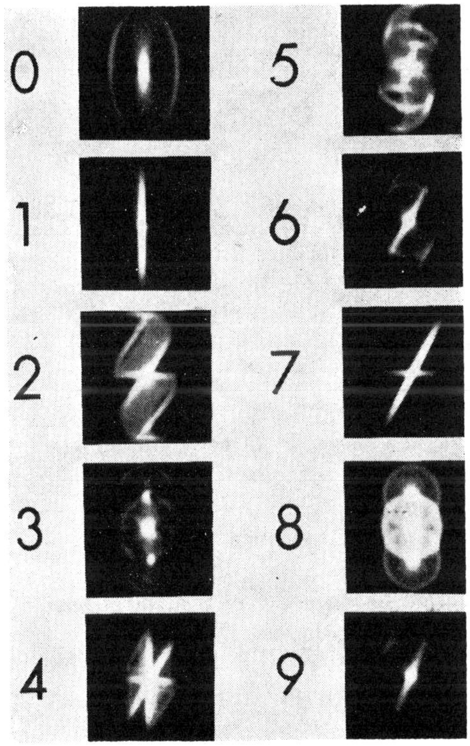

Fig. 13—The autocorrelation functions of a set of numerals generated by the device shown in Fig. 11.

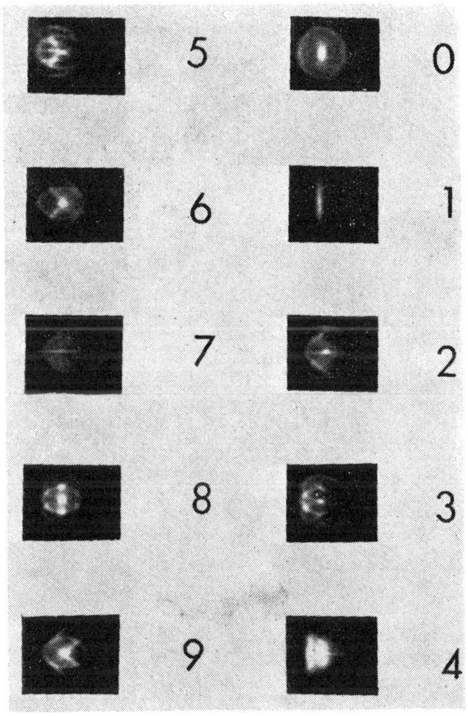

Fig. 15—The autocorrelation functions of a set of numerals generated by the device shown in Fig. 14.

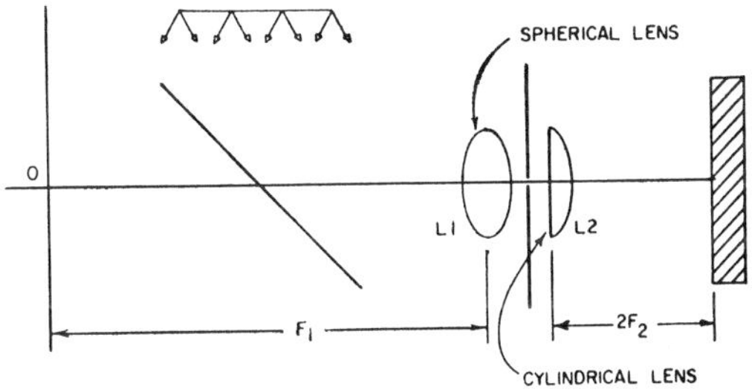

Fig. 14—Autocorrelation function generator modified to eliminate 180° ambiguity problem.

The autocorrelation function can now be used in a recognition system in exactly the same manner that the wave number spectrum was employed. Fig. 13 is a set of autocorrelation functions for the same set of numerals used previously. It was pointed out earlier that there is a 180° ambiguity problem in both the wave number spectrum and the autocorrelation function. The autocorrelation function generator can be modified as shown in Fig. 14 to eliminate this ambiguity. In this arrangement, registration invariance is maintained along the generators of the cylindrical lens, but lost in the normal direction. The intensity in the output plane is the result of autocorrelating in the direction of the generators of the cylinder and convolving in the direction normal to the generators. That is,

$$I(\mu, \nu) = \int\int f(x, y) \cdot f(x - \mu, \nu - y) dx dy$$

or

$$I(\mu, \nu) = \int\int f(x, y) \cdot f(\mu - x, y - \nu) dx dy.$$

One may regard this physically as being the result of introducing a phase shift on the system which is constant along one coordinate and variable along the other coordinate. The particular arrangement illustrated in Fig. 14 gives the autoconvolution; however, any other phase plate could be introduced, but if registration invariance is to be retained in one dimension, the phase shift in this dimension must be constant. The autoconvolution of the set of numerals is shown in Fig. 15.

The two optical systems that have been described have certain practical limitations. These are:

1) The operation of these systems demands thta a transparency of the character be supplied as an input.
2) Low-level light intensities are inherent to both modes of operation.

If the optical realizations are to become applicable to a large class of recognition problems, some practical method of converting, in real time, a projected image of a character from a conventional document to a transmitting input must be developed, and some method of supplying image intensification in either the Fraunhofer plane or the correlation plane must be supplied. Optical pumping techniques in solid-state materials are being investigated as possible means of solving the first problem, and phosphor technology is being investigated as a means of solving the latter problem.

C. Electronic Autocorrelation Model

In view of the outstanding problems in the optical realizations, it became apparent that an electronic realization would be desirable to handle the more general problems. In discussing the electronic realization, a binary matrix representation of the input character will be used, and the sequential digital signals will be those generated by a vertical scan moving from left to right.

The geometrical interpretation of the autocorrelation function is that the value of the autocorrelation function at any specified point is proportional to the number of pairs of occupied points having a given displacement and direction. Physically, this is equivalent to taking the same character on two matrices and shifting one with respect to the other through all possible translations and counting the number of coincidences for each relative translation. Fig. 16(a) illustrates this procedure, and shows one-half of the autocorrelation function. Making use of the symmetric properties of the autocorrelation function, the complete function is constructed and shown in Fig. 16(b).

It would be desirable to define an operation on the sequential scan pattern which is the equivalent of the two-dimensional shifting operation. If one considers a skewed character matrix, as shown in Fig. 17(a), wrapped around a cylinder such that the rows are parallel to the generators of the cylinder and the upper and lower rows are adjacent members, it becomes clear that the columns of the matrix form a continuous spiral, and that the sequence of bits is the same as that formed by connecting, from left to right, the lower end of one column to the upper end of the adjacent column. A translation of the spiral along the helix retains the spatial relation between the bits, but causes a movement of the pattern both along and around the cylinder. Thus, the pattern may be placed in any desired location on the cylinder by translating the spiral. If two like matrices are wrapped around the cylinder and one held stationary while the other is moved by translating the spiral, the complete two-dimensional shifting operation described previously will be carried out. Fig. 17(b) shows the autocorrelation generated in this way, and the sequence of shifts is described by the spiral of the correlation matrix. After some reflection one can realize that if the maximum character height is N bits tall, the minimum column height must be $(2N-1)$ bits tall to prevent overlaps. The two spirals may now be considered as one-dimensional words which preserve the spatial relations under shifting operations.

The autocorrelation function of an input pattern can therefore be generated by putting the same bit

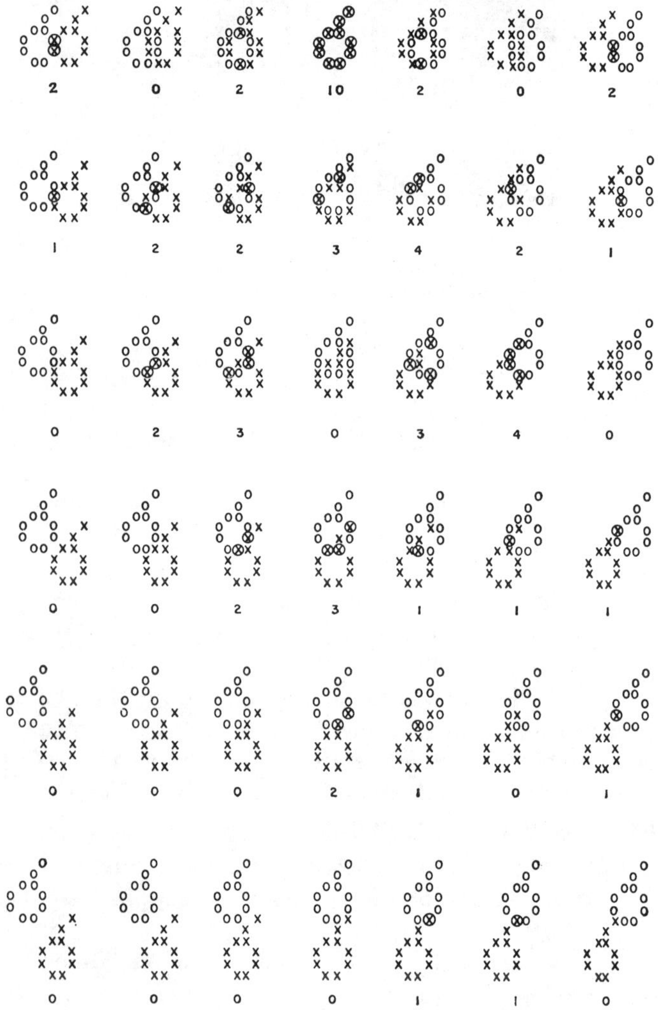

(a)

0	1	1	0	0	0	0
1	0	1	2	0	0	0
1	1	1	3	2	0	0
0	4	3	0	3	2	0
1	2	4	3	2	2	1
2	0	2	10	2	0	2
1	2	2	3	4	2	1
0	2	3	0	3	4	0
0	0	2	3	1	1	1
0	0	0	2	1	0	1
0	0	0	0	1	1	0

(b)

Fig. 16—(a) One-half of the autocorrelation function. (b) The complete function.

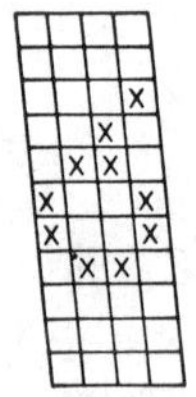

(a)

0	1	1	0	0	0	0
1	0	1	2	0	0	0
1	1	1	3	2	0	0
0	4	3	0	3	2	0
1	2	4	3	2	2	1
2	0	2	10	2	0	2
1	2	2	3	4	2	1
0	2	3	0	3	4	0
0	0	2	3	1	1	1
0	0	0	2	1	0	1
0	0	0	0	1	1	0

(b)

Fig. 17—(a) Two-dimensional shifting operation. (b) Where the sequence of shifts is described by the spiral of the correlation matrix.

pattern in two shift registers and counting the number of coincidences each time one register is shifted with respect to the other. Fig. 18 shows the generation of an autocorrelation function using this method. The result of not having at least $(2N-1)$ column positions is also shown in this figure. This mode of operation is not the most desirable, since there must be at least a full word separation between incoming patterns.

A more useful method, developed by McDermid, Petersen, and Shelton, of obtaining the autocorrelation function is illustrated in Fig. 19. In this structure, the first position in the register is ANDED independently to all other register positions, and the outputs of the ANDS each fed into a counter. Reading from left to right, the counter positions and the equivalent shifts are as follows:

1) Total number of bits in the pattern
2) Unit shift in $+y$ direction
3) 2-unit shift in $+y$ direction
4) 3-unit shift in $+y$ direction
5) 4-unit shift in $+y$ direction
6) 5-unit shift in $+y$ direction
7) 1-unit shift in x direction and 5-unit shift in $-y$ direction
8) 1-unit shift in x direction and 4-unit shift in $-y$ direction
9) And so forth, with every twelfth position corresponding to an additional x shift.

The values in the counters at any particular time represent the autocorrelation function of that part of the word that is in the register. After the last bit of the word moves into the register, the values in the counters become and remain stationary until the first bit in some new word enters the register.

The appropriate time to compare the function in the counters to the comparison functions is that time when the values in the counters become stationary for some specific number of vertical scans. The number chosen would depend on the nominal separation of the input.

The comparison is made by taking the scalar product of the function in the counters with each normalized comparison function. This is equivalent to multiplying each number in the counter by the corresponding comparison number in each comparison channel and summing the products in each channel. The channel having the largest sum identifies the input character. Fig. 19 shows a block diagram of this operation.

Earlier it was noted that two symbols which differ from each other only by a 180° rotation result in an ambiguous result. This ambiguity can be resolved by making use of the partial autocorrelation functions that exist in the counters as the input enters the register. In general, at corresponding times, these partial autocorrelation functions are characteristically different even for the ambiguous case. Fig. 20 is an illustration of the generation of the autocorrelation functions of a six and a nine, and it is clear that the partial functions are significantly different.

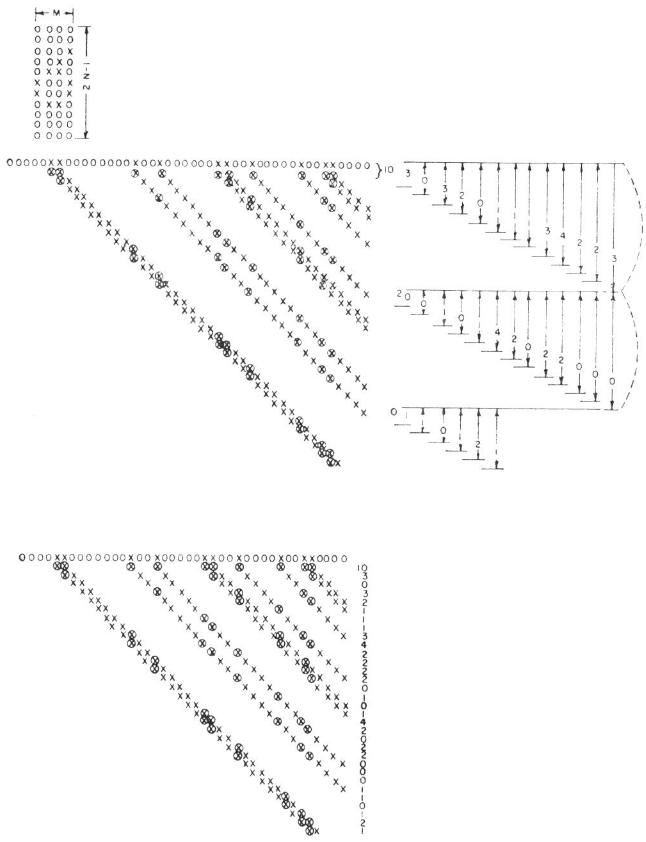

Fig. 18—Generation of an autocorrelation function by putting the same bit pattern in two shift registers and counting the coincidences each time one register is shifted with respect to the other.

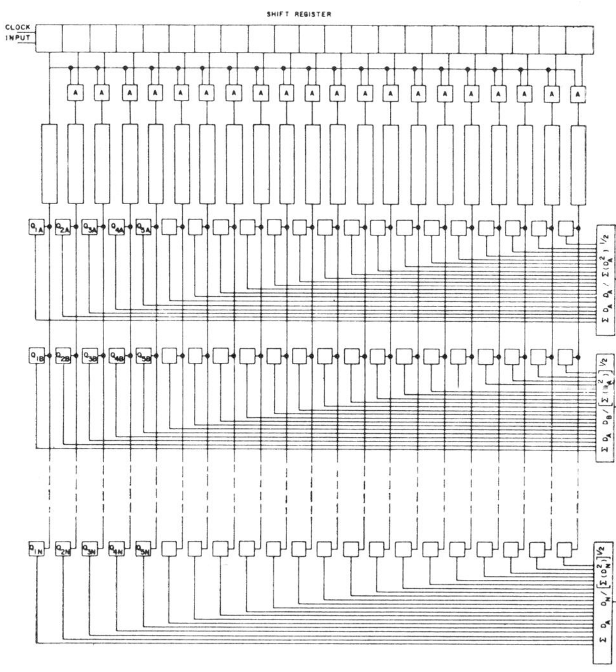

Fig. 19—Block diagram of operation. $A=$AND; $Q_{in}=$normalized reference multiplied by counter value; $E=N$th channel output voltage.

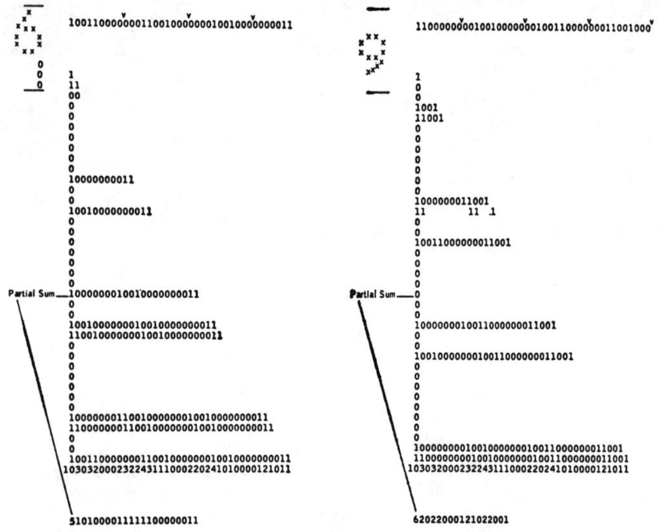

Fig. 20—Autocorrelation of the numbers six and nine.

Simulated results of the system discussed indicate a good capability to handle printed characters of a large alphabet. A continuous trade between discrimination and stability can be made by use of the M operators described in Section I. Thus, multifont recognition can be realized at the cost of discrimination. There are indications that extensions of the notions described will result in improved multifont recognition and in gains toward the isolation of the character separation problem.

III. Acknowledgment

The authors wish to thank Mrs. L. Selfridge and Miss F. Reilly for carrying out the calculations referred to above.

Bibliography

[1] E. L. O'Neill, "Spatial filtering in optics," IRE Trans. on Information Theory, vol. IT-2, pp. 56–65; June, 1956.
[2] T. P. Cheatham, Jr. and A. Kohlenberg, "Optical filters: their equivalence to and difference from electrical networks," 1954 IRE Convention Record, pt. 4, pp. 6–12.
[3] P. Elias, D. S. Gray, and D. Z. Robinson. "Fourier treatment of optical processes," *J. Opt. Soc. Amer.*, vol. 42, pp. 127–174; February, 1952.
[4] L. S. G. Kovásznay and H. M. Joseph, "Image processing," Proc. IRE, vol. 43, pp. 560–470; May, 1955.
[5] D. Z. Robinson, "Methods of background description and their utility," Proc. IRE, vol. 47, pp. 1554–1561; September, 1959.
[6] G. F. Aroyan, "The technique of spatial filtering," Proc. IRE, vol. 47, pp. 1561–1568; September, 1959.

A Class of Algorithms for Fast Digital Image Registration

DANIEL I. BARNEA, MEMBER, IEEE, AND HARVEY F. SILVERMAN, MEMBER, IEEE

Abstract—The automatic determination of local similarity between two structured data sets is fundamental to the disciplines of pattern recognition and image processing. A class of algorithms, which may be used to determine similarity in a far more efficient manner than methods currently in use, is introduced in this paper. There may be a saving of computation time of two orders of magnitude or more by adopting this new approach.

The problem of translational image registration, used for an example throughout, is discussed and the problems with the most widely used method-correlation explained. Simple implementations of the new algorithms are introduced to motivate the basic idea of their structure. Real data from ITOS-1 satellites are presented to give meaningful empirical justification for theoretical predictions.

Index Terms—Registration efficiency, sequential similarity detection algorithms, spatial cross correlation, spatial registration of digital images.

I. Introduction

THE AUTOMATIC determination of local similarity between two structured data sets is fundamental to the disciplines of pattern recognition and image processing. A class of algorithms, which may be used to determine similarity in a far more efficient manner than methods currently in use, is introduced in this paper. By adopting this new approach there may be a saving of computation time of two orders of magnitude or more.

The method most widely used for similarity detection is correlation. In fact the similarity detection problem itself is generally called "correlation." In this paper, however, the distinction between the mathematical method, correlation, and the class of similarity detection problems will be maintained; for it will be shown that procedures other than those currently in practice will yield efficient and accurate results.

The class of sequential similarity detection algorithms (SSDAs) introduced here may be applied to the entire spectrum of similarity detection problems. A broad theory relating general properties of these algorithms can be developed. In this paper, however, the utility of SSDAs will be illustrated by their application to a specific problem, translational registration, which is basic to image processing.

In Section II, the translational registration problem will be introduced. Section III will describe the computational efficiency of correlation methods. In Section IV some examples of SSDAs are presented in the order they were chronologically investigated. It is intended that this section impart an intuitive feeling about SSDAs to the reader. Section V generalizes this class of algorithms and presents, in light of this general concept, some ideas for further development of the given examples.

Manuscript received May 13, 1971; revised August 6, 1971.

D. I. Barnea was with the IBM T. J. Watson Research Center, Yorktown Heights, N. Y. 10598. He is now with Eljim, Holon, Israel.

H. F. Silverman is with the IBM T. J. Watson Research Center, Yorktown Heights, N. Y. 10598.

II. The Translational Registration Problem

Registration is inherently basic to any image processing system. When it is desired to detect changes or perform a mapping of two similar images, it is necessary for meaningful results to have the images registered. If the pictures do not differ in magnification and rotation, then the best translational fit will yield the required registration. (The problems arising from magnification and rotation will not be considered here for the sake of brevity. However, the methods are applicable when proper modifications are introduced.)

Let two images, S the *search area* and W the *window* be defined as shown in Fig. 1. S is taken as an $L \times L$ array of digital picture elements which may assume one of K grey levels; i.e.,

$$0 \le S(i, j) \le K - 1$$
$$1 \le i, j \le L.$$

W is considered to be an $M \times M$, M smaller than L array of digital picture elements having the same grey scale range; i.e.,

$$0 \le W(l, m) \le K - 1$$
$$1 \le l, m \le M.$$

It will be convenient to introduce a notation for $M \times M$ wholly contained subimages.

$$S_M{}^{i,j}(l, m) \equiv S(i + l - 1, j + m - 1), \quad \begin{cases} 1 \le l, m \le M \\ 1 \le i, j \le L - M + 1. \end{cases} \tag{1}$$

Each $M \times M$ subimage of S can be uniquely referenced by the specification of its upper left corner's coordinates (i, j). These will be used to define *reference points*. It will be assumed that enough *a priori* information is known about the dislocation between the window and search area so that the parameters L and M may be selected with the virtual guarantee that, at registration, a complete subimage is contained in the search area as shown in Fig. 1.

Translational registration, therefore, is a search over some subset of the allowed range of reference points to find a point (i^*, j^*) which indicates a subimage that is most similar to the given window.

Reprinted from *IEEE Trans. Comput.*, vol. C-21 pp. 179–186, Feb. 1972.

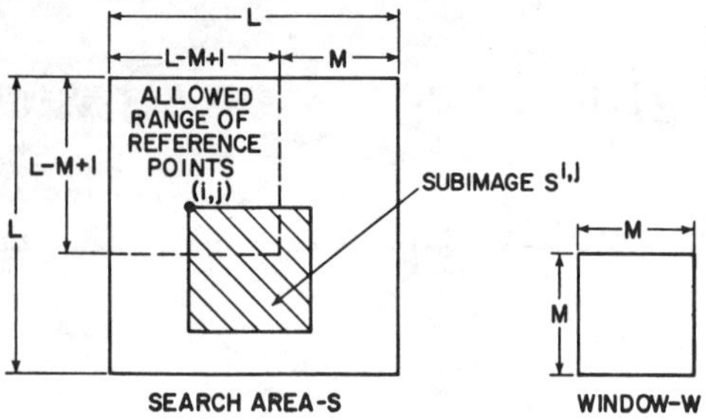

Fig. 1. Search space.

III. Computational Efficiency of Correlation Methods

A. Correlation Method

The method most widely used for the automatic determination of translation is correlation [1], [2]. The elements of the unnormalized cross-correlation surface $R(i, j)$ are defined to be

$$R(i, j) \equiv \sum_{l=1}^{M} \sum_{m=1}^{M} W(l, m) S_M{}^{i,j}(l, m), \qquad 1 \leq i, j \leq L - M + 1. \quad (2)$$

In the correlation scheme a representative output surface such as $R(i, j)$ is searched for a maximum $(\hat{\imath}, \hat{\jmath})$. The procedure is successful if $(\hat{\imath}, \hat{\jmath})$ and (i^*, j^*) are equivalent. As a counterexample, however, consider the unnormalized cross correlation of (2) even in the *ideal case* where W exactly matches some subimage; i.e., $W = S_M{}^{i^*,j^*}$. Then

$$R(i^*, j^*) = \sum_{l=1}^{M} \sum_{m=1}^{M} W^2(l, m). \quad (3)$$

Also, for this ideal case, consider the nonmatching point $(\hat{\imath}, \hat{\jmath})$ where

$$S_M{}^{\hat{\imath},\hat{\jmath}}(l, m) = \max_{l,m} W(l, m) = W_M, \qquad \text{for all } (l, m). \quad (4)$$

Clearly

$$R(\hat{\imath}, \hat{\jmath}) = W_M \sum_{l=1}^{M} \sum_{m=1}^{M}, \qquad W(l, m) \geq R(i^*, j^*). \quad (5)$$

Therefore even in the ideal case, a search for a maximum over $R(i, j)$ does not necessarily yield the registration point. Normalization is therefore necessary in even the simplest of cases! For completeness, the usual normalized correlation surface is defined in (6).

$$R_N^2(i, j) = \frac{\left(\sum_{l=1}^{M} \sum_{m=1}^{M} W(l, m) S_M{}^{i,j}(l, m) \right)^2}{\left[\sum_{l=1}^{M} \sum_{m=1}^{M} W^2(l, m) \right] \left[\sum_{l=1}^{M} \sum_{m=1}^{M} S_M^2{}^{i,j}(l, m) \right]}, \qquad 1 \leq i, j \leq L - M + 1. \quad (6)$$

The choice of a similarity detection algorithm should be justified by its probability for error and its computational complexity rather than by tradition or expediency. Perhaps the two reasons which are generally given for using the correlation method are that: 1) correlation appears to be a natural solution for the mean-square-error criteria [3]; and 2) that analog–optical methods implement correlation easily [4].

However, there is no guarantee for *any* method that a solution is correct or unique. There seems to be, therefore, no adequate justification for the use of correlation to solve all digital registration problems. Algorithms, such as those presented in this paper, which have selectable distance measure properties and lower computational complexity, appear to be a more fitting choice.

B. The Cost of Correlation

The normalized correlation surface may be calculated by direct means or by fast Fourier transform (FFT) methods [5]. In each case the amount of computation required for normalization is the same. The numerator of (6) is all that may be treated by FFT.

In Table I the approximate number of calculations for each procedure are shown. One should note that although the FFT method requires fewer operations for L and M large, this method also requires a memory capacity of $2L^2$ real words which may be infeasible for an L larger than 256.

IV. Some Examples of SSDA Implementation

A. The Basic Concept

For a particular reference point (i, j), there are M^2 points of the subimage $S_M{}^{i,j}$ which may be compared with the M^2 corresponding points in W. (Each set of points for comparison (e.g., $\langle S_M{}^{i,j}(l, m), W(l, m) \rangle$) will be called a *windowing pair.*) In correlation the maximum number of windowing pairs or $M^2(L-M+1)^2$ are compared. Each reference point, regardless of content, is therefore processed with very high precision. However, accuracy is required only for those relatively few points near surface maxima. Hence, there is considerable waste in performing high-accuracy calculations at a vast majority of points.

SSDA reduces this redundancy by performing a sequential search which may be terminated before all M^2 windowing pairs for a particular reference point are tested. Furthermore, the algorithms do not implicitly contain any fixed error measure or measure evaluation method.

B. Constant Threshold Algorithm

A simple but important SSDA has been extensively studied and will be used as a vehicle to introduce the concept. Here, a search over each of the $(L-M+1)^2$ reference points is performed, as in correlation. However, the criteria for similarity at each reference point is significantly different from correlation.

In the constant threshold algorithm, windowing pairs are selected for comparison in a random order so that, in general, a great deal of new information is considered in each test; i.e., a random nonrepeating sequence of the integers $1, 2, \cdots, M^2$ is generated and used to yield the random, nonrepeating sequence of coordinates (l_n, m_n),

TABLE I

COST OF NORMALIZED CORRELATION

Direct Correlation	FFT Correlation	Ordinary Normalization	"Fast" Normalization
$M^2(L-M+1)$-Mults.	$6L^2(\log_2 L)$-Complex Mult-Adds	L^2+M^2-Mults.	L^2+M^2-Mults.
$M^2(L-M+1)^2$-Adds	FFT of S,W and IFFT of Product	$M^2(L-M+1)^2$-Adds	$4(L-M+1)^2$-Adds
$(L-M+1)^2$-Squarings	L^2-Complex Mults.	$(L-M+1)^2$-Mults & Divides	$(L-M+1)^2$-Mults & Divides
	$(L-M+1)^2$-Squarings		

$n=1, 2, \cdots, M^2$. Thus the windowing pairs $\langle S_M^{i,j}(l_n, m_n), W(l_n, m_n)\rangle$ are compared in random order as n increases.

Unnormalized or normalized measures for evaluating the error between windowing pairs may be defined, respectively, as

$$\epsilon'(i, j, l_n, m_n) \equiv | S_M^{i,j}(l_n, m_n) - W(l_n, m_n) | \quad (7)$$

$$\epsilon(i, j, l_n, m_n) \equiv | S_M^{i,j}(l_n, m_n) - \hat{S}(i, j) - W(l_n, m_n) + \hat{W} | \quad (8)$$

where

$$\hat{W} \equiv \frac{1}{M^2} \sum_{l=1}^{M} \sum_{m=1}^{M} W(l, m) \quad (9)$$

and

$$\hat{S}(i, j) \equiv \frac{1}{M^2} \sum_{l=1}^{M} \sum_{m=1}^{M} S_M^{i,j}(l, m). \quad (10)$$

Unlike the correlation methods cited previously, in the *ideal case*, where $W = S_M^{i^*,j^*}$, a minimum of zero is guaranteed for the nonnormalized case, i.e., for

$$\|E(i, j)\| \equiv \sum_{l=1}^{M} \sum_{m=1}^{M} | S_M^{i,j}(l, m) - W(l, m) | \quad (11)$$

$$0 = \|E(i^*, j^*)\| \leq \|E(i, j)\|. \quad (12)$$

Thus, in this ideal case no normalization is necessary and, obviously, a comparison of very few points will yield the answer. The error measure, based upon the L_1 norm between two images, is also computationally simpler than the multiplicative measure of correlation.

In this SSDA implementation, a constant threshold T is introduced. As the error for randomly selected windowing pairs is accumulated, a test is made against T. When the accumulated error exceeds T at test N, operations cease for reference point (i, j) and the value N is recorded. The SSDA surface $I(i, j)$ is therefore defined as

$$I(i, j) = \left\{ r \,\middle|\, \min_{1 \leq r \leq M^2} \left\{ \sum_{n=1}^{r} \epsilon(i, j, l_n, m_n) \geq T \right\} \right\}. \quad (13)$$

Reference points where $I(i, j)$ is large (those which require many windowing pair tests to exceed T) are considered points of similarity.

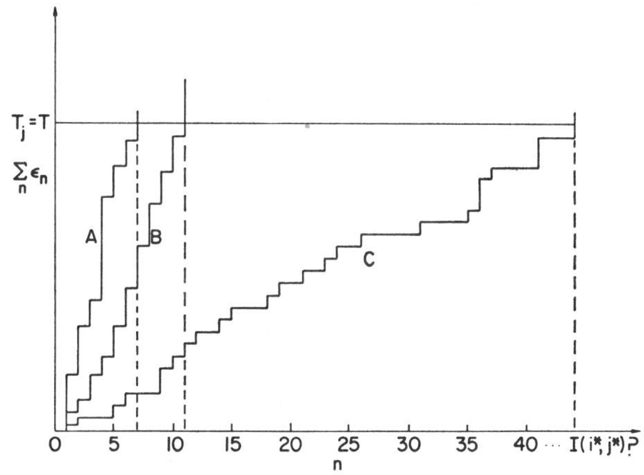

Fig. 2. Growth curves with T=const.

It is clear that if a suitable value for T is selected, many fewer than M^2 tests will be required for reference points which rapidly accumulate error. It is this property which significantly reduces the computational complexity of an SSDA! This fact is made clear by Fig. 2. Curves A, B, and C depict the cumulative error for three different reference points as a function of test. A and B accumulate error rapidly and operations for their reference points terminate early with $I(i, j)$ obtaining values of 7 and 11, respectively. Curve C, however, accumulates error more slowly. It is, therefore, much more likely to be a candidate for registration, and will accordingly have a value of 44 assigned to $I(i, j)$.

Data from the NOAA weather satellite ITOS-1 have been used to test this algorithm. Fig. 3 is an example of a typical data set. Images (a) and (b) are segments of uncorrected vidicon output taken over Baja California on August 9, and 11, 1970, respectively. In this application, one is interested in registering land masses. Therefore, in addition to the noise due to optics and scanners, there is a great deal of intense noise due to clouds and cloud shadow, as well as from the fiducial mark (the ×) in (b). Fig. 3(c) shows the window function W taken from the upper left corner of (b). In Fig. 3(d), the answer is displayed. The window of Fig. 3(c) is put into the search area [Fig. 3(a)] at the position indicated by the algorithm.

The fit is visually accurate, but as real rather than contrived data were tested, there is virtually no way to absolutely check this fit. However, the same data were run in an FFT

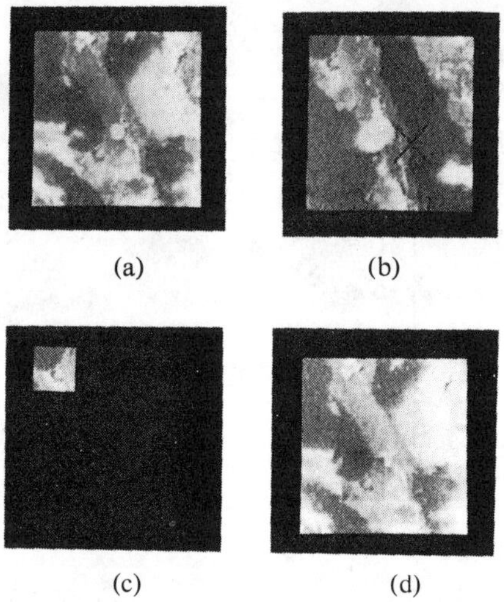

Fig. 3. Example of registered images taken from NOAA data. (a) Search area. (b) Picture 2. (c) Window area taken from (b). (d) Window inserted into search area at (i^*, j^*).

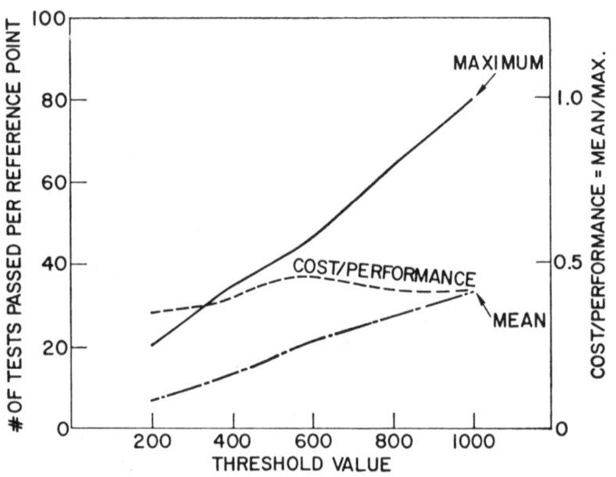

Fig. 4. Statistical results for constant threshold algorithm.

correlation-registration scheme, for the purposes of a relative comparison. As expected, there were some complications involved in the application of this more traditional method. Only after clouds had been detected and their bright surface values replaced by random noise would correlation yield a meaningful peak. (No corrections whatsoever were necessary for SSDA!) For this data, correlation gave the registration point as one picture element higher than did this SSDA. As correct registration might easily lie between two points, there is little to argue!

The data in Fig. 3 have the following attributes:

1) $L=128$;
2) $M=32$;
3) $0 \leq S(i,j) \leq 255$;
4) $E[S(i,j)] \approx 75$ (expected value).

For this data set, tests were made for various values of T. The important results from these tests are given in Fig. 4. The average value of a point in the surface $I(i,j)$—and hence the number of calculations required—grows linearly with threshold. The maximum value in the surface $I(i,j)$, an indicator of the accuracy of the method, also grows linearly with threshold value, although at a somewhat higher rate. The ratio of these two items can be considered as a cost/performance measure. For the case of a constant threshold, this measure is shown to be relatively constant. This implies that accuracy may be achieved only by an associated increase in computation.

C. Monotonic-Increasing Threshold Sequence Algorithm—(Algorithm A)

The growth curve for a particular reference point (three of which are shown in Fig. 2), is a monotonically increasing function. It is the *average slope of the growth curve* that is important in the determination of a threshold crossing. It therefore seems reasonable that the replacement of the constant threshold T by a threshold T_j, increasing monotonically with test, would improve performance when the following criteria are considered.

1) The sequence T_j should have "shape" approximating that of the growth curve for $I(i^*, j^*)$, but should bound this growth curve from above for an arbitrarily large n.

2) The T_j sequence should have initial values high enough so that a trend might be established even for reference points far from registration.

An example of a monotonically increasing threshold sequence is shown in Fig. 5. Growth curves A and B, for reference points far from registration, are eliminated earlier than in Fig. 2, at 4 and 7, respectively. As most reference points do exhibit rapid growth, the total number of tests will be diminished significantly. Growth curve C, however, which appears to be a strong candidate for registration, will undergo a larger number of tests than before. Therefore, a high degree of accuracy will be achieved for those few points which have growth curves with low average slope.

Monotonically increasing threshold curves were generated by several methods. A stochastic approach yielded excellent analytic and experimental results. The distances at registration $\epsilon(i^*, j^*, l, m)$ may be considered a random variable with an exponential distribution

$$f_\alpha(x) = \begin{cases} \dfrac{1}{\lambda} e^{-x/\lambda}, & 0 \leq x \leq \infty \\ 0, & \text{otherwise.} \end{cases} \tag{14}$$

At the point of registration, the sum of "ϵ terms" is a cumulative measure of the noise between two different pictures. At any other reference point $\epsilon(i, j, l, m)$ must be considered as the sum of the noise due to the fact that the pictures are different, *image noise*, and that the pictures are out of registration, *registration noise*. One assumes, in order to say a correct answer has been achieved, that the total noise is minimal at the point of registration.

Optimally, one desires to find a threshold sequence which does the following.

1) It minimizes the probability that a point other than the registration point will remain beneath the threshold as long as the "true" registration point.

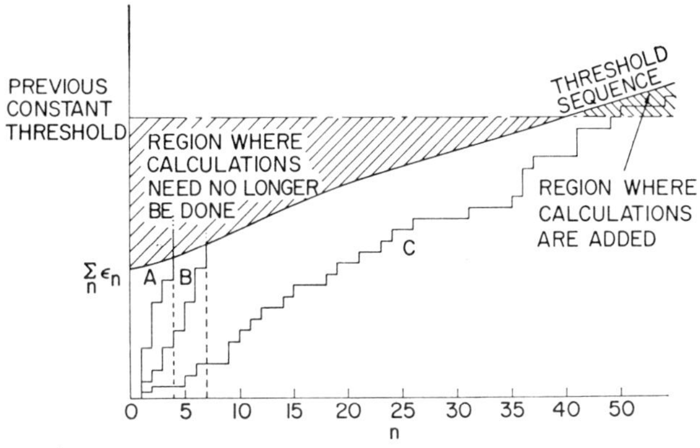

Fig. 5. Growth curve for algorithm Q with T_j a monotonically increasing function.

2) It maximizes the probability that the registration point remains below threshold.

3) It allows the total number of operations to be small.

Some analytic difficulty developed when this total optimization problem was considered. However, suboptimal threshold sequences may be found based upon the image noise alone which appear to satisfy the two hypotheses at the beginning of this section. Empirical results also tend to justify their contention for near optimality.

Fig. 6 shows a set of curves derived simply from the mean and variance of the image noise's statistics. If one has a measure of the mean of this noise λ and a bound to the deviation from the mean that he is willing to allow, he may select a threshold sequence. This method of design, however, leads to results which are somewhat less satisfying than those obtained by more careful derivation.

The curves shown in Fig. 7 are drawn from a computer solution of an analytic system of recursive equations. Here, one considers the probability P_k, which is the probability for the cumulative error at the registration point $I(i^*, j^*)$ to exceed T_k at test k, given that the cumulative error has stayed beneath the threshold for all tests $n = 1, 2, \cdots, k-1$. The useful set of design curves of Fig. 7 may be derived by restricting P_k to be a constant q for all k.

While the curves in Figs. 6 and 7 appear to have the same shape, the equiprobable curves, which take into account the results from past as well as present tests, are closer to the optimal. Any particular one of these curves may be generated by an exact set of equations or by an approximate model. Once an estimate to the mean of the image noise λ has been obtained, a reasonable design follows.

Several tests were conducted on the data set shown in Fig. 8. Fig. 8(a) and (b) are each taken from the edge of a large, unnormalized NOAA vidicon output. As before the images were taken two days apart each of the Gulf coast of Texas. The data have many distortions due to sun angle, spacecraft attitude, and optical edge effects. There are also very few prominent features in evidence. The insertion in Fig. 8(d) appears to be correct upon visual inspection.

For this set of data, *a priori* knowledge of the approximate registration point allowed one to find the parameter λ

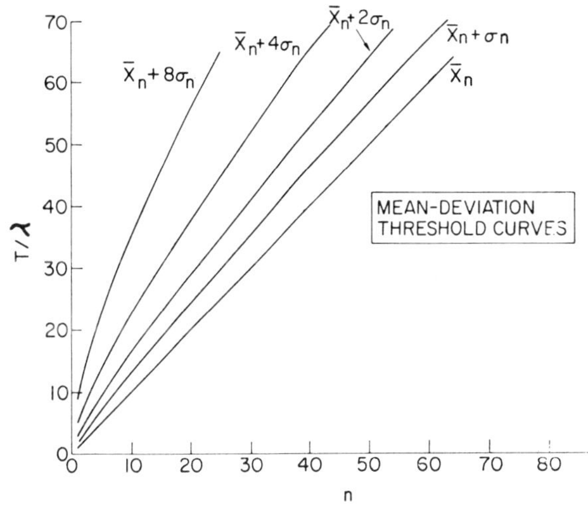

Fig. 6. Mean deviation threshold curves.

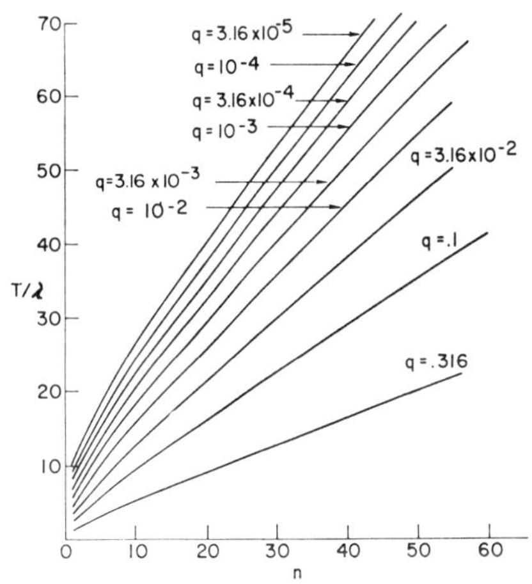

Fig. 7. Calculated equiprobable threshold curves.

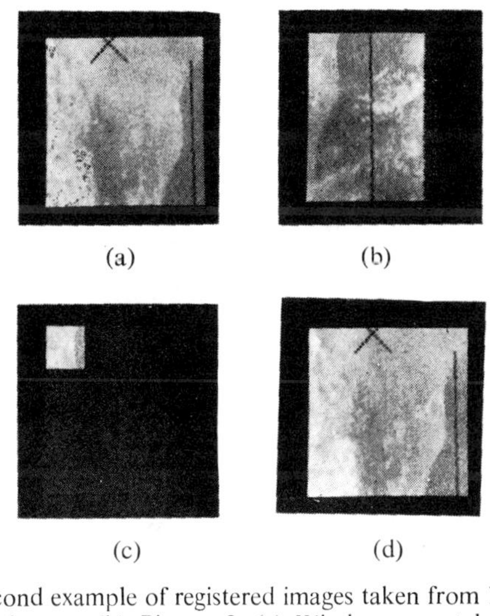

Fig. 8. Second example of registered images taken from NOAA data. (a) Search area. (b) Picture 2. (c) Window area taken from (b). (d) Window inserted into search area at (i^*, j^*).

empirically. Fig. 9 shows the sample distribution for the data of Fig. 8 for the absolute value of the image noise. The exponential distribution assumed in (14) appears to have some justification, with a value of 9.089 for λ. Based upon this data, quantized versions of threshold functions for several values of q were constructed. These are shown in Fig. 10. A quantized approximation was used so that thresholds might be stored in a lookup table and the speed of calculation increased. It was found that a coarse quantization yielded accurate results.

Curves based upon the results of tests on the data of Fig. 8, using the constant q design criteria, are given in Fig. 11. One immediately can see the benefit of using the monotone-increasing threshold sequence by looking at the cost/performance ratio. As one would expect, there is a region where this ratio has minimal—implying optimal—value(s). In the tests conducted, the optimal value(s) for q lie in the range $10^{-2} \leq q \leq 10^{-3}$. Hence, knowledge of the parameter λ and an optimal value for q can lead to a highly efficient design.

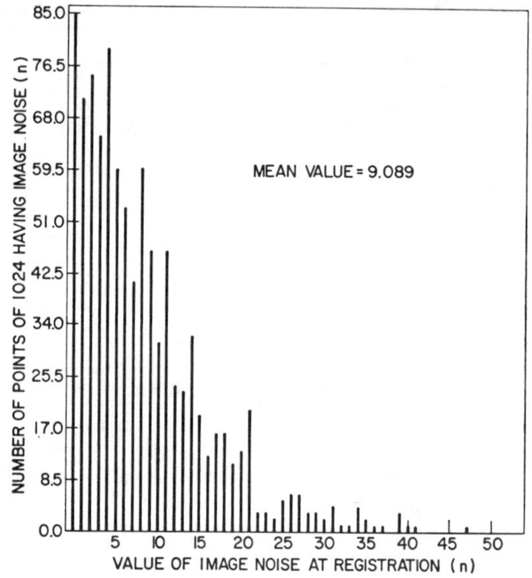

Fig. 9. Sample distribution for image noise for data.

D. Computational Aspects

The efficiency of SSDA has been demonstrated in the given examples. The saving which results can be illustrated by relating numbers of operations required for various implementations. Time ratios relating arithmetic operations for the IBM 360/65 will be used. Namely,

1) real multiply time/read add time = 3;
2) real add time/integer add time = 2;
3) integer multiply time/integer add time = 3.5;
4) compare time = integer add time;
5) complex add = 2 real adds;
6) complex multiply = 3 real multiplies + 5 real adds.

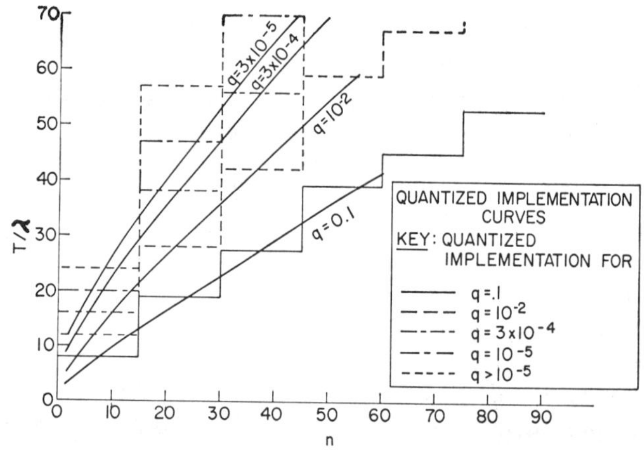

Fig. 10. Quantized implementation of equiprobable curves, where q is the probability for the growth curve at (i^*, j^*) to exceed T_k at any k given that the growth curve is below $T_{k'}$ for all $k' = 1, 2, \cdots, k-1$.

From Table I the direct cross-correlation method, *not including normalization,* reduces to

number (direct)

$$= 4.5M^2(L - M + 1)^2 \text{ equivalent integer adds} \quad (15)$$

and the FFT cross-correlation method, *not including normalization,* reduces to

number (FFT)

$$= 200L^2 \log_2 L \text{ equivalent integer adds.} \quad (16)$$

The monotone increasing threshold SSDA, *using the normalized distance measure,* requires:

1) 4 adds/reference point to obtain normalization;
2) 2 adds + 1 compare/windowing point to get the error measure;
3) 1 compare/windowing point;
4) An average of $\bar{n}$ windowing tests/reference point.

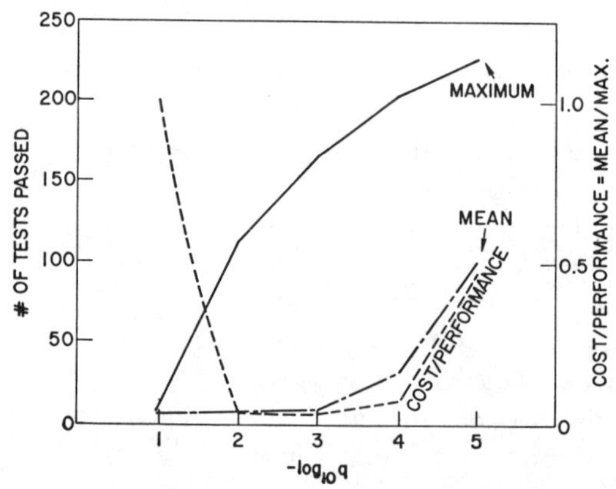

Fig. 11. Statistical results for constant q algorithm.

Thus

number (SSDA)

$$= (4 + 4\bar{n})(L - M + 1)^2 \text{ equivalent integer adds.} \quad (17)$$

For M of 32, $\bar{n}$ should be about 10–15. However, as M increases, more points should be checked to establish conclusively that misregistration has occurred. A figure of $\bar{n}=10(M/32)^{1/2}$ is reasonable. Thus

$$\text{number (SSDA)} = 4(1 + 10(M/32)^{1/2}) \cdot (L - M + 1)^2 \text{ equivalent integer adds.} \quad (18)$$

Table II lists corresponding calculational costs for the several methods for various values of M and L. Values which are power of two are selected so that best figures for the FFT method might be achieved. One should note, however, that no other algorithm requires L and M to be a power of two. (It is recognized that this condition might be relaxed somewhat for the FFT, but most all available routines maintain this requirement.)

Table II shows that the given example of an SSDA is about a factor of 50 faster than the FFT correlation method. (This is a quite conservative estimate.) The order of magnitude saving by use of SSDAs is very important to production digital image processing.

V. Generalization and Conclusion

A. The General Concept of SSDA

A structure for the class of SSDAs may now be stated in a general form so that the versatility of the method is clear. There are many implementations other than those of the examples, each of which contains the following basic elements.

1) Ordering Algorithm O_1: O_1 orders the $(L-M+1)^2$ reference points. The window is successively compared to subimages which are selected by O_1. This ordering need not be fixed in advance, and may vary in a way that depends upon events occurring during the execution. In these cases, the total number of reference points which are processed is less than $(L-M+1)^2$.

2) Ordering Algorithm O_2: O_2 orders the M^2 (or fewer) windowing pairs to be compared at each reference point. O_2 may be fixed—the same for all reference points—or it may adapt in a data-dependent fashion.

3) Distance Measure (Norm) $\|x\|$: $\|x\|$ is used as a measure of error when windowing pairs are compared.

4) Sequential Measuring Algorithm Q: Q is a mapping from a subset of the M^2 possible distance measures for a particular reference point (i, j) into an element in the inspection surface $I(i, j)$.

$I(i, j)$ is directly analogous to the correlation surface described in Section III. Q operates upon distances $\|x\|$ for windowing pairs in the sequential manner specified by O_2. This sequence of operations continues until an event (like the passing of a threshold) occurs. At that instant $I(i, j)$ is evaluated on the basis of the measurements taken.

Each of the four basic properties may be tuned to fit a specific similarity detection problem so that a proper balance between accuracy and efficiency is realized. There are many variations that are far different from the examples presented. In the next few paragraphs, reasonable extensions to the monotonic-increasing threshold algorithm are given to show just how the general formulation can be used.

The reference points, in the examples above, were exhaustively searched. The following list presents some alternative approaches for O_1, if it is desired to sacrifice some accuracy for computation speed.

1) Apply a two-step coarse–fine uniform search, i.e., a first search on a coarse grid at every mth point and then do a full search in local regions about the maxima. (This procedure was implemented on several data sets, and a tradeoff, in fact, does exist. The degradation in accuracy is highly dependent upon the smoothness of the fully sampled $I(i, j)$ surface.)

2) Apply other multistep open-loop search criteria such as a coarse random search.

3) Apply, assuming that the surface $I(i, j)$ is monotone in some sense and has a global maximum, closed-loop hill climbing or gradient techniques.

The order for the windowing pairs O_2 might be easily changed to yield rapid elimination of misregistration points. Rather than selecting a completely random sequence, one might, for example, select a pseudorandom sequence based upon prominent features of W. That is, the first points in the sequence would be as random as possible while representing the key points for registration. There is a host of methods for the legitimate selection of feature points.

Each data structure has its own problem of normalization. SSDA may adapt to any necessary condition by the selection of suitable $\|x\|$. Most measures and process normalizations are acceptable. However, complex measures will somewhat degrade the ratio of saving shown in Table II.

It is apparent that there is a wide area yet unexplored for determining Q. Other mappings might be tried. It is possible, for example, to set a fixed threshold and test each of the windowing points against it independently. The surface value $I(i, j)$ might then be the number of comparisons prior to the occurrence of a certain number of threshold crossings.

Perhaps of great value to the monotonic-increasing threshold algorithm would be threshold adaptation. A conservative guess to the optimal threshold sequence might be used at first. On the basis of measurements made at succeeding reference points, the threshold sequence might be adaptively lowered. This procedure could allow an efficient algorithm to develop with very little *a priori* design on the part of the user. The algorithm would also gain a certain amount of independence from data.

B. Conclusion

A general class of sequential algorithms for similarity detection has been introduced, referencing the specific problem of translational image registration. Experimental and analytic results have been presented to show orders of magnitude improvement in efficiency. Several ideas for further improvement of experimentally implemented algorithms have also been presented.

The structure of the new algorithms is ideally suited for digital similarity detections. There is a time saving of at least 50 for typical problems on a representative medium-size computer—a prediction substantiated by experiment.

TABLE II

EQUIVALENT INTEGER ADDS FOR VARIOUS ALGORITHMS FOR SEVERAL VALUES OF L AND M

		Direct Method	FFT Correlation	Algorithm A
		$4.5M^2(L-M+1)^2$	$200\ L^2 \log_2 L$	$4(1+10(M/32))^{1/2}(L-M+1)^2$
L	M			
128	32	4.4×10^7	2.25×10^7	4.2×10^5
256	32	2.57×10^8	1×10^8	2.2×10^6
512	32	1.1×10^9	4.6×10^8*	1.05×10^7
1024	32	4.5×10^9	2×10^9*	4.35×10^7
2048	32	1.85×10^{10}	8.8×10^9*	1.75×10^8
128	64	8.15×10^7	2.25×10^7	2.5×10^5
256	64	6.9×10^8	1×10^8	2.2×10^6
512	64	3.7×10^9	4.6×10^8*	1.2×10^7
1024	64	1.7×10^{10}	2×10^9*	5.5×10^7
2048	64	7.4×10^{10}	8.8×10^9*	2.4×10^8
256	128	1.15×10^9	1×10^8	1.37×10^6
512	128	1.1×10^{10}	4.6×10^8*	1.25×10^7
1024	128	5.8×10^{10}	2×10^9*	6.7×10^7
2048	128	2.5×10^{11}	8.8×10^9*	2.9×10^8
512	256	2×10^{10}	4.6×10^8*	7.5×10^6
1024	256	1.8×10^{11}	2×10^9*	7×10^7
2048	256	1×10^{12}	8.8×10^9*	3.7×10^8
1024	512	2.7×10^{11}	2×10^9*	4.1×10^7
2048	512	2.6×10^{12}	8.8×10^9*	4×10^8
2048	1024	4.5×10^{12}	8.8×10^9*	5.7×10^8

* FFT's not possible on most machines without much disk accessing.

ACKNOWLEDGMENT

The authors wish to thank H. Kobayashi, J. Mommens, D. Grossman, and P. Franaszek for their help in many fruitful discussions.

REFERENCES

[1] P. E. Anuta, "Spatial registration of multispectral and multitemporal digital imagery using fast Fourier transform techniques," *IEEE Trans. Geosci. Electron.*, vol. GE-8, pp. 353–368, Oct. 1970.

[2] J. A. Leese, C. S. Novak, and B. B. Clark, "An automated technique for obtaining cloud motion from geosynchronous satellite data using cross-correlation," *J. Appl. Meteorol.*, vol. 10, pp. 110–132, Feb. 1971.

[3] A. Papoulis, *Probability, Random Variables and Stochastic Processes.* New York: McGraw-Hill, 1965.

[4] J. W. Goodman, *Introduction to Fourier Optics.* San Francisco: McGraw-Hill, 1968.

[5] J. W. Cooley, P. A. W. Lewis, and P. D. Welch, "Application of fast Fourier transform to computation of Fourier integrals, Fourier series, and convolution integrals," *IEEE Trans. Audio Electroacoust.*, vol. AU-15, pp. 79–84, June 1967.

The Representation and Matching of Pictorial Structures

MARTIN A. FISCHLER AND ROBERT A. ELSCHLAGER

Abstract—The primary problem dealt with in this paper is the following. Given some description of a visual object, find that object in an actual photograph. Part of the solution to this problem is the specification of a descriptive scheme, and a metric on which to base the decision of "goodness" of matching or detection.

We offer a combined descriptive scheme and decision metric which is general, intuitively satisfying, and which has led to promising experimental results. We also present an algorithm which takes the above descriptions, together with a matrix representing the intensities of the actual photograph, and then finds the described object in the matrix. The algorithm uses a procedure similar to dynamic programming in order to cut down on the vast amount of computation otherwise necessary.

One desirable feature of the approach is its generality. A new programming system does not need to be written for every new description; instead, one just specifies descriptions in terms of a certain set of primitives and parameters.

There are many areas of application: scene analysis and description, map matching for navigation and guidance, optical tracking, stereo compilation, and image change detection. In fact, the ability to describe, match, and register scenes is basic for almost any image processing task.

Index Terms—Dynamic programming, heuristic optimization, picture description, picture matching, picture processing, representation.

Manuscript received November 30, 1971; revised May 22, 1972, and August 21, 1972.

The authors are with the Lockheed Palo Alto Research Laboratory, Lockheed Missiles & Space Company, Inc., Palo Alto, Calif. 94304.

INTRODUCTION

THE PRIMARY PROBLEM dealt with in this paper is the following. Given some description of a visual object, find that object in an actual photograph. The object might be simple, such as a line, or complicated, such as an ocean wave, and the description can be linguistic, pictorial, procedural, etc. The actual photograph will be called the "sensed scene," a two-dimensional array of gray-level values, while the object being sought is called the "reference."

This ability to find a reference in a sensed scene, or, equivalently, to match or register the images of two scenes, is basic for almost any image processing task. Application to such areas as scene analysis and description, map matching for navigation and guidance, optical

Reprinted from *IEEE Trans. Comput.*, vol. C-22, pp. 67–92, Jan. 1973.

tracking, stereo compilation, and image change detection is direct and obvious.

There are two basic approaches to solving the image-matching problem as defined above.

If we possess a precise description of the noise and distortion process which defines the mapping between the reference and its image in the sensed scene, we can employ statistical decision theory techniques to derive an image-matching procedure which optimizes some objective criterion (e.g., minimum error, or minimum risk, in determining the best embedding of the reference in the sensed scene). A typical outcome of such an analysis is the use of a correlation-like matching procedure. However, in most practical problem situations, the required noise and distortion model is not available, nor is it feasible to attempt to construct one. For example, we might consider all human faces to be perturbed versions of some single ideal or reference face. However, to attempt to define completely a valid noise and distortion model for this situation would be a hopeless task.

A second and more general approach to the image-matching problem bypasses the need for a noise and distortion model by accepting an embedding metric without requiring its justification as being equivalent to a minimum error (or minimum risk) procedure. In fact, without a noise and distortion model, there is no theoretically valid way to derive or predict the error performance of a selected procedure prior to its actual application. Our primary concern in this paper is with this latter case.

We offer the following two sets of criteria for an embedding metric not theoretically derived on the basis of error performance. First, it must be successful in application (i.e., its observed error performance must be acceptable), intuitively satisfying (so we can have some confidence in its ability to deal with as yet untried applications), and general enough so that it can be employed over a wide range of problems without significant modification. Second, it must be possible to specify a computationally feasible decision algorithm which selects a suitable embedding based on the given embedding metric. (The combination of embedding metric and corresponding decision algorithm will be called an embedding model.) In the remainder of this paper, we will present an embedding metric and corresponding decision algorithm, present examples of the application of this embedding model to a number of distinct problem areas, and show how this embedding model is relevant to the problem of scene representation as well as to image matching.

An Embedding Metric

To introduce the generic form of our embedding metric, and establish its intuitive validity, let us first consider the following process. Assume that the reference is an image on a transparent rubber sheet. We move this sheet over the sensed image and, at each possible placement, we pull or push on the rubber sheet to get the best possible alignment between the reference image on the sheet and the underlying sensed image. We evaluate each such embedding both by how good a correspondence we were able to obtain and by how much pushing and pulling we had to exert to obtain it.

Let us now consider a discrete version of the above process which is both more precise and more reasonable from an implementation standpoint. In a specific application, we might have some information on the range of permissible distortions that can occur between the reference and sensed images. For instance, some subset of the items appearing in the sensed image might always retain their internal shape even though their relative positions might be subject to change with respect to their locations in the reference scene. Further, where change of relative position is possible, we might be able to bound the extent of such change; and, finally, we might like to assign variable "costs" to the different types of change of relative position or relative change in some nongeometric attribute.

To achieve these capabilities, we replace the rubber sheet by a reference image which is composed of a number of rigid pieces (components) held together by "springs." A rigid piece of the reference image can be as small as a single resolution cell, or as large as the entire reference image, and corresponds to a single coherent entity in the reference image. The springs joining the rigid pieces serve both to constrain relative movement and to measure the "cost" of the movement by how much they are "stretched." (Typically, the springs will be highly nonlinear in their behavior.) In determining the cost of an embedding, we measure the "tension" on each spring (the tension can be a function of direction as well as stretch or even a relative change in some locally defined attribute), and also make a local evaluation of how well each coherent piece is embedded as an independent entity.

The above model permits two interesting dichotomies. The first dichotomy is the separation of "syntactic" and "semantic" information. The semantic information, which is application dependent, is embodied in the specific partitioning of the reference into coherent pieces, the placement and cost functions assigned to the springs, and the cost functions associated with the independent embedding of the coherent pieces. The syntactic information, which is relatively independent of the particular application, defines the class of descriptions which the algorithm can process. These data are embodied in the limits set on reference decomposition (e.g., number and maximum size of pieces, etc.); in the formats which must be employed to specify the global constraints and costs; and in the form of the embedding metric which evaluates "global" fit. The separation of semantic and syntactic information is essential to permit application of the model to a broad range of problem areas without the necessity of making significant changes in the implementation.

The second dichotomy is the separation between the

local and global evaluation functions. The global evaluation function, associated with the relative positioning of the coherent pieces as described previously, has strong syntactic controls on its form to permit its integration directly into the decision algorithm. This is important because the global evaluation produces the most severe combinatorial problems. A local evaluation function, associated with how well a given coherent piece is independently embedded, is easily changed from problem to problem (based on problem-dependent considerations) without requiring any change in the core algorithms. Thus, the form of a local evaluation function can be a (conventional) correlation function together with a pictorial reference component, or a procedure based on linguistic concepts together with a formal description of a reference component,[1] or even a series of guesses inserted interactively by a human evaluator. The decoupling of the local evaluation functions from the core algorithms provides a great deal of flexibility in making changes or improvements in the evaluation functions for a given problem, as well as when switching from problem to problem. Further, because of the above separation, the performance of the algorithms (both local and global) can be independently evaluated in a direct and intuitively obvious manner. Such an evaluation then permits iterative improvement in performance by selective alteration in the problem-dependent options.

We are now in a position to present formally the proposed embedding metric. Let the reference be composed of p components (i.e., p coherent, or primitive, pieces). For $1 \leq i \leq p$, let x_i be a variable ranging over the set of all locations of the sensed scene. x_i is defined to be the postion of the ith component. Suppose there is a mechanism, either a computer program, or possibly a person, or some mechanical device, which, for location x_i of the ith component, outputs a numerical value $l_i(x_i)$ that indicates how strongly the ith component fits at location x_i of the sensed scene. The smaller $l_i(x_i)$, the better the fit.

While not formally required, the intent is that $l_i(x_i)$ measure the presence of the ith component at a location in the sensed scene independent of any knowledge of the locations of the other components. That is, $l_i(x_i)$ is a purely local and possibly imprecise measure of the presence of the ith component at location x_i.

In addition to the purely local measure l_i, $1 \leq i \leq p$, there are the following considerations: 1) how well the different components are situated in the required spatial relations to each other; and 2) how *relative* values of attributes of the components compare with the corresponding measured values in the sensed image (e.g., we might want to specify that the ith component be thicker and more greenish than the jth component). The extent to which the above specifications are not satisfied is reflected in the "stretching" of the springs between the corresponding components.

[1] Note that we are now further generalizing the concept of "component." It no longer has to be a rigid entity defined pictorially, but rather may be any information structure or decision procedure which can be used to define a real-valued function whose domain of definition is the set of all locations in the sensed image.

Each location in the sensed image can be associated with a two-dimensional vector (e.g., the components of the vector can be the row and column number of the location in the sensed scene). In that case, $x_i - x_j$ (usual vector subtraction) is a vector pointing from x_j to x_i. We can now let $g_{ij}(x_i, x_j) = g_{ij}(x_i - x_j)$ be the cost associated with the spring joining the ith and jth components. If there is no spring between these components, then g_{ij} is identically zero.

If we set $g_{ij}(x_i, x_j) = l_i(x_i)$ when $i = j$; and let $X_i = \{x_i, x_2, \cdots, x_i\}$, then the total cost of embedding p components at locations X_p is $G(X_p)$.

$$G(X_p) = \sum_{i=1}^{p} \sum_{j=1}^{i} g_{ij}(x_i, x_j). \tag{1}$$

Expression (1) can also be written as

$$G(X_p) = \sum_{i=1}^{p} h_i(X_i) \tag{2}$$

where

$$h_i(X_i) = \sum_{j=1}^{i} g_{ij}(x_i, x_j).$$

$h_i(X_i)$ can be thought of as the cost of embedding the ith component at location x_i, given that the previous $i-1$ components are at the locations specified by X_i.

Computational Procedures

In this section of the paper, we will present computational procedures for locating a suitable embedding of one image in another, based on the embedding metric just presented. A discussion of dynamic programing (DP) is included to place our proposed algorithm [the "linear embedding algorithm" (LEA)] in proper perspective. In particular, a generic (but computationally impractical) approach to solving the embedding problem is some form of DP. The specific form of our embedding metric permits a simplification of the general DP formulation, and the LEA is offered as a computationally feasible approximation to this restricted DP formulation. A graph theoretic interpretation is included to provide a better intuitive appreciation of the LEA in relation to DP.

Let us assume that the sensed image, designated by the abbreviation SM, is composed of M resolution elements; while the reference, designated by the abbreviation RM, is composed of P pictorially defined components (coherent pieces) with a total of $N = \sum_{i=1}^{P} n_i$ resolution elements, n_i being the number of resolution elements in the ith component.

The most direct procedure for locating a best embedding is to select combinationally N resolution elements at a time from the SM, determine if each such selection satisfies the coherent (intracomponent) and global (intercomponent) constraints, and, if acceptable, then evaluate the embedding metric for the given selection. Obviously, such an approach is completely impractical even for small pictures. For example, a 50×30 SM ($M=1500$) and a 5×5 RM ($N=25$) would require more than 10^{54} selections and evaluations, a hopelessly large number. If we assume that the coherent and relational constraints provide us with $P=6$ nonoverlapping components, each component sequentially constrained to stay within $w=10$ locations referenced to the location of the previously placed component, then we would still be required to perform on the order of $1500\times 10^5 = 1.5\times 10^8$ evaluations. Assuming 10^{-3} s per evaluation, we would require 1.5×10^5 s or approximately two days of computation time. It is thus obvious that a more effective technique is required.

Dynamic Programming

Expressed in formal terms, the evaluation of the embedding metric for a typical picture results in a nonlinear, integer programming problem with local optima different from the global optimum (i.e., no particular regularity, such as unimodality). The only available class of computational procedures for finding the global optimum under the above conditions (other than the exhaustive search techniques discussed earlier) is usually designated by the generic name dynamic programming.

DP is a multistage or iterative optimization procedure which can be described in general terms as follows (see [6] and [7]). We wish to find

$$\min_X G(X) = \sum_{i\in I} h_i(X^i) \tag{3}$$

where $X=\{x_1, x_2, \cdots, x_p\}$, each x_i, $1\le i\le p$, ranges over a set of vectors with discrete components, $I=\{1, 2, \cdots, p\}$, and X^i is the set of those variables (among X) upon which h_i depends. $\{h_i\}$, $1\le i\le p$, are a given set of real-valued functions.[2]

Let $|X^i|$ be the number of variables in X^i, and let each x_i, $1\le i\le p$, range over M values. Then each component $h_i(X^i)$ of the cost function is specified by means of a table with $|X^i|+1$ columns and $M^{|X^i|}$ rows.

The solution proceeds as follows. We select a variable $y_1\in X$ and compute the following expressions (this gives us the minimization of G with respect to y_1):

$$f_1(\Gamma(y_1)) = \min_{y_1} \sum_{i\in I_1} h_i(X^i) \tag{4}$$

$$y_1^*(\Gamma(y_1)) = \text{the value of } y_1 \text{ which minimizes expression (4)} \tag{5}$$

where $\Gamma(y_1)$ is the set of all those variables (except y_1) which occur in any one of those X^i which contains y_1. In other words, $\Gamma(y_1)$ is the set of variables which interacts with y_1. I_1 is the set of those i such that X^i contains y_1. y_1^* is the optimizing assignment for y_1 as a function of the variables of $\Gamma(y_1)$.

The operation described by (4) is called the elimination of the variable y_1 and results in the following transformation of (3):

$$\min_X G(X) = \min_{X-\{y_1\}} \left[f_1(\Gamma(y_1)) + \sum_{i\in[I-I_1]} h_i(X^i) \right]. \tag{6}$$

Expression (6) has the same form as (3), but does not contain y_1. Thus, we can find an optimal assignment for X be sequentially "eliminating" all of the variables, and then tracing back through the stored tables of $y_i^*(\Gamma(y_i))$, where $\Gamma(y_i)$ can only contain y_j such that $j>i$. That is, we must eventually reach a point where, for some s,

$$\bigcup_{j=1}^{s} I_j = I$$

and expression (6) has the form

$$\min_X G(X) = \min_{\{y_{s+1},\cdots,y_p\}} [f_s(\overbrace{y_{s+1}, \cdots, y_p}^{\Gamma(y_s)})]. \tag{7}$$

From (7) we can directly determine the global minimum cost for G and also the optimizing values for $Y=(y_s, y_{s+1}, \cdots, y_p)$. Given the value for Y, we can determine the value of y_{s-1}^* from the stored table for $y_{s-1}^*(\Gamma(y_{s-1}))$, as indicated in expression (5). This "backward" recursive process is continued to provide us with the complete optimizing assignment for X.

The computational feasibility of the DP approach depends on storage and computing time requirements. For a given objective function (in our case, the embedding metric), storage and computing time requirements are a function of the order in which the variables are eliminated [3], [7] and the dimensionality of each of the eliminated variables. We will say that variable y_i has dimensionality $|\Gamma(y_i)|$, where $|\Gamma(y_i)|$ is the number of variables in the set $\Gamma(y_i)$ as defined following (4) and (5). For many of the problems we shall be concerned with, the dimensionality will be essentially constant over all variables (i.e., a constant number of springs attached to each component and a symmetric interconnection topology) and relatively independent of order of elimination. Let us associate the dimensionality of the variables with the embedding function itself employing the designation $D(G)$ to denote the maximum dimensionality of any of the variables to be eliminated for a given order of elimination.

Where the dimensionality of all variables is constant for a given order of elimination, the complete embedding procedure will involve the iterative application of (4) and (5) $[p-D(G)]$ times, to evaluate the p arguments of G corresponding to the embedding of the p components. In the kth iteration,[3] we compute and store a

[2] The $\{h_i\}$ defined in (2) are independent of all x_j for $j>i$; the $\{h_i\}$ defined in the general DP formulation can be a function of any x_j.

[3] The kth iteration evaluates the "cost" of embedding the kth component at y_k^*, given some specific embedding of the components associated with the variables in $\Gamma(y_k)$. This evaluation corresponds to the elimination of y_k.

table for f_k and $y_k{}^*$. The number of lookups required for the construction of this kth table will be proportional to the number[4] of its rows $[M^{D(G)}]$ and (for each row) the constrained number of feasible locations of the variable being eliminated (denoted w_k, where $w_k \leq M$). Thus we have a storage requirement of $2 \cdot M^{D(G)}$ elements per table (the two entries stored in each row of the table are values of f_k and $y_k{}^*$), and a computation requirement of up to $M^{D(G)+1}$ lookups (with one or more additions and comparisons per lookup).

If $w_k = w$ for all k, and all variables have dimensionality $D(G)$, the complete embedding procedure then has a computation requirement proportional to

$$[p - D(G)](w)[M^{D(G)}] \quad \text{lookups} \tag{8}$$

transient (fast store) storage requirement proportional to

$$2 \cdot [M^{D(G)}] \quad \text{entries} \tag{9}$$

and a static (secondary store) storage requirement (for the "backward" selection of the y_i after obtaining the minimum global cost) proportional to

$$[p - D(G)][M^{D(G)}] \quad \text{entries.} \tag{10}$$

For the earlier example where $M = 50 \times 30 = 1500$ locations, (for $1 < k \leq P$) $w_k = w = 10$ locations, and $p = 6$ components, if these components are connected in a linear sequence where each interior component has only one spring attached to each of its two immediate neighbors and the end components only one spring each, the number of computations[5] would involve 75×10^3 lookups and evaluations, or 75 s of computing time at 10^{-3} s per computation. This is certainly a much more reasonable requirement than the two days of computing time for a direct evaluation. We pay for this speedup[6] by having a fast storage requirement of 3×10^3 entries (or words), and backup storage requirement of 7.5×10^3 entries, versus a storage requirement of only a few entries for the direct evaluation.

Now, however, note that, if we permit just one additional spring per component (or even a single spring linking the first and last components), $D(G) = 2$ and the number of computations increase by a factor of almost $M = 1500$ to 9×10^7 lookups and evaluations, or 9×10^4 s $= 25$ h. The fast storage requirement increases by a factor of $M = 1500$ to 4.5×10^6 entries, and the backup storage requirement increases to 9×10^6 entries.

This exercise demonstrates that, for even a small increase over unit dimensionality, the utility of dynamic programming as a computational technique is questionable for the embedding task. The next subsection introduces a heuristic modification to the DP type of sequential optimization which eliminates the growth of dimensionality as more global constraints (springs) are permitted.

[4] Relational constraints can reduce the number of feasible rows to a value considerably below the unconstrained case. However, the computational problems, in attempting to take "advantage" of this reduction, may be prohibitive.

[5] Assuming the variables are eliminated in the order in which they appear in the linear sequence, then $D(G) = 1$.

[6] Dynamic programming can be considered to be a way of trading storage for computation time.

Linear Embedding

The sequential embedding technique which we present in this section will be called the linear embedding algorithm (LEA). The essential property of this algorithm is its ability to locate a suitable embedding with a linear, rather than exponential, growth of storage and computing time requirements as a function of the number of components in the reference. The algorithm is formally described as follows.

Given a reference with p components, for $1 \leq i \leq p$, $l_i(x_i)$ is the externally supplied local evaluation array for the ith component as a function of x_i, its embedded location. Given that the ith component is embedded in location x_i, $w_i(x_i)$ is the constrained set of feasible locations for x_{i-1}.

If y is the sequence (1, 3, 2, 5), then $y * 8$ will be another way of writing (1, 3, 2, 5, 8). $S_i(x_1, \cdots, x_i) = \sum_{j=1}^{i-1} g_{ij}(x_i, x_j)$, where each g_{ij} is an externally supplied spring array, specified as a function of the relative embeddings of the ith and jth components. This decomposition of S_i into a sum of two-place functions (springs) is not required in the following LEA. Thus, if desired, we could extend the scope of the embedding metric to include more complex relational forms without increasing the computational complexity of the embedding algorithm (LEA).

The LEA is the computation, in order, of the following sequence of $2p+2$ equations. (Note that $h_i = s_i + l_i$.)

$$g_1(x_1) = l_1(x_1) \tag{11}$$

$$y_1(x_1) = x_1 \tag{12}$$

$$g_2(x_2) = \min_{x_1 \in w_2(x_2)} [S_2(x_1, x_2) + l_2(x_2) + g_1(x_1)] \tag{13}$$

$$y_2(x_2) = y_1(x_1) * x_2, \quad \text{where } x_1 \text{ is that value which minimizes the previous equation.} \tag{14}$$

$$\vdots$$

$$g_i(x_i) = \min_{x_{i-1} \in w_i(x_i)} [S_i(y_{i-1}(x_{i-1}) + x_i) * l_i(x_i) + g_{i-1}(x_{i-1})] \tag{15}$$

$$y_i(x_i) = y_{i-1}(x_{i-1}) * x_i, \quad \text{where } x_{i-1} \text{ is that value which minimizes the previous equation.} \tag{16}$$

$$\vdots$$

$$g_p(x_p) = \cdots \tag{17}$$

$$y_p(x_p) = \cdots \tag{18}$$

$$G = \min_{x_p} g_p(x_p). \tag{19}$$

$$Y = y_p(x_p), \quad \text{where } x_p \text{ is that value which minimizes the previous equation.} \tag{20}$$

As in the subsection on dynamic programming, G [see (19)] is the total embedding cost, and the corresponding locations of the embedded components are determined from Y [see (20)].[7]

Given the restriction that the h_i in (3) are independent of all x_j for $j>i$ [this is consistent with the embedding metric as presented in (2)], then dynamic programming can be viewed as a procedure for finding the shortest path through a graph. We define this graph in the following way. The nodes of the graph are arranged in p columns, where each node in the ith column is labeled by the values of the variables corresponding to a unique set of embeddings for the first i components; there are as many nodes in the ith column as there are unique embeddings of the first i components. The length of a branch between a node in column $i-1$ (with label X_{i-1}) and a node in column i (with label X_i) is $h_i(X_i)$. We delete branches with infinite length.

To determine the shortest path from the root node (a single node placed in column zero) to some node in column p, we can proceed as follows. In the ith column, for each node, sum all the branch lengths corresponding to the label of the node. We now determine that set of variables (x_i) which do not appear in any h_j for $j>i$, and call this set of variables Z_i (the variables in Z_i correspond to components in columns $j\leq i$ which do not have spring connections to components in columns $j>i$). We now place the nodes in the ith column into sets, such that, for each set, the nodes are identical in their labels except for the variables in Z_i. For each such set, we retain the node with the shortest path length from the root node, and delete all the other nodes in the set as well as those nodes in columns $j>i$ which branch from deleted nodes. It is this pruning process which gives DP its computational advantage over complete enumeration. After the above set of operations is carried out through the Pth column, we select the node defining the shortest path through the tree, and thus the lowest cost embedding for the P components. The LEA differs from DP in that in the sequential determination of the shortest path, a maximum of only m_i nodes will be retained in the ith column (m_i is the number of permissible locations for embedding the ith component). In processing the ith column, the nodes are grouped into sets such that, for each set, the nodes are identical in their labels for x_i. For each such set we then proceed as in the DP case. Thus, at the ith iteration we save only the m_i "best" current embeddings, such that every possible positioning of the ith component occurs in one of the embeddings. This approximation technique may fail to find the best embedding (shortest path) if the components with low indicies (small i) incur a high embedding cost when placed in their optimal locations.

If the components of the reference are linked by a single chain of springs (which are then called primary springs), DP and the LEA provide identical solutions. When additional springs are present, the LEA no longer assures the optimal embedding, and these additional springs are called hueristic springs (hueristic in the sense that while these additional springs provide more information and thus give an intuitively better match than in the single chain case, the best possible use of this additional information is not always assured). The operation of the LEA is illustrated by the example given in Fig. 1.

Additional Computation Speedup and Storage Reduction Techniques

By slowing the growth of the computation and storage requirements to a linear function of the size (M) of the pictures, the LEA establishes itself as a feasible embedding procedure. However, because of the large proportionality constants, the practicality of employing the LEA will probably depend on the efficiency of the associated computer programing, and on the employment of additional (second-order) speedup and storage reduction techniques. Two of the more important speedup techniques, applicable to both DP and the LEA, are discussed below.

The dynamic programming and LEA formalisms presented earlier do not explicitly consider constraints on the variables (x_i). The simplest way of treating such constraints is to introduce into the objective function (embedding metric) cost terms which become infinite if the relational constraints are violated. This approach handles the problem by increasing the computation time needed to evaluate the additional cost terms. A much more desirable technique is to employ the explicit relational constraints (the w_i) to limit both the table size and search requirements by ignoring or eliminating infeasible variable combinations. This technique is illustrated in [1]–[3], and was considered in deriving expression (8). When the relational constraints are implicit (i.e., must be computed from the given data), it is not clear whether any advantage can be gained from first converting them to explicit form, and then applying the above technique.

Dynamic programming and the linear embedding algorithm previously described can both be speeded up by the following (type of branch-and-bound) technique. We examine a number of complete embeddings, and select the lowest cost corresponding to any one of these trial embeddings. (The embeddings themselves could have been obtained by random placements of the com-

[7] The y_i defined in (16) clarify the presentation of the LEA. However, in a computer implementation one need not calculate these y_i, which are arrays, each element of which is a sequence of locations. Rather, it is only necessary to compute a more restricted function z_i defined by

$$z_i(x_i) = x_{i-1}$$

where x_{i-1} is that value which minimizes the previous equation

These z_i are arrays, each element of which is a single location rather than a sequence of locations. Then in the computations involving the S_i, if S_i depends on more than the two rightmost locations of $y_{i-1}(x_{i-1})*x_i$, the additional locations may be retrieved from the z_k, $1\leq k\leq i-1$.

SM

y				
4	5	2	8	8
3	7	5	1	3
2	8	1	5	7
1	4	3	2	4
	1	2	3	4 → z

x = (z, y)	SM (z, y)
1,4	5
2,4	2
3,4	8
4,4	8
1,3	7
2,3	5
3,3	1
4,3	3
1,2	8
2,2	1
3,2	5
4,2	7
1,1	4
2,1	3
3,1	2
4,1	4

(a)

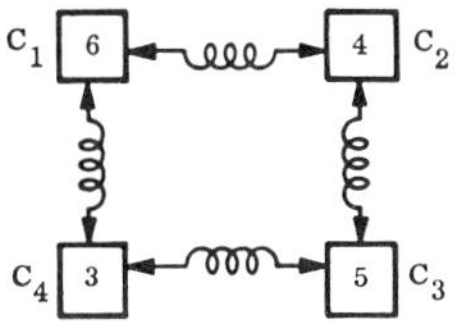

$C_1(1,1) = C_1 = 6$

$C_2(1,1) = C_2 = 4$

$C_3(1,1) = C_3 = 5$

$C_4(1,1) = C_4 = 3$

$\ell_i(z,y) = |SM(z,y) - C_i|$ for $1 \le i \le 4$

Spring definition when (i, j) = (2, 1) or (i, j) = (4, 3)

$x_i - x_j = (z_i - z_j,\ y_i - y_j)$	$g_{ij}(x_i - x_j)$
1,0	0
2,0	1
otherwise	∞

Spring definition when (i, j) = (4, 1) or (i, j) = (3, 2)

$x_i - x_j = (z_i - z_j,\ y_i - y_j)$	$g_{ij}(x_i - x_j)$
0,1	0
0,2	1
otherwise	∞

(b)

Evaluation of g_2

x_2 $z_2\ y_2$	x_1 $z_1\ y_1$	g_1 ℓ_1	ℓ_2	S_2 g_{21}	g_2
2 4	1 4	1	2	0	3
3 4	1 4	1	4	1	6
	2 4	4	4	0	
4 4	2 4	4	4	1	
	3 4	2	4	0	6
2 3	1 3	1	1	0	2
3 3	1 3	1	3	1	
	2 3	1	3	0	4
4 3	2 3	1	1	1	3
	3 3	5	1	0	
2 2	1 2	2	3	0	5
3 2	1 2	2	1	1	4
	2 2	5	1	0	
4 2	2 2	5	3	1	
	3 2	1	3	0	4

Evaluation of g_3

x_3 $z_3\ y_3$	x_2 $z_2\ y_2$	x_1 $z_1 y_1$	ℓ_3	S_3 g_{32}	g_2	g_3
2 3	2 4	1 4	0	0	3	3
3 3	3 4	1 4	4	0	6	10
4 3	4 4	3 4	2	0	6	8
2 2	2 3	1 3	4	0	2	6
	2 4		4	1	3	
3 2	3 3	2 3	0	0	4	4
	3 4		0	1	6	
4 2	4 3	2 3	2	0	3	5
	4 4		2	1	6	
2 1	2 2		2	0	5	
	2 3	1 3	2	1	2	5
3 1	3 2	1 2	3	0	4	7
	3 3		3	1	4	
4 1	4 2	3 2	1	0	4	5
	4 3		1	1	3	

Evaluation of g_4 = G

x_4 $z_4\ y_4$	x_3 $z_3\ y_3$	x1 $z_1\ y_1$	g_{43}	g_{41}	S_4	ℓ_4	g_3	g_4 G	
1 3	2 3	1 4	0	0	0	4	3	7	
	3 3	1 4	1	0	1	4	10	15	
2 3	3 3	1 4	0			2			
	4 3	3 4	1			2			
3 3	4 3	3 4	0	0	0	2	8	10	
1 2	2 2	1 3	0	0	0	5	6	11	
	3 2	2 3	1			5			
2 2	3 2	2 3	0	0	0	2	4	6	← [(2, 3), (3, 3), (3, 2), (2, 2)]
	4 2	2 3	1	0	1	2	5	8	
3 2	4 2	2 3	0			2			
1 1	2 1	1 3	0	1	1	1	5	7	
	3 1	1 2	1	0	1	1	7	9	
2 1	3 1	1 2	0			0			
	4 1	3 2	1			0			
3 1	4 1	3 2	0	0	0	1	5	6	← [(3, 2), (4, 2), (4, 1)(3, 1)]

(c)

Fig. 1. An example illustrating the operation of the linear embedding algorithm. The definitions of x, g_{ij}, l_i are given on pages 69 and 71. z and y are the components of x; that is, $x = (z, y)$. (a) The sensed image. (b) The reference description. (c) Linear embedding algorithm.

ponents, or perhaps by someone guessing at what a "good" embedding might actually look like.) Now, employing the LEA (or DP) to place and evaluate the cumulative cost of placement of the components sequentially, we can eliminate from further consideration any of those placements whose costs exceed the bound established by our best trial embedding. It should be noted that this technique is valid only if the cost associated with the embedding of each component is nonnegative. However, this can almost always be the case for the class of problems we are discussing in this paper.

A heuristic embellishment of the above branch-and-bound technique would be to use some fraction (say $[k/N]^{\frac{1}{2}}$) of the bound at the kth stage of an N-stage process as the threshold for eliminating a possible sequence of embeddings.

Scale and Rotation (S&R) Considerations

In attempting to match or register two images, we frequently are faced with the problem of unknown relative scale and orientation. While such variations are conceptually indistinct from any of a host of unwanted variations between the reference and the image, they (S&R) can serve as a vehicle for clarifying some important issues pertaining to the way the LEA is employed.

As noted in earlier sections, the embedding process is carried out at two levels. First, the components of the reference are searched for as independent entities. The particular processes by which these searches are executed are not a direct issue of concern here; the important point is that, regardless of the search mechanism, the outcome of the search for any individual component is presented to the LEA by a tabulation called the local evaluation array [L(EV)A]. Each entry in the L(EV)A corresponds to a possible embedding in the image of the associated reference component, indexed by the variables used to define the embedding. The entry consists of a number related to the probability that the component is actually present at the "location" specified by the indexing variables, and each entry can also contain the values of attributes of the component as measured at the indexed location. The LEA has no knowledge of the component beyond what is presented to it in the L(EV)A for that component. The purpose of the LEA is to integrate global or structural knowledge with the information provided in the L(EV)A's to find the best overall embedding (or embeddings) of the reference in the image. The acceptability of the final embedding selected by the LEA will thus be dependent on the quality of the information presented in the L(EV)A's, where the extent of this dependence is related to the relative importance of local (component definition) versus global (intercomponent) information for the particular problem. Thus, it is the responsibility of the local evaluation function, in attempting to gather evidence about the presence of some given reference component, to be able to deal with the various noise and distortion processes (such as S&R) which might be encountered. To the extent that these same noise and distortion processes affect the global or structural relationships between the reference components, the LEA provides the machinery necessary to deal with the resulting problems.

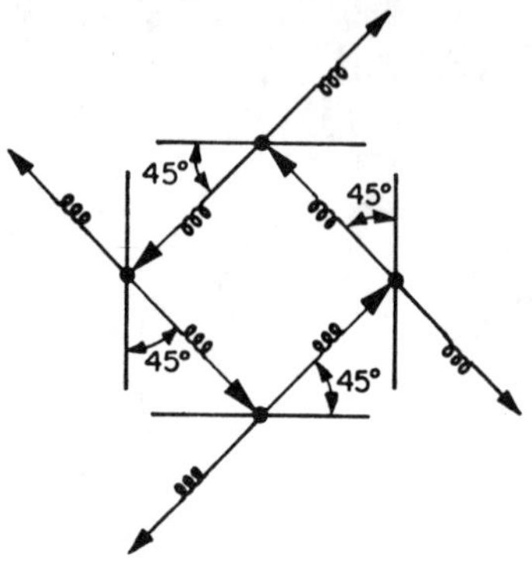

Fig. 2. Reference description of a square.

Ability to deal with variations at the global level is accomplished by defining "attributes" which measure (or estimate) these variations, and then making the "spring" parameters functions of these attributes.

Thus, in the case of S&R, if a component P_1 has scale and rotation attributes S_1 and θ_1, the springs (vectors) attached to P_1 would be (conceptually) scaled as a function of S_1, and rotated as a function of θ_1. The following example illustrates some of the above comments.

Problem

Given a two-dimensional region in which there are k randomly oriented and positioned line segments, find the four-line segments which best approximate a square. Each line is specified by a four-tuple of the type (x, y, θ, l) where the x, y coordinates locate the center of the segment, θ specifies the orientation, and l the length of the segment. To simplify this example, we will ignore the detection problem and assume that the given values for each segment are known with probability one. We assign a cost C_1 for each unit of positional disparity between the sides of a candidate square, and a cost C_2 for each degree of rotational disparity between the sides (i.e., sides should meet at right angles).

Solution

Consider a single "local evaluation array" consisting of the given list of four-tuples (x, y, θ, l). We can consider x, y to be the "location" indexing variables, and θ, l to be attributes. All entries have unit probability, entries with zero probability are deleted. The "description" of a square is shown schematically in Fig. 2. Each (spring) vector is rigidly attached to its line segment at a fixed angle of 45°. Costs (C_1) associated with (x, y) disparity between the end of a vector and the center of the next line segment are circularly symmetric (i.e., the "cost" function increases with distance from the tip of the vector) about the tip of the vector. We also assess a cost (C_2) proportional to the difference in measured (attribute value) orientation of more or less than 90° for sequential line segments.

The general form of the LEA, with orientation adjusted springs, and spring costs augmented by attribute (S&R) differences, is adequate to deal with the problem as posed. A minor difficulty arises from the fact that the orientation of a line segment is actually two valued (i.e., θ and $\theta+180°$). If we list each line segment twice in the local evaluation array, once for each of its two orientations, then the LEA can be applied without modification.

We can handle squares of differing size by using the line-segment-length attribute in the same manner as the orientation attribute (except that double entries are not required here). Spring stretching is augmented by a cost proportional to the difference in length attribute for sequential line segments. That is, when the LEA examines a new line segment as a possible additional side for a square already partially formed, it compares the length of the new line segment with the length of the line segment in the partial square to which the new segment will be attached. The spring between the new and the old line segments then is stretched (over and above any stretching due to angular and positional disparity) by an amount proportional to the difference in these lengths.

In the above example note that, because each entry in the local evaluation array had either zero (∞ cost) or unit (0 cost) probability associated with its occurrence, the size of this array could be reduced to listing only those few coordinate combinations associated with the feasible (nonzero probability) occurrence of a component (line segment). This procedure can be used in other situations where it is reasonable to reduce all low probability entries in the local evaluation array to zero.

Pictorial Representation

A central problem in much of the work concerned with the computer processing of pictorial data is that of representation. Since we cannot manipulate the real world object (itself) within the computer, we attempt to construct a representation (or model) which can be used in place of the actual object and which has the following (somewhat overlapping) properties.

Complete: Any question of interest which could be resolved by reference to the actual object should also be capable of being resolved by reference to the representation.

Compact: The representation should be free of information redundant to the purposes for which it will be used. This is necessary to minimize computer storage requirements.

Transformable: Much of the information contained in a representation will be implicit rather than explicit in form. The ability to manipulate easily the representation to extract required information is essential. For example, if we represent a picture by an intensity matrix or raster, then a count of the number of isolated objects appearing in the picture would be implicit information which could be extracted from the representation after considerable processing. However, if the representation consisted of the contours of the object appearing in the picture, then the required count could be obtained rather simply.

Incrementally Changeable: If we observe a slight change in the real world object, it should be a relatively simple and straightforward task to alter the representation. Further, from the standpoint of image matching, a small change in the real world should require only a small change in the representation.

Accuracy and Simplicity of Translation: Given a real world object, it should be relatively simple to derive an accurate representation of the object.

Over the past ten years or so, much of the work concerned with pictorial representation has been restricted to the domain of line type drawings, and the use of formal linguistic methods (see [8]–[10]). Very little success has been achieved in attempts to extend this work to scenes of terrains, cloud covers, human faces, etc., which can only be described meaningfully in terms of picture components which are not line elements, but which are regions with colors, textures, shadings, etc.[8]

Perhaps the most serious failing of the linguistic (and similar) techniques occurs with respect to the "translation" property. These techniques build a representation by constructing a hierarchy based on picture primitives, assembled into linear expressions employing specified relational forms, and satisfying a set of syntax rules. The problem arises from the fact that (usually) the only direct correspondence between the actual object and its representation occurs at the level of the primitive elements (typically points, intensities, and lines), while in practice there are pieces of a picture that are too involved or complicated to describe in terms of these primitives. Theoretically, the description is possible since the matrix representing the picture is finite. However, such a description would be so complicated that, aside from the difficulty of composing it, there would be a considerable likelihood of error and inaccurate representation.

In the previous portions of this paper we have presented a representational scheme for pictures, and were primarily concerned with its application to image matching. We will now show that the representational scheme has wide general applicability, and avoids many of the problems of the linguistic approaches. First we note that it is a hybrid type of representation in that it invokes symbolic (numerical) elements as well as allowing actual picture segments to be part of the representation. The ability to intermix picture segments and symbolic data in the same representation greatly simplifies

[8] Specialized systems have been developed, highly tailored to specific problems, which are exceptions to this assertion; e.g., see Kelly [11]. A number of papers, including Bledsoe [15], [16] and Goldstein and Harmon [17], effectively consider the problem of face identification based on feature measurements obtained manually.

the translation problem. Where a pictorial concept is difficult to describe symbolically, we can use an actual piece of the picture as part of the description. A second aspect of our representation that simplifies the translation problem is the fact that the components (picture pieces, local evaluation arrays, etc.) and the relational forms (springs) are two-dimensional rather than one-dimensional entities. We thus avoid the problem of having to construct a one-dimensional model for a two-dimensional structure.

Let us now examine some of the other representational attributes. Incremental changeability follows directly from ease of translation (although the inverse relationship would not necessarily hold). Transformability with respect to image matching has certainly been established; the fact that the representation is already in two-dimensional form, with metric, geometrical, and topological relationships explicitly and quantitatively expressed, implies that transformability for many other applications is more than adequate.

In many respects, compactness and transformability are antithetical since data in explicit form are usually more extensive than the equivalent implicit information. This is the case with our representation. It requires considerably more storage than might be required for a linguistic representation.

The one area where the linguistic approach has an obvious advantage is with respect to completeness. A linguistic representation can treat nonpictorial information (e.g., relations between items in a picture and other items not visible, but perhaps implied or normally associated with the pictured items) in a way that would be extremely difficult in our representation. This additional capability could, of course, be achieved by appending the necessary linguistic machinery to our current scheme, although the final result might well be a mixture of two representations rather than one integrated representation.

Experimental Results

In order to evaluate the practical implications of the techniques presented in this paper, we have initiated a program involving experiments on a variety of line type drawings and gray-level imagery. Over 400 experiments have already been performed with the following general results.

1) On well-defined imagery (i.e., relatively noise-free and unblurred pictures), the embeddings produced by the LEA were almost always in agreement with the best embedding as predetermined by human evaluators. Where the few deviations did occur, they were reasonable, and usually related to the crude component descriptions employed.

2) On noisy imagery (course resolution, additive random and coherent noise), the fall in performance paralleled the difficulty human evaluators had in locating suitable embeddings. Where the components were discernible in the image, the embeddings were usually correct; those components which were significantly altered by the noise were sometimes missed, but the substitution error was usually in close proximity to the correct embedding location, and, even in error, the correct embedding almost always had a score close to the best score.

Since the programmed version of the algorithm is still evolving, and some of the discussed features have not yet been implemented (e.g., the "attribute" feature is not operational as yet), most of the experiments were informal in nature. However, for the purposes of this paper, two sets of controlled experiments were run and are described below.

Image-Matching Experiments Using Faces

The majority of the experiments we have run to date had human faces as their subject. Reasons for this selection include the following.

1) The availability of a set of digitized gray-scale pictures containing faces.

2) A single reference (face) could be tested on all the faces in the data set. In the case of, say, terrain pictures, a separate reference (or, at best, a unique composition of standard reference components) is necessary for each picture.

3) Our familiarity with faces and their components (eyes, nose, mouth, etc.) facilitates evaluation of performance as noise and distortion are introduced.

The data set used in the face experiments consisted of 15 human faces,[9] both men (some with beards) and women, digitized to approximately 16 true gray levels, and each face typically was contained in a picture field of from 2000 to 3000 resolution elements. Using a reference as shown schematically in Fig. 3(a), with components as described in Fig. 3(b) and (c), almost 300 formal and informal experiments were performed. In each experiment, additive (truncated) Gaussian random noise with zero-mean and standard deviation of either 0, 10, or 15 units was added to each resolution element (relative to a pseudogray scale of 64 units for the noise-free pictures). In some of the informal experiments, coherent noise consisting of randomly placed lines was also inserted [see Fig. 4(a)].

With no more than two or three exceptions, when the reference was restricted to hair, eyes, and sides of face, correct embedding was achieved. These results are gratifying in view of the simple component descriptions employed, and the equivocation displayed by the resulting L(EV)A's. (See examples shown in Fig. 4.)

Two series of formal experiments were run on the

[9] These data were obtained from the Stanford Artificial Intelligence Laboratory and are a subset of the data employed in the experiments described by Kelly [11].

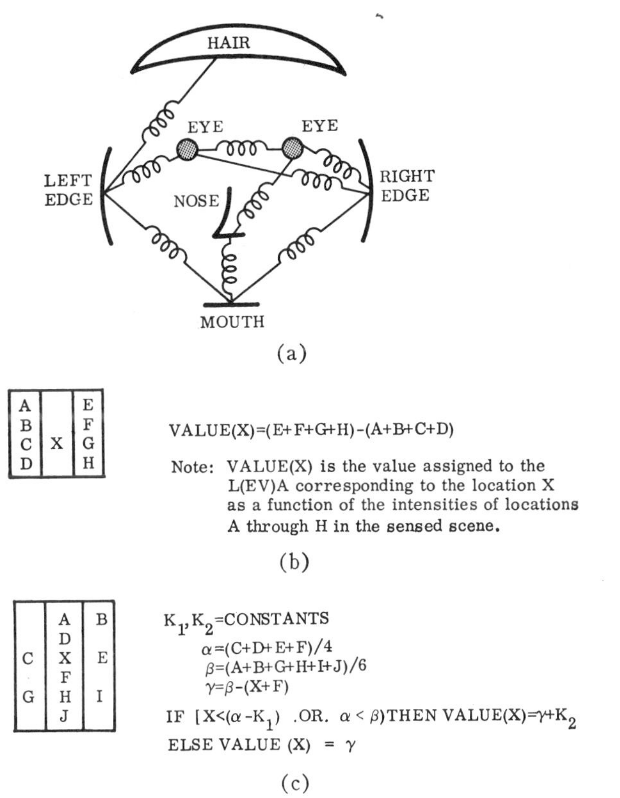

Fig. 3. Reference description of a face. (a) Schematic representation of face reference, indicating components and their linkages. (b) Reference description for left edge of face. (c) Reference description for eye.

(noisy) face pictures using two references which included, but differed in, the nose/mouth definitions. In the first series, consisting of 90 experiments, there were 83 completely correct embeddings, and 7 partially incorrect embeddings. The errors involved six experiments in which the nose/mouth complex was offset by three to four resolution cells from its ideal location, and one experiment in which both the eyes and the nose/mouth complex were improperly placed. In the second series, consisting of 45 experiments, the placement of the nose/mouth complex was judged incorrect in 3 experiments, while all the other components were always correctly embedded.

Analysis of the face experiments led to the following conclusions. In spite of almost perfect performance in embedding the hair, eyes, and sides of the face, precise placement of the nose/mouth complex based on strictly local evaluation was almost impossible in some of the noisy pictures due to loss of detail [e.g., see Fig. 4(b)]. With the attribute feature of the LEA not yet operational, and with the arbitrary decision to use binary (rather than multivalued) weights in the spring arrays for these experiments, the LEA restricted the feasible region over which an optimum value could be selected for embedding the nose/mouth complex, but did not bias the selection as would generally be the case. In the presence of heavy noise, the simple nose/mouth descriptions used in these experiments were not always adequate to produce a local optimum in the L(EV)A at or near the ideal embedding location. (A three-resolution cell deviation was considered an error.)

Image-Matching Experiments Using Terrain Scenes

Approximately 40 experiments have been performed using terrain scenes (including both aerial and ground scenes). The object in each case was to create a relatively simple description of some portion of the scene and then attempt to find the proper embedding of the description in the image (or some distorted or alternate view of the image).

The descriptions employed two basic types of components: 1) *texture components*, in which the "texture value" of a point was defined as a crude statistical function of the intensity values and gradients in some local region surrounding the point; and 2) *shape components*, which were defined by collections of "edge" points having specified gradients.

Fig. 5(a) shows an example of a terrain (reference) description. Fig. 5(b) shows its successful embedding relative to the computer-stored version of the photograph of the actual terrain segment as shown in Fig. 5(c). Each coherent piece in reference 5(a) is represented by several points enclosed by a dotted line. In this example, the points of each enclosure of the reference com-

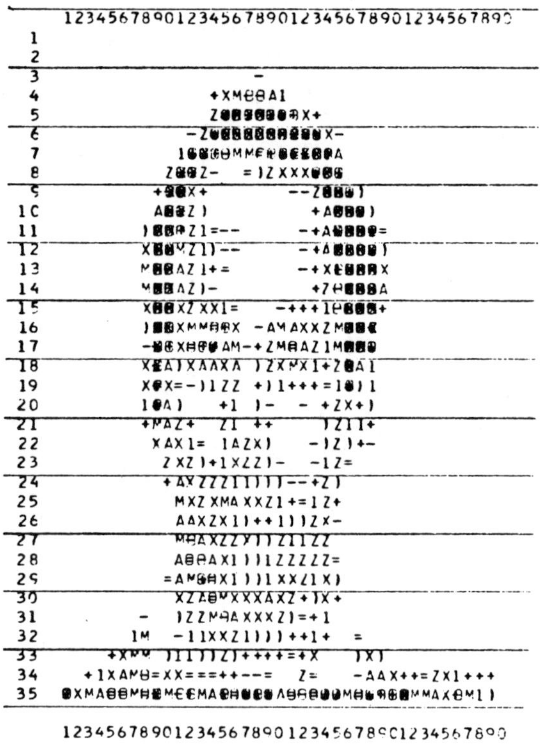

Original picture.

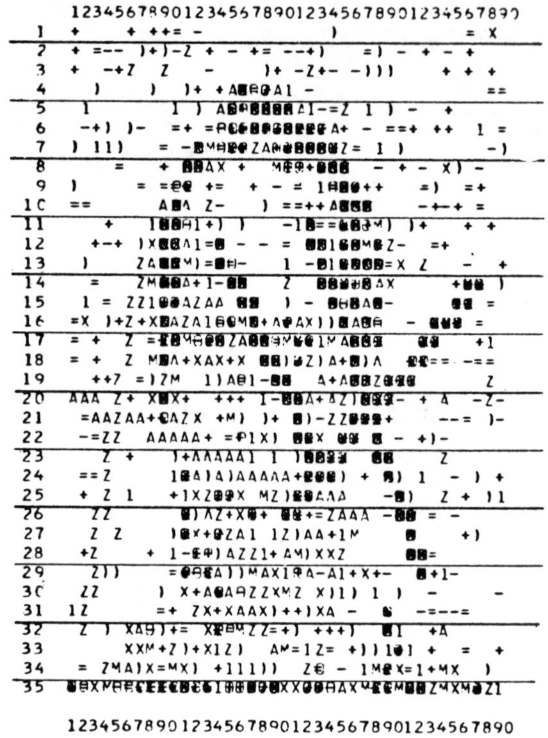

Noisy picture (sensed scene) as used in experiment.

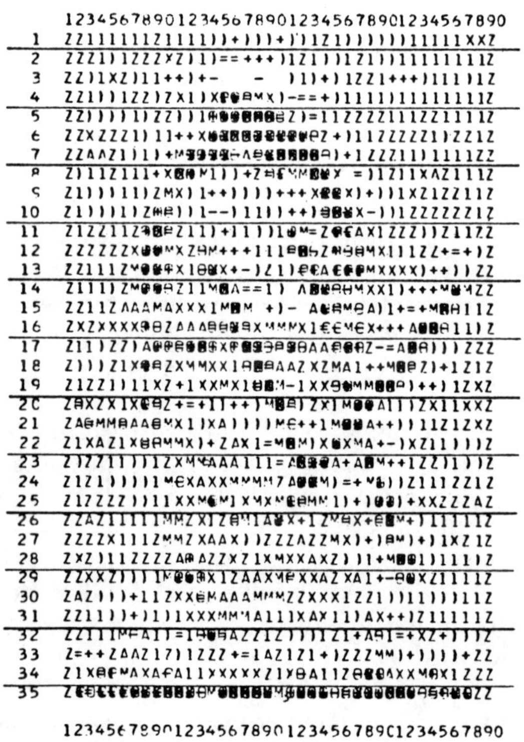

L(EV)A for eye. (Density at a point is proportional to probability that an eye is present at that location.)

HAIR WAS LOCATED AT (6, 18)
L/EDGE WAS LOCATED AT (18, 10)
R/EDGE WAS LOCATED AT (18, 25)
L/EYE WAS LOCATED AT (17, 13)
R/EYE WAS LOCATED AT (17, 21)
NOSE WAS LOCATED AT (22, 18)
MOUTH WAS LOCATED AT (24, 17)

(a)

Fig. 4. Examples of image-matching experiments using faces. (a) Successful embedding under coherent noise.

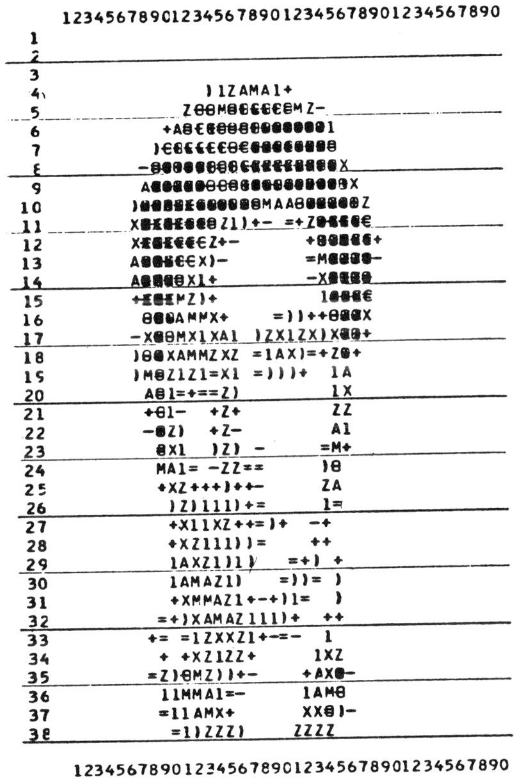

Original picture.

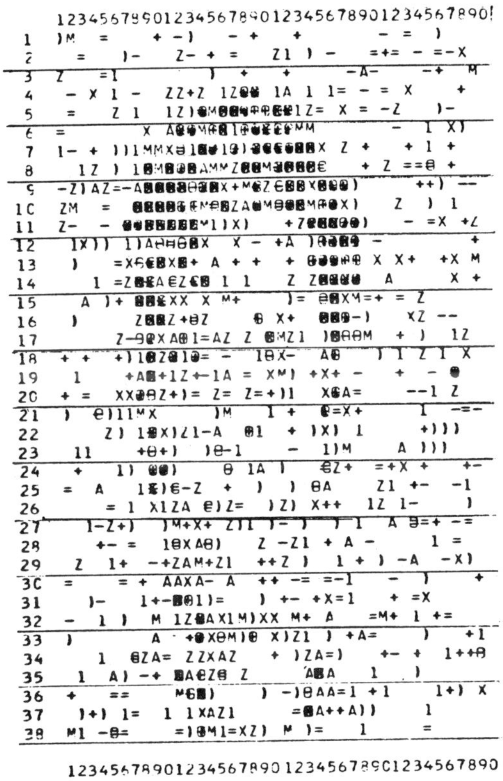

Noisy picture (sensed scene) as used in experiment.

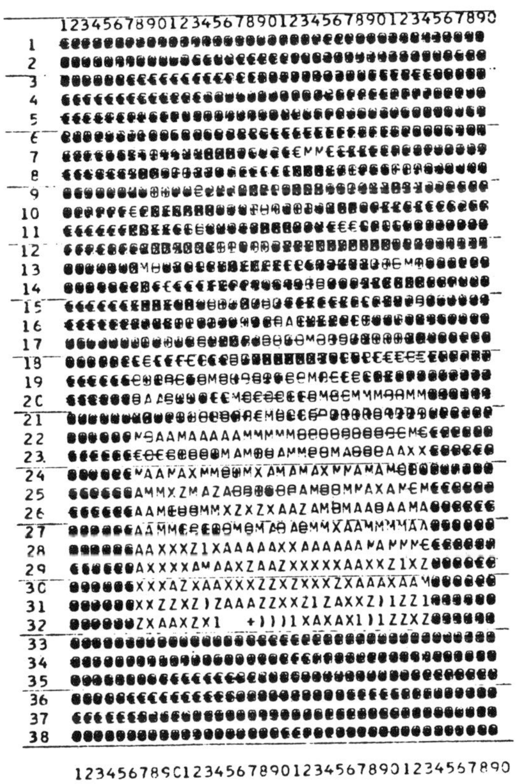

L(EV)A for nose. (Density at a point is proportional to probability that nose is present at that location.)

HAIR WAS LOCATED AT (8, 21)
L/EDGE WAS LOCATED AT (17, 11)
R/EDGE WAS LOCATED AT (17, 25)
L/EYE WAS LOCATED AT (17, 14)
R/EYE WAS LOCATED AT (17, 20)
NOSE WAS LOCATED AT (21, 16)
MOUTH WAS LOCATED AT (23, 16)

(b)

Fig. 4 (continued). (b) Incorrect embedding of nose under random noise.

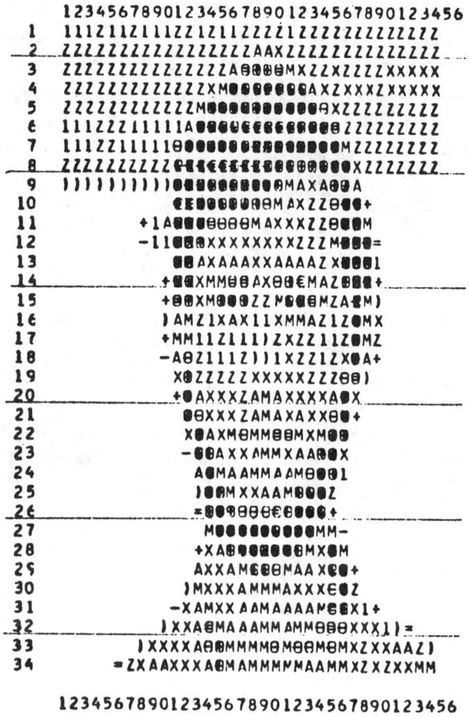

Original picture.

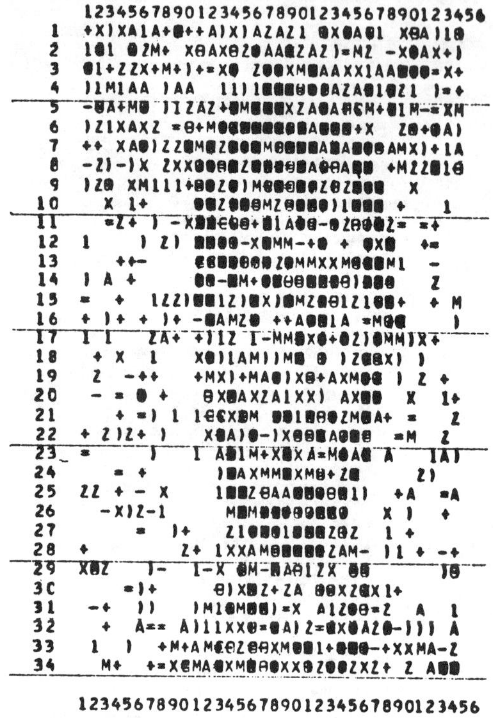

Noisy picture (sensed scene) as used in experiment.

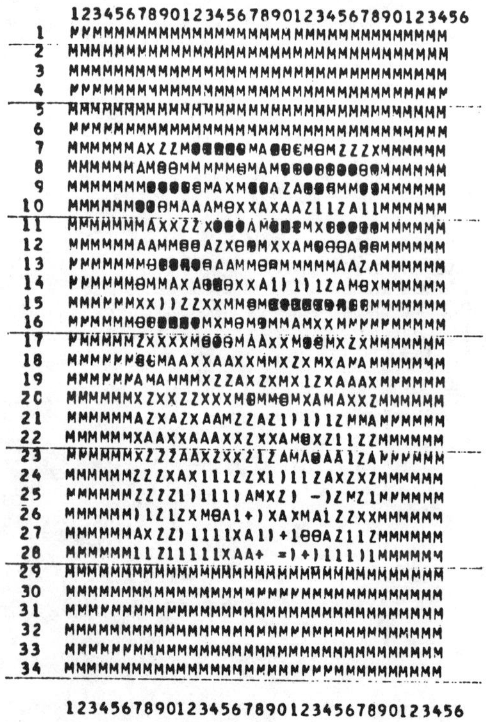

L(EV)A for nose. (Density at a point is proportional to probability that nose is present at that location.)

HAIR WAS LOCATED AT (7, 23)
L/EDGE WAS LOCATED AT (17, 13)
R/EDGE WAS LOCATED AT (17, 26)
L/EYE WAS LOCATED AT (14, 17)
R/EYE WAS LOCATED AT (14, 23)
NOSE WAS LOCATED AT (20, 20)
MOUTH WAS LOCATED AT (22, 20)

(c)

Fig. 4 (continued). (c) Successful embedding under random noise.

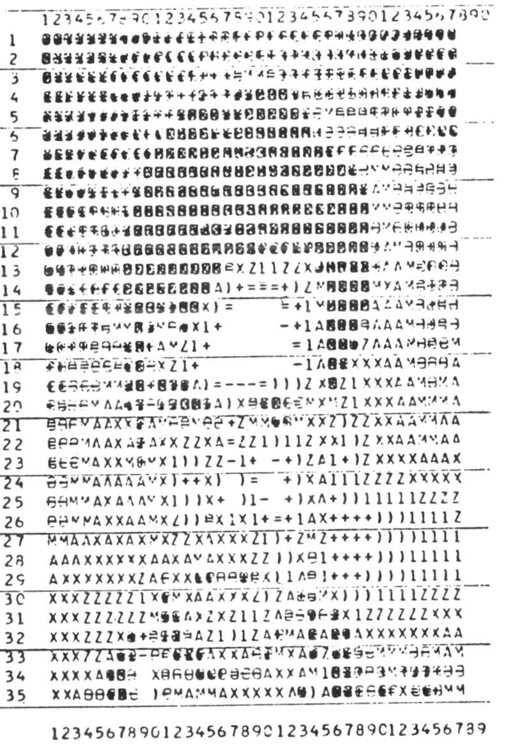

Original picture.

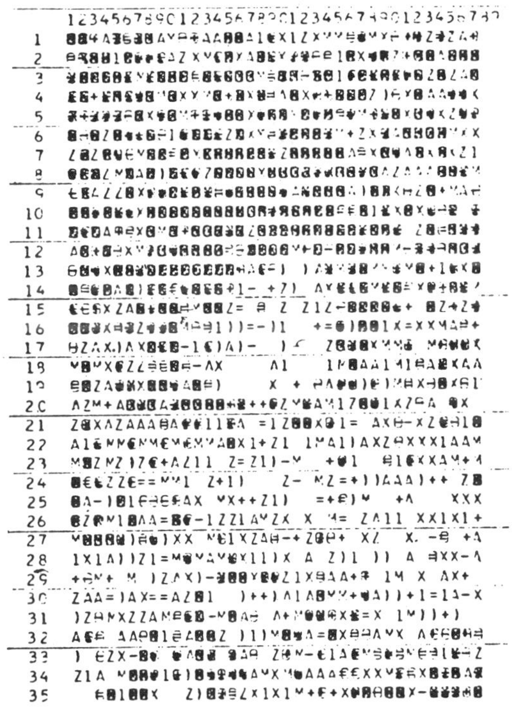

Noisy picture (sensed scene) as used in experiment.

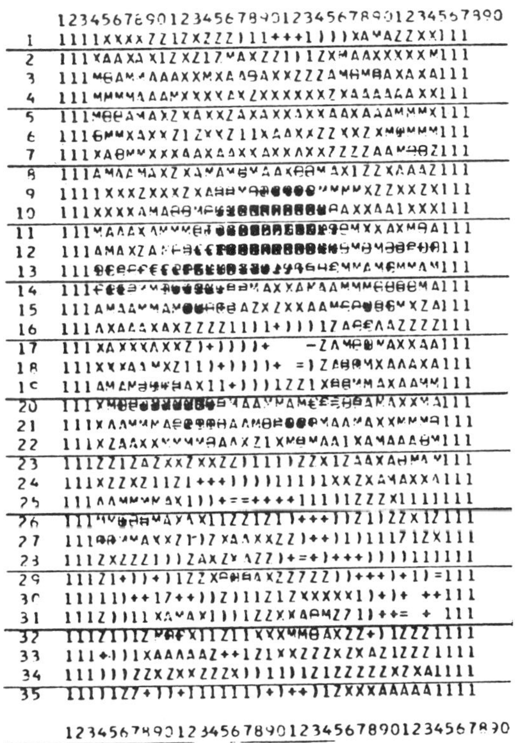

L(EV)A for hair. (Density at a point is proportional to probability that hair is present at that location.)

HAIR WAS LOCATED AT (11, 21)
L/EDGE WAS LOCATED AT (25, 11)
R/EDGE WAS LOCATED AT (25, 24)
L/EYE WAS LOCATED AT (21, 15)
R/EYE WAS LOCATED AT (21, 21)
NOSE WAS LOCATED AT (26, 18)
MOUTH WAS LOCATED AT (29, 17)

(d)

Fig. 4 (continued). (d) Successful embedding under random noise.

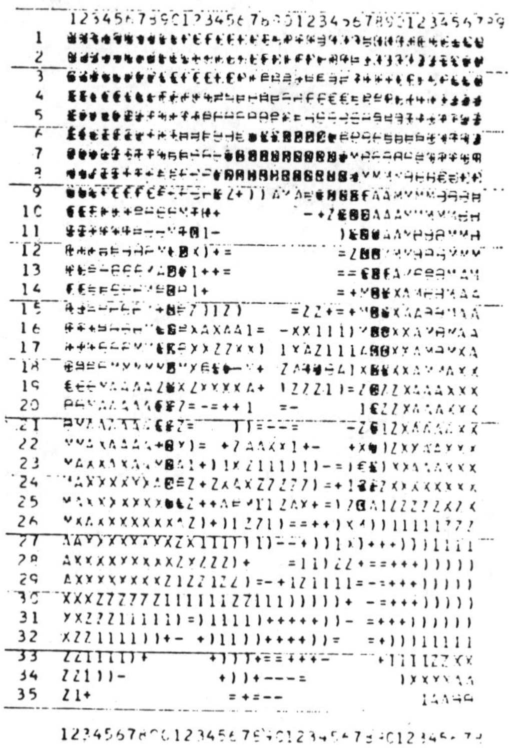

Original picture.

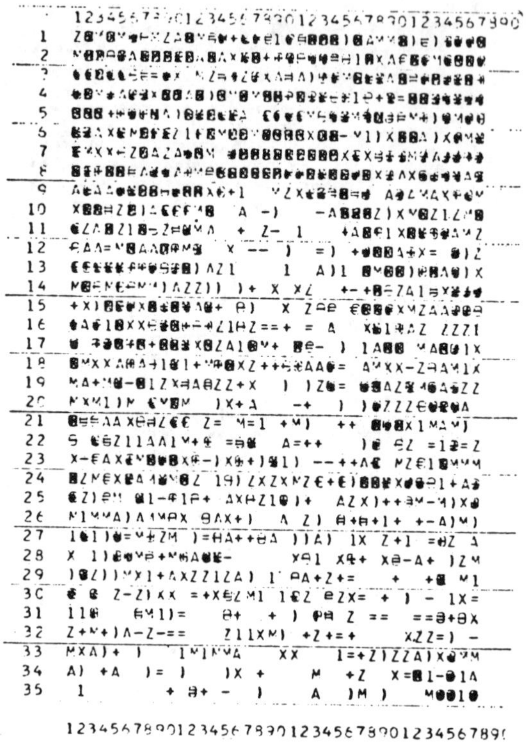

Noisy picture (sensed scene) as used in experiment.

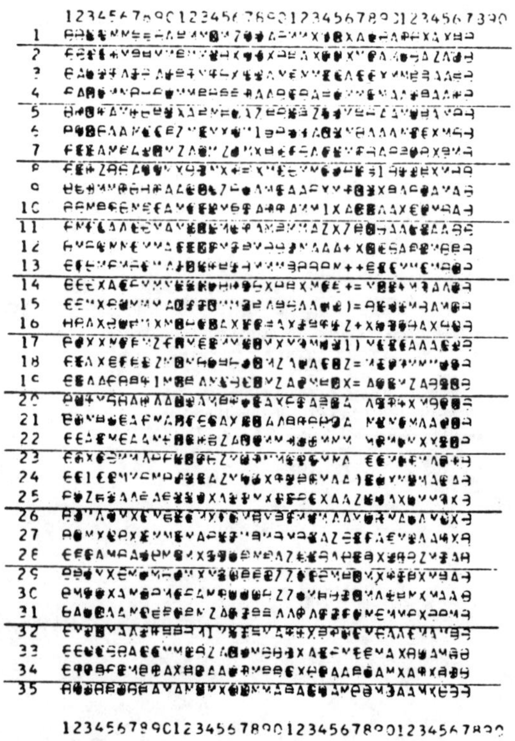

L(EV)A for L/EDGE. (Density at a point is proportional to probability that L/EDGE is present at that location.)

HAIR WAS LOCATED AT (8, 20)
L/EDGE WAS LOCATED AT (20, 11)
R/EDGE WAS LOCATED AT (20, 27)
L/EYE WAS LOCATED AT (18, 15)
R/EYE WAS LOCATED AT (18, 23)
NOSE WAS LOCATED AT (23, 18)
MOUTH WAS LOCATED AT (25, 18)

(e)

Fig. 4 (continued). (e) Successful embedding under random noise.

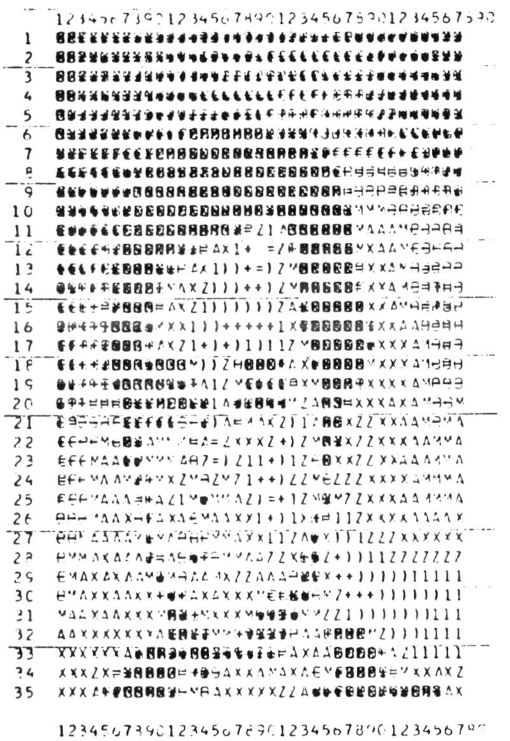

Original picture.

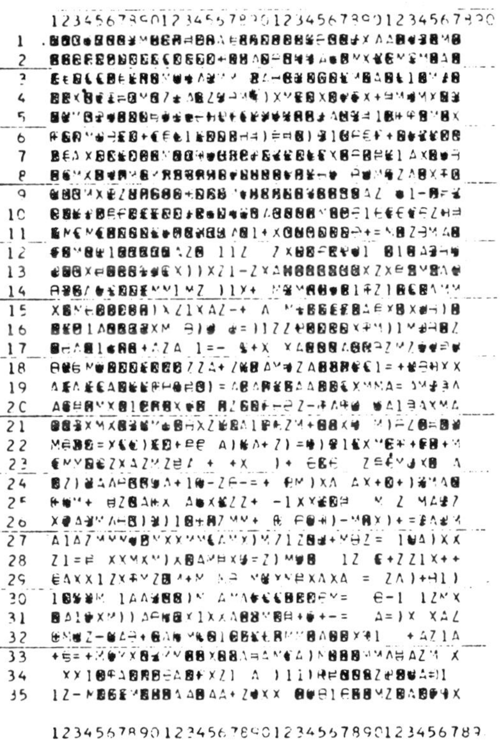

Noisy picture (sensed scene) as used in experiment.

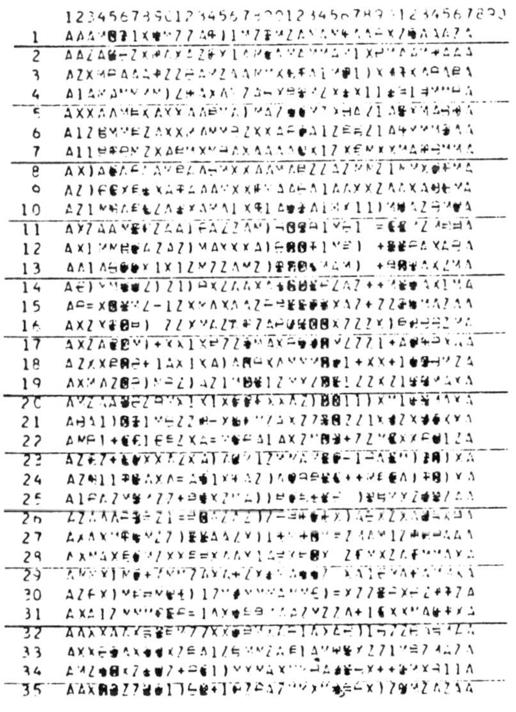

L(EV)A for R/EDGE. (Density at a point is proportional to probability that R/EDGE is present at that location.)

HAIR WAS LOCATED AT (11, 15)
L/EDGE WAS LOCATED AT (19, 8)
R/EDGE WAS LOCATED AT (19, 24)
L/EYE WAS LOCATED AT (19, 12)
R/EYE WAS LOCATED AT (19, 20)
NOSE WAS LOCATED AT (25, 15)
MOUTH WAS LOCATED AT (28, 15)

(f)

Fig. 4 (continued). (f) Successful embedding under random noise.

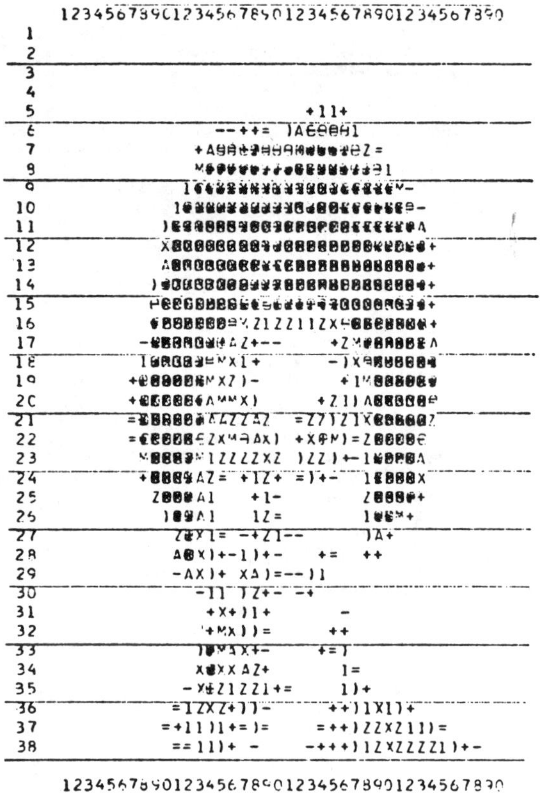

Original picture.

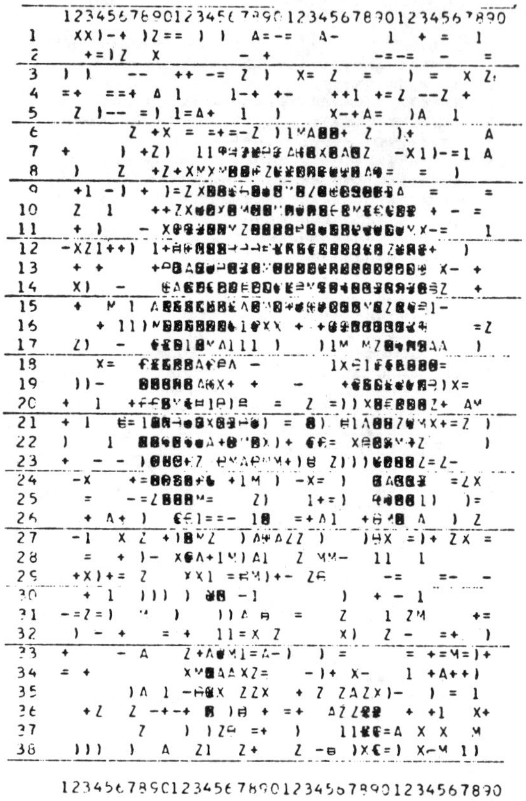

Noisy picture (sensed scene) as used in experiment.

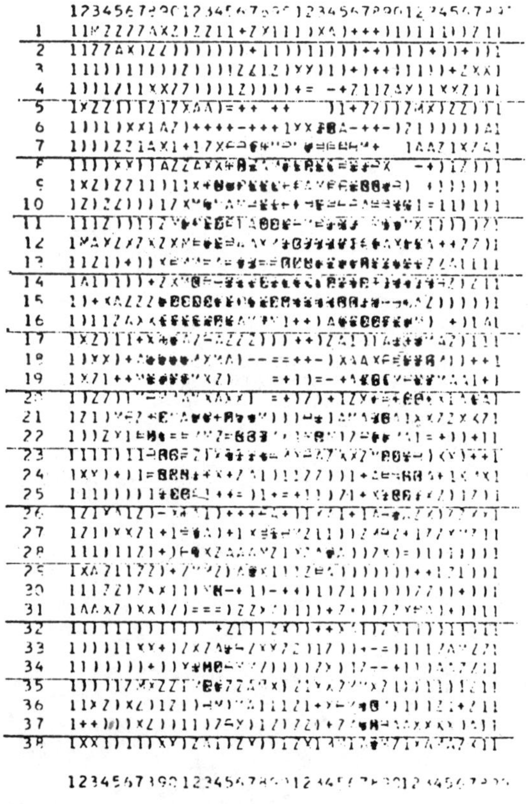

L(EV)A for eye. (Density at a point is proportional to probability that eye is present at that location.)

HAIR WAS LOCATED AT (13, 23)
L/EDGE WAS LOCATED AT (25, 13)
R/EDGE WAS LOCATED AT (25, 28)
L/EYE WAS LOCATED AT (22, 16)
R/EYE WAS LOCATED AT (22, 23)
NOSE WAS LOCATED AT (27, 20)
MOUTH WAS LOCATED AT (29, 19)

(g)

Fig. 4 (continued). (g) Successful embedding under random noise.

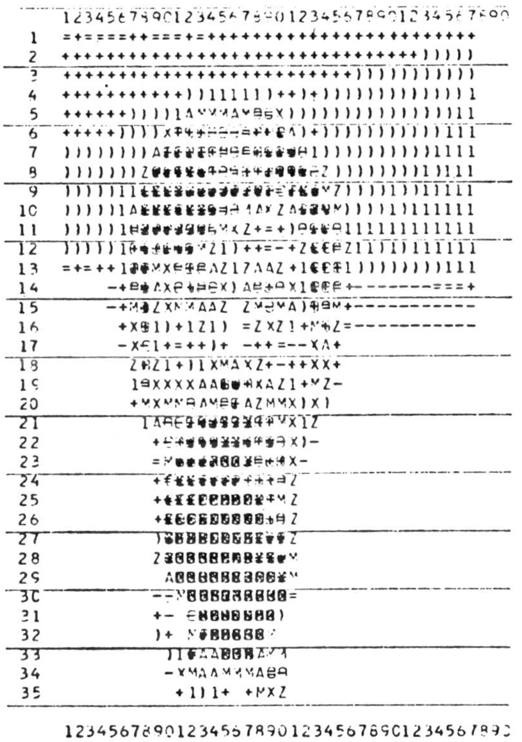

Original picture.

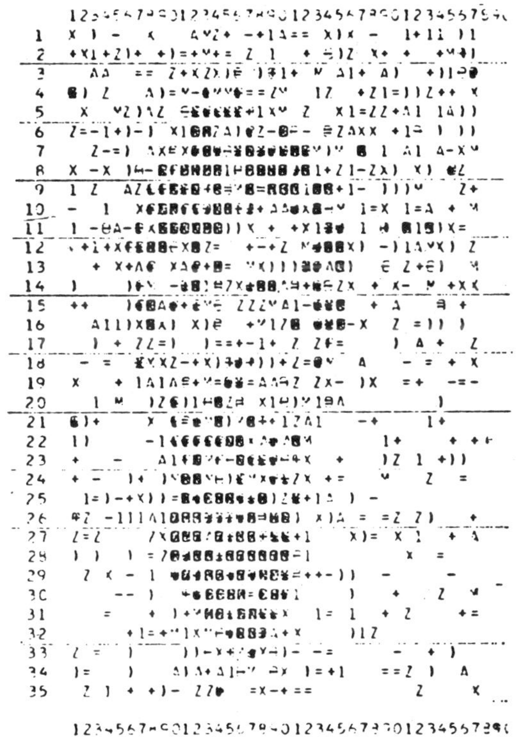

Noisy picture (sensed scene) as used in experiment.

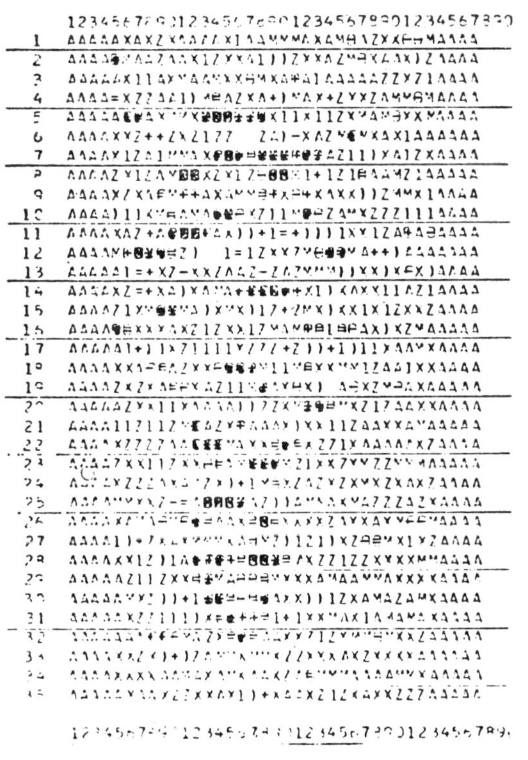

L(EV)A for mouth. (Density at a point is proportional to probability that mouth is present at that location.)

HAIR WAS LOCATED AT (8, 19)
L/EDGE WAS LOCATED AT (16, 9)
R/EDGE WAS LOCATED AT (16, 23)
L/EYE WAS LOCATED AT (14, 12)
R/EYE WAS LOCATED AT (14, 18)
NOSE WAS LOCATED AT (18, 16)
MOUTH WAS LOCATED AT (21, 16)

(h)

Fig. 4 (continued). (h) Successful embedding under random noise.

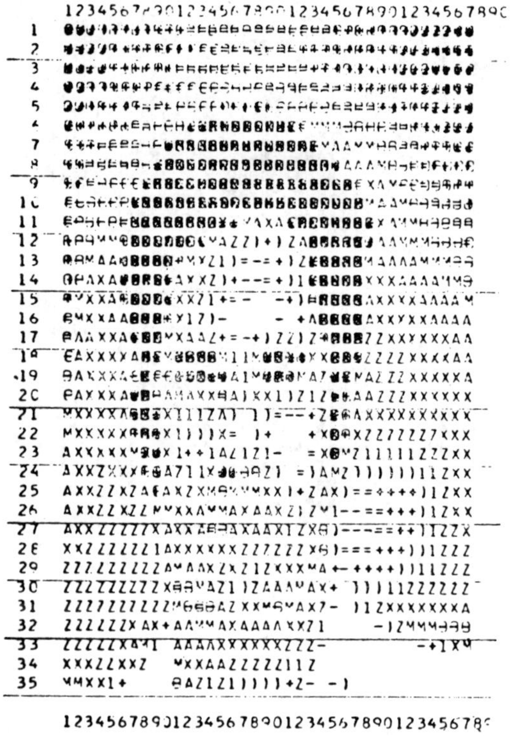

Original picture.

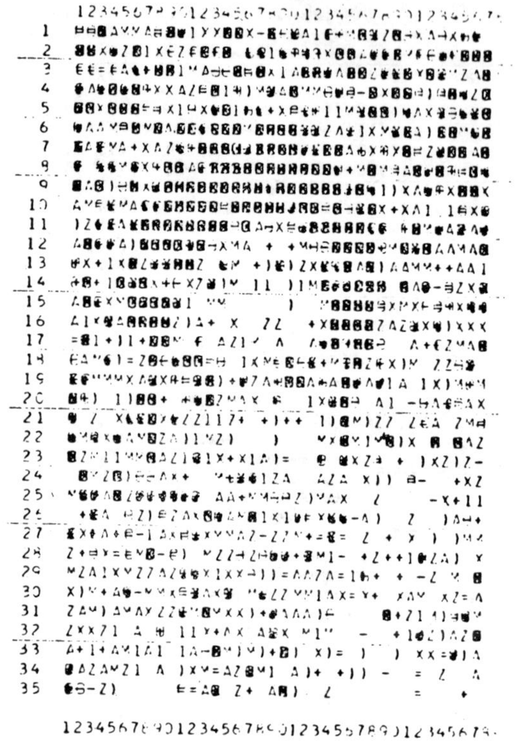

Noisy picture (sensed scene) as used in experiment.

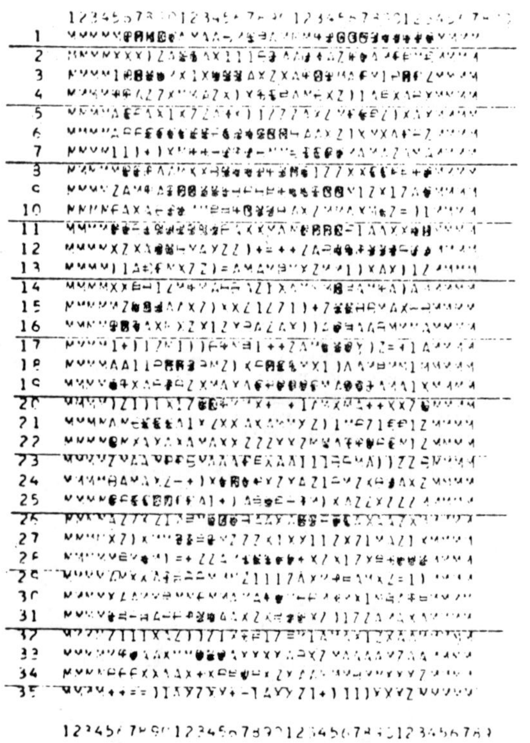

L(EV)A for nose. (Density at a point is proportional to probability that nose is present at that location.)

HAIR WAS LOCATED AT (9,19)
L/EDGE WAS LOCATED AT (22,9)
R/EDGE WAS LOCATED AT (22, 22)
L/EYE WAS LOCATED AT (19, 12)
R/EYE WAS LOCATED AT (19, 19)
NOSE WAS LOCATED AT (24, 16)
MOUTH WAS LOCATED AT (26,15)

(i)

Fig. 4 (continued). (i) Successful embedding under random noise.

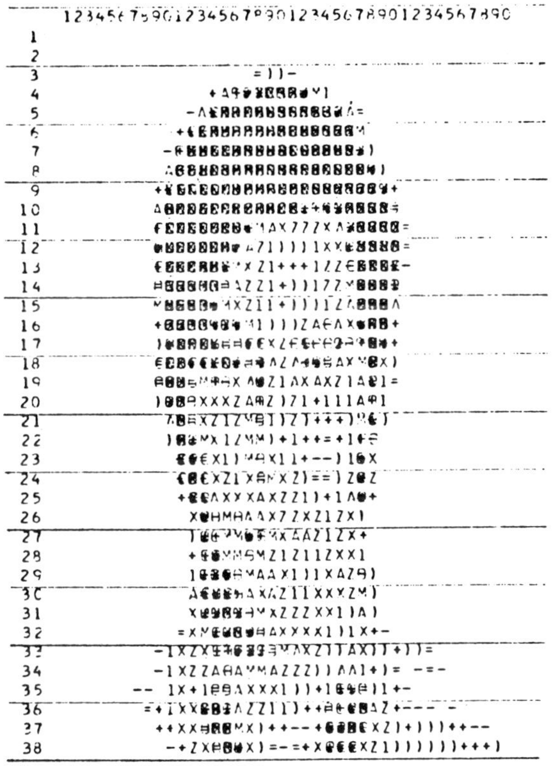

Original picture.

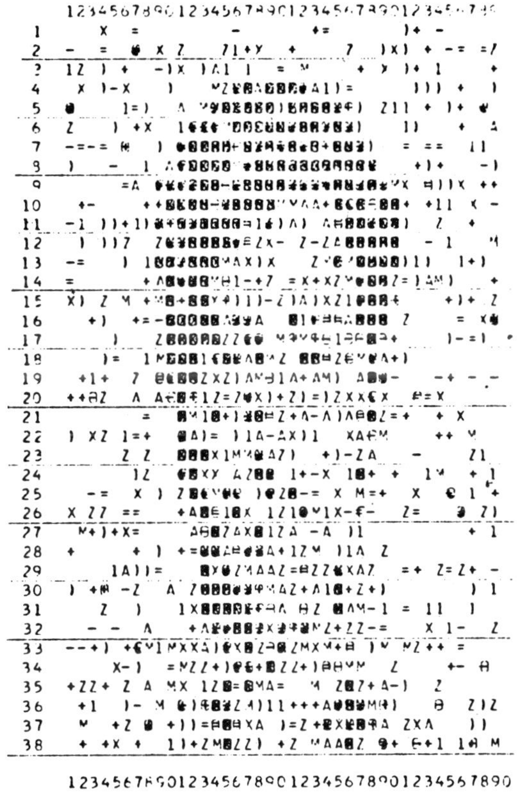

Noisy picture (sensed scene) as used in experiment.

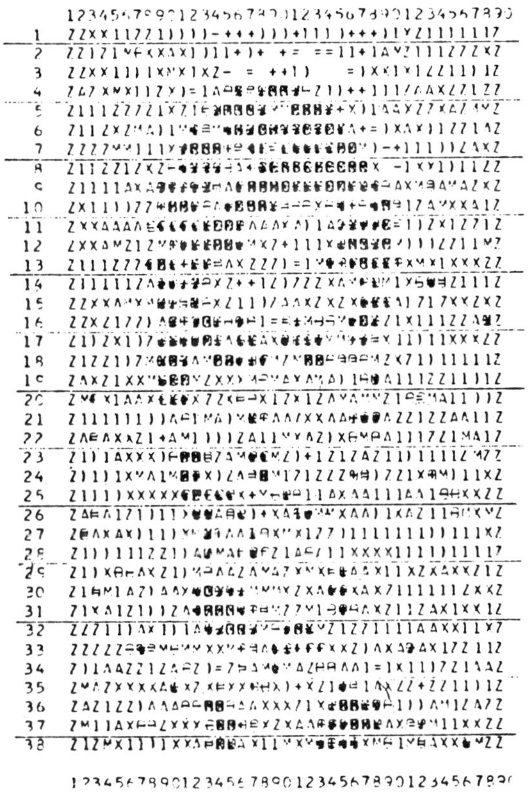

L(EV)A for eye. (Density at a point is proportional to probability that eye is present at that location.)

HAIR WAS LOCATED AT (9, 20)
L/EDGE WAS LOCATED AT (20,12)
R/EDGE WAS LOCATED AT (20,27)
L/EYE WAS LOCATED AT (18,15)
R/EYE WAS LOCATED AT (18, 22)
NOSE WAS LOCATED AT (24, 18)
MOUTH WAS LOCATED AT (27, 19)

(j)

Fig. 4 (continued). (j) Successful embedding under random noise.

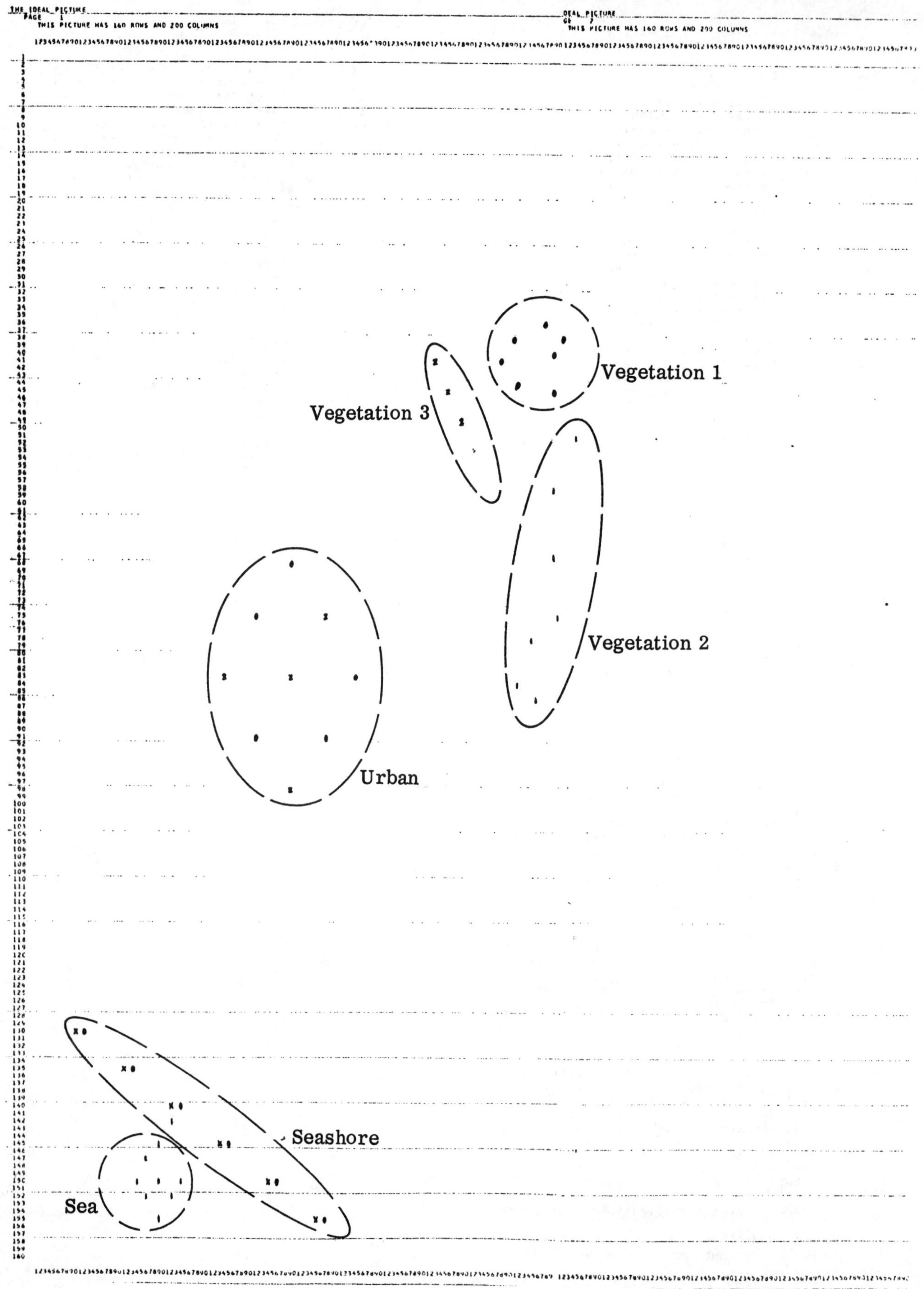

(a)

Fig. 5. Example of image-matching experiment using a terrain scene. (a) Reference for terrain.

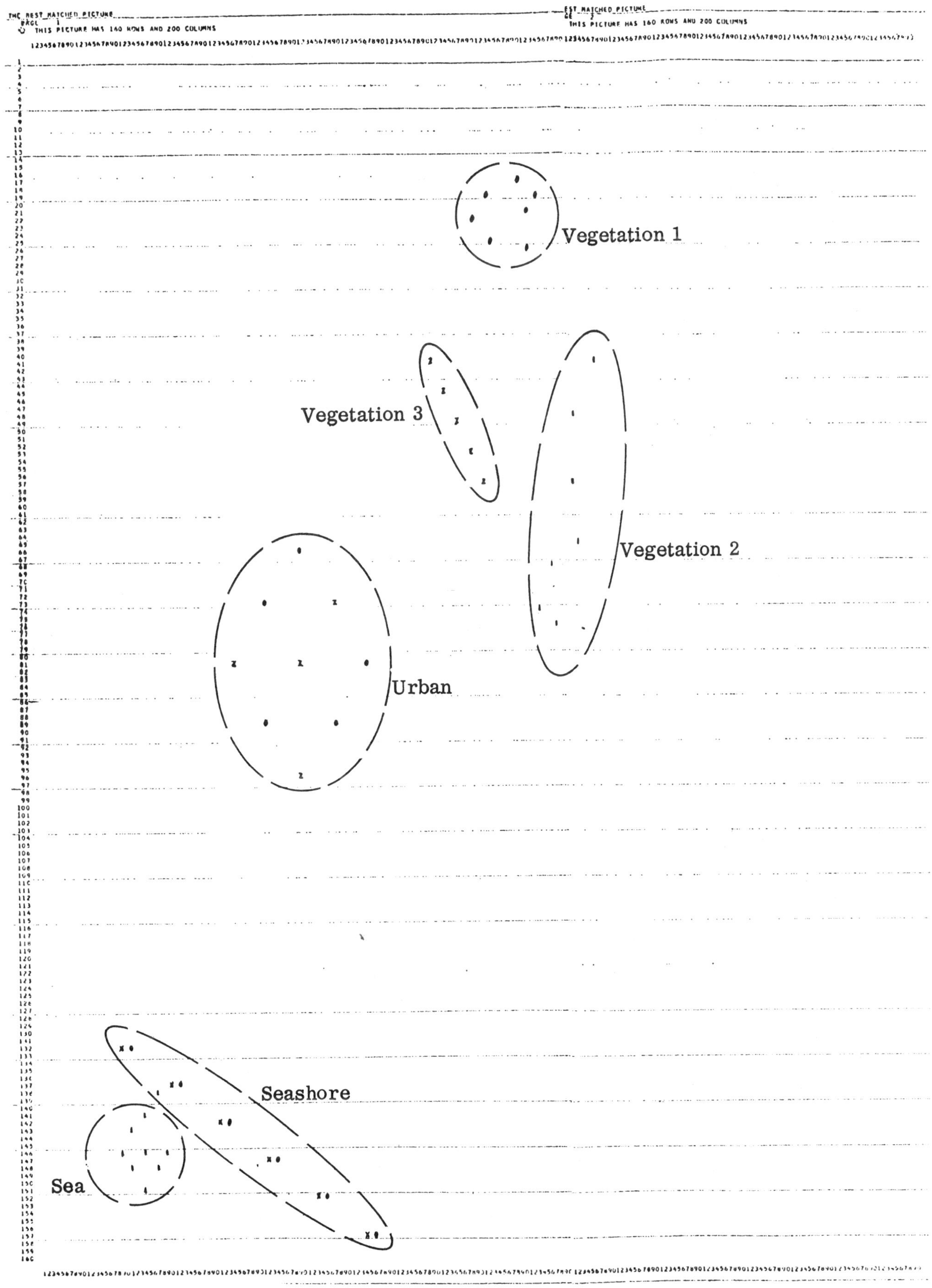

(b)

Fig. 5 (continued). (b) Embedding of reference in sensed image for terrain.

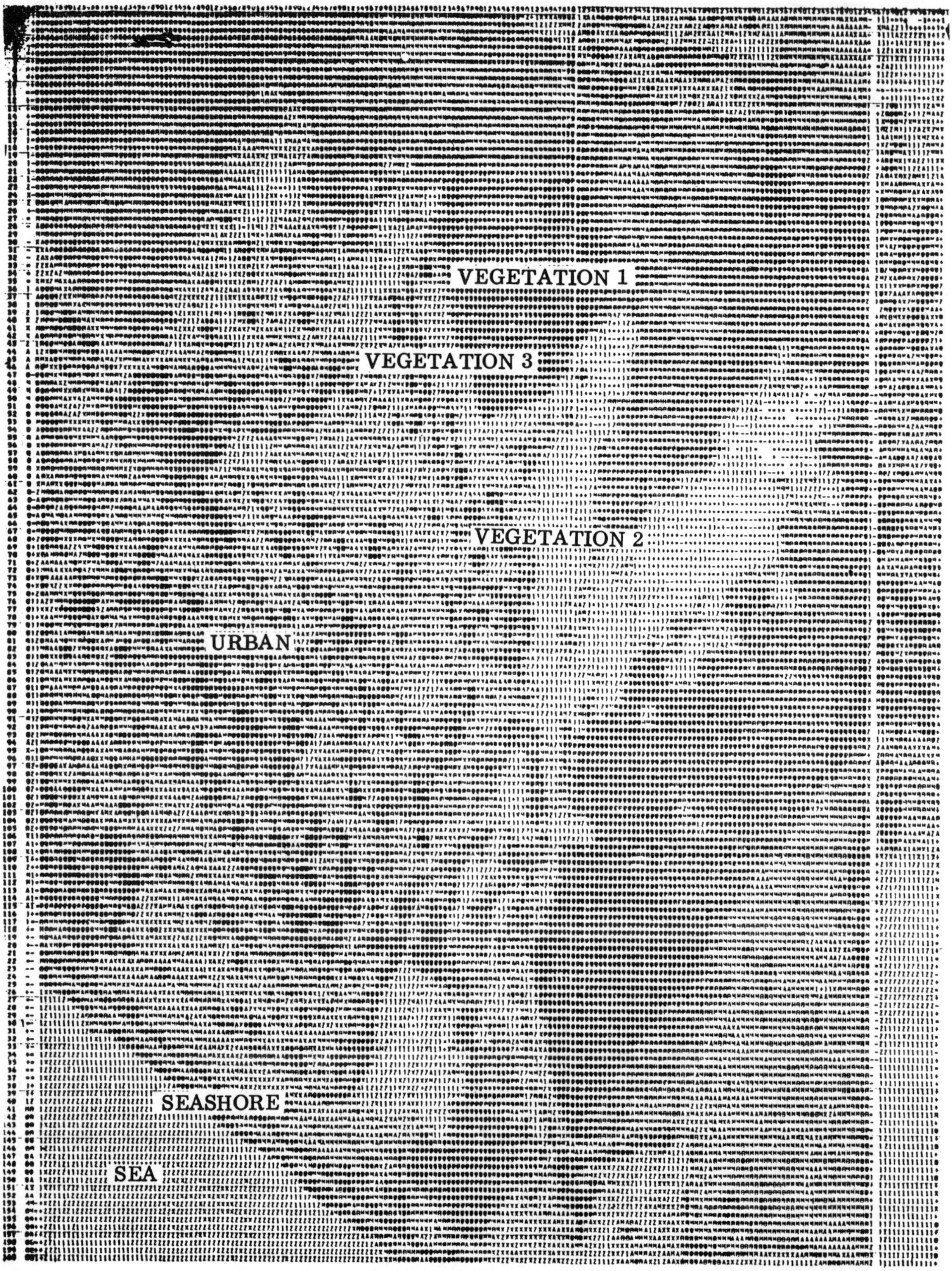

(c)

Fig. 5 (continued). (c) Sensed image for terrain.

posed a rigid template which is correlated at the different positions of the sensed scene in order to get the local evaluation array. The correlation is performed only at the indicated points of the template. Reproduction difficulties make it hard to see the intensities of these enclosed points. For "Vegetation 1," all the points are of intensity 40, for "Vegetation 2," 20. For "Urban," the points were randomly assigned intensities of 25 and 35, and so on for the other pieces.

The terrain-matching experiments represent a continuing area of investigation. Our current efforts are primarily directed toward obtaining a satisfactory set of primitives which can be used as the basis for reference component description. In low-noise experiments, with limited geometric distortion, the simple descriptions (i.e., ad hoc shape and texture components linked by springs) produced correct embeddings. Experiments under more severe conditions remain to be performed.

Implementation Details

In all of the face and terrain experiments presented in this paper, the springs were assigned the values

$$g_{ij}(x_j - x_i) = \begin{cases} 0, & \text{if } A \leq \text{row}\,(x_j - x_i) \leq B \text{ and} \\ & C \leq \text{column}\,(x_j - x_i) \leq D \\ \infty, & \text{otherwise.} \end{cases}$$

The values A, B, C, and D were typically set by taking a number of sample pictures and determining the smallest box encompassing the variation in the relative position between the components i and j. A subroutine was written to perform this task and automatically set the derived parameters into the LEA.

In our current implementation, for a typical 35×39 (face) picture, we require 13 s to compute all the L(EV)A's, and 35 s to execute the LEA (these times include picture input from a disk library, and storage of results on the disk). Most of the programming is in Fortran, and the computer is an IBM/360/40 (the addition instruction on the 360/40 executes in 11.88 μs). Total core and disk storage in bytes for all arrays is $M(4f+2h)$ and $NM(f+2h)$, respectively, where M, N, f, h are the number of resolution cells in the sensed image, the number of L(EV)A's, the amount of core needed for a floating point number, and the amount of core required for a (small) integer. For a typical picture in the face experiments ($M=1600$, $N=8$, $f=4$, $h=2$), these requirements work out to approximately 32 K bytes of core, and 100K bytes of disk storage.

Discussion

Many, though by no means all, visual objects can be described by breaking down the object into a number of more "primitive parts," and by specifying an allowable range of spatial relations which these "primitive parts" must satisfy for the object to be present.

As an example, suppose we want to describe a frontal view of a standing person. This visual object could be decomposed into six primitive pieces: a head, two arms, a torso, and two legs. For this visual object to be present in an actual picture, it is required that these six primitives occur (or at least that some significant subset of them occurs), and also that they occur within a certain spatial relationship one to the other—that is, the legs should be next to each other, and below the torso; the torso should be between and below the tops of the two arms; and the head should be on top of the torso.

It may be noticed that in the previous two paragraphs, we implicitly separated the local aspects from the global aspects of the description; the local aspects are the primitive parts of the picture, and the global aspects are the spatial relations between these parts. While at first glance it does not seem unnatural to make this separation, in practice there is a frequently encountered difficulty, that is, the feedback between the local and the global.

To illustrate this difficulty, let us go back to the example of the view of a standing person. Any method which detects torsos on a local level (that is, a method which detects torsos without using any knowledge of the positions of nearby arms and legs) might very well detect several torsos in a picture. In fact, the actual or "true" torso may be one of the weaker of the torsos detected by the method; it may even happen that the true torso is not detected at all. What does determine the position of the true torso is the position of the true arms, legs, and head. But, unfortunately, the reverse is true. The positions of, say, the true arms depends on the position of the true torso. Thus, the possible position of each piece affects the possible position of each other piece, making for a circular type of dependency.

It seems that whatever the visual object is, whenever we try to separate the global and the local, the same circular dependency occurs. Many times attempts to recognize visual objects described in such a way involve alternating between local and global analysis, heuristics, backup procedures, etc.

One conceptual way of avoiding this circularity is to evaluate simultaneously a complete interpretation of the picture; e.g., in the example given above, we would look at, and evaluate, complete configurations of head, arms, legs, torso, etc. The best complete interpretation could then be chosen. This approach, however, requires that we make an infeasibly large number of evaluations. It was just this computational problem that in the first place led to the decomposition approach.

The implication of the above discussion is that, in general, we cannot hope to decompose the global evaluation problem into a number of smaller *independent* problems, but rather must use something akin to the simultaneous evaluation, taking advantage of any reduction in total variable interdependency to reduce the required number of such evaluations.

In this paper, we accomplish this through the following machinery. First, an embedding metric is presented which sets the framework for evaluating how well any composition of primitive picture pieces (parts of the decomposed picture) matches the desired composite picture. Second, a sequential optimization (dynamic programing-type) algorithm is developed which takes advantage of the decomposition to reduce drastically the computational requirements (our computational requirements grow linearly with the size of the picture, rather than exponentially). The contribution of this paper is the simultaneous offering of the above two components and their suitability for application to a wide class of pictorial objects.

In addition to the image-matching application, which was the center of most of the development in this paper, we have also attempted to establish the utility of the representational aspects of the embedding metric for general picture description applications.

The work presented here is a continuation of the investigation described in [1] and [2], where Fischler uses sequential optimization for matching two-dimensional scenes, introduces the generic form of the embedding metric elaborated on here, and presents the concepts of coherent segmentation, arbitrary serialization, and sequential constraints. The relation of the heuristic embedding problem to formal decision theory is also discussed. The only other paper in which sequential optimization is applied to a broad *class* of problems involving two-dimensional scenes is where Martelli and Montanari [4] present a metric and matched algorithm for smoothing pictures. Kovalewsky [5] and Montanari [3] have applied dynamic programing to the detection of (one-dimensional) line-like pictures. Reference [3] is an outstanding paper, in which Montanari provides much insight into the characteristics of sequential optimization. Both Kovalewsky [5] and Montanari [3] comment on the representational aspects associated with their optimization procedures. An embedding metric conceptually very similar to the one given in [1], [2], and this paper is discussed in a broad and interesting work by Bremermann[10] [12] with respect to its potential use in character recognition, speech recognition, and control of effectors (e.g., manipulators).

[10] Other relevant publications by Bremermann include [13] and [14].

Acknowledgment

The authors wish to thank O. Firschein and J. Tenenbaum for many constructive suggestions relative to the organization of this paper.

References

[1] M. A. Fischler, "The detection of scene congruence," Lockheed Missiles & Space Company, Inc., Palo Alto, Calif., Rep. 6-83-71-2, Jan. 1971.
[2] M. A. Fischler, "Aspects of the detection of scene congruence," in *Proc. 2nd Int. Joint Conf. Artificial Intelligence* (Advance Paper), Sept. 1971, pp. 88–100.
[3] U. Montanari, "On the optimal detection of curves in noisy pictures," *Commun. Ass. Comput. Mach.*, vol. 14, pp. 335–345, May 1971.
[4] A. Martelli and U. Montanari, "Optimal smoothing in picture processing," in *Proc. IFIP Congr.* Amsterdam, The Netherlands: North-Holland, 1971.
[5] V. A. Kovalewsky, "Sequential optimization in pattern recognition and pattern description," in *Proc. IFIP Congr.* Amsterdam, The Netherlands: North-Holland, 1968, pp. 146–151.
[6] R. Bellman and S. Dreyfus, *Applied Dynamic Programming.* Princeton, N. J.: Princeton Univ. Press, 1962.
[7] U. Bertelé and F. Brioschi, "A new algorithm for the solution of the secondary optimization problem in nonserial dynamic programming," *J. Math. Anal. Appl.*, vol. 27, no. 3, pp. 565–574, 1969.
[8] O. Firschein and M. A. Fischler, "Describing and abstracting pictorial structures," *Pattern Recognition*, vol. 3, pp. 421–444, Nov. 1971.
[9] A. Rosenfeld, *Picture Processing by Computer.* New York: Academic, 1969.
[10] W. F. Miller and A. S. Shaw, "Linguistic methods in picture processing—A survey," in *1968 Fall Joint Comput. Conf., AFIPS Conf. Proc.*, vol. 33. Washington, D. C.: Thompson, 1968, pp. 279–290.
[11] M. D. Kelly, "Edge detection in pictures by computer using planning," in *Machine Intelligence 6*, B. Meltzer and D. Michie, Ed. New York: Elsevier, 1971, pp. 397–410.
[12] H. J. Bremermann, "Cybernetic functionals and fuzzy sets," in *Ann. Symp. Rec. 1971 IEEE Syst., Man Cybern. Group*, Oct. 1971, pp. 248–253.
[13] ——, "Pattern recognition, functionals, and entropy," *IEEE Trans. Bio-Med. Eng.*, vol. BME-15, pp. 201–207, July 1968.
[14] ——, "What mathematics can and cannot do for pattern recognition," in *Pattern Recognition in Biological and Technical Systems*, O. J. Grüsser, Ed. Heidelberg, Germany: Springer, 1971, pp. 31–45.
[15] W. W. Bledsoe, "The model method in facial recognition," Panoramic Research, Inc., Palo Alto, Calif., Rep. PRI:15, Aug. 1966.
[16] ——, "Man-machine facial recognition," Panoramic Research, Inc., Palo Alto, Calif., Rep. PRI:22, Aug. 1966.
[17] A. J. Goldstein, L. D. Harmon, and A. B. Lesk, "Identification of human faces," *Proc. IEEE*, vol. 59, pp. 748–760, May 1971.

Part II
Segmentation: Regions

A major step in the process of classifying and describing objects is the isolation of the objects from the background. One widely applicable procedure for object isolation is thresholding. Consider, for example, the isolation of black characters printed on white paper. When the print contrast is sufficiently high, one can find a threshold such that intensities below that threshold correspond to character points, while intensities above threshold correspond to background points. Once the characters are isolated, they can be individually analyzed in a variety of ways.

Even when thresholding is in principle an appropriate technique for object isolation, simple fixed thresholding is rarely adequate. The figure/background decision is in fact a pattern recognition problem of sometimes surprising subtlety. The first paper, by Bartz, illustrates the degree of sophistication needed to implement an adaptive thresholding system that can cope with the kind of printing variability encountered in practical, multifont character recognition. [For other thresholding techniques for character recognition and graphical data processing, see Ullmann (1973; cited in Part I).]

Thresholding is one of several techniques for producing a binary image from a gray-scale image. Binary images play a central role in image analysis, and their properties have been intensively investigated [for example, see Gray (1971)]. Perhaps most fundamental are the topological properties, such as connectedness. The connected components of a binary image are typically the figures or parts of the figures of interest.

Several algorithms have been developed for finding connected components. For example, Rosenfeld and Pfaltz (1966) give a basic sequential procedure that labels all components, Shelman (1972) gives a convenient list-processing technique for extracting any particular component, and Preston and Norgren (1972; cited in Part III) give a procedure suitable for parallel hardware based on the operations described by Golay in Part III. The second paper, by Rosenfeld [see also Rosenfeld (1970)], gives mathematical proofs of the validity of such algorithms and related border-following algorithms. The structural relations among components receive particular attention, which relates this paper to the paper by Zahn in Part V.

Returning to the topic of region formation, we note that the thresholding procedure described by Bartz takes advantage of the fact that a relatively small percentage of a printed page contains ink, so that the local white level can be estimated from the maximum intensity in a local region. More generally, dark figures on light backgrounds give rise to a bimodal distribution of intensity values in regions containing both the figure and the background. [One can generalize this notion to produce segmentation based on bimodal distributions of other properties, such as texture; for examples, see Tsuji and Tomita (1973), Davis, Rosenfeld, and Weszka (1975), and Zucker, Rosenfeld, and Davis (1975).]

A sophisticated procedure for locating a threshold between these peaks is described by Chow and Kaneko (1972). This approach can be generalized to multilevel slicing by placing thresholds at all the significant minima of the intensity histogram. Prewitt and Mendelsohn (1966) used this technique to separate the background, cytoplasm, and nucleus in photomicrographs of white blood cells, while Ohlander (1975) applied such ideas recursively to segment color photographs of natural scenes.

While the peaks in a histogram disclose dominant intensity values that may not be apparent locally, if the area histogrammed is too large, gradual shadings will broaden and blur these peaks. An alternative approach to finding uniform regions is to start with a partitioning of the image into a very large number of small uniform regions, and to grow them by merging similar neighbors. The third paper, by Brice and Fennema, shows how two simple heuristics can produce good segmentations of images of blocks world objects. (This paper also gives a useful method for boundary description, which relates it to Part V, and uses the techniques described by Guzmán in Part VI.) A related region-growing procedure, given by Muerle and Allen (1968), uses a statistical test for merging regions, and the third paper in Part VII, by Yakimovsky and Feldman, examines such approaches from the viewpoints of decision theory and artificial intelligence.

A major problem with growing small regions is that it is computationally very expensive for high-resolution pictures. The other alternative of starting with the complete picture and forming regions by repeated subdivision can be even more expensive: although there are fewer regions to consider, more sophisticated computation is needed to decide how to split a region. [On this approach, see, for example, Gupta and Wintz (1975).] In the fourth paper, Horowitz and Pavlidis present a natural compromise—a split-and-merge procedure that starts with an intermediate partition and can move in either direction. This is not only computationally more efficient, but also an aesthetically more appealing approach to region formation.

M. R. Bartz

The IBM 1975 Optical Page Reader

Part II: Video Thresholding System

Abstract: An adaptive video thresholding system is used in the Page Reader to minimize recognition failures due to contrast and line width variation. The main threshold operator is a linear function of the average contrast over a specified area. Adjustments to the threshold level are made by circuits that compute the average line width in a character, and by circuits that filter out spatial noise in the vicinity of the character. The different types of print quality distortions predominant in typewriter printing and business machine printing are handled by switching between different sets of threshold operators.

Introduction

The IBM 1975 Optical Page Reader is a multifont optical character recognition system designed to read the quarterly earnings reports submitted to the Social Security Administration. R. B. Hennis has described the organization and function of this system.[1] The present paper discusses the video thresholding portion of the 1975 system. The purpose of the video thresholder is to provide a means of determining whether or not an area on the document being read contains a portion of a character.

It accomplishes this purpose by measuring characteristics of the analog video signal obtained from a cathode ray tube flying-spot scanner and then setting a video threshold based on this measurement. A decision as to whether the video signal is to be classified as "black" or "white" (i.e., whether or not part of a character is in the area scanned) is made by comparing the amplitude of analog video to the threshold. The threshold system output containing the black-white information is then sent to a recognition unit for analysis and character identification.[2]

The engineering problem discussed in this paper is a version of the classical communications problem of separating a signal from noise. In this case there is no known model of the complex noise process, making it difficult to use statistical decision methods. The method adopted was to classify the types of noise that caused recognition errors with typical machine input documents, to design a set of thresholding operators to handle each of the error classes, and to adjust the parameters of the operators to minimize the number of errors.

The complexity of the video thresholding problem can be illustrated by contrasting the ideal case with the situation actually observed. In the ideal situation, where a character is printed with the best printer available on a very clean sheet of paper, the analog video signal would have only two, well-separated voltage levels corresponding to the white and black areas on the paper (if one ignores the sloping signal edges caused by a finite scanning aperture). The only noise signals would be caused by scanner electronics and paper surface variations, and they would be relatively small. Signal waveforms could be observed on an oscilloscope and a video threshold set half way between the white and black levels.

Under the above conditions a black vs. white decision would be very easy to make. Any number of different characters with equally good print quality could be scanned without readjusting the video threshold. Recognition errors would be extremely rare.

However, if print quality is variable, video waveforms have many signal levels and the threshold must be continually adjusted to prevent excessive recognition errors. Because of wide variations in print quality encountered on the documents that are read by the 1975 system, a means of automatically adjusting the threshold is necessary. Furthermore, the particular type of print quality distortion

The author is at The Systems Development Division Laboratory in Rochester Minnesota.

presented to the scanner determines the type of threshold adjustment that must be used.

Design of the video thresholder for the 1975 was preceded by a study of the distortions that could be expected to occur on the input documents. The basic structure of the thresholder was then developed to handle the predominant distortions, and techniques were selected for adjusting the threshold in such a way as to maximize the information content in the data flow from the scanner to recognition unit. This was done by evaluating the dependence of each threshold parameter on recognition performance since the recognition unit provides a useful means of measuring information content.

The bulk of this paper is a discussion of the engineering considerations that led to the selection of threshold operators. System schematics and an evaluation of the thresholder performance are also given. As a background to the discussion, the important print quality distortions are described.

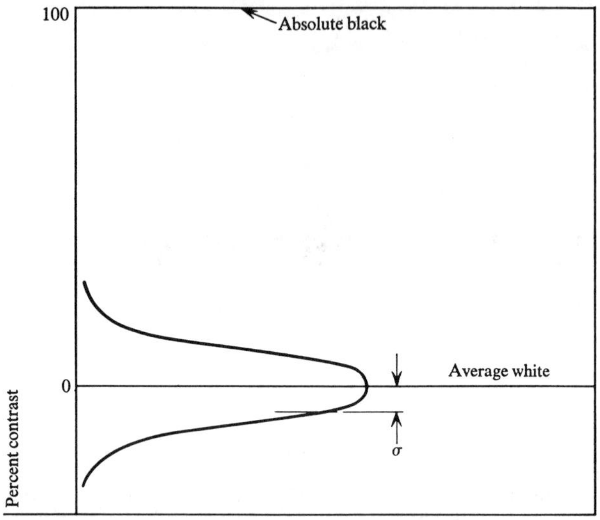

Figure 1 Definition of video signal levels. All values of contrast between 0 and 100 percent are considered "gray levels."

Print quality

Input to the IBM 1975 is printed on a standard document form (see Hennis[1]). The characters are printed with typewriters or machine printers and the type font is not specified. Upper and lower case characters are acceptable. Since no constraints are imposed on print quality, a very wide range of character distortions is present at the input to the system.

The predominant distortions observed on the input documents were classified as contrast variations, line width variations, and background noise. Each of these types of distortion has a particular influence on the design of the threshold system.

Contrast, C, is defined by the relation $C = [(R_P - R_I)/R_P] \times 100$, where R_P and R_I are percent paper reflectance and percent ink reflectance, respectively. Since contrast decreases with ribbon usage, large variations in contrast occur among documents printed with different machines. Contrast gradients (shading) may occur within a field of printed characters or within a single character. Field shading is a horizontal gradient across a line of characters that is produced when both ends of a printing mechanism, such as a chain-printer chain, are not equidistant from the surface of the paper. Character shading is usually a vertical gradient that occurs when a single type element does not strike flat on the surface of the paper. The contrast between two adjacent characters may differ if the type keys strike the paper with different impact forces. When contrast drops off radically within a small area of a character, the area is referred to as a void. Voids can be caused by defective ribbons or defective type keys.

Background noise consists of all the contrast variations surrounding or adjacent to the printed character which would not be present if the printing were of high quality. For example, new ribbons often cause some degree of ink splatter around the character. Some type mechanisms move horizontally across the paper and after the primary type element strikes the paper, an adjacent element may have enough bounce to cause a secondary background image to be superimposed on the primary image. Another type of background noise results when small openings begin to fill in. This can occur with new ribbons or when type element openings become filled with debris from the ribbon. Considerable background noise (e.g., dirt) is caused by the document handling process.

Character line widths vary widely among documents which must be read by the 1975 system because of the many different type fonts which are used to print the documents. Line widths also decrease with ribbon usage.

Objectives of threshold design

The input signal to the thresholding system is called "corrected video";[1] it can be described by an average white level, a white level noise distribution curve with standard deviation σ, the intermediate gray levels, and an absolute black level (Fig. 1). The noise distribution curve results from noise frequencies (primarily from shot noise) contained in the information bandwidth (25.6 to 500 kHz), and must therefore be tolerated by the video thresholder. Average white is defined as the mean of the noise distribution curve and is assumed to be the zero level of the corrected video signal. Absolute black corresponds to 100% contrast, or ink reflectance, R_I, equal to zero. Video levels between average white and absolute black are called gray levels, and include all contrast values between 0 and 100%.

The standard deviation, σ, of the noise distribution curve (RMS noise) is about 0.025 (2.5%) for this scanner.

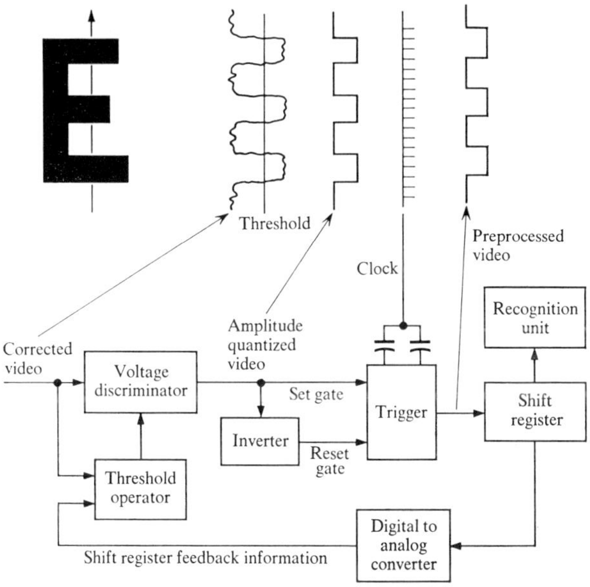

Figure 2 Video signal processing.

Figure 3 Notation for different signal-level ranges. Variation within ΔI is due to distortions of printed characters and within ΔN is due to background noise. Thresholds should be set within ΔT to achieve best patterns to present to the recognition logic. The constant K depends on the empirically determined difference between the peak value and the average value of the signal.

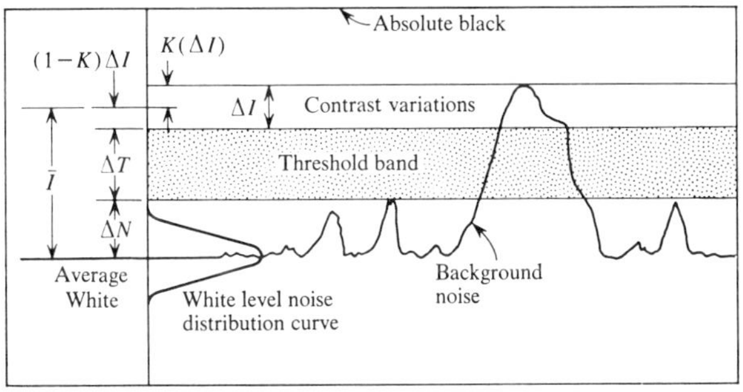

The lower limit of the information bandwidth (25.6 kHz) is determined by the fundamental raster frequency while the upper limit (500 kHz) is determined by the highest spatial frequency of the printing.

Fig. 2 illustrates the operation of the thresholding portion of the 1975 system. As the CRT spot scans vertically through a character, the corrected video signal varies according to the contrast along the line of scan. This signal is compared with a threshold by a voltage discriminator, and the discriminator's output is sampled by a 1-MHz system clock. The sampled video signal is then entered into a shift register in preparation for character recognition. Each shift register position represents a 5-mil square area on the document.

The problem of determining the level at which the threshold should be set can be described in terms of certain characteristics of the video signal resulting from a single scan through a character (Fig. 3). Let $\bar{I}$ represent the average value of the corrected video signal, ΔI the range of contrast variations due to distortions of the printed character, and ΔN the range of background noise above the average white level. Then, a threshold T can be set within a range ΔT, which falls between ΔN and ΔI, so that neither background noise or contrast variations will be reflected in the voltage discriminator output.

Furthermore, T can be adjusted within ΔT to give different pulse widths W at the voltage discriminator output. For T within ΔT, and assuming linear slopes on the pulse edge, the pulse width is defined by

$$W = W_{1/2} - \frac{200}{S}(T - T_{1/2}), \tag{1}$$

where $T_{1/2} = \frac{1}{2}\bar{I}$, $W_{1/2}$ is the pulse width at $T_{1/2}$, and S is the slope of the pulse edge. The units of W are inches and of S, percent contrast change per inch.

Although the edges of the ΔI and ΔN bands as defined above could be detected during each scan through a character, their values can vary widely from scan to scan, and hence, it would not be practical or very meaningful to use such measurements as the basis for setting threshold levels. Instead, the ΔN and ΔI bands were regarded as loosely defined ranges having conceptual, rather than practical, value in designing the threshold operators.

As an illustration of how the threshold level affects character recognition, Fig. 4 shows the bit-pattern representations of the same letter "O" generated when the threshold was set within the ΔI, ΔT, and ΔN bands, respectively. In Fig. 4a, with the threshold in the ΔI band, the pattern may be rejected by a recognition system because of difficulty in distinguishing it from a "C" pattern with a low-contrast right side. If such a noise situation had appeared on the left side of the pattern, however, the rejection probably would not occur. In Fig. 4c, where the threshold is in the ΔN band, the additional noise has made the "O" look like a "Q" to the recognition system. If a similar group of bits appeared disconnected and to the left of the pattern, correct recognition would be much easier. Another feature of the bit patterns produced by the different threshold levels is that they all have different average line widths. Since recognition logic may be sensitive to variations in line width, this feature must also be considered in choosing the threshold level.

The many causes of variation within the ΔI and ΔN bands make it impossible to set T at a level that is acceptable for every character. In the most difficult situation, where ΔN and ΔI are so large as to make $\Delta T = 0$, it is impossible to prevent rejections and substitutions from occurring at the output of the recognition unit. What is

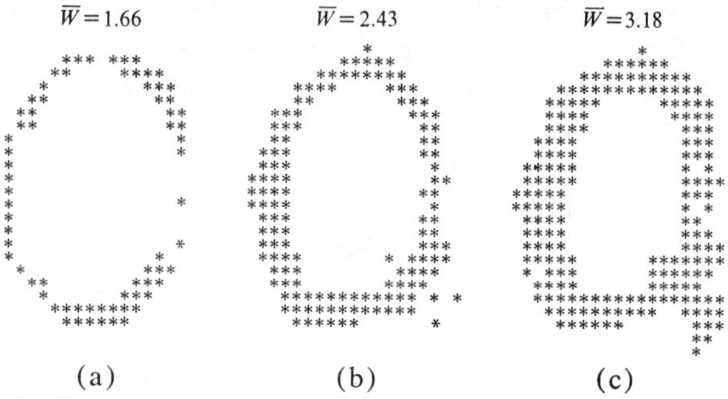

Figure 4 Effect of threshold setting on bit pattern; (a) threshold set within ΔI band; (b) within ΔT; and (c) within ΔN. The $\bar{W}$ values are the average line widths computed for each pattern.

possible, however, is to determine which types of noise occur most often and make recognition most difficult, and then set T at a level that will give recognition results that are statistically the best. The studies of noise sources and their influence on recognition were accomplished by visual examination of Social Security Administration documents and by analytical techniques involving computer processing of data.

The most significant conclusion of these studies was that threshold operator requirements were considerably different for documents printed with typewriters than for those printed with business machines. Background noise was a serious problem in both instances, but its nature differed in each. In business machine printing, most of the noise was due to ink splatter and background images. In typewriter printing fill-in noise was predominent because of the small openings in lower case letters. Line width variations were far more extreme in machine printing than in typewriter printing. These differences were used to advantage in designing the threshold operators because input documents to the system can be hand-sorted to separate machine-printed documents from the others. Hence, it is practical to call different threshold operators into use in the system by a manual switch.

Threshold operator for machine printing

The threshold operator for machine printing was designed to minimize shift register bit pattern distortions resulting from the two predominant noise sources; background noise due to ink splatter and background images, and line width variations.

Background noise is not confined to areas adjacent to the printed character, but can occur throughout the entire white area between characters. The threshold must remain above ΔN when scanning through these white areas, but above ΔN and below the ΔI band when scanning through characters.

Two types of threshold operators were considered, peak detection and video averaging. Peak detection involves measuring the highest signal amplitude, storing it on a capacitor, and then allowing it to decay at a rate slow enough to prevent appreciable errors in the stored value while scanning through areas between characters. This approach was rejected since the peak value depends on the highest contrast video amplitudes and ignores the fact that information is carried by video amplitudes covering a contrast range ΔI.

The averaging technique was adopted instead. A minimum threshold level $T_{\min}$ was defined as a threshold level between video amplitudes which have an extremely low probability of representing information and amplitudes which have a very high probability of being part of the ΔI band. $T_{\min}$ was set as close to the white level noise distribution curve as possible.

A term, $\bar{V}$, was defined as the average of all video samples greater than $T_{\min}$ within a predetermined area such that

$$\bar{V} = \frac{1}{N} \sum_{i=1}^{M_x} \sum_{j=1}^{M_y} V(i, j), \qquad V(i, j) > T_{\min} \tag{2}$$

where $\bar{V}(i, j)$ is the jth sample of the ith scan, N is the total number of video samples with $\bar{V}(i, j) > T_{\min}$ and M_x and M_y define the area over which $\bar{V}$ is evaluated.

The value for $\bar{V}$ is established during the left-to-right scan that precedes the right-to-left recognition scan of each line; it is evaluated over the space of several characters. During operation, the machine averages continuously, even during the recognition scan.

A threshold, $T(\bar{V})_M$, was defined to be a linear function of $\bar{V}$ such that,

$$T(\bar{V})_M = g\,\bar{V} + C_M. \tag{3}$$

The final output of the threshold operator, $(TO)_1$, is either $T_{\min}$ or $T(\bar{V})_M$ as shown in Fig. 5. A "black analog" OR circuit, which generates an output equal to whichever input is nearer black, prevents the output $(TO)_1$ from falling below $T_{\min}$. The transfer function of this operator is also shown in Fig. 5 along with measured maximum and minimum values of $\bar{V}$.

There was not a great deal of freedom in choosing the constants of Eq. (3). $\bar{V}$ was subject to a certain amount of error caused by the analog integrating circuit used to generate it. This, in turn, caused an error in $T(\bar{V})_M$ which was intolerable if g was greater than 2. Of course, g must be large enough to provide an adequate dynamic range for $T(V)_M$ as $\bar{V}$ varies from one extreme to another. On the basis of these considerations, $g = 2$ was chosen as the best compromise.

The value of C_M represents the quiescent point of $T(\bar{V})$; it was determined experimentally as -0.53 by evaluating its dependence on the recognition reject and substitution

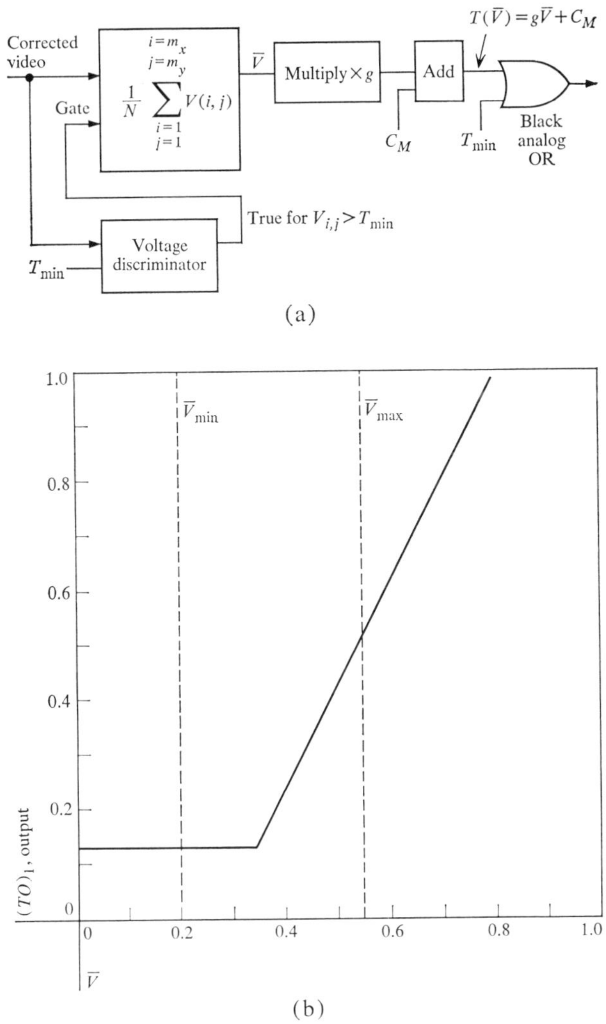

Figure 5 (a) Threshold operator for machine-printed documents; (b) graph of threshold as function of averaged video signal, $\bar{V}$.

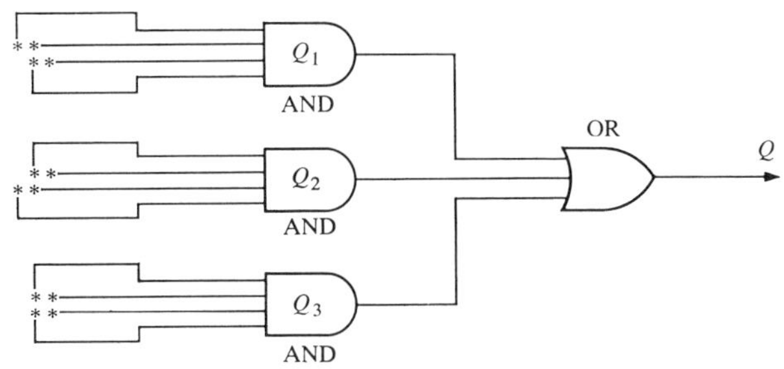

Figure 6 Logic for line-width measurement. Whenever certain shift register positions are detected as containing one of the three bit configurations shown, the output, Q, causes a counter to be incremented.

rates. To determine T_{min}, the very lowest contrast printing available was scanned. Under these conditions $T(\bar{V})_M < T_{min}$, so the output of $(TO)_1$ was essentially equal to T_{min}. Several settings of T_{min} ranging from 5% to 20% contrast were then tried. It was found that $T_{min} = 0.13$ (13% contrast) was the best setting because the number of rejects and substitutions were minimal for this setting.

Character line width measurement

Examination of Eq. (1) shows that the output pulse width obtained from a single scan through a character can be adjusted about a value $W_{1/2}$. The value of $W_{1/2}$ is more or less a characteristic of the type font but it also depends on $\bar{V}$. Control of signal pulse width is limited by ΔN and ΔI, which vary significantly from document to document. Therefore, the main purpose of this control is to normalize the effect of variations in $W_{1/2}$ on the generation of character bit patterns.

However, to enhance the recognition capability of the system, it is necessary to have a method of measuring and normalizing the line width of characters in terms of the number of "bits" that result from scanning. A method has been developed[3] for measuring character line width in which bits passing through the preprocessor output shift register are examined to see how often they occur in certain patterns. This method uses three logic ANDs (Q_1, Q_2, and Q_3) which can be satisfied only when the bit configurations shown in Fig. 6 are present in the pattern represented in the shift register. Each time one or more of the three operators is satisfied as the pattern passes through the shift register, a counter is incremented. The total number of these counts for a pattern is called Q. In addition, the total number of bits in a pattern is counted and labeled A. The ratio of Q to A is then used to compute an "average line width" for the pattern. The circuits Q_1 and Q_2 are used for measuring the width of diagonal pattern segments, and Q_3 measures vertical and horizontal segments.

Consider, for example, the pattern of a vertical line W bits wide and L bits long. It is easy to convince oneself that the operator Q_3 will be satisfied $(L - 1) \times (W - 1)$ times as this pattern goes through the shift register. It can also be seen that although Q_1 and Q_2 will each be satisfied $(L - 1) \times (W - 2)$ times, none of these will be a time that Q_3 is not also satisfied; if $L \gg 1$,

$$\frac{Q_3}{A} = \frac{L(W-1)}{LW} = \frac{W-1}{W}. \tag{4}$$

Rearranging terms and replacing Q_3 by Q and W by $\bar{W}$ gives an expression that is used for computing the average line width of any pattern:

$$\bar{W} = \frac{1}{1-(Q/A)}. \tag{5}$$

Threshold adjustments based on a measured value of $\bar{W}$ can be made after a character is scanned. A line width

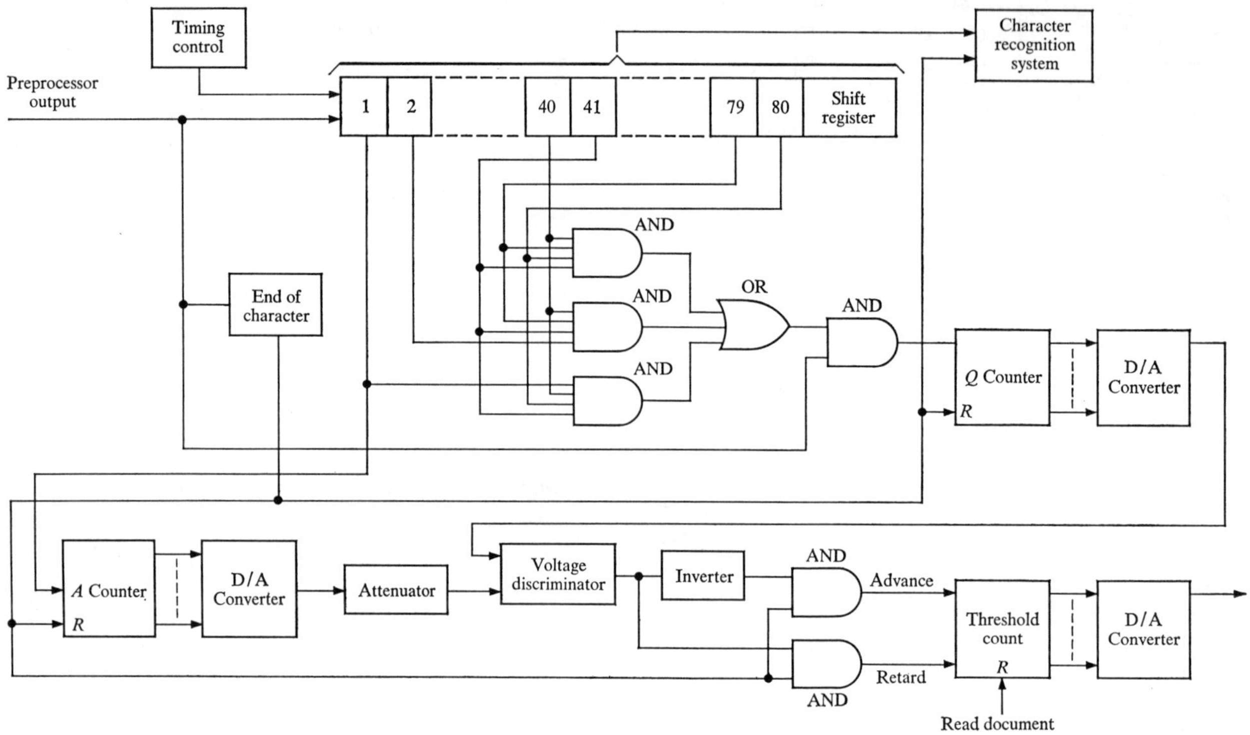

Figure 7 Circuit for generating threshold adjustment based on bit pattern line width. The output is the function $(TC)C_w$ included in Eq. (6).

servomechanism (Fig. 7) evaluates Q/A after each character is scanned, compares it to a predetermined constant q corresponding to some nominal average line width $\bar{W}_n$ and then adjusts the threshold in a direction which will make Q/A on the next character fall closer to q (or in other words, make $\bar{W}$ on the next character fall closer to $\bar{W}_n$).

The "end of character" signal (in Fig. 7) defines the beginning and end of a character and the area over which Q and A are determined. Q and A are converted to analog signals with digital-to-analog (D/A) convertors. A is then multiplied by q and compared to Q. If $Q > qA$, a four-bit threshold counter is advanced; if $Q < qA$, the threshold counter is retarded. The threshold count, (TC), is then converted to one of sixteen possible levels $(TC) \times C_w$, where C_w is a constant whose value is determined experimentally. $(TC)\,C_w$ can then be added to Eq. (3) to provide a threshold adjustment based on line width:

$$T(\bar{V}, \bar{W})_M = 2\bar{V} - 0.53 + \frac{15C_w}{2} - (TC)C_w\,. \qquad (6)$$

The threshold $T(\bar{V}, \bar{W})_M$ now depends on both contrast and line width. The contrast information is used to determine the mid-point of the ΔT range and then another adjustment is made within ΔT based on line width information. $\bar{W}_n$ and C_w were calculated experimentally to be 4.0 and 0.013, respectively.

Threshold operator design for typewriter printing

Line width variations are not a major factor in typewriter printing. Background noise of the variety present in machine printed documents was not predominant, but enough was present to warrant use of a threshold operator based on contrast variations over the entire scanning area:

$$T(\bar{V})_T = 2(\bar{V}) + C_T. \qquad (7)$$

C_T was determined experimentally to be -0.63.

Background noise due to "fill-in" was predominant in typewritten documents because of the small openings in lower case letters. This type of noise is adjacent to the character, so a threshold operator can be based on contrast information from a localized area surrounding the video sample on which a decision is to be made.

To decide on the size and shape of the area, some practical as well as theoretical considerations were made. The biggest limitation on the size of the area lies in the amount of analog storage required. If an analog delay line is to be large enough to store a complete scan, its length, τ, must be

$$\tau = (N - 1)\,RS + NS, \qquad (8)$$

where S is the sample time (1 μsec); N, the width of the localized area, a, in number of samples; and R, the number of samples in one complete scan, including retrace time ($R = 39$).

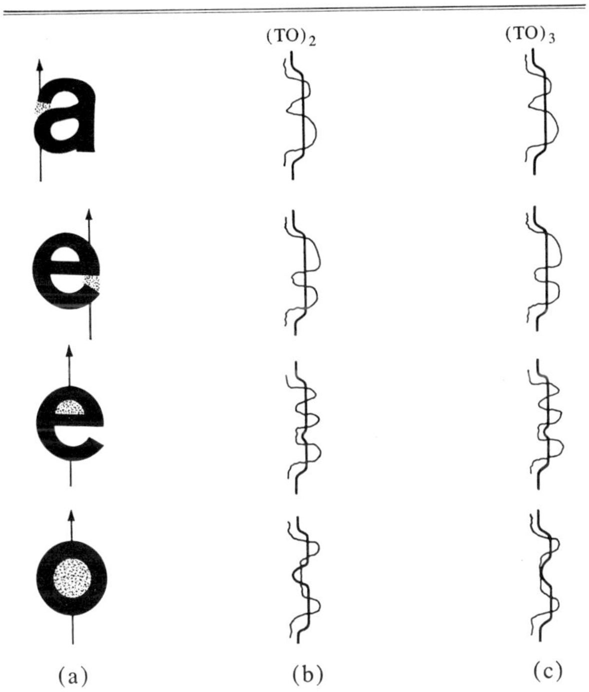

Figure 8 Fill-in noise in printed characters; (a) line of scan; (b) video signal (light line) and threshold generated by $(TO)_2$ operator (heavy line); (c) video signal and threshold generated by operator that includes spatial noise rejection feature.

For an N of three, $\tau = 81$ sec. The delay line bandwidth would have to be 500 kHz, making this approach rather impractical with presently available lines.

The IBM 1428 Optical Reader[4] uses a delay line which stores information from a 0.015 in. × 0.030 in. document area. A two dimensional raster is used to eliminate the need for an excessively long delay line. However, a two dimensional raster could not be considered in the IBM 1975 without drastically altering the system design. The approach taken was to use a delay line long enough to store only five samples within a single scan and then re-evaluate the situation after experimental results were available to see if a larger line could be justified. The delay line had five taps (including input and output) spaced 1 μsec (one sample time) apart. The center tap video was compared to a threshold to determine whether the sample was black or white. This is a very important feature because it enables a decision to be based on two samples ahead and two samples behind the sample in question.

Design of the threshold operator was accomplished by considering several special situations of fill-in noise (Fig. 8a). In many of the special cases illustrated, a situation often arises where only one or two of the five samples are information samples and the others are lower amplitude noise samples which should be called white bits.

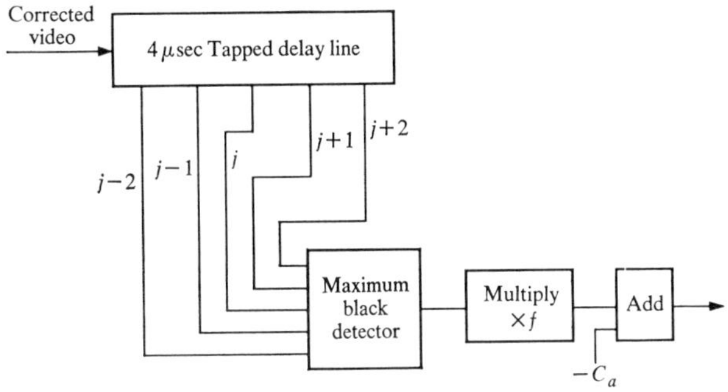

Figure 9 $(TO)_2$ operator.

The quantity V_{max} can be measured by determining the highest video level in the delay line. A threshold, $(TO)_2$, can then be made a linear function of V_{max}:

$$(TO)_2 = f[V_{max}] + C_a. \tag{9}$$

Proper choice of f and C_a depends almost entirely on the printing noise levels. The best performance was achieved with $f = 1.23$ and $C_a = -0.38$.

Waveforms showing $(TO)_2$ relative to the analog video are shown in Fig. 8b. Examination of these waveforms shows that satisfactory results were obtained in the first three cases but not the fourth. The situation represented by the filled-in o occurred often enough to warrant combining the $(TO)_2$ operator (Fig. 9) with a spatial noise rejection feature.

The new operator, $(TO)_3$, is shown in Fig. 10. Here, the capacitor circuit stores a voltage, T_c, with negligible leakage over a period of several scans. T_c can be increased (moved towards black) by $(TO)_2$ until $T_c = (TO)_2$. T_c can be decreased to T_{avg} or zero depending on which level is nearest white when $(TO)_2 < T_{avg}$.

$$T_{avg} = \frac{1}{5} \sum_{n=j-2}^{n=j+2} V_n + C_{avg}, \tag{10}$$

where V_n is one of the delay line samples and j is the index of the sample for which a decision is to be made.

The analog OR circuits generate an output that is equal to whichever input is nearest white or black as indicated in Fig. 10. The trigger is set equal to one on every black-to-white transition. It is reset to zero during retrace or after the video has been black for 3 μsec.

Fig. 8c shows $(TO)_3$ relative to analog video. Notice that the noise rejection feature has not changed the threshold setting in the top three cases, but the case with the relatively large filled-in area has been improved considerably.

T_{avg} must be capable of discharging the capacitor storage circuit to a level such that $T_{avg} < T_{min}$ when the delay line is filled with white video. The value chosen for C_{avg} should be less than T_{min}, which equals 0.13, and it should be large

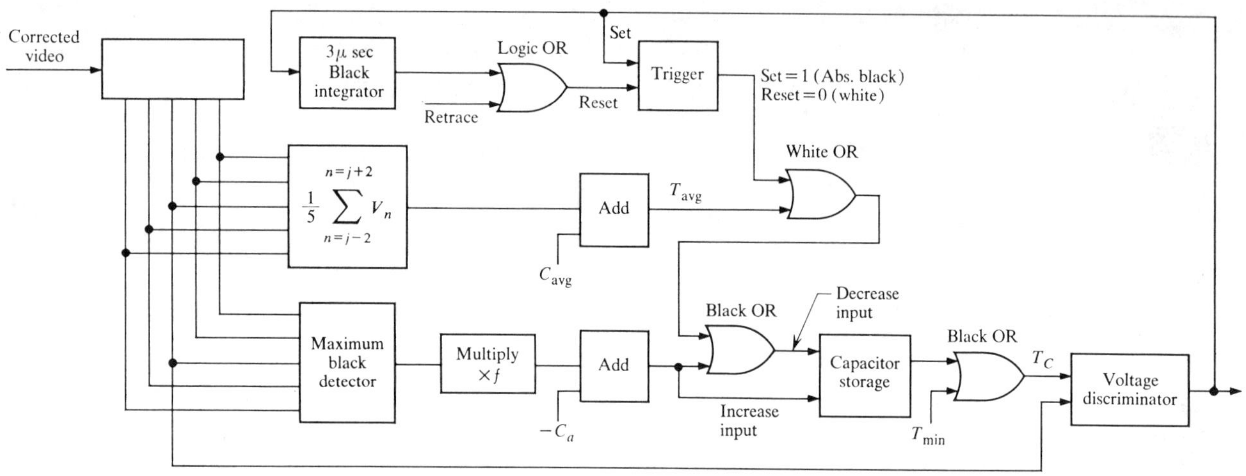

Figure 10 $(TO)_3$ operator.

enough to hold T_c above the level of fill-in noise, as illustrated in the bottom row of Fig. 8. A value of $C_{avg} = 0.1$ was chosen as a reasonable compromise.

Using $(TO)_2$ in the form defined by Eq. (9) presents an objectionable pattern characteristic for the recognition logic. When the input consists of high contrast printing with narrow line widths ($W_{1/2} < 2.5$ bits), the $(TO)_2$ operator sometimes produces patterns with $W = 1$ bit. To prevent patterns of this type from occurring, the threshold operator was modified to include a measurement derived from the dynamic line-width circuit. The modified operator is given by

$$(TO)_2' = (TO)_2 - 0.20 + 0.013\,(TC). \qquad (11)$$

Here, the line width circuit senses the presence of narrow line width patterns and, in effect, moves the threshold closer to white.

Evaluation methods

As an initial step in the evaluation of the thresholder design, documents with a variety of print quality distortions were scanned and computer print-outs of the character bit patterns were produced. Adjustments of the threshold operator constants were then made on the basis of a visual inspection of the bit patterns. This procedure was continued until the preprocessor could produce enough patterns that were judged by subjective criteria as satisfactory input for the recognition unit. This sort of adjustment is necessary, but limited in effect, primarily because the number of documents that can be analyzed in a reasonable time is small and because the decision on whether a pattern is acceptable or not is completely subjective. Hence, more elaborate techniques involving computer processing were devised to refine the design.

The effect on recognition reject rates of varying the operator constants was used as a means of selecting the best values. Reject rates were determined for several values of each constant and the value that minimized the reject rate was selected for use in the threshold operator. In order to obtain reliable reject rates about 35,000 characters from 350 documents were sampled for each calculation. The bit patterns of the characters rejected by the recognition system simulator were then visually examined to determine what kind of distortion was responsible for rejection. If one particular type of distortion were the cause of a large number of rejects, it would be seen as an indication that major redesign of one or more threshold operator was necessary.

Another, less successful, computer simulation technique for selecting threshold operator values involved converting the analog video signal into digital form and storing the gray-level information on magnetic tape. This tape served as input for use with a computer program that simulated threshold operators and generated character bit patterns. Since the simulation did not include an entire recognition logic, it was necessary to decide whether patterns were acceptable or not by a manual process. Although this technique had serious limitations on its effectiveness, it was valuable in studying video characteristics in the early development of the threshold operators.

Evaluation of the component parts of the thresholder enabled us to conclude that the circuits for averaging the video signal over the area occupied by several characters and for dynamically adjusting character line-widths have the most significant effect in improving character recognition rates. It was obvious from visual inspection of character bit patterns that the local contrast operator, which uses only five samples of video information, is inadequate

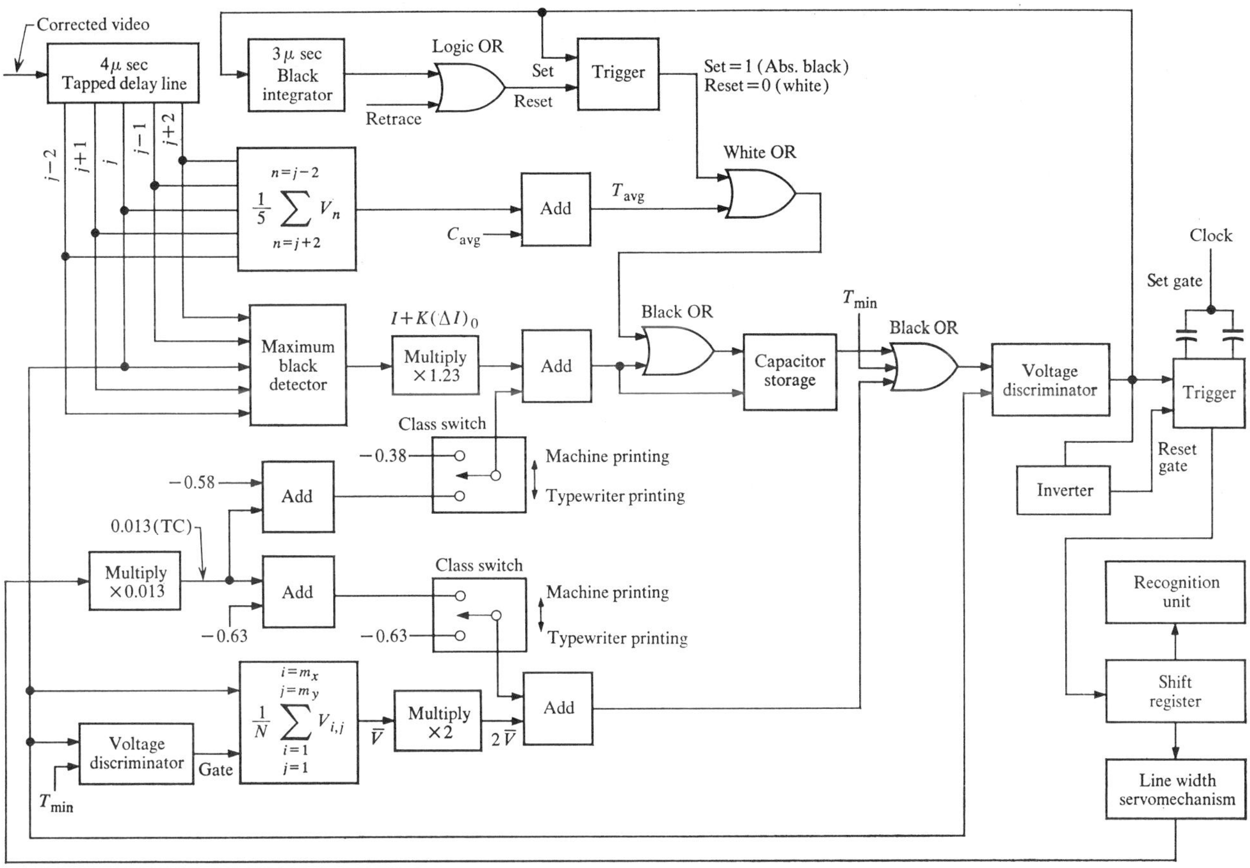

Figure 11 Final configuration of thresholding system.

Figure 12 Some examples showing effect of adaptive thresholding. The left bit pattern of each pair was obtained with a threshold fixed at 20% contrast.

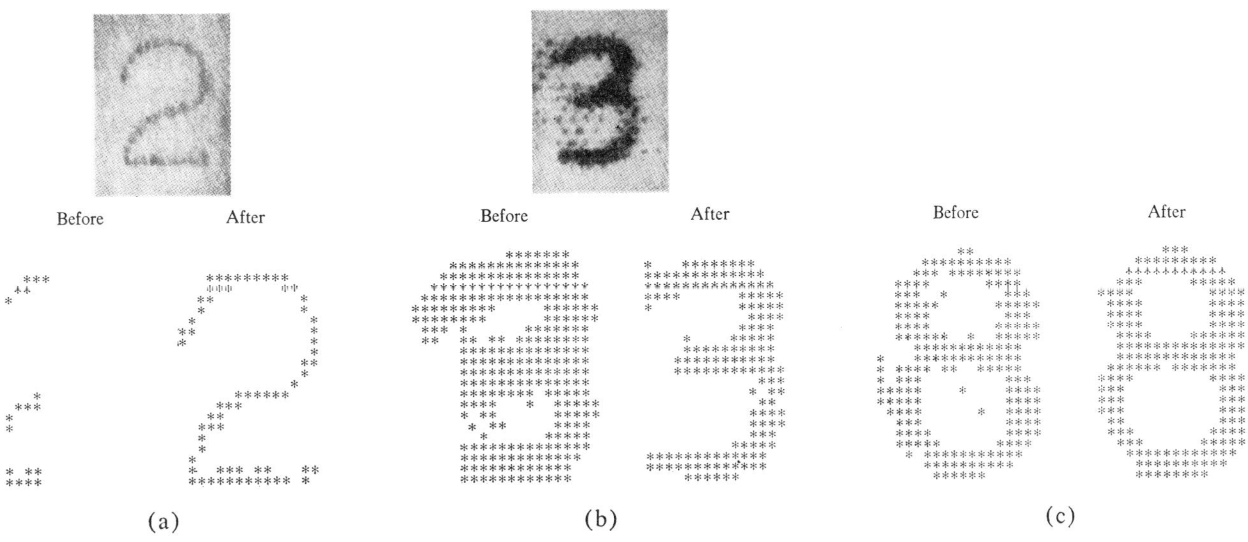

when used by itself, but it did make a significant improvement in recognition rates of typewritten characters when it was used in conjunction with the large-area operator. Use of the dynamic line-width circuit reduced the reject rate of machine-printed characters to about half of that observed when the circuit was not used.

Fig. 11 shows the final configuration and operator values selected for the thresholder. It combines the large-area contrast operator, the dynamic line-width circuit, the local contrast operator, the spatial noise filter, and the switches used to select the operator constants appropriate for machine-printed and typewritten documents. It can be seen how the switches allow the large-area contrast operator to be dominant for machine-printed documents and the local contrast operator to be dominant for typewritten documents.

Figure 12 shows some character samples and the bit patterns generated with a threshold fixed at 20% contrast, and with thresholds determined adaptively by the preprocessor system described in this paper.

References

1. R. B. Hennis, "The IBM 1975 Optical Page Reader, Part I: System Design," *IBM J. Res. Develop.* **12,** No. 5, 346–353 (1968), this issue.
2. D. R. Andrews, A. J. Atrubin and K. C. Hu, "The IBM 1975 Optical Page Reader, Part III: Recognition Logic Development," *IBM J. Res. Develop.* **12,** No. 5, 364–371 (1968), this issue.
3. The line width measurement technique was developed by G. L. Shelton, Thomas J. Watson Research Center, Yorktown Heights, N. Y.
4. "IBM Field Engineering Manual of Instruction, 1428 Alphameric Optical Reader," Form No. 225-1725-1, December, 1964.

Received February 1, 1967.

Adjacency in Digital Pictures*

AZRIEL ROSENFELD

Computer Science Center, University of Maryland, College Park, Maryland 20742

Let S be a subset of a digital picture, and let $\bar{S}$ be the complement of S. It is well known that the connected components of S and $\bar{S}$, under thc relation "is adjacent to," form a tree, and algorithms for constructing this tree have been devised. The main purpose of this paper is to prove that the components do form a tree, and in the process, to provide a basis for proving the validity of the tree-constructing algorithms.

1. BASIC CONCEPTS

Let $I \times I$ be the set of pairs of integers, and let $x = (i, j)$ be in $I \times I$. The four pairs $(i \pm 1, j)$ and $(i, j \pm 1)$ are called the 4-*neighbors* of x, and are said to be 4-*adjacent* to x. These same pairs, together with the four pairs $(i \pm 1, j \pm 1)$, are called the 8-neighbors of x (8-*adjacent* to x). More generally, if S, T are subsets of $I \times I$, we say that S is (4- or 8-) adjacent to T if some $x \in S$ is (4- or 8-) adjacent to some $y \in T$. From now on, elements of $I \times I$ will be called *points*.

A *path* from x to y is a sequence of distinct points $x = x_0, x_1, \ldots, x_{n-1}, x_n = y$ such that x_i is adjacent to x_{i-1}, $1 \leqslant i \leqslant n$. There are two versions of this definition, "4-path" and "8-path", depending on which type of adjacency is used. Similarly, there are two versions of most of the definitions which follow.

If $x, y \in S \subseteq I \times I$, we say that x is *connected* to y in S if there is a path from x to y consisting entirely of points of S. Readily, "is connected to" is an equivalence relation; the equivalence classes under this relation are called the connected *components* of S. If S has only one component, it is called *connected*.

PROPOSITION 1. *If S contains T, every component of T is contained in a unique component of S.*

* The support of the Office of Naval Research, under Contract N00014-67-A-0239-0012, is gratefully acknowledged. The author is also indebted to Anupam N. Shah for several helpful discussions, and to Mrs. Eleanor B. Waters for her help in preparing the manuscript.

Reprinted with permission from *Inform. Contr.*, vol. 26, pp. 24–33, Sept. 1974. Copyright © 1974 by Academic Press, New York, NY.

Let $\bar{S}$ be the complement of S. If S is finite, all points of $\bar{S}$ that are sufficiently far from the origin are evidently connected in $\bar{S}$. It follows that exactly one component of $\bar{S}$ is infinite; we call this component the *background* of S. All other components of $\bar{S}$, if any, are called *holes* in S. If S is connected and has no holes, it is called *simply-connected.* We assume from now on that S is finite. We shall sometimes refer to points of S as 1's, and to points of $\bar{S}$ as 0's; this convention is suggested by the definition of the "characteristic function" K_S of a set S: $K_S(x) = 1$ if $x \in S$, $K_S(x) = 0$ if $x \in \bar{S}$.

Many of the theorems that can be proved about connectedness hold only if opposite definitions are used for S and $\bar{S}$; i.e., if we use 4-connectedness for 1's and 8-connectedness for 0's, or vice versa. We shall do so in the remainder of this paper. Thus, for example, we now have only two types of simply-connected S's: a 4-connected S that has no 8-holes, and an 8-connected S that has no 4-holes.

PROPOSITION 2. *If $\bar{S}$ is connected, every component of S is simply-connected. If $\bar{S}$ has $k + 1$ components, no component of S has more than k holes.*

It should be pointed out that if we use a hexagonal grid of points in place of $I \times I$, all of the results of this paper remain valid, with the simplification that only one type ("6-") of adjacency, connectedness, etc., is needed. Generalizations to other tessellations of the plane are also possible, but would seem to be of little practical interest. For surfaces more complicated than the plane, e.g., surfaces such as the torus, many of our results are no longer valid.

2. BORDERS

By the *border* of S we mean the set of points of S that are adjacent to points of $\bar{S}$. More generally, if C is a component of S, and D is a component of $\bar{S}$, by the D-border of C we mean the set of points of C that are adjacent to points of D. From now on, we shall use only the versions of these definitions in which "adjacent" means "4-adjacent".

It is known (Rosenfeld, 1970) that if S is simply-connected, a certain algorithm will find all border points of S, given an initial border point. In the following paragraphs we state a slightly modified version of this algorithm and prove that, for any C and D, it can be used to find all points of the D-border of C, given an initial pair of 4-adjacent points c, d with $c \in C$, $d \in D$. We assume below that C has more than one point; otherwise its sole point c is also its D-border, and there is nothing to do.

Our algorithm, which we shall refer to as BF, proceeds as follows.

(a) Change the values of c and d to 3 and 2, respectively.

(b) Let the 8-neighbors of c in (say) clockwise order, starting with d and ending with the first occurrence of 1, 3, or 4, be $e_1, \ldots, e_k$. If $c = 3$, $e_k = 4$, and $e_h = 2$ for some $h < k$, change the 3 to 4, the 2 to 0, and stop. Otherwise, change the value of c to 4 (if it was 1). Take e_k as the new c and e_{k-1} as the new d, and return to step (b).

We claim that when BF stops, the 4's are exactly the points of the D-border of C. [The version of BF just given assumes that C is 8-connected and D 4-connected. If the reverse is true, we need only modify step (b) as follows: Let e_k be the first 4-neighbor that has value 1, 3 or 4. If e_{k-1} does not have one of these values, take e_k as the new c and e_{k-1} as the new d. If it does have one of these values, take e_{k-1} as the new c and e_{k-2} as the new d.]

Rosenfeld (1970) showed, by induction on the number of points of S, that BF (or rather, an algorithm closely similar to it) works in the case where $S = C$ is simply-connected. This proof depends on the existence of "deletable points" in any simply-connected C (provided C has more than one point), and so does not generalize directly to an arbitrary C. We can, however, develop a general proof by using the simply-connected case together with a result in (Rosenfeld, 1973) about components that have exactly one hole.

Note first that the operation of BF remains the same if all 0's except those in D and those in the background are changed to 1's. This is because the c's, d's, and e's examined by BF are all in C and D, so that changes to points that lie in neither C nor D can have no effect on BF. (In the case where C is 4-connected and D 8-connected, some of the e's may lie in neither C nor D; but readily, the operation of BF is independent of the values of these e's.) By Proposition 2, it thus suffices to prove that BF works provided C has at most one hole. Note that since the background still consists of 0's, C is still finite.

If C has no holes, it is simply-connected, and the proof in (Rosenfeld, 1970) applies. If C has exactly one hole, by Theorem 4.4 of (Rosenfeld, 1973), it has deletable points unless it is a simple curve (see there for the definitions). But readily, BF works for simple curves; hence here again we can use an induction argument involving deletable points, analogous to the argument used in (Rosenfeld, 1970).

3. THE ADJACENCY TREE

Let B be the background component of $\bar{S}$, and let F, G be distinct components of S or $\bar{S}$. We say that G *surrounds* F, or that F is *inside* G, if any 4-path from any point of F to any point of B must meet G. Readily, B surrounds every other component of S or $\bar{S}$. It is also easily shown that "surrounds" and "inside" are strict partial order relations, i.e.,

(1) if F is inside G, then G is not inside F;

(2) if F is inside G, and G is inside H, then F is inside H.

Our first goal in this section is to prove

THEOREM 3. *Let C be a component of S, and D a component of $\bar{S}$ that is adjacent to C. Then either C surrounds D or D surrounds C; moreover, exactly one such D surrounds C.*

Proof. Let D^* be the component of $\bar{C}$ that contains D (see Proposition 1). Let c^*, d^*, and c, d be pairs of 4-adjacent points in C, D^* and C, D, respectively. If we use BF starting from c^*, d^*, we must eventually obtain c as one of the c's and d as one of the e's, since BF finds the entire D^*-border of C. But readily, the e's having value 0 that are examined by BF are all 4-connected (or, in the modified version of BF, 8-connected); hence d^* is connected to d, so that d^* is in D.

Let D, E be distinct components of $\bar{S}$ that are adjacent to C. By the preceding paragraph, D and E cannot be in the same component of $\bar{C}$. In particular, at most one D_0 can be in the background component of $\bar{C}$; thus every D except for D_0 is inside C. On the other hand, let d_0 be the point to the right of a rightmost point of C; thus d_0 has value 0, and so lies in some D_0, and the right half-line emanating from d_0, which eventually reaches B, never meets C, so that D_0 cannot be inside C. Let ρ be any 4-path from C to B; then ρ meets at least one component of $\bar{S}$ that is adjacent to C, since it must leave C at some border of C. Let D_1 be the last such component that ρ meets; then ρ cannot meet C after leaving D_1, so that D_1 cannot be inside C, which proves that $D_1 = D_0$, and this shows that D_0 surrounds C. ▌

It follows that any C has exactly one *outer border*, namely its D_0-border, where D_0 is the unique adjacent component of $\bar{S}$ that surrounds C; the other D-borders of C, if any, will be called its *inner borders* or *hole borders*. Analogous results hold with S and $\bar{S}$ interchanged, except that the component B of $\bar{S}$ has no outer border.

Let T be the graph whose nodes are the components of S and $\bar{S}$, and in which two nodes are joined by an arc if the corresponding components are adjacent. As shown above, if D, E are adjacent to C, they cannot be in the same component of $\bar{C}$, i.e., any path joining them must meet C. Thus deletion of node C from T disconnects any two of the neighboring nodes. It follows that T is acyclic, hence is a tree. We shall call T the *adjacency tree* of S. It is convenient to regard node B as the root of T. We can then represent T by a well-formed parenthesis string, with B corresponding to the outermost pair of parentheses.

It should be pointed out that a number of algorithms for finding and marking borders (e.g., Minsky and Papert (1969, pp. 136–139); Rosenfeld (1969, pp. 137–8)) depend on the above results. Specifically, these algorithms make use of the fact that in a TV raster scan of a binary-valued digital picture (i.e., a rectangular array of 0's and 1's, with no 1's on its edges), the outer border of any C is always first encountered as a left–right transition from 0 to 1, while inner borders are first encountered as left–right transitions from 1 to 0. To prove this, let D_0 surround C, and let c be the leftmost of the uppermost points of C; thus the left-hand neighbor d of c is 0. Since the left half-line emanating from d does not meet C, d cannot be inside C; hence d is in D_0. Similarly, let C surround D, and let d be the leftmost of the uppermost points of D; thus the left-hand neighbor c of d is 1. Since the left half-line emanating from c does not meet D, c cannot be inside D; hence c must be in the unique adjacent component of 1's that surround D, i.e., in C.

4. Bicurves

The results in Section 3 were based on the validity of the general BF algorithm of Section 2, which in turn depended on the results of (Rosenfeld, 1970, 1973). In this section we present a more direct method of establishing these results. In what follows, a *pair* (x, y) means a pair of 4-adjacent points, where x has value 1 and y has value 0; we assume, as usual, that the set S of 1's is finite.

We say that two pairs (x, y), (u, v) are *adjacent* if any of the following conditions holds (compare Section 6 of (Rosenfeld, 1970)).

(a) x is 4-adjacent to u and y is 4-adjacent to v.

(b) $v = y$ and x is 8-adjacent to u.

(c) $u = x$, and y is 8-adjacent to v, and the other common 4-neighbor of y and v has value 0.

Thus (x, y) and (u, v) are adjacent in the following cases.

$$\begin{array}{ll} x & u; \\ y & v; \end{array} \qquad \begin{array}{lll} x & & w \\ y = v & & u \end{array} \text{ whether } w = 0 \text{ or } 1; \qquad \begin{array}{ll} x = u & v. \\ y & 0. \end{array}$$

This definition is designed for the case where we use 8-connectedness for the 1's and 4-connectedness for the 0's. To handle the opposite case, we need only assume that first terms of pairs are 0's and second terms 1's, and replace "0" by "1" in (c).

PROPOSITION 4. *Any pair has exactly two adjacent pairs, one in each of the two-by-two squares that contain the given pair.*

A sequence of distinct pairs $(x_1, y_1), \ldots, (x_n, y_n)$ will be called a *bipath* if (x_i, y_i) is adjacent to (x_{i+1}, y_{i+1}), $1 \leqslant i < n$. A bipath will be called a *bicurve* if $n > 2$ and (x_n, y_n) is adjacent to (x_1, y_1). By Proposition 4, in any bicurve, the only pairs adjacent to (x_i, y_i) are the two pairs $(x_{i\pm 1}, y_{i\pm 1})$(modulo n). Note also that the 1's in two adjacent pairs are always 8-connected, and the 0's always 4-connected; hence the first terms of the pairs in any bipath are 8-connected, and the second terms 4-connected.

PROPOSITION 5. *Distinct bicurves are disjoint.*

PROPOSITION 6. *Let P be a set of pairs such that, for all $(x, y) \in P$, exactly two pairs of P are adjacent to (x, y). Then P is a union of bicurves.*

Let β be a bicurve, p any point, and let H_p be the right half-line emanating from p. We say that H_p *crosses* β at (x_i, y_i), where (x_i, y_i) is a pair in β, if x_i and y_i both lie on H_p. We say that p is *inside* β if the number of times H_p crosses β is odd; *outside*, if the number is even.

PROPOSITION 7. *If p and q are 4-adjacent and are not a pair in β, or if they are 8-adjacent 1's, they are either both inside or both outside β. If they are a pair in β, one of them is inside and the other outside.*

Proof. This is clear if p and q are horizontally adjacent. Suppose they are vertically adjacent and are not a pair in β. Let $p = (x_r, y_r), \ldots, (x_s, y_s)$ be a run of pairs in β both of whose terms lie on $H_p \cup H_q$; thus both terms of (x_{r-1}, y_{r-1}) and (x_{s+1}, y_{s+1}) do not lie on $H_p \cup H_q$. It is easily verified that for any such run, either H_p crosses β twice and H_q does not cross it; or vice versa; or both H_p and H_q cross β once. Thus in any case, the difference between the numbers of crossings is even, and since this is true for any run, p and q are either both inside or both outside β.

A similar argument applies if p and q are diagonally adjacent 1's; one must verify here that no trouble can arise in the case of a run just at the beginning of $H_p \cup H_q$. Note that if p and q are diagonally adjacent 0's, a run can cross H_p at its beginning without crossing H_q, e.g., for $\begin{smallmatrix} p & u & z \\ x & q & w \end{smallmatrix}$, where $x = u = 1$, $z = w = 0$, we can have the pairs (x, q), (u, q), (u, z).

Finally, if p and q are a pair in β, there is in fact a run just at the beginning of $H_p \cup H_q$, namely the run containing (p, q); and this run crosses either H_p or H_q, but not both. All other runs still give even differences in the numbers of crossings; hence the total numbers have opposite parity, so that p, q cannot be both inside or both outside. ∎

Corollary 8. *If p and q are 4-connected 0's or 8-connected 1's, they are either both inside or both outside β.*

Corollary 9. *The background 4-component B of 0's is outside β.*

Corollary 10. *Let p be inside β, and let $t_1, \ldots, t_n$ be a 4-path from p to a point of B; then (t_i, t_{i+1}) is a pair in β for some $1 \leqslant i < n$.*

Let C be any component of S, and D any component of $\bar{S}$ that is adjacent to C. Let (C, D) be the set of pairs (x, y) with $x \in C$, $y \in D$. It is easily verified that (C, D) has the property of Proposition 6, hence is a union of bicurves. Let (u, v) be a rightmost pair in (C, D), i.e., no term of any pair in (C, D) is further to the right than u or v. Suppose that (u, v) is a pair in the bicurve β of (C, D). Readily, u and v must be horizontally adjacent. Thus if u is to the left of v, C is inside β and D outside, and if v is to the left of u, the reverse is true. Moreover, let β_1 be any other bicurve of (C, D), and let (c_1, d_1) be a pair in β_1. Then c_1 is inside β and d_1 outside (or vice versa); but (c_1, d_1) isn't a pair in β (Proposition 5), which contradicts Proposition 7. We have thus shown that (C, D) is a single bicurve, with C inside it and D outside it (or vice versa). It follows from Corollary 10 that C is inside D (or vice versa).

Finally, let C be inside both D and E; let (c', d) and (c'', e) be pairs in (C, D) and (C, E), respectively. Evidently these bicurves are distinct, hence disjoint; thus by Proposition 7, d is inside (C, E) and e inside (C, D), so that by Corollaries 8 and 10, D is inside E and E inside D, contradiction. Hence at most one D adjacent to C can have C inside it; all other such D's must be inside C.

Evidently, if D is inside C and C is inside E, any path from D to E must meet C. Let D and E both be inside C; then E is outside (C, D), while D is inside it, so that any path from D to E must cross (C, D)(in the sense of

Corollary 10), hence must meet C. Using these results together with the analogous results with 1's and 0's interchanged, we thus have once again proved that the adjacency graph T is a tree.

5. Construction of the Adjacency Tree

Suppose that we divide the background component B into two connected parts B_1 and B_2, where B_1 is finite and surrounds the other components to which B was adjacent, and B_2 surrounds B_1. The adjacency tree of components, with B_1 replacing B and B_2 excluded, is unchanged; and we can now describe algorithms for constructing this tree by scanning the finite set $I \times I - B_2$. Note that in particular, if we take B_2 to be the outside of a large rectangle, $I \times I - B_2$ is just a binary-valued digital picture (i.e., a rectangular array of 0's and 1's) with 0's outside it.

Our first (well-known) method of constructing the tree depends for its validity on the following observations: If C is inside D and has no holes, for any x in C there exists y on the D-border of C, to the left of x, with nothing but 1's between y and x; for, as we move left from x, the 1 that we pass just before we first hit a 0 is a border point of C, and this can only be on the D-border. Conversely, if C has holes, there exists a point z on the C-border of some hole, and y on the D-border of C to the left of z, with nothing but 1's between y and z. Indeed, the lefthand neighbor z of a leftmost hole point is in C, and there can be nothing but 1's between y and z. The above assumes that C is a component of 1's; the analogous observations also hold with 1's and 0's interchanged.

To construct the tree, we start at a point of the outer border of B_1. We uniquely mark this starting point, and use BF to mark the entire border, say with stars. We then use BF again, but this time, after we move to each border point b, we also scan to the right through 0's or primed points inside B_1, marking any nonprimed 0's with primes as we go, until we hit a starred point. We then move leftward back to b (which is identifiable, since it is starred) and resume BF. If we can finish this process and return to the starting point, we can conclude by the remarks in the preceding paragraph that B_1 has no holes. Thus in this case, the tree consists of a single pair of parentheses. On the other hand, if B_1 has holes, we will hit a 1 at some stage before hitting a starred point; this 1 is on the B_1-border of some component C of 1's adjacent to B_1.

Suppose, in general, that we have already constructed the tree for all components outside C, and that we have just found C by hitting a point of its

643/26/1-3

outside border, as in the preceding paragraph. The "current" pair of parentheses π in the string being constructed represented the component (D, say) adjacent to and surrounding C; when we hit C, we add a new pair π' inside π, representing C itself, and π' becomes the new current pair. We then treat C exactly the way we treated B_1 in the preceding paragraph (but with 1's and 0's interchanged, if C is a component of 1's). If we find a hole E in C, we add a new current pair of parentheses inside π', and repeat the process of the preceding paragraph. If not, we can finish the scan of C, so that it has no holes, and π' is an innermost pair of parentheses in the string. In this case we once again use BF to traverse the outer border of C and turn all stars into primes; thus every point of C has now been primed. We now revert to π as the current pair of parentheses, and resume scanning D. If D has other holes that have not yet been primed, this scan will find them; if not, we will be able to finish scanning D and resume scanning its surrounding component. Ultimately, this process will enable us to finish scanning B_1, at which point the procedure terminates.

We conclude by describing a second method of constructing the adjacency tree; this is a simplified version of a procedure due to (Bunemann, 1969). Here we shall assume that we are given a rectangular array of 0's and 1's with no 1's at its edges; i.e., the top and bottom rows, and leftmost and rightmost columns, are all 0's, all of which belong to the background component B. In the process of constructing a parenthesis-string representation of the adjacency tree, we will use brackets instead of parentheses to represent components that are currently being tracked. We assume that we have some way of telling, for any run of 1's or 0's on the current row of the array, to which pair of brackets it corresponds.

We scan the top row of the array and construct an initial string consisting of a single pair of brackets, corresponding to the background region. Suppose that we have scanned the first k rows, and have constructed a string of brackets and parentheses representing the adjacency tree for the subarray consisting of these k rows; we can certainly do so for $k = 1$. We now scan the $(k + 1)$ st row and update the string as follows.

(a) If a component C, say of 1's, that met the kth row does not meet the $(k + 1)$ st row, i.e., none of C's runs on the kth row is adjacent to any run of 1's on the $(k + 1)$ st row, we turn C's brackets into parentheses. In particular, if the kth row was the bottom row, the outermost pair of brackets turns into parentheses; by our assumption that only the background component meets the bottom row, this is the only pair of brackets in the string at that stage.

(b) If a component C, say of 1's, that met the kth row, splits on the $(k + 1)$ st row, i.e., one of C's runs is adjacent to two or more runs of 1's on the $(k + 1)$ st row, we insert one or more new pairs of brackets inside C's brackets, corresponding to the new runs (of 0's) on the $(k + 1)$ st row, each of which may be the beginning of a hole in C.

(c) If two (or more) components D, E, say of 1's, that met the kth row merge on the $(k + 1)$ st row, i.e., a run of 1's on the $(k + 1)$ st row is adjacent to runs belonging to both D and E, we combine their brackets into a single pair; the substrings (if any) that were inside their brackets are concatenated and are put immediately inside this new pair of brackets.

It is not hard to see that the resulting updated string correctly represents the adjacency tree for the subarray consisting of the first $k + 1$ rows.

RECEIVED: December 4, 1973

REFERENCES

BUNEMAN, O. P. (1969), A grammar for the topological analysis of plane figures, *in* "Machine Intelligence 5" (B. Meltzer and D. Michie, Eds.), pp. 383–393, Edinburgh University Press.

MINSKY, M. AND PAPERT, S. (1969), "Perceptrons," M.I.T. Press, Cambridge, Mass.

ROSENFELD, A. (1969), "Picture Processing by Computer," Academic Press, New York.

ROSENFELD, A. (1970), Connectivity in digital pictures, *J. Assoc. Comput. Mach.* **17**, 146–160.

ROSENFELD, A. (1973), Arcs and curves in digital pictures, *J. Assoc. Comput. Mach.* **20**, 81–87.

Scene Analysis Using Regions

Claude R. Brice and
Claude L. Fennema

Stanford Research Institute, Menlo Park, California

Recommended by D. Michie

ABSTRACT

One of the vision projects of the Stanford Research Institute Artificial Intelligence Group is described. The method employed uses regions as basic data and progresses by successive partitioning of the picture toward an interpretable "goal partition," which is then explored by a heuristic decision tree. A general structure is discussed and an example problem is shown in detail.

[1] The research reported here was supported by the Advanced Research Projects Agency and the Rome Air Development Center under Contract F30602–69–C–0056, and is continuing under Contract NAS 12–2221 with the National Aeronautics and Space Administration and the Advanced Research Projects Agency.

1. Introduction

One important component of the SRI automation project is a set of programs that provide the automaton with a means of interpreting visual data. This automaton, or robot, described in detail by Nilsson [1] and by Munson [2], is equipped with touch sensors, a range finder, and a standard vidicon camera. The output of the camera is transmitted by microwave to an analog-to-digital converter which produces a 120 × 120 digitized picture with 16 levels of gray.

In the early stages of the robot project a prime concern was the simple task of navigation, for which the vision programs needed only to find areas of environment through which the robot could freely move. More recently, however, interest has turned to the more advanced tasks of collecting various objects and using tools to manipulate these objects. For these tasks it has become necessary for the programs *to isolate and to recognize* objects.

The classical paper of Roberts [3] describes a method for recognizing objects by transforming a digitized gray scale picture into a line drawing, which he then analyzes using mathematical models and projective transformations. His gray scale-to-line drawing transformation applied a local edge finder to the picture to find points through which edges probably pass, and then fit straight lines to these points. His mathematical models were then

Reprinted with permission from *Artificial Intelligence*, vol. 1, pp. 205–226, Fall 1970.

transformed by projection and rotation to find the model that best explained (or fit) the line drawing.

Roberts' treatment of objects was elegant, and many who followed sought to improve the technique with better edge finders and line detectors. Many of these results are described and referenced in the work of A. Rosenfeld [4]; other interesting approaches are described by Duda and Hart [5] and by Feldman [6].

In 1968 Guzman's thesis (Guzman [8]) described a way to isolate objects from a line drawing by observing the vertex configurations of the regions of the picture. This led to efforts to find regions as well as their boundaries. The Guzman techniques would then group these regions into objects that can be recognized either by Roberts' model matching or by simpler means.

Marvin Minsky and Seymour Papert [7] are perhaps the first persons who seriously investigated the direct transformation of a gray scale picture to regions, bypassing the edge-finding, line-fitting procedures. The method they describe constructs *regions that are the union of squares* whose corners have the same or nearly the same gray scale.

The process described in this paper breaks the digitized picture into atomic regions of uniform gray scale. Then a pair of heuristics is used to join these regions in such a way as to obtain regions whose boundaries are determined more by the natural lines of the scene than by the artificial ones introduced by quantization and noise. Next, a simple line-fitting technique is used to approximate the region boundaries by straight lines, and finally the scene analyzer interprets the picture using some simple tests on object groups generated by a Guzman-like procedure.

In Section 2, the region-oriented structure is described, together with its associated operators and information retrieval functions. Section 3 then describes the preprocessing heuristics and the line-fitting procedure, and Section 4 describes the scene analyzer.

2. The Data Representation and Related Operations

While it is possible to represent the information in a scene accurately by a digitized gray scale picture, this representation alone does not put emphasis on the interesting properties of the scene. It would be more convenient to have a description in terms of the "natural" elements of the picture, such as regions and lines. For example, if we had an ideal picture of a cube, a natural description would be in terms of regions corresponding to the quadrilaterals and the lines corresponding to the boundaries of the quadrilaterals. In reality, pictures are far from ideal, and many steps of processing are necessary; however, by using regions and lines to describe the data of each step we will be working in terms of the elements of the picture. The boundaries, regions, and their properties will be reorganized, but they will be represented in a uniform manner. Furthermore, describing pictures in terms of regions provides easy access to some useful global information.

2.1. Preliminary Definitions

A picture is first stored in the computer as an array P on an $n \times m$ grid G. The array P is a function on G, whose value on each point is the gray scale of that point. (More generally, $P(i, j)$ could also include other properties such as range, color, and texture.) Each pair (i, j) is called a *picture element*, whose gray scale is the value of $P(i, j)$ and whose coordinates are i and j; thus a picture may be thought of as a set of picture elements (provided we remember the topological structure imposed by grid G).

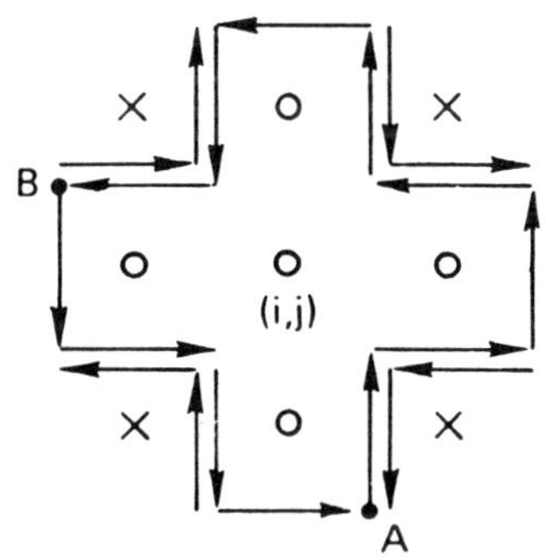

FIG. 1. A region made up of grind point (i, j) and its four neighbors.

In what follows we often refer to *neighbors* of a point (i, j); by this we mean any of the four points that are nondiagonally adjacent (see Fig. 1). We say that two points p_1 and p_2 belonging to a subset R of G are connected with respect to the set R, provided that there exists a sequence of points from R, the first of which is p_1 and the last of which is p_2, and such that consecutive points are neighbors. With this definition in mind, we then define a *region* as a set of $R \subseteq G$ in which any pair of points is connected with respect to R. These regions are the basic elements of the picture representation.

A *partition* of a set X is any collection of sets $\{R_1, R_2, \ldots, R_n\}$ such that the union of the R_k is exactly X and the pairwise intersection of the R_k is nil unless the two sets are identical.

If we define some equivalence relation on the array P—say, as a trivial example, $P(i, j)$ is equivalent to $P(k, l)$ if their values are equal—then this in turn induces a natural equivalence relation on G given by: (i, j) is equivalent to (k, l) is and only if $P(i, j)$ is equivalent to $P(k, l)$.

Any equivalence relation on G yields a partition of G into equivalence classes. These classes can be further broken down into maximally connected subsets called connected *components*; we call these homogeneous connected

Artificial Intelligence **1** (1970), 205–226

components *atomic regions*. Using the equivalence relation induced by the equality of the gray scale, the atomic regions are obtained by the first step of our technique. They are the building blocks that, if properly joined, will give us a representation of the picture in a form usable in the interpretation methods.

2.2. Region Representation

The process described in this paper makes extensive use not only of the region boundaries, but also of the local properties of the points they surround. Furthermore, during the analysis, regions are constantly being joined with other regions, and sometimes a single region is split into two. This makes it difficult, if not impossible, to use the region representation schemes of Cook [9] and Pfaltz *et al.* [10], whose main concern is with the boundaries themselves and the relationship of the regions with the regions with which they share a common boundary.

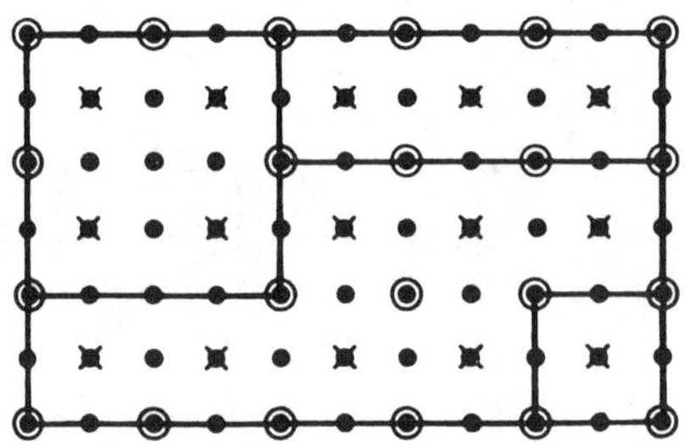

FIG. 2. Four regions represented in the super grid S. The region points are marked with a cross, and the boundary points with a circle.

One may think of the picture grid G as a subgrid of some supergrid S. If G is an $n \times m$ grid, then S is $(2n + 1) \times (2m + 1)$, and the points (i, j) of G are those where i and j are both odd (see Fig. 2). The points of the regions here are represented by appropriate points in G, and the boundaries of these regions are closed curves made up of horizontal and vertical line segments whose endpoints are points in S that have even coordinates (see Fig. 2). These boundaries are then between the points they separate. The points of the subgrid B of boundary segment endpoints are of some importance later; we call them boundary points.

Representing regions in this manner admits a simple algorithm for finding the atomic regions of a picture. Each point of G (except for edge effects) is compared with the one above it and the one to its right and, if a difference in gray scale is encountered, the boundary segment is inserted between them (Fig. 3). When each point has been considered, the grid is partitioned into regions.

Artificial Intelligence **1** (1970), 205–226

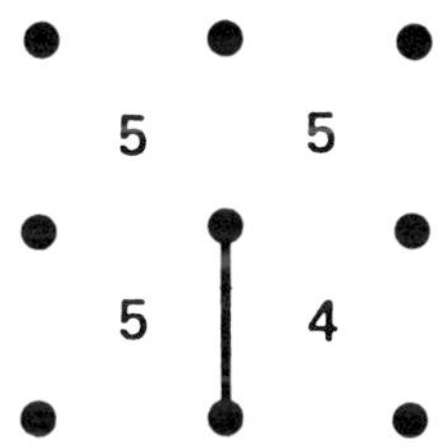

FIG. 3. Insertion of boundary segment. When the lower left-hand point (each point of G is replaced by a number representing its gray scale) is compared with the point to its right, a difference is encountered and the boundary segment is inserted as shown; a second boundary segment will be inserted when the "4" is compared with the point above it.

2.3. Important Operations

At all times the region boundaries are assumed to be oriented in such a way that the region lies to the left of the boundary. This allows a simple means of performing the join operation MERGE. This operation merges two regions into one by adding their boundaries as in Fig. 4.

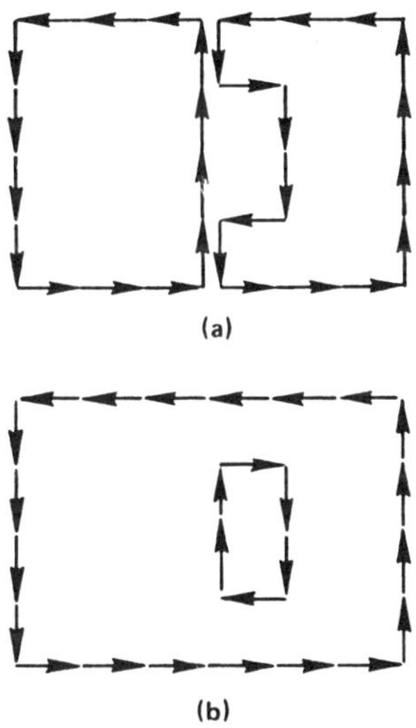

FIG. 4. Addition of the boundaries of two regions, as shown in (a), results in cancellation of the common part, as shown in (b).

A second operation, CUT, completes the list of operations that affect the picture partition. This operation splits a region along a straight line (see Fig. 5). A more general operation can be defined to cut along arbitrary curves, but straight lines are sufficient in our current problem domain.

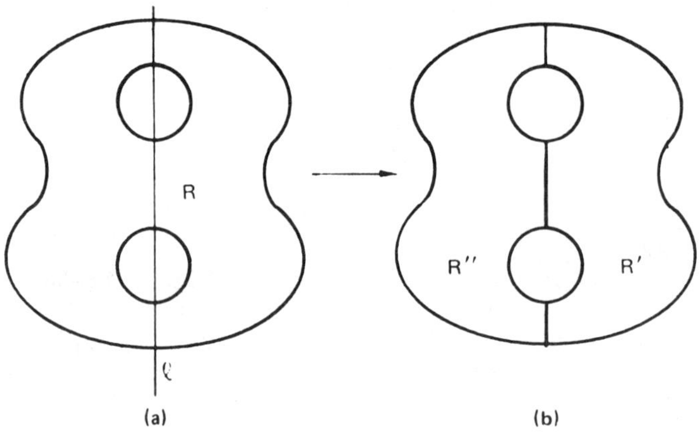

FIG. 5. The operator *cut* divides the region R into two parts, R' and R'', along the line l.

FIG. 6. Original optical photographs of the examples used in this paper.

Starting with the original partition into atomic regions, two heuristics are used to guide these operations in their repartitioning of the picture. We next describe these heuristics.

3. Initial Processing

The TV pictures (as in Fig. 6) are first digitized (Fig. 7), then partitioned into homogeneous connected components. As can be seen from Fig. 8, this first partition does not typically permit simple interpretation within the context of the problem domain. Many false boundaries are created by lens distortion, uneven illumination, shadows, reflections, distance from the light source, noise, and nonuniformities in the composition of the surfaces of the objects.

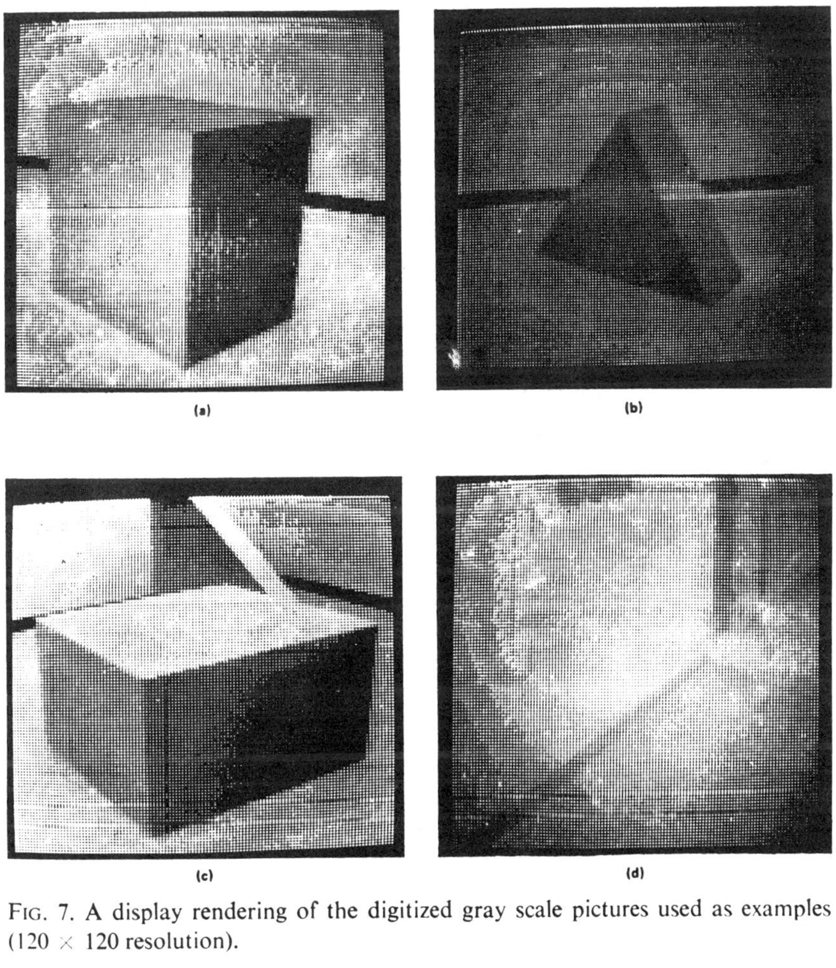

(a) (b) (c) (d)

FIG. 7. A display rendering of the digitized gray scale pictures used as examples (120 × 120 resolution).

Furthermore, the straight lines of the pictures are not easily extracted from their quantized representations.

In this section we describe two heuristics that effect a repartition of a picture, and a process for fitting lines to the boundaries.

FIG. 8. The initial partition into atomic regions of the gray scale pictures of Fig. 6 (60 × 60 resolution).

3.1. Repartitioning

One approach to obtaining the desired regions of the picture is to group points if certain of their properties are not too different. It is difficult, however, to use such a method without producing regions that extend across the natural lines of the picture. If, for example, two natural regions are even only locally—not too different across the boundary that separates them—they will be considered as one. The problem becomes, then, one of breaking regions into the natural regions they contain. Then a picture partition will often consist of only one region with many boundaries. This presents problems quite similar to those of analyzing a gradient picture.

The procedure chosen here is to start with the atomic regions and use more global criteria to join them. Because these criteria consider the entire boundaries of two regions to be joined, the regions are not so often erroneously joined.

The assumption used to further process such a picture is that boundaries between regions belonging to the same surface generally are not so "strong" as boundaries between regions from different surfaces. On the other hand, while it is often true that intersections of two different surfaces produce the strong boundary segments, this criterion of boundary strength is not sufficiently reliable. More sophisticated criteria are needed to form significant regions. The heuristics described below guide the merging of regions to yield a new partition of the picture that respects the natural lines of the picture.

3.1.1. THE PHAGOCYTE HEURISTIC. The first heuristic tends to guide the merging of regions in such a way as to smooth or shorten the resulting boundary. Two regions that differ strongly along their common boundary are never joined; but, even if this boundary is weak, they are joined only if the resulting boundary does not grow too fast.

More precisely, the *strength* of each elementary vector of a boundary is defined as the magnitude of the difference between the properties of the picture elements on the right and on the left of this vector—in our case, the absolute value of the difference in gray scale of the two elements. The strength of a curve is the average of the strength of each vector of the curve. We are particularly interested in those parts of boundaries where the strength is small, since they are potentially places where regions can be joined. We define the length W of the *weak* part of the boundary between two regions as the number of boundary vectors having a strength less than some threshold σ. Then the phagocyte heuristic is to merge adjacent regions R_i and R_j if $W/PM > \theta_1$, where $PM = \min(P_i, P_j)$, P_i is the perimeter of R_i, P_j is the perimeter of R_j, and θ_1 is a threshold. The threshold σ is hardware-dependent, and grows with the number of levels of gray and the dynamic range of the picture. The threshold θ_1 is important. If θ_1 is small, the criterion is weak and many regions may be joined. On the other hand, if θ_1 is large, two regions are joined only if one of the regions almost surrounds the other.

Another way of analyzing this heuristic is to observe the growth of the boundary of the resulting region as new regions are joined. If I is the length of the boundary between R_i and R_j, the perimeter P_R of the union of R_i and R_j is

$$P_R = P_i + P_j - 2I$$

and

$$I/\theta_1 > W/\theta_1 > PM = P_j$$

say. Then

$$P_R < P_i + I\left(\frac{1}{\theta_1} - 2\right)$$

This formula shows how the growth of the perimeter P_R is related to θ_1. Note that the value $\theta_1 = \frac{1}{2}$ is significant. For $\theta_1 > \frac{1}{2}$ the boundary must shrink, and for $\theta_1 < \frac{1}{2}$ it is allowed to grow. In Fig. 9 the regions in part (*a*) would be joined if $\theta_1 > \frac{1}{2}$, but those in (b) would not. If θ_1 is sufficiently small, however, both configurations could be joined.

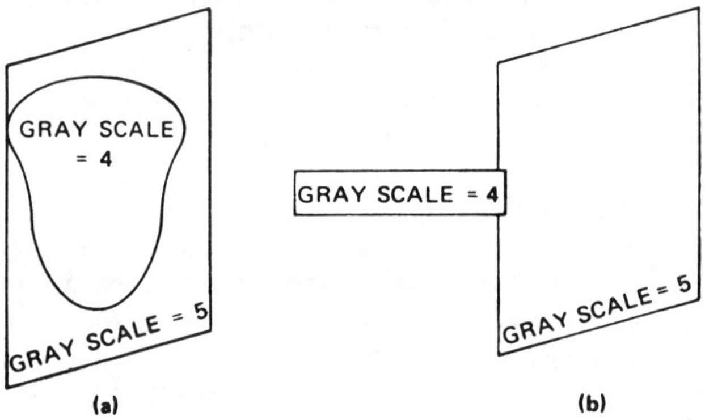

FIG. 9. The phagocyte heuristic will for $\theta_1 = \frac{1}{2}$ join the regions of (a) but not those of (b).

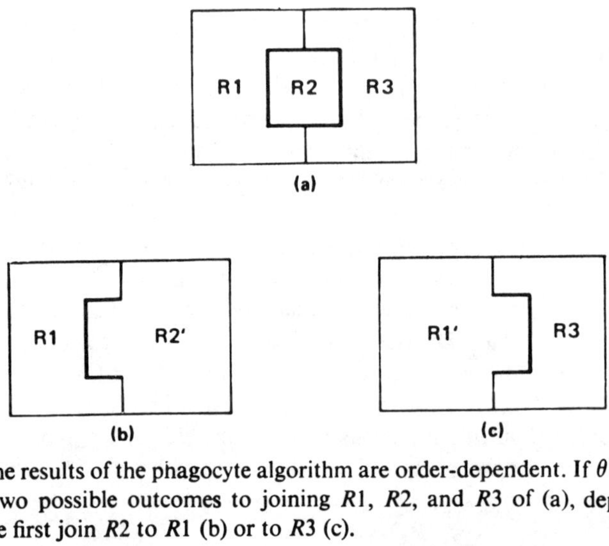

FIG. 10. The results of the phagocyte algorithm are order-dependent. If $\theta_1 = \frac{1}{2}$, then there are two possible outcomes to joining *R*1, *R*2, and *R*3 of (a), depending on whether we first join *R*2 to *R*1 (b) or to *R*3 (c).

Artificial Intelligence **1** (1970), 205–226

Theoretically, this criterion is order-dependent. For example, three regions of Fig. 10 will be joined differently if we first join R_1 and R_2 from the way they will be joined if we first join R_2 and R_3 (let $\theta_1 = 0.5$). However, in our experience this effect has not proved to be very important. This criterion

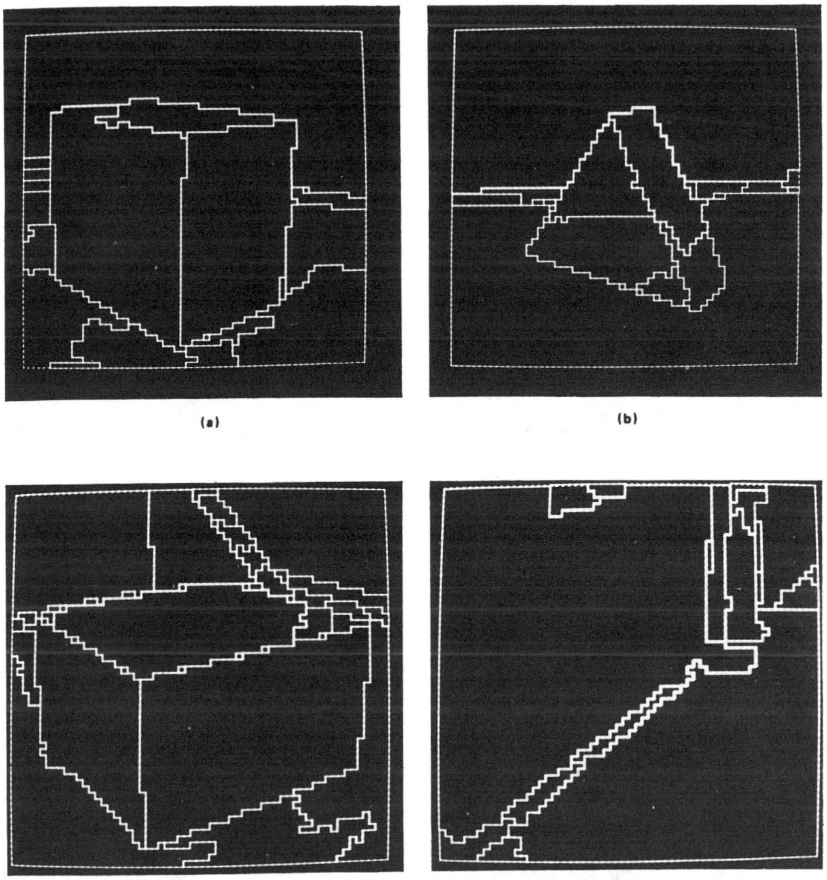

(a) (b) (c) (d)

FIG. 11. Results of the phagocyte heuristic applied to the partitioned pictures (60 × 60 resolution).

is applied to pictures until it is no longer satisfied. Results of this heuristic on real examples have not been extensively tested, but some typical results are shown in Fig. 11. The "smoothing" character allows the complete joining of the cube in Fig. 11a, but will not allow joining of parts of the face of the wedge (Fig. 11b). The thresholds here were $\sigma = 2$ and $\theta_1 = 0.45$.

3.1.2. THE WEAKNESS HEURISTIC. We have seen that the phagocyte heuristic cleans up the pictures considerably, but it is clear that many false separations still exist, and more processing is needed.

The second heuristic, the weakness heuristic, is more natural than the first in the sense that it joins regions solely on the basis of the strength of the boundary that separates them.

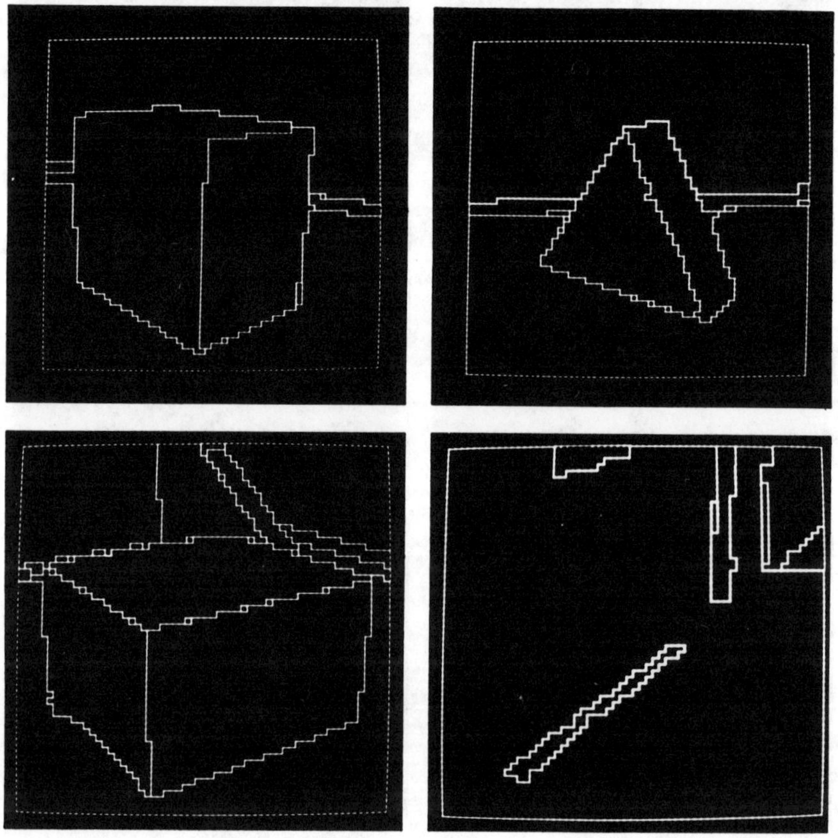

FIG. 12. Weakness heuristic applied to results of the phagocyte heuristic (60 × 60 resolution).

As before, W is the weak portion of the intersection I of two regions R_1 and R_2. Then we join R_1 and R_2 if $W/I > \theta_2$ (i.e. if the weak part is at least a certain percentage of the intersection). This heuristic is applied as before until no further merges are possible. Figure 12 shows the results (with $\theta_2 = 0.75$) of applying this criterion to the pictures of Fig. 11.

This heuristic is more natural than the first, and one might ask why not

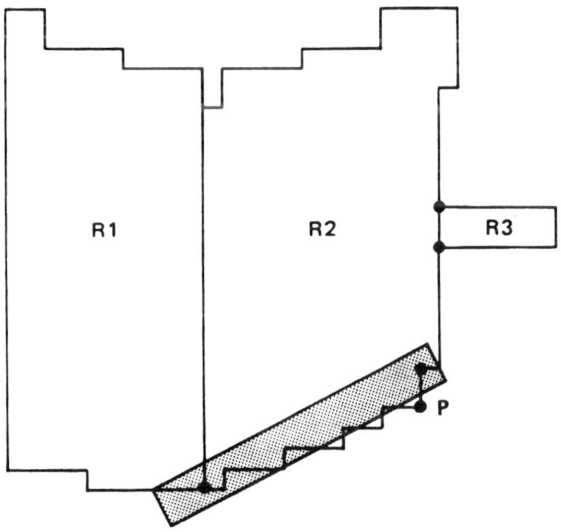

(a) POINT P LIES OUTSIDE THE MASK

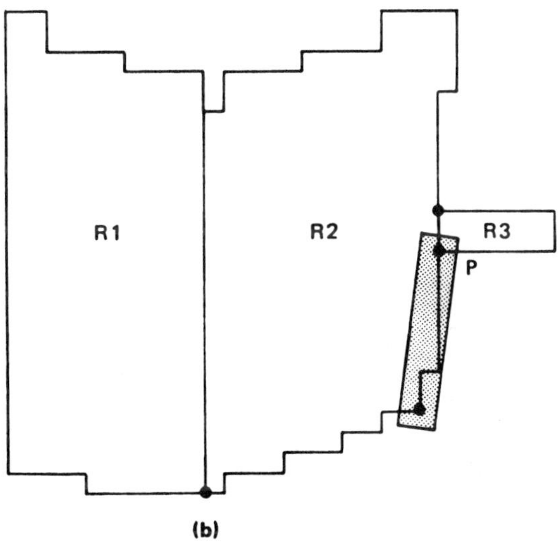

(b) VERTEX ENCOUNTERED AT P

FIG. 13. The use of a sliding mask to get an approximate line drawing, by using straight lines to represent curves.

simply apply it alone. The answer is that it is too local. The results are to wipe out almost all the regions. The phagocyte heuristic must be applied first.

These two heuristics yield a partition that admits a more simple interpretation than that of the original. It is also certainly possible that other heuristics, using more contextual information, could be applied. For example, a restrictive condition to guide the first two heuristics could be that merging should not be allowed if it tends to break long, straight lines. Better partitions could be obtained at greater computation cost, but it is felt here that further perfection lies at a higher level. Application of the two heuristics above yields a considerably cleaner picture which can be used for the purpose of scene analysis.

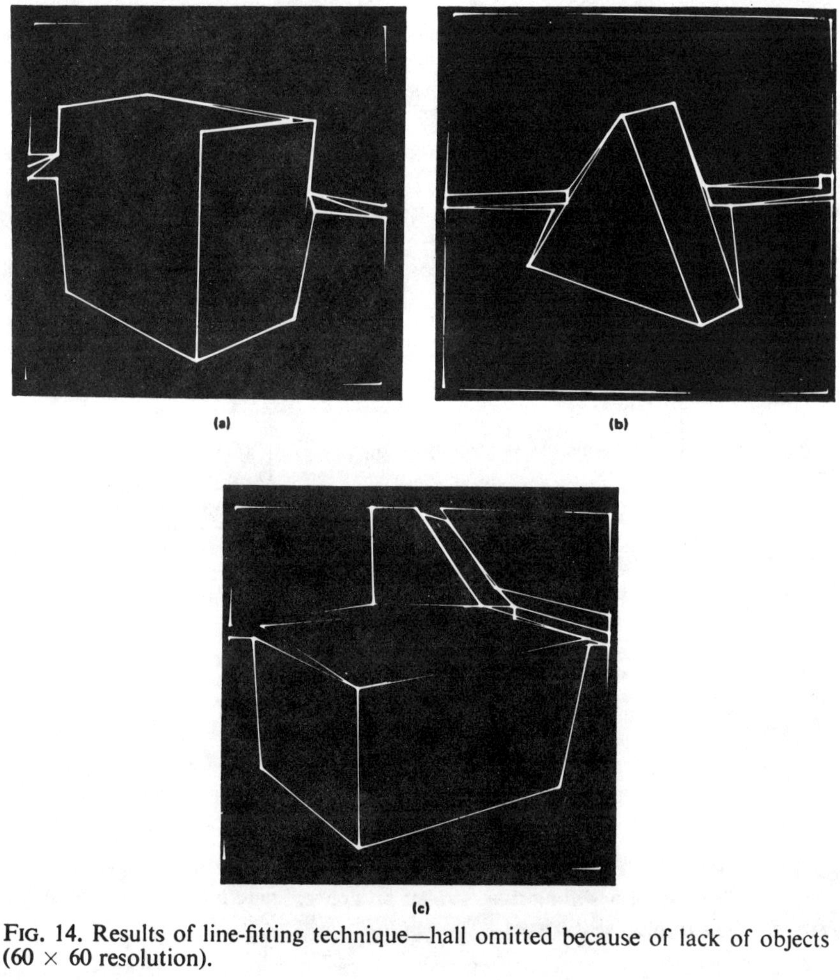

(a) (b) (c)

FIG. 14. Results of line-fitting technique—hall omitted because of lack of objects (60 × 60 resolution).

3.2. Line Representation

Even after the final partition is obtained, the boundaries of the regions are still represented as a list of small unit vectors. This representation is not easily used for shape analysis. We describe in this section a simple straight line-fitting program to represent the boundary in terms of long, straight lines.

The literature abounds with methods for fitting straight lines to curves, but we have chosen a simple sliding mask of fixed width together with some global criteria to get an approximate line drawing. This by no means gives a perfect line drawing but globalizes the information fed to the scene analyzer.

The operation consists of three passes. The first pass has as input data the low-level data structure of elementary vector curves. Starting with the lowest vertex (a point where three regions come together) on the boundary of a region, the algorithm applies a mask between successive points on the boundary until either a vertex is encountered or the mask is so situated that at least one intermediate point lies outside the mask (see Fig. 13). When a vertex is encountered, the line approximation is from the starting point to the vertex. Otherwise the endpoints of the line approximation are the starting point and the last point for which the mask covered the boundary. In either case, the procedure continues from the new endpoint. The second pass does the same thing on the output of the first with a more generous mask.

The third pass is similar, but the mask is still larger and vertices are ignored. This technique is somewhat crude, and Fig. 14 shows that the resulting line drawing is not perfect, but already there is enough information for the scene analyzer, which we describe next.

4. The Scene Analyzer

This scene analyzer, although simple in form, completes the present system. Its goal is to interpret the data in terms of the following object classes:

1. Wedges.
2. Cubes.
3. Wall.
4. Floor.

This problem domain is simple, but the problem is one of working with imperfect data; we deal here with missing lines, broken lines, and occlusion.

The well-known work of Roberts [3] shows well what can be done with projective transformations; but we have decided here to explore the use of clues combined with a two-dimensional description of the objects, a method that humans use heavily (Gibson [11]; Yarbus [12]).

As we said before, the goal of the scene analyzer is to give an interpretation of the picture data. Such an interpretation is, say, a list

((OB1 FLOOR) (OB2 WALL) (OB3 WEDGE))

where each OBN is a collection or group of interpreted regions. By this we mean that each region in this group is labeled as one of

(QUAD, TRIANGLE, PART OF FLOOR, PART OF WALL)

For each of the examples above we might well have

OB1 = ((R17 PART OF FLOOR) (R135 PART OF FLOOR))
OB2 = ((R22 PART OF WALL))
OB3 = ((R1 QUAD) (R2 TRIANGLE))

Basically, the scene analyzer extracts easily recognized regions first (here the floor and wall), groups the regions, using a simplified Guzman technique, into "objects," and then tries to recognize the faces of the objects. If recognition fails, it proposes lines, regroups regions, and begins again.

4.1. Extracting Easily Recognized Objects

We begin by extracting the floor and walls by using several strong hints provided by the nature of this particular environment.

1. Floor and wall are separated by a black baseboard of known height.
2. Floor and walls are light in intensity.
3. Wall is high in the picture.
4. Floor is low on the picture.

In many cases this is sufficient information to extract floor and walls readily—especially if any of the baseboard is visible. If none of these clues is available, the program has to go into a default condition and try to recognize some object in the scene.

4.2. Grouping Regions

The remaining regions are grouped by a simple Guzman-like technique. Two regions are linked at a vertex if they follow one another going clockwise around that vertex and if both have angles less than 90° at that vertex. We exclude vertices where any angle is 180°, and we do not treat objects that are occluded enough to be divided. These region groups, then, are tentative objects, and a figure recognition program tries to interpret the regions of this object.

4.3. Region Interpretation

The first region of an object is examined to see if it:

(1) has exactly four sides of reasonable length or

(2) has exactly three sides of reasonable length.

If it satisfies one of these two criteria it is labeled appropriately QUAD or

TRIANGLE. Otherwise we analyze the region boundary to find clues that might suggest missing information.

Clues are numerous, but essentially they come in two types:

1. The region boundary touches the edge of the picture frame, or some object that is not the floor or wall.
2. The region boundary contains some syntactic information such as collinear segments.

Clues of the first type indicate that this object may be occluded and that extending the broken lines into the occluding body may provide the needed information. This is illustrated by the wedge in the background of Fig. 14c. Here, extending the lines beyond the frame of the picture resolves the interpretation question—this is indeed a triangle.

Clues of type 2 indicate that there are missing lines that may be proposed in a way suggested by Minsky [7] if low-level information is present or simply inserted when the evidence is already strong enough. We often choose the latter approach, since the low-level information is sometimes absent.

4.4. Object Recognition

The object recognition program in its present state is quite unsophisticated. Its task is to control the grouping and region interpretation.

First, the regions remaining after the floor and wall have been extracted are grouped into tentative objects, and a region interpretation is attempted on the regions of the first of these objects. If all the regions of that object are interpretable, the object is marked as recognized and is given an interpretation. This object interpretation is simple:

Wedge = TRIANGLE OR TRIANGLE + QUAD

Cube = QUAD + QUAD + QUAD

Unknown = ANY OTHER COMBINATION

The unknown class has been invented to take care of ambiguous cases. Thus only two QUADS may be visible, and it is impossible by any means to say whether it is a cube or a wedge with triangular side up.

The region interpreter is not always successful, however. As in Fig. 14a, line extension is not enough and new regions are formed by inserting the missing lines. In this case the object recognizer makes the appropriate changes, regroups the regions, and begins again.

4.5. Example Runs

Figures 15a and 15b show the results of the program on two single scenes, one containing only a cube, the other containing only a wedge. It is worth noting that in Fig. 15a the missing lines (see Fig. 14a) have been entered exactly as described above.

Figure 16 summarizes the results of the program on the scene of Fig. 6c. The floor baseboard and wall provide no problems, and the regrouping gives us an easily recognized cube. One of the groups (see Fig. 16h) contains only one QUAD that did not satisfy any of the object criteria, and this was left as object unknown. It was, in fact, one reflection on the floor of the baseboard.

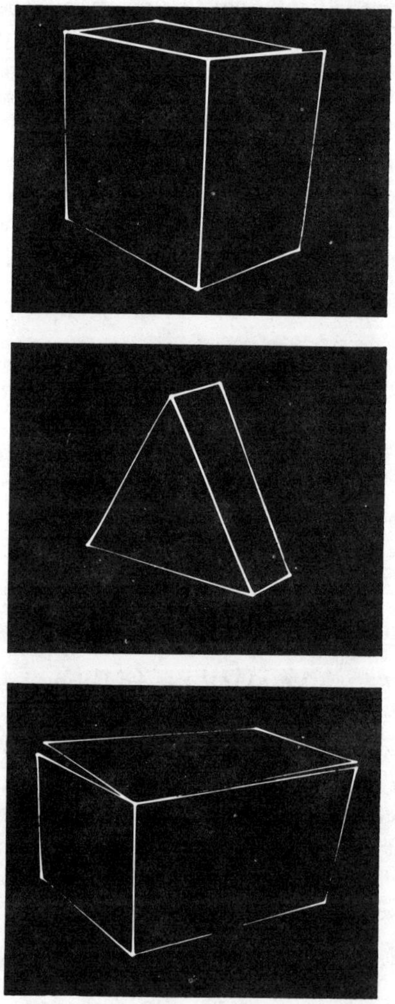

FIG. 15. Objects isolated, hall omitted owing to lack of objects (60 × 60 resolution).

Figure 16i shows the beginning of the wedge recognition. This triangle was occluded in the picture frame, as well as the cube, but recognition was accomplished by extending the lines above the picture frame. Because this extrapolated information is less reliable than that obtained from recognition of unoccluded objects, the occlusion is remembered.

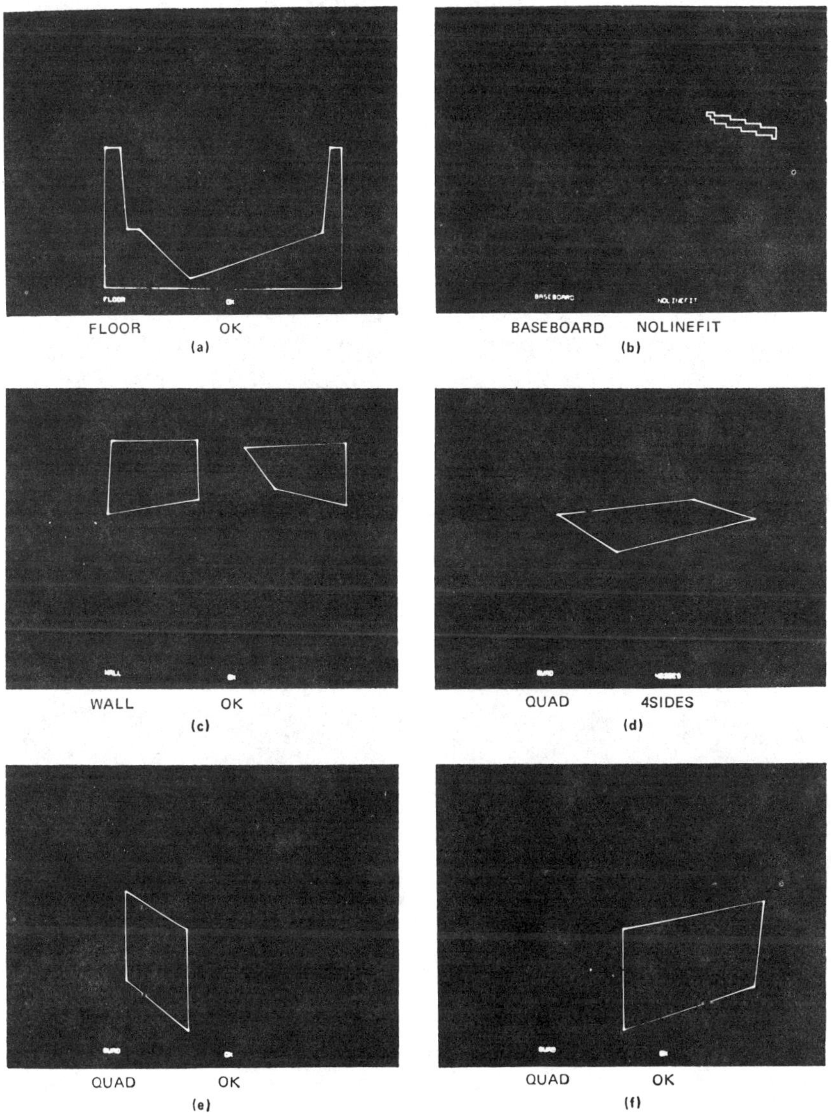

FIG. 16. A sample run of the scene analysis program showing intermediate steps.

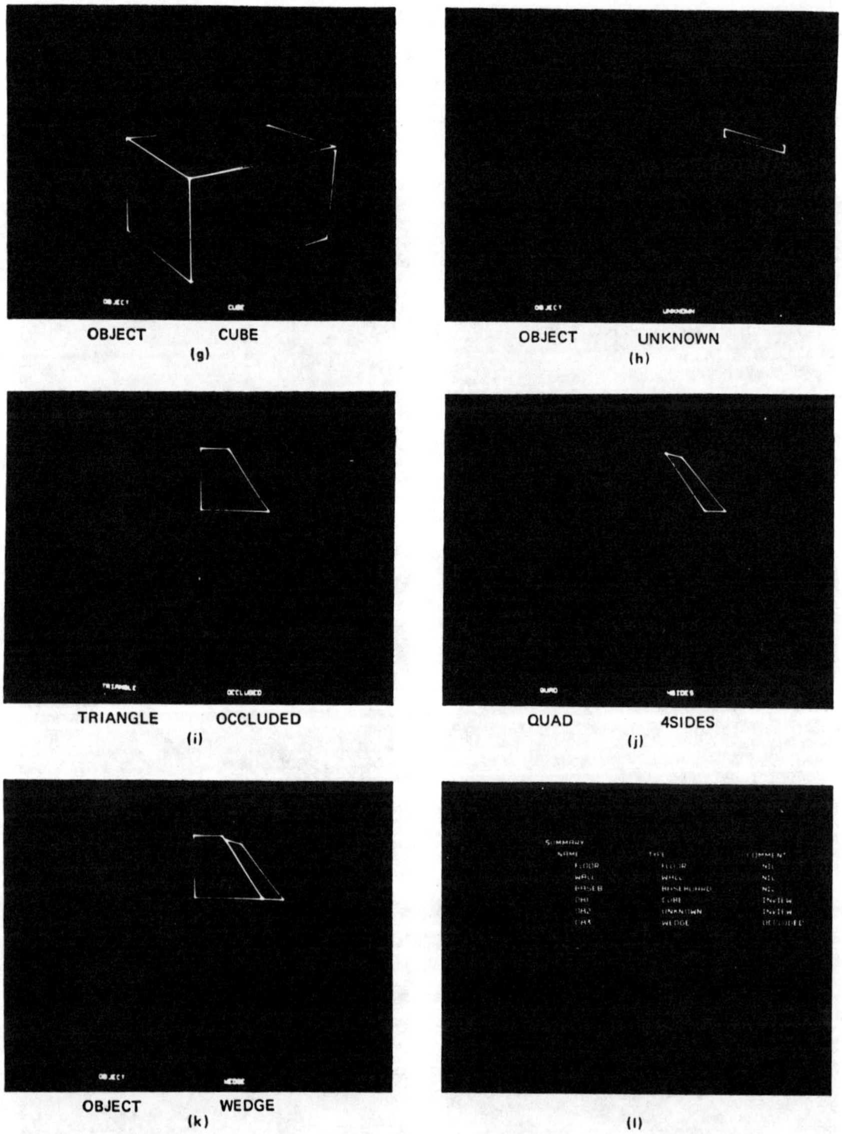

FIG. 16 (*continued*).

5. Conclusion

The advantages of a uniform data type are perhaps obvious. At any step of processing, information can be extracted in the same way and the same operations can be used throughout. In the discussion above it suffices to have the operations MERGE and CUT, and the local data can always be extracted in the same way—lines, gray scale, neighbors, area, perimeter, etc. Our data type was regions; the original picture was broken into regions of an atomic nature, and this type was kept throughout.

The choice of regions as a data type has proved quite successful. The global information available has led to the two heuristics, phagocyte and weakness, which apply global information at the very outset. They use information about the whole region to determine what happens locally. As has been the conjecture of many for some time, global information proves to be quite fruitful when compared with the blind application of local operators. (The application of a local criterion too often blindly destroys needed information.)

The construction of an entire system has permitted the experimentation with various heuristics for repartitioning boundary descriptions, etc. The output of the phagocyte and weakness heuristics has been quite impressive and seems to be relatively independent of context. The results have been to give "clean" pictures in various situations.

One of the weaker links in the analysis chain is the line-fitting technique. It is not so faithful as it possibly could be. Neighboring regions are often separated, as in Fig. 14, and T connections are not even so faithfully represented as in the original data. A new technique might make better use of the local and global information available.

The scene analyzer described here is very simple and is one of the prime targets for improvement. The next scene analyzer will have to include a new class of objects called "doors" and will make a richer use of the robot's world model to reduce unnecessary analysis of "known" areas. The use of clues together with such a model can make analysis easier.

This new scene analyzer will incorporate the ideas of our co-workers Duda and Hart [5], which provide several useful tests derived from projective geometry, and the ideas presented above.

REFERENCES

1. Nilsson, N. J. A mobile automaton: an application of artificial intelligence techniques. *Proc. Internat. Joint Conf. on Artificial Intelligence*, Washington, D.C. (May 1969), 509–520.
2. Munson, J. H. The SRI intelligent automaton program. *Proc. First Symp. on Industrial Robots*, Chicago, to be published.
3. Roberts, L. G. Machine Perception of Three Dimensional Solids. *Optical and Electro-Optical Information Processing*, in Tippett *et al.* (Eds.). MIT Press, Cambridge, Massachusetts, 1965.
4. Rosenfeld, A. *Picture Processing by Computer.* Academic Press, New York, 1969.
5. Duda, R. O., and Hart, P. E. Experiments in scene analysis. *First Nat. Symp. on Industrial Robots*, Chicago, to be published.

Artificial Intelligence **1** (1970), 205–226

6. Feldman, J. A., *et al.* The Stanford hand-eye project. *Proc. Internat. Joint Conf. on Artificial Intelligence*, Washington, D.C., May 1969.
7. Minsky, M. A., and Papert, S. *Project MAC Progress Report IV*, MIT Press, Cambridge, Massachusetts, 1967.
8. Guzman, A. Decomposition of scenes into bodies. *AFIPS Conf. Proc.* 33, Part 1 (Dec. 1968), 291–304.
9. Cook, B. G. A computer representation of plane region boundaries. *Australian Computer J.* (Nov. 1967), 44–50.
10. Pfaltz, J. L., Snively, J. W., Jr., and Rosenfeld, A. Local and global picture processing by computer. *Pictorial Pattern Recognition.* Thompson Book Co., Washington D.C., 1968, pp. 353–371.
11. Gibson, J. *The Senses Considered as Perceptual Systems.* Houghton and Mifflin, Boston, 1966.
12. Yarbus, A. L. *Eye Movement and Vision.* Plenum Press, New York, 1967.

Received December, 1969.

PICTURE SEGMENTATION BY A DIRECTED SPLIT-AND-MERGE PROCEDURE*

by

Steven L. Horowitz

and

Theodosios Pavlidis

Department of Electrical Engineering
Princeton University
Princeton, N. J. 08540

In the past, picture processing has been performed by merging small primitive regions, or by recursively splitting the whole picture. This paper combines the two approaches with significant increase in processing speed while maintaining small memory requirements. The data structure is described in detail and examples of implementations are given.

I. INTRODUCTION

The picture segmentation problem has been treated most often in one of two ways:

(1) Merging or Bottom-up: The picture is divided into a large number of small regions (possibly coinciding with single pixels) which are then merged to form larger regions. The "phagocyte" algorithm of Brice and Fennema [1] is typical of this kind. Algorithms which scan a waveform sequentially from left to right to determine the longest interval such that an approximation is below a given tolerance [2] also belong to this class. There have been a number of merging criteria for different types of primitive regions which are discussed in the literature [3-9].

(2) Splitting or Top-down: The picture is successively divided into smaller and smaller regions until certain criteria are satisfied. There have been very few applications of this scheme in picture processing until recently [10,11], but it has been a well-known method for polygonal approximation of boundaries [5].

It is possible to devise a mixed <u>split-and-merge</u> scheme, and this has been carried out in the case of waveforms and boundaries with encouraging results [12-14]. Extension of this strategy is proposed here for pictures. The advantages are:

(1) an increase in computation speed, and

(2) an additional degree of freedom allowing improvement in segmentation quality, both without expanding memory requirements.

Below is an abstract formulation of the problem:

Let X be the domain of a picture (e.g. a square). Let f(x,y) be the brightness function defined on X. A logical predicate P is defined on subsets S of X as follows:

$$P(S) = \begin{cases} \text{true if there exists a constant } a \\ \quad \text{such that } |f(x,y) - a| \le e\ * \\ \quad \text{for all points } (x,y) \in S. \\ \text{false otherwise.} \end{cases}$$

where e is a prescribed error tolerance.

A segmentation of X is a partition of X into subsets S_i i=1,...,m for some m such that:

(a) $X = \bigcup_{i=1}^{m} S_i$

(b) $S_i \cap S_j = \emptyset$ for all $i \neq j$

(c) $P(S_i)$ = true for all i

(d) $P(S_i \cup S_j)$ = false for all $i \neq j$

provided S_i and S_j are adjacent in X.

Note that the number m is not unique, nor must it be the minimum under which conditions (a) - (d) hold. In many practical cases, however, imposing these conditions yields an m close to the minimum.

A <u>merging</u> scheme starts with a partition satisfying (c) and proceeds to fulfill (d); a <u>splitting</u> scheme starts with a partition satisfying (d) and proceeds to fulfill (c). A <u>split-and-merge</u> procedure begins with an arbitrary partition satisfying neither condition and produces a partition satisfying both.

The above formulation can be expressed in a graph theoretical framework by a tree whose nodes correspond to the subsets of X such that:

(i) X is the root of the tree.

(ii) The successors of the node corresponding to a subset S of X are the nodes corresponding to proper, disjoint, collectively exhaustive subsets of S.

*This research was supported by NSF Grant GK-36180.

*equivalent to $\left[\max_S (f(x,y)) - \min_S (f(x,y))\right] \le 2e$

Reprinted from *Proc. 2nd Int. Joint Conf. on Pattern Recognition*, Aug. 13-15, 1974, pp. 424-433.

(iii) The leaves of the tree represent the smallest subsets of X under consideration (e.g. single pixels).

Figure 1 shows an example of such a segmentation tree construction. It can be shown that a segmentation corresponds to a node cutset which is the minimal set of nodes separating the root from all the leaves. The root R may be separated from a leaf L in one of three ways:

(i) R is in the cutset.

(ii) L is in the cutset.

(iii) There exists one and only one node N in the cutset such that R is an ancestor of N and L is a descendent of N.

Thus there is no other cutset closer to the root with nodes where P(S) is true. The terms top-down and bottom-up used before can be visualized in terms of moves along the segmentation tree and the corresponding alterations on the original node cutset.

Another operation known as sidewise merging or grouping [6] is accomplished either by transforming the tree structure, or by partitioning the final cutset into equivalence classes, based on node maxima and minima with respect to the prescribed error tolerance as discussed earlier. Thus adjacent nodes with different predecessors and belonging to different levels may be merged or grouped together to form larger irregular subsets S (possibly of genus greater than zero) provided P(S) holds true. Figure 2 shows an example of grouping and the resulting modifications in tree structure.

II. COMPUTATIONAL CONSIDERATIONS

In most practical applications the computational efforts involved in splitting and merging are not the same. Merging d subsets into a single set requires two groups of d comparisons (updating the maximum and minimum values of f(x,y) on the domain of the subsets). On the other hand, splitting a single set into d subsets necessitates a reevaluation of the maximum and minimum of f(x,y) on the domain of each of the d subsets.

Let: cost of merging $\{S_1,\ldots,S_d\}$ into $S_0 = C_m$

cost of splitting S_0 into $\{S_1,\ldots,S_d\} = C_s$

Note that C_m is independent of $|S_0|$ but dependent on d (i.e. $C_m = 2 \cdot d \cdot \delta$) while C_s is dependent on $|S_0|$ but independent of d (i.e. $C_s = 2 \cdot |S_0| \cdot \delta$) where δ is the cost of a single comparison.

Common as the above model might be, there may be cases where a more symmetrical situation exists, especially in uniform approximation by high order polynomials; also, the dependence on size need not be linear [15,16].

The choice between a bottom-up or top-down precedure should depend on the level of the tree at which the majority of the nodes in the cutset corresponding to the final segmentation occurs. Asymptotically the top-down procedure is of order $|X| \log_d |X|$ while the bottom-up is of order $|X|$, where d as before is also the node degree of the tree. The difference in asymptotic behavior explains the popularity of bottom-up methods in the literature as compared to top-down methods.

Theoretically and experimentally, savings in computation can be achieved if an intermediate level of the tree is chosen as a starting cutset. Such a choice should attempt to minimize the number of splits and merges needed to reach the final cutset, which usually encompasses many levels (some nodes close to the root and others close to or coinciding with leaves) due to the non-uniformity in the distribution of detail over a typical picture. Figure 3a illustrates the computational effort yielded by applying the two methods.

If ℓ_s and ℓ_m are such initial levels for top-down and bottom-up methods respectively, a combined split-and-merge procedure, starting at a middle level ℓ_0 such that $\ell_m < \ell_0 < \ell_s$, can move in both directions with an overall economy as a result. Figure 3b illustrates the smaller computational effort associated with a split-and-merge approach. That this is indeed the case has been verified for waveforms and boundaries [12,13], and holds true for the picture examples presented in this paper.

III. IMPLEMENTATION OF SPLIT-AND-MERGE FOLLOWED BY GROUPING

In order to process an n-by-n picture P (where $n=2^{\ell_n}$), the concepts of "segmentation tree" and "node cutset" must be utilized. The segmentation tree is implemented as follows:

Each node corresponds to a square picture region segment (block). Leaves of the tree are one-by-one blocks (i.e. single pixels) at level $\log_2 n = \ell_n$, and all other nodes are defined recursively: a level ℓ node b (where $0 \le \ell < \ell_n$) which is located (taken to mean the index of the upper left corner pixel of the corresponding block) at (x,y) has sides z of length $n/2^\ell$ and has four successors with sides of length z/2 located at (x,y), (x+z/2,y), (x,y+z/2), and (x+z/2, y+z/2). Note that the above definition has the root node (level 0) located at (1,1). This is, of course, the complete picture with sides of length n. Aside from specifying x_k, y_k and z_k, each node b_k has associated values M_k and m_k equal to the maximum and minimum of the brightness function $f(x,y)$ on the corresponding block.

Rather than storing the complete tree in memory (necessitating $5 \sum_{i=0}^{\ell_n} 4^i = 5(4n^2 - 1)/3$ integers), only the cutset is stored as 5 arrays $(x_k, y_k, z_k, M_k, m_k)$ of size n^2. This is achieved by a suitable encoding for the position of b_k in the array given (x_k, y_k) which enables us to retrieve the tree structure, i.e. subsets of four nodes and their common predecessor. Initialization of the array at level ℓ_0 where $s_0 = n/2^{\ell_0}$ is performed as follows:

```
k←0;
DO i =1 TO n BY s0;
 DO j =1 TO n BY s0;
  k←k+1; xk←i; yk←j; zk←s0;
  Mk←MAX (pi,j:i=xk,...,xk+s0, j=yk,...,yk+s0);
  mk←MIN (pi,j:i=xk,...,xk+s0, j=yk,...,yk+s0);
  END;
 END;
```

Note that k contains the value $c_0=4^{\ell_0}$ after the initialization step (the number of nodes in the initial cutset).

Now nodes $b_{k1}, b_{k2}, b_{k3}, b_{k4}$ have a common predecessor b_{k0} located at (x_{k0}, y_{k0}) if:

$$k1 = 2^{\ell_0} \cdot [(x_{k0}-1) + 2^{\ell_0}(y_{k0}-1)]/n + 1$$

$$k2 = k1 + 2^{\ell_0 - \ell}$$

$$k3 = k1 + 2^{2\ell_0 - \ell}$$

$$k4 = k3 + 2^{\ell_0 - \ell}$$

for a given level ℓ (where $0 < \ell \le \ell_0$). Figure 4a illustrates the relationship between the five nodes and represents some of the associated variables.

Merging nodes $b_{k1}, b_{k2}, b_{k3}, b_{k4}$ is a process that removes the four nodes from the cutset and replaces them with node b_{k0} (actually b_{k1} updated):

```
ℓ←ℓ0; s←s0;
DO WHILE ℓ > 0;
 DO FOR ALL bk1 ε level ℓ;
  IF zk1=zk2=zk3=zk4=s THEN DO;
   g←MAX (Mk1, Mk2, Mk3, Mk4);
   h←MIN (mk1, mk2, mk3, mk4);
   IF g-h ≤ 2e* THEN DO;
    zk2←0; zk3←0; zk4←0;
    zk1←2s; Mk1←g; mk1←h;
    END;
   END;
  END;
 ℓ←ℓ-1; s←2s;
 END;
```

The cutset still has c_0 nodes, but merged nodes either have been updated or have had z set to zero.

Splitting a node b_{k0} (actually b_{k1} before updating) is a process that removes the node from the cutset and replaces it with nodes $b_{k1}, b_{k2}, b_{k3}, b_{k4}$.

```
c←c0;
 DO k1=1 TO c;
  IF 0 < zk1 ≤ s0 THEN DO;
   ℓ←ℓ0; s←s0;
   DO WHILE ℓ < ℓn;
    IF Mk1-mk1 > 2e* THEN DO;
     k2←c+1; k3←c+2; k4←c+3;
     zk1←s/2; zk2←zk1; zk3←zk1; zk4←zk1;
     xk2←xk1; xk3←xk1+zk1; xk4←xk3;
     yk2←yk1+zk1; yk3←yz1; yk4←yk2;
     [Mk1,..,Mk4 and mk1,..,mk4 are calculated
     using values of pi,j as demonstrated
     previously];
     ℓ←ℓ+1; c←c+3;
     END;
    ELSE ℓ←ℓn;
    END;
   END;
  END;
```

This yields the final cutset satisfying properties (a) - (d) in section I.

It is easy to see that both processes terminate. Merging can proceed until the cutset consists solely of the root node (e.g. a one-color picture), and splitting can proceed until the cutset consists of the $4^{\ell_n}=n^2$ leaf nodes only (e.g. an alternating two-color picture). Furthermore, the two processes are mutually exclusive (i.e. all of the merging operations are followed by all of the splitting operations without violating final cutset properties), since once four nodes have been merged, they cannot be split. Also, nodes that have been split cannot be merged back to their original configuration, although merging with other adjacent nodes is possible

* as defined in Section I

through grouping, which is the third step in the processing sequence.

For grouping, an auxilliary adjacency matrix A is initialized as follows:

```
DO k=1 TO c SUCH THAT z_k≠0;
 DO i=x_k TO x_k+s_k;
  DO j=y_k TO y_k+s_k;
   a_i,j←k;
   END;
  END;
 END;
```

Thus if a point $p_{i,j}$ in the picture is a member of the block specified by node b_k, the corresponding entry $a_{i,j}$ in the adjacency matrix contains k. Grouping is then executed by a labeling algorithm. A block is either unlabeled and unscanned ($z_k > 0$) or labeled and unscanned ($z_k < 0$ and b_k is on the stack S), or labeled and scanned ($z_k < 0$ and b_k is not on the stack S). Nodes that are labeled during one application of the algorithm are placed consecutively on a region list R, and after each application, a special endmarker cell (e.g. containing zero) is added to the region list:

```
s←empty;
DO k=1 TO c;
 IF z_k > 0 THEN DO;
  z_k← - z_k; PUSH(b_k)→S; u←M_k; v←m_k;
  DO WHILE S ¬empty;
   POP (b_i)←S; PUSH (b_i)→R;
   DO FOR ALL b_j ε A SUCH THAT b_j@‡ b_i;
     g←MAX (u,M_j); h←MIN (v,m_j);
     IF g-h≤2e*   THEN DO;
      PUSH (b_j)→S; u←g; v←h;
      END;
    END;
   END;
  END;
 PUSH (zero) → R;
 END;
```

Figure 4b shows an example of the grouping procedure outlined above. The resulting partitioning of the cutset into equivalence classes determining the tree structure transformations is obviously a segmentation satisfying properties (a) - (d) in section II.

‡is adjacent to

*as defined in section I

IV. EXAMPLES OF IMPLEMENTATION

Tests of the three-step algorithm presented in section III were run on an IBM 360/91 computer using the PL/I optimizing compiler. Samples were 7-bit, 64-by-64 pictures ($0 \le f(x,y) \le 127$, ℓ_n=6, n^2=4096). The total processing time was about two seconds per picture, most of which was spent performing I/O.

Figure 5 is an original digital TV photograph and figures 6 and 7 are reconstructions with different prescribed error tolerances. Fig. 8 shows the region boundaries for the reconstruction of fig. 6. The initial segmentation produces a large number of segments whose size has a bimodal distribution [17]. This allows the use of simple editing algorithms [17,18] for eliminating a large number of regions of small size. All figures in this paper show the region boundaries after the editing process. The statistics of the configuration are shown in Table I where R_i and R_f denote the number of regions before and after editing. ℓ_0, c_0 and e are as defined in section II while the rest of the symbols have the following meaning:

M number of merges

S number of splits

B number of nodes in the cutset after split-and-merge step

Fig. 9 corresponds to fig. 7 but during the editing regions of larger size than in the example of fig. 8 have been eliminated. It is important to point out that no use of semantics has been made up to this point. They can be introduced after this stage for further analysis of the data.

Fig. 10 shows a printer-plotter representation of a chromosome picture and fig. 11 the corresponding boundaries. Note that they have been obtained without choosing a threshold level but only an "error" tolerance. The latter essentially corresponds to the number of insignificant bits and thus it is much easier determined then the threshold [19].

Fig. 12 and 13 are the original and reconstruction of a scanning electron microscope picture of an Al-CdS junction. In this case segmentation is used primarily for picture enhancement by filtering out high frequency noise while preserving sharp boundaries.

In all reconstructions the brightness value used is not the constant a as defined in section I, but the average of f(x,y) within each region. This was found to result in better picture contrast.

REFERENCES

1. Brice, C.R. and C.L. Fennema, "Scene Analysis Using Regions", *Artificial Intelligence Journal*, Vol. 1, No. 3 (Fall 1970).

2. Pavlidis, T., "Linguistic Analysis of Waveforms", *Software Engineering*, (J. Tou, editor), Academic Press, 1971, pp. 203-205.

3. Rosenfeld, A., *Picture Processing by Computer*, Academic Press, 1969.

4. Meisel, W.S., *Computer-Oriented Approaches to Pattern Recognition*, Academic Press, 1972.

5. Duda, R.O. and P.E. Hart, *Pattern Classification and Scene Analysis*, J. Wiley, 1973.

6. Pavlidis, T., "Segmentation of Pictures and Maps through Functional Approximation", *Comp. Graphics and Image Processing*, Vol. 1 (1972), pp. 360-372.

7. Strong, T.P. and A. Rosenfeld, "A Region Coloring Technique for Scene Analysis", *Com. ACM*, Vol. 6 (April 1973), pp. 237-246.

8. Yakimovsky, Y., "Scene Analysis Using a Semantic Base for Region Growing", *CS-73-380*, Comp. Science Dept., Stanford University, June 1973.

9. Harlow, C.A. and S.A. Eisenbeis, "The Analysis of Radiographic Images", *IEEE Trans. on Computers*, Vol. C-22 (July 1973), pp. 678-689.

10. Robertson, T.V., P.H. Swain and K.S. Fu, "Multispectral Image Partitioning", *TR-EE73-26*, Purdue University, Aug. 1973.

11. Klinger, A., "Data Structures and Pattern Recognition", *Proc. First Int. Joint Conference on Pattern Recognition*, (Oct. 1973), pp. 497-498.

12. Pavlidis, T. and S.L. Horowitz, "Piecewise Approximation of Plane Curves", *Proc. First Int. Joint Conference on Pattern Recognition*, (October 1973), pp. 396-405.

13. Pavlidis, T., and S.L. Horowitz, "Segmentation of Plane Curves", *IEEE Trans. Computers*, Vol. C-23 (Aug. 1974).

14. Horowitz, S.L., "A General Peak Detection Algorithm with Applications in the Computer Analysis of Electrocardiograms", This Conference. Also available as *TR-129*, Computer Science Lab., Princeton University, August, 1973.

15. Rice, J.R., *The Approximation of Functions*, Vol. I, (1964), and Vol. II (1969), Addison-Wesley.

16. Esch, R.E. and W.L. Eastman, "Computational Methods for Best Approximation", *Tech. Report SEG-TR-67-30*, (Dec. 1967), Sperry Rand, Sudbury, Mass.

17. Feng, H.Y. and T. Pavlidis, "The Generation of Polygonal Outlines of Objects from Gray Level Pictures," *TR-150*, Computer Science Lab., Princeton University, April 1974.

18. Horowitz, S.L. and T. Pavlidis, in preparation.

19. Nagel, R.N. and A. Rosenfeld, "Steps Toward Handwritten Signature Verification", *Proc. First Inst. Joint Conference on Pattern Recognition*, (October 1973), pp. 59-66.

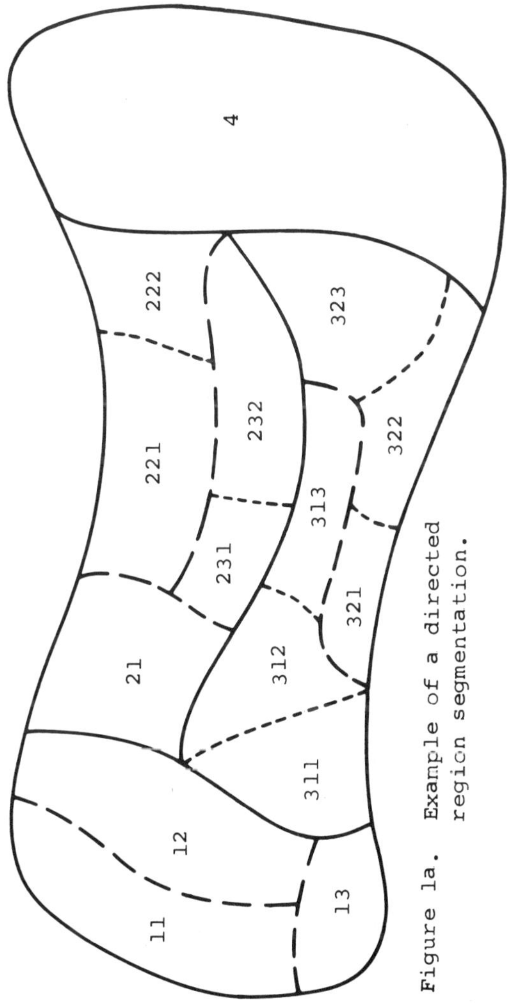

Figure 1a. Example of a directed region segmentation.

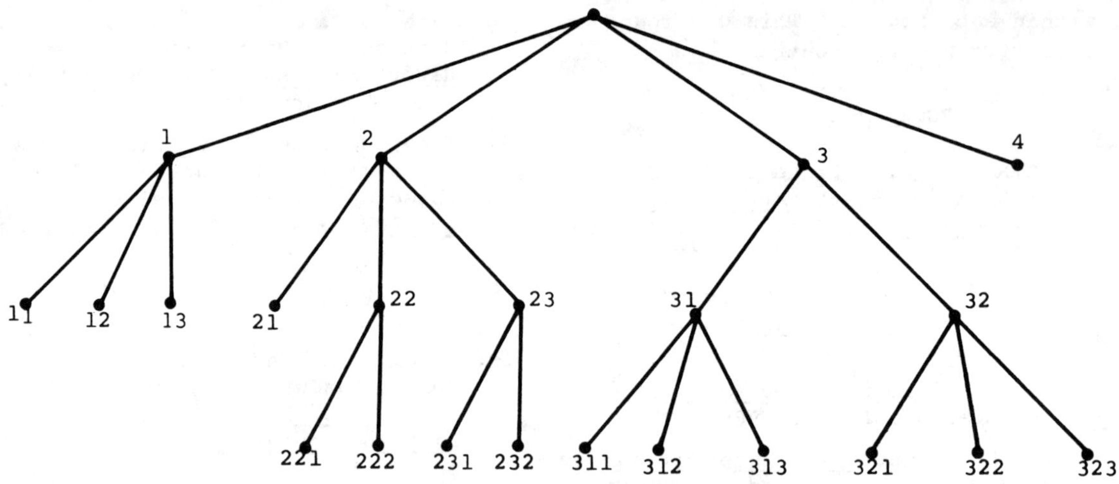

Figure 1b. Tree representing the segmentation of Figure 1a.

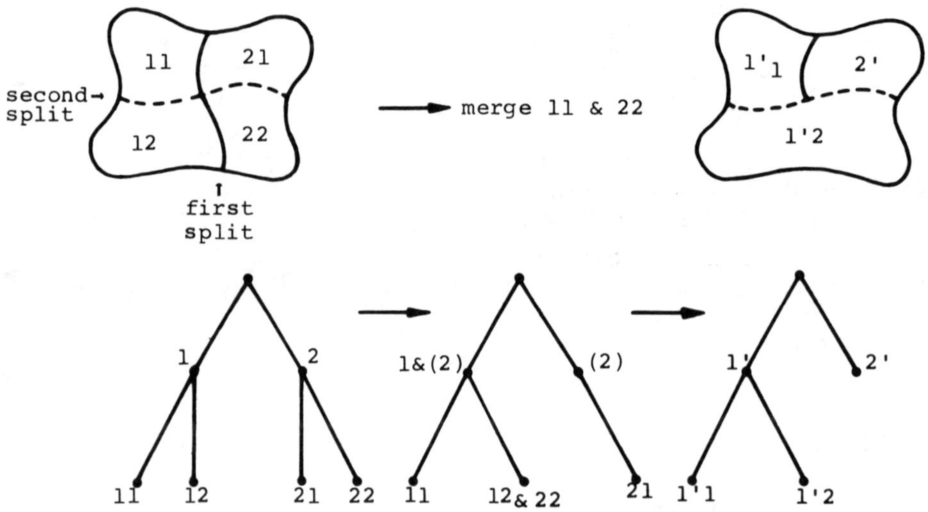

Figure 2. Illustration of the split-and-merge procedure.

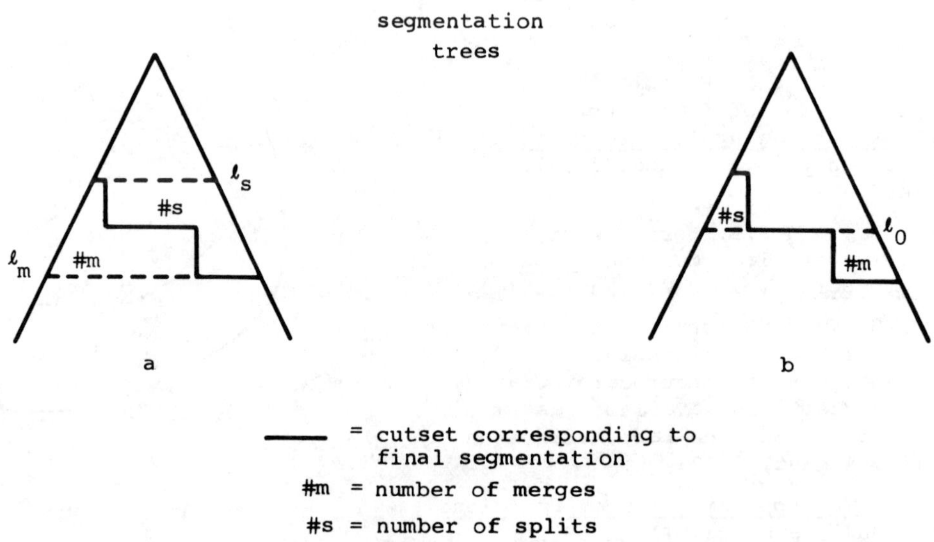

Figure 3. Illustration of the computational economy achieved by the split-and-merge method.

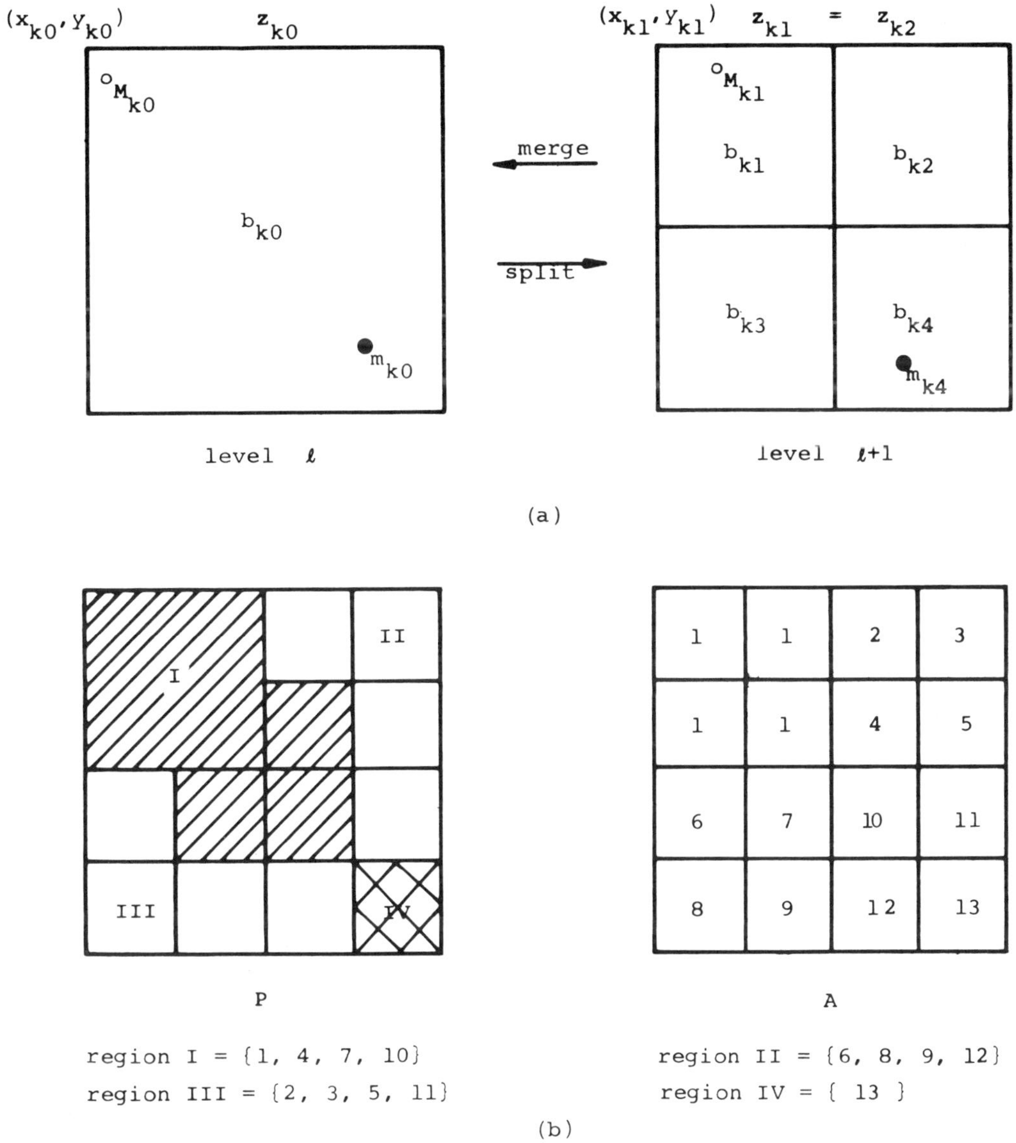

Figure 4. Illustration of the effects of the algorithm of Section III.

Figure 5. Original digitized picture of a face.

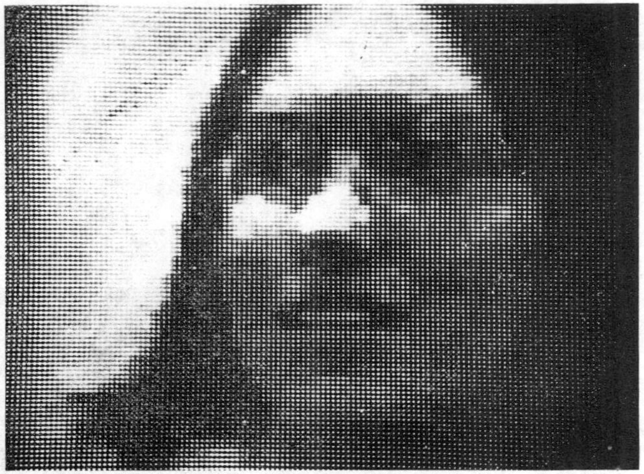

Figure 6. Reconstruction of Figure 5.

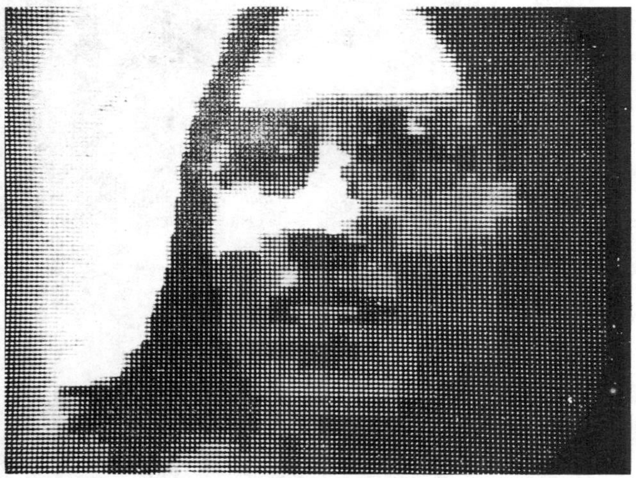

Figure 7. Reconstruction of Figure 5.

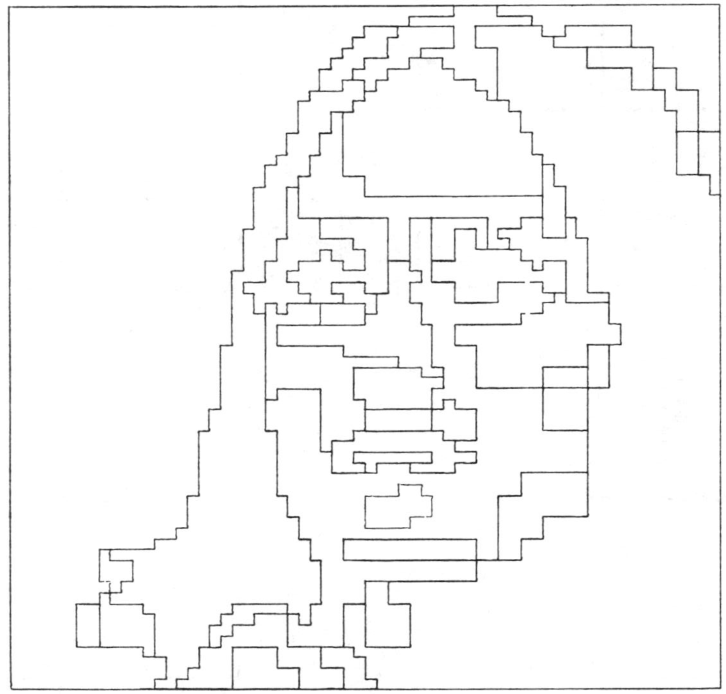

Figure 8. Region boundaries for segmentation of Figure 6.

Figure 10. Original chromosome picture produced by an overprinting routine.

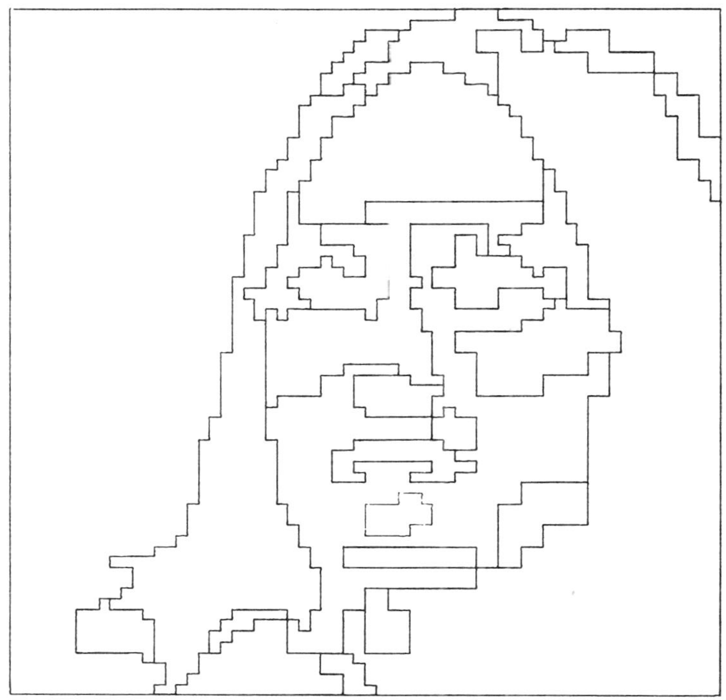

Figure 9. Region boundaries for segmentation of Figure 7.

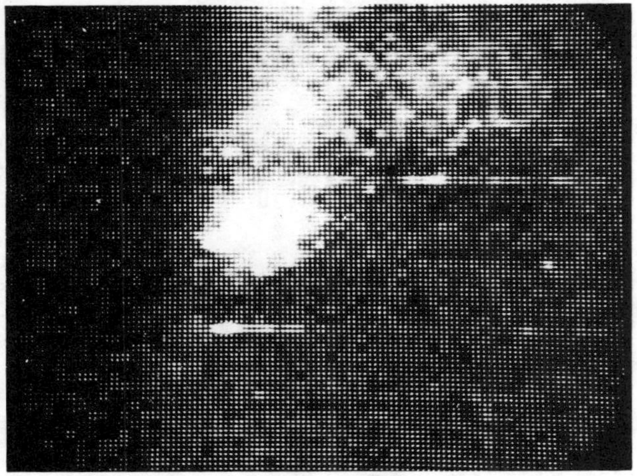

Figure 12. Original of a scanning electron microscope picture of an Al-CdS junction.

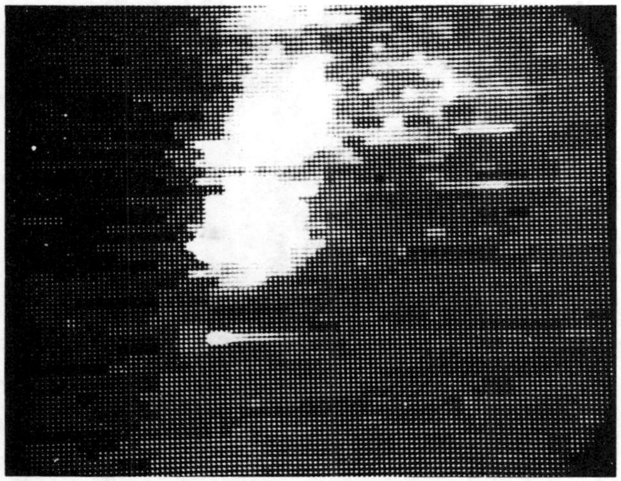

Figure 13. Reconstruction of Figure 12 for noise removal.

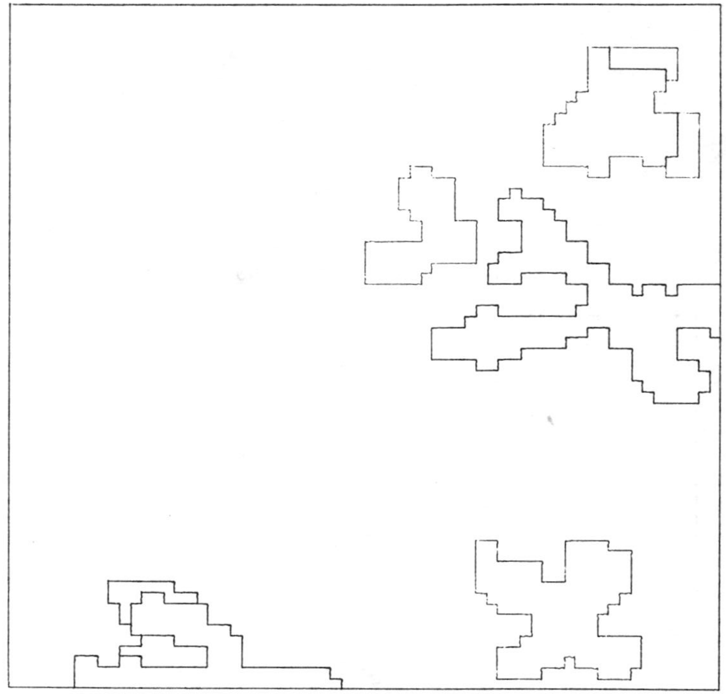

Figure 11. Boundaries of segmentation outlining chromosomes.

Table I

Fig	ℓ_0	c_0	e	M	S	B	R_i	R_f
8(6)	5	1024	6	143	225	1270	300	29
9(7)	5	1024	9	169	131	910	184	45
13(12)	5	1024	12	223	80	595	115	10

Part III
Region Representation and Description

Once an interesting region in a picture has been defined, it can be represented and described in a variety of ways. Some representations preserve all of the information about the region, while other representations—and virtually all descriptions—intentionally lose information, preserving only what is of interest for a particular task. Since the ultimate goal of image analysis is always the extraction of just the desired information, the main reason for considering any alternative representation is that it may simplify the subsequent analysis.

One popular, mathematically complete representation is provided by two-dimensional moments. Used in many early pattern recognition systems, moments have the advantage that they can be applied to a region containing an object of interest without first having to isolate the object. The lower-order moments, which are insensitive to noise and quantization error, provide an incomplete, but useful, gross representation of the region. [For other early papers on the use of moments, see Giuliano *et al.* (1961, 1962) and Alt (1962).]

Suppose that the region contains one major object of interest, but that the object can vary in location, orientation, and perhaps scale. Then certain moments are very convenient for normalization. For example, the first moments give the centroid, which is a natural origin of coordinates for translation invariant measurements. Similarly, the eigenvectors of the matrix of second central moments establish the directions of the principal axes, whose use leads to rotation invariant measurements. The first paper, by Hu [see also Hu (1961)], develops these ideas systematically. In particular, Hu shows how one can avoid explicit determination of the principal axes, and gives a general procedure for developing functions of the central moments that are invariant to general geometrical transformations. [When rotation is the chief problem, an interesting alternative is to exploit moments in polar coordinates; see Smith and Wright (1971).]

The moment representation is related through the moment generating function to another popular, mathematically complete representation—the two-dimensional Fourier transform. The main virtue of the Fourier domain—aside from its ability to be implemented optically—is the way in which it simplifies the basic linear, shift-invariant operations of convolution and cross correlation (recall the paper by Vander Lugt in Part I). The Fourier domain also provides a natural way to obtain measurements of the textural properties of regions, particularly when the texture is periodic. [For a review of early work on other kinds of texture features, see Hawkins (1970).]

The second paper, by Lendaris and Stanley, describes how one can derive such texture measurements from the power spectrum, which is basically the squared magnitude of the Fourier transform. Except for edge effects, the power spectrum is automatically invariant to translation, and Lendaris and Stanley show how one can obtain signatures that are invariant to rotation or to scale as well. In addition, they describe the SARF system, an interactive facility for designing Fourier-based features for pattern recognition. [For an alternative approach to using Fourier transforms for texture measurement, see Bajcsy (1973).]

The power spectrum is the Fourier transform of the autocorrelation function. Interpreted statistically, the autocorrelation function is just a particular moment of the joint distribution of intensities at two different points. It is natural to view visual texture in terms of certain kinds of statistical uniformity, and this joint distribution gives a general measure of spatial gray-level interdependence. The third paper, by Haralick, Shanmugam, and Dinstein, describes a variety of different texture measurements that can be extracted from the joint distribution. In particular, they include measurements that are invariant under monotonic gray-scale transformations, which is an important property in many applications. Analogous measures can be extracted from first-order distributions of gray level differences; see Weszka, Dyer, and Rosenfeld (1976). Another approach to texture analysis, based on time-series concepts, can be found in McCormick and Jayaramamurthy (1975).

In the important case of binary images, the significant region properties concern topology (which was treated in Part II) and shape. As with texture, there are a variety of techniques that can be used to measure shape. In principle, both moments and the power spectrum provide perfectly good shape features. However, such features often lack simple, readily understandable interpretations, making it difficult to decide which of the many features to use for a particular application.

One class of intuitively appealing shape features can be derived from the boundary of the region; such features are the main topic of Part V. Another useful class of shape features can be obtained by shrinking or thinning the region to some kind of skeleton. The fourth paper, by Blum, presents a completely general, skeletal representation of regions—the medial-axis (symmetric-axis, grassfire) transformation. With any information-preserving transformation, one should determine the class of problems for which it is particularly well suited. The medial-axis description is simplest for a circular disk, very simple for "worm-shaped" regions, and basically attractive for any shape that has a simply described skeleton. While one can view the medial-axis representation as a generalization of the center-radius description of a circular disk to arbitrary point sets, which makes it applicable to any region, it is these "biological" shapes for which this representation is most convenient.

Blum's work stimulated a number of investigations of skeletal representations. His own extensions of these ideas are lucidly presented in Blum (1973). The mathematical properties of the medial-axis transformation were investigated by Calabi and Hartnett (1968). Algorithms for digital skeletons were developed by Rosenfeld and Pfaltz (1966; cited in Part II) and examined further by Pfaltz and Rosenfeld (1967); for the related topic of digital distance functions, see Rosenfeld and Pfaltz (1968). Extensions to skeletons of gray-scale images were made by Hilditch (1969) and by Levi and Montanari (1970).

Many of the operations involved in skeletonizing binary images, as well as in related, more elementary processes such as thinning, end-point detecting, growing, smoothing, and gap filling, are basically local operations. The last paper in this part, by Golay, presents a set of basic local operations for transforming one binary image into another one. With these so-called Golay transforms, the state of an output pixel is a simple function of the state of the input pixel and the states of its immediate neighbors. By choosing a hexagonal sampling pattern, Golay obtains simple procedures for operations such as thinning without disconnecting. These Golay transforms have been developed further and applied by Preston (1971) and Preston and Norgren (1972), and have formed the basis for a parallel hardware implementation. [For other parallel thinning algorithms, see Stefanelli and Rosenfeld (1971) and Rosenfeld (1975).]

Visual Pattern Recognition by Moment Invariants*

MING-KUEI HU† SENIOR MEMBER, IRE

Summary—In this paper a theory of two-dimensional moment invariants for planar geometric figures is presented. A fundamental theorem is established to relate such moment invariants to the well-known algebraic invariants. Complete systems of moment invariants under translation, similitude and orthogonal transformations are derived. Some moment invariants under general two-dimensional linear transformations are also included.

Both theoretical formulation and practical models of visual pattern recognition based upon these moment invariants are discussed. A simple simulation program together with its performance are also presented. It is shown that recognition of geometrical patterns and alphabetical characters independently of position, size and orientation can be accomplished. It is also indicated that generalization is possible to include invariance with parallel projection.

I. Introduction

RECOGNITION of visual patterns and characters independent of position, size, and orientation in the visual field has been a goal of much recent research. To achieve maximum utility and flexibility, the methods used should be insensitive to variations in shape and should provide for improved performance with repeated trials. The method presented in this paper meets all these conditions to some degree.

Of the many ingeneious and interesting methods so far devised, only two main categories will be mentioned here: 1) The property-list approach, and 2) The statistical approach, including both the decision theory and random net approaches.[1] The property-list method works very well when the list is designed for a particular set of patterns. In theory, it is truly position, size, and orientation independent, and may also allow for other variations. Its severe limitation is that it becomes quite useless, if a different set of patterns is presented to it. There is no known method which can generate automatically a new property-list. On the other hand, the statistical approach is capable of handling new sets of patterns with little difficulty, but it is limited in its ability to recognize patterns independently of position, size and orientation.

This paper reports the mathematical foundation of two-dimensional moment invariants and their applications to visual information processing.[2] The results show that recognition schemes based on these invariants could be truly position, size and orientation independent, and also flexible enough to learn almost any set of patterns.

In classical mechanics and statistical theory, the concept of moments is used extensively; central moments, size normalization, and principal axes are also used. To the author's knowledge, the two-dimensional moment invariants, absolute as well as relative, that are to be presented have not been studied. In the pattern recognition field, centroid and size normalization have been exploited[3-5] for "preprocessing." Orientation normalization has also been attempted.[5] The method presented here achieves orientation independence without ambiguity by using either absolute or relative orthogonal moment invariants. The method further uses "moment invariants" (to be described in III) or invariant moments (moments referred to a pair of uniquely determined principal axes) to characterize each pattern for recognition.

Section II gives definitions and properties of two-dimensional moments and algebraic invariants. The moment invariants under translation, similitude, orthogonal transformations and also under the general linear transformations are developed in Section III. Two specific methods of using moment invariants for pattern recognition are described in IV. A simulation program of a simple model (programmed for an LGP-30), the performance of the program, and some possible generalizations are described in Section V.

II. Moments and Algebraic Invariants

A. A Uniqueness Theorem Concerning Moments

In this paper, the two-dimensional $(p + q)$th order moments of a density distribution function $\rho(x, y)$ are defined in terms of Riemann integrals as

$$m_{pq} = \int_{-\infty}^{\infty} \int_{-\infty}^{\infty} x^p y^q \rho(x, y)\, dx\, dy, \qquad p, q = 0, 1, 2, \cdots . \tag{1}$$

If it is assumed that $\rho(x, y)$ is a piecewise continuous therefore bounded function, and that it can have nonzero values only in the finite part of the xy plane; then moments of all orders exist and the following uniqueness theorem can be proved.

Uniqueness Theorem: The double moment sequence $\{m_{pq}\}$ is uniquely determined by $\rho(x, y)$; and conversely, $\rho(x, y)$ is uniquely determined by $\{m_{pq}\}$.

It should be noted that the finiteness assumption is important; otherwise, the above uniqueness theorem might not hold.

* Received by the PGIT, August 1, 1961.
† Electrical Engineering Department, Syracuse University, Syracuse, N. Y.

[1] M. Minsky, "Steps toward artificial intelligence," Proc. IRE, vol. 49, pp. 8–30; January, 1961. Many references to these methods can be found in the Bibliography of M. Minsky's article.

[2] M-K. Hu, Pattern recognition by moment invariants," Proc. IRE (Correspondence), vol. 49, p. 1428; September, 1961.

[3] W. Pitts and W. S. McCulloch, "How to know universals," *Bull. Math. Biophys.*, vol. 9, pp. 127–147; September, 1947.

[4] L. G. Roberts, "Pattern recognition with an adaptive network," 1960 IRE International Convention Record, pt. 2, pp. 66–70.

[5] Minsky, *op. cit.*, pp. 11–12.

Reprinted from *IRE Trans. Inform. Theory*, vol. IT-8, pp. 179–187, Feb. 1962.

B. Characteristic Function and Moment Generating Function

The characteristic function and moment generating function of $\rho(x, y)$ may be defined, respectively, as

$$\phi(u, v) = \int_{-\infty}^{\infty}\int_{-\infty}^{\infty} \exp(iux + ivy)\rho(x, y)\, dx\, dy, \tag{2}$$

$$M(u, v) = \int_{-\infty}^{\infty}\int_{-\infty}^{\infty} \exp(ux + vy)\rho(x, y)\, dx\, dy. \tag{3}$$

In both cases, u and v are assumed to be real. If moments of all orders exist, then both functions can be expanded into power series in terms of the moments m_{pq} as follows:

$$\phi(u, v) = \sum_{p=0}^{\infty}\sum_{q=0}^{\infty} m_{pq} \frac{(iu)^p}{p!}\frac{(iv)^q}{q!}, \tag{4}$$

$$M(u, v) = \sum_{p=0}^{\infty}\sum_{q=0}^{\infty} m_{pq} \frac{u^p}{p!}\frac{v^q}{q!}. \tag{5}$$

Both functions are widely used in statistical theory. If the characteristic function $\phi(u, v)$ which is essentially the Fourier transform of $\rho(x, y)$ is known, $\rho(x, y)$ may be obtained from the following inverse Fourier transform,

$$\rho(x, y) = \frac{1}{(2\pi)^2}\int_{-\infty}^{\infty}\int_{-\infty}^{\infty} \exp(-iux - ivy)\phi(u, v)\, du\, dv. \tag{6}$$

The moment generating function $M(u, v)$ is not as useful in this respect, but it is convenient for the discussion in Section III. The close relationships and differences between $\phi(u, v)$ and $M(u, v)$ may be seen much more clearly, if we consider both as special cases of the two-sided Laplace transform of $\rho(x, y)$,

$$\mathfrak{L}[\rho(x, y)] = \int_{-\infty}^{\infty}\int_{-\infty}^{\infty} \exp(-sx - ty)\rho(x, y)\, dx\, dy, \tag{7}$$

where s and t are now considered as complex variables.

C. Central Moments

The central moments μ_{pq} are defined as

$$\mu_{pq} = \int_{-\infty}^{\infty}\int_{-\infty}^{\infty} (x - \bar{x})^p(y - \bar{y})^q\rho(x, y)\, d(x - \bar{x})\, d(y - \bar{y}), \tag{8}$$

where

$$\bar{x} = m_{10}/m_{00}, \qquad \bar{y} = m_{01}/m_{00}. \tag{9}$$

It is well known that under the translation of coordinates,

$$x' = x + \alpha, \quad y' = y + \beta, \qquad \alpha, \beta - \text{constants} \tag{10}$$

the central moments do not change; therefore, we have the following theorem.

Theorem: The central moments are invariants under translation.

From (8), it is quite easy to express the central moments in terms of the ordinary moments. For the first four orders, we have

$$\begin{aligned}
&\mu_{00} = m_{00} \equiv \mu, \qquad \mu_{10} = \mu_{01} = 0,\\
&\mu_{20} = m_{20} - \mu\bar{x}^2, \qquad \mu_{11} = m_{11} - \mu\bar{x}\bar{y},\\
&\mu_{02} = m_{02} - \mu\bar{y}^2,\\
&\mu_{30} = m_{30} - 3m_{20}\bar{x} + 2\mu\bar{x}^3,\\
&\mu_{21} = m_{21} - m_{20}\bar{y} - 2m_{11}\bar{x} + 2\mu\bar{x}^2\bar{y},\\
&\mu_{12} = m_{12} - m_{02}\bar{x} - 2m_{11}\bar{y} + 2\mu\bar{x}\bar{y}^2,\\
&\mu_{03} = m_{03} - 3m_{02}\bar{y} + 2\mu\bar{y}^3.
\end{aligned} \tag{11}$$

From here on, for the simplicity of description, all moments referred to are central moments, and μ_{pq} will be simply expressed as

$$\mu_{pq} = \int_{-\infty}^{\infty}\int_{-\infty}^{\infty} x^p y^q \rho(x, y)\, dx\, dy, \tag{12}$$

and the moment generating function $M(u, v)$ will also be referred to central moments.

D. Algebraic Forms and Invariants

The following homogeneous polynomial of two variables u and v,

$$f = a_{p0}u^p + \binom{p}{1}a_{p-1,1}u^{p-1}v + \binom{p}{2}a_{p-2,2}u^{p-2}v^2 + \cdots + \binom{p}{p-1}a_{1,p-1}uv^{p-1} + a_{0p}v^p, \tag{13}$$

is called a binary algebraic form, or simply a binary form, of order p. Using a notation, introduced by Cayley, the above form may be written as

$$f \equiv (a_{p0}; a_{p-1,1}; \cdots; a_{1,p-1}; a_{0p})(u, v)^p. \tag{14}$$

A homogeneous polynomial $I(a)$ of the coefficients $a_{p0}, \cdots, a_{0p}$ is an algebraic invariant of weight w, if

$$I(a'_{p0}, \cdots, a'_{0p}) = \Delta^w I(a_{p0}, \cdots, a_{0p}), \tag{15}$$

where $a'_{p0}, \cdots, a'_{0p}$ are the new coefficients obtained from substituting the following general linear transformation into the original form (14).

$$\begin{bmatrix} u \\ v \end{bmatrix} = \begin{bmatrix} \alpha & \gamma \\ \beta & \delta \end{bmatrix}\begin{bmatrix} u' \\ v' \end{bmatrix}, \qquad \Delta = \begin{vmatrix} \alpha & \gamma \\ \beta & \delta \end{vmatrix} \neq 0. \tag{16}$$

If $w = 0$, the invariant is called an absolute invariant; if $w \neq 0$ it is called a relative invariant. The invariant defined above may depend upon the coefficients of more than one form. Under special linear transformations to be discussed in Section III, Δ may not be limited to the determinant of the transformation. By eliminating Δ between two relative invariants, a nonintegral absolute invariant can always be obtained.

In the study of invariants, it is helpful to introduce another pair of variables x and y, whose transformation with respect to (16) is as follows:

$$\begin{bmatrix} x' \\ y' \end{bmatrix} = \begin{bmatrix} \alpha & \beta \\ \gamma & \delta \end{bmatrix} \begin{bmatrix} x \\ y \end{bmatrix}. \tag{17}$$

The transformation (17) is referred to as a cogredient transformation, and (16) is referred to as a contragredient transformation. The variable x, y are referred to as covariant variables, and u, v as contravariant variables. They satisfy the following invariant relation

$$ux + vy = u'x' + v'y'. \tag{18}$$

The study of algebraic invariants was started by Boole, Cayley and Sylvester more than a century ago, and followed vigorously by others, but interest has gradually declined since the early part of this century. The moment invariants to be discussed in Section III will draw heavily on the results of algebraic invariants. To the authors knowledge, there was no systematic study of the moment invariants in the sense to be described.

III. Moment Invariants

A. A Fundamental Theorem of Moment Invariants

The moment generating function with the exponential factor expanded into series form is

$$M(u, v) = \int_{-\infty}^{\infty} \int_{-\infty}^{\infty} \sum_{p=0}^{\infty} \frac{1}{p!} (ux + vy)^p \rho(x, y)\, dx\, dy. \tag{19}$$

Interchanging the integration and summation processes, we have

$$M(u, v) = \sum_{p=0}^{\infty} \frac{1}{p!} (\mu_{p0}, \cdots, \mu_{0p})(u, v)^p. \tag{20}$$

By applying the transformation (17) to (19), and denoting the coefficients of x' and y' in the transformed factor $(ux + vy)$ by u' and v', respectively, or equivalently by applying both (16) and (17) simultaneously to (19), we obtain

$$M_1(u', v') = \int_{-\infty}^{\infty} \int_{-\infty}^{\infty} \sum_{p=0}^{\infty} \frac{1}{p!} \cdot (u'x' + v'y')^p \rho'(x', y') \frac{1}{|J|}\, dx'\, dy' \tag{21}$$

where $\rho'(x', y') \equiv \rho(x, y)$, $|J|$ is the absolute value of the Jacobian of the transformation (17), and $M_1(u', v')$ is the moment generating function after the transformation. If the transformed central moments μ'_{pq} are defined as

$$\mu'_{pq} = \int_{-\infty}^{\infty} \int_{-\infty}^{\infty} (x')^p (y')^q \rho'(x', y')\, dx'\, dy', \qquad p, q = 0, 1, 2, \cdots \tag{22}$$

then we have

$$M_1(u', v') = \frac{1}{|J|} \sum_{p=0}^{\infty} \frac{1}{p!} (\mu'_{p0}, \cdots, \mu'_{0p})(u', v')^p. \tag{23}$$

In the theory of algebraic invariants, it is well known that the transformation law for the a coefficients in the algebraic form (14) is the same as that for the monomials, $x^{p-r}y^r$, in the following expression:

$$(ux + vy)^p = (x^p, x^{p-1}y, \cdots, y^p)(u, v)^p. \tag{24}$$

From (19), (20), (21) and (23), it can be seen clearly that the same relationship also holds between the pth order moments and the monomials except for the additional factor $1/|J|$. Therefore, the following fundamental theorem is established.

Fundamental Theorem of Moment Invariants: If the algebraic form of order p has an algebraic invariant,

$$I(a'_{p0}, \cdots, a'_{0p}) = \Delta^w I(a_{p0}, \cdots, a_{0p}), \tag{25}$$

then the moments of order p have the same invariant but with the additional factor $|J|$,

$$I(\mu'_{p0}, \cdots, \mu'_{0p}) = |J| \Delta^w I(\mu_{p0}, \cdots, \mu_{0p}). \tag{26}$$

This theorem holds also between algebraic invariants containing coefficients from two or more forms of different orders and moment invariants containing moments of the corresponding orders.

B. Similitude Moment Invariants

Under the similitude transformation, *i.e.*, the change of size,

$$\begin{bmatrix} x' \\ y' \end{bmatrix} = \begin{bmatrix} \alpha & 0 \\ 0 & \alpha \end{bmatrix} \begin{bmatrix} x \\ y \end{bmatrix}, \qquad \alpha - \text{constant}, \tag{27}$$

each coefficient of any algebraic form is an invariant

$$a'_{pq} = \alpha^{p+q} a_{pq}, \tag{28}$$

where α is not the determinant. For moment invariants we have

$$\mu'_{pq} = \alpha^{p+q+2} \mu_{pq}. \tag{29}$$

By eliminating α between the zeroth order relation,

$$\mu' = \alpha^2 \mu \tag{30}$$

and the remaining ones, we have the following absolute similitude moment invariants:

$$\frac{\mu'_{pq}}{(\mu')^{(p+q)/2} + 1} = \frac{\mu_{pq}}{\mu^{(p+q)/2} + 1}, \quad p + q = 2, 3, \cdots \tag{31}$$

and $\mu'_{10} = \mu'_{01} \equiv 0$.

C. Orthogonal Moment Invariants

Under the following proper orthogonal transformation or rotation:

$$\begin{bmatrix} x' \\ y' \end{bmatrix} = \begin{bmatrix} \cos\theta & \sin\theta \\ -\sin\theta & \cos\theta \end{bmatrix} \begin{bmatrix} x \\ y \end{bmatrix} \tag{32}$$

we have

$$J = \begin{vmatrix} \cos\theta & \sin\theta \\ -\sin\theta & \cos\theta \end{vmatrix} = +1. \tag{33}$$

Therefore, the moment invariants are exactly the same as the algebraic invariants. If we treat the moments as the coefficients of an algebraic form

$$(\mu_{p0}, \cdots, \mu_{0p})(u, v)^p \tag{34}$$

under the following contragredient transformation:

$$\begin{bmatrix} u \\ v \end{bmatrix} = \begin{bmatrix} \cos\theta & -\sin\theta \\ \sin\theta & \cos\theta \end{bmatrix} \begin{bmatrix} u' \\ v' \end{bmatrix}, \tag{35}$$

then we can derive the moment invariants by the following algebraic method. If we subject both u, v and u', v' to the following transformation:

$$\begin{bmatrix} U \\ V \end{bmatrix} = \frac{1}{2}\begin{bmatrix} 1 & i \\ 1 & -i \end{bmatrix} \begin{bmatrix} u \\ v \end{bmatrix}, \quad \begin{bmatrix} U' \\ V' \end{bmatrix} = \frac{1}{2}\begin{bmatrix} 1 & i \\ 1 & -i \end{bmatrix} \begin{bmatrix} u' \\ v' \end{bmatrix}, \tag{36}$$

then the orthogonal transformation is converted into the following simple relations,

$$U' = Ue^{-i\theta}, \qquad V' = Ve^{i\theta}. \tag{37}$$

By substituting (36) and (37) into (34), we have the following identities:

$$\begin{aligned} (I_{p0}, \cdots, I_{0p})(U, V)^p &\equiv (\mu_{p0}, \cdots, \mu_{0p})(u, v)^p \\ &\equiv (\mu'_{p0}, \cdots, \mu'_{0p})(u', v')^p \\ &\equiv (I'_{p0}, \cdots, I'_{0p})(Ue^{-i\theta}, Ve^{i\theta})^p, \end{aligned} \tag{38}$$

where $I_{p0}, \cdots, I_{0p}$ and $I'_{p0}, \cdots, I'_{0p}$ are the corresponding coefficients after the substitutions. From the identity in U and V, the coefficients of the various monomials $U^{p-r}V^r$ on the two sides must be the same. Therefore,

$$\begin{aligned} &I'_{p0} = e^{ip\theta} I_{p0}; \qquad I'_{p-1,1} = e^{i(p-2)\theta} I_{p-1,1}; \cdots; \\ &I'_{1,p-1} = e^{-i(p-2)\theta} I_{1,p-1}; \; I'_{0p} = e^{-ip\theta} I_{0p}. \end{aligned} \tag{39}$$

These are $(p + 1)$ linearly independent moment invariants under proper orthogonal transformations, and $\Delta = e^{i\theta}$ which is not the determinant of the transformation. From the identity of first two expressions in (38), it can be seen that $I_{r,p-r}$ is the complex conjugate of $I_{p-r,r}$,

$$\begin{aligned} I_{p0} &= \mu_{p0} - i\binom{p}{1}\mu_{p-1,1} - \binom{p}{2}\mu_{p-2,2} \\ &\quad + i\binom{p}{3}\mu_{p-3,3} + \cdots + (-i)^p \mu_{0p}, \\ I_{p-1,1} &= (\mu_{p0} + \mu_{p-2,2}) - i(p-2)(\mu_{p-1,1} + \mu_{p-3,3}) \\ &\quad + \cdots + (-i)^{p-2}(\mu_{2,p-2} + \mu_{0p}), \\ I_{p-2,2} &= (\mu_{p0} + 2\mu_{p-2,2} + \mu_{p-4,4}) \\ &\quad - i(p-4)(\mu_{p-1,1} + 2\mu_{p-3,3} + \mu_{p-5,5}) \\ &\quad + \cdots + (-i)^{p-4}(\mu_{4,p-4} + 2\mu_{2,p-2} + \mu_{0p}), \\ &\cdots\cdots\cdots \\ I_{p-r,r} &= [(\mu_{p0}; \mu_{p-2,2}; \cdots; \mu_{p-2r,2r})(1, 1)^r; \\ &\quad (\mu_{p-1,1}; \mu_{p-3,3}; \cdots; \mu_{p-2r-1,2r+1})(1, 1)^r; \cdots; \\ &\quad (\mu_{2r,p-2r}; \mu_{2r+2,p-2r-2}; \cdots; \mu_{0p})(1, 1)^r](1, -i)^{p-2r}, \\ &\qquad p - 2r > 0. \\ &\cdots\cdots\cdots \end{aligned} \tag{40}$$

and

$$I_{p/2,p/2} = \mu_{p0} + \binom{p/2}{1}\mu_{p-2,2} + \binom{p/2}{2}\mu_{p-4,4} + \cdots + \mu_{0p}, \qquad p = \text{even}.$$

It may be noted that these $(p + 1)$ I's are linearly independent linear functions of the μ's, and vice versa.

For the following improper orthogonal transformation, *i.e.*, rotation and reflection:

$$\begin{bmatrix} x' \\ y' \end{bmatrix} = \begin{bmatrix} \cos\theta & \sin\theta \\ \sin\theta & -\cos\theta \end{bmatrix} \begin{bmatrix} x \\ y \end{bmatrix}, \quad \begin{vmatrix} \cos\theta & \sin\theta \\ \sin\theta & -\cos\theta \end{vmatrix} = -1. \tag{41}$$

Similarly, we have

$$U' = Ve^{i\theta} \qquad V' = Ue^{-i\theta} \tag{42}$$

and

$$\begin{aligned} &I'_{p0} = e^{-ip\theta} I_{0p}; \qquad I'_{p-1,1} = e^{-i(p-2)\theta} I_{1,p-1}; \cdots; \\ &I'_{1,p-1} = e^{i(p-2)\theta} I_{p-1,1}; \; I'_{0p} = e^{ip\theta} I_{p0}, \end{aligned} \tag{43}$$

where $I_{p0}, \cdots, I_{0p}$ and $I'_{p0}, \cdots, I'_{0p}$ are the same as those given by (40).

The orthogonal invariants were first studied by Boole, and the above method was due to Sylvester.[6]

[6] E. B. Elliott, "Algebra of Quantics," Oxford Univ. Press, New York, N. Y., 2nd ed., ch. 15; 1913.

D. A Complete System of Absolute Orthogonal Moment Invariants

From (39) and (43), we may derive the following system of moment invariants by eliminating the factor $e^{i\theta}$:

For the second-order moments, the two independent invariants are

$$I_{11}, \qquad I_{20}I_{02}. \tag{44}$$

For the third-order moments, the three independent invariants are

$$\begin{aligned} &I_{30}I_{03}, \qquad I_{21}I_{12}, \\ &(I_{30}I_{12}^3 + I_{03}I_{21}^3). \end{aligned} \tag{45}$$

A fourth one depending also on the third-order moments only is

$$\frac{1}{i}(I_{30}I_{12}^3 - I_{03}I_{21}^3) \tag{46}$$

There exists an algebraic relation between the above four invariants given in (45) and (46). The first three given by (45) are absolute invariants for both proper and improper rotations but the last one given by (46) is invariant only under proper rotation, and changes sign under improper rotation. This will be called a skew invariant. Therefore it is useful for distinguishing "mirror images."

One more independent absolute invariant may be formed from second and third order moments as follows:

$$(I_{20}I_{12}^2 + I_{02}I_{21}^2). \tag{47}$$

For pth order moments, $p \geq 4$ we have $[p/2]$, the integral part of $p/2$, invariants

$$I_{p0}I_{0p}; \qquad I_{p-1,1}I_{1,p-1}; \cdots ; \qquad I_{p-r,r}I_{r,p-r}; \cdots . \tag{48}$$

If p is even, we also have

$$I_{p/2,p/2}. \tag{49}$$

And also combined with $(p-2)$th order moments, we have $[p/2 - 1]$ invariants

$$\begin{aligned} &(I_{p-1,1}I_{0,p-2} + I_{1,p-1}I_{p-2,0}), \\ &(I_{p-2,2}I_{1,p-3} + I_{2,p-2}I_{p-3,1}), \\ &\cdots\cdots\cdots\cdots \\ &(I_{p-r,r}I_{r-1,p-r+1} + I_{r,p-r}I_{p-r+1,r-1}), \quad p - 2r > 0 \\ &\cdots\cdots\cdots\cdots \end{aligned} \tag{50}$$

combined with second-order moments, if $p =$ odd we have

$$(I^2_{[p/2],[p/2]+1}I_{20} + I^2_{[p/2]+1,[p/2]}I_{02}), \tag{51}$$

if $p =$ even,[7]

$$(I_{p/2-1,p/2+1}I_{20} + I_{p/2+1,p/2-1}I_{02}). \tag{52}$$

Therefore we always have $(p + 1)$ independent absolute invariants. By changing the above sums into differences, we can also have the skew invariants.

All the independent moment invariants together form a complete system, *i.e.*, for any given invariant; it is always possible to express it in terms of the above invariants. The proof is omitted here.

E. Moment Invariants Under General Linear Transformations

From the theory of algebraic invariants under the general linear transformations (17), it is known that the factor Δ is the determinant of the transformation. For linear transformations, J is also the determinant. For simplicity, let A, B, C and a, b, c, d denote the second and third order moments, then we may write the following two binary forms in terms of these moments as

$$\begin{aligned} &(A, B, C)(u, v)^2 \\ &(a, b, c, d)(u, v)^3. \end{aligned} \tag{53}$$

From the theory of algebraic invariants, we have the following four algebraically independent invariants,

$$\begin{aligned} I_1 &\equiv AC - B^2, \\ I_2 &\equiv (ad - bc)^2 - 4(ac - b^2)(bd - c^2), \\ I_3 &\equiv A(bd - c^2) - B(ad - bc) + C(ac - b^2), \\ I_4 &\equiv a^2C^3 - 6abBC^2 + 6acC(2B^2 - AC) \\ &\quad + ad(6ABC - 8B^3) \\ &\quad + 9b^2AC^2 - 18bcABC + 6bdA(2B^2 - AC) \\ &\quad + 9c^2A^2C - 6cdBA^2 + d^2A^3, \end{aligned} \tag{54}$$

of weight $w = 2$, 6, 4 and 6, respectively.

For the zeroth order moment, we have

$$\mu' = |J| \mu. \tag{55}$$

With the understanding that $\Delta^2 = |J|^2$, the following four absolute moment invariants are obtained,

$$\frac{I_1}{\mu^3}, \quad \frac{I_2}{\mu^7}, \quad \frac{I_3}{\mu^5}, \quad \text{and} \quad \frac{I_4}{\mu^7}. \tag{56}$$

There also exists a skew invariant,[8] I_5, of weight 9 depending on the moments A, B, C and a, b, c, d. This also may be normalized as

$$\frac{\Delta}{|J|}\left(\frac{I_5}{\mu^{10}}\right), \tag{57}$$

where $\Delta/|J|$ indicates the sign of the determinant. This invariant contains thirty monomial terms, and it is not algebraically independent.

By counting the number of relations among the moments and the number of parameters involved, it can be shown

[7] For $p = 4$, (52) is the same as the one given by (50); instead of (52), $(I_{40}I_{02}^2 + I_{04}I_{20}^2)$ may be used.

[8] G. Salmon, "Modern Higher Algebra," Stechert, New York, N. Y., 4th ed., p. 188; 1885. (Reprinted 1924.)

that four is the largest number of independent invariants possible for this case. Various methods have been developed in finding algebraic invariants, and many invariants have been worked out in detail. In case extension to higher moment invariants are required, the known results for algebraic invariants will be of great help.

IV. Visual Information Processing and Recognition

A. Pattern Characterization and Recognition

Any geometrical pattern or alphabetical character can always be represented by a density distribution function $\rho(x, y)$, with respect to a pair of axes fixed in the visual field. Clearly, the pattern can also be represented by its two-dimensional moments, m_{pq}, with respect to the pair of fixed axes. Such moments of any order can be obtained by a number of methods. Using the relations between central moments and ordinary moments, the central moments can also be obtained. Furthermore, if these central moments are normalized in size by using the similitude moment invariants, then the set of moment invariants can still be used to characterize the particular pattern. Obviously, these are independent of the pattern position in the visual field and also independent of the pattern size.

Two different ways will be described in the next two sections to accomplish orientation independence. In these cases, theoretically, there exist either infinitely many absolute moment invariants or infinitely many normalized moments with respect to the principal axes. For the purpose of machine recognition, it is obvious that only a finite number of them can be used. In fact, it is believed only a few of these invariants are necessary for many applications. To illustrate this point, a simple simulation program, using only two absolute moment invariants, and its performance will be described in Section V.

B. The Method of Principal Axes

In (39) and (40), let $p = 2$, we have the following moment invariants,

$$\begin{aligned}(\mu_{20}' - \mu_{02}') - i2\mu_{11}' &= e^{i2\theta}[(\mu_{20} - \mu_{02}) - i2\mu_{11}],\\ (\mu_{20}' - \mu_{02}') + i2\mu_{11}' &= e^{-i2\theta}[(\mu_{20} - \mu_{02}) + i2\mu_{11}],\\ \mu_{20}' + \mu_{02}' &= \mu_{20} + \mu_{02}.\end{aligned} \tag{58}$$

If the angle θ is determined from the first equation in (58), to make $\mu_{11}' = 0$, then we have

$$\tan 2\theta = \frac{+2\mu_{11}}{\mu_{20} - \mu_{02}}. \tag{59}$$

The x', y' axes determined by any particular values of θ satisfying (59) are called the principal axes of the pattern. With added restrictions, such as $\mu_{20}' > \mu_{02}'$ and $\mu_{30}' > 0$, θ can be determined uniquely. Moments determined for such a pair of principal axes are independent of orientation. If this is used in addition to the method described in the last section, pattern identification can be made independently of position, size and orientation.

The discrimination property of the patterns is increased if higher moments are also used. The higher moments with respect to the principal axes can be determined with ease, if the invariants given by (39) and (40) are used. These relations are also useful in other ways. As an illustration, for $p = 3$ we have

$$\begin{aligned}&(\mu_{30}' - 3\mu_{12}') - i(3\mu_{21}' - \mu_{03}')\\ &\qquad = e^{i3\theta}[(\mu_{30} - 3\mu_{12}) - i(3\mu_{21} - \mu_{03})],\\ &(\mu_{30}' + \mu_{12}') - i(\mu_{21}' + \mu_{03}')\\ &\qquad = e^{i\theta}[(\mu_{30} + \mu_{12}) - i(\mu_{21} + \mu_{03})].\end{aligned} \tag{60}$$

The two remaining relations, which are the complex conjugate of these two, are omitted here. If θ and the four third moments are known, the same moments with respect to the principal axes can be computed easily by using the above relations. There is no need of transforming the input pattern here.

In the above method, because of the complete orientation independence property, it is obvious that the numerals "6" and "9" can not be distinguished. If the method is modified slightly as follows, it can differentiate "6" from "9" while retaining the orientation independence property to a limited extent. The value of θ is still determined by (59), but it is also required to satisfy the condition $|\theta| < 45$ degrees. The use of third order moments in this case is also essential.

If the given pattern is of circular symmetry or of n-fold rotational symmetry, then the determination of θ by (59) breaks down. This is due to the fact that both the numerator and the denominator are zero for such patterns. As an example, assume that the pattern is of 3-fold rotational symmetry, *i.e.*, if the pattern is turned $2\pi/3$ radians about its centroid, it is identically the same as the original. In the first equation in (58), there are only two possible values for θ to make the imaginary part of $I_{20}' = 0$, *i.e.*, to make $\mu_{11}' = 0$. Under this symmetry requirement, there are more than two possible values to make the imaginary part of $I_{20}' = 0$; therefore, the only possibility is to have $I_{20}' \equiv 0$, and also $I_{20} \equiv 0$. In this 3-fold rotational symmetry case, the first equation in (60) can be used to determine the θ and the principal axes by requiring $3\mu_{21}' - \mu_{03}' = 0$. Based upon this example, we may state the following theorem.

Theorem: If a pattern is of n-fold rotational symmetry, than all the orthogonal invariants, I's, with the factor $e^{\pm iw\theta}$ and $w/n \neq$ integer must be identically equal to zero. For the limiting case of circular summetry, only I_{11}, I_{22}, $\cdots$ are not zero.

For patterns with mirror symmetry, a similar theorem may be derived.

C. The Method of Absolute Moment Invariants

The absolute orthogonal moment invariant described in Section III-D can be used directly for orientation independent pattern identification. If these invariants are combined with the similitude invariants of central moments, then pattern identification can be made independently of position, size and orientation. A specific example is given in Section V-A.

For the second and third order moments, we have the following six absolute orthogonal invariants:

$$\begin{aligned}
&\mu_{20} + \mu_{02},\\
&(\mu_{20} - \mu_{02})^2 + 4\mu_{11}^2,\\
&(\mu_{30} - 3\mu_{12})^2 + (3\mu_{21} - \mu_{03})^2,\\
&(\mu_{30} + \mu_{12})^2 + (\mu_{21} + \mu_{03})^2,\\
&(\mu_{30} - 3\mu_{12})(\mu_{30} + \mu_{12})[(\mu_{30} + \mu_{12})^2 - 3(\mu_{21} + \mu_{03})^2]\\
&\qquad + (3\mu_{21} - \mu_{03})(\mu_{21} + \mu_{03})\\
&\qquad \cdot[3(\mu_{30} + \mu_{12})^2 - (\mu_{21} + \mu_{03})^2],\\
&(\mu_{20} - \mu_{02})[(\mu_{30} + \mu_{12})^2 - (\mu_{21} + \mu_{03})^2]\\
&\qquad + 4\mu_{11}(\mu_{30} + \mu_{12})(\mu_{21} + \mu_{03}),
\end{aligned} \tag{61}$$

and one skew orthogonal invariants,

$$\begin{aligned}
&(3\mu_{21} - \mu_{03})(\mu_{30} + \mu_{12})[(\mu_{30} + \mu_{12})^2 - 3(\mu_{21} + \mu_{03})^2]\\
&\quad - (\mu_{30} - 3\mu_{12})(\mu_{21} + \mu_{03})[3(\mu_{30} + \mu_{12})^2 - (\mu_{21} + \mu_{03})^2].
\end{aligned} \tag{62}$$

This skew invariant is useful in distinguishing mirror images.

This method can be generalized to accomplish pattern identification not only independently of position, size and orientation but also independently of parallel projection. In this generalization the general moment invariants are used instead of the orthogonal and similitude moment invariants.

V. Visual Pattern Recognition Models

A. The Simulation of a Simple Model

A simulation program of a simple pattern recognition model, using only two moment invariants, has been written for an LGP-30 computer. No information, properties, or features about the patterns to be recognized are contained in the simulation program itself; it learns. The visual field is a 16×16 matrix of small squares. A pattern is first projected onto the matrix and then each small square is digitalized to the values, 0, 2, 4, 6, or 8. After loading each pattern, the following two moment invariants[9]

$$\begin{aligned}
X &= \mu_{20} + \mu_{02}\\
Y &= \sqrt{(\mu_{20} - \mu_{02})^2 + 4\mu_{11}^2}
\end{aligned} \tag{63}$$

are computed. The central moments, μ_{20}, μ_{11}, μ_{02} used above (normalized with respect to size) are obtained from the ordinary moments by (11). This point (X, Y) in a two dimensional space is used as a representation of the pattern.

[9] X and Y are, respectively, the sum and difference of the two second moments with respect to the principal axes, and may be interpreted as "spread" and "slenderness."

Assume the program or model has already learned a number of patterns, represented by (X_i, Y_i), $i = 1, 2, \cdots, n$, together with their names. If a new pattern is presented to the model, a new point (X, Y) and the distances

$$d_i = \sqrt{(X - X_i)^2 + (Y - Y_i)^2}, \quad i = 1, 2, \cdots, n \tag{64}$$

between (X, Y) and (X_i, Y_i) are computed. Let the minimum distance, $d_{\min}$, be defined as

$$d_{\min} = \min_i d_i. \tag{65}$$

The distance d_k satisfying $d_k = d_{\min}$ is selected (if more than one of the distances satisfying the condition, one d_k is selected at random). Then d_k is compared with a preselected recognition level L. 1) If $d_k > L$, the computer will type out "I do not know", then wait to learn the name of the new pattern. If a name is now entered, the computer then stores (X, Y) as (X_{n+1}, Y_{n+1}) together with the assigned name. Hence, a new pattern is learned by the program. 2) If $d_k \leq L$, the computer will identify the pattern with (X_k, Y_k) by typing out the name associated.

A very simple performance-improving program is also incorporated. When this program is used, it replaces the values of X_i, Y_i corresponding to the name now told, by

$$\frac{1}{\alpha}[(\alpha - 1)X_i + X], \quad \frac{1}{\alpha}[(\alpha - 1)Y_i + Y] \quad \alpha > 1. \tag{66}$$

This operation moves the point (X_i, Y_i) toward (X, Y).

B. Performance of the Simulation Program

Several experiments have been tried on the simulation program. For the convenience of description, two patterns are described as strictly similar, if one pattern can be transformed exactly into the other by a combination of translation, rotation and similitude transformations. In one experiment, patterns which are strictly similar after digitalization were fed to the program. If any one of such patterns is taught to the program just once, then it can identify correctly any other pattern of the same class. The number of different pattern classes capable of being learned is quite large, even with this simple program. There is no wrong identification except for specially constructed patterns.

Another experiment dealt with character recognition. A set of twenty-six capital letters from a 2-inch Duro Lettering Stencils were copied onto the 16×16 matrix and digitalized as inputs to the program. The values of X and Y, in arbitrary units, are given in Table I and Fig. 1, and two samples of the digitalized inputs for the letter M and W are shown in Fig. 2. The following may be noted:

1) Fig. 1 shows that the points for all the twenty-six letters are separated.

2) If inputs, prepared by using the same stencils but not strictly similar after digitalization are used, the corresponding points are not the same as those shown in Fig. 1. For a limited number of cases tried, the maximum

TABLE I

	X	Y		X	Y
A	6.2020	2.4986	N	5.7885	1.7933
B	6.1104	2.0853	O	8.2829	2.6246
C	10.4136	4.1818	P	7.0329	2.6456
D	8.2045	3.0911	Q	6.7674	1.9611
E	8.2147	4.3144	R	6.2707	1.9149
F	8.0390	4.5017	S	7.7501	3.3660
G	8.6096	3.0127	T	10.6216	7.1239
H	7.6243	1.1825	U	9.1728	2.1383
I	11.9780	11.2824	V	6.8761	3.2715
J	10.4118	6.6854	W	5.6667	0.1893
K	7.3278	2.5620	X	7.5714	3.5651
L	12.0662	8.3889	Y	8.3538	3.8612
M	5.7356	0.0540	Z	8.8843	5.1580

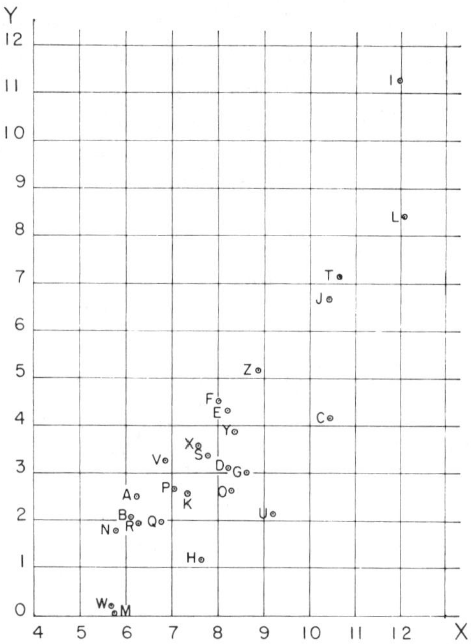

Fig. 1—Point representation of the twenty-six capital letters.

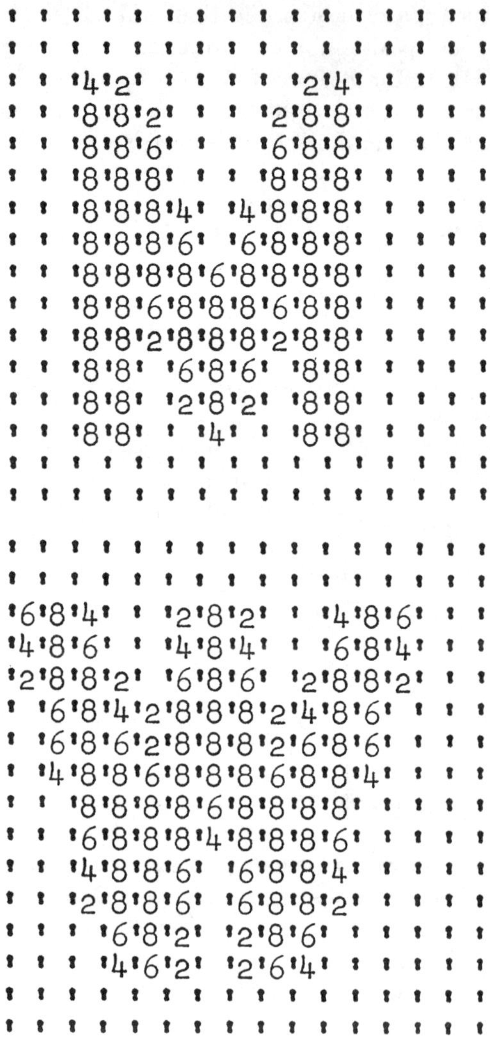

Fig. 2—Two samples of the digitalized inputs for the letters M and W.

variation in terms of distance between two points representing the same letter is of the order of 0.5. Compared with Fig. 1, it is obvious that overlapping of some classes will occur. If the resolution of the visual field is increased, the performance will definitely be improved.

3) In Fig. 1, it can be seen that some letters which are close to each other are of considerable difference in shape. A typical case is shown in Fig. 2, it is not difficult to conclude that the third order moments for the M and W examples shown will be considerably different.

From these results, it is clear that both the resolution and the number of invariants used should be increased but probably not greatly.

One additional experiment concerned the simple learning program. In this experiment, patterns belonging to the same class were generally represented by different points, clustered together, in the plane. As already described, a class represented by such a cluster was represented by a single point in this program, but this point together with the recognition level really form a circular recognition region for the class. For good performance, this region should be centered over the cluster of points representing the class. The point for the first sample of a class is not necessarily at the center of this region. Because of this fact, incorrect identifications may occur. The simple learning program, sometimes, is useful for such cases. If the clusters of points of different classes do not 'overlap,' generally, the program will improve the performance; otherwise, the performance may become worse. Another learning program will be described in the next section.

C. Other Visual Pattern Recognition Models

From the simulation program and the theoretical considerations described in IV, a considerably improved pattern recognition model is as follows: P absolute moment invariants or P normalized moments with respect to the principal axes, denoted by $X^1, X^2, \cdots, X^P$, are used; and the point $(X) \equiv (X^1, X^2, \cdots, X^P)$ in a P dimensional space is used as the representation for a pattern. It is believed that $P = 6$, (*i.e.*, using four more invariants related to the third order moments) and a 32×32 or

50 × 50 matrix as the visual field will be adequate for many purposes.

Let (X_i), $i = 1, 2, \cdots, n$ be the points representing the patterns already learned, and N_i be the number of samples of the ith pattern already learned. After each learning process for the ith pattern, N_i is replaced by $(N_i + 1)$, and (X_i) by

$$\left(\frac{N_i X_i + X}{N_i + 1}\right) \equiv \bar{X}_i \tag{67}$$

where (X) is the representation of the new sample. This new (X_i) is obviously equal to the average of all the $(N_i + 1)$ samples learned.

Instead of using a common recognition level, L, a separate one is determined for each pattern class in the learning process. After each sample is learned, L_i is replaced by the larger one of

$$L_i \quad \text{and} \quad \sqrt{\sum_{p=1}^{P} (\bar{X}_i^p - X^p)^2}. \tag{68}$$

The L_i thus determined, as the sample number increases, approaches the minimum radius of a hypersphere which includes most if not all the sample points in its interior. The center of the hypersphere is located at $(\bar{X}_i)$.

In this model, the following are stored for each class of patterns learned,

$$\text{Name}, \quad (X_i), L_i, N_i. \tag{69}$$

(X_i) and L_i form a spherical recognition region for the ith pattern. When a new pattern represented by (X) is entered, the distances

$$d_i = \sqrt{\sum_{p=1}^{P} (X_i^p - X^p)^2} \qquad i = 1, 2, \cdots, n \tag{70}$$

are computed. The distances d_j satisfying

$$d_j \leq L_j \tag{71}$$

are then selected. If no d_j is obtained, the pattern is considered as not yet learned, otherwise

$$D_j = \frac{d_j}{N_j} \tag{72}$$

is computed and

$$D_k = \min_j (D_j) \tag{73}$$

is selected, as in Section V-A, to identify the pattern. The use of N_i in the identification is believed to be useful when overlapping occurs.

If automatic input and digitalization equipment is used, there may be other types of noise introduced in addition to that due to digitalization. The well known local averaging process[10,11] can be used to reduce some of such noise, but the potential for discrimination possessed by such models is useful to combat whatever remains. In this connection, it seems worthwhile to point out the following two facts. 1) If two classes are separated, say, in two dimensions; they can never overlap when additional dimensions are introduced. 2) The use of moment invariants makes possible the derivation of models which may automatically generate additional dimensions—moment invariants—for the purpose of discrimination or combating noise.

The representation of a pattern by a point in a P dimensional space converts the problem of pattern recognition into a problem of statistical decision theory. Depending upon the particular decision method used, different statistical models may be devised. The work done by Sebestyen[12] is an example, his method can be used here directly.

The method of principal axes developed here has another application in connection with the statistical approaches mentioned at the beginning of this paper. It may be used as a preprocessor to normalize the inputs before the main processer is used. All the parameters necessary for translation, size and orientation normalizations can be obtained from some of the relations used in the method of principal axes. Such a preprocessor undoubtedly will increase the ability of the models based upon the statistical approach.

ACKNOWLEDGMENT

The author would like to express his deep appreciation to Dr. W. R. LePage for his constructive criticism and invaluable help during the preparation of this paper.

[10] G. P. Dineen, "Programming pattern recognition," *Proc. Western Joint Computer Conf.*, pp. 94–100; March, 1955.

[11] J. S. Bomba, "Alpha-numeric character recognition using local operations," *Proc. Eastern Joint Computer Conf.*, pp. 218–224; December, 1959.

[12] G. S. Sebestyen, "Recognition of membership in classes," IRE TRANS. ON INFORMATION THEORY, vol. IT-7, pp. 44-50; January, 1961.

Diffraction-Pattern Sampling for Automatic Pattern Recognition

GEORGE G. LENDARIS, MEMBER, IEEE,
AND GORDON L. STANLEY

Abstract—This paper describes diffraction-pattern sampling as a basis for automatic pattern recognition in photographic imagery; it covers: diffraction-pattern generation, diffraction-pattern/image-area relationships, diffraction-pattern sampling, algorithm development (using an interactive computer-graphic based facility), facility description, and experimental results which have been obtained over the last few years at General Motors' AC Electronics-Defense Research Laboratories, Santa Barbara, Calif.

Sampling the diffraction pattern results in a sample signature—a different one for each sampling geometry. The kinds of information obtainable from sample signatures are described, and considerations for developing algorithms based on such information are discussed.

A tutorial section is included for the purpose of giving the reader an intuitive feeling for the kinds of information contained in a diffraction pattern and how it relates to the original photographic imagery.

INTRODUCTION

THE PURPOSE of this paper is to demonstrate the utility of diffraction-pattern sampling for automatically classifying patterns in photographic transparencies.

An important step in the design of automatic recognition devices using diffraction-pattern sampling is for the designer to become intimately familiar with diffraction patterns and their relationships to the original photographs. Accordingly, an allied purpose of this paper is to help the reader develop an intuitive feeling for the information available in diffraction patterns and its relationship to patterns in the original photographs.

Once the designer has developed this familiarity, he can determine the features of the diffraction pattern which correspond to the features of the original photographic images which are important to the recognition task at hand. On the basis of this, he can design a sampling scheme which will preserve this information. His task then is to study the resulting sampled information (called a sample signature), and become familiar with its properties and how they relate to the diffraction pattern (and, ultimately, to the original photograph). With this as a basis, he can proceed to develop means for automatically performing the recognition task on the basis of the sampled information.

The basic pattern-recognition system configuration against which the discussions in this paper are made is shown in Fig. 1. The preprocessor processes the photographic input, extracts what information it can, and passes this on to the decision processor (this could be a computer); the output of the decision processor is the desired classification.

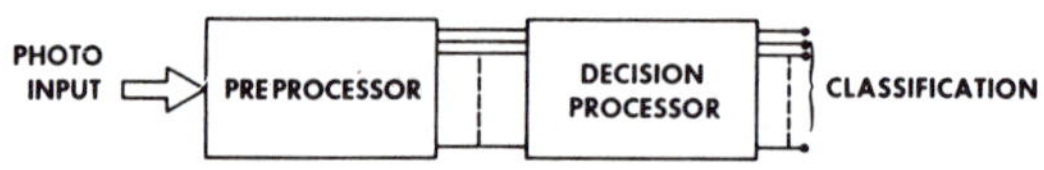

Fig. 1. Pattern-recognition system configuration assumed in this paper.

Diffraction-pattern sampling is intended to serve as a preprocessor for such a system. The decision processor must then use this sampled information to arrive at the desired classification. We discuss certain aspects of the decision processor design, and cite some experimental results which demonstrate the utility of this approach.

This paper has two main sections. Section I is concerned with helping the reader develop an intuitive feeling for the information contained in a diffraction pattern and how it relates to the original photographic imagery. Section II-A describes various aspects of the diffraction-pattern sampling process, and Section II-B cites some experimental results.

I. DEVELOPING A FEELING FOR DIFFRACTION PATTERNS

As mentioned in the Introduction, one of our purposes here is to give the reader an intuitive feeling for interpreting diffraction patterns. Here we show the diffraction patterns of some basic input patterns and discuss them with this aim in mind. Accordingly, these discussions will not be rigorous; for details, the reader is referred to texts such as [1]–[3].

Diffraction Pattern Basics

As in [4],[1] we start the discussion by stating two well-known facts.

1) The Fraunhofer diffraction pattern of an aperture is the two-dimensional Fourier transform of the transmission function of that aperture.

[1] This was the first paper in which we described our use of diffraction-pattern sampling for automatic pattern recognition.

Manuscript received June 18, 1969; revised December 5, 1969. Essentially, this paper is the same as one presented, by invitation, at the Pattern Recognition Studies, Seminar-in-Depth, sponsored by the Pattern Recognition Society and the Society of Photo-Optical Instrumentation Engineers on June 10, 1969, New York, N. Y., and appears under the same title in the proceedings of that Seminar.

G. G. Lendaris was with AC Electronics-Defense Research Laboratories, General Motors Corporation, Santa Barbara, Calif. He is now with the Oregon Graduate Center, Beaverton, Ore. 97005.

G. L. Stanley is with the AC Electronics-Defense Research Laboratories, General Motors Corporation, Santa Barbara, Calif. 93015.

Reprinted from *Proc. IEEE*, vol. 58, pp. 198–216, Feb. 1970.

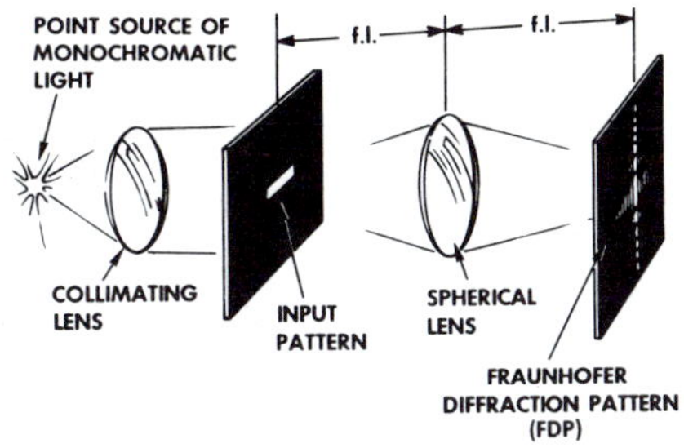

Fig. 2. FDP generation system.

2) Except for a linear phase factor, the Fourier transform of any function is independent of a translation of that function.

An optical configuration for generating the Fraunhofer diffraction pattern (FDP) of an input pattern is shown in Fig. 2.

Since the Fourier transform preserves rotation, if the input pattern is rotated, the FDP (which is centered on the optical axis of the system) is rotated about the optical axis but remains otherwise unchanged. We assume that the pattern given for identification is in the form of a photographic transparency, and call each portion of the transparency which is being "looked at" for identification a scan area (usually circular).

From 1) it follows that the FDP of a scan area is a spatial-frequency representation of the scan area. Since, however, physical sensors (such as the eye, film, photoelectric cells, etc.) sense the intensity (hence the phase factor of 2) disappears) and not simply the amplitude of a light distribution, the FDP as observed by these sensors is equivalent to the power spectrum[2] of the scan area's transmission function. To keep this distinction clear, in the remainder of this paper we will use FDPMS (FDP modulus squared) when referring to the FDP as it is photographed, as it is measured, or as we see it.

The center of the FDP corresponds to a spatial frequency of zero. Any other point in the plane of the FDP corresponds to a spatial frequency proportional to the points' distance from the origin and to the direction given by drawing a radius vector from the origin through that point. An intense high spatial-frequency indication along a given direction in the FDP corresponds to an object in the scan area with a small dimension in that direction. Similarly, an intense low-frequency indication along a given direction corresponds to an object with a large dimension in that direction. Further, the FDPMS is symmetric with respect to the origin; that is, every intensity indication in the FDP has an identical counterpart which is located on the same direction line, on the other side of and the same distance from the origin.

A linear object in the original imagery gives rise to a distribution of light along a line centered on the optical axis. Similarly, a circular object gives rise to a distribution of light in concentric annular rings centered on the optical axis. Finally, a lattice-like object gives rise to lattice-like distribution of light in the diffraction pattern.

Single Simple Shapes

In Fig. 3(a) we show a simple rectangular aperture, and in Fig. 3(b), its FDPMS.

The intensity of the FDP along the direction corresponding to the width of the rectangle has the form $(\sin x/x)^2$, where $x=(\pi D/\lambda) \sin \alpha$, D is the width of the slit, λ is the wavelength of the (monochromatic) light used, and α is the angle between the optical axis and the line drawn from the lens center to the point x (in the focal plane). It follows that $x=0$ at the point where the optical axis intercepts the focal plane, and is the "center" of the FDP.

A semilogarithmic plot of the function $(\sin x/x)^2$ is shown in Fig. 4. The logarithm was chosen because it yields a better representation of what the eye perceives, e.g., Fig. 3(b). We note that the $(\sin x/x)$ function has its zeros at $x=\pi, 2\pi, 3\pi \cdots$; and from its definition, x takes on these values when $\sin \alpha$ takes on the values λ/D, $2\lambda/D$, $3\lambda/D \cdots$. Thus, the narrower the rectangle (smaller D), the wider the central lobe of the $(\sin x/x)^2$ curve and hence the FDPMS. Similarly, the wider the slit, the narrower the center portion of the FDPMS.

In Fig. 3(a), the length L of the rectangle is equal to $10D$. Therefore, the center lobe (and proportionately the rest) of the FDPMS (Fig. 3(b)) in the direction corresponding to the length of the rectangle is 1/10 as large as in the direction corresponding to the width of the rectangle.

We next look at a circular aperture, Fig. 5(a). Fig. 5(b) is the FDPMS of a circular aperture of radius R, and Fig. 5(c) is the FDPMS of a circular aperture of radius $2R$ (these are called Airy disks).

Inasmuch as the 2-dimensional Fourier transform preserves circular symmetry, this FDPMS is circularly symmetric. The intensity as a function of radius r, measured from the optical axis, has the form $(J_1(r)/r)^2$ where $J_1(r)$ is a first-order Bessel function.

[2] During review it was suggested that the generality of the approach described in this paper might better be conveyed with the words "spatial-frequency power-spectrum sampling" rather than the words "diffraction-pattern sampling" since the fact that an optical system can be arranged to provide a Fourier transform is really incidental to the concepts involved. Consideration showed it would have been a major undertaking to make the necessary attendant changes throughout the paper. However, it is pointed out that the basic ideas expressed in this paper concerning FDPMSs apply to spatial-frequency power-spectra *in general*, no matter how they are obtained.

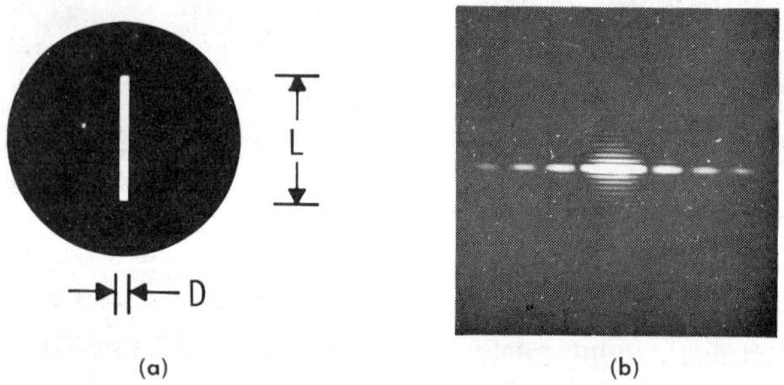

Fig. 3. (a) Rectangular aperture. (b) FDPMS of (a).

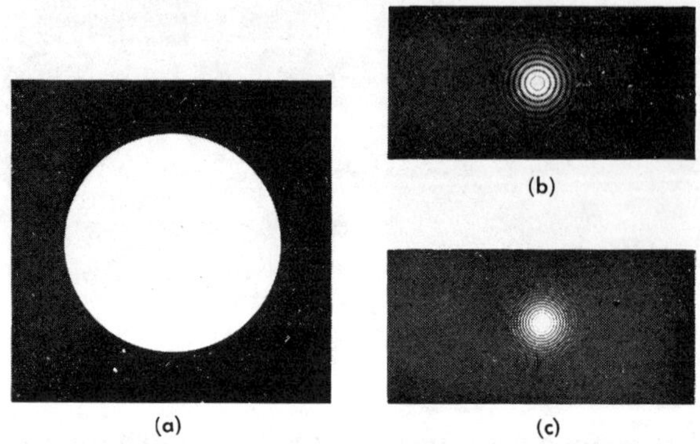

Fig. 5. (a) Circular aperture. (b) FDPMS of (a) with radius R. (c) FDPMS of (a) with radius $2R$.

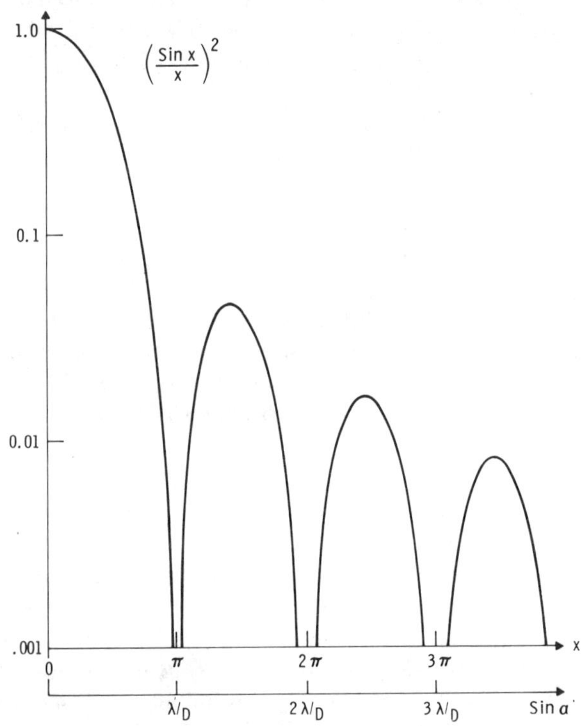

Fig. 4. Semilog plot of $(\sin x/x)^2$.

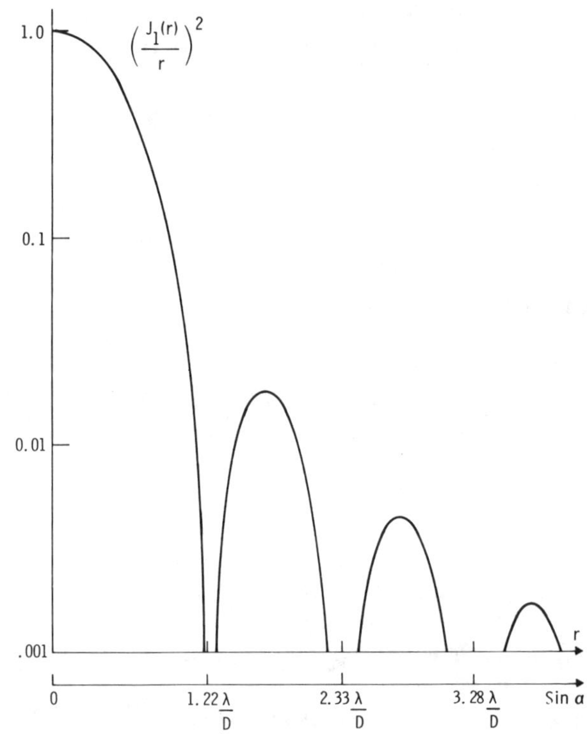

Fig. 6. Semilog plot of $(J_1(r)/r)^2$.

A semilogarithmic plot of the function $(J_1(r)/r)^2$ is shown in Fig. 6.

The relationships are not as straightforward as in the rectangular case, but it turns out that the zeros of this function occur at $1.22\ \lambda/D$, $2.33\ \lambda/D$, $3.238\ \lambda/D \cdots$. As in the rectangular case, the scaling of this function is inversely proportional to the diameter D of the circular aperture.

Multiple Simple Shapes

As a convenience in this paper, we will utilize the notions of impulse-sheets and two-dimensional impulses. An impulse-sheet has an infinite length in one direction (we call this the sheets' "direction"); its cross section, which remains the same along the sheets' entire length, has the usual delta-function properties. Multiplication of two intersecting impulse-sheets results in a single (two-dimensional) impulse located at the intersection of the two impulse-sheets.

The impulse-sheets and two-dimensional impulses have the following properties. (Since they are equivalent, we will use the term FDP instead of "two-dimensional Fourier transform.")

1) The FDP of an impulse-sheet is an impulse-sheet, centered at the origin and in a direction orthogonal to the direction of the original impulse-sheet.
2) The FDP of an infinite array of uniformly spaced parallel impulse-sheets is an infinite string of impulses along the direction orthogonal to the impulse-sheets' direction, with a spacing inversely proportional to the impulse-sheet separation, and with one of the impulses located at the origin.
3) Conversely, the FDP of a string of uniformly spaced impulses is an array of parallel impulse-sheets whose direction is orthogonal to the impulse string, whose separation is inversely proportional to the impulse

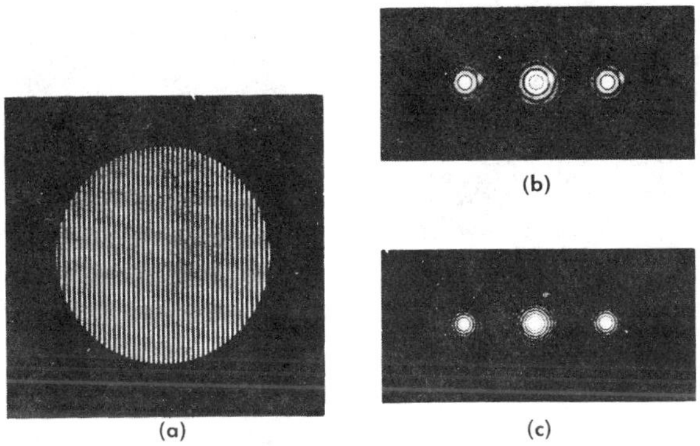

Fig. 7. (a) Parallel lines through a circular aperture. (b) FDPMS of a grating through radius R. (c) FDPMS of a grating through radius $2R$.

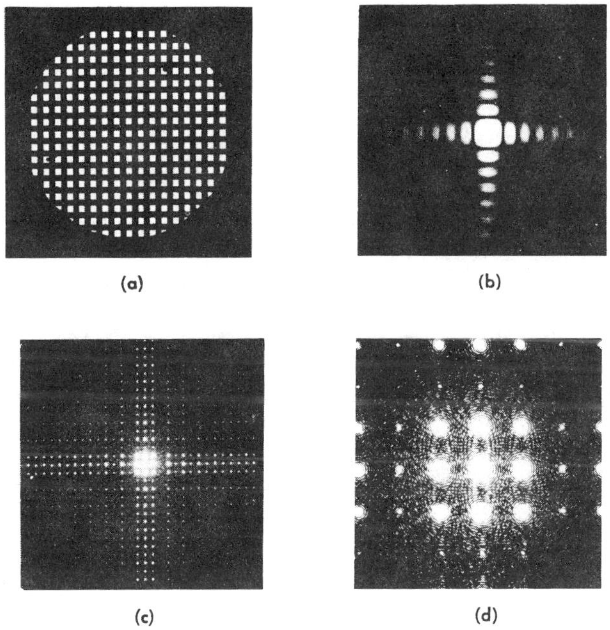

Fig. 8. (a) Array of rectangles through a circular aperture. (b) FDPMS of one of the rectangles of (a). (c) FDPMS of (a). (d) Center portion of (c).

separation, and one of whose impulse-sheets goes through the origin.

4) The FDP of an infinite lattice-like array of impulses is an infinite lattice-like array of impulses whose dimensions are inversely related to those of the original lattice, with an impulse at the origin.
5) When convolving a function, call it y, with an array of impulses, the function y is simply replicated at the locations of each of the impulses.

Further, we will use the fact that the Fourier transform of a product of two functions is equal to the convolution of the Fourier transforms of the two functions; and, conversely, the Fourier transform of the convolution of two functions is equal to the product of the Fourier transforms of the two functions.

As an example, consider the scan area of Fig. 7(a). This is an array of uniformly spaced parallel straight lines as "seen" through a circular aperture. This scan area can be viewed as the result of multiplying an infinite array of parallel lines with the transmission function of the circular aperture. Therefore, to get the FDP of the combination, we convolve their two individual FDPs. The parallel lines are analogous to the parallel impulse-sheets noted earlier; therefore, by property 2), their FDP will be a string of impulses, with separation inversely related to the spacing of the lines and in a direction orthogonal to the lines. As previously noted, the FDP of the circular aperture is an Airy disk. Therefore, by property 5), the resulting FDP should consist of an Airy disk at the origin and at the locations of each of the remaining impulses (corresponding to the fundamental spatial frequency and all the harmonics which are present).

Fig. 7(b) and 7(c) is the FDPMSs of a grating as seen through the apertures which generated, respectively, Fig. 5(b) and 5(c). Note that in the FDPMSs, the replicated Airy disks have different sizes, corresponding inversely to the diameter of their respective circular apertures, while the replication positions, corresponding to the grating spacing, remain the same.

From a slightly different viewpoint, this can be seen as an analog of a sampled-data system. For this case, the parallel equally spaced linear apertures perform a "sampling" of the circular aperture function. As shown in sampling theory, in the frequency domain, the sampling operation introduces replications of the functions' spectrum at locations determined by the sampling frequency. Fig. 7(b) and 7(c) shows just exactly this situation.

As an example of multiple simple shapes, consider the configuration shown in Fig. 8(a). This is a two-dimensional array of identical rectangles as "seen" through a circular aperture. Three aspects of this array contribute to its FDP: the individual rectangle, the array geometry, and the scan aperture.

This array of rectangles can be viewed as the result of convolving one rectangle with an array of impulses (whose relative locations are the same as those of the array of Fig. 8(a)). Accordingly, the FDP of this array will be the product of the FDP of the rectangle and the FDP of the array of impulses. (Assuming this array to be infinite, then, by property 4), its FDP will also be an infinite array of impulses.)

The FDP of one of the rectangles of this array is shown in Fig. 8(b). Multiplying the FDP of the rectangle with the array of impulses yields what we may describe as a "sampled" FDP of the rectangle (where the "sampling" points are determined by the original array geometry and its dimensions). This is shown in Fig. 8(c).

As with the previous example, Fig. 8(a) can be viewed as the product of the scan-area transmission function with that of an infinite array of rectangles. Again, this multiplication is equivalent to a convolution of the respective FDPs. Therefore, at each "sample point" of the FDP in Fig. 8(c) (up to this point, each of these is supposed to be an impulse) we should find replicated the FDP of the circular aperture. Since the aperture is much larger than the rectangles or the array dimensions, its FDP will be much smaller. The center

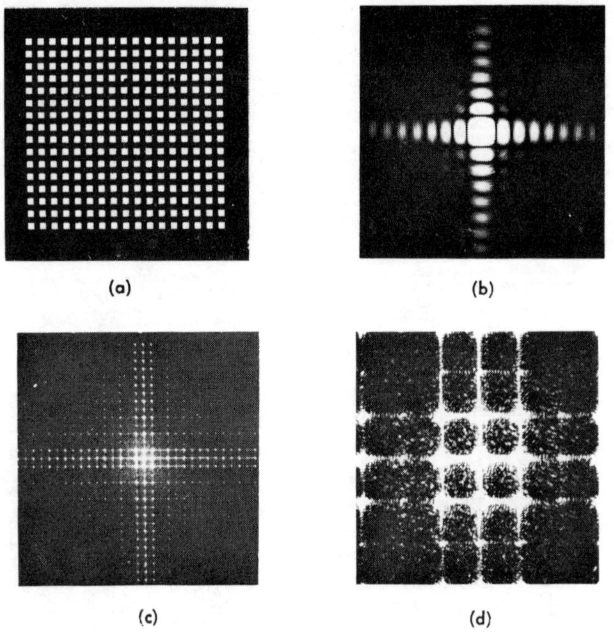

Fig. 9. (a) Array of rectangles through a rectangular aperture. (b) FDPMS of one of the rectangles of (a). (c) FDPMS of (a). (d) Center portion of (c).

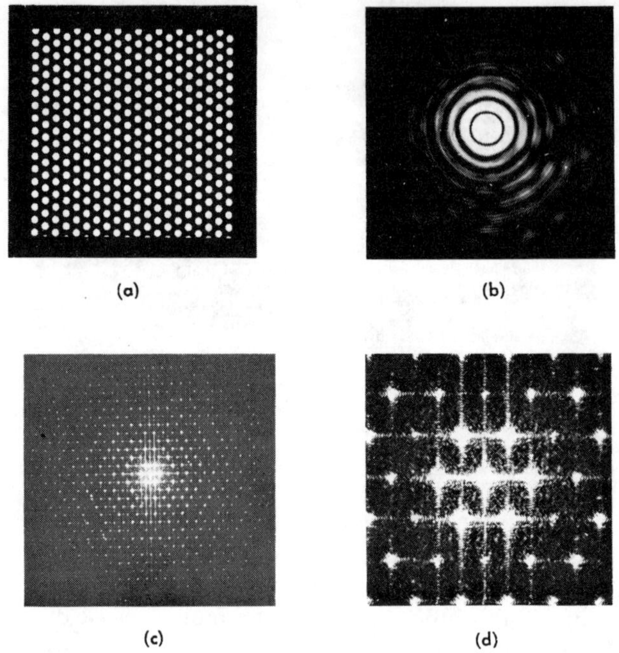

Fig. 11. (a) Array of circles through rectangular aperture. (b) FDPMS of one of the circles of (a). (c) FDPMS of (a). (d) Center portion of (c).

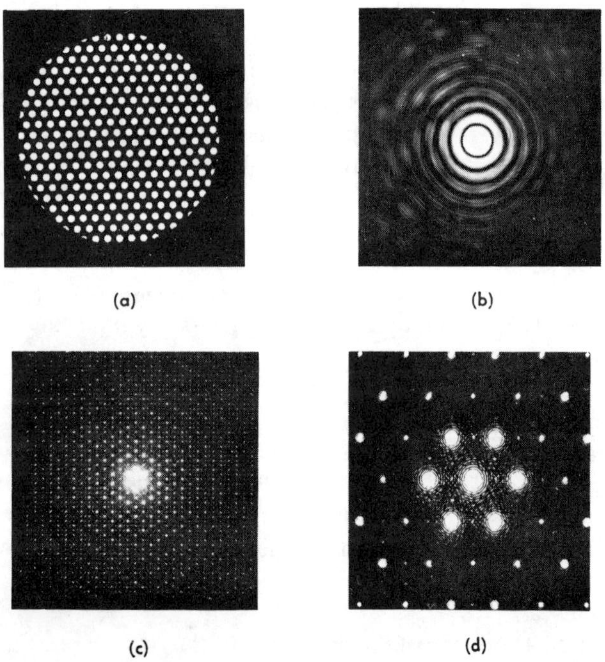

Fig. 10. (a) Array of circles through a circular aperture. (b) FDPMS of one of the circles of (a). (c) FDPMS of (a). (d) Center portion of (c).

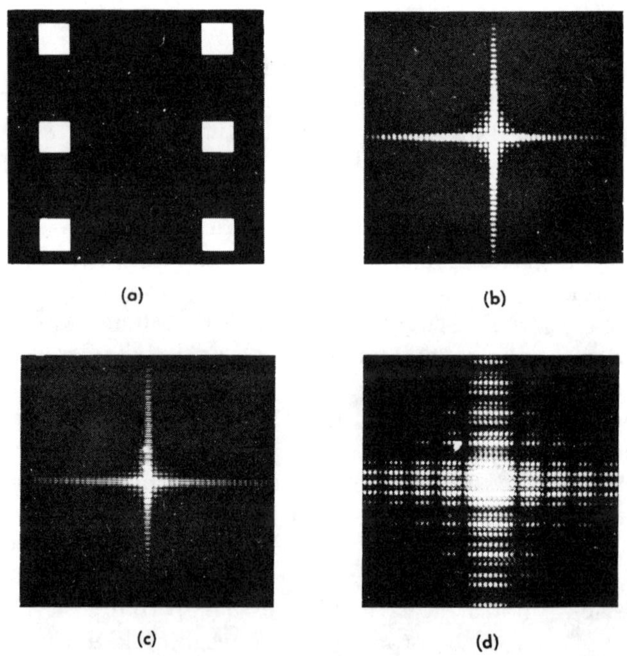

Fig. 12. (a) Array of six squares. (b) FDPMS of one of the squares of (a). (c) FDPMS of (a). (d) Center portion of (c).

of the Fig. 8(c) FDP is magnified in Fig. 8(d); we see there that the aperture FDP is in fact replicated at each "sample point."

This same discussion applies to Figs. 9 through 11, where in Fig. 9 we use a rectangular aperture; and in Figs. 10 and 11 we use an array of circles instead of rectangles.

In comparing these figures, we note that the circular aperture provides less visual contamination of the properties of the lattice than does the rectangular aperture (this would be even more obvious if the rectangular aperture were rotated, say 45°). This is true in general, and for this reason, when scanning photographic images the circular aperture is generally preferred to a rectangular (or square) one.

As a less symmetrical example, consider the array of Fig. 12(a). This array can be viewed as the result of convolving one rectangle with an infinite array of impulses (whose vertical and horizontal spacings are as in Fig. 12(a)) and the result multiplied by an aperture to allow only the six squares

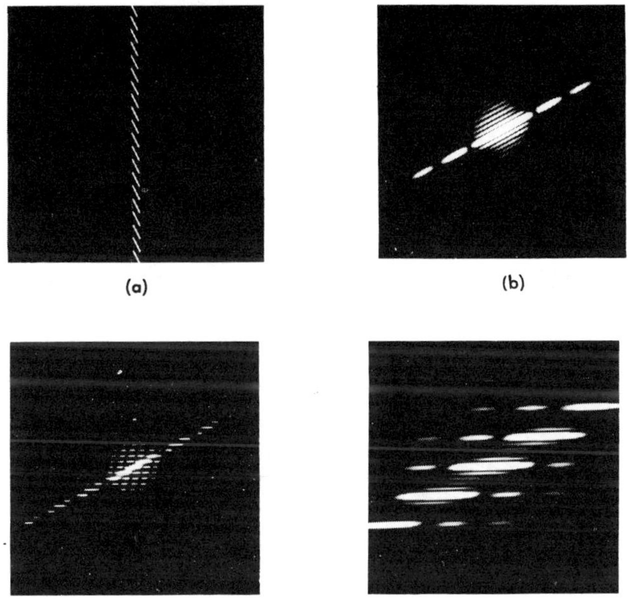

Fig. 13. (a) Staggered rectangular objects (similar to parked cars). (b) FDPMS of one of the rectangles of (a). (c) FDPMS of (a). (d) Center portion of (c).

of Fig. 12(a) to appear. The FDPMS of one of the rectangles is shown in Fig. 12(b). The result of multiplying this FDP with that of the impulse array is shown in Fig. 12(c) (again we have a "sampled" version of the Fig. 12(b) FDPMS). Note that the "sampling" points have a different spacing in the horizontal and vertical directions, corresponding to the differences in the respective spacings in the original array [Fig. 12(a)].

Whereas in the examples of Figs. 8 through 11 the aperture dimensions were considerably larger than the array dimensions, and therefore the FDPs of the aperture were considerably smaller than the spacings in the FDP array, in this example the aperture dimensions are nearly the same as those of the array. Therefore, the aperture FDP will be very noticeable as it is replicated at each of the array impulse locations. This is particularly noticeable in the vertical direction of Fig. 12(c), and as shown in the 5× magnification, Fig. 12(d). The lighter indications in the FDP are the second lobe (refer to Fig. 4) of the aperture FDP.

As a final example, consider the array of Fig. 13(a). This can be viewed as the result of convolving a small rectangle (oriented at approximately 60°) with a vertically oriented infinite string of uniformly spaced impulses, and the result multiplied by an aperture to limit its vertical extent. The FDPMS of one of the small rectangles is shown in Fig. 13(b). By property 3) the FDP of a vertically oriented string of uniformly spaced impulses is an infinite array of parallel horizontal impulse-sheets. Multiplying these two FDPs, we again obtain a "sampled" version of Fig. 13(b), shown in Fig. 13(c); only this time we have "sampling lines" instead of "sampling points." Further, the vertical component of the aperture's FDP will be replicated at each of the impulse-sheets (since the cross section of each of these impulse-sheets has the usual impulse properties). These can be seen in Fig. 13(d), which is a 5× magnification of the center portion of Fig. 13(c).

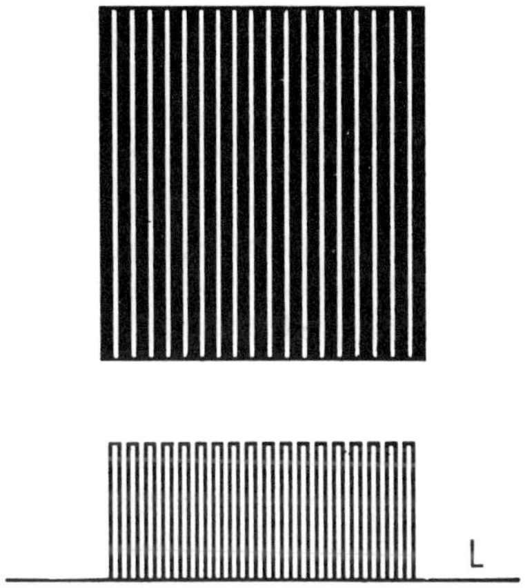

Fig. 14. Example of using projections to obtain FDP profile (along projection direction).

Interpreting the FDP in Terms of Projections

Another convenient tool for looking at a scan area and getting a gross feeling for what its FDP will look like is briefly discussed next.

Assume that the pattern of interest has a transmission value of 1, and that of the background is 0. Strike a line L in any direction, but outside the scan area. Now perform a vertical projection (via a line integral) of the pattern onto the given line (as illustrated in Fig. 14). Take the (one-dimensional) Fourier transform of the resulting "signal" on the projection line, and this gives the profile of the FDP along that direction. (In the FDP, the direction line corresponding to each projection line passes through the origin.)

As a simple example, consider the scan area of Fig. 14. We see that the projection onto L is a "signal" consisting of a number of equally spaced pulses; in no other direction will the "signal" be periodic. Along the (horizontal) L direction then, we expect to see contributions at the locations corresponding to the period of the "signal" and at the locations corresponding to the harmonics of this (spatial) frequency that are present in the "signal." Along the other directions, no such components should be expected. This is a "quick and dirty" way of predicting the gross characteristics of the FDP we obtained in Fig. 7.

Scan-Area/FDPMSs from Aerial Photographs

An important fact to extract from all of the above examples is that every pattern in a scan area will have its own diffraction pattern, and each of these will be superimposed, centered on the optical axis. Accordingly, a building-block approach to interpreting the FDPs is useful, and the foregoing patterns can be considered as basic building blocks.

Figs. 15 through 26 are scan-area/FDPMS pairs taken from aerial photographs. The reader is invited to study

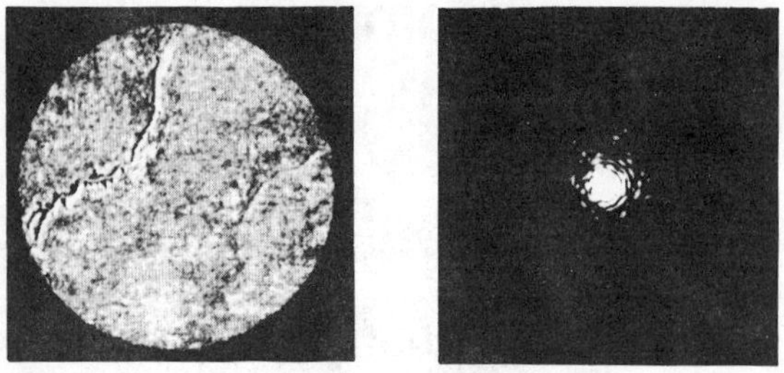

(a) (b)

Fig. 15. (a) Natural terrain. (b) FDPMS of (a).

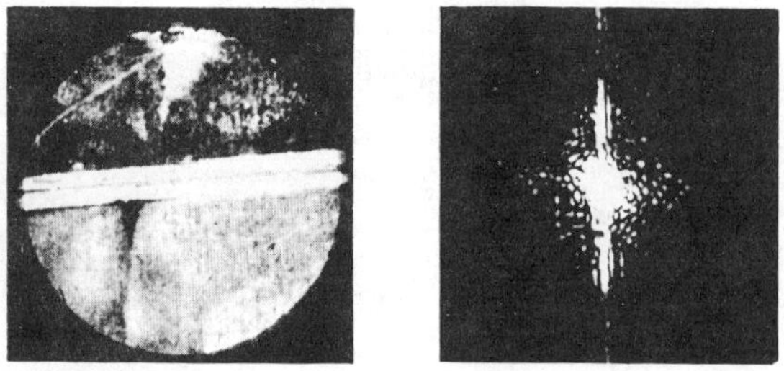

(a) (b)

Fig. 16. (a) Roadway with natural terrain background. (b) FDPMS of (a).

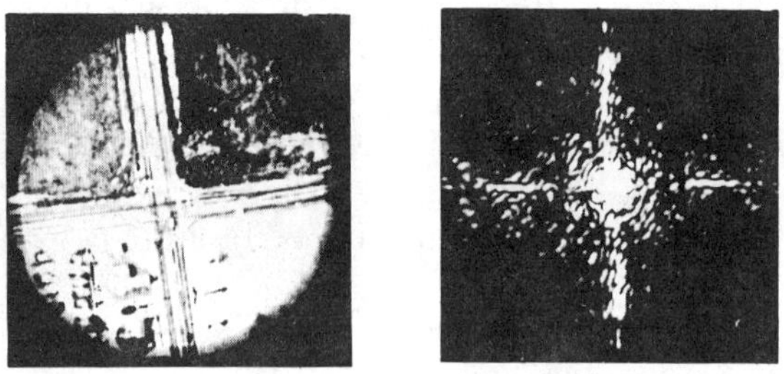

(a) (b)

Fig. 17. (a) Road intersection. (b) FDPMS of (a).

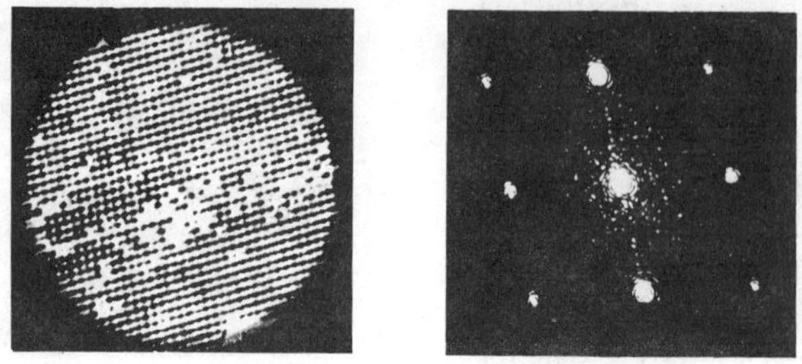

(a) (b)

Fig. 18. (a) Orchard. (b) FDPMS of (a).

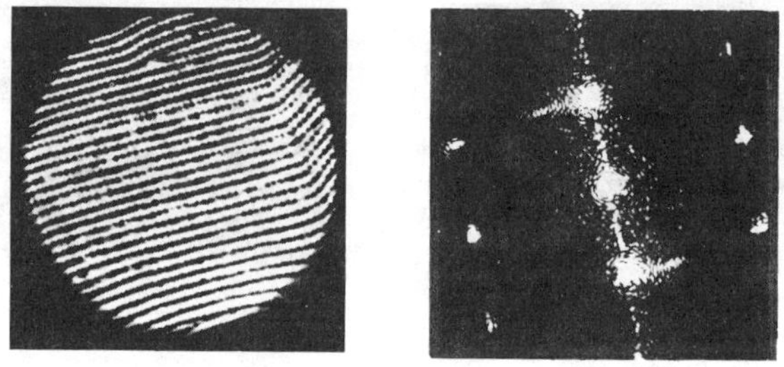

(a) (b)

Fig. 19. (a) Orchard on hillside. (b) FDPMS of (a).

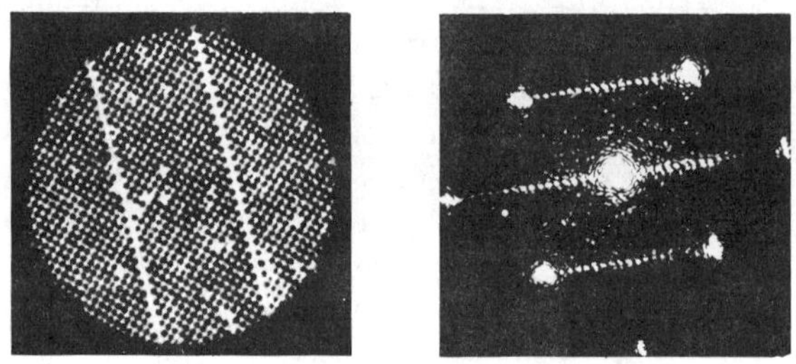

(a) (b)

Fig. 20. (a) Orchard with service roads. (b) FDPMS of (a).

them, keeping in mind the basic information cited above, and the importance, mentioned earlier, of the designer developing an ability to interpret the FDP in terms of content of the scan area, for this will lead to an understanding of significant features in the sample signatures. The latter will be discussed in the next section.

II. Sampling Diffraction Patterns

A. *Description*

Reason for Sampling

Although the process of going from a scan area to its FDP considerably reduces the complexity of the pattern recognition task, we are still dealing with a two-dimensional distribution of light. This will have to be converted to a form more convenient for subsequent processing.

As we noted earlier, the transformation of the scan area to its FDP is done in the preprocessor portion of the system (refer to Fig. 1). To perform its appropriate task, the preprocessor should reduce as much as possible the amount of information the decision processor has to handle.

Sampling is a good tool for further reducing the amount of information to be sent to the decision processor. The basic task is to design the sampling properly so that the resulting sampled data contain sufficient information for the subsequent classification to be effected. The FDP is particularly suited to sampling, because, as noted earlier, the scan area is decomposed into its basic building-block patterns, and the contribution of each of these to the FDP is centered on the optical axis of the system no matter where the pattern is located within the scan area. The optical axis, therefore, provides a reference point from which all sampling can be done. (No such reference point exists in the original scan area.) Due to this building block and centering property of the FDP, sampling geometries can be configured which are rotation invariant, size invariant, or both (depending on the subsequent processing). These are attractive features for a preprocessor to have.

The purpose, then, of sampling the FDP is to reduce this two-dimensional distribution of light to a relatively small set of numbers—called a sample signature.

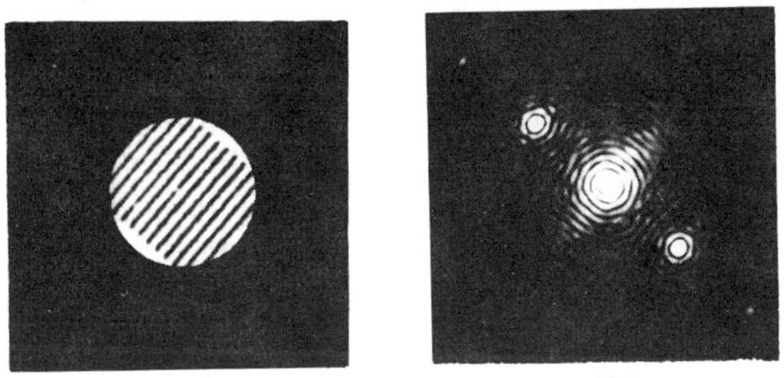

(a) (b)

Fig. 21. (a) Vine crop. (b) FDPMS of (a).

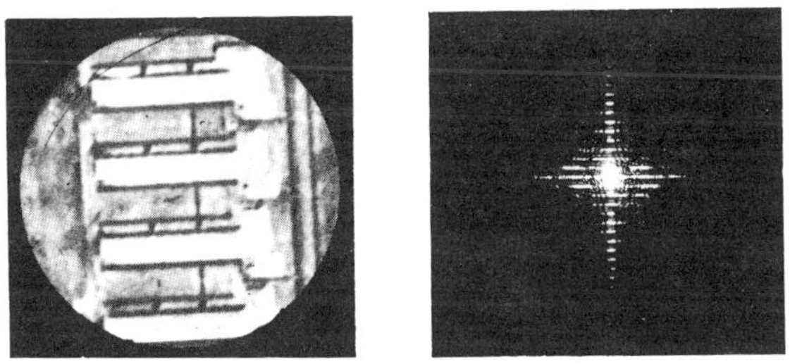

(a) (b)

Fig. 22. (a) Complex of large buildings. (b) FDPMS of (a).

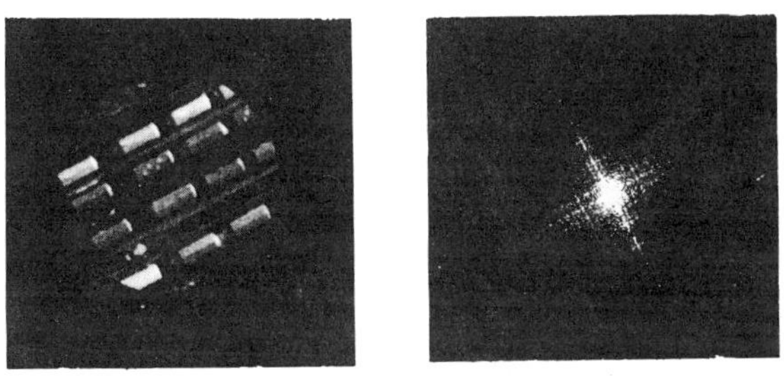

(a) (b)

Fig. 23. (a) Complex of smaller buildings. (b) FDPMS of (a).

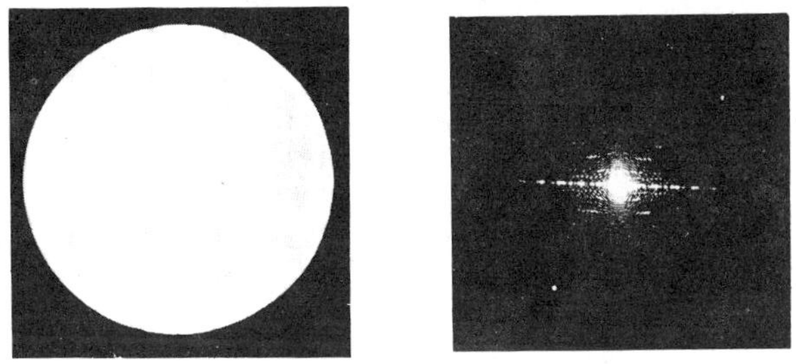

(a) (b)

Fig. 24. (a) Runway strip with chevrons. (b) FDPMS of (a).

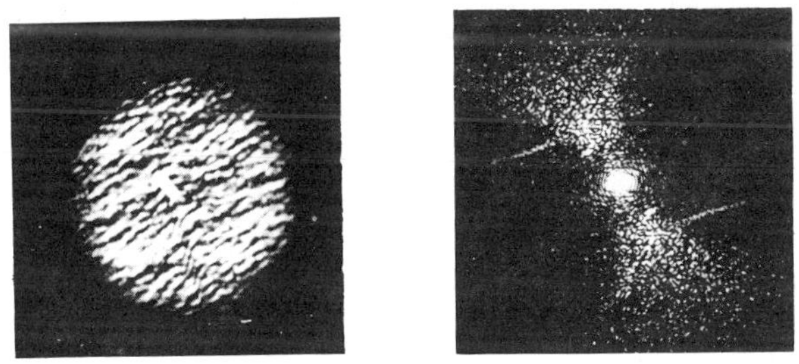

(a) (b)

Fig. 25. (a) Ship at sea and wake. (b) FDPMS of (a).

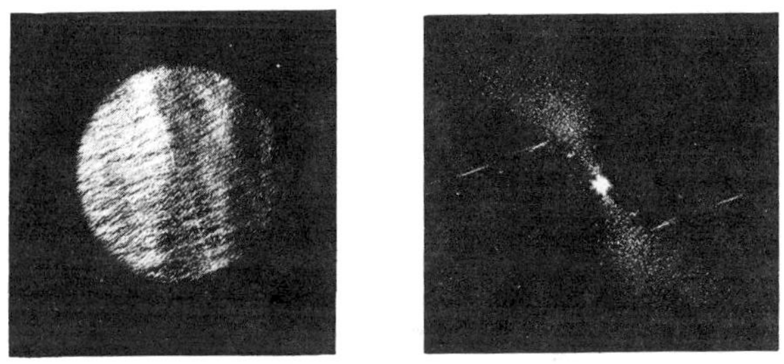

(a) (b)

Fig. 26. (a) Ship wake and ocean surface. (b) FDPMS of (a).

Sampling Geometries

The process of sampling an FDP is basically one of measuring the amount of light energy falling within specified areas of the FDP. To do this, we define a set of areas, which, taken together, normally cover the FDP; we call this set of areas a "sampling geometry." Then we measure the total light energy falling within each area; we call the resulting set of measurements a sample signature. In principle, this sampling can be performed by constructing a physical set of sampling windows, one window for each of the defined areas. Each window is transparent over its corresponding sampling area, and opaque elsewhere. Each window is successively placed at the diffraction-pattern plane, and the light transmitted through the clear portion of the window is measured and converted to a number. The resulting numbers form the sample signature.

Because the sampling is done over two dimensions, there is a large variety of possible sampling geometries. A few of these possibilities intuitively hold promise for being especially useful in those cases where regularity of any kind is present in the scan area (e.g., linear features, one- or two-dimensional periodicity, etc.). This is important, because man-made objects, in aerial photographs, for example, normally exhibit some kind of regularity.

For one of these possibilities, we recall that any point in the FDP plane corresponds to a frequency determined by its radial distance from the origin, and note that a circle in the FDP plane centered on the optical axis corresponds to one frequency in all directions. Therefore, by integrating the light energy of the diffraction pattern along a circle centered on the optical axis, we obtain the total contribution of one frequency component, independent of direction. In practice, an annular-ring sampling window will have a finite width [see Fig. 27(a)]; accordingly, each sampling window will obtain the contribution of a small number of adjacent frequencies. Having a finite width is advantageous, both because it helps to reduce the number of samples we have to take, and because it helps to make the sample signature insensitive to minute variations within the scan area (e.g., for noise considerations). A set of annular-ring

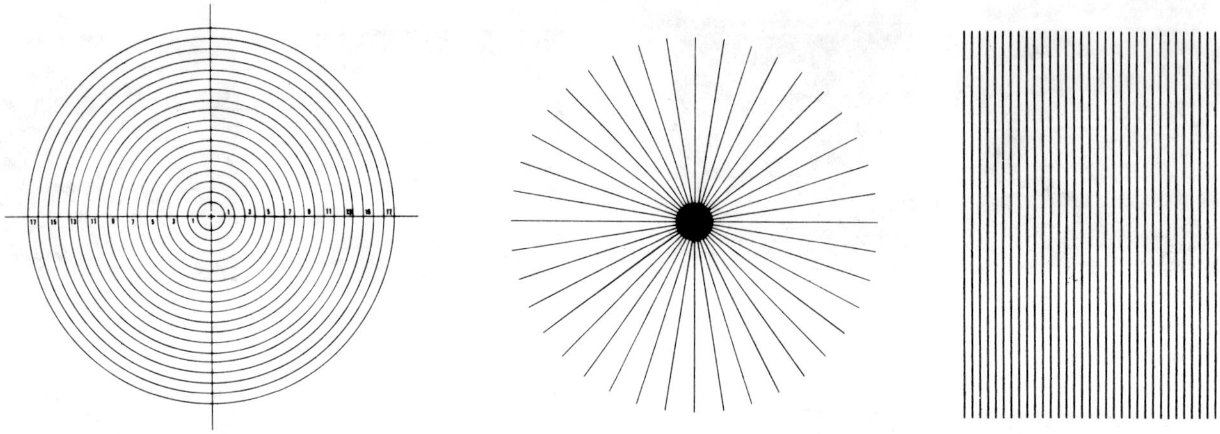

(a) (b) (c)

Fig. 27. (a) Annular-ring sampling geometry. (b) 9° wedge sampling geometry. (c) Parallel-slit sampling geometry.

sampling windows obtains a spatial-frequency profile of the contents of the scan area, and takes into account all directions simultaneously. (See Fig. 28(e) and 28(f) for examples of annular-ring sample signatures.) This technique of sampling provides a means for detecting one- or two-dimensional spatial periodicity within the scan area and, further, it is insensitive to rotation of the scan area.

For another sampling geometry, we recall that any point in the FDP plane corresponds to a direction (in the scan area) determined by drawing a radius vector from the origin through that point. Note that a radial line in the FDP plane corresponds to a single direction in the scan area, which includes all frequencies. Therefore, by integrating the light energy of the diffraction pattern along a radial line, we obtain the total contribution of one direction, independent of frequency (and independent of location in the scan area). In practice, a wedge-shaped sampling window is used [see Fig. 27(b)]. This obtains the contributions of a small number of adjacent directions, and has the advantages of reducing the number of samples required and reducing the effects of small variations. A set of wedge-shaped sampling windows obtains a direction profile of the contents of the scan area, and takes into account all frequencies simultaneously. (See Fig. 28(g) and 28(h) for examples of wedge sample signatures.) This technique of sampling is, grossly speaking, insensitive to scale.

Another possibility is what we call a parallel-slit sampling geometry. In this case, after rotating the FDP to a specified orientation, we measure the light energy transmitted through a slot of a specified width and long enough to cover the FDP. This sampling geometry is useful for distinguishing one-dimensional regularity from two-dimensional regularity. (The annular-ring geometry is used to detect the regularity and the wedges to find the principal directions. The slit is taken along these directions to distinguish the one-dimensional from the two-dimensional regularity.)

Finally, a frequency profile can be obtained along any desired direction by placing a very small aperture at, or near, the origin and moving it along the desired direction radius; taking sufficient measurements to cover the line. (See Fig. 39 for example.)

A mathematical description of these sampling geometries is given in Appendix I.

Sample Signature Examples

Examples of sample signatures obtained via the annular-ring and wedge sampling geometries are given in Fig. 28.

FDP Sampling System

The implementation of diffraction-pattern sampling at General Motors' AC Defense Research Labs. (DRL) has undergone a two-step evolution from that described in [4]. The first of these implementations was used for two basic studies. The first study [5], performed on an aerial photograph of the Goleta Valley, Calif., was used to determine the feasibility of utilizing diffraction-pattern sampling as an aid to classifying areas in aerial photographs. The second study [6] was performed to determine to what extent dimensional information concerning the objects to be classified could be extracted from diffraction patterns. The results of these two studies were very encouraging; and in order to perform more in-depth studies of diffraction-pattern sampling, a fully automated Diffraction-Pattern Sampling System (DPSS) was built. The DPSS (shown in Fig. 29) provides for a wide variety of sampling geometries (important in a research environment), and for greater precision and higher sampling rates than either of our previous systems.

The DPSS is directly coupled to a large-scale computer and is controllable via a control station ⑤ or the General Motors developed Signature Analysis Research Facility (SARF) design console ⑥. The system basically operates as follows. A laser beam illuminates a selectable area of a photographic transparency mounted on an X–Y table ①, the diffraction pattern is formed optically, and can be sampled by the computer-controlled mechanisms at ②. In an alternate path, the diffraction pattern and/or the scan area of the photograph is observable by the analyst at either the viewing screen at ③ or the one at ④. When at ③, the analyst

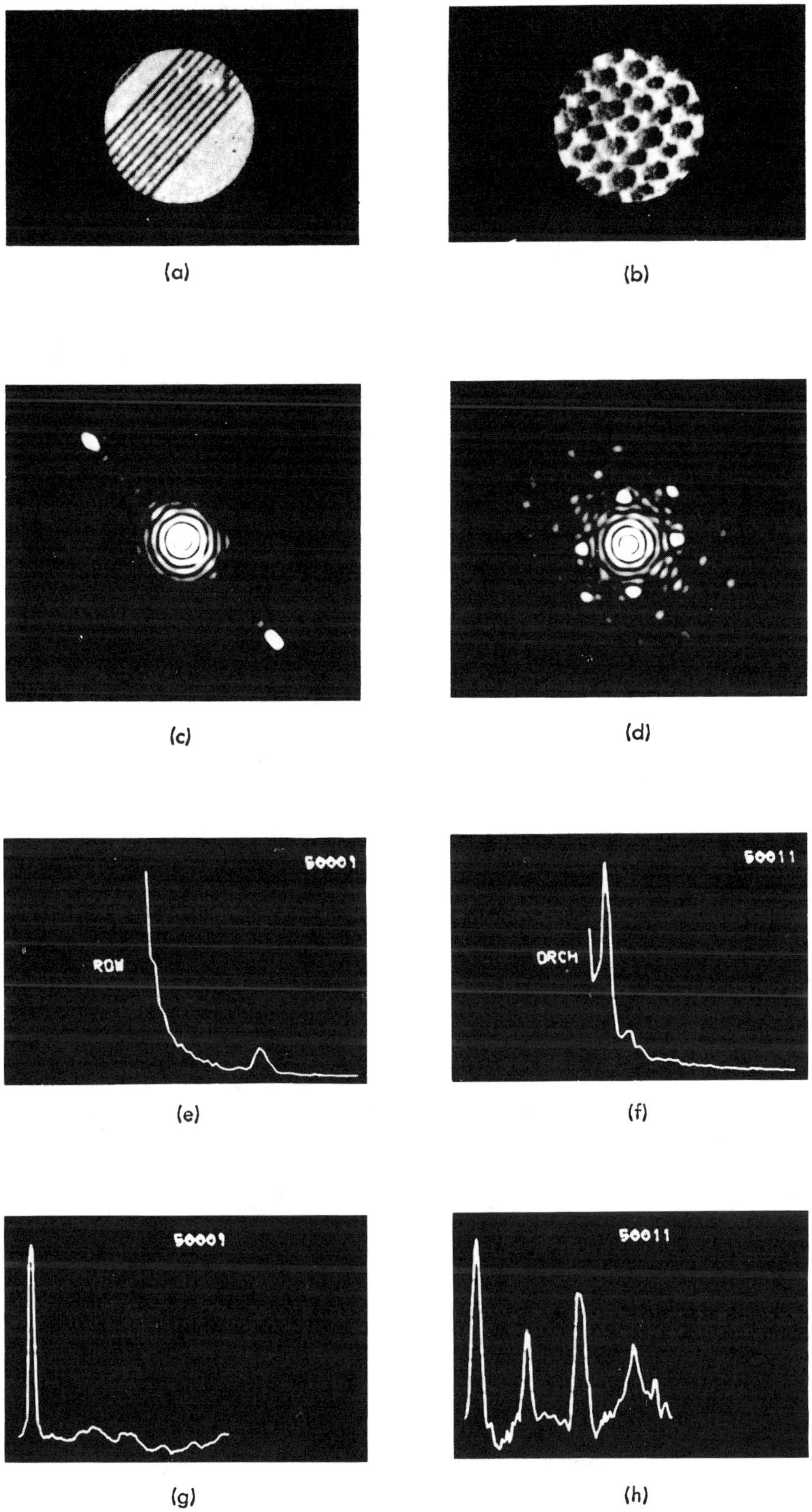

Fig. 28. (a) Agriculture one-dimensional periodicity. (b) Agriculture two-dimensional periodicity. (c) FDPMS of (a). (d) FDPMS of (b). (e) Annular-ring sample signature of (c). (f) Annular-ring sample signature of (d). (g) 1.8° wedge sample signature of (c). (h) 1.8° wedge sample signature of (d).

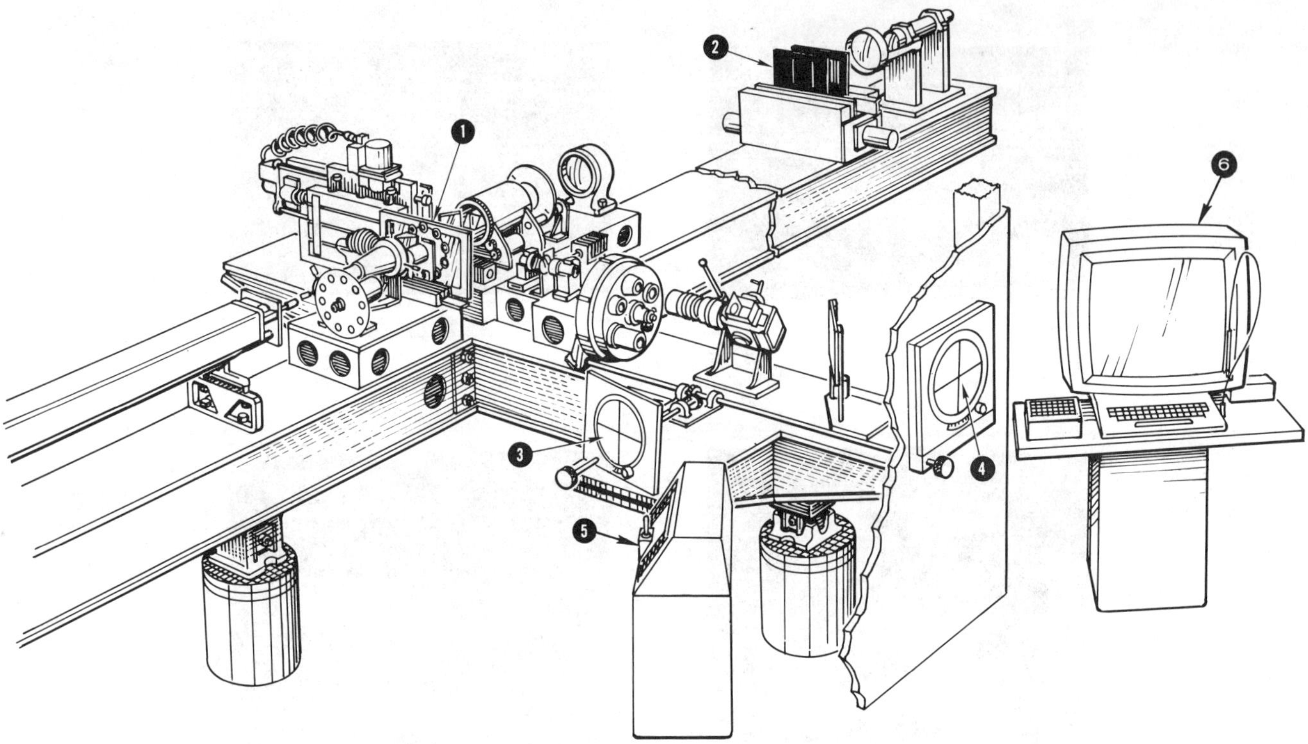

Fig. 29. Diffraction-pattern sampling system (DPSS).

can control the coordinates of the scan aperture using the joy stick at the control station ⑤ and can perform non-computer experiments. When at ④, the analyst can view the diffraction patterns and/or the scan areas, and can simultaneously view on the SARF graphics console ⑥ the sample signatures and any processing of the signatures he wishes to correlate with the diffraction pattern and/or the scan area while developing automatic classification algorithms.

A more detailed description of a DPSS is given in Appendix II.

Utilizing FDP Sample Signatures

As noted earlier, the purpose of sampling the FDP is to reduce the FDP to a representation consisting of a relatively small set of numbers. The next question is, how do we best use the resulting sample signatures?

In contrast to various "template matching" approaches, we treat the collection of numbers (which is the sample signature) as a "signal," or "signature" and plot it graphically. The human algorithm designer studies the signatures, discovers their distinguishing features, and devises schemes (algorithms) for machines to perform the classification task based on these features.

We noted earlier that the algorithm designer must become intimately familiar with the FDP and its relationships to the scan area so he can interpret the FDP in terms of important features of the scan area. Similarly, the designer must now study the sample signatures and become intimately familiar with their characteristics and how they relate to features of the FDP so that he will be able to arrive at the processing steps necessary to perform the pattern recognition task.

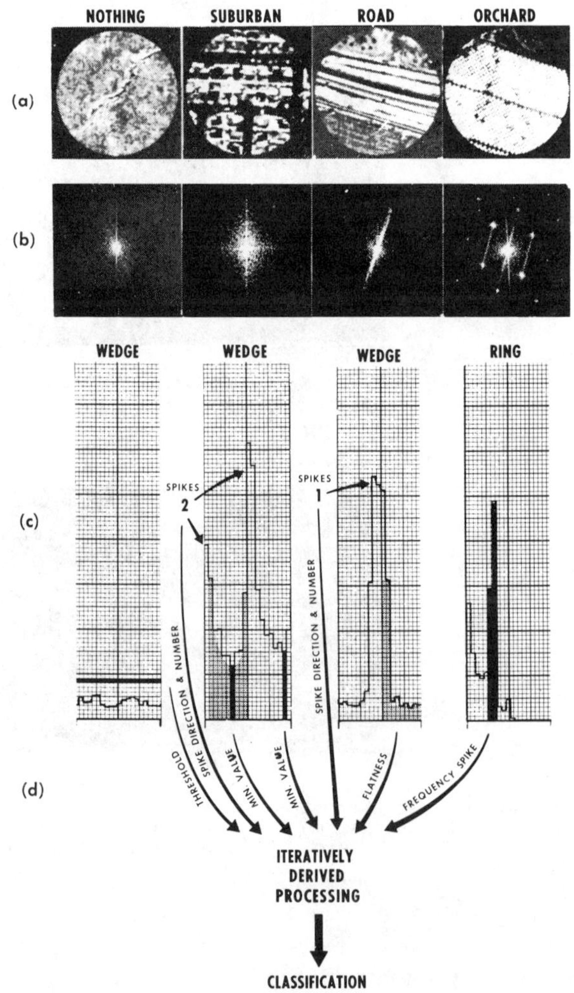

Fig. 30. Classification process.

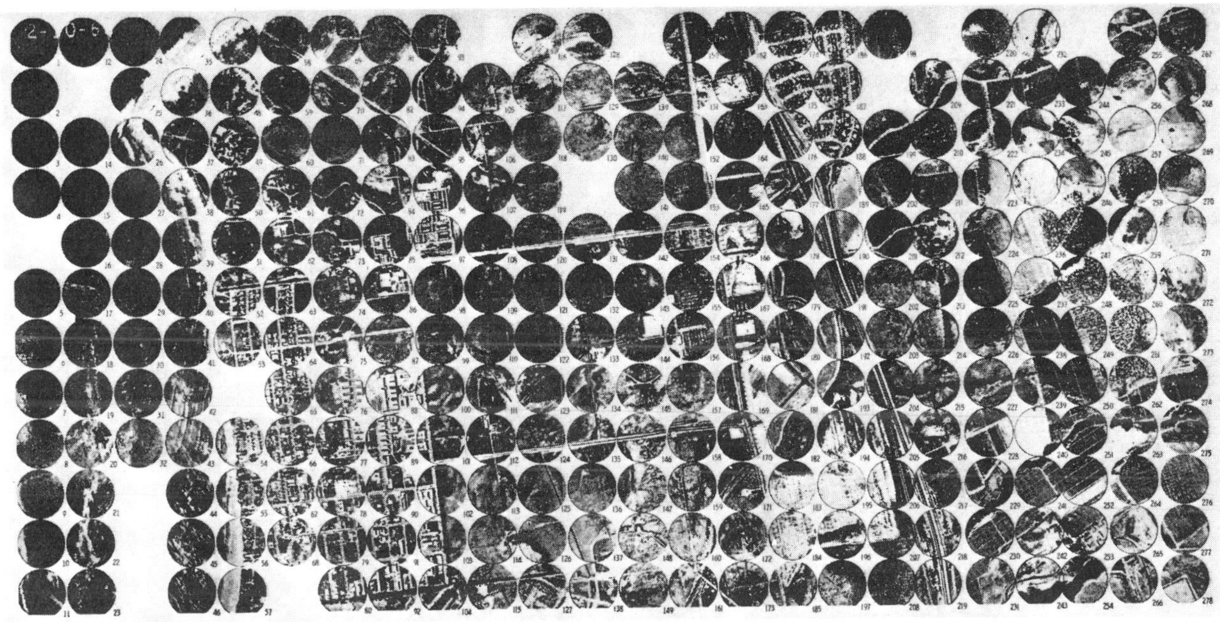

Fig. 31. Aerial photograph of Goleta Valley with scan-area selections indicated.

With reference to Fig. 30, our overall design approach can be stated as follows.

1) Studying (a) and (b), discover pertinent features in the FDP as related to content in the scan areas, and determine appropriate sampling geometries for the given task(s).

2) Studying (b) and (c), discover features in the sample signatures (and/or transformed versions of the signatures) which correspond to the FDP features discovered in 1), and determine appropriate data processing schemes for extracting measures of these features.

3) Iteratively arrive at decision procedures (d) utilizing these feature measures to perform the classification.

In practice, this design process has been found to be a highly intuitive and iterative process, and the "iteratively derived processing" represents the algorithm designer adapting as he interacts with the data and its processing.

SARF was developed by General Motors at AC-DRL to facilitate this "adaptive" design process (where we emphasize that the adaption is done by the algorithm designer). It is based on interactive computer graphics, and provides for real-time interaction between the designer, his data, and its processing, and is designed to take advantage of human intuition, judgment, and pattern discovery capabilities. It allows the designer to formulate hypotheses and to quickly test them, and, in particular, to utilize the exceptions or unexpected results of these tests. (SARF is described in [7]–[9].)

B. Experimental Results

The experimental results of this section are offered as examples of the utility of diffraction-pattern sampling for performing various aspects of automatic pattern recognition. The two algorithms described were developed only far enough to show a point, and have not been incorporated in any operational system.

When setting out to design an automatic pattern recognizer, we must first define the task to be performed. We define the work 'task' as looking for a specified object or class of objects in the given imagery. Associated with the objects of interest are certain physical constraints such as size and shape. These constraints determine the ground-distance diameter (GDD) of the scan-area aperture to be used for performing the task. In general, the smaller the object to be classified, the smaller the GDD required.

Goleta Valley Study

The main study performed on the second of the three implementations of a diffraction-pattern sampling system at AC-DRL was based on an aerial photograph of the Goleta Valley, Calif. [5]. This photograph included a ground area of roughly 1.7 by 3.4 miles, and, as shown in Fig. 31, was sectioned off into disjoint circular areas of 750 feet ground-distance diameter (these were the scan areas). The task was to classify each of these scan areas according to whether or not they displayed any gross man-made characteristics. Accordingly, two major classifications were defined: 1) gross man-made, and 2) nothing man-made. The gross man-made category was further divided into a) roads, b) intersections of roads, c) buildings, and d) orchards.

Since gross man-made characteristics were sought rather than small man-made objects—that is, for example, freeways and suburbia rather than small dirt roads and single dwellings—we choose a GDD of 750 feet. With this GDD, a road less than 20 feet wide could not be detected, for instance, (unless it was in strong contrast to the rest of the scan-area background and traversed approximately the center of the scan area) and a very large building, say 1000 feet on a side, would probably be misclassified.

Scan areas that contained more than one category were classified in the category whose features were most pronounced. In this manner, no scan area was classified into more than one category.

The FDP of each of these scan areas was sampled via two sets of sampling windows. The first set consisted of 20 annular rings, and the second consisted of 20 nine-degree wedges. Classification parameters were derived from each of the resulting sample signatures; and these parameters (rather than the original sample signatures) were used in the subsequent classification process.

The initial set of classification parameters derived from the sample signatures were, in a sense, equivalent to the differences observed visually in the diffraction patterns. This first set provided a usable basis for deriving other parameters; and it was these that were used to effect the final classification algorithm of this study.

Features Utilized: The annular-ring sample signature for a scan area containing no regularity will fall off exponentially as a function of radius. (This statement assumes that the width of the annular-ring window is such that it cannot resolve the variations of the scan-aperture Airy disk.) If the scan area contains regularity, the energy in the FDP is redistributed and yields "bright" spots in certain areas of the FDP. The reading from the annular-ring window that coincides with such a bright spot will be higher than expected from the exponential fall-off model cited above. It turns out, then, that if an annular-ring window reading is equal to or larger than its neighbor on the low-frequency side, then the scan area can be assumed to have some kind of regularity. In particular, since orchard scan areas give rise to a lattice-like diffraction pattern, their annular sample signatures show sudden changes in the sampled value. This property alone allowed us to correctly classify the orchard scan areas in the Goleta Valley photograph with 100 percent accuracy.

The wedge sample signature is a good representation of those FDPs whose corresponding scan area contains linear images. Linear images in the scan area give rise to energy distributions along certain radii in the diffraction pattern, and these bright radii give rise to spikes in the wedge sample signature plot. The term "spike" refers to an abrupt rise in the sample signature plot, which in turn corresponds to a high intensity portion of the FDP impinging on a sample window [see Fig. 28(g) and (h) and Fig. (30)]. The number of these spikes, their angular relationship, and the distribution of energy between them, formed the basis upon which the remaining discriminations in this study were made.

Further details of the algorithm arrived at in this study are given in [5]. The classification results of the final algorithm for that study are given in Fig. 32.

	Actual	Hit	False Alarm	Hit (%)
NOTHING	108	93	2	86.1
MANMADE	167	165	15	98.8

False Rejection Rate
$\frac{2}{167} = 1.2\%$

Subclassification of MANMADE

Buildings	37	36	10 (7 I, 3N)	97.2
Roads	50	47	9 (9N, 1B)	94.0
Inter-sections	35	28	5 (1B, 1R, 3N)	80.0
Orchards	45	45	0	100
TOTAL	167	156	24	93.4

Fig. 32. Summary of machine classification results for Goleta Valley study.

San Francisco Bay Area Algorithm

Upon completion of the SARF and DPSS facilities described earlier, a demonstration algorithm was developed for distinguishing residential, agricultural, and natural areas in an aerial photograph of a portion of the San Francisco Bay Area [7]–[9].

This algorithm again demonstrated the utility of diffraction-pattern sampling for performing this kind of classification. More importantly, however, the development of this algorithm (and subsequent work done on the facility) demonstrated the rapidity with which such algorithms can be developed using a facility like SARF/DPSS.

For this algorithm, annular-ring and wedge sample signatures were again used—only this time, 100-point signatures instead of 20-point signatures were used.

These sample signatures were perused, and after class-distinguishing features were discovered in them, parameters for measuring them were formulated and implemented, and the resulting parameter values used for subsequent classification.

Three of these parameters were used as coordinates in a three-dimensional space. Each scan area was plotted in this three-dimensional parameter (or feature) space using the three parameter values, respectively, as the x, y, and z coordinates. Each scan area was assigned a letter, A, R, or N according to its classification as an agricultural, residential, or natural area. (Note, during this design phase, the scan areas were classified by us ahead of time, and the data for each area were so tagged within the computer.)

This three-dimensional space was explored to determine the distribution of the scan areas within it, the object being to establish criteria for separating the different categories. Fig. 33 shows a typical feature space plot as seen on the SARF console, and Fig. 34 shows some geometric decision boundaries injected into the plot. These boundaries were manipulated until separation of the categories were achieved.

To preserve the feature-extraction processing and the arrived-at decision boundaries, a logic decision tree is specified at the SARF console, as shown in Fig. 35.

To test the algorithm (now represented by the tree) all the data is "fed" into the tree, and the results collected in the form of a scoreboard, Fig. 36. The rows represent the actual categories, and the columns represent the tree output branches. All the agricultural and residential examples are

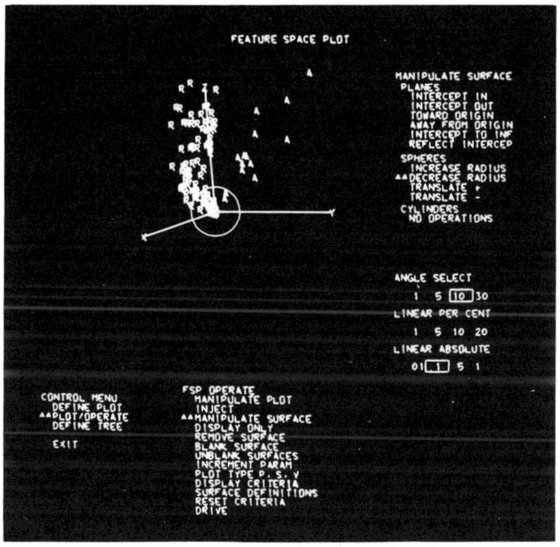

Fig. 33. Feature space plot on SARF console.

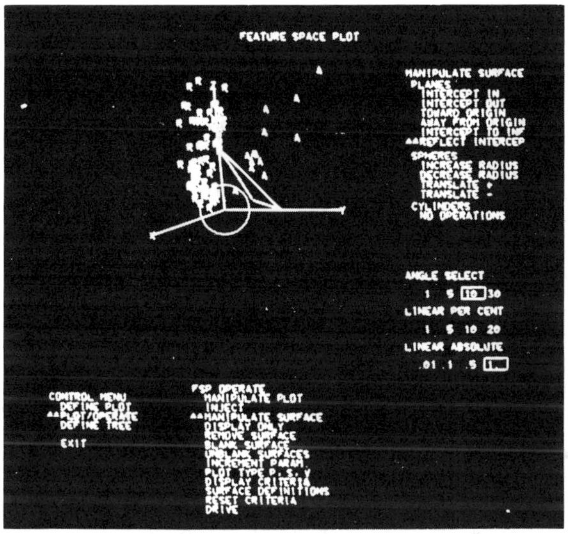

Fig. 34. Feature space plot with decision boundaries injected.

correctly classified, and all but one of the naturals are correctly classified.

For more details on this process, see [7]–[9].

Dimensional Information from Diffraction Patterns

In another of our studies [6], we found that the FDP not only provides an excellent means for detecting regularity in a scan area, but it also provides dimensional information concerning both the object to be identified and, if the scan area contains an array of such objects, the lattice-like properties of the array. This kind of information concerning the pattern within the scan area can be important in an automatic pattern-recognition system in which clues are being assembled for pattern identification. This statement is based on the notion that, given the scale of the photograph and the general context of the area being scanned, the existence of and/or the dimension of regularity within the scan area is a very telling feature.

As an example of this notion, consider the collection of parked aircraft in the scan area shown in Fig. 37. The annular-ring signatures of these scan areas display a number of local peaks. Analysis of these signatures shows that the relative spacing of these peaks are directly related to the following:

1) nacelle separation,
2) wing-to-stabilizer separation,
3) wing and stabilizer lengths.

Further, because this FDP sampling was done with annular rings, this dimensional information is independent of the orientation or position of the parked aircraft.

In this example we note that the diffraction-pattern sampling technique serves for *both* object detection and ob-

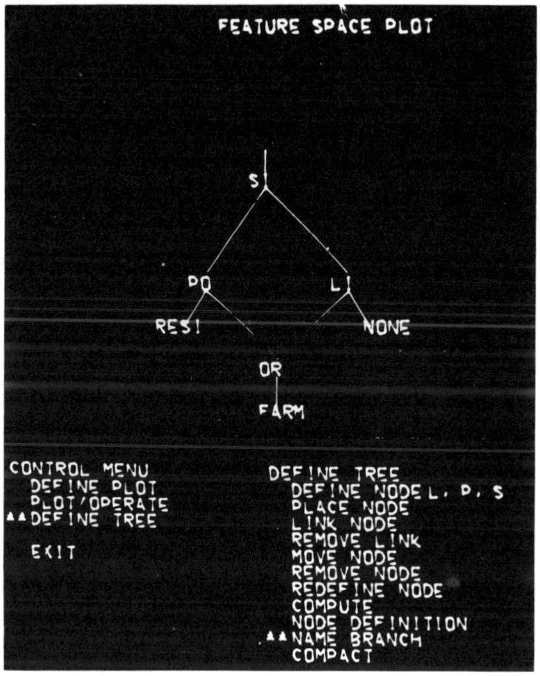

Fig. 35. Logic decision tree.

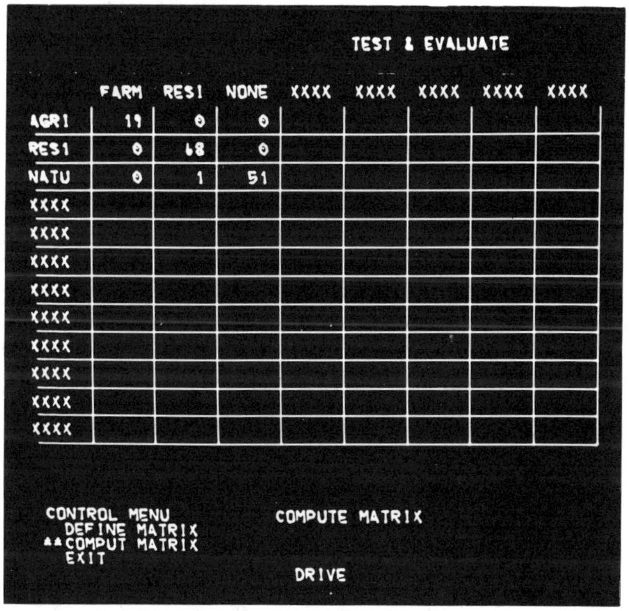

Fig. 36. Algorithm results.

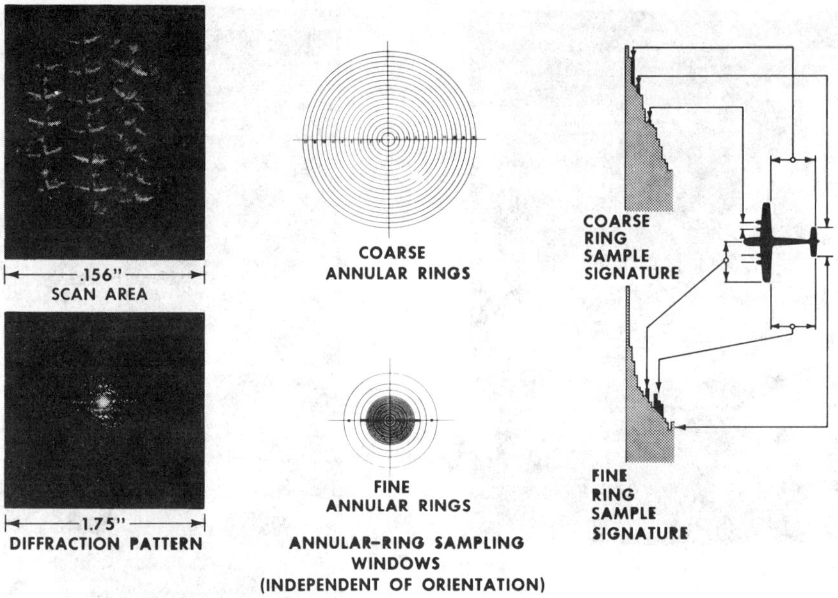

Fig. 37. Identification dimensions for aircraft extracted from sample signatures.

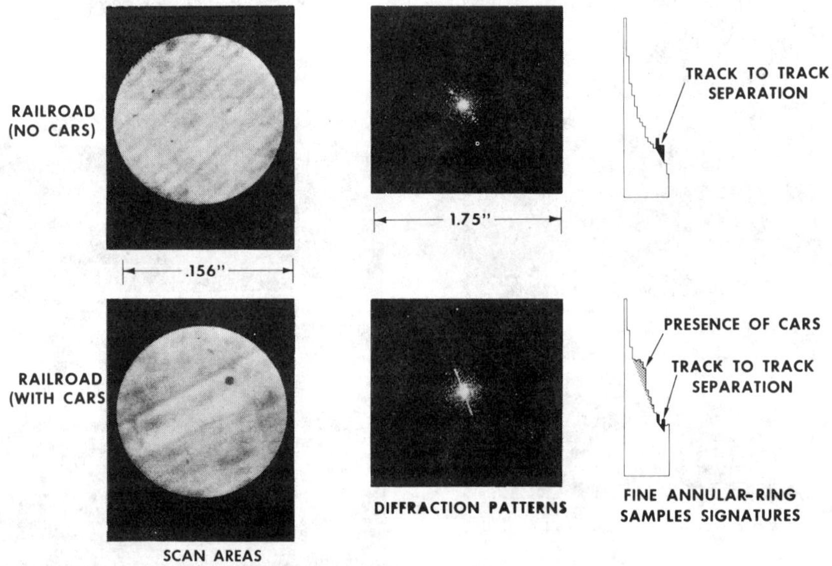

Fig. 38. Identification clues for railroads obvious in sample signatures.

ject identification. In addition, this example shows that even for a large collection of a given object (e.g., parked aircraft), identification can be made with only one set of measurements taken in parallel, in contrast to collating multiple measurements taken on each object in the group.

In addition to the annular-ring and wedge sample signatures, a technique was developed for measuring and recording a continuous intensity profile of an FDP in any direction and location. This was done by recording the light energy passing through a small aperture (called a fine probe) which traversed the FDP in the specified direction and location. In this way, having established the principal directions of information in an FDP (e.g., via a wedge sample signature) an intensity profile in that direction can be taken.

These were used to determine how well the annular-ring sample signatures captured the pertinent information, and/or to obtain a higher resolution means of obtaining dimensional information concerning the object in the scan area. As an example, we compare Figs. 38 and 39.

Fig. 38 shows scan areas, diffraction patterns, and annular-ring sample signatures for railroad tracks with and without cars. Fig. 39 shows the frequency profiles of these FDPs along the direction indicated by line A. These frequency profiles clearly provide a higher resolution means of analyzing the FDP, and therefore provide more precise dimensional information concerning the objects in the scan area.

In the data acquired in this study, a high degree of correlation was achieved between object dimensions computed from the sample signatures and the corresponding dimensions measured directly from the photographs. It is significant that much of these data were derived from the annular-ring signatures, because this sampling is independent of

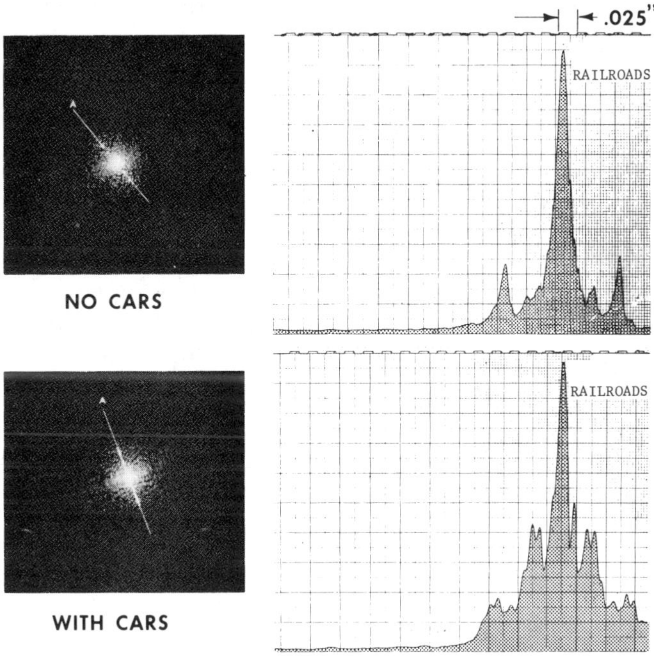

Fig. 39. Constant-direction intensity profiles for Fig. 38 scan-area diffraction patterns.

orientation (of either a single object, or relative orientation and location among a set of similar objects).

The experience acquired in this study suggested the following algorithm for extracting pertinent object signatures: 1) use annular-ring sampling for general frequency characteristics (especially for collections of objects); 2) use wedge sampling for determining angular relations and for selecting directions in which to obtain detailed intensity profiles; and 3) if appropriate, obtain the indicated profile plot or plots (particularly useful for single-object identification).

Many clues can be obtained by harmonic analysis of this frequency information, and these clues could be quite characteristic of specific object types, because they are directly related to the dimensions of the object. In addition, since the various dimensions of any given object (or object class) form certain fixed ratios, the derived characteristic frequency/harmonic relationships will be independent of scale. Adding this to the other properties of this technique results in a method that is basically independent of scale, translation, and rotation. Furthermore, the technique is one that could be implemented for extremely rapid parallel processing.

Conclusion

As mentioned several times in this paper, our investigations have shown that diffraction-pattern sampling can be a useful tool in effecting automatic pattern recognition in photographic transparencies. We do not mean to imply, however, that diffraction-pattern sampling can answer all the questions and problems associated with automatic pattern recognition.

Generally speaking, the design of an automatic pattern recognizer will depend very much upon the task to be performed. We do not believe that there is a general-purpose automatic pattern recognizer (other than the human) in the offing for some time to come.

However, we do feel that we can learn a lot from the way humans do the job, and inject this into the design of automatic pattern-recognition algorithms. First of all, the human performs pattern recognition in a dynamic way. General information is sought first, and as information is obtained, or context is established, different types of information are sought, at each step more detailed, until the desired identification is made.

For example, in searching for a specific type of aircraft on the ground in an aerial photograph, a human, having first established the approximate scale, would scan the photograph very quickly to see if there are any linear features in it, and in particular those linear features which exhibit characteristics of airfields. Having found such an airfield, he will focus his attention (this time looking for details of a smaller scale) to looking for linear features characteristic of aircraft. Once he finds one or more aircraft, he then focuses his attention to looking for the specific kind he *had in mind*. If he finds one meeting this requirement, then his pattern-recognition task is completed.

For another example of the dynamic aspects of this process, consider the game of Twenty Questions. The contestant may ask any question of the type "is it x or not?" After each question is answered, the contestant formulates a new question and in this way proceeds, with up to 20 questions, to gather information to arrive at an answer. Those familiar with this game are well aware that it would be next to impossible to win this game if all 20 of the questions had to be asked before receiving any answers.

The human is able to win this game only because his experience in our environment has allowed him to build up a good model of that environment, and thus, by proper

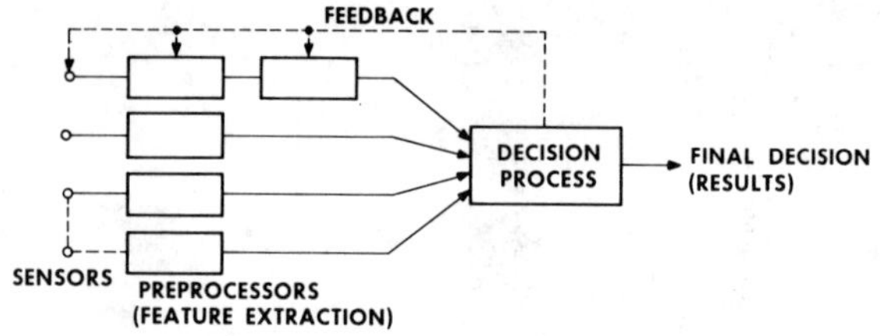

Fig. 40. General structure for an automatic pattern-recognition system.

questioning, can guide himself through the combinatorial tree of possibilities.

We believe that this kind of dynamic approach should be taken in the design of automatic pattern-recognition devices. The word 'dynamic' here is meant to imply the use of feedback, wherein the decision part of the device uses feedback to change the source and/or kind of information it receives (i.e., change the questions it is asking) as a function of its current state of knowledge. This, of course, implies that we must build into the decision maker a good model of the "environment" of the task it is to perform. A system structure of such a recognizer is shown in Fig. 40.

We feel that diffraction-pattern sampling can serve well as one of the preprocessors in such a system.

Appendix I

Mathematical Description of Some Sampling Geometries

If the amplitude transmission of the scan area is described by $t(x, y)$, and if the scan area is uniformly illuminated by a normally incident monochromatic plane wave, of amplitude A and wavelength λ, then the FDP formed by a spherical lens of focal length F is described by

$$T(\mu, v) = \frac{A}{\lambda F} \int_{-\infty}^{\infty} \int_{-\infty}^{\infty} e^{-2\pi i(\mu x + vy)} t(x, y) dx dy$$

where $\mu = (x_t/\lambda F)$, $v = (y_t/\lambda F)$, and x_t and y_t are coordinates in the transform (FDP) plane.

We recall that our sensors record the FDPMS, not the FDP. The FDPMS is described by

$$I(\mu, v) = T(\mu, v)T^*(\mu, v) = |T(\mu, v)|^2$$

where * denotes complex conjugate.

The sample signatures consist of a set of measurements $(m_1, m_2, \cdots, m_n)$, where n is the number of sampling areas defined for the given sampling geometry.

For an annular-ring sampling geometry, the measurements are described by

$$m_j = \int_0^{2\pi} \int_{\rho_j}^{\rho_j + \Delta\rho} I(\rho, \theta) \rho d\rho d\theta, \qquad j = 1, 2, \cdots, n$$

where

$$\rho = \sqrt{\mu^2 + v^2} \quad \text{and} \quad \theta = \tan^{-1} \frac{v}{\mu}.$$

Each of the m_j represent the total power in the band of spatial frequencies defined by

$$(\rho_j, \rho_j + \Delta\rho).$$

For a wedge sampling geometry, the measurements are described by

$$m_j = \int_{\rho_{\min}}^{\rho_{\max}} \int_{\theta_j}^{\theta_j + \Delta\theta} I(\rho, \theta) \rho d\theta d\rho, \qquad j = 1, 2, \cdots, n.$$

Each of the m_j represent the total power in the band of directions defined by $(\theta_j, \theta_j + \Delta\theta)$, for all frequencies larger than $\rho_{\min}$ out to the limits of the measuring system ($\rho_{\max}$). To eliminate large zero-frequency contributions, $\rho_{\min}$ is normally not set to zero.

For a slit sampling geometry, the measurements are described by

$$m_j(\theta) = \int_{-v_{\max}}^{v_{\max}} \int_{\mu_j}^{\mu_j + \Delta\mu} I(\mu, v) d\mu dv, \qquad j = 1, 2, \cdots, n$$

where the FDP is first rotated to a desired direction (an angle θ from some reference direction) before the measurements are taken. The height of the slit window ($2v_{\max}$) is arbitrary, but is normally determined by the extent of the FDP or the limits of the measuring system; the width of the slit window is $\Delta\mu$.

A frequency profile can be obtained along any desired direction (an angle θ from some reference direction) as follows.

$$m_j(\theta) = \int_{-v_a}^{v_a} \int_{\mu_j}^{\mu_j + \Delta\mu} I(\mu, v) d\mu dv, \qquad j = 1, 2, \cdots, n$$

where $2v_a$ is the height of the sampling window and $\Delta\mu$ is its width. For this case v_a and $\Delta\mu$ are generally small, while n is large, and the μ_j are selected for slight overlap of the samples.

This profile sampling can be generalized to perform a raster-scan sampling of the FDP. These measurements are described by

$$m_{jk} = \int_{v_k}^{v_k + \Delta v} \int_{\mu_j}^{\mu_j + \Delta\mu} I(\mu, v) d\mu dv \qquad j = 1, 2, \cdots, n_j \quad k = 1, 2, \cdots, n_k.$$

Appendix II

Functional Description of General Motors' AC-DRL DPSS

As mentioned in the Introduction, the implementation of FDP sampling at AC-DRL has undergone a two-step evolution from that described in [4], the last step being an automated Diffraction Pattern Sampling System (DPSS).

The following is a functional description of the DPSS. DPSS can be controlled manually by way of a control panel provided for the operator, or it can be operated in a completely hands-off mode by the computer system to which it is attached.

In the following description, reference is made to Fig. 41.

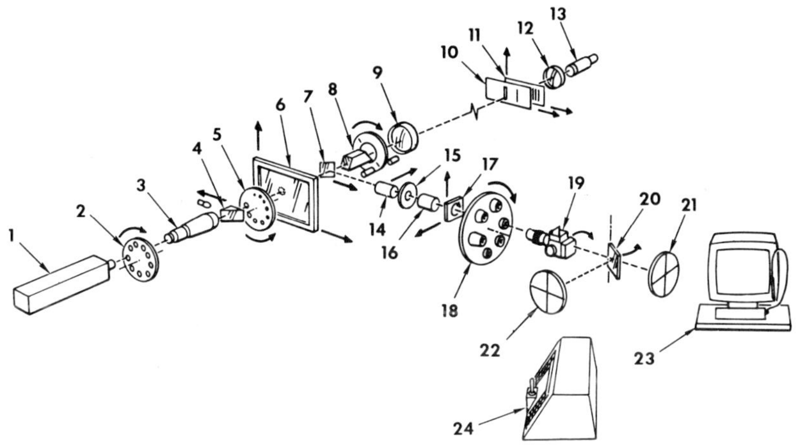

Fig. 41. Schematic of General Motors' AC-DRL diffraction-pattern sampling system.

Item Number	Item Name	Description and/or Function
1	Laser	Provides monochromatic, coherent light for generating the diffraction pattern.
2*	Intensity control disk	Contains neutral density filters of different densities. Used to control the laser light intensity.
3	Beam expanding and collimating optics	Used to generate a 20-mm or 50-mm collimated light beam. Includes a pin hole spatial filter to "clean up" the beam.
4	Prism	This is moved in and out of the optical path. When out of the path, the laser illuminates the film 6; when in the path, a white light source illuminates the film.
5*	Aperture disk	Provides variable scan-aperture capability. Contains a sequence of different apertures ranging in size from 0.040 inch diameter to 0.596 inch diameter used when sampling, and one aperture of 1.00 inch diameter for context viewing. These apertures determine the size of the laser beam illuminating the film plate—hence, they determine the size of the scan area.
6*	Film holder	The film is mounted on a digitally driven, two-axis table. Any location on a 9 by 9 inch film can be positioned on the optical axis. This determines the center of the scan area; the diameter of the scan area is determined by the aperture selected 5. The position of the table can be controlled either manually via the joy stick, or directly by the digital computer.
7	Mirror	This is moved in and out of the optical path. When in the path, the viewing portion of the system is selected; when out, the sampling portion of the system is selected.
8*	Dove prism	The dove prism rotates the diffraction pattern for the purpose of generating certain of the sample window sets. It is driven, selectively, by either a digital stepping motor or a continuous motor.
9	Lens	This forms the Fourier transform of the scan area at the plane defined by 11.

* These units are under computer control, and can be utilized as part of a dynamic pattern-recognition algorithm.

Item Number	Item Name	Description and/or Function
10*	Sample window selector plate	This is moved in relation to the plate at 11 to select one of the windows on that plate. While sampling, the two plates are moved together. They are both under computer control, and the stepping increments are nominally 0.001 inch.
11*	Sample window plate	A photographic plate, all opaque with the exception of various sample windows. A window is selected by 10, and both windows are moved to the optical axis; then appropriate motions are made according to the sampling desired.
12	Collection lens	Collects all the light energy coming through the sampling window and focuses it onto the photo detector 13.
13	Photo detector	A photomultiplier tube with an S-20 photocathode. The output is fed to an extended range integrator which has a linear range of nearly 5 decades. The output of the integrator is converted and sent to the computer via a data control unit.
14	Relay lens 1	This is moved in and out of the optical path. When in the path, the scan area is projected onto 19, 21, or 22; when out, the diffraction pattern is projected onto 19, 21, or 22.
15	Iris	Controls the intensity of the projected images at 21 and 22.
16	Relay lens 2	This is part of the lens system to accomplish the projections described in 14.
17	Spatial filter holder	Has both rotational and X, Y motion capability (under manual control) and is used for spatial filtering experiments.
18	Lens turret	Brings (under manual control) the appropriate optics into the optical path to make the projections (of the scan area or diffraction pattern) at 21 or 22 either $10\times$, $20\times$, $50\times$, or $100\times$.
19	Camera	Records scan areas and/or their corresponding diffraction patterns. Can be moved in or out of the optical path.

(continued)

Item Number	Item Name	Description and/or Function
20	Mirror	This is moved in and out of the optical path. When in the path, the projection is at screen 22; when out, the projection is at screen 21.
21, 22	Projection screens	Rotatable projection screens with crosshairs, and a ruling engraved on one axis.
23	Computer graphics console	DPSS can be controlled from the SARF graphics console, via the computer. The analyst "flies" over the film (using the joystick of 24), views it on screen 21, selects the scan-areas he wishes to sample, and by light-pen selecting appropriate instructions on the display, initiates automatic transfer of their coordinates on the film plate into the computer. He then classifies the scan areas by light-pen selecting the appropriate name from a list of categories which appears on the display console. The selected scan areas with their corresponding identification tags are compiled into a list in the computer. This list is used later by the computer in effecting an automatic sampling of the FDPs of the selected scan areas. Proximity of the console to screen 21 allows the analyst to relate the contents of the scan area and its FDP (viewed on screen 21) to the sample signatures, to the features derived therefrom, and to any other processing displayed on the graphics console during his algorithm design sessions on SARF.
24	Control panel	This control panel allows the operator to operate DPSS for noncomputer experiments.

References

[1] J. W. Goodman, *Introduction to Fourier Optics*. San Francisco, Calif.: McGraw-Hill, 1968.

[2] M. Born and E. Wolf, *Principles of Optics*. New York: Pergamon, 1959.

[3] F. W. Sears, *Optics*. Cambridge, Mass.: Addison-Wesley, 1949.

[4] G. G. Lendaris and G. L. Stanley, "An opticalogical self-organizing recognition system," in *Optical and Electro-Optical Information Processing*. Cambridge, Mass.: M.I.T. Press, 1965, ch. 29.

[5] ——, "Interpreting aerial photography with the GM-DRL pattern recognition system," General Motors AC-Defense Res. Labs., Santa Barbara, Calif., Tech. Rept. TR65-91, January 1966.

[6] ——, "Applications of diffraction-pattern sampling to automatic target recognition," General Motors AC-Defense Res. Labs., Santa Barbara, Calif., Tech. Rept. TR67-06, January 1967.

[7] "The application of SARF/DPSS to the development of an ATR system," General Motors AC-Defense Res. Labs., Santa Barbara, Calif. Publ. no. S68-10, July 1968.

[8] "SARF, signature analysis research facility," General Motors AC-Defense Res. Labs., Santa Barbara, Calif., Publ. no. S68-11A, October 1968.

[9] G. L. Stanley, W. C. Nienow, and G. G. Lendaris, "SARF, an interactive signature analysis research facility," General Motors AC-Defense Res. Labs., Santa Barbara, Calif., Tech. Rept. TR69-15, March 1969; also *Proc. Purdue University Centennial Year Symp. on Information Processing*, 1969.

For further information the following were suggested by the reviewers.

A. Papoulis, *Systems and Transforms with Applications in Optics*. New York: McGraw-Hill, 1968.

C. A. Taylor and H. Lipson, *Optical Transformation*. Ithaca, N. Y.: Cornell University Press, 1964.

"Fourier-transform methods for pattern recognition," Air Force Cambridge Res. Lab., Cambridge, Mass., July 1962.

Asendorf, "Pictorial pattern recognition," in *Proc. Symp. on Automatic Photo Interpretation*, Washington, D. C.: Thompson Books, 1969, ch. 10.

Textural Features for Image Classification

ROBERT M. HARALICK, K. SHANMUGAM, AND ITS'HAK DINSTEIN

Abstract—**Texture is one of the important characteristics used in identifying objects or regions of interest in an image, whether the image be a photomicrograph, an aerial photograph, or a satellite image. This paper describes some easily computable textural features based on gray-tone spatial dependancies, and illustrates their application in category-identification tasks of three different kinds of image data: photomicrographs of five kinds of sandstones, 1:20 000 panchromatic aerial photographs of eight land-use categories, and Earth Resources Technology Satellite (ERTS) multispecial imagery containing seven land-use categories. We use two kinds of decision rules: one for which the decision regions are convex polyhedra (a piecewise linear decision rule), and one for which the decision regions are rectangular parallelpipeds (a min–max decision rule). In each experiment the data set was divided into two parts, a training set and a test set. Test set identification accuracy is 89 percent for the photomicrographs, 82 percent for the aerial photographic imagery, and 83 percent for the satellite imagery. These results indicate that the easily computable textural features probably have a general applicability for a wide variety of image-classification applications.**

I. Introduction

With the advent of high-speed general-purpose digital computers it is becoming possible to perform mathematical or algorithmic processes on pictorial data from images of photographic quality. In most of these processes, the pictorial information is represented as a function of two variables (x,y). The image in its digital form is usually stored in the computer as a two-dimensional array. If $L_x = \{1,2,\cdots,N_x\}$ and $L_y = \{1,2,\cdots,N_y\}$ are the X and Y spatial domains, then $L_x \times L_y$ is the set of resolution cells and the digital image I is a function which assigns some gray-tone value $G \in \{1,2,\cdots,N_g\}$ to each and every resolution cell; I: $L_x \times L_y \rightarrow G$. Various two-dimensional analyses are performed on I to achieve specific image-processing tasks such as coding, restoration, enhancement, and classification. In recent years a tremendous amount of computer processing of photographs has occurred, with facilities having been developed to process anything from aerial photographs to photomicrographs [1], [2].

In this paper we are concerned with the task of developing a set of features for classifying or categorizing pictorial data. The classification of pictorial data can be done on a resolution cell basis (such as in identifying the crop category of a resolution cell on satellite imagery) or on a block of contiguous resolution cells (such as in identifying the crop category of an entire agricultural field extending over a large number of resolution cells). The most difficult step in categorizing pictorial information from a large block of resolution cells is that of defining a set of meaningful features to describe the pictorial information from the block of resolution cells. Once these features are defined, image blocks can be categorized using any one of a multitude of pattern-recognition techniques.

In a search for meaningful features for describing pictorial information, it is only natural to look toward the types of features which human beings use in interpreting pictorial information. Spectral, textural, and contextual features are three fundamental pattern elements used in human interpretation of color photographs. Spectral features describe the average tonal variations in various bands of the visible and/or infrared portion of an electromagnetic spectrum, whereas textural features contain information

Manuscript received April 25, 1973; revised June 12, 1973. This work was supported by the NASA Goddard Space Flight Center under Contract NAS5-21822, and the Geographic Sciences Laboratory of the U.S. Army Engineer Topographic Laboratories under Contract DAAK02-70-C-0388.

R. M. Haralick and I. Dinstein are with the School of Electrical Engineering and the Center for Research, Inc., University of Kansas, Lawrence, Kans. 66044.

K. Shanmugam was with the Center for Research, Inc., University of Kansas, Lawrence, Kans. He is now with the Department of Electrical Engineering, Wichita State University, Wichita, Kans. 67218.

Reprinted from *IEEE Trans. Syst., Man, Cybern.*, vol. SMC-3, pp. 610–621, Nov. 1973.

about the spatial distribution of tonal variations within a band. Contextual features contain information derived from blocks of pictorial data surrounding the area being analyzed. When small image areas from black and white photographs are independently processed by a machine, then texture and tone are most important.

The concept of tone is based on the varying shades of gray of resolution cells in a photographic image, while texture is concerned with the spatial (statistical) distribution of gray tones. Texture and tone are not independent concepts; rather, they bear an inextricable relationship to one another very much like the relationship between a particle and a wave. Context, texture, and tone are always present in the image, although at times one property can dominate the other.

Texture can be evaluated as being fine, coarse, or smooth; rippled, molled, irregular, or lineated. For example, in the humid tropics, fine texture on radar imagery can be indicative of nonresistant fine-grain sedimentary rocks and unconsolidated sediments, while a coarse texture can be indicative of coarser grained sedimentary rocks. A massive texture with high-contrast components may be indicative of igneous rocks. A hummocky texture can be indicative of eroded igneous rocks.

Texture is an innate property of virtually all surfaces—the grain of wood, the weave of a fabric, the pattern of crops in a field, etc. It contains important information about the structural arrangement of surfaces and their relationship to the surrounding environment. Although it is quite easy for human observers to recognize and describe in empirical terms, texture has been extremely refractory to precise definition and to analysis by digital computers. Since the textural properties of images appear to carry useful information for discrimination purposes, it is important to develop features for texture. We present in this paper a computationally quick procedure for extracting textural features of images and discuss the usefulness of these features for discriminating between different kinds of image data.

Early image texture studies have employed autocorrelation functions [3], power spectra [4], restricted first- and second-order Markov meshes [5], and relative frequencies of various gray levels on the unnormalized image [6]. These had some degree of success, but we know little more about texture after finding the results of these experiments than before because they did not try to specifically define, characterize, or model texture. They only used some general mathematical transformation which assigns numbers to the transformed image in a nonspecific way.

Recent attempts to extract textural features have been limited to developing algorithms for extracting specific image properties such as coarseness and presence of edges. Many such algorithms have been developed and tried on special imagery. The subjective parameters (such as the selection of thresholds) associated with the techniques do not enable them to be generalized to imagery other than that processed by the authors. Recently, Rosenfeld and his coinvestigators presented a set of procedures for extracting some textural properties of pictorial data [7]–[9]. In [7] Rosenfeld and Troy described a procedure for obtaining a measure of the texture "coarseness" of images. Their procedures were based on the differences between the gray-tone values of adjacent image elements and on the autocorrelation of the image gray-tone values. In [8] Rosenfeld and Thurston gave a procedure for detecting boundaries separating regions which differ in texture coarseness. In [9] Troy *et al.* described gray-level manipulation procedures which can be used for preprocessing of pictures before applying the algorithms given in [7] and [8]. Procedures for detecting textural properties such as lines and dots have also been suggested by other investigators (see, for example, [10]–[12]). Before applying these procedures to pictures other than the ones processed by the authors of the respective papers, the investigator has to make a choice as to the method appropriate to the picture in question as well as the selection of parameters for the particular method.

We are presenting in this paper a general procedure for extracting textural properties of blocks of image data. These features are calculated in the spatial domain, and the statistical nature of texture is taken into account in our procedure, which is based on the assumption that the texture information in an image I is contained in the overall or "average" spatial relationship which the gray tones in the image have to one another. We compute a set of gray-tone spatial-dependence probability-distribution matrices for a given image block and suggest a set of 14 textural features which can be extracted from each of these matrices. These features contain information about such image textural characteristics as homogeneity, gray-tone linear dependencies (linear structure), contrast, number and nature of boundaries present, and the complexity of the image. It is important to note that the number of operations required to compute any one of these features is proportional to the number of resolution cells in the image block. It is for this reason that we call these features quickly computable.

We also investigate the usefulness of textural features for categorizing or classifying image blocks from three different data sets ranging from high-resolution photomicrographs to low-resolution satellite imagery. The accuracy of classification on multiclass categorization using textural features of these data sets was in the range of 80–90 percent.

II. Textural Features

Our initial perspective of texture and tone is based on the concept that texture and tone bear an inextricable relationship to one another. Tone and texture are always present in an image, although one property can dominate the other at times. The basic intuitively perceived relationships between tone and texture are the following. When a small-area patch of an image has little variation—i.e., little variation of features of discrete gray tone—the dominant property of that area is tone. When a small-area patch has a wide variation of features of discrete gray tone, the dominant property of that area is texture. Crucial to this distinction are the size of the small-area patch, the relative sizes of the discrete features, and the number of distinguishable discrete features. As the number of distinguishable tonal discrete features decreases, the tonal properties will predominate. In fact, when the small-area patch is only the size of one resolution cell, so that there is only one discrete feature, the only property present is tone. As the number of distinguishable features of discrete gray tone increases within the small-area patch, the texture property will dominate.

One important property of tone–texture is the spatial pattern of the resolution cells composing each discrete tonal feature. When there is no spatial pattern and the gray-tone variation between features is wide, a fine texture results. As the spatial pattern becomes more definite and involves more and more resolution cells, a coarser texture results. An excellent set of photographs of different types of texture may be found in [13].

The preceding description of texture is, of course, a gross simplification and idealization of what actually occurs. Discrete tonal features are really quite fuzzy in that they do not necessarily stand out as entities by themselves. Therefore the texture analysis we suggest is concerned with more general or macroscopic concepts than discrete tonal features.

The procedure we suggest for obtaining the textural features of an image is based on the assumption that the texture information on an image I is contained in the overall or "average" spatial relationship which the gray tones in the image I have to one another. More specifically, we shall assume that this texture information is adequately specified by a set of gray-tone spatial-dependence matrices which are computed for various angular relationships and distances between neighboring resolution cell pairs on the image. All of our textural features are derived from these angular nearest-neighbor gray-tone spatial-dependence matrices.

Gray-Tone Spatial-Dependence Matrices

Suppose an image to be analyzed is rectangular and has N_x resolution cells in the horizontal direction and N_y resolution cells in the vertical direction. Suppose that the gray tone appearing in each resolution cell is quantized[1] to N_g levels. Let $L_x = \{1,2,\cdots,N_x\}$ be the horizontal spatial domain, $L_y = \{1,2,\cdots,N_y\}$ be the vertical spatial domain, and $G = \{1,2,\cdots,N_g\}$ be the set of N_g quantized gray tones. The set $L_y \times L_x$ is the set of resolution cells of the image ordered by their row–column designations. The image I can be represented as a function which assigns some gray tone in G to each resolution cell or pair of coordinates in $L_y \times L_x$; $I: L_y \times L_x \to G$.

An essential component of our conceptual framework of texture is a measure, or more precisely, four closely related measures from which all of our texture features are derived. These measures are arrays termed angular nearest-neighbor gray-tone spatial-dependence matrices, and to describe these arrays we must emphasize our notion of adjacent or

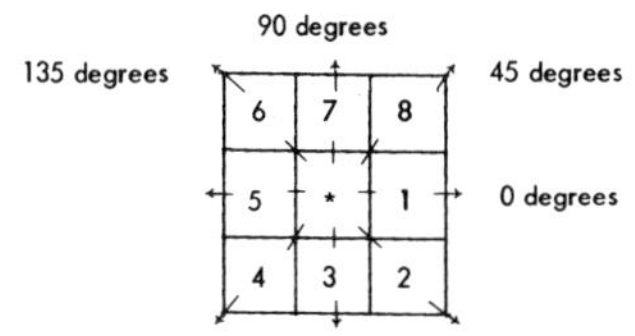

Fig. 1. Resolution cells 1 and 5 are 0° (horizontal) nearest neighbors to resolution cell $*$; resolution cells 2 and 6 are 135° nearest neighbors; resolution cells 3 and 7 are 90° nearest neighbors; and resolution cells 4 and 8 are 45° nearest neighbors to $*$. (Note this information is purely spatial, and has nothing to do with gray-tone values.)

nearest-neighbor resolution cells themselves. We consider a resolution cell—excluding those on the periphery of an image, etc.—to have eight nearest-neighbor resolution cells as in Fig. 1.

We assume that the texture-context information in an image I is contained in the overall or "average" spatial relationship which the gray tones in image I have to one another. More specifically, we shall assume that this texture-context information is adequately specified by the matrix of relative frequencies P_{ij} with which two neighboring resolution cells separated by distance d occur on the image, one with gray tone i and the other with gray tone j. Such matrices of gray-tone spatial-dependence frequencies are a function of the angular relationship between the neighboring resolution cells as well as a function of the distance between them. Fig. 2 illustrates the set of all horizontal neighboring resolution cells separated by distance 1. This set, along with the image gray tones, would be used to calculate a distance 1 horizontal gray-tone spatial-dependence matrix. Formally, for angles quantized to 45° intervals the unnormalized frequencies are defined by

$$\begin{aligned}
P(i,j,d,0^\circ) &= \#\{((k,l),(m,n)) \in (L_y \times L_x) \\
&\qquad \times (L_y \times L_x) \mid k-m=0, |l-n|=d, \\
&\qquad I(k,l)=i, I(m,n)=j\} \\
P(i,j,d,45^\circ) &= \#\{((k,l),(m,n)) \in (L_y \times L_x) \\
&\qquad \times (L_y \times L_x) \mid (k-m=d, l-n=-d) \\
&\qquad \text{or } (k-m=-d, l-n=d), \\
&\qquad I(k,l)=i, I(m,n)=j\} \\
P(i,j,d,90^\circ) &= \#\{((k,l),(m,n)) \in (L_y \times L_x) \\
&\qquad \times (L_y \times L_x) \mid |k-m|=d, \\
&\qquad l-n=0, I(k,l)=i, I(m,n)=j\} \\
P(i,j,d,135^\circ) &= \#\{((k,l),(m,n)) \in (L_y \times L_x) \\
&\qquad \times (L_y \times L_x) \mid (k-m=d, l-n=d) \\
&\qquad \text{or } (k-m=-d, l-n=-d), \\
&\qquad I(k,l)=i, I(m,n)=j\} \qquad (1)
\end{aligned}$$

where $\#$ denotes the number of elements in the set.

Note that these matrices are symmetric; $P(i,j;d,a) = P(j,i;d,a)$. The distance metric ρ implicit in the preceding

[1] Variations in lighting, lens, film, developer, and digitizer usually introduce monotonic transformations of the "true" image gray-tone values. Under these conditions, we would want two images of the same scene, one being a gray-tone monotonic transformation of the other to produce the same features. Image normalization by equal-probability quantizing guarantees that images which are monotonic transformations of one another produce the same results [14], [15]. An equal-probability quantizing algorithm is given in Appendix III.

(1,1)	(1,2)	(1,3)	(1,4)
(2,1)	(2,2)	(2,3)	(2,4)
(3,1)	(3,2)	(3,3)	(3,4)
(4,1)	(4,2)	(4,3)	(4,4)

$L_y = \{1,2,3,4\}$

$L_x = \{1,2,3,4\}$

$$
\begin{aligned}
R_H = & \{((k,l),(m,n)) \in (L_y \times L_x) \times (L_y \times L_x) \mid k-m = 0, |l-n| = 1\} \\
= & \{((1,1),(1,2)), ((1,2),(1,1)), ((1,2),(1,3)), ((1,3),(1,2)), \\
& ((1,3),(1,4)), ((1,4),(1,3)), ((2,1),(2,2)), ((2,2),(2,1)), \\
& ((2,2),(2,3)), ((2,3),(2,2)), ((2,3),(2,4)), ((2,4),(2,3)), \\
& ((3,1),(3,2)), ((3,2),(3,1)), ((3,2),(3,3)), ((3,3),(3,2)), \\
& ((3,3),(3,4)), ((3,4),(3,3)), ((4,1),(4,2)), ((4,2),(4,1)), \\
& ((4,2),(4,3)), ((4,3),(4,2)), ((4,3),(4,4)), ((4,4),(4,3))\}
\end{aligned}
$$

Fig. 2. Set of all distance 1 horizontal neighboring resolution cells on 4 × 4 image.

0	0	1	1
0	0	1	1
0	2	2	2
2	2	3	3

(a)

		Grey Tone 0	1	2	3
Grey Tone	0	#(0,0)	#(0,1)	#(0,2)	#(0,3)
	1	#(1,0)	#(1,1)	#(1,2)	#(1,3)
	2	#(2,0)	#(2,1)	#(2,2)	#(2,3)
	3	#(3,0)	#(3,1)	#(3,2)	#(3,3)

(b)

0°
$$P_H = \begin{pmatrix} 4 & 2 & 1 & 0 \\ 2 & 4 & 0 & 0 \\ 1 & 0 & 6 & 1 \\ 0 & 0 & 1 & 2 \end{pmatrix}$$

(c)

90°
$$P_V = \begin{pmatrix} 6 & 0 & 2 & 0 \\ 0 & 4 & 2 & 0 \\ 2 & 2 & 2 & 2 \\ 0 & 0 & 2 & 0 \end{pmatrix}$$

(d)

135°
$$P_{LD} = \begin{pmatrix} 2 & 1 & 3 & 0 \\ 1 & 2 & 1 & 0 \\ 3 & 1 & 0 & 2 \\ 0 & 0 & 2 & 0 \end{pmatrix}$$

(e)

45°
$$P_{RD} = \begin{pmatrix} 4 & 1 & 0 & 0 \\ 1 & 2 & 2 & 0 \\ 0 & 2 & 4 & 1 \\ 0 & 0 & 1 & 0 \end{pmatrix}$$

(f)

Fig. 3. (a) 4 × 4 image with four gray-tone values 0–3. (b) General form of any gray-tone spatial-dependence matrix for image with gray-tone values 0–3. $\#(i,j,)$ stands for number of times gray tones i and j have been neighbors. (c)–(f) Calculation of all four distance 1 gray-tone spatial-dependence matrices.

equations can be explicitly defined by

$$\rho((k,l),(m,n)) = \max \{|k - m|, |l - n|\}.$$

Consider Fig. 3(a), which represents a 4 × 4 image with four gray tones, ranging from 0 to 3. Fig. 3(b) shows the general form of any gray-tone spatial-dependence matrix. For example, the element in the (2,1) position of the distance 1 horizontal P_H matrix is the total number of times two gray tones of value 2 and 1 occurred horizontally adjacent to each other. To determine this number, we count the number of pairs of resolution cells in R_H such that the first resolution cell of the pair has gray tone 2 and the second resolution cell of the pair has gray tone 1. In Fig. 3(c)–3(f) we calculate all four distance 1 gray-tone spatial-dependence matrices.

If needed, the appropriate frequency normalizations for the matrices are easily computed. When the relationship is nearest horizontal neighbor ($d = 1$, $a = 0°$), there will be $2(N_x - 1)$ neighboring resolution cell pairs on each row and there are N_y rows, providing a total of $2N_y(N_x - 1)$ nearest horizontal neighbor pairs (see Fig. 3). When the relationship is nearest right-diagonal neighbor ($d = 1$, $a = 45°$) there will be $2(N_x - 1)$ 45° neighboring resolution cell pairs for each row except the first, for which there are none, and there are N_y rows. This provides a total of $2(N_y - 1)(N_x - 1)$ nearest right-diagonal neighbor pairs. By symmetry there will be $2N_x(N_y - 1)$ nearest vertical neighbor pairs and $2(N_x - 1)(N_y - 1)$ nearest left-diagonal neighbor pairs. After the number of neighboring resolution cell pairs R used in computing a particular gray-tone spatial-dependence matrix is obtained, the matrix is normalized by dividing each entry in the matrix by R.

It is appropriate at this point to comment on the computational aspects of obtaining the gray-tone spatial-dependence matrices. The number of operations required to process an image using our procedure is directly proportional to the number of resolution cells n present in the image. In comparison, the number of operations are of the order of $n \log n$ if one wishes to use Fourier or Hadamard transform to extract texture information. Also, to compute the entries in the gray-tone spatial-dependence matrices, one needs to keep only two lines of image data in core at a time. Thus no severe storage constraints are imposed. For most of the images we have processed, the computations have been carried out on a small digital computer (PDP 15/20 with 12 kwords of core and two DEC tape drives).

Textural Features Extracted from Gray-Tone Spatial-Dependence Matrices

Our initial assumption in characterizing image texture is that all the texture information is contained in the gray-tone spatial-dependence matrices. Hence all the textural features we suggest are extracted from these gray-tone spatial-dependence matrices. The equations which define a set of 14 measures of textural features are given in Appendix I. Some of these measures relate to specific textural characteristics of the image such as homogeneity, contrast, and the presence of organized structure within the image. Other measures characterize the complexity and nature of gray-tone transitions which occur in the image. Even though these features contain information about the textural characteristics of the image, it is hard to identify which specific textural characteristic is represented by each of these features.

For illustrative purposes, we will define 3 of the 14 textural features in this section and explain the significance of these features in terms of the kind of values they take on for two images of distinctly different textural characteristics The features we consider are

$$f_1 = \sum_{i=1}^{N_g} \sum_{j=1}^{N_g} \left(\frac{P(i,j)}{R}\right)^2 \tag{2}$$

$$f_2 = \sum_{n=0}^{N_g-1} n^2 \left\{ \sum_{|i-j|=n} \left(\frac{P(i,j)}{R}\right) \right\} \tag{3}$$

$$f_3 = \frac{\sum_{i=1}^{N_g} \sum_{j=1}^{N_g} [ijP(i,j)/R] - \mu_x\mu_y}{\sigma_x\sigma_y} \tag{4}$$

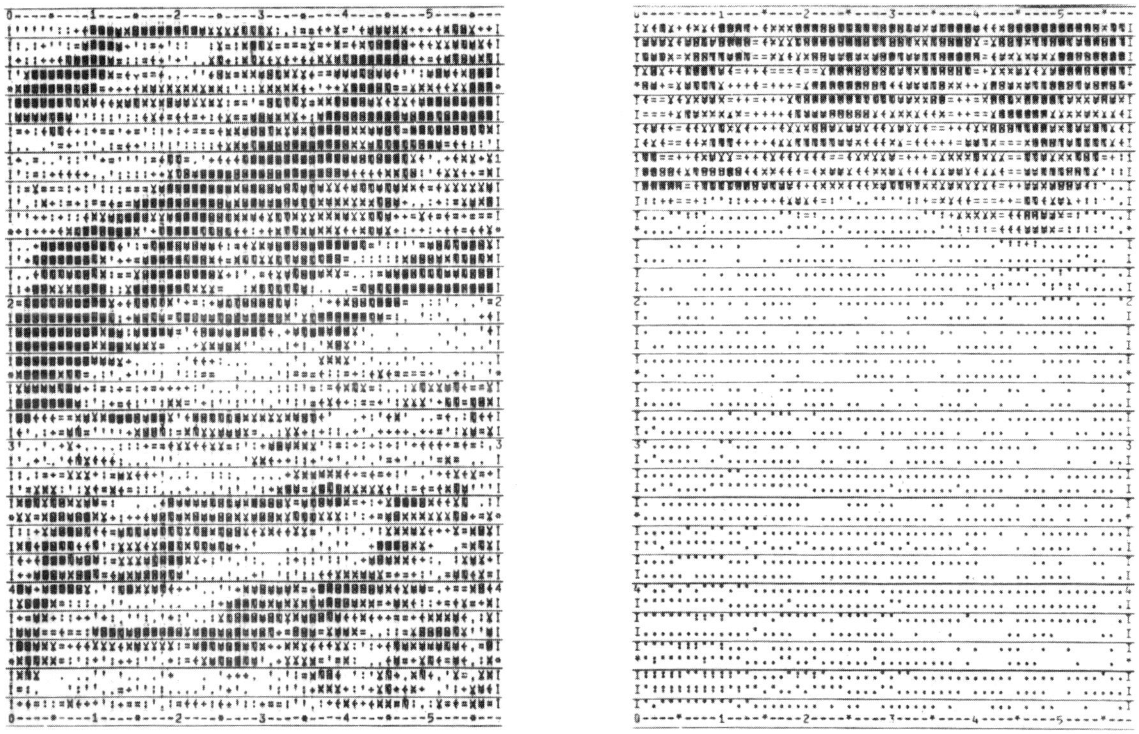

	Grassland			Water Body		
Angle	ASM	Contrast	Correlation	ASM	Contrast	Correlation
0°	.0128	3.048	.8075	.1016	2.153	.7254
45°	.0080	4.011	.6366	.0771	3.057	.4768
90°	.0077	4.014	.5987	.0762	3.113	.4646
135°	.0064	4.709	.4610	.0741	3.129	.4650
Avg.	.0087	3.945	.6259	.0822	2.863	.5327

(a) (b)

Fig. 4. Textural features for two different land-use category images.

where μ_x, μ_y, σ_x, and σ_y are the means and standard deviations of the marginal distributions associated with $P(i,j)/R$, and R is a normalizing constant.

Fig. 4 shows the digital printout of two 64×64 image blocks taken from a satellite picture over the California coastline (NASA ERTS Image no. 1002-18134). The image shown in Fig. 4(a) is a representative sample of grasslands and Fig. 4(b) is a representative sample of water bodies in the area. The values of the features f_1, f_2, and f_3 obtained from gray-tone spatial-dependence matrices for distance $d = 1$, are shown below the images in Fig. 4.

The angular second-moment feature (ASM) f_1 is a measure of homogeneity of the image. In a homogeneous image, such as shown in Fig. 4(b), there are very few dominant gray-tone transitions. Hence the P matrix for this image will have fewer entries of large magnitude. For an image like the one shown in Fig. 2(a), the P matrix will have a large number of small entries, and hence the ASM feature (which is the sum of squares of the entries) in the P matrix will be smaller. A comparison of the ASM values given below the images in Fig. 4 shows the usefulness of the ASM feature as a measure of the homogeneity of the image.

The contrast feature f_2 is a difference moment of the P matrix and is a measure of the contrast or the amount of local variations present in an image. Since there is a large amount of local variation present in the image of Fig. 4(a) compared to the image shown in Fig. 4(b), the contrast feature for the grassland image has consistently higher values compared to the water-body image.

The correlation feature f_3 is a measure of gray-tone linear-dependencies in the image. For both the images shown in Fig. 4, the correlation feature is somewhat higher in the horizontal (0°) direction, along the line of scan. The water-body image consists mostly of a constant gray-tone value for the water plus some additive noise. Since the noise samples are mostly uncorrelated, the correlation features for the water-body image have lower values compared to the grassland image. Also, the grassland image has a considerable amount of linear structure along 45° lines across the image, and hence the value of the correlation feature is higher along this direction compared to the values for 90° and 135° directions. More examples of the significance of some of the textural features are presented in Appendix II.

The various features which we suggest are all functions of distance and angle. The angular dependencies present a special problem. Suppose image A has features a, b, c, and d for angles 0°, 45°, 90°, and 135°, respectively, and image B is identical to A except that B is rotated 90° with respect to A. Then B will have features c, d, a, and b for angles 0°, 45°, 90°, and 135°, respectively. Since the texture context of A is the same as the texture context of B, any decision rule using the angular features a,b,c,d must produce the same results for c,d,a,b. To guarantee this, we suggest that the angularly dependent features not be used directly. Instead we suggest that two functions of a, b, c, and d, their average and range (which are invariant under rotation), be used as inputs to the classifier.

III. Applications of Textural Features for Image Classification

In this section we present the results of our studies on the usefulness of the textural features for categorizing images. Three data sets were used in our study. These data sets were extracted from photomicrographs of different rocks, from aerial photographs of man-made and natural scenes, and from high-altitude satellite pictures of the earth. A brief description of the data sets and classification algorithms used and the results of classification experiments will be presented. For further details the interested reader is referred to [16]–[18].

Data Set Description

1) *Photomicrographs of Sandstones:* Identification of the types of rocks (sandstones) present in an underground reservoir of crude oil is important in petroleum production studies; hence we undertook a study to identify different types of rocks from the textural characteristics of their photomicrographs. The data set consisted of 243 image blocks of size 64 × 64 and the gray tones of the images were equal-probability quantized into 16 levels. Details of the algorithm used to perform this quantization are given in Appendix III. There were five sandstone categories in the data set, and samples of the photomicrographs are shown in Fig. 5. The textural features for the 243 samples were calculated from distance 1 gray-tone spatial-dependence matrices.

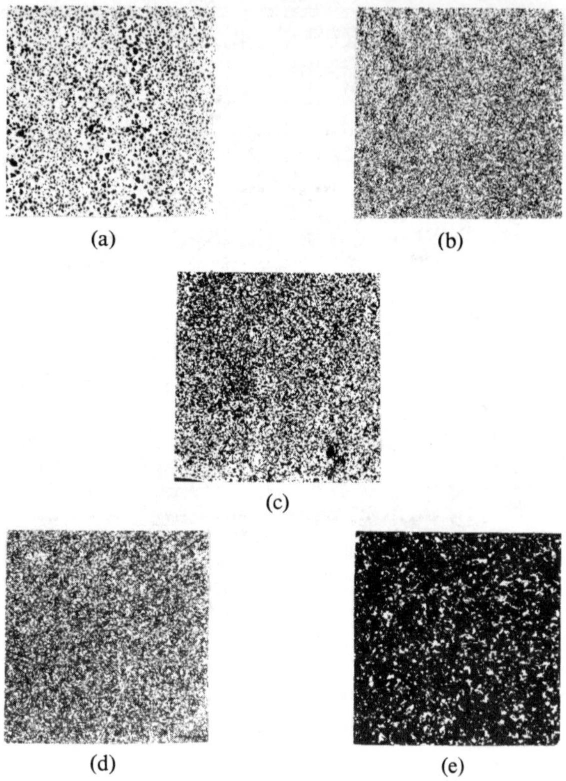

Fig. 5. Samples of photomicrographs of sandstones. (a) Dexter-L. (b) Dexter-H. (c) St. Peter. (d) Upper Muddy. (e) Gaskel.

2) *Aerial Photographic Data Set:* This data set consisted of 170 images belonging to eight categories. These 170 images were manually sampled from 9 × 9-in 1:20 000 aerial photography negatives provided by the Army Environmental Topographic Laboratories, Fort Belvoir, Va. [17]. The digitization of the sampled images was done on the Image Discrimination, Enhancement, and Combination System (IDECS)/PDP. The IDECS is an analog–digital near real-time processing system that has been developed at the University of Kansas Center for Research, Inc. A PDP 15/20 computer is interfaced to the IDECS and the system is capable of performing various functions of image processing [25]. The IDECS is capable of processing up to four image inputs. Three flying spot scanners can input transparencies up to a maximum size of 4 × 5 in. A vidicon scanner accepts larger transparencies as well as printed images. Multi-image inputs are electronically registered and combined into a format more suitable for interpretation. Processed images are displayed on a color display unit and/or a black and white monitor. A 24-channel disk is synchronized with the system and used for storage of processed digital images. Either manually or under computer control, the IDECS/PDP is able to perform image enhancement, video level slicing, area integration, category discrimination, display of histograms and scatograms, quantizing, and similar image-processing operations. The algorithm used for quantizing the aerial photographic data set digitized the images into eight levels in such a way that the areas of the different levels (in each image) were equal. Each image was digitized into a 20 × 50 array. The eight categories were residential old (RSOLD), residential new (RESNU), LAKE, SWAMP, MARSH, URBAN, RAIL, and SCRUB or WOOD (SCROD). The first six categories contained 20 samples each, the RAIL and SCROD categories contained 10 and 40 samples, respectively. Samples of the images are shown in Fig. 6.

3) *Data Set Derived from Satellite Imagery:* A variety of satellite imagery is currently used for the remote sensing of Earth resources. The most commonly used satellite

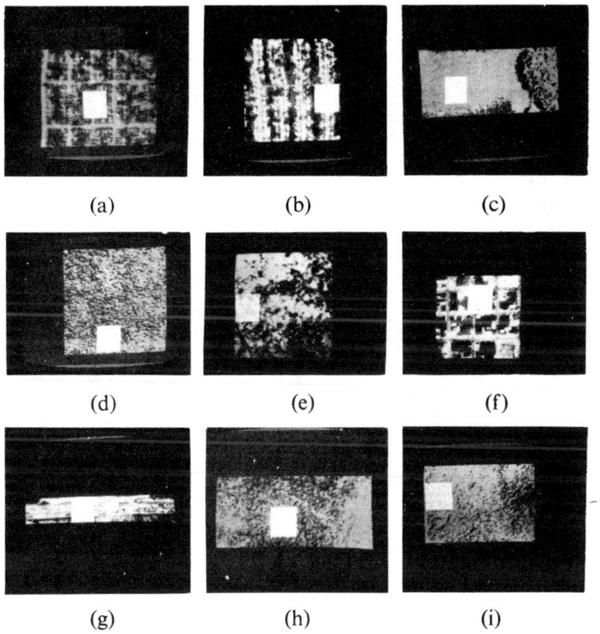

Fig. 6. Samples of aerial photographic data set. (a) RSOLD. (b) RESNU. (c) LAKE. (d) SWAMP. (e) MARSH. (f) URBAN. (g) RAIL. (h) SCRUB. (i) WOOD. (SCRUB and WOOD are classified together as SCROD.)

Fig. 7. Satellite picture of Monterey Bay, Calif.

imagery is the multispectral scanner (MSS) imagery, which consists of a set of images of a scene, where each image in the set is created by scanning the radiance of the scene from the satellite in a narrow band of wavelengths. The MSS operates in wavelengths ranging from 0.3 to 16 μm, and up to 24 scanners have been used in the past.

The data set used in our study was derived from a high-altitude four-channel multispectral satellite image taken over the California coastline. The image in one of the four MSS bands is shown in Fig. 7. The ground area covered by the image is about 14 400 mi^2 and the size of the digital image is 2340 $\times$ 3200. Out of this large image, a subset of 624 contiguous image blocks of size 64 $\times$ 64 was taken and used in our study. There were seven land-use categories in the image; these are coastal forest, woodlands, annual grasslands, urban areas, small and large irrigated fields, and water bodies. The land-use category for each 64 $\times$ 64 image block was obtained by interpreting the color composite of the multispectral image set. The textural features were computed from the distance 1 gray-tone spatial-dependence matrix of the second MSS band image in the multispectral image set.

Since the ground truth category information was derived from the color composite image, we used a set of spectral (color) features in addition to the textural features for categorizing these images. The spectral features consisted of the mean and standard deviation of the gray-tone values of the 64 $\times$ 64 image blocks in each of the four spectral bands.

Classification Algorithms

1) *Piecewise Linear Discriminant Function Method:* For categorizing image blocks in data sets 1 and 3, which have a reasonably large number of samples, we used a piecewise linear-discriminant function method. The data sets were partitioned arbitrarily into training and test sets. The classification algorithm was developed using the samples in the training set and it was tested on the samples in the test set.

A variety of algorithms have been discussed in the pattern-recognition literature for obtaining piecewise linear-discriminant functions for pattern classification [19]–[22]. In a widely used algorithm, the pattern space is partitioned into a number of regions using a set of hyperplanes (decision boundaries) whose locations are determined by the sample patterns. Each region is dominated by sample patterns of a particular category. When a new pattern is presented for identification, it is assigned a category depending on the region to which it belongs. If the new pattern X is located in a region dominated by sample patterns of category c_j, then X is classified as coming from category c_j.

For the multicategory problem involving N_R categories, a total of $N_R(N_R - 1)/2$ hyperplanes are used to partition the pattern space. These hyperplanes are defined by a set of weight vectors W_{ij}, $i = 1,2,\cdots,N_R$, $j = 1,2,\cdots,N_R$, $j > i$, which separates the sample patterns belonging to the ith and jth categories. A regression type algorithm given in [19, ch. 4] was used to obtain the weight vectors. After the location of the hyperplanes are determined, the classification of new patterns is done as follows. For each category c_i, the number of hyperplanes V_i which give a positive response when the new pattern X is presented are determined using

$$V_i = \sum_{\substack{j=1 \\ j \neq i}}^{N_R} \frac{|W_{ij}{}^T Z| + W_{ij}{}^T Z}{2|W_{ij}{}^T Z|}, \qquad i = 1,2,\cdots,N_R \quad (5)$$

where Z is the augmented pattern vector obtained by adding a component of value 1 to X, i.e.,

$$Z = \begin{bmatrix} 1 \\ X \end{bmatrix}. \tag{6}$$

X is assigned to category c_j if $V_j = \max \{V_i\}$. If there is a tie between categories c_m and c_n, then X is assigned to c_m if $W_{mn}{}^T Z \geq 0$, or to c_n if $W_{mn}{}^T Z < 0$. Several modifications of the linear-discriminant function method and a multitude of other classification procedures may be found in the references cited.

2) *Min–Max Decision Rule:* For the aerial photograph data set which had a small number of samples per category we used a min–max decision rule for categorizing the images based on their textural features. The decision boundaries were obtained by using all but one of the samples in the data set, and the sample which was left out was then classified. The procedure was repeated for all the samples in the data set to obtain the overall accuracy of the classification scheme. This method of "leaving one sample out" is widely used by investigators in the pattern-recognition field when the data set has a small number of samples. The decision rule is described as follows.

Decision rule: If b_{n_k} and a_{n_k} define the minimum and maximum values of the uniform distribution, $(x_1, x_2, \cdots, x_N)$ is assigned to category k if and only if

$$b_{n_k} \leq x_n \leq a_{n_k}, \qquad n = 1, 2, \cdots, N$$

$$\prod_{n=1}^{N} \frac{1}{(a_{n_k} - b_{n_k})} \geq \prod_{n=1}^{N} \frac{1}{(a_{n_j} - b_{n_j})} \tag{7}$$

for all j such that $b_{n_j} \leq x_n \leq a_{n_j}$, $n = 1, 2, \cdots, N$. If there exists no k such that $b_{n_k} \leq x_n \leq a_{n_k}$, $n = 1, 2, \cdots, N$, then $(x_1, x_2, \cdots, x_n)$ is assigned to category k if and only if

$$\sum_{n=1}^{N} \min \frac{\{|x_n - a_{n_k}|, |x_n - b_{n_k}|\}}{(a_{n_k} - b_{n_k})}$$

$$\geq \sum_{n=1}^{N} \min \frac{\{|x_n - a_{n_j}|, |x_n - b_{n_j}|\}}{(a_{n_j} - b_{n_j})}, \qquad j = 1, 2, \cdots, K \tag{8}$$

where K is the number of categories.

Results of Image Classification Studies

1) *Photomicrograph Data Set:* A set of 8 variables, comprising the mean and variance of the textural features f_1, f_2, f_3, and f_9 (for definitions of these features, see Appendix I), computed from distance 1 gray-tone spatial-dependence matrices, was used as an input to the classifier. The data set was arbitrarily divided into a training set of 143 samples and a test set of 100 samples. Piecewise linear-discriminant functions for separating the patterns belonging to the five sandstone categories were derived using the samples in the training set. The contingency table for classifying the samples in the test set is shown in Table I. The overall accuracy of the classification scheme on the test set (obtained by summing the diagonal elements in the contingency table and dividing by the total number of samples in the test set) was 89 percent.

TABLE I
CONTINGENCY TABLE FOR THE CLASSIFICATION OF PHOTOMICROGRAPHS OF SANDSTONES

		ASSIGNED CATEGORY					Total
		Dexter-L	Dexter-H	St. Peter	Upper Muddy	Gaskel	
TRUE CATEGORY	Dexter-L	29	0	1	0	0	30
	Dexter-H	0	15	0	0	0	15
	St. Peter	2	0	22	4	0	28
	Upper Muddy	0	0	4	11	0	15
	Gaskel	0	0	0	0	12	12
	TOTAL	31	15	27	15	12	100

Number of samples in test set = 100; number of samples in training set = 143; overall accuracy of classification of test set = 89 percent.

TABLE II
CONTINGENCY TABLE FOR THE CLASSIFICATION OF THE AERIAL PHOTOGRAPHIC DATA SET

		ASSIGNED CATEGORY								TOTAL
		RSOLD	RESNU	LAKE	SWAMP	MARSH	URBAN	RAIL	SCROD	
TRUE CATEGORY	RSOLD	17	0	0	0	1	1	0	1	20
	RESNU	1	15	0	0	1	2	1	0	20
	LAKE	0	0	19	0	0	0	0	1	20
	SWAMP	0	0	0	19	1	0	0	0	20
	MARSH	0	0	0	0	12	0	0	8	20
	URBAN	2	3	0	0	0	15	0	0	20
	RAIL	0	1	0	0	0	1	5	3	10
	SCROD	0	0	0	0	2	0	0	38	40
	TOTAL	20	19	19	19	17	19	6	51	170

140 out of 170 images or 82.3 percent of the images were correctly classified.

2) *Aerial Photographic Data Set:* A min–max decision rule was used for the classification of 170 images into eight categories. The processing was done as follows. Four gray-tone spatial-dependencies (for four directions) were computed for each image. Eleven textural features (f_1–f_{11} as defined in Appendix I) were computed, yielding a measurement vector of $4 \times 11 = 44$ components for each image. Computation of the mean, range, and deviation of each feature over the four directions reduced the dimensionality to 33. The minimum and maximum statistics were computed for 169 images and the min–max decision rule was used to classify the image which was left out. This procedure was repeated 170 times to classify all the images. Out of these 170 images, 140, or 82.3 percent of the images, were correctly classified. The contingency table is given in Table II.

3) *Satellite Imagery Data Set:* The piecewise linear-discriminant function method was used for identifying the land-use category of each image block in this data set. The input variables to the classifier consisted of the mean variance of the four textural features (f_1, f_2, f_3, and f_9 obtained from the distance 1 gray-tone spatial-dependence matrices) and eight spectral features (comprised of the mean variance of the image gray-tone values) in each of the four spectral bands. The 624 samples in the data set were divided arbitrarily into a training set of 314 samples and a test set of 310 samples. The classifier was trained on the 314 samples in the training set, and each sample in the test set was assigned to one of seven possible land-use categories. The

TABLE III
CONTINGENCY TABLE FOR LAND-USE CLASSIFICATION OF SATELLITE IMAGERY

		ASSIGNED CATEGORY							
		Coastal Forest	Woodlands	Annual Grass-Lands	Urban Area	Large Irrigated Fields	Small Irrigated Fields	Water	Total
TRUE CATEGORY	Coastal Forest	23	1	2	0	0	0	1	27
	Woodlands	0	17	10	0	1	0	0	28
	Annual Grasslands	1	3	109	1	1	0	0	115
	Urban Area	0	3	10	13	0	0	0	26
	Large Irrigated Fields	1	2	6	0	37	2	0	48
	Small Irrigated Fields	0	0	4	0	3	24	0	31
	Water	0	0	0	0	0	0	35	35
	Total	25	26	141	14	42	26	36	310

Number of training samples = 314; number of test samples = 310; accuracy of classification on training set = 84.0 percent; accuracy of classification on test set = 83.5 percent.

contingency table for the classification of the test samples is given in Table III. The overall accuracy of the classifier on the test samples was found to be 83.5 percent.

Classification of multispectral data such as the satellite imagery we have processed is usually carried out using only the spectral features. We attempted a spectral classification on the satellite imagery data and achieved a classification accuracy of 74–77 percent on the test set. This result, compared with the 83.5-percent classification accuracy achieved using a combination of spectral and textural features, shows that a significant improvement in the classification accuracy might result if the textural features are used as additional inputs to the classifier.

IV. CONCLUDING REMARKS

We have described a class of quickly computable textural features which seem to have general applicability to many kinds of image data. The textural features are based on statistics which summarize the relative frequency distribution (which describes how often one gray tone will appear in a specified spatial relationship to another gray tone on the image). We have used these features in category-identification tasks of three different kinds of image data. Identification accuracy on independent test sets are 89 percent for the photomicrograph image set (five categories of sandstones), 82 percent for the aerial photographs (eight land-use categories), and 83 percent for the satellite imagery (seven land-use categories).

These initial experimental results are promising. Much work needs to be done, however, on gray-tone normalization on the imagery and the use of features which are invariant under monotonic gray-tone transformations. The reason is that in one important sense, texture is independent of tone. Two people examining photographs of the same texture may actually be seeing two different, though related, kinds of tones in the texture. One photograph may have been developed such that its tones are light and thin, and the other photograph may have been developed such that its tones are dark and heavy. Most people could easily make the observation that the texture on the two images is the same. For a machine to find that the textures are the same, either the images must be probability quantized and the features computed from the probability quantized images (which are invariant under monotonic gray-tone transformations), or the features themselves must be invariant under monotonic gray-tone transformations. Of the textural features described in Appendix I, the angular second-moment, the entropy, the sum entropy, the difference entropy, the information measure of correlation, and the maximal-correlation features have the invariance property. We intend to repeat the experiments reported here using these kinds of features. We expect that these features will provide more generalized results.

Additional investigations are necessary to determine the size of the subimage region and the distances which should be used in computing the gray-tone dependence matrices. Too small a subimage region will not have enough textural information to separate image categories of interest, while a large subimage region may have objects belonging to several different categories. Also, a large subimage region will increase the storage requirements. The distance which must be used in computing the gray-tone spatial-dependencies may be obtained from the autocorrelation function of the image. The distance at which the normalized autocorrelation function of the image becomes too small can serve as an upper bound on the distance which may be used for computing the gray-tone spatial-dependence matrices.

APPENDIX I
TEXTURAL FEATURES

We suggest a set of 28 textural features which can be extracted from each of the gray-tone spatial-dependence matrices. The following equations define these features.

Notation

$p(i,j)$ (i,j)th entry in a normalized gray-tone spatial-dependence matrix, $= P(i,j)/R$.

$p_x(i)$ ith entry in the marginal-probability matrix obtained by summing the rows of $p(i,j)$, $= \sum_{j=1}^{N_g} P(i,j)$.

N_g Number of distinct gray levels in the quantized image.

$\sum_i$ and $\sum_j$ $\sum_{i=1}^{N_g}$ and $\sum_{j=1}^{N_g}$, respectively.

$$p_y(j) = \sum_{i=1}^{N_g} p(i,j).$$

$$p_{x+y}(k) = \sum_{\substack{i=1 \\ i+j=k}}^{N_g} \sum_{j=1}^{N_g} p(i,j), \qquad k = 2,3,\cdots,2N_g.$$

$$p_{x-y}(k) = \sum_{\substack{i=1 \\ |i-j|=k}}^{N_g} \sum_{j=1}^{N_g} p(i,j), \qquad k = 0,1,\cdots, N_g - 1.$$

Textural Features

1) Angular Second Moment:

$$f_1 = \sum_i \sum_j \{p(i,j)\}^2.$$

2) Contrast:

$$f_2 = \sum_{n=0}^{N_g-1} n^2 \left\{ \sum_{\substack{i=1 \\ |i-j|=n}}^{N_g} \sum_{j=1}^{N_g} p(i,j) \right\}.$$

3) Correlation:

$$f_3 = \frac{\sum_i \sum_j (ij)p(i,j) - \mu_x \mu_y}{\sigma_x \sigma_y}$$

where μ_x, μ_y, σ_x, and σ_y are the means and standard deviations of p_x and p_y.

4) Sum of Squares: Variance

$$f_4 = \sum_i \sum_j (i - \mu)^2 p(i,j).$$

5) Inverse Difference Moment:

$$f_5 = \sum_i \sum_j \frac{1}{1 + (i - j)^2} p(i,j).$$

6) Sum Average:

$$f_6 = \sum_{i=2}^{2N_g} i p_{x+y}(i).$$

7) Sum Variance:

$$f_7 = \sum_{i=2}^{2N_g} (i - f_8)^2 p_{x+y}(i).$$

8) Sum Entropy:[2]

$$f_8 = -\sum_{i=2}^{2N_g} p_{x+y}(i) \log \{p_{x+y}(i)\}.$$

9) Entropy:

$$f_9 = -\sum_i \sum_j p(i,j) \log (p(i,j)).$$

10) Difference Variance:

$$f_{10} = \text{variance of } p_{x-y}.$$

11) Difference Entropy:

$$f_{11} = -\sum_{i=0}^{N_g-1} p_{x-y}(i) \log \{p_{x-y}(i)\}.$$

12), 13) Information Measures of Correlation:

$$f_{12} = \frac{HXY - HXY1}{\max \{HX, HY\}}$$

$$f_{13} = (1 - \exp[-2.0(HXY2 - HXY)])^{1/2}$$

$$HXY = -\sum_i \sum_j p(i,j) \log (p(i,j))$$

where HX and HY are entropies of p_x and p_y, and

$$HXY1 = -\sum_i \sum_j p(i,j) \log \{p_x(i)p_y(j)\}$$

$$HXY2 = -\sum_i \sum_j p_x(i)p_y(j) \log \{p_x(i)p_y(j)\}.$$

14) Maximal Correlation Coefficient:

$$f_{14} = (\text{Second largest eigenvalue of } Q)^{1/2}$$

where

$$Q(i,j) = \sum_k \frac{p(i,k)p(j,k)}{p_x(i)p_y(k)}.$$

These measures of correlation have some desirable properties which are not brought out in the rectangular correlation measure f_3. For details, see [23], [24].

Referring to (1) in the text, for a chosen distance d we have four angular gray-tone spatial-dependency matrices. Hence we obtain a set of four values for each of the preceding 14 measures. The mean and range of each of these 14 measures, averaged over the four values, comprise the set of 28 features which can be used as inputs to the classifier. In this set of 28 features some of the features are strongly correlated with each other. A feature-selection procedure may be applied to select a subset or linear combinations of the 28 features.

Appendix II
Examples of the Significance of Some Textural Features

The problem "What do the textural features represent?" from a human perception point of view can be a subject for a thorough experiment. There are some intuitive expectations as to properties represented by some features. For example, one might expect the entropy feature to take higher values for more complex images. One might expect to be able to notice some linear dependencies in images with high values of the correlation feature. As a beginning for an investigation of this problem we used an aerial photography data set that consisted of 12 categories, 6 samples in each [17].

We ordered the categories in the following manner. We computed the range and the mode of values that each feature takes for each category. The categories were then ordered for each feature according to the modes, and the ranges were plotted. The prospective of this Appendix is to comment on the ordering of categories by some of the features. Figs. 8–10 show plots of the ranges of three of the features, each accompanied by three pictures. These pictures are of units that have small, medium, and large values of the respective feature. The order of categories of the average angular moment feature (ANGMOM) in Fig. 8 is almost the reverse of the order of categories of the average entropy (ENTROP) in Fig. 9. The three representative pictures show that in both figures the units were ordered according to the complexity of the image. Images with more gray levels have lower average ANGMOM and higher

[2] Since some of the probabilities may be zero, and log (0) is not defined, it is recommended that the term log $(p + \varepsilon)$ (ε an arbitrarily small positive constant) be used in place of log (p) in entropy computations.

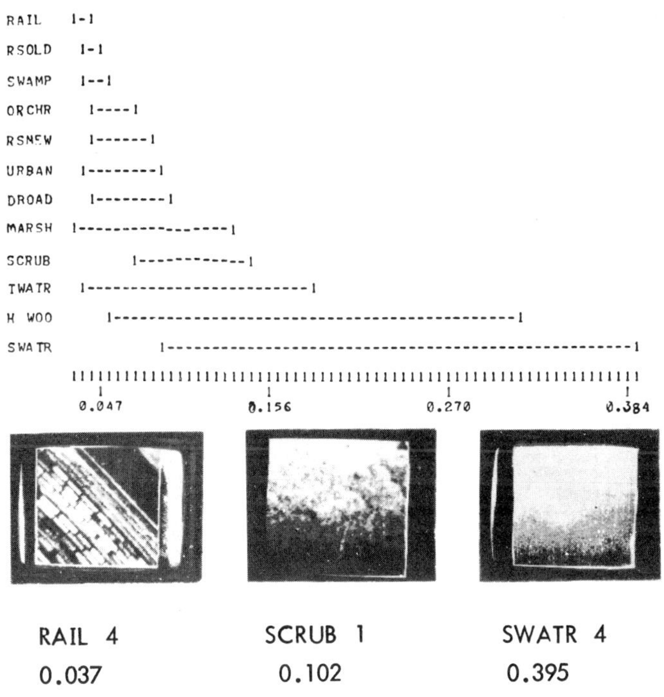

Fig. 8. Range plots and representative pictures for average second angular moment.

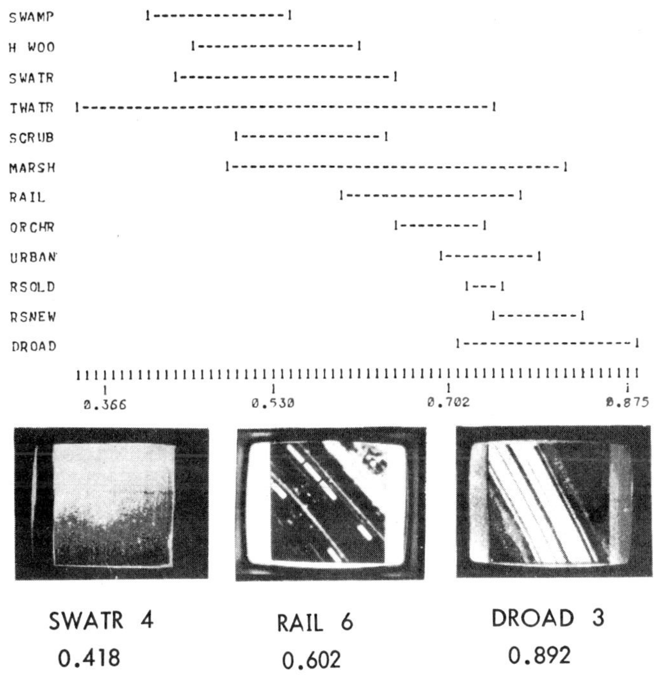

Fig. 10. Range plots and representative pictures for average correlation.

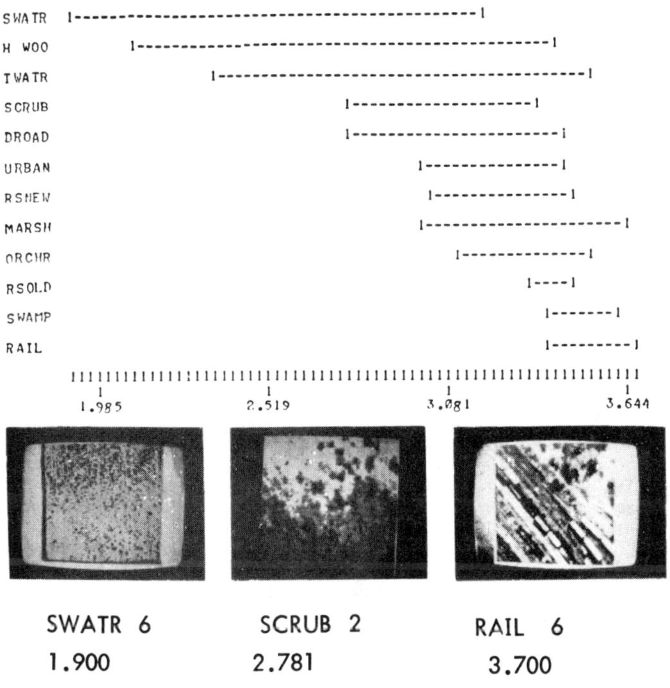

Fig. 9. Range plots and representative pictures for average entropy.

average ENTROP. The average ratio (RATIO) in Fig. 10 is lower for natural scenes and higher for man-made scenes. The three representative pictures in Fig. 10 show correspondence between the average RATIO values and the linear dependency of gray levels in the images. These comments reflect our subjective perception. Much more experimentation and analysis should be done in this subject.

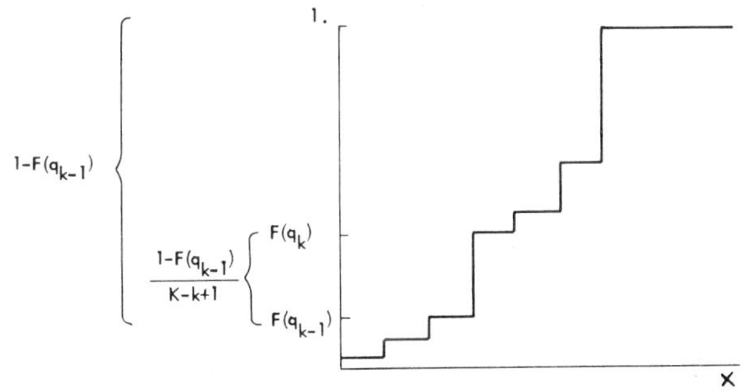

Fig. 11. Quantizing algorithm. At kth iteration, $F(q_k - 1)$ probability has already been allocated to $k - 1$ levels and $1 - F(q_k - 1)$ probability remains to be allocated. If $1 - F(q_k - 1)$ probability is split up equally among remaining $K - k + 1$ quantizing levels to be allocated, each level would get $(1 - F(q_{k-1}))/(K - k + 1)$. Since F is a step function, there is no guarantee that a q_k can be found and that $F(q_k) = F(q_{k-1}) + (1 - F(q_{k-1}))/(K - k + 1)$. Hence we look for a q_k which is closest to satisfying the equality.

Appendix III
Equal-Probability Quantizing

Let X be a random variable with cumulative probability function F_X. Let Q_1, the K-level equal-probability quantizing function for X, be defined by

$Q_1(x) = k$, if and only if

$$\text{lub}\left\{w \mid F_X(w) = \frac{k-1}{K}\right\} \le x < \text{lub}\left\{w \mid F_X(w) = \frac{k}{K}\right\}$$

where lub denotes least upper bound.

For any strictly monotonic function g, define the random variable Y by $Y = g(X)$. Let Q_2, the K-level equal-probability quantizing function for Y, be defined by

$Q_2(y) = k$, if and only if

$$\text{lub}\left\{w \mid F_Y(w) = \frac{k-1}{K}\right\} \le y < \text{lub}\left\{w \mid F_Y(w) = \frac{k}{K}\right\}.$$

The following lemma states that the equal-probability quantization of X produces a random variable which is identical to the equal-probability quantization of Y.

Lemma: $Q_1(X) = Q_2(Y)$.

Proof: $Q_1(X) = k$, if and only if

$$\text{lub}\left\{x \mid F_X(x) = \frac{k-1}{K}\right\} \le X < \text{lub}\left\{x \mid F_X(x) = \frac{k}{K}\right\}. \quad (9)$$

Since g is strictly monotonic, (9) holds if and only if

$$g\left(\text{lub}\left\{x \mid F_X(x) = \frac{k-1}{K}\right\}\right) \le g(X) < g\left(\text{lub}\left\{x \mid F_X(x) = \frac{k}{K}\right\}\right). \quad (10)$$

Since g is strictly monotonic, we may reverse the order of g and lub so that (10) holds if and only if

$$\text{lub}\left\{g(x) \mid F_X(x) = \frac{k-1}{K}\right\} \le g(X) < \text{lub}\left\{g(x) \mid F_X(x) = \frac{k}{K}\right\}. \quad (11)$$

However,

$$F_X(x) = P(X \le x) = P(g(X) \le g(x)) = P(Y \le g(x)) = F_Y(g(x)).$$

Hence, (11) holds if and only if

$$\text{lub}\left\{g(x) \mid F_Y(g(x)) = \frac{k-1}{K}\right\} \le g(X) < \text{lub}\left\{g(x) \mid F_Y(g(x)) = \frac{k}{K}\right\}. \quad (12)$$

Finally, (12) is equivalent to

$$\text{lub}\left\{y \mid F_Y(y) = \frac{k-1}{K}\right\} \le Y < \text{lub}\left\{y \mid F_Y(y) = \frac{k}{K}\right\} \quad (13)$$

so that $Q_2(Y) = k$.

Equal Probability Quantizing Algorithm: Let X be a nonnegative random variable with cumulative probability function F_X. Let Q, the K-level equal-probability quantizing function for X, be defined by $Q(x) = k$ if and only if $q_{k-1} \le x < q_k$. We define $q_0, q_1, q_2, \cdots, q_k$ in an iterative manner. Let $q_0 = 0$. Suppose q_{k-1} has been defined. Then let q_k be the smallest number such that

$$\left|\frac{1 - F_X(q_{k-1})}{K - k + 1} + F_X(q_{k-1}) - F_X(q_k)\right| \le \left|\frac{1 - F_X(q_{k-1})}{K - k + 1} + F_X(q_{k-1}) - F(q)\right|, \qquad \text{for all real } q.$$

Fig. 11 illustrates the equal-probability quantizing algorithm.

REFERENCES

[1] *Special Issue on Digital Picture Processing*, *Proc. IEEE*, vol. 60, pp. 763–922, July 1972.
[2] H. C. Andrews, A. G. Tescher, and R. P. Kruger, "Image processing by digital computer," *IEEE Spectrum*, vol. 9, pp. 20–32, July 1972.
[3] H. Kaizer, "A quantification of textures on aerial photographs," Boston Univ. Res. Lab., Boston, Mass., Tech. Note 121, 1955, AD 69484.
[4] R. Chevallier, A. Fontanel, G. Grau, and M. Guy, "Application du filtrage optique a l'etude des photographies aeriennes," presented at the 11th Congr. Int. Photogrammetric, July 1968.
[5] R. Bixby, G. Elerding, V. Fish, J. Hawkins, and R. Loewe, "Natural image computer," Aeronutronic Division, Philco-Ford Corp., Newport Beach, Calif., Final Tech. Rep., vol. 1, Publ. C-4035, May 1967.
[6] E. M. Darling and R. D. Joseph, "Pattern recognition from satellite altitudes," *IEEE Trans. Syst. Sci. Cybern.*, vol. SSC-4, pp. 38–47, Mar. 1968.
[7] A. Rosenfeld and E. Troy, "Visual texture analysis," Comput. Sci. Cent., Univ. Maryland, College Park, Tech. Rep. 70-116, June 1970.
[8] A. Rosenfeld and M. Thurston, "Visual texture analysis, 2," Comput. Sci. Cent., Univ. Maryland, College Park, Tech. Rep. 70-129, Sept. 1970.
[9] E. B. Troy, E. S. Deutsch, and A. Rosenfeld, "Gray-level manipulation experiments for texture analysis," *IEEE Trans. Syst., Man, Cybern.*, vol. SMC-3, pp. 91–98, Jan. 1973.
[10] *Proc. Symp. Computer Image Processing and Recognition*, Univ. Missouri, Columbia, Aug. 24–26, 1972.
[11] R. M. Pickett, "Visual analysis of texture in the detection and recognition of objects," in *Picture Processing and Psychopictorics*, B. S. Lipkin and A. Rosenfeld, Eds. New York: Academic Press, 1970, pp. 289–308.
[12] A. Rosenfeld and B. S. Lipkin, "Texture synthesis," in *Picture Processing and Psychopictorics*, B. S. Lipkin and A. Rosenfeld, Eds. New York: Academic Press, 1970, pp. 309–345.
[13] P. Brodatz, *Textures*. New York: Dover, 1966.
[14] R. M. Haralick, K. Shanmugam, and I. Dinstein, "On some quickly computable features for texture," in *Proc. Symp. Computer Image Processing and Recognition*, Univ. Missouri, Columbia, vol. 2, pp. 12-2-1–12-2-8, Aug. 1972.
[15] E. E. Hall *et al.*, "A survey of preprocessing and feature extraction techniques for radiographic images," *IEEE Trans. Comput.*, vol. C-20, pp. 1032–1044, Sept. 1971.
[16] R. M. Haralick and D. Anderson, "Texture–tone study with applications to digitized imagery," Univ. Kansas Center for Research, Inc., Lawrence, CRES Tech. Rep. 182–2, Nov. 1971.
[17] R. M. Haralick, I. Dinstein, K. Shanmugam, and D. Goel, "Texture–tone study with applications to digitized imagery," Univ. Kansas Center for Research, Inc., Lawrence, CRES Tech. Rep. 182-4, Dec. 1972.
[18] R. M. Haralick and K. Shanmugam, "Computer classification of reservoir sandstones," *IEEE Trans. Geosci. Electron.*, vol. GE-11, pp. 171–177, Oct. 1973.
[19] K. Fukunaga, *Introduction to Statistical Pattern Recognition*. New York: Academic Press, 1972.
[20] K. S. Fu and J. M. Mendel, *Adaptive Learning and Pattern Recognition Systems*. New York: Academic Press, 1972.
[21] W. Miesel, *Computer Oriented Approaches to Pattern Recognition*. New York: Academic Press, 1972.
[22] N. J. Nilsson, *Learning Machines*. New York: McGraw-Hill, 1965.
[23] E. H. Linfoot, "An information measure of correlation," *Inform. Contr.*, vol. 1, pp. 85–89, 1957.
[24] C. B. Bell, "Mutual information and maximal correlation measures of dependence," *Ann. Math. Statist.*, vol. 43, pp. 587–595, 1962.
[25] P. N. Anderson *et al.*, "An automated image processing system—IDECS," Ph.D. project, Univ. Kansas, Lawrence, 1972.

A TRANSFORMATION FOR EXTRACTING NEW DESCRIPTORS OF SHAPE

Harry Blum *
Data Sciences Laboratory
Air Force Cambridge Research Laboratories
Bedford, Massachusetts

Introduction

I have approached the problem of shape by assuming that the current mathematical tools were somehow missing essential elements of the problem. For despite more than two millennia of geometry, no formulation which appears natural for the biological problem has emerged. This is not surprising perhaps when one recognizes that geometry has been born of surveying and has grown in close collaboration with physical science and its mensuration problems. A corollary to this position is that there is some central difference between the biological problem that we are trying to deal with and the physical problem that we have been dealing with. Consequently, such an approach requires a restudy of visual function to assess what such a geometry should indeed try to accomplish. Unfortunately, the problem of exploring function is not easy to do in isolation, since the visual world is extremely rich, and hypotheses about visual shapes and their functional value to an organism may reflect the cultural bias of the experimenter to a large degree. I have chosen to enter the problem from the middle by hypothesizing simple shape processing mechanisms, and then exploring together the geometry and visual function that result. One such mechanism is presented in this paper. Since it leads to a drastic reformulation of a number of notions of visual shape, it may be useful to review briefly some of the notions implicit in our views and experiments.

While local attributes of the visual field (intensity, color, edge, angle, motion) and average global properties such as total luminous flux have been subject to progressive experimentation and understanding, the attributes of shape, global properties in which the particular distribution is precisely relevant, are still largely unkown. Yet, shape has been shown to be a particularly stable organization of our visual stimulus by the degree and variety of distortion that are the stock in trade of artists and cartoonists – even being used to enhance recognition. Why, then, has it been so difficult to tease out the properties

*Now with the Laboratory for Statistical and Mathematical Methodology, Division of Computer Research, National Institutes of Health, Bethesda, MD 20014.

of shape? A primary reason is that the number and variety of shapes are enormous, so that it is impossible within the limits of experiment to explore more than a small sampling of the variables. Consequently, one assumes a theory of shape, at least implicitly, in the selection of experimental stimuli. This reason is compounded by the degree to which geometry and reading, both synthetic processes, imposed on us culturally and with considerable difficulty, have biased our view of shape. As a result, stimuli in even the best animal experiments have been largely rectilinear or other unnatural geometric shapes presented in isolation on a uniform field (for example, see Sutherland, 1960). Except for a limited number of terminated experiments (for example, Attneave and Arnoult, 1956), quantitative work with humans has been subject to similar limitations and greater confusion due to the intimate interaction with cultural artifacts. Consequently, the most primitive problems of shape vision, segmentation of the field into objects, defining locations and recognition of biologically relevant amorphous shapes, are implicitly ignored. It is precisely toward these problems that the process considered here is useful.

If it cannot be assumed that a shape can be isolated and normalized a priori, or that the wide variety of perturbations that a shape may take and still be identified, can be exhaustively enumerated, it is necessary to develop some way of obtaining shape properties without becoming enmeshed in the huge combinatorics of simple congruence views. A number of people have proposed to have the shape interact with itself to accomplish this. The Gestalt school (Koffka, 1935, for example) used field theoretic notions. Deutsch (1955), and Bitterman, Krauskopf and Hochberg (1954), have proposed propagation or diffusion models of interaction. Kazmierczak (1960) has used an electric field analogy and Kirsch, Cahn, Ray and Urban (1958) have used computer rules on a rectangular grid to allow a wide variety of self-interactions, including propagation. However, these models failed to arrive at a set of incisive shape attributes. The present model develops these notions somewhat differently and arrives at a number of distinct attributes which are useful in two ways: (a) in developing a physchology of shape for everyday life, and (b) in focusing attention on some of our biases. The model can be developed in a number of ways. The propagation scheme used has been chosen for its tutorial ease. Another view of it, perhaps having more physiological relevance, is introduced later. A word of warning is important. The model does not intend to deal with the central perceptual problem in which the entire history of the organism, its set, and its total stimulus input must be considered. The output of the transformation merely gives a set of shape attributes on which the organism can perform selection, using criteria of usefulness and relevance similar to those used in other sensory processes.

Definition Using Time

Consider a continuous isotropic plane (an idealization of an active granular material or a rudimentary neural net) that has the following properties at each point: (a) excitation – each point can have a value of either 0 or 1; (b) propagation – each excited point excites an adjacent point with a delay proportional to the distance; and (c) refractory or dead time – once fired, an excited point is not affected by a second firing for some arbitrary interval of time. A visual stimulus from which the contours or edges have been extracted, impinges on such a plane at some fixed time and excites the plane at those points. This excitation spreads uniformly in all directions but in such a way that the waves generated do not flow through each other. Figure 1 shows this process occurring for an input pattern of two points. The waves start as two expanding circles and merge into a single contour when these circles intersect. (One can visualize the contours as the front of a grass fire.) Figure 2 shows the generated wave for a number of other simple shapes. This wave passes by each point of the space once and only once. For most points of the space the wave flows by in a good fashion, each point on the wave front generating a new point on a new wave front. This is not the case, however, for those places where corners appear in the wave, or where waves collide with each other in a frontal or circular manner. The wave front undergoes cancellation at these places. These places are shown by the dotted lines in Figures 1 and 2. The two dots of Figure 1 give rise to a line which starts with an infinite velocity and becomes asymptotic to the space velocity. From Figure 2, it can be seen that such points occur only on the inside of an angle or an arc. The corner appears only after the center of curvature is reached. In the triangle three corners start propagating and disappear at the center of the largest inscribed circle. In the case of the ellipse, these points start at the shortest radius of curvature and disappear at the center of the largest inscribed circle. Since both of these shapes are convex, they generate no such

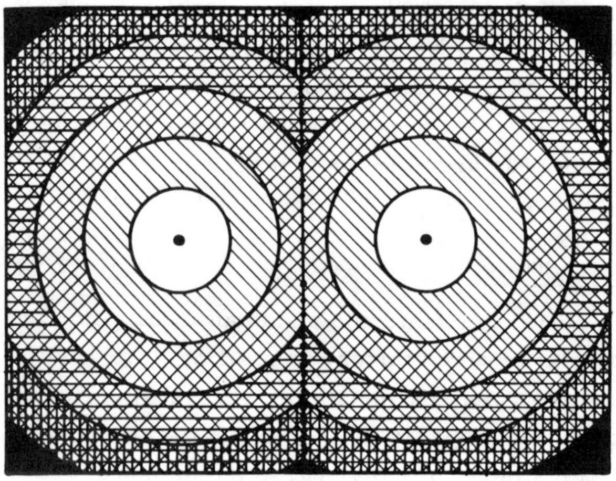

Fig. 1. Wave fronts generated by a two-point grass fire excitation. (Note the absence of flow-through associated with propagation of waves in a pond. The mid-line represents the locus of corners in this process.)

Fig. 2. Wave fronts and "corners" generated by some simple excitations (dotted lines represent locus of "corners").

points outside of themselves. A straight line and a point, since they are convex and have no inside, have no such points unless a special definition is used to include them. The circle is a special case whose description is a single point. The series of dots in a circle is included to show a closure property of the process. By setting a velocity threshold (this is equivalent to a smoothness criterion of the wave front), one sees that it is possible to have these points disappear and reappear at the center. In this way, a circle and a series of dots in a circle can be equated. Clearly, under these conditions, the process is not dimension preserving. Let us examine the process more carefully since these points along with the times or distances associated with them form an alternate description of a shape.

Figure 3 consists of 2 circular arcs and the tangents connecting them. A corner does not immediately appear in the wave front since it is a smooth figure. A corner appears at time t_1, propagates to the right and disappears at time t_2. Each stage of these expanding waves represents a figure parallel to the original one. The appearance of a single corner occurs at the minimum radius of curvature of the figure. The dissappearance of a corner represents a largest circle that can be drawn in the figure. All the corner points have a velocity associated with them that is determined by the angle formed – the sharper the angle, the faster the velocity. Since the parallel figures generated in Figure 3 all have the same angle, the velocity along this central line is a constant. An accelerating point would result from curvature toward this locus; a decelerating point would result from curvature away from this locus. The locus of corners will be called the "medial axis" (MA). When the times of corner occurrence on the medial axis is included, it will be referred to as the "medial axis function" (MAF).[1] As long as the MA is a straight line, the figure is symmetrical. Clearly an arbitrary parameter, such as parallel distance, could be substituted for time.

The transformation from the original pattern to the medial axis function description is unique and hence invertible. It can be seen that the original pattern can be described by the envelope of circles of proper radius (MAF time divided by the space velocity) associated with each point. This can be done more elegantly by the following inversion procedure. Excite the space along the MA at a time defined by the negative of the MAF. Thus, the later points of the function are pre-excited and allowed to expand by the proper amount before the earlier

---

[1] This locus has also been called the "skeleton" and the function on the skeleton, the "skeletal function."

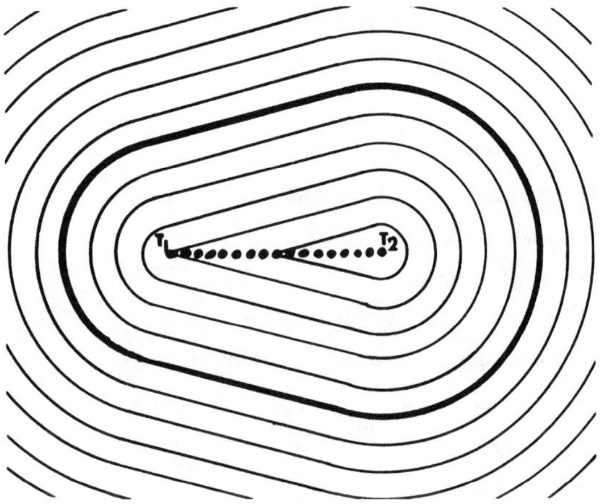

Fig. 3. Generation of new contour description (medial axis function) centered on enclosed space. (t_1 is appearance of corner, t_2 is disappearance. The locus of points and their times are required.)

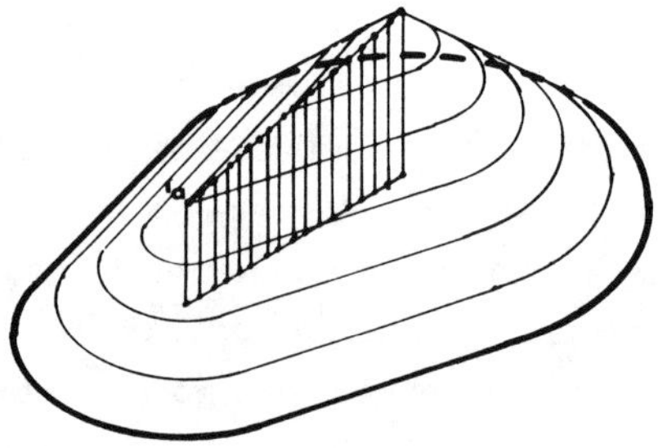

Fig. 4. A three-dimensional static alternative to the two-dimensional kinetic view of the process. (The MAF is the ridge formed where the union of cones on the input contour intersect each other.)

points of the function are excited. This procedure uses a reciprocity property like that of electromagnetic theory.

Alternate Definitions

There are a number of alternate ways of looking at the transformation, each of which sheds a somewhat different light on the process. Consequently, it is worth describing them. Figure 4 shows a formulation proposed by Kotelly (1963) in applying an earlier version of this process to the problem of pattern masking (metacontrast). Here, vertical distance is substituted for time. Instead of a propagating wave, one obtains a representation as a surface which is the envelope of cones whose apexes sit on the original contour. The MAF is then the locus of discontinuities in this surface. Still another way of looking at the process is from the viewpoint of field theory. The wave can be considered as a "nearest distance" field. The MAF is then the locus of discontinuities in the derivative of that field. Such a field departs sharply and in important ways from the conventional fields of physics and psychology, in which the value at a point is an integral of a distance function to the total input. Still one more, and I feel particularly interesting definition of it, is gotten from an equidistance viewpoint. Consider the (minimum) distance from each point of the plane to the stimulus shape. For most points this distance is unique to one point on the shape. For points on the MA, however, this distance is not unique to one point on the stimulus. Consequently, the MA is the locus of points equidistant from the pattern and in this sense represents a line of symmetry of the pattern. The MAF can be considered a symmetrical central description of the space whose boundary is the stimulus contour. It has been particularly satisfying to me to find that the process can be expressed in so simple a geometric notion. If one considers contours as coming from a world of objects, then these original contours are extremities of objects or transition points between objects. For even so simple a notion as location of an object, the contour or perimeter description is particularly poor.

Properties

Let us see what kind of categorizations are natural to such a process by considering a number of smooth, simply connected closed curves. The MAF of the circle and the ellipse were shown in Figure 2. Figure 5 shows the MAFs for a number of more undulating curves. The "pinch" in the upper left figure generates a double set of MA points which propagate in opposite directions. The pinches in the lower left "3-bulb" do the same. The outward moving points disappear, indicating closure; the inward points form a triple branch indicative of the three-sided structure of the figure. That the "3-star" on the lower right has such a triple structure is again shown by this three branch. At this point, it can be seen heuristically that by use of the directed graph of the MAF, it is possible to specify a generic 3-bulb, for example. This leads to a geometry whose character lies between topology, which is too general for shape categorization, and the "hard" congruent geometries such as Euclidean and projective geometry. Figure 6 shows such a geometry at work in describing an anthropomorphic outline and a gross distortion of it. It should be pointed out that since these figures are not convex, they have MAF points which lie outside themselves. The outside MAFs are generated only by the points of the contours that are <u>not</u> on the convex hull of the contours.

Since the contour and the MAF description are equivalent and invertible there is no information lost or gained in the transformation. It is clear, then, that this categorization could have been done on the contour itself. What then is the virtue of the new description? I believe

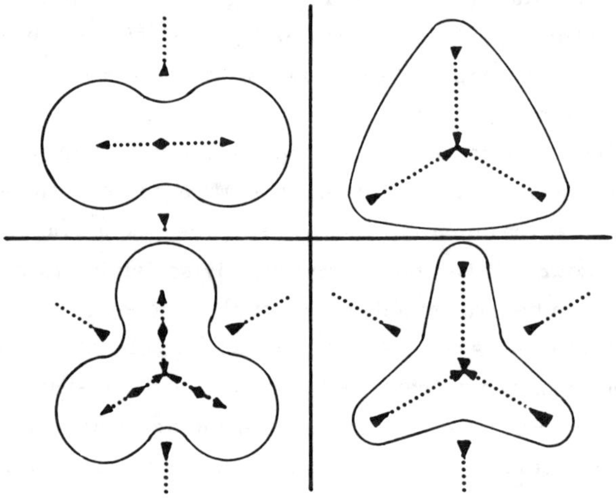

Fig. 5. Representation of simple shape properties by directed graph structure of the MAF.

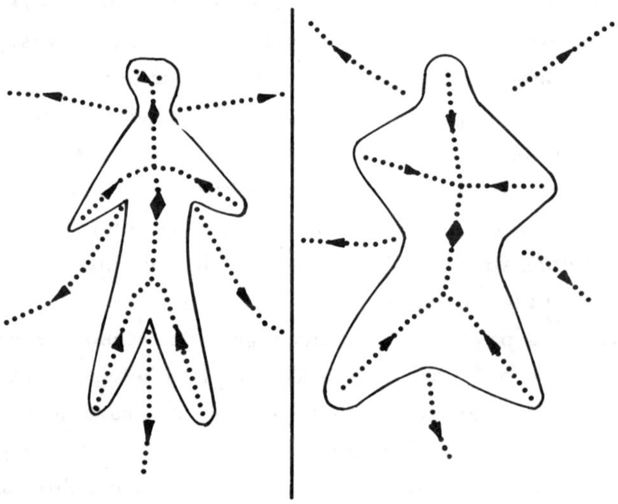

Fig. 6. Two anthropomorphs and their MAFs.

it to be in defining natural shape properties on the new primitive descriptors. It could be argued that the properties I have discussed in the last paragraph could be gotten by such simple contour properties as counting the inflection points and the total curvature of the shape.

Figure 7 shows a case where simple equivalent descriptors cannot be found on the boundary. The upper and lower figures can be arranged to have the same number of inflexion points and the same total curvature; yet they are quite different visual figures. What is different about the figures has to do with the distance across the enclosed space. Carrying along such a distance function to all sorts of other points across the contour as one moves along the contour is a formidable task indeed. To try to deal with such properties from a boundary description such as the intrinsic co-ordinates, for example, implies that one cannot describe a swan's neck until one has gone completely around the swan. This point is illustrated in a different way in Figure 8. The upper figure shows a rectangle and its MAF. The heavy dots indicate points which represent parallel lines and so occur at the same time (infinite velocity). The middle figure shows this rectangle distorted perspectively. It can be seen that the velocity of the central element, a local property, becomes the cue to perspective angle if one assumes one is looking at a rectangle. The lowest figure has been included to emphasize that all is not roses with this theory. For extreme distortions, the MAF structure stays the same but the perspective cue changes. Note that in going from the top to the bottom figure, one passes through a degenerate state when the figure can be circumscribed about a circle. It is an interesting conjecture, that with proper constraints, the MAF structure cannot change without going through such a degenerate state.

An interesting question of continuity is raised in Figure 9. In both parts of this figure the outside shape is continuously transformed to the circle, but by different paths. The intermediate shapes lead to distinctly different MAs. In the upper figure, the MA is continuously deformed from that of the original figure to that of the circle, the central dot. In the lower figure, however, the limiting MA is a radius of the circle. The velocity of the lower MAF, however, becomes asymptotic to the space velocity. Consequently, a viable notion of

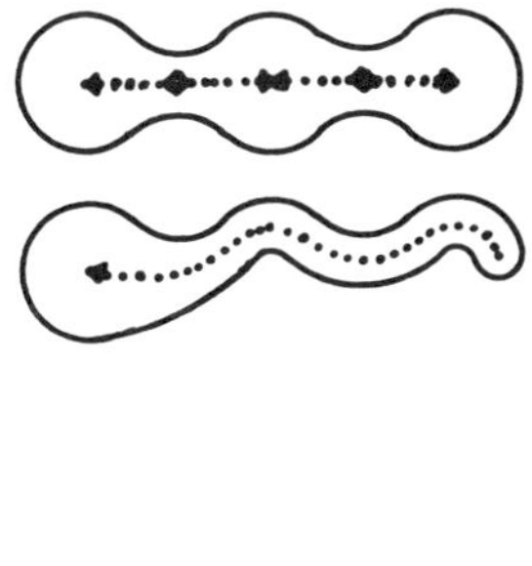

Fig. 7. The width measure of the shape as a primitive property.

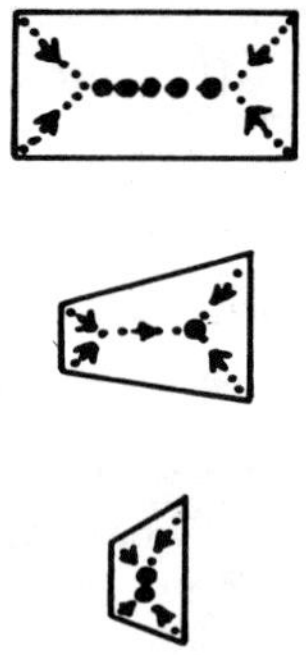

Fig. 8. The MAF for a rectangle "full face" and in perspective.

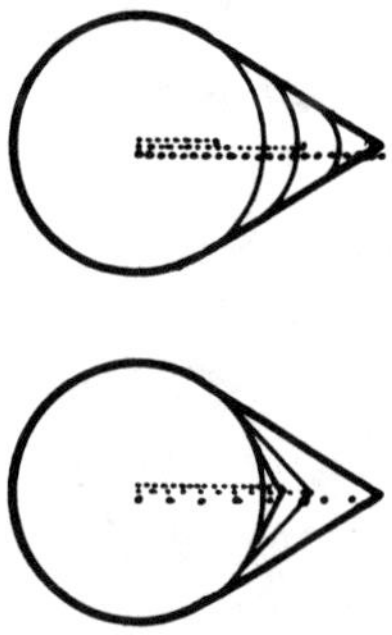

Fig. 9. Dependency of MAF limit on path of curve perturbation.

continuity must include the velocity dimension. This dimension is also important for the closure property introduced in the circle of dots in Figure 2 and in the completing of masked figures discussed later.

Figure 10 illustrates the process considered as a co-ordinate transformation. The upper and middle figures depict symmetrical contours and so have a straight-line MA. The new description consists of the location of the MA point as a function of time (or length of the normal) to the MA. Acceleration represents curvature to the axis; deceleration, away from the axis. An explicit formulation can be given for curvature in terms of velocity and acceleration of the MAF point. The condition that no branching occurs at a point on the MA generated by two contours is that the center of curvature of the points mapping into it do not lie between the MA and those points. (The condition of two contours is inserted since a third contour may enter for curves with their centers of curvature on the side opposite the MA such as is shown in the case of the 3-bulb of Figure 5. The lowest figure shows a curved MAF. In this case the contours are described by two components, the MAF curvature and the acceleration curvature – one curve

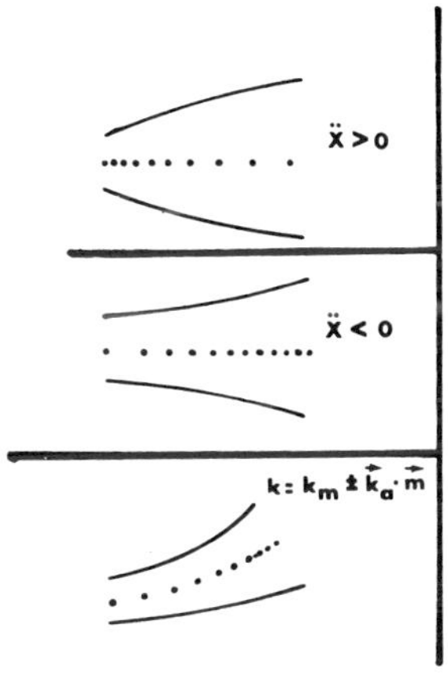

Fig. 10. The MAF as a co-ordinate transformation.

being the sum and the other being the difference when proper components of these curvatures have been taken. Consequently, a curve has an infinite number of descriptions since it can be described from a wide variety of MAs, and the location of the MAs is determined by what else is in the input field. Figure 11 illustrates this. The upper left figure is a semiellipse, its MA being its evolute continued by the perpendicular bisector of its end points. The upper right figure shows the change in form of description gotten by completing the ellipse. The lower left figure shows a descriptor appearing outside the elliptic arc when space is enclosed there, and the altered description obtained by putting a barrier line underneath the elliptic arc. These illustrate clearly the barrier effect of the process. Only the closest contour can affect the MAF. This guarantees that a description of a shape represents the bounding contour alone. Conventional field theories alter the relevant description by modifications due to outside contours both far and near. As such, these and other complete interaction processes, such as two-dimensional autocorrelation, are certain to become unwieldy as the complexity of the input goes up. I feel that this type of segmentation property is essential to a realistic theory of shape.

Figure 12 shows an abstract generalized input. Except for special cases, which complicate the description, but not the properties, the transformation maps each point on the contour into a 2 MAF points, and each MAF point results from 2 contour points. (To make the statement general, it is necessary to consider ray paths rather than contours and MAFs.) Gross properties of the MAF are equivalent to gross properties of the pattern. Each connected space is represented

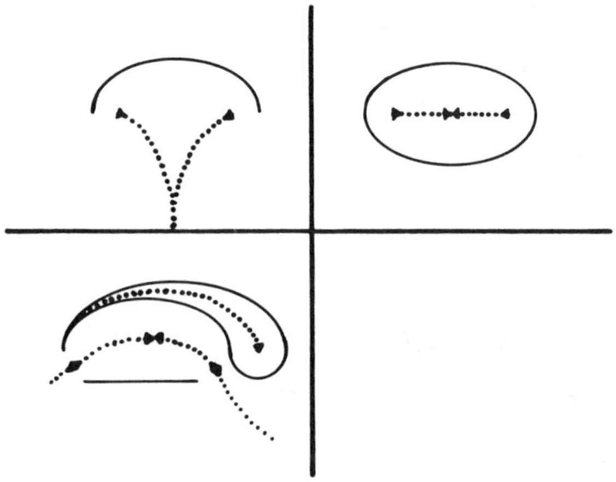

Fig. 11 . Nonsuperposability of partial descriptions and barrier effect.

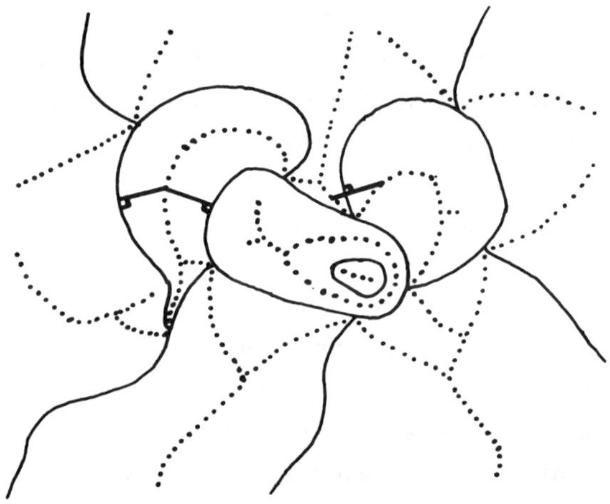

Fig. 12. A generalized input showing MAF as limit of space seen from contour.

by a separate graph piece. A connected space with no holes in it is represented by a "tree," each hole in a space being represented by a loop. (Topological properties of the space are thus available as primitive properties of the graph.) The order and connection of the expanding spaces is represented by the structure of the directed graph. Since the MA represents discontinuities in the wave, the discontinuities of the MAF (appearance, disappearance, impulse, and branch points) can be considered as second-order discontinuities of the process. These can be categorized by a number pair in which the first number is the number of MAF points immediately prior to the discontinuity and the second, the number immediately after the discontinuity. That each of these represents a particular natural figure property is shown by Figure 13. A 0,0 represents a circle; a 0,1 a circular arc of less than 180 degrees; a 1,0 a circular arc of greater than 180 degrees; a 2,0 a double closure such as in an ellipse; a 0,2 a pinch in the contour; a 1,1 a case where the center of curvature lies on the MAF, etc. Simple constraints of geometry, such as the closest distance between three points being between two of them, impose a number of constraints on the possibility of such pairs, which leads to forbidden pairs and pairs possible only at t = 0. Figure 14 illustrates the equivalent matrix of the graphical description. Graph discontinuity properties go into the diagonal, continuous properties into the off diagonal. Each term in the matrix is a vector, in general. Finding good vectors cannot in general be done theoretically but must be decided by new experiments on visual shape. This may be a fruitful area of exploration.

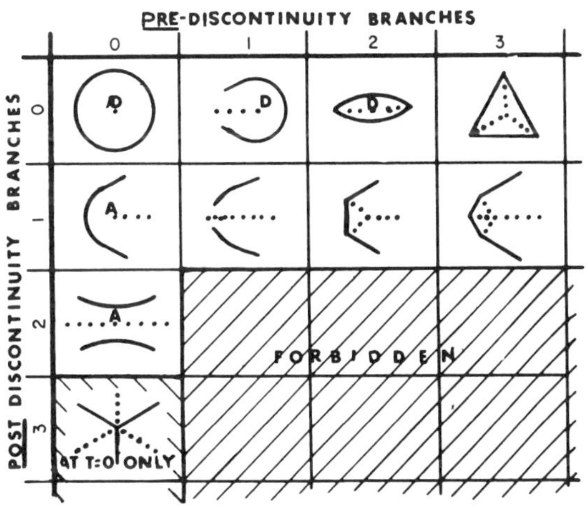

Fig. 13. Shape equivalents of discontinuities on MAF.

Compound Shapes

Thus far we have been considering shapes merely as spaces on a plane. Figure 15 contains an illustrative example for showing an extension of this process for shapes which are masked by others. In the

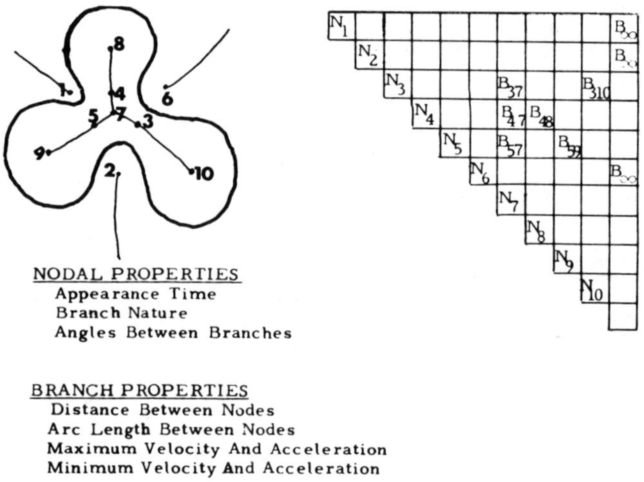

Fig. 14. MAF description allows analysis by matrix of graph.

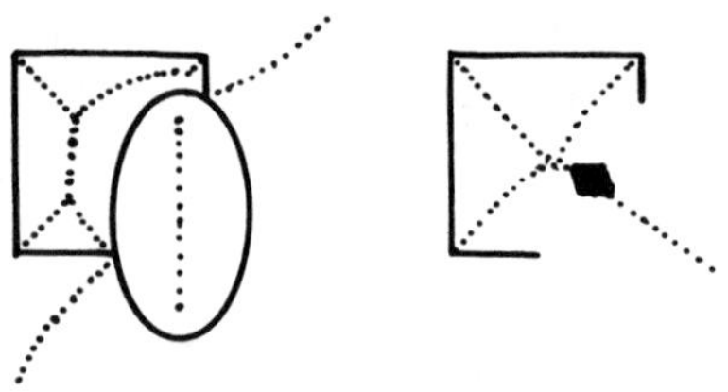

Fig. 15. Extension of MAF to masked shapes.

left-hand figure, an ellipse and partially masked square are shown, the masking considerably complicating the MAF of the square. If the MAF of the ellipse is recognized and its contour removed after its recognition, as shown in the right figure, the representation is considerably simplified. (This can be done from the MAF function by presenting it in negative time so that the ellipse is formed at t = 0. If further propagation is stopped, the input can be made refractory at the locations of the ellipse.) The remaining square contour can easily be recognized. In this way, one can attempt to cancel sequentially the entire stimulus contour and equate this cancellation with an acceptable interpretation of it. Clearly, alternate sequences are possible. A number of properties of such a sequential process are particularly interesting. (a) Masking cues are primary outputs of the process. (b) There is an inherent figure-ground effect that comes from matching MAF descriptions rather than the contour itself. (c) There is an asymmetry of the process with respect to additive and subtractive noise. Additive noise, such as a masking object, must be explicitly removed. Subtractive noise, such as a missing contour, is implicitly filled in. Figure 16 shows a stylized animal behind a tree. If the tree is recognized by any cues, such as the green of its leaves and brown of its trunk, its contour can be removed. This facilitates making a single object of the animal. Such an object completion property may be an extremely important one to vision; a cat must see through grass, a monkey through leaves.

Fig. 16. Making a contiguous world of masked objects.

Psychology and Physiology

Although the transformation has been developed to deal with the visual problems of everyday life, it is interesting to see how it checks with the anomalies of vision, the illusions. In Figure 17, I have shown a number of standard illusions and the MAFs associated with these. In the right-hand figures, I have accented the faster velocity MAFs. If one assumes that the visual process tries to line up MAFs, (these represent the spaces or objects) rather than the contours themselves, the illusions have a natural meaning. The figures on the left are self-explanatory, I feel. The MAF opens a new range of processes for explaining the standard illusions. I feel that it goes much deeper than that, however. The process allows us to take a much broader view of shape vision. We have been looking at vision as a recognition process, rather than as a segmentation process. The closure properties suggest that the appearance of a dotted figure as a continuous contoured one (Figure 2) is an illusion. The process suggests that the whole of our sensitivity to circles, parallels, and symmetric figures may be an illusion. If one asks what life and death problem such sensitivities have emerged to solve, one finds no ready answer. It is suggested here that these are by-products of an inverting process of the visual co-ordinates useful for coping with the problems of localization, segmentation, and organization of a world of visual objects.

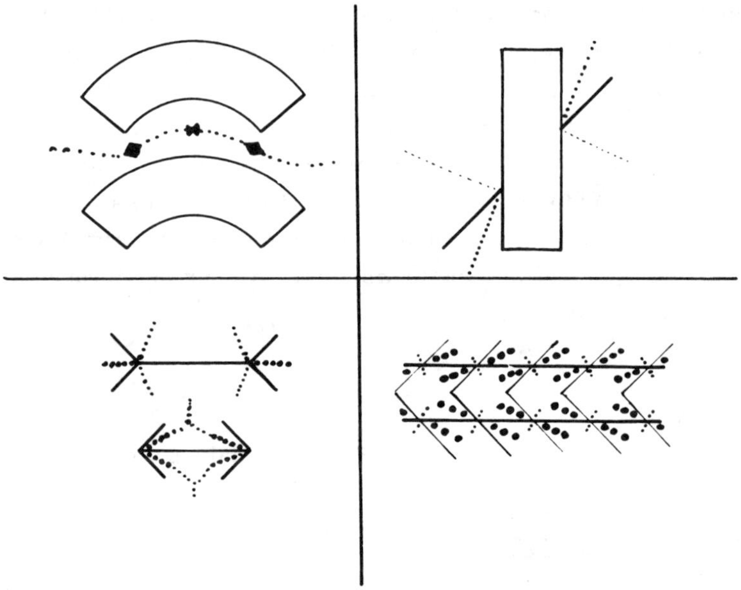

Fig. 17. The MAF as a source of new explanations for optical illusions.

A word must be said about how such a process relates to the results of Hubel and Wiesel (1962, 1965) on cortical cell responses. This has been abstracted and stylized in Figure 18. The upper-left figure shows the receptive fields of the "simple cells" that would be fired by a particular contour. The upper right shows the receptive fields of their "complex cells." Whereas Hubel and Wiesel have looked at these responses as detectors of lines at particular angles, I am suggesting that they can be looked at as generators of parallel shapes. The lower left shows the kinds of corners that could be formed from the interaction of the output of such cells. The lower right shows a selection of proper "symmetrical corners" obtained by having a cell that responds to a corner at a particular location, but not in an adjacent location. Such a sequence can perform the transformation without the barrier property. This requires a further condition. Unfortunately no one has looked for such a property, as far as I know.

General Comments and Conclusion

A word should also be said about the mathematical implication of the transformation. (It should be clear that I speak for mathematicians

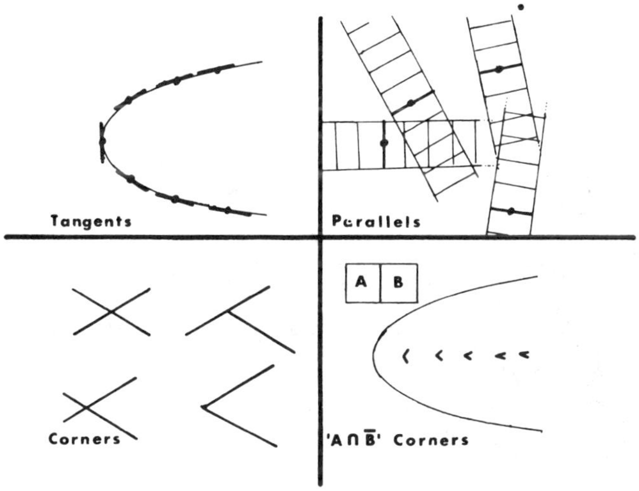

Fig. 18. MAF extraction by idealized properties of visual cortical cell receptive fields.

here without their permission.) The general Cartesian view of geometry metricizes a space (builds a co-ordinate system) and then describes the curve in that metric in some functional form. Unfortunately, the transformation from geometry to analysis is highly discontinuous and gives preference to shapes which have easy functional rather than geometric meaning. The approach introduced here suggests that the co-ordinate system be built on the curve and space be explored from the curve. The MA is then the boundary of the space (in some Riemann sense) that must exist if one is to have a good single valued co-ordinate system. The process indicates that the curve is then describable from the boundary as the envelope of circles on that boundary. I have talked hitherto about a Euclidean orthogonal process, but there is no need to impose such a limitation. This transformation can be done with non-Euclidean processes, as the spherical visual mapping may well demand, or it can be done in higher dimensions, as a three- or four-dimensional world may demand.

Figure 19 shows my first physical embodiment of the process. It uses a movie projector and camera with high contrast film. These are symmetrically driven apart from the lens in such a way as to keep a one to one magnification, but to increase the circle of confusion (defocussing). The lower right inset shows how this dilates the contour. Corner detection is done separately by a subsequent process. I am presently building a closed loop electronic system to do both the wave

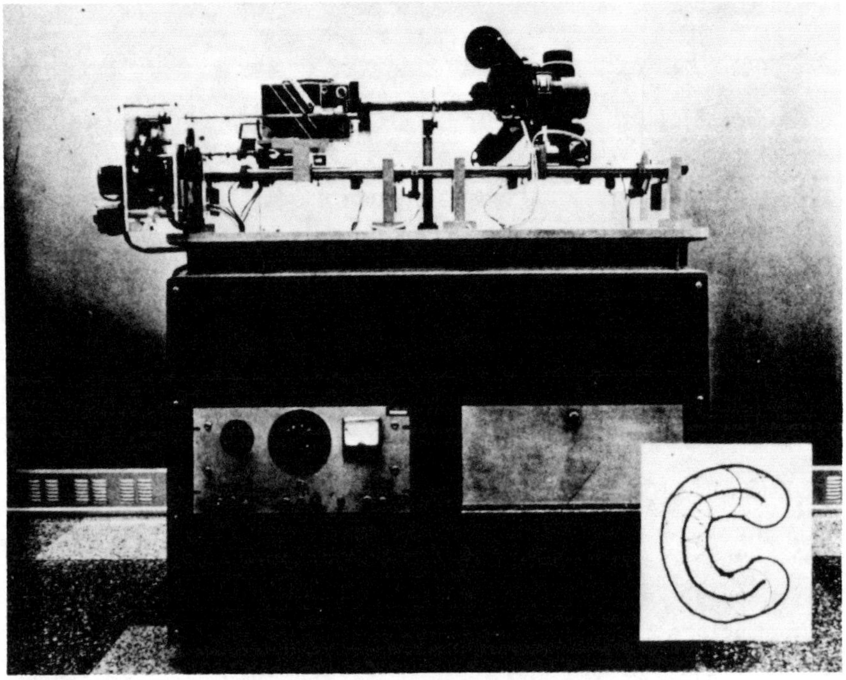

Fig. 19. An early opticomechanical device for obtaining grassfire propagation using photographic defocussing.

generation and corner detection.

In conclusion, I would like to emphasize that our view of vision has been constrained for too long by problems of physical geometry. Successful methods there have looked at shape as a superposable process where the total collection of elements is equal to the sum of the parts. I have tried to depart from this by looking for processes in which the interaction is precisely the item of interest, without getting into a conceptual space so rich in possibilities that all things are encompassable in it.

References

Attneave, F., and Arnoult, M. D. The quantitative study of shape and pattern perception. Psychol. Bull., 1956, 53, 452-471.

Bitterman, M. E., Krauskopf, J., and Hochberg, J. Threshold for visual form: A diffusion model. Amer. J. Psychol., 1954, 67, 205-219.

Blum, H. An associative machine for dealing with the visual field and some of its biological implications. In E. E. Bernard and M. R. Kare (Eds.), Biological prototypes and synthetic systems. Vol. 1. New York: Plenum, 1962.

Deutsch, J. A. A theory of shape recognition. Brit. J. Psychol., 1955, 46, 30-37.

Hubel, D. H., and Wiesel, T. N. Receptive fields, binocular interaction and functional architecture in the cat's visual cortex. J. Physiol., 1962, 160, 106-164.

Hubel, D. H., and Wiesel, T. N. Receptive fields and functional architecture in two non-striate visual areas (18 and 19) of the cat. J. Neurophysiol., 1965, 28, 229-289.

Kazmierczak, H. The potential field as an aid to character recognition. In Proceedings international conference on information processing. London: Butterworth, 1960.

Kirsch, R. A., Cahn, L., Ray, C., and Urban, G. H. Experiments in processing pictorial information with a digital computer. East. joint Comp. Conf., 1958, 12, 221-229.

Koffka, K. Principles of Gestalt psychology. New York: Harcourt Brace, 1935.

Kotelly, J. C. A mathematical model of Blum's theory of pattern recognition. AFCRL-TR-63-164, USAF Camb. Res. Labs., 1963.

Sutherland, N. S. The methods and findings of experiments on the visual discrimination of shape by animals. Exp. Psychol. Monogr. No. 1. Cambridge, Eng.: Heffer, 1960.

Hexagonal Parallel Pattern Transformations

MARCEL J. E. GOLAY, FELLOW, IEEE

Abstract—**The concept of the two-dimensional (2-D) parallel computer with square module arrays was first introduced by Unger. It is the purpose of this paper to discuss the relative merits of square and hexagonal module arrays, to propose an operational symbolism for the various basic hexagonal modular transformations which may be performed by these comupters, to illustrate some logical circuit implementation, and to describe a few elementary applications.**

Index Terms—**Blowing up and shrinking, hexagonal, modular fields, nearest neighbor logic, parallel computer.**

Introduction

THE 2-D parallel computer of Unger[1] was based on a square array of modules, each module having four immediate or primary neighbors at the distance unity, four secondary neighbors at the distance $\sqrt{2}$, four tertiary ones at the distance 2, eight quarternary ones at the distance $\sqrt{5}$, etc.

Such computers may be well suited for programming the studies of games such as checkers, chess, and Go, or for performing operations specified with definite reference to two mutually perpendicular directions, but are more cumbersome than needed for general problems of pattern transformation and classification which are not firmly tied to a preassigned system of cartesian coordinates.

An important logical difficulty is that created by the concept of connectivity which is fundamental to several pattern transformations, and which is based on nearest neighbors logic. For instance, in Fig. 1, let the modules at P, A, and B be in the state **1**. The binit **1** indicates that these modules coincide with elements of the image of an object which constitutes an input in the computer, and the connectivity between A and B is established by the primary neighborhood of the **1**'s in P and A, and P and B. The question is now raised whether this connectivity may be affected by a transformation which replaces the **1** at P by a 0. This question is obviously unanswerable unless it is known that E is in the state **1**, or unless it can be stipulated that the secondary neighborhood of the **1**'s at A and B maintains this connectivity. In either case this implies that P also should be aware of the states of its secondary neighbors at E, F, G, and H. Yet, these points being more distant than the immediate neighbors, their importance should somehow be of a lower order. Thus, a logical decision taken about P should consider the states at eight neighboring points,

```
E       B       F

A       P       C

H       D       G
```

Fig. 1. Square module array.

```
    B       C

A       P       D

    F       E
```

Fig. 2. Hexagonal module array

four of a first order of hierarchy, and four of a second order. This, of itself, implies formidable logical complications.

Consider, by contrast, the greater simplicity of the hexagonal array[2] illustrated in Fig. 2 in which P is surrounded by six equally distant neighbors at A, B, C, D, E, and F. Let the modules at P, A, and B be in state **1** which indicates the presence of image elements, while C, D, E, and F are in the state 0. It is immediately apparent that the connectivity of A and B is not impaired by an operation which would replace the **1** of P by a 0. On the other hand, had the P, A, and C modules been in state **1**, while B, D, E, and F were in state 0, the connectivity of A and C through P may obviously be affected by the placement of P in state 0. Thus, the knowledge of the states of six immediate neighbors is sufficient to make decisions about connectivity for which the knowledge of four nearest and also four more distant neighbors is required in a square array.

It can be noted also that a greater angular resolution is offered by the hexagonal array. Thus, from any one module there are six directions to the nearest neighbors, spaced 60° apart, and these directions are bisected by the six directions to the next nearest neighbors, after which unequal angular divisions are effected by the

Manuscript received February 9, 1968; revised June 24, 1968 and April 17, 1969.

The author is with the Perkin-Elmer Corporation, Norwalk, Conn.

[1] S. H. Unger, "A computer oriented toward spatial problems," *Proc. IRE*, vol. 46, pp. 1744–1750, 1958, and S. H. Unger, "Pattern detection and recognition," *Proc. IRE*, vol. 47, pp. 1737–1752, 1959.

[2] The hexagonal array has been proposed briefly by B. H. McCormick in "The Illinois pattern recognition computer—ILLIAC III," *IEEE Trans. Electronic Computers*, vol. EC-12, pp. 791–813, December 1963, where it is referred to as the rhombic array.

Reprinted from *IEEE Trans. Comput.*, vol. C-18, pp. 733–740, Aug. 1969.

TABLE I

Pattern	0 0 0 + 0 0 0	0 1 0 + 0 0 0	1 1 0 + 0 0 0	1 1 1 + 0 0 0	1 1 1 + 0 1 0	1 1 1 + 0 1 1	1 1 1 + 1 1 1	
Index	0	1	2	3	4	5	6	
Weight	1	6	6	6	6	6	1	Σ = 32
Pattern	0 1 1 + 0 0 1	1 1 0 + 0 1 0	1 1 0 + 0 0 1	1 1 1 + 0 0 1	0 0 0 + 1 1 0	0 1 0 + 0 1 0	1 0 1 + 1 0 1	
Index	7	8	9	10	11	12	13	
Weight	2	6	6	6	6	3	3	Σ = 32

directions to the modules at distances $\sqrt{7}$, $\sqrt{13}$, etc. This must be contrasted with the four directions to the nearest neighbors of module in a square array which are bisected by the four directions to the next nearest neighbors, after which unequal angular divisions are effected by the directions to the modules at distances $\sqrt{5}$, $\sqrt{10}$, etc.

It should be pointed out that only three types of planar, symmetric, isotropic point arrays exist, namely the square and hexagonal arrays referred to above, and a third one, the triangular array of points in a field which can be formed by removing one-third of the points in a hexagonal field, such as the **1**'s in the array of Fig. 3. Such an array appears to be the one which is the least suited to nearest neighbor logic transformations, for in order to determine whether the removal of, e.g., a 2 destroys the connectivity of two neighboring 3's, it is necessary to know the states of that 2's three nearest neighbors, six neighbors at distance $\sqrt{3}$, and three neighbors at distance 2.

```
  3   1   2   3   1   2
1   2   3   1   2   3   1
  3   1   2   3   1   2
1   2   3   1   2   3   1
  3   1   2   3   1   2
```

Fig. 3. Subdivision of a hexagonal field of modules into three subfields of nonneighboring modules.

Modular Operations and Symbols in Hexagonal 2-D Parallel Computers

A module can be in the state 0 or **1**, and given the state of any one module of a hexagonal group and the states of its six nearest neighbors, a simple modular operation may serve either to change the state or not change the state of this module as a function of its state and of the state of its six neighbors. It will be considered, in general, that a module in state **1** represents one element of some image or pattern, e.g., a well-defined line to be thinned or swollen, a formless "blob" to be counted or inspected for holes, etc., and the interest will be in that image rather than in the emptiness with which it is surrounded. Simple modular operations may serve to swell this image without permitting it to become connected to others, or to fill holes it may have, or again they may serve to shrink a "holeless" image to an open-ended chain, or to a point, or to shrink images with holes to networks. These simple operations will be classified tentatively as shrinking or swelling operations pending the introduction of more general ones.

The $2^6 = 64$ possible patterns of 0's or **1**'s forming the immediate neighborhood or surround of a module may be divided into a first set of 32 patterns in which there is at most a single group of adjacent **1**'s and a second set of 32 patterns in which there are more than one group of adjacent **1**'s, as shown in Table I.

The weights indicate how many distinct similar patterns can be formed by merely rotating the pattern shown, and it will be noted that the sum of the weights of all patterns is $64 = 2^6$, i.e., the total number of possible patterns. The patterns 8 and 9 are the only asymmetric patterns. They can be obtained from each other by mirror imaging and are kept as separate patterns in order to permit the distinction of clockwise or counterclockwise rotation. It is noteworthy, and in a sense gratifying, that six is the minimum number for which a cyclical sense can be specified by means of binits on a circle which read differently whether traversed clockwise or counterclockwise, this cyclical sense being specifiable by a choice between one or the other of the single pair of mirror imaged patterns, 8 and 9.

The two sets of patterns are sharply separated by the concept of connectivity, for in the first set the addition or suppression of a **1** in the center cannot alter the connectivity of the **1**'s of its surround, whereas it can in the second set. Therefore, it can be anticipated that where connectivity invariant operations are concerned, shrinking or swelling operations will be of interest in connection with the first set only, whereas more general operations may be of interest in connection with both the first and second set.

It must be noted also that insofar as this study is concerned, all patterns which can be derived from one of any index by rotation will be considered as equivalent and completely characterized by their common index. This means that all the operations to be examined in what follows will be isotropic in the sense that they will be invariant against rotation of the entire field by 60°, 120°, etc.

It is necessary that all operation in 2-D parallel computers be static in the sense that two neighboring modules, A and B, should never be operated on simultaneously as a function of each other, as this could introduce the logical difficulty that the computer would not be able to decide, e.g., whether to use the old state of A when determining the surround of B, and then the new state of B when determining the surround of A, or vice versa. This difficulty is resolved by assigning an ordering sequence to A and B, i.e., by letting the computer use, say, the first of the two alternatives just stated. A ready mechanism for carrying out this ordering and maintaining the "staticity" of the operation consists of subdividing the hexagonal field into three subfields as illustrated in Fig 3 (no two modules of any one subfield being ever adjacent), and then operating on all the modules of each subfield simultaneously, taking one subfield after the other.

There will be cases when a hexagonal field of modules should be even more finely divided, and Figs. 4 and 5 show how such a field can also be subdivided symmetrically into four and seven subfields, respectively, each subfield always having the same relationship to every other.

1 2 1 2 1 2 1

3 4 3 4 3 4

2 1 2 1 2 1 2

4 3 4 3 4 3

1 2 1 2 1 2 1

Fig. 4. Subdivision of a hexagonal field of modules into four subfields of nonneighboring modules.

1 2 3 4 5 6 7 1 2

4 5 6 7 1 2 3 4 5

6 7 1 2 3 4 5 6 7

2 3 4 5 6 7 1 2 3

4 5 6 7 1 2 3 4 5

7 1 2 3 4 5 6 7 1

2 3 4 5 6 7 1 2 3

5 6 7 1 2 3 4 5 6

Fig. 5. Subdivision of a hexagonal field of modules into seven subfields of modules in which each is surrounded by the six others.

Consider now the simple shrinking (or emptying, or skeletonizing) operation defined by the symbol

$$M^{3}_{1-3,n}[k = \overline{a_{i(k)}} \cdot k].$$

The subscript of a, $i(k)$, is the index of the surround of any element of the field k about to be operated on, and the first subscript of M indicates that whenever the index is 1, 2, or 3 as defined in Table I above, we have

$$a_{i(k)} = 1 \quad \text{and} \quad \overline{a_{i(k)}} = 0,$$

whereas, whenever that element of the k field is surrounded by a pattern of index 0 or 4, 5, $\cdots$, 13, we have

$$a_{i(k)} = 0 \quad \text{and} \quad \overline{a_{i(k)}} = 1.$$

The second subscript n indicates the number of times the complete operation is performed. The superscript 3 indicates that each complete operation consists of the three partial operations performed successively on the 3 subfields of k shown in Fig. 3 for the sake of staticity, as just explained. The equation inside the bracket indicates that whenever $a_{i(k)} = 0$ for any module of the subfield operated on, that module is placed in state 0; otherwise, when $a_{i(k)} = 1$, that module is left unchanged. Thus, k before the = sign refers to the next state of a module, whereas, after the = sign, it refers to its former state. If that operation is repeated until no further changes occur in k, that is, until $a_{i(k)} = 0$ for all modules, which is indicated by making $n = \infty$, all simple blobs of k will have been shrunk to a single 1, while blobs with holes such as rings, figure eights, rings within rings, etc., will have been reduced to one or more loops of thinness unity. Finally, the semicolon separates the operation just defined from a next operation.

It will be noted that the letter k is used alternately to denote the field or to represent the 0 or 1 state of any module of that field, as no ambiguity has been found to result from this double use.

Consider next the simple swelling (or filling, or blowing up) operation defined by the symbol

$$M^{3}_{3-5,\infty}[k = a_{i(k)} + \overline{a_{i(k)}} \cdot k].$$

The first subscript of M indicates that whenever the index of the surround of any module of the field k is 3, 4, or 5, $a_{i(k)} = 1$; otherwise, $a_{i(k)} = 0$. The second subscript of M, ∞, indicates that the complete operation is repeated until there are no surrounds left of index 3, 4, or 5 in the k field. The superscript 3 indicates as before that the parallel operation is performed for one of three subfields at a time. And the presence of the $a_{i(k)}$ term in the left-hand side of the equation within the brackets indicates that whenever a module of the subfield of k operated on has a surround for which $a_{i(k)} = 1$, that module is placed in state **1**; otherwise, it is left unchanged. The result of this operation is to cause all the blobs of field k to smooth their outsides until they are convex everywhere, or until the proximity of another blob prevents any further swelling, while any simple hole in a blob shrinks to a single 0.

It is appropriate to indicate at this point that any field must be assumed framed by 0's which serve to define the surrounds of the modules on the edge of the field, but which are never operated on themselves; they remain 0's at all times.

With this preliminary we may now generalize the concept of modules and consider more general operations which may be performed with these. For this purpose, we shall consider the hexagonal field to be made up by the overlay of a multiplicity of layers, $k, p, q, \cdots, r, s$, each with the same number of what will now be termed submodules, and all overlayable on each other. Then, on the one hand, all homologous submodules located along one vertical line together with certain additional logical elements will constitute one module of the hexagonal field. On the other hand, all the submodules located at the same height will constitute one of the layers.

The entire computer will consist of:

1) the modules, organized in fields, all the submodules forming part of one module being interconnected and connected to the neighboring modules, as well as to certain busses;

2) the "brain" of the computer which will send over busses the commands required by the various submodules for this or that modular operation; and

3) input and output organs, the existence of which will be merely stipulated in what follows.

The general modular operation will be designated by

$$M_{a_0a_1\cdots a_{13},n}[k = a_{i(L_1)}\cdot L_2\cdot L_3 + \overline{a_{i(L_1)}\cdot L_2}\, k] \qquad (1)$$

where M stands for the general modular operation. The a_0, a_1, etc., are unity whenever their suffixes 0, 1, etc., are also the indices of those of the 14 surrounds listed in Table I for which a modular operation will be performed; otherwise, they are zero. Thus, in the case of the shrinking operation considered earlier, M would carry the suffixes $M_{01110000000000,n}$, and may be written simply $M_{1-3,n}$ for short. n stands for the number of times a given operation is to be repeated, and the symbol ∞ in place of n indicates that the modular operation should be repeated until there are no more changes in the field operated on. When $n=1$ it will merely be left out. Most operations examined later will be with either $n=1$ or $n=\infty$.

Inside the square brackets in (1), k represents the module of the layer being operated on, the new value of k being on the left-hand side of the $=$ sign, and the old one on the right-hand side. L_1 stands for a logical function or one or more of the submodules constituting the surround of the submodule being operated on. Thus, we may have $L_1 = k$, as in the case of the shrinking or swelling operations examined earlier. Alternately, we may have $L_1 = kr$, i.e., L_1 is the output of an AND gate to which are fed the states of homologous submodules in the k and r layers. More generally, we shall have $L_1 = L_1(k, p, q, \cdots, r, s, \cdots)$. $i(L_1)$ is the index of the L_1 surround of the module being operated on, so that $a_{i(L_1)} = 1$ if the L_1 surround is one of those whose index is listed in the subscript of M; otherwise, we have $a_{i(L_1)} = 0$.

L_2 corresponds to one of the three, four, or seven subfields into which the entire field is subdivided. Thus, L_2 is a wired-in command communicated by the brain to all modules sequentially, having $L_2 = 1$ for, e.g., one-third of all modules at one time, while having $L_2 = 0$ for the other two-thirds. This serves to preserve staticity or may serve other purposes. L_2 will be indicated as 1/3, 2/3, or 3/3 when the vertical subdivision in three's is utilized, and similarly for the subdivisions in fours or sevens, and a superscript of M, M^3, M^4, or M^7 will be utilized instead of L_2 to indicate that the entire modular operation must be performed once for each of the 3, 4, or $7L_2$ subfields when $n=1$, and for each in turn until there is no more change in the subfield operated on when $n=\infty$. L_3 stands for either a central command or a logical function of certain layers such as $L_3 = L_3(p, q, \cdots)$.

The requirement of staticity demands that $p, q, \cdots, \neq k$. Furthermore, if L_1 also is not a function of k, the modular operation becomes a mere "marking" operation, and can be performed simultaneously for all modules of the k field without the need for vertical subfields of 3, 4, or 7.

IMPLEMENTATION

The visualization of the general modular operation described above may be helped by figures showing the logical circuits required for a single composite module of the field, which consists of several submodules and some auxiliary logical functions.

Fig. 6 illustrates a submodule of the k layer and its inputs and outputs. The state of the submodule is stored in the double stability circuit shown at the top of the figure. This double stability circuit consists of two NAND gates and two auxiliary OR gates. It has two outputs, k and $\bar{k}$, which are fed to other parts of the module, as well as to the output organs of the computer which are not discussed here.

The inputs into the double stability circuit comprise the preset inputs shown as dotted lines, which can be used initially to place an input image into the kth layer by means not discussed here, as well as other inputs organized to perform the logical function defined in the right-hand side of [1], which are as follows.

The d_k input is fed by the computer brain, and is a **1** whenever the k layer is selected for a modular operation. Thus, d_k is fed in parallel to all submodules of the k layer.

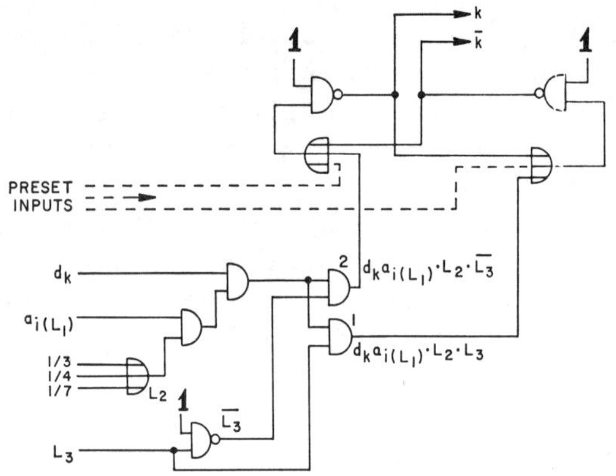

Fig. 6. Submodule of kth field.

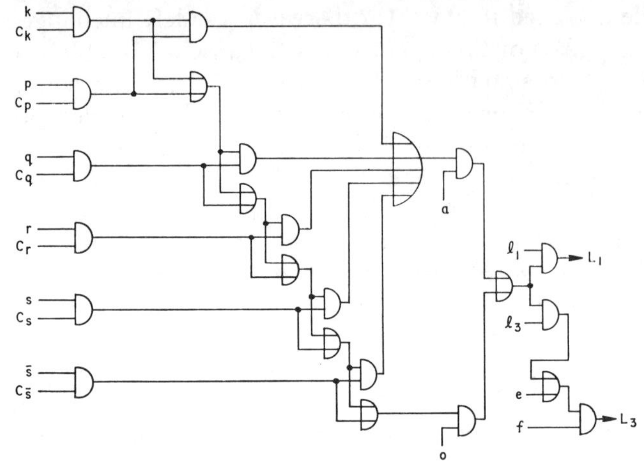

Fig. 7. Formation of L_1 or L_3.

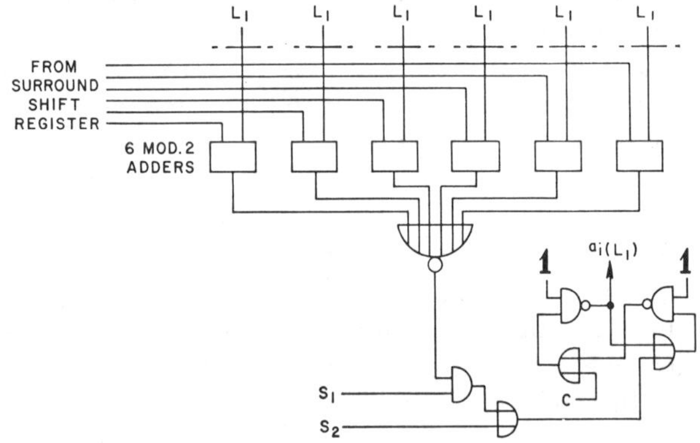

Fig. 8. Formation of $a_i(L_1)$.

The $a_{i(L_1)}$ input is a local input formed within the module itself, as will be described below.

The L_2 input is the output of an OR gate to the inputs of which are given the overall command of the brain relative to the one out of three, four, or seven, subfields. Thus, for example, the one out of three consists of one out of three busses which is fed in parallel to all the submodules of the modules which form the one out of three subfields.

Whenever $d_k a_{i(L_1)} L_2 = 0$, the outputs of both gates 1 and 2 are 0 and the particular k submodule concerned is left in its original state. On the other hand, whenever $d_k a_{i(L_1)} L_2 = \mathbf{1}$, the outputs of gates 1 and 2 will be **1** and 0, respectively, whenever L_3 is **1**, and 0 and **1**, respectively, whenever L_3 is 0, so that the state of L_3 is placed in the double stability circuit of the k submodule concerned through the auxiliary OR gates shown.

Fig. 7 illustrates the modular logical circuit for forming L_1 or L_3. The same circuit can be utilized for both, since, in general, for each new modular operation, L_1 is formed first, then $a_{i(L_1)}$ is formed, and finally L_3 is formed for the modular operation, the only exceptions being special operations for which all a_i's are **1**'s.

The inputs from just five layers have been illustrated: the k, p, q, r, and s layers. c_k, c_p, etc., a, o, l_1, l_3, e, and f are commands from the computer's brain. For instance, if we wish to form

$$L_1 = p \cdot q \qquad (p \text{ AND } q),$$

a **1** is placed on c_p, c_q, a, and l_1. Alternately, for

$$L_1 = k \wedge r \qquad (k \text{ OR } r),$$

a **1** is placed on c_k, c_r, o, and l_1, etc.

As a further example, the function

$$L_3 = r \cdot \bar{s}$$

is formed by placing a **1** on c_r, $c_{\bar{s}}$, a, l_3, and f.

Alternately, to have simply

$$L_3 = \mathbf{1},$$

a **1** is placed on e and f.

The L_1's are fanned out to the $a_{i(L_1)}$ logical circuits of the six neighboring modules. Likewise, the $a_{i(L_1)}$ logical circuit illustrated by Fig. 8 receives its L_1 (surround) inputs from its six neighboring modules.

When an M_{1-3} operation is desired, the pattern of index 1 is first placed in a shift register in the brain and fed together with the neighboring L_1's to six mod 2 adders. If the pattern of index 1 so placed agrees with the L_1 surround, the output of the 6-fold NOR gate shown will be a **1**, and a clock impulse fed to s_1 will place that **1** in the double stability circuit for $a_{i(L_1)}$; otherwise, nothing happens. The shift register then operates five times and rotates the pattern of index 1 around. The patterns of indices 2 and 3 are then placed in their turn on the shift register and rotated. Meanwhile, s_1 will have been pulsed 18 times, to establish whether the L_1 pattern is of index 1, 2, or 3.

If the logical operation desired should take place for any surround, i.e., for an M_{0-13} operation, a **1** is simply placed on S_2 and the shift register is not operated. The $a_{i(L_1)}$ output is, of course, fed to all the logical circuits for the submodules of the composite module concerned. The c command serves to reset the $a_{i(L_1)}$ double stability circuits.

Elementary Applications

While input and output organs will be left out of this discussion, we will assume that as the computer operates, it may be occasionally interfered with by a "deus ex machina" (DEM) which may ask the question: "Is the x-layer empty?" The question and an affirmative answer will be written as

$$\text{DEM}\ (0,\ x) = 0.$$

This question may be implied in all operations for which $n = \infty$ in the subscript of M, for we have the choice of either making n so large that for any conceivable input, this particular operation performed for the nth time would not change the pattern any more if applied for the $(n+1)$st time, or making n smaller, utilizing auxiliary fields x and y with the following operators for the particular case of 1 being the only field operated on. Thus, we may have, for instance:

Symbols	*In Words*
$M_{0-13}[x=1]$	Layer 1 is marked in x.
$M_{a_0{}^3, a_1, \cdots, a_{13}}[1 = a_{i(L)} \cdot L_2 + \overline{a_{i(L_1)} \cdot 1}]$	The M^3 operation is performed once more.
$M_{0-13}[y = 1 + x - 2(1 \cdot x)]$	The new layer 1 is added to x (mod 2) and marked in y.
$\text{DEM}(O, y) = ?$	Is y empty?

If $\text{DEM}(O,\ y) = 0$, the operation M^3 on the 1 layer introduces no more changes and is discontinued.

1) *Counting of Blobs with at Most Simple Holes:* The operator $M^3{}_{3-6,\infty}[1 = a_{i(1)} + \overline{a_{i(1)}} \cdot 1]$ is first applied. This serves to fill all the simple holes in any hollow blob. The outer part of the blob swells until the outline is entirely convex, except where the nearness of another blob prevents further swelling in order not to alter the connectivity. The operator $M^3{}_{1-5,\infty}[1 = a_{i(1)} \cdot 1]$ is then applied. This serves to shrink every blob to a single **1**, after which the **1**'s can be counted by means of one of the output organs which are not discussed here.

2) *Blob Smoothing:* The operator $M^3{}_{4-6,\infty}[1 = a_{i(1)} + a_{i(1)} \cdot 1]$ is first applied. This serves to fill in all small cavities inside or outside the blob until the zero-centered 4 to 6 patterns are no longer present. The $M_{0-2,\infty}[1 = a_{i(1)} \cdot 1]$ operator then serves to cut off appendages. This latter operator is one of the exceptional operators alluded to earlier for which the use of the subfields is not required, even though $a_{i(1)}$ is a function of the subfield operated on.

3) *Skeletonizing of Closed or Open Blobs:* The application of the operator $M^3{}_{2-3,\infty}[1 = a_{i(1)} \cdot 1]$ will, in general, yield a one-element thick skeleton of structure. However, there are structures, such as the one shown in Fig. 9, in which the skeletonizing operator may just nibble away two-element thick limbs without leaving a skeleton behind. A way out of this difficulty consists of changing at random the order of the three vertical subfields, i.e., after each subfield operation the brain selects at random which of the two others will be used next. Alternately, the subfield order may be changed periodically.

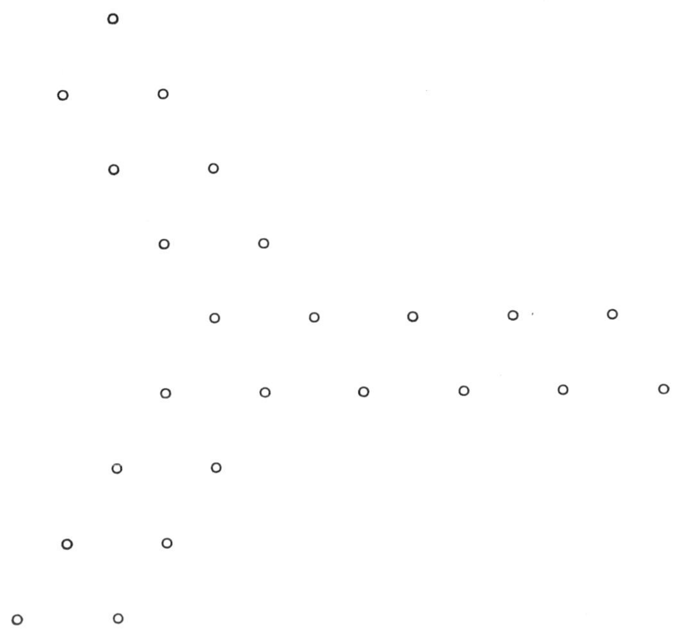

Fig. 9. Structure presenting skeletonizing difficulties.

4) *Determination of Line Ends:* The operator $M_1[2 = a_{i(1)} \cdot 1]$ serves to mark in layer 2 the line ends of layer 1.

5) *Determination of Junction Points:* Whenever layer 1 contains lines only, i.e., contains no **1** centered patterns of index 2–5, the following successive operations serve to mark in layer 2 the line junctions of 1 while leaving 1 unchanged:

$$M_7[2 = a_{i(1)} \cdot 1];\ M_{10}[2 = a_{i(1)} \cdot 1/3 \cdot 1 + \overline{a_{i(1)} \cdot 1/3} \cdot 2];$$
$$M_{10}[1 = \overline{a_{i(1)} \cdot 1/3} \cdot 1];\ M_{10}[2 = a_{i(1)} \cdot 2/3 \cdot 1 + \overline{a_{i(1)} \cdot 2/3} \cdot 2];$$
$$M_{10}[1 = \overline{a_{i(1)} \cdot 2/3} \cdot 1];\ M_{10}[2 = a_{i(1)} \cdot 3/3 \cdot 1 + \overline{a_{i(1)} \cdot 3/3} \cdot 2];$$
$$M_{10}[1 = \overline{a_{i(1)} \cdot 3/3} \cdot 1];\ M_7[2 = a_{i(1)} \cdot 1 + \overline{a_{i(1)}} \cdot 2];$$
$$M_{8-9}[2 = a_{i(1)} \cdot 1/3 \cdot 1 + \overline{a_{i(1)} \cdot 1/3} \cdot 2];$$
$$M_{8-9}[1 = \overline{a_{i(1)} \cdot 1/3} \cdot 1];\ M_{8-9}[2 = a_{i(1)} \cdot 2/3 \cdot 1 + \overline{a_{i(1)} \cdot 2/3} \cdot 2];$$
$$M_{8-9}[1 = \overline{a_{i(1)} \cdot 2/3} \cdot 1];\ M_{8-9}[2 = a_{i(1)} \cdot 3/3 \cdot 1 + \overline{a_{i(1)} \cdot 3/3} \cdot 2];$$
$$M_{8-9}[1 = \overline{a_{i(1)} \cdot 3/3} \cdot 1];\ M_7[2 = a_{i(1)} \cdot 1 + \overline{a_{i(1)}} \cdot 2];$$
$$M_{0-13}[1 = 1 \wedge 2].$$

The actual position of the junctions marked in 2 may vary slightly with the order of the 3 vertical subfields, but the number of junctions marked is invariant against this order.

6) *Blob Filling and Counting of Regions:* The example will be taken of a layer a in which there are blobs, holes in the blobs, secondary blobs in the holes, and secondary holes in the secondary blobs. The problem is to count the number of connected regions and to fill all the primary blobs. Two auxiliary layers b and c and 3 ancillary layers 1, 2, and 3 are used. The successive operators are written at the left and the operations they perform are written in words at the right.

Symbols	*In Words*
DEM $(O, a)\neq 0$	The a layer is not empty.
$M_{0-13}[b=a]$;	The a layer is marked in the b layer.
$M^3{}_{1-3,\infty}[a=\overline{a_{i(a)}}\cdot a]$;	The simple blobs of a are shrunk to nuclei.
$M_0[1=a_{i(a)}\cdot a]$;	The nuclei of a are marked in 1.
$M^3{}_{1-13,\infty}[1=a_{i(1\cdot b)}\cdot b+\overline{a_{i(1\cdot b)}}\cdot 1]$;	The simple blobs of b are restored in 1.
$M_{0-13}[a=b-1]$;	The simple blobs of b are removed and the remainder is marked in a.
DEM $(O, a)\neq 0$	The a layer is not empty.
$M_{0-13}[b=a]$;	The a layer is marked in b.
$M^3{}_{3-5,\infty}[a=a_{i(a)}+\overline{a_{i(a)}}\cdot a]$;	The simple cavities of a are filled except for 0 nuclei.
$M_6[c=a_{i(a)}\cdot\bar{a}]$;	The 0 nuclei of a are marked as 1 nuclei in c.
$M^3{}_{1-13,\infty}[c=a_{i(c)}\cdot\bar{b}+\overline{a_{i(c)}}\cdot c]$;	The simple cavities of b are restored as blobs in c.
$M_{0-13}[b=b+c]$;	The simple cavities of b are filled. This serves, inter alia, to remove the secondary cavities of the original a layer.
$M_{0-13}[a=b]$;	The b layer is marked in the a layer.
$M^3{}_{1-3,\infty}[a=\overline{a_{i(a)}}\cdot a]$;	The simple blobs of a are shrunk to nuclei.
$M_0[2=a_{i(a)}\cdot a]$;	The nuclei of a are marked in 2.
$M^3{}_{1-13,\infty}[2=a_{i(2)}\cdot b+\overline{a_{i(2)}}\cdot 2]$;	The simple blobs of b are restored in 2.
$M_{0-13}[a=b-2]$;	The simple blobs of b are removed and the remainder is marked in a. This serves to remove the secondary blobs.
DEM $(O, a)\neq 0$	The a layer is not empty.
$M_{0-13}[_{b=a}]$;	The a layer is marked in b.
$M^3{}_{3-5,\infty}[a=a_{i(a)}+\overline{a_{i(a)}}\cdot a]$;	The simple cavities of a are filled except for 0 nuclei.
$M_6[c=a_{i(a)}\cdot\bar{a}]$;	The 0 nuclei of a are marked as 1 nuclei in c.
$M^3{}_{1-13,\infty}[c=a_{i(c)}\cdot\bar{b}+\overline{a_{i(c)}}\cdot c]$;	The simple cavities of b are restored as blobs in c.
$M_{0-13}[b=b+c]$;	The simple cavities of b are filled. This serves to remove the remaining primary cavities.
$M_{0-13}[a=b]$;	The b layer is marked in the a layer.
$M^3{}_{1-3,\infty}[a=\overline{a_{i(a)}}\cdot a]$;	The simple blobs of a are shrunk to nuclei.
$M_0[3=a_{i(a)}\cdot a]$;	The nuclei of a are marked in 3.
$M^3{}_{1-13,\infty}[3=a_{i(3)}\cdot b+\overline{a_{i(e)}}\cdot 3]$;	The simple blobs of b are restored in 3.
$M_{0-13}[a=b-3]$;	The simple blobs of b are removed and the remainder is marked in a. This serves to remove the remaining primary blobs.
DEM $(O, a)=0$	The a layer is empty. This signals the end of the processing operations.
$M_{0-13}[a=1\wedge 2\wedge 3]$.	The Boolean superposition (OR) of the 1, 2, and 3 field in a gives the original a layer in which all blobs are filled.

The number of original regions in a is equal to the total number of blobs in the 1, 2, and 3 layers. The blobs in the 1, 2, and 3 layers are all simple and this number is obtained by shrinking them to nuclei and by counting these nuclei with one of the output organs not discussed here.

7) *Straight-Line Prolongation:* Let there be two adjacent **1**'s in a layer which it is desired to prolong into a straight line. It is necessary to maintain a sense of succession and for this purpose the first **1** is printed in empty layer 1 and the second **1** in empty layer 2. The task consists in marking in 1 a third point, just beyond the **1** in 2, making a straight line with the two first **1**'s while erasing the original **1** in 1, after which the process can be repeated with interversion of the role of the 1 and 2 layers. This will be accomplished by the following set of operations, in which two auxiliary empty layers a and b are utilized:

$$M_2[a = a_{i(1+2)}]; \qquad M_2[b = a_{i(2+a)}]; \qquad M_3[1 = a_{i(2+b)}].$$

This serves to place a **1** in 1 beyond the **1** in 2 from the original **1** in 1 which is removed, after which the same operators are applied to 1 and 2 interverted, in order to proceed in the same direction, etc.

Traveling Operators

Traveling operators may be useful for various tasks such as those performed by the eye following the pattern of a line or a blob contour. These tasks may consist in integrating successive bends in a line to ascertain whether it is an outside or an inside border, or to determine its curvature, etc. Three types will be illustrated.

8) *Traveling Operators on Lines Without Junctions:* Let there be a subfield 1 in which there are lines without junctions, i.e., in which any **1** has at most two neighboring **1**'s, and let there be one or more pairs of two consecutive, ordered **1**'s picked out of one or more lines and printed in otherwise empty auxiliary subfields a and b. More than one pair may be picked out of one line provided: a) the line is closed, b) the traveling operators are going in the same direction on it, and c) no two **1**'s belonging to different pairs are adjacent. For each pair a third point beyond b can be printed on a third auxiliary subfield c by the operator: $M_1[c=a_{i(1\cdot b)}\cdot 1\cdot\bar{a}]$, and the a, b, c, subfields can be used in rotation for the traveling point by the operator just written and the other two operators:

$$M_1[a = a_{i(1\cdot c)}\cdot 1\cdot\bar{b}];$$
$$M_1[b = a_{i(1\cdot a)}\cdot 1\cdot\bar{c}].$$

9) *Traveling Operators on the Border of Blobs:* Let there be in a subfield 1 simply or multiply connected blobs which are at least two **1**'s thick everywhere, so that all their **1**'s have surrounds which are patterns of index 2 to 6, and let one border element or several nonadjacent border elements of that blob be defined by as many **1**'s in otherwise empty subfield t. The task considered is that of a robot which, standing on any first **1** of 1 which is also marked in t with the blob on its right

and the void on its left, selects out of the two **1**'s which are border neighbors of the **1** on which it is standing, the second **1** which is in front of it. This is equivalent to selecting that second **1** of 1 which, if traveled to from the first **1** with return through the common neighboring third **1** belonging to the blob, will define a clockwise travel. This in turn is equivalent to asking that if the third **1** is imaged into a symmetric **1** with respect to the second **1**, the surround thus formed around the second **1** and consisting of the first **1** and the third **1** and its image, with the three other elements being zeroes, be of index 8.

10) Traveling Operators Along Lines with Junctions: Directivity and handedness are both required whenever a traveling operator reaches a line junction in order to determine its next position when that operator is required to take, say, the left fork. Let *a*, *b*, *c*, *a*, etc., be the order in which successive **1**'s in field 1 are marked in the *a*, *b*, and *c* subfields *a* and *b* at the start, on positions homologous to two adjacent **1**'s on lines of the 1 subfield. The problem consists of examining the surround of *b*, and starting and proceeding clockwise, determine the next **1** of 1 encountered and marking it in *c*, after which the process is repeated.

Conclusion

A few other applications may be mentioned briefly. The detection of a large "gulf" in a blob, which is accomplished by swelling the blob with an $M_{3-6,\infty}$ operator. This produces an irregular hexagon which is tangent to the original blob at six points. Thus, the added **1**'s may be separated into six regions and the presence of a region much larger than the other five is indicative of the filling of a gulf.

The detection of the roughness of outline of a blob may be likewise obtained by first applying an $M_{3-6,\infty}$ swelling operator, and by applying a limited number of times a thinning M_{1-13} operator to the added regions. The roughness of the original blob will be manifested by the number of smaller regions into which these added regions will be broken up by the thinning operator, and this in turn is given by the number of times the surround of the removed **1**'s are of index 8 to 13.

There are applications such as the analysis of bubble chamber photographs in which it may be desired to prolongate a given curved and thinned line segment into a longer line of the same curvature. This operation involves considerable complexity and a sketch of the principle employed is as follows.

A parallelogram of points is formed in which the ends of the line segment coincide with two opposite curves of the parallelogram. A measure of the line segment curvature *C* is given by the expression

$$C = \frac{m_1 - m_2}{N^3},$$

which is calculated by means of an auxiliary computer, and where m_1 and m_2 are the number of **1**'s in the two respective halves portion of the parallelogram into which it is divided by the line segment, and N is the number of **1**'s contained in the thinned line segment. The three forward points which are neighbors of the penult point are tried successively and the value of C for the line prolongated by each is calculated, and that point is selected for which the calculated curvature is the closest to the original curvature.

It will be noted that all the applications mentioned above are intrinsically isotropic in character because no preferred orientations were introduced in addition to those naturally given by the hexagonal pattern itself. Unless and until such preferred orientations are introduced, it can be concluded tentatively that hexagonal pattern processing is well suited to the studies of patterns which can be presented to the computer in any arbitrary orientation, such as blood cell photographs, bubble chamber photographs, or aerial photographs for which a histogram is made of the various connected regions to detect man-made elongated structure, but which are less well orientated for the recognition of letters, numbers, or other patterns with predetermined preferred orientations.

Part IV
Segmentation: Boundaries

Another basic approach to object isolation involves edge or boundary detection. When complete, closed boundaries can be produced, this approach can be viewed as an alternative to the previously described methods of region formation. However, for images in which only fragments of boundaries can be found, or for basically curvilinear objects such as line drawings or particle chamber tracks, edges and curves provide the most appropriate approach to segmentation.

Considerable effort has been devoted to the problem of developing effective edge detectors. [For a good survey of edge detection techniques, see Davis (1975).] The simplest model of an edge is that of a step discontinuity in the picture function, perhaps obscured by noise, reflectance variability, and blurring. For sharp pictures with little noise, the edge is disclosed by a steep change in the picture function, i.e., by large derivative values. As Kovasznay and Joseph (1955) showed, either first derivatives (the gradient) or second derivatives (the Laplacian) can be used to detect such edges, although the Laplacian tends to be more of a spot detector than an edge detector. By replacing derivatives by finite differences, one can obtain discrete analogs of these operators. (The paper by Roberts in Part VI introduced a simple and popular technique for approximate calculation of the magnitude of the gradient vector that has come to be known as the Roberts-cross operator.)

At higher noise levels and with greater blurring, first or second differences are not effective edge detectors. Instead of working very locally, one must difference some kind of weighted average taken over windows on either side of the edge. The choice of the window size presents a problem. A small window may locate the edge accurately, but will be susceptible to noise. A large window will average out the noise, but will respond everywhere in the general area of the edge; in addition, if the window is so large that the region covered is not homogeneous, then the averaging operation is no longer valid. The first paper, by Rosenfeld and Thurston, addresses this problem by employing a family of windows of different sizes, using a simple criterion to select the largest allowable window size. In addition, related nonlinear methods are described for detecting spots, thin curves, and thick curves or streaks. [Further development of these ideas is described in Rosenfeld, Thurston, and Lee (1972).]

A formal approach to edge detection can be based on a signal-plus-noise model and used to derive statistically optimal edge detectors. However, as can be appreciated from the work of Griffith (1973), realistic models for edges can become rather complicated. A natural alternative—reminiscent of the use of least squares fitting even in the absence of a data model—is to attempt to adjust the parameters of a simple edge model to best fit the picture function locally. This is the approach taken in the second paper, by Hueckel. Hueckel's edge operator attempts to find the best location for a single edge inside a moderately large, roughly circular window. Fitting the edge involves selecting values for four independent parameters so as to minimize a sum-of-squared-error criterion. Hueckel's analysis yields analytical solutions for three of the parameters in terms of the fourth, which must be found by a search procedure. [A modified version of this procedure—extended to detect lines as well as edges—is given in Hueckel (1973).]

When the edges in the image are known to form long straight lines (or, more generally, curves of known shape), more global procedures can be employed effectively. One interesting approach—described originally by Hough—involves the use of a transformation that maps edge points into curves in such a way that the line-finding problem is converted into a rather simple clustering problem. The third paper, by Duda and Hart, develops the theory of Hough's point-to-curve transformation and shows how it can be applied to find edges that form simple parametric curves. [This approach has since been extended further by several authors, including Kimme, Ballard, and Sklansky (1975), Shapiro (1975), and O'Gorman and Clowes (1976).]

A simple and popular way to link edges into curves is to do

contour following. The fourth paper, by Greanias, Meagher, Norman, and Essinger, describes a system for recognizing hand-printed characters that uses contour following to isolate the characters in real time. In addition, simple slope features extracted during a second traversal allow classification by hardwired sequential logic circuits. While this particular curve follower uses an analog flying-spot scanner, it is easy to develop similar contour following algorithms for digital pictures. In particular, Ledley, Jacobsen, and Belson (1966) describe a general system based on contour following, and Pingle and Tenenbaum (1971) describe an interesting accommodating edge follower. (McKee and Aggarwal (1975) describe a more parallel approach based on finding the skeleton of an array of edge points.)

As with edge detection, at higher noise levels simple contour followers become confused by noise streaks and by gaps in the contour. One general approach to finding noisy contours is to define a cost function or figure of merit function and to formulate the problem as an optimization problem. (Recall the analogous approach of Fischler and Elschlager in Part I.) Montanari (1971) showed how the curve detection problem could be formulated as a dynamic programming problem. The fifth paper, by Martelli, generalizes this procedure, formulating the problem as a graph searching problem [see also Martelli (1972)]. This provides a convenient way to introduce various kinds of special knowledge about a particular problem to speed up the optimization process.

The use of search techniques and special domain knowledge are characteristic of artificial intelligence techniques for scene analysis. The final paper, by Kelly, introduces another artificial intelligence technique—planning. Kelly uses a small, low-resolution image and a simple edge detection technique to provide a crude approximation to the desired contour. This approximation then serves as a "plan" to guide a search for the contour in the full-resolution picture. In general, by exploiting specific knowledge about the problem domain, one can improve the performance of such low-level operators as edge detectors and curve followers, as we shall see in Part VI.

Edge and Curve Detection for Visual Scene Analysis

AZRIEL ROSENFELD, MEMBER, IEEE, AND MARK THURSTON

Abstract—Simple sets of parallel operations are described which can be used to detect texture edges, "spots," and "streaks" in digitized pictures. It is shown that, by comparing the outputs of the operations corresponding to (e.g.,) edges of different sizes, one can construct a composite output in which edges between differently textured regions are detected, and isolated objects are also detected, but the objects composing the textures are ignored. Relationships between this class of picture processing operations and the Gestalt psychologists' laws of pictorial pattern organization are also discussed.

Index Terms—Curve detection, edge detection, Gestalt psychology, scene analysis, spot detection, streak detection, texture discrimination, visual perception.

I. Introduction

OVER the past decade there has been increasing acceptance of the notion that the human visual system performs a variety of parallel operations on an input scene—in particular, operations that are sensitive to the presence of dots or "spots," lines or streaks, and edges in the scene. In practical automatic scene analysis systems such operations also play important roles, but they are generally implemented using sequential rather than parallel processing techniques. This is done not only for reasons of expediency, but also because the parallel detection operations that are conventionally used typically produce rather noisy output, which can be improved if sequential edge following is used. Moreover, the conventional techniques, whether parallel or sequential, can be used only to detect edges or curves defined by simple gray level differences, but not the more general class of those defined by textural differences.

The purpose of this paper is to demonstrate that classes of very simple parallel detection operations can be designed which yield relatively noise-free output, and which are capable of handling textural differences. At the same time, it will be shown that by combining the outputs of the operations in a given class, one can extract useful structural information about the given scene. For example, by combining in a straightforward way the outputs of operations that detect edges of different widths, one can obtain a composite output which retains only the "conspicuous" edges in the scene—in particular, it retains edges between textured regions, as well as the edges of isolated objects, while suppressing the edges of the objects that compose the textures. Many of the Gestalt laws of psychologists' pictorial pattern organization can also be "predicted" if such operations are postulated. The fact that it is easy to formulate notions of "conspicuousness" and goodness of pattern organization in terms of the proposed operations suggests that some of the features of this approach may be relevant to the modeling of visual systems.

Manuscript received September 1, 1970; revised October 16, 1970. The work reported here was supported in part by the U. S. Atomic Energy Commission under Contract AT-(40-1)-3662, and in part by the U. S. Air Force under Contract AF-19-628-70-C-0208.

The authors are with the Computer Science Center, University of Maryland, College Park, Md.

II. Edge Detection

Gray Level Edges

Visible "edges" in a scene are often associated with abrupt changes in average gray level. Such changes can vary greatly in degree of localness ("size"); at one extreme, two adjacent points having different gray levels define a microedge, while at the other extreme, two large coarsely textured regions having different average gray levels can also define a sharp edge if the transition between them is abrupt.

Edges through a point can be detected by comparing the average gray levels in pairs of nonoverlapping neighborhoods that meet at the point. Here the relative orientation of the neighborhoods determines the direction of the edges that will be detected, and the size of the neighborhoods determines the widths of the edges that will be detected. To detect microedges, small neighborhoods must be used, but detecting edges between textured regions requires neighborhoods large enough for gross averaging over the detail in the textures. Thus a complete edge detection system should employ, at each point, pairs of neighborhoods of various sizes and at various orientations; the largest of these should have size comparable to that of the entire picture.

The implementation of a system of edge detectors such as just described can be quite practical computationally, if one uses a suitably chosen discrete set of neighborhood sizes. (Experiments indicate [1] that the results obtained are not sensitive to the choice of sizes.) For digital pictures on a square grid, square neighborhoods whose sizes are powers of 2 are especially convenient. In fact, if we shift the picture, say, to the right by the successive amounts 1, 2, 4, 8, $\cdots$, 2^{k-1}, and pointwise add these shifted pictures, the result is an array whose value at each point (i, j) is the sum of the gray levels in the original picture at the points (i, j), $(i, j-1)$, $\cdots$, $(i, j-2^k+1)$; in other words, k shifts and addition operations suffice to add up the 2^k gray levels to the left of each point.

Note that this result is not correct for the portion of the picture near its right edge.

If we alternate the rightward shifts with, say, downward shifts—so that the total number of shifts and adds is $2k$—

Reprinted from *IEEE Trans. Comput.*, vol. C-20, pp. 562–569, May 1971.

then we have at each point the sum of the gray levels in a 2^k by 2^k square having the point as lower right-hand corner.

Moreover, as intermediate steps we have also computed the gray level sums for all such 2^h by 2^h squares, $0 \leq h \leq k$ (here $h=0$ is the original picture itself). These sums can be regarded as averages by simply shifting the binary point, since the number of terms in each sum is a power of 2. Finally, the difference between the averages for any given pair of these neighborhoods can be computed at every point by a single shift and pointwise subtraction operation (for example, the difference between the averages in horizontally adjacent 2^k by 2^k squares requires a horizontal shift by 2^k).[1] All of these operations could be performed very efficiently in parallel using a computer such as Illiac III [2], and can be performed with fair efficiency on a conventional computer using the Illiac III simulation language PAX [3], [4]; it would not be difficult to design special-purpose layered networks to implement them.

The outputs of vertical edge detectors of sizes 8, 16, and 32 for the central portion of a probabilistically generated picture (108 by 108 elements) are shown in Fig. 1.

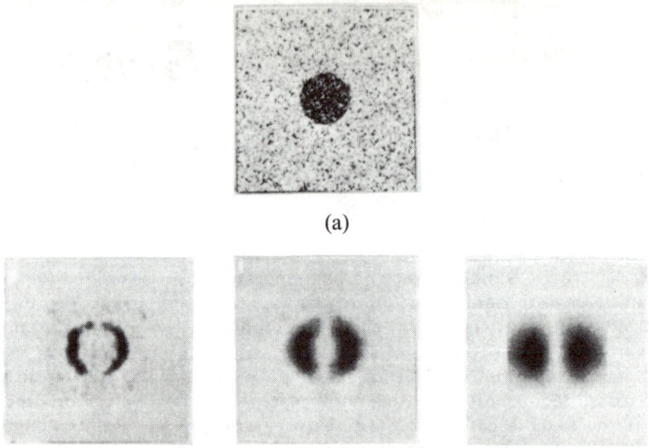

(b) (c) (d)

Fig. 1. Outputs of vertical edge detectors for a probabilistically generated binary-valued picture (probability of a "1" inside the circle 0.8; outside, 0.2) shown in (a). The edge detector sizes are (b) 8, (c) 16, (d) 32.

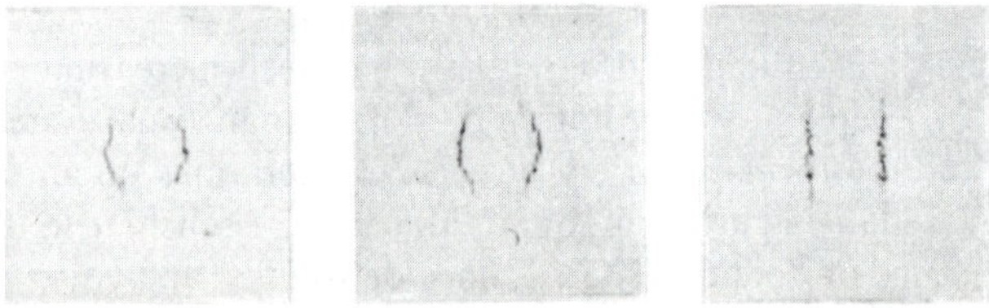

Fig. 2. Maxima (in the direction across the edge) for the outputs of Fig. 1.

Edge Sharpening

When large neighborhoods are used to detect a "large" edge, they do it very redundantly; neighborhood pairs which just straddle the edge are likely to have the highest differences in average gray level, but other pairs at nearby positions (and orientations) will have differences that are nearly as high, even though the location and direction of the edge are sharply defined to the eye. However, one can obtain a sharp response to a sharply defined edge by comparing difference values at nearby points to one another, and suppressing all but the largest. For example, if the neighborhoods used have diameter d, and the neighborhood pairs have relative orientation θ (i.e., if the aim is to detect edges of size d and direction perpendicular to θ), we can compare the value at each point to the values at the points within, say, distance $d/2$ of it in direction θ, and suppress it if any of these values is larger. (A more complicated method of sharpening large edge detector output, by multiplying it by the outputs of the smaller detectors at the same point, was described in [5]–[7].) If $d=2^k$, the required comparison can be carried out using only k shifts and comparison operations, by methods analogous to that described above for the averaging operation. The results for the outputs of Fig. 1 are shown in Fig. 2.

Conspicuous Edges: The "Best" Edge Size at a Given Point

When a scene is very briefly viewed, some of the edges in it are more likely to be noticed than others. In general, large edges (i.e., edges between large regions separated by an abrupt change in average gray level) are more conspicuous than small ones; but an *isolated* small edge may be more conspicuous than are larger ones belonging to the objects in a cluttered textured region. In the following paragraphs it will be shown that, interestingly, if simple comparisons are made between the outputs of edge detectors of different sizes at each point of a picture, it is easy to construct a composite output which retains only such conspicuous edges.

For simplicity, we first consider edges in a single direction only, say of sizes $1, 2, 4, \cdots, 2^k$. Let $e_{ij}^{(h)}$ be the output of the detector of size 2^h at the point (i, j). Let $f_{ij}^{(h)}$ be the maximum of $e_{ij}^{(h)}$ over all points within distance 2^{h-1} of (i, j) in the direction perpendicular to the edge; it will be recalled that this maximum was computed in connection with the edge sharpening problem. Let $m = m(i, j)$ be the largest h such that

$$f_{ij}^{(k)} < f_{ij}^{(k-1)} < \cdots < f_{ij}^{(h)}$$

but

$$f_{ij}^{(h)} \nless f_{ij}^{(h-1)}$$

where $<$ means "is significantly smaller than" (in the program used to obtain the figures shown below, it meant "is less than three-quarters of"). To see the significance of m as a "best" edge size at the point (i, j), let us consider four examples.

Suppose first that (i, j) is just on an edge between two large uniformly textured regions, each of size 2^k or larger. Since 2^k is comparable with the picture size, it is reasonable to assume that both 2^k and 2^{k-1} are large compared to the size of the *detail* in the textured regions. Thus both $e^{(k)}$ and

[1] Experiments indicate that horizontally adjacent squares and vertically adjacent squares provide useful edge detection power for edges at all orientations. However, there is a tendency for objects to be "squared off" (see Figs. 2 and 5); this could be eliminated if other orientations (and more circlelike neighborhoods) were used.

$e^{(k-1)}$ are differences of averages taken over substantial samples of the textures, so that they will ordinarily be approximately equal. Moreover, $f^{(k)}=e^{(k)}$ and $f^{(k-1)}=e^{(k-1)}$ (approximately), since e will be a maximum just on an edge; thus $f^{(k)} \nless f^{(k-1)}$, and by definition of m, we have $m=k$. Thus when we are actually at an edge between large textured regions, the largest edge detector is chosen as "best."

Next, suppose that (i, j) is *near* such an edge, say within distance 2^k of it. By our assumption that 2^{k-1} is large compared to the texture detail size, $e^{(k)}$ and $e^{(k-1)}$ will almost certainly be smaller at an interior point than at a point just on or closer to the edge. Thus $f^{(k)}$ and $f^{(k-1)}$ will correspond to points on, or as close as possible to, the edge; and since the 2^k size is larger, $f^{(k)}$ will come from a point at least as close to the edge as $f^{(k-1)}$. Hence we are highly unlikely to have $f^{(k)}<f^{(k-1)}$, and so here again by definition of m we have $m=k$. In other words, near an edge between large regions, the largest detector is still the best. But since its value away from the edge will be smaller than its values closer to the edge, our edge sharpening procedure will suppress it, thus giving a zero output at (i, j).

Suppose next that (i,j) is inside a uniform textured region, far from its edge. Here both $f^{(k-1)}$ and $f^{(k)}$ should have near zero values, since each is a difference between averages over two large samples of the same uniform texture. Thus we will not have $f^{(k)}<f^{(k-1)}$, and once again this implies $m=k$. In other words, in the interior of a textured region the "best" edge output will be that from the largest detector, which will be near zero even at its maxima.

Finally, suppose that (i,j) is on the edge of a small isolated object, e.g., a small black disk on a white background, where the disk diameter d is smaller than 2^k and the background is more than 2^k wide. In this case it is evident that we will have $f^{(h+1)}=e^{(h+1)}<e^{(h)}=f^{(h)}$ (indeed, $e^{(h)}$ will be about four times $e^{(h+1)}$) as long as 2^h is larger than d; but $f^{(h+1)}=e^{(h+1)}$ will be approximately equal to $f^{(h)}=e^{(h)}$ when 2^h is d or smaller. Thus here the "best" edge size m is that for which 2^m just exceeds d. In other words, even if d is very small, as long as the disk is isolated, our method will select a "best" edge output that detects the disk edge.

Further examples could be given, e.g., of small objects which are near larger objects; a set of such examples, in one dimension, is shown in Fig. 3. Fig. 4 shows some further one-dimensional results, using single TV lines from a TIROS cloud cover picture as input. Fig. 5 shows results for pictures similar to Fig. 1; the "best" size used at each point (as predicted in the discussion of the first three examples above) is evidently nearly always the largest size. Some additional two-dimensional results, for a TIROS cloud cover picture, are shown in Fig. 6. These results use horizontal and vertical detectors only; the output at each point is the larger of the best horizontal and best vertical detectors.

Other Types of Texture Edges

The edge detection scheme described up to now can only detect differences in average gray level; this makes it blind

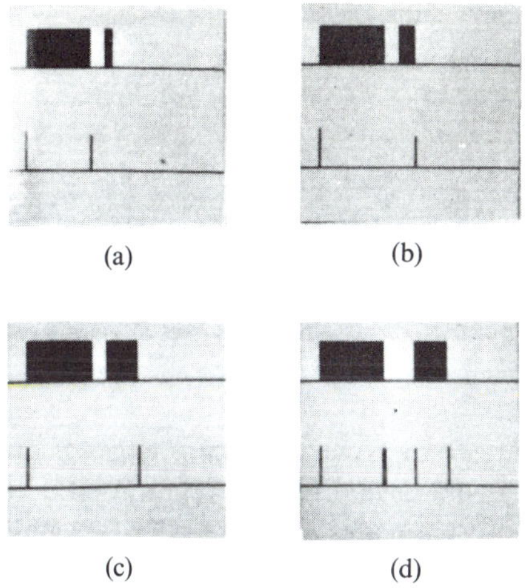

Fig. 3. One-dimensional examples of conspicuous edge detection. (a) Objects of length 32 and 4, separated by gap of length 7. (b) 32 and 8, gap of 7. (c) 32 and 16, gap of 7. (d) 32 and 16, gap of 15. The edge detector sizes are powers of 2 up to 32.

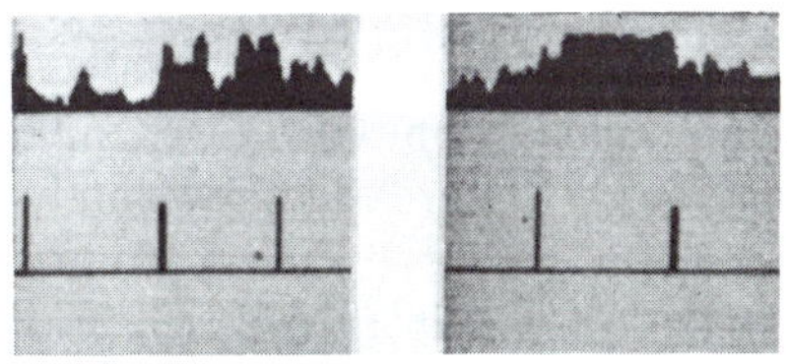

Fig. 4. One-dimensional examples of conspicuous edge detection using two single TV lines from a TIROS cloud cover picture as inputs.

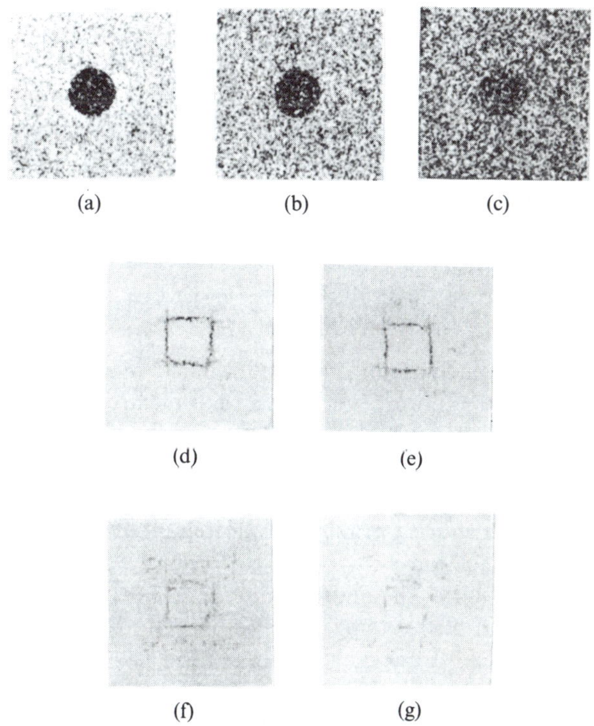

Fig. 5. Conspicuous edge detection results for three pictures (a)–(c) analogous to Fig. 1(a) (probabilities 0.8–0.2, 0.7–0.3, 0.6–0.4). Results for (a)–(c) are shown in (d)–(f). In (g), points of (f) with values below 6 have been suppressed.

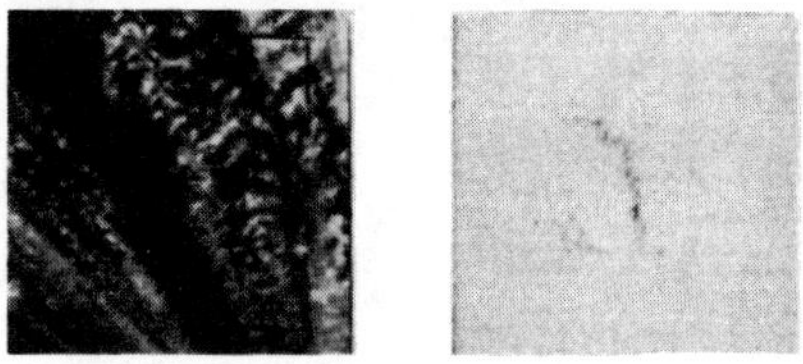

Fig. 6. Conspicuous edge detection results for a TIROS cloud cover picture. (a) Input. (b) Results for points over 32 away from the picture edge.

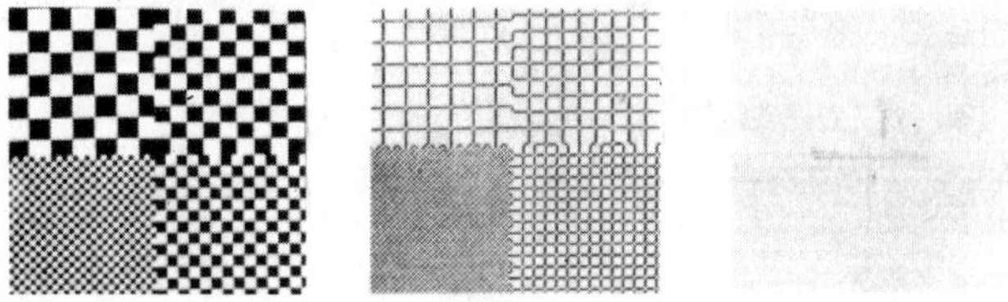

Fig. 7. Detection of coarseness edges. (a) Visual cliff picture. (b) Roberts gradient. (c) Coarseness edges in (a) obtained by applying conspicuous edge detectors to (b).

to certain important types of texture edges. For example, suppose that two adjacent regions contain the same texture seen at two different scales; the abrupt "coarseness" change at the edge between the regions will not be detected, since the average gray level is the same on both sides. (We are assuming here that the largest detectors are large when compared even to the coarser of the two detail sizes.) Similarly, suppose that the regions contain a strongly directional texture (parallel streaks or the like) seen at two different orientations; then the abrupt "directionality" change at the edge will not be detected.

It should be quite simple, however, to extend our scheme to handle such types of texture edges, by simply first "preprocessing" them so as to convert them into gray level edges [1], [5]–[7]. For example, coarseness is related (inversely) to the amount of (micro-) edge per unit area [8]; if we apply a small-edge detection operation to the picture, differences in average coarseness thus turn into differences in average gray level. If we then apply our edge detection scheme to the results, we should be able to detect the original coarseness edges, and analogously for directionality edges.

Examples of this type of edge detection are shown in Figs. 7–9. Fig. 7(a) shows a "visual cliff" picture composed of checkerboard patterns having four different coarsenesses (the squares have sizes 2×2, 4×4, 6×6, and 8×8 in the four quadrants). Fig. 7(b) is the result of applying the "Roberts gradient"

$$|a_{ij} - a_{i+1,j+1}| + |a_{i,j+1} - a_{i+1,j}|$$

to Fig. 7(a); the quadrants now differ in average gray level, since there are more edge points per unit area in a fine checkerboard than in a coarse one. Fig. 7(c) shows the results of applying the conspicuous edge detectors described earlier to Fig. 7(b). Fig. 8 is analogous to Fig. 7, but using an input picture which shows terrazzo tile photographed at four different magnifications.

In Fig. 9, half of the input picture [Fig. 9(a)] is streaked vertically, half horizontally. Fig. 9(b) was produced by applying a vertical streak detection operation (see Section III) to Fig. 9(a); this preserved the vertical streaks but suppressed the horizontal ones, so that the two halves of the picture now differ greatly in average gray level. Finally, Fig. 9(c) was obtained by applying the conspicuous edge detectors to Fig. 9(b).

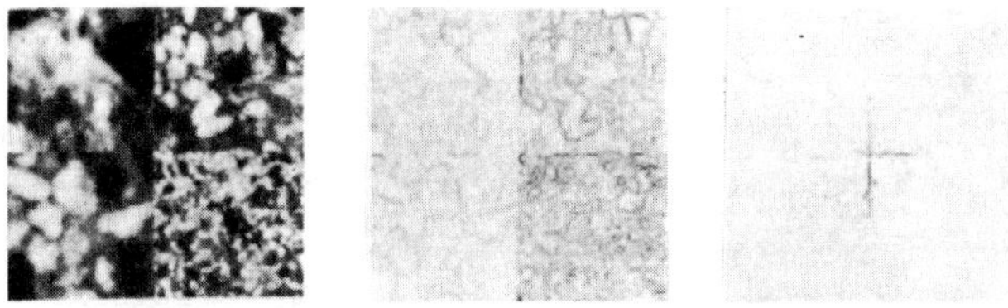

Fig. 8. Analogous to Fig. 7, using a composite picture of terrazzo tile photographed at four different magnifications.

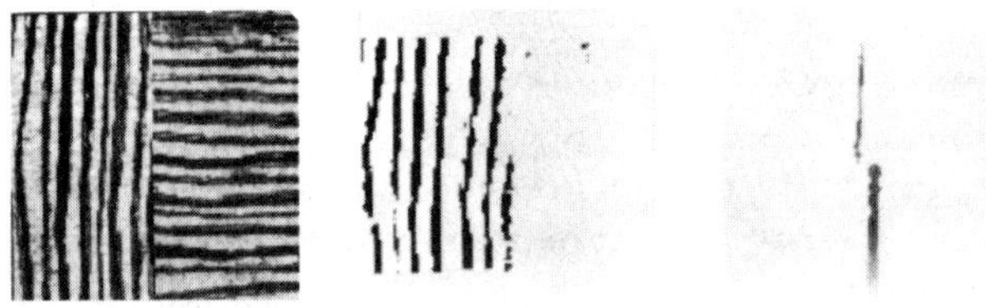

Fig. 9. Detection of directionality edges. (a) Edge due to difference in streak direction. (b) Results of applying a vertical streak detection operation (see Fig. 13). (c) Directionality edge in (a) obtained by applying conspicuous edge detectors to (b).

III. Spot, Curve, and Streak Detection

Spot Detection

In Section II, edge detectors of various sizes and orientations were obtained by shifting and pointwise subtracting blurred versions of a picture from themselves. If other simple shift-and-compare operations are applied to the same blurred pictures, "spots" of various sizes, and curves or streaks of various widths and directions, can be detected. For example, if the picture has been blurred by averaging over 2^k by 2^k squares, we can detect high-value spots of size $\leq 2^k$ on a lower value background by simply looking for points at which the (averaged) gray level is significantly greater than at, say, the four points 2^k to the right, left, above, and below.[2] The "spotness" at a point can be measured by subtracting the (averaged) gray levels at those four points from that at the point, and taking, e.g., the smallest of these differences as the spotness value. All of this can be done in parallel using only four shifts, subtractions, and comparisons. (The detection of curves and streaks will be described below.)

[2] Throughout this section we shall consider only high-value spots, curves, or streaks on a lower valued background; the modifications required to detect the "negatives" of such objects are obvious.

The approach just described will detect a large spot (or wide streak) in a wide range of positions; but the output can be "sharpened" by suppressing all spotness values which have larger values within a distance equal to the spot size (in any direction, for spots; in the direction across the streak, for streaks), as was done in the edge detection case. Moreover, many different sized detectors may detect the same spot; thus here again, as for edge detection, one would want to determine a "best" spot size (or streak width) at each point. This can be done in exactly the same manner as was done for edges. (For a more elaborate method of "size detection," based on edge separation, see [9].) Fig. 10 shows the results of this approach for some one-dimensional inputs (where spotness of size 2^k at a point is defined as the smaller of the differences between the value of the blurred string at the point and its values 2^k to the right and left). All of the foregoing can evidently be extended to "texture spots" by using suitable preprocessing before the detectors are applied to the picture.

Curve Detection

In a digital picture, a point which lies on a thin curve must satisfy two conditions.

Condition 1: It has a pair of lower valued neighbors on opposite sides of it (in the direction across the curve).

Condition 2: It has two other neighbors (in the direction along the curve) which satisfy condition 1.[3]

Since these conditions are quite local, the number of shifts and comparison (or logical) operations which are required in order to apply them is not very large.

In the examples to be given below, only two versions of condition 1 are used: 1a) involves the point's two vertical neighbors, 1b) involves the two horizontal neighbors, so that only four shifts-and-compares are needed. A smooth curve[4] whose slope is between $-45°$ and $45°$ should satisfy condition 1a), while one whose slope lies between $45°$ and $135°$ should satisfy condition 1b).

Condition 2 can be more complicated, since many different pairs of neighbors can satisfy it even in a smooth curve—at least seven such pairs in each of cases 1a) and 1b) [5], [7]. If one wants to allow neighbors that are not immediately adjacent to the point, so as to permit broken or dotted curves to be detected, the number of neighbor pairs to be tested becomes even larger. In [5], a total of 54 pairs of nearest and next nearest neighbors was suggested. However, this can be simplified without much loss in detection performance. Specifically, let the 25 neighbors of point m (which satisfies condition 1) be

[3] A *curve end* point will not in general satisfy condition 2. However, operations analogous to those described in this section can easily be designed that will detect curve ends and streak ends.

[4] If curves are allowed to make sharp turns, other versions of condition 1 may be necessary; alternatively, curve points where sharp turns occur can be handled by special "angle detector" operations. Such operations could easily be defined, e.g., by ANDing the results of curve end detections in pairs of directions.

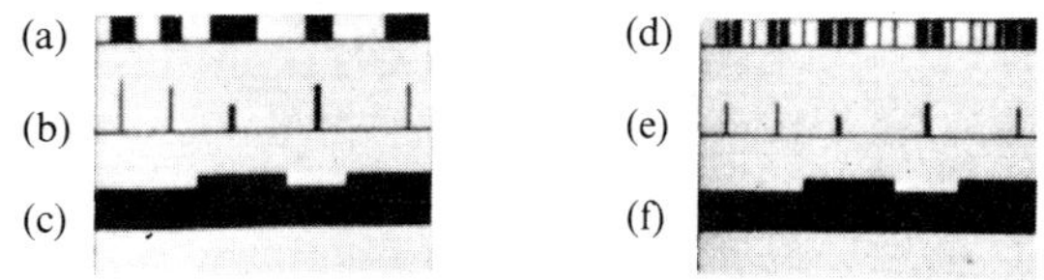

Fig. 10. One-dimensional examples of best spot size detection. (a) Input string containing object of size 8, gap of 8, object of size 7, gap of 9, object of size 16, gap of 16, object of size 9, gap of 16, object of size 14. (b) Spotness maxima. (c) Best size at each point (higher values are 16s, lower are 8s). (d)–(f) Same for input string with 15 percent random noise.

$$\begin{matrix} a & b & c & d & e \\ f & g & h & i & j \\ k & l & m & n & o \\ p & q & r & s & t \\ u & v & w & x & y. \end{matrix}$$

Let

$$A = a \vee b \vee c \vee g \vee h$$
$$C = c \vee d \vee e \vee h \vee i$$
$$E = e \vee i \vee j \vee n \vee o$$
$$O = o \vee n \vee t \vee s \vee y$$
$$Y = y \vee x \vee w \vee s \vee r$$
$$W = w \vee v \vee u \vee r \vee q$$
$$U = u \vee p \vee k \vee q \vee l$$
$$K = k \vee f \vee a \vee l \vee g.$$

In other words, we say that A "satisfies condition 1" if any of a, b, c, g, h does and similarly for the others. Then to impose condition 2 it suffices to require that one of the four pairs A and Y, C and W, E and U, or O and K "satisfy condition 1."

If one needs to detect curves in heavy noise, condition 2 may be too weak, since three consecutive points satisfying condition 1 may occur frequently in the noise; one may want to require more than three such points before a curve is detected. One simple way to impose such a stronger requirement is to *iterate* condition 2, i.e., to apply it again to the points which survive the first application, and so on. Fig. 11 shows the results of iterating the simplified condition 2 on an arbitrary dotted curve in noise; these results should be compared with Fig. 4 in [7], which used a more complicated condition 2. As also pointed out in [5], [7], curve detectors can be used to clean up the output of edge detectors; Fig. 12 shows the results of applying the curve detectors to one of the outputs in Fig. 5.

It is worth stressing the differences between the foregoing scheme and the linear filtering approach to curve detection which has been pursued by some investigators [10], [11]. Suppose that we attempt to detect curves by convolving the picture with templates matched to short line segments at various orientations, e.g., the vertical line template might have the form

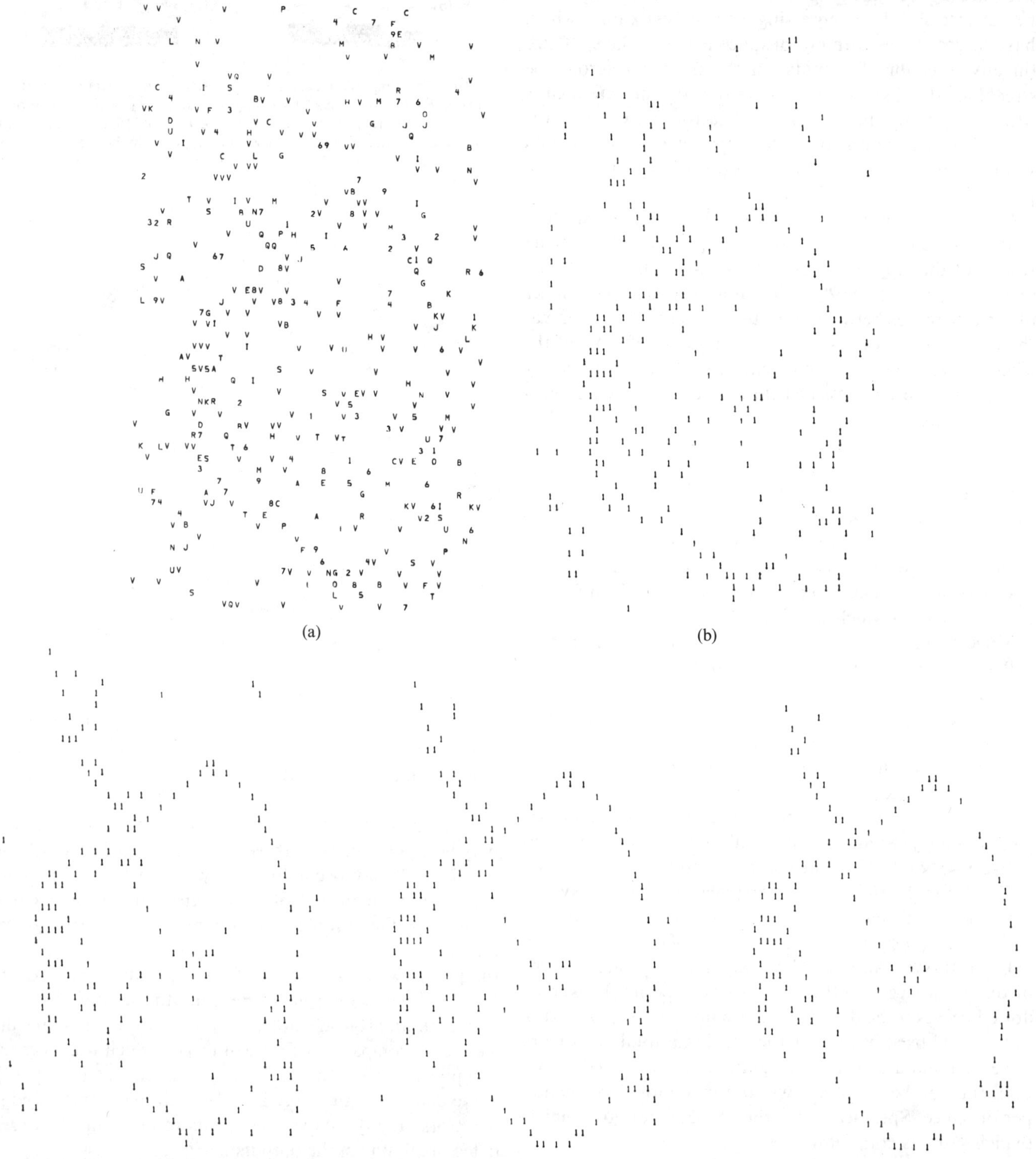

Fig. 11. Dotted curve detection. (a) Input, curve in 12 percent noise. (b)–(e) Four iterations of simplified dotted curve detectors. In input, 1, · · · , V represent gray levels 1, · · · , 31; level 0 is represented by a blank.

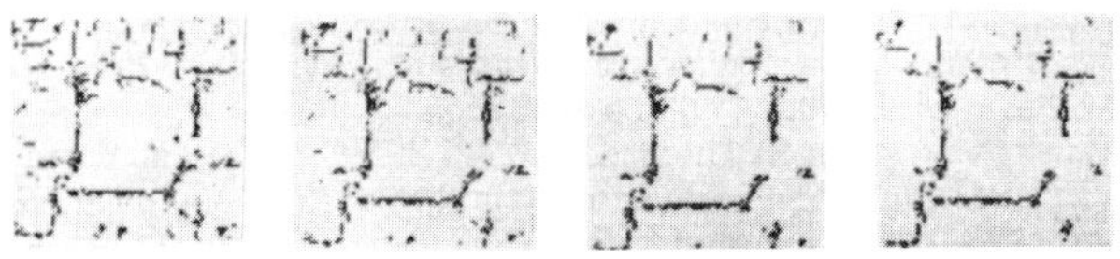

(a) (b) (c) (d)

Fig. 12. Results of applying the simplified dotted curve detectors to Fig. 5(f). (a)–(d) Four iterations.

$$\begin{array}{ccc} -\frac{1}{2} & 1 & -\frac{1}{2} \\ -\frac{1}{2} & 1 & -\frac{1}{2} \\ -\frac{1}{2} & 1 & -\frac{1}{2}. \end{array}$$

This will have output $3k$ when in register with a vertical line segment whose gray level is k greater than that of the background; but it will have the same output for a single point whose gray level is $3k$ greater than the background level, or for a vertical edge between gray levels which differ by $2k$. Our scheme, by breaking the detection operation up into stages (1 and 2), avoids these difficulties.

Streak Detection

The above approach to curve detection can be immediately generalized to detecting thick streaks rather than thin curves: we simply apply conditions 1 and 2 to a blurred version of the picture, with "neighbors" spaced sufficiently far apart that the points being examined represent averages over nonoverlapping neighborhoods on the original picture. Fig. 13 shows streak detection results, using 4×4 neighborhoods, for a bubble chamber picture. At step 2, we accept any pair of "neighbor" points lying in two nonoverlapping 4×4 squares on opposite sides of the 4×4 square containing the given point; three such pairs of squares are used in conjunction with condition 1a), and three with condition 1b).

As indicated at the beginning of this section, a general streak detection system would use neighborhoods of many sizes, would sharpen the results by suppressing all but maxima (in the direction across the streak), and would pick a "best" streak width at each point by comparing the outputs obtained using the various neighborhood sizes. Such a general scheme could also be applied to the detection of elongated parts of objects; for a more complicated approach to this problem see [12].

Application to "Grouping" of Picture Parts

If a picture processing system employs spot and streak detection operations such as those described above, it can be thought of as imposing an "organization," or "grouping" structure, on complex patterns when it is applied to them. This organization seems to resemble in many respects that described by the Gestalt psychologists (e.g., [13]).

Each spot or streak detection operation can be regarded as "accepting" a spotlike or streaklike patch of the picture, provided certain conditions are met (the average gray level in the patch is greater than each of the averages in certain surrounding patches), as described above. The patch accepted by a given detector will be called its *field.* Let us make the following basic assumption: *A portion of the picture is a "grouping" if it is the field of a single detector, or consists of the fields of all detectors of a given type.*

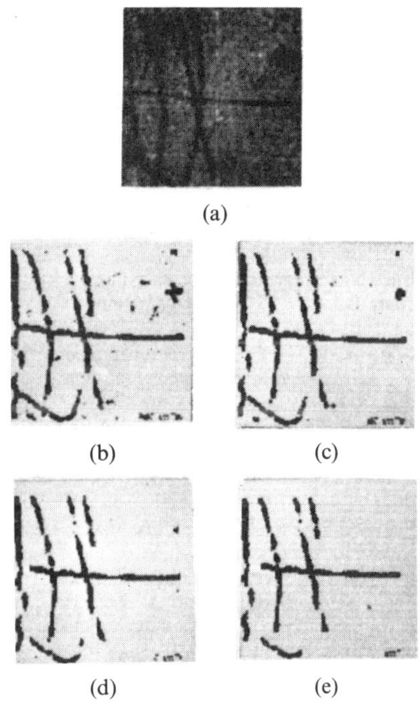

Fig. 13. Streak detection results for a portion of a bubble chamber picture. (a) Input. (b)–(e) Four iterations.

Example 1: Suppose that a picture contains simple objects of two (or more) types—spots and streaks, small spots and large spots, horizontal streaks and vertical streaks, etc. Then the objects of each type constitute a grouping defined by the fields of all the detectors of that type. This class of examples is essentially that referred to by the Gestaltists as illustrating the "law of similarity."

Example 2: A dense (spotlike or streaklike) cluster of objects on a sparser background is a grouping, since it is the field of a single detector. This class of examples can be thought of as illustrating the Gestaltists' "law of proximity."

Example 3: Objects which lie along a line group together, since they are the field of a single streak detector, but lines which make a sharp angle with each other do not group together; this suggests the Gestalt "law of good continuation."

Example 4: Conversely, if a portion of the picture is a grouping according to these principles, then it is very hard to see a proper part of this portion as making up a grouping in conjunction with other parts of the picture. In particular, a figure can be "hidden" by extending parts of it so that the extensions constitute groupings.

Example 5: If one portion of the picture becomes a grouping when one type of detector is used, and another portion becomes a grouping when another type is used, the two portions are hard to see as groupings at once. This is true, in particular, when the two portions are an object and its complement.

It should be stressed that our basic assumption certainly does not account for all of the natural ways in which parts of a picture can be grouped. For example, it takes no account of such fundamental pattern properties as closure, periodicity, symmetry, familiarity, etc. However, it is of interest to see that many natural types of grouping do appear to follow from the simple assumption, now very commonly made by modelers, that the visual system incorporates spot and streak detectors.

References

[1] M. Abbamonte, E. G. Johnston, Y. H. Lee, R. Nagel, A. Rosenfeld, and M. Thurston, "Edge and curve enhancement in digital pictures, 2," Computer Sci. Ctr., Univ. of Maryland, College Park, Tech. Rep. 70-103, Jan. 1970.

[2] B. H. McCormick, "The Illinois pattern recognition computer—ILLIAC III," *IEEE Trans. Electronic Computers*, vol. EC-12, Dec. 1963, pp. 791–813.

[3] E. B. Butt and J. W. Snively, Jr., "The PAX II picture processing system," Computer Sci. Ctr., Univ. of Maryland, College Park, Tech. Rep. 68-67, May 1968 (revised Sept. and Dec. 1969; revised version edited by E. G. Johnston and R. Lipsett).

[4] E. G. Johnston, "The PAX picture processing system," in *Picture Processing and Psychopictorics*, B. S. Lipkin and A. Rosenfeld, Eds. New York: Academic Press, 1970, pp. 427–512.

[5] A. Rosenfeld, R. B. Thomas, and Y. H. Lee, "Edge and curve enhancement in digital pictures," Computer Sci. Ctr., Univ. of Maryland, College Park, Tech. Rep. 69-93, May 1969.

[6] A. Rosenfeld, "A nonlinear edge detection technique," *Proc. IEEE* (Lett.), vol. 58, May 1970, pp. 814–816.

[7] A. Rosenfeld, R. B. Thomas, and Y. H. Lee, "Edge and curve detection for texture discrimination," in *Picture Processing and Psychopictorics*, B. S. Lipkin and A. Rosenfeld, Eds. New York: Academic Press, 1970, pp. 381–393.

[8] A. Rosenfeld and E. B. Troy, "Visual texture analysis," Computer Sci. Ctr., Univ. of Maryland, College Park, Tech. Rep. 70-116, June 1970; abridged version in *Conf. Rec., 1970 Symp. Feature Extraction and Selection in Pattern Recognition*, IEEE Publ. 70C-51C, Oct. 1970, pp. 115–124.

[9] C. M. Cook, "Size detectors," Computer Sci. Ctr., Univ. of Maryland, College Park, Tech. Rep. 70-127, Aug. 1970.
C. M. Cook and A. Rosenfeld, "Size detectors," *Proc. IEEE* (Lett.), vol. 58, Dec. 1970, pp. 1956–1957 (abridged version).

[10] S. Deutsch, "Conjectures on mammalian neuron networks for visual pattern recognition," *IEEE Trans. Syst. Sci. Cybern.*, vol. SSC-2, Dec. 1966, pp. 81–85.

[11] S. Deutsch, *Models of the Nervous System*. New York: Wiley, 1967.

[12] A. Rosenfeld, C. M. Park, and J. P. Strong, "Noise cleaning in binary-valued digital pictures using 'propagation' processes," Computer Sci. Ctr., Univ. of Maryland, College Park, Tech. Rep. 69-91, May 1969; *1969 EASCON Conv. Rec.*, IEEE Publ. 69C31-AES, pp. 264–273 (abridged version).

[13] M. Wertheimer, "Investigations on the Gestalt theory," *Psychol. Forsch.*, vol. 4, 1923, pp. 301–350.

An Operator Which Locates Edges in Digitized Pictures

MANFRED H. HUECKEL

Stanford University,* Stanford, California

ABSTRACT. Because of the fundamental importance of edges as primitives of pictures, automatic edge finding is set as goal. A set of requirements which should be met by a local edge recognizer is formulated. Their main concerns are fast and reliable recognition in the presence of noise. A unique optimal solution for an edge operator results. The operator obtains the best fit of an ideal edge element to any empirically obtained edge element. Proof of this is given. A reliability assessment accompanies every recognition process.

KEY WORDS AND PHRASES: artificial intelligence, picture processing, pattern recognition, edge recognition, edge operator, primitives of pictures, computer vision, scene analysis, feature extracting, information reduction, noise elimination, heuristic process, Hilbert space

CR CATEGORIES: 3.63, 3.65, 5.10

1. *Introductory Remarks*

Natural scenes usually consist of objects with sharply defined surfaces.[1] Therefore 2-dimensional projections of natural scenes often consist of domains of different brightness separated by well-defined edges. Thus the edge is a fundamental primitive in pictures. This paper is concerned with locating edges.

Edge finding may seem easy to the reader because human vision extracts useful edge information in the presence of a variety of disturbances. However, it seems from the work of Hubel and Wiesel [1] that the retina may do the work of edge abstraction before the visual information is even presented to the brain. So it may be that the problem of edge recognition eludes human awareness: Because the edges already seem "to be there," we may be led to underestimate the difficulties of the problem.

As part of the Stanford Artificial Intelligence Project an edge operator was mathematically derived and then programmed as a subroutine for a large picture processing system.

To show an example, Picture 1 (below) was passed through a vidicon camera, a digitizer, the picture processing system, and a plotter. Then the linear edge elements were linked to form edge curves. Picture 2 (below) resulted. Picture data structures on this level of abstraction are only an intermediate step, for a scene needs ultimately to be reconstructed as a composition of objects [2].

An extensive and recent literature survey on picture processing in general is given by Rosenfeld [3].

The sections of this paper are ordered according to the steps which led to the de-

The research reported here was supported in part by the Advanced Research Projects Agency of the Office of the Secretary of Defense (SD-183).

* Artificial Intelligence Project.

[1] Exceptions may occur in a scene containing patches of fog or in a picture containing objects out of focus.

Reprinted with permission from *J. Ass. Comput. Mach.*, vol. 18, pp. 113–125, Jan. 1971.

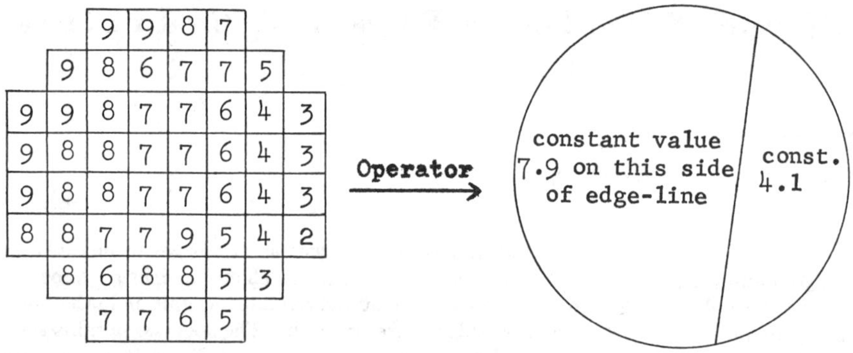

(a) (b)

FIG. 1. (*a*) Empirically obtained edge-element $\mathcal{E}(x, y)$; (*b*) idealization $\mathfrak{F}(x, y)$ of (*a*)

velopment of the operator. Section 2 intuitively introduces the notion of edge operators and then establishes five requirements for reliable local edge recognition under noisy conditions. Section 3 presents the formal definition of the unique operator which meets the five requirements. Section 4 sets forth a theorem showing that the operator has a practical solution, and Section 5 proves the theorem. Section 6 provides experimental demonstration of the operator's performance. Section 7 is the summary and conclusion.

A note on terminology: The term "edge data" specifies the linear border between two domains plus the brightness values left and right of the border. A short section of an edge will be called an "edge element." The term "ideal edge element" will be used to indicate a precise brightness step-function. "Picture" will mean a 2-dimensional array of discrete grey values. (No color information was recorded.) "Empirical edge element" refers to a part of a "picture" containing a brightness step within a scene.

2. *Difficulties With Which the Operator Has to Cope*

This section includes subsections corresponding to five different difficulties with which the operator must cope. For the understanding of this section only the following sketch of the operator is needed. The exact operation of the operator will be presented in Sections 3 and 4.

Define D to be a set of grid squares, arranged so as to best approximate a disk. An example for a 52-element D is the grid pattern of Figure 1(a). D will be called the input disk. Imagine D to be a subgrid of a digitized picture. With $p \in D$, let E_p be the brightness of the picture at point p. If an edge passes over the input disk, the disk cuts out an edge element. Such is the case for the function E_p over D in Figure 1(a). It is a part of an actual television picture which was digitized into a 231×297 matrix with 16 grey values possible for each matrix element.

This diagram demonstrates how an edge in a digitized picture may differ from a clear-cut step function due to blurring and white noise. Compared with most other empirically obtained edge elements, Figure 1(a) is rather distinct.

Figure 1(b) shows the ideal edge which best approximates Figure 1(a). Figure 1(b) was produced by one operator application on a PDP-10 in 10^{-2} seconds. It is the task of the operator to model an empirical edge element by means of the ideal edge element which fits best. While it takes 52 numbers to describe Figure 1(a), it takes only four numbers to describe Figure 1(b), namely, two for the borderline and two for the adjacent average brightnesses. However, this data reduction is not what primarily motivates the application of the operator; rather it is the need to know if there is an edge element in a given input disk, where it is (if applicable), and how reliably it may be distinguished from noise.

The operator accepts any $\{E_p \mid p \in D\}$ as input. As the first computational step the operator calculates eight intermediate data a_i as follows:

$$a_i = \sum_{p \in D} H_{ip} E_p \qquad (i = 0, \cdots, 7). \tag{1}$$

The second computational step maps these eight a_i in a nonlinear manner into six final edge data (c, s, ρ, d, b, k) which will be defined later. The first step (1) consumes about 80 percent, the second about 20 percent of the operator's total computation time.[2] The H_{ip} form a matrix of constants which were obtained from continuous counterparts $H_i(x, y)$ $(i = 0, \cdots, 7)$, as will be described in Section 2.1. $\{H_0(x, y), \cdots, H_7(x, y)\}$ is the solution of a set of lengthy functional equations which may be presented in detail at a later time. However, the requirements which determine these functional equations, as well as the H_{ip}, will be presented in intuitive form in Sections 2.1–2.5. The functional equations have a unique solution, and it is in this sense that the operator is of optimal design.

2.1. Nullifying Rounding Errors on the Periphery of the Disk. Although the problem consists of finding the edge in a discrete disk D as in Figure 1(a), the operator was derived using a continuous circular disk $\mathfrak{D}$ as in Figure 1(b). Define

$$\mathfrak{D} = \{(x, y) \mid x^2 + y^2 \leq 1\}.$$

While formula (1) is applicable to the discrete case, the corresponding formula for the continuous case is

$$a_i = \int_{\mathfrak{D}} H_i(x, y) \mathcal{E}(x, y) dx\, dy \qquad (i = 0, \cdots, 7). \tag{2}$$

When superimposing the disks D and $\mathfrak{D}$ [compare Figures 1(a) and 1(b)],

some grid squares p of D have corners which fall outside $\mathfrak{D}$ (2′)

and

some squares just outside of D have corners which fall inside $\mathfrak{D}$. (2″)

Thus, in a naive transition [such as $H_{ip} = H_i(x, y)$ with (x, y) = center of p] from the continuous to the discrete situation, errors result. Accomplishing this transition from the discrete to the continuous case with minimum error establishes the first requirement.

Three measures were taken which together virtually nullify these errors and thus "round" D: First, the size of the "offending" corners was minimized by choosing particular disk sizes. An investigation showed that the disks which contain 32, 52, 69, 88, and 137 grid squares are the most favorable choices within the range of interest. The corresponding radii of $\mathfrak{D}$ are given by

$$r = [(\text{area of } D)/\pi]^{\frac{1}{2}}.$$

[2] For this reason, and because of its simplicity, (1) suggests itself for hardware implementation, which would speed up the operator by a factor of 5.

Second, the transition from formula (2) to formula (1) was made via

$$H_{ip} = \int_p H_i(x, y)dx\,dy.$$

Now, an error which otherwise would result from an excess corner (2′) disappears by simply setting $H_i(x, y) = 0$ for $x^2 + y^2 > 1$. Each error due to (2″) was compensated for by distributing the integral over the missing corner over the adjacent grid squares in D. This numerical effort resulted in a one-time tabulation of H_{ip} for $i = 0, \cdots, 7$, $p \in D$, and (size of D) $\in \{32, 52, 69, 88, 137\}$. A third measure results from the low weight given to the peripheral points as will be described in Section 2.2.

2.2. The Weight of the Input Data Decreases toward the Disk's Periphery. $\{H_i(x, y) \mid i = 0, \cdots, 7\}$ has been designed so as to ascribe varying weight to points within $\mathfrak{D}$. For reasons of continuity the weight function must approach zero on the periphery of the disk because this is the weight which is assigned to the outside of the disk. Furthermore, disturbances such as other edges appear predominately at the periphery of the disk. Thus, it would be desirable to have a weight high at the center and diminishing continuously toward the periphery. Finally, if data close to the periphery have little effect on the results, then errors near the periphery will also have little effect.

The weight function was chosen to be $[1 - (x^2 + y^2)]^{\frac{1}{2}}$ for a disk whose periphery is $x^2 + y^2 = 1$.

2.3 An Operator Which Locates Edges Need Not be Sensitive to Noise of High Spatial Frequencies. The term "spatial frequencies" refers to the components of a Fourier spectrum of $\mathcal{E}(x, y)$. What is meant by the Fourier spectrum of a function over a 2-dimensional disk $\mathfrak{D}$? The set of all[3] continuous functions over $\mathfrak{D}$ is a Hilbert space. A complete and orthonormalized sequence $\{\mathcal{H}_i \mid i = 0, \cdots, \infty\}$ of functions over $\mathfrak{D}$ defines a Cartesian basis for the Hilbert space. The Hilbert vector $\mathcal{E}(x, y)$ when represented in this basis is $(\alpha_0, \alpha_1, \alpha_2, \alpha_3, \cdots)$, with

$$\alpha_i = \int_{\mathfrak{D}} \mathcal{H}_i(x, y)\mathcal{E}(x, y)dx\,dy \qquad (i = 0, \cdots, \infty). \tag{3}$$

This vector is a Fourier spectrum if the $\mathcal{H}_i$ are spatial wave functions and if they are monotonically ordered by increasing frequencies.

The requirements of Sections 2.2–2.5 lead to the choice of a particular basis sequence $\{\mathcal{H}_i \mid i = 0, \cdots, \infty\}$ whose basis functions $\mathcal{H}_i$ separate into products of radial and angular wave functions, i.e. a Fourier analysis in polar coordinates.

The frequency behavior of the first eight[4] of the $\mathcal{H}_i$ can be observed from the zero crossings shown below. All the $\mathcal{H}_i$ are zero on the periphery of the disk and over any additional circle or line which is shown below. Define $\mathcal{H}_i$ to be of low or high spatial frequency depending on whether $i \leq 7$ or $i > 7$. Only the amplitudes of low frequencies proved useful for the computations of the operator since

(1) empirically observed edges are blurred, and blurring removes high-spatial-frequency information [Figure 1(a) is a well-focused image of an edge of frequently encountered clarity], and

(2) the television camera amplifier fills up the high-frequency range with white noise.

[3] Actually we are only interested in brightness functions which are zero on the periphery of $\mathfrak{D}$. This is because of the weighing according to Section 2.2.

[4] For the expressions of these $\mathcal{H}_i$, see Section 3.

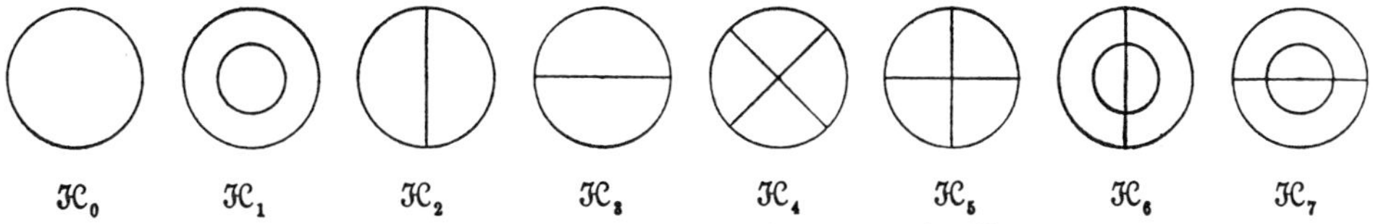

FIG. 2. Zero crossings of the first eight $\mathcal{H}$'s

Therefore, it was decided to low-pass filter the input edge element $\mathcal{E}(x, y)$ by replacing its equivalent Hilbert representation $(\alpha_0, \alpha_1, \alpha_2, \cdots, \alpha_\infty)$ by the finite[5] vector $(\alpha_0, \alpha_1, \cdots, \alpha_7)$. High frequencies do not pass the filter and formula (3) becomes formula (2). This explains the meaning of the first (linear) computational step of the operator. The meaning of the second (nonlinear) step will be explained in Section 4.

2.4. An Operator Which Locates Edges Is by Its Nature Sensitive to Noise of Low Spatial Frequencies. Usually a low-frequency disturbance in an input disk comes in the form of another edge element. The extra edge may belong to a surface irregularity, to a corner, to the background, to a shadow, or it may originate in the digitization[6] or in faulty hardware.[7] Even if the disturbance is not part of an edge, its low-frequency character is likely to simulate one.

A second edge in the input disk with all its characteristics of relevant information is a serious kind of disturbance. Therefore, the operator is designed to be able to simultaneously recognize two edge elements within the input disk and describe each by the four parameters of edge data. On the basis of the edge signal-to-noise ratio (see Section 4), it then decides which edge is more dominant and reports it as its output. It was for this reason that the eight $\mathcal{H}_i$ were chosen as the cut-off for the low-pass filter, for the two sets of interwoven edge data can be conveyed in the eight calculated a_i in formulas (1) or (2).

In experiments, simultaneous recognition of two similarly dominant edge elements was successful, but the less dominant edge element is likely to come out inaccurate. Because there are only eight intermediate data, any low-frequency disturbance added to two superimposed edge elements will seriously disturb the data. Therefore it must be stressed that the operator is not appropriate for simultaneous location of two edge elements in general, nor can it be used as a corner finder. However, a single edge does enjoy partial protection from low-frequency noise.

One could use a larger H_{ip} for more intermediate data a_i, but this would tremendously lengthen the second computation step of the operator.

2.5. The Computing Time for the Operator Should be Minimal. At the beginning of Section 2 it was stated that the operator proceeds in two successive computational steps. The computation time of the second step depends on the choice of the H_{ip}. Minimizing the latter time establishes the requirement of this Section.

Let (4) denote

(a set of simultaneous functional equations which formalize
the requirements of Sections 2.2–2.5.) (4)

[5] The choice of eight for the dimensionality of this vector will be explained in Section 2.4.

[6] A gradual change of brightness in the analog picture before digitization may come out as several step changes afterwards.

[7] For example, patchlike faults sometimes appeared in pictures as a result of comparator oscillations in the digitizer.

The actual derivation of (4) will not be given in this paper. A set $\{\mathcal{H}_i(x, y) \mid i = 0, \cdots, 7\}$ emerged as the sole[8] solution of (4). With the resulting H_{ip}, the second computational step of the operator can be performed in one-fourth the time of the first step. The solution $\{\mathcal{H}_i(x, y) \mid i = 0, \cdots, 7\}$ of (4) will be presented in Section 3.

3. *Definition of the Operator*

Let $\mathcal{E}(x, y)$ and $\mathcal{F}(x, y)$ be functions of the kinds shown in Figures 1(a) and 1(b), respectively. $\mathcal{E}$ is continuous and constant within each grid square, and arbitrary otherwise. $\mathcal{F}$ is continuous and constant except over the edge line $c \cdot x + s \cdot y = \rho$. We normalize this equation by setting $c^2 + s^2 = 1$. With b as the lower and $b + d$ the higher of the two brightnesses of the step function $\mathcal{F}$, the exact definition of the latter is

$$\mathcal{F}(x, y, c, s, \rho, d, b) = b \qquad \text{if } c \cdot x + s \cdot y \leq \rho,$$

$$\mathcal{F}(x, y, c, s, \rho, d, b) = b + d \qquad \text{if } c \cdot x + s \cdot y > \rho.$$

There is a one-to-one correspondence between ideal edges $\mathcal{F}(x, y)$ and quintuples (c, s, ρ, d, b). Actually, the latter contain only four independent parameters because $c^2 + s^2 = 1$. There is no need to define $\mathcal{E}$ and $\mathcal{F}$ outside $\mathfrak{D}$.

The task of the operator is to best approximate a given empirical edge element $\mathcal{E}$ by an ideal edge element $\mathcal{F}$. As the measure of closeness the square of the Hilbert distance between $\mathcal{E}$ and $\mathcal{F}$ was chosen, namely,

$$\Lambda(c, s, \rho, d, b) = \int_{\mathfrak{D}} [\mathcal{E}(x, y) - \mathcal{F}(x, y, c, s, \rho, d, b)]^2 dx\, dy. \tag{5}$$

Now our goal is to find an efficient algorithm which selects (c, s, ρ, d, b) so as to minimize $\Lambda(c, s, \rho, d, b)$.

Let the set $\{\mathcal{H}_i \mid i = 0, \cdots, \infty\}$ be a complete and orthonormalized basis of the Hilbert space of functions over $\mathfrak{D}$. Orthonormal means

$$\int_{\mathfrak{D}} \mathcal{H}_i(x, y)\mathcal{H}_j(x, y) dx\, dy = \begin{cases} 1 & \text{if } i = j, \\ 0 & \text{if } i \neq j. \end{cases}$$

Define

$$\alpha_i = \int_{\mathfrak{D}} \mathcal{H}_i(x, y)\mathcal{E}(x, y) dx\, dy \tag{6}$$

and

$$f_i(c, s, \rho, d, b) = \int_{\mathfrak{D}} \mathcal{H}_i(x, y)\mathcal{F}(x, y, c, s, \rho, d, b) dx\, dy. \tag{7}$$

Expression (5) then takes on the form

$$\Lambda(c, s, \rho, d, b) = \sum_{i=0}^{\infty} [\alpha_i - f_i(c, s, \rho, d, b)]^2.$$

Any process which exactly locates the minimum of Λ requires an unacceptably large amount of computation time. Instead, we permit one approximation. It changes Λ into $\bar{\Lambda}$, which can be minimized by means of the solution theorem of

[8] It seems likely that the solution would remain the same unique solution if the requirement of Section 2.2 were dropped; however, no proof of this was sought.

Section 4. We weigh the terms of the sum of Λ as follows:

$$\bar{\Lambda}(c, s, \rho, d, b) = \sum_{i=0}^{\infty} \{w_i[\alpha_i - f_i(c, s, \rho, d, b)]\}^2, \tag{8}$$

with

$$\begin{aligned}
w_0 &= 1 &&= 1.000, & w_4 &= (4/3)^{\frac{1}{2}} = 1.155,\\
w_1 &= (4/3)^{\frac{1}{2}} &&= 1.155, & w_5 &= (4/3)^{\frac{1}{2}} = 1.155,\\
w_2 &= 1 &&= 1.000, & w_6 &= (4/5)^{\frac{1}{2}} = 0.895,\\
w_3 &= 1 &&= 1.000, & w_7 &= (4/5)^{\frac{1}{2}} = 0.895,\\
& && & w_i &= 0 \text{ for all } i \geq 8.
\end{aligned}$$

These weights were chosen by reference to the functional equations (4) with the additional provision that their values lie close to 1.0.

Finally, we present $\{H_i \mid i = 0, \cdots, 7\}$ and $\{\mathcal{H}_i \mid i = 0, \cdots, 7\}$.

Let

$$r^2 = x^2 + y^2, \qquad Q(r) = (1 - r^2)^{\frac{1}{2}}.$$

Then

$$\begin{aligned}
H_0(x, y) &= (5/6\pi)Q(r)\cdot(1 + 2r^2),\\
H_1(x, y) &= (8/27\pi)^{\frac{1}{2}}Q(r)\cdot(5r^2 - 2),\\
H_2(x, y) &= (3/\pi)^{\frac{1}{2}}Q(r)\cdot x\cdot(4r^2 - 1),\\
H_3(x, y) &= (3/\pi)^{\frac{1}{2}}Q(r)\cdot y\cdot(4r^2 - 1),\\
H_4(x, y) &= (3/\pi)^{\frac{1}{2}}Q(r)\cdot x\cdot(3 - 4r^2),\\
H_5(x, y) &= (3/\pi)^{\frac{1}{2}}Q(r)\cdot y\cdot(3 - 4r^2),\\
H_6(x, y) &= (32/3\pi)^{\frac{1}{2}}Q(r)\cdot(x^2 - y^2),\\
H_7(x, y) &= (32/3\pi)^{\frac{1}{2}}Q(r)\cdot 2xy;
\end{aligned}$$

$$\mathcal{H}_0 = (24\pi/25)^{\frac{1}{2}}H_0, \tag{9}$$

$$\mathcal{H}_1 = (9/2)^{\frac{1}{2}}H_1, \tag{10}$$

$$\mathcal{H}_2 = H_2 + H_4, \tag{11}$$

$$\mathcal{H}_3 = H_3 + H_5, \tag{12}$$

$$\mathcal{H}_4 = (3/2)H_6, \tag{13}$$

$$\mathcal{H}_5 = (3/2)H_7, \tag{14}$$

$$\mathcal{H}_6 = (5/4)^{\frac{1}{2}}(H_2 - H_4), \tag{15}$$

$$\mathcal{H}_7 = (5/4)^{\frac{1}{2}}(H_3 - H_5). \tag{16}$$

$\{\mathcal{H}_i \mid i = 0, \cdots, 7\}$ is the orthonormal set of Hilbert vectors which was the original form of the solution of (4). $\{H_i \mid i = 0, \cdots, 7\}$ is used in the actual computations because this form saves a few arithmetic operations in the operator program. Each is a set of linearly independent vectors which define the same 8-dimensional subspace onto which $\mathcal{E}$ is projected by (2).

Now it is possible to define the exact function of the operator: For given input $\mathcal{E}(x, y)$ find (c, s, ρ, d, b) so that $\bar{\Lambda}(c, s, \rho, d, b)$ is minimal, and output that (c, s, ρ, d, b).

4. *Presentation of the Solution Theorem*

The theorem below reduces minimization of (8) to extremization[9] of (22). This means a decisive speed gain for the operator. Before the theorem is presented we need the following six definitions:

$$a_i = \int_{\mathfrak{D}} H_i(x, y)\mathcal{E}(x, y)dx\,dy \qquad (i = 0, \cdots, 7), \tag{17}$$

$$e_0(c, s) = a_2c + a_3s, \tag{18}$$

$$e_1(c, s) = a_4c + a_5s, \tag{19}$$

$$e_2(c, s) = a_1 + a_6(c^2 - s^2) + a_7 2sc, \tag{20}$$

$$U(c, s) = [e_0(c, s)/|\,e_0(c, s)\,|]\cdot[e_1^2(c, s) + e_2^2(c, s)]^{\frac{1}{2}}, \tag{21}$$

$$A(c, s) = e_0(c, s) + U(c, s). \tag{22}$$

The Solution Theorem. The locations of the global and local minima of $\bar{\Lambda}(c, s, \rho, d, b)$[10] coincide with the locations of the global and local extrema of $A(c, s)$, respectively. At these locations,

$$\rho = e_2(c, s)/\{2^{\frac{1}{2}}[U(c, s) + e_1(c, s)]\}, \tag{23}$$

$$d = 4A(c, s)/[(3\pi)^{\frac{1}{2}}(1 - \rho^2)^2(1 + 2\rho^2)], \tag{24}$$

and

$$b = a_0 - d\,\{4 + \rho[3 + \rho(2 + \rho)]\}\,(1 - \rho)^2/8. \tag{25}$$

The solution theorem will be proved in Section 5.

Thus, the operator program first locates the global extremum of $A(c, s)$. It then enters (23)–(25) to obtain (c, s, ρ, d, b). Finally, it decides if the edge data it generates are sufficiently noise free to warrant output. It does this by comparing the ideal edge element with the original data as follows.

Let $\mathcal{E}'$ and $\mathcal{F}'$ be the projections of the Hilbert vectors $\mathcal{E}$ and $\mathcal{F}$ onto the linear subspace spanned by $\{\mathcal{H}_i \mid i = 1, \cdots, 7\}$. Also let $\mathcal{E}$ and $\mathcal{F}$ be in input-output relation relative to the operator. Then the program calculates the cosine k between $\mathcal{E}'$ and $\mathcal{F}'$ using the formula

$$\frac{\mathcal{E}'\mathcal{F}'}{|\,\mathcal{E}'\,|\cdot|\,\mathcal{F}'\,|} = \left|\frac{A(c, s)}{[6a_1^2 + 2(a_2^2 + a_3^2 + a_4^2 + a_5^2) + 3(a_6^2 + a_7^2)]^{\frac{1}{2}}}\right|. \tag{26}$$

If $\mathcal{E}$ is already an ideal edge element, then $k = 1$. If $k < 0.9$, then the operator discards $\mathcal{E}$ as being too noisy.

To give an example, in the case of Figure 1(a), the following output quintuple

[9] "Extremization" means "maximization or minimization."
[10] That $c^2 + s^2 = 1$ is always assumed.

resulted:

$$(c, s, \rho, d, b) = (-0.99, 0.14, -0.28, 3.80, 4.08),$$

with $k = 0.95$.

Actually, it is desirable to make the acceptance of an edge somewhat dependent on the brightness step across the edge. This is handled by setting two thresholds (conf and diff) and then allowing an edge to be reported only if

$$1 - k < (1 - \text{conf})d^2/(\text{diff}^2 + d^2).$$

For 4-bit television pictures (16 grey shades differentiated), choices of thresholds in the range

$$0.86 < \text{conf} < 0.9, \qquad 1.50 < \text{diff} < 4.0$$

turned out to be most successful.

5. *Proof of the Solution Theorem*

Define λ and ω as follows:

$$\lambda = (3\pi/16)^{\frac{1}{2}} d(1 - \rho^2)^2(1 + 2\rho^2), \tag{27}$$

$$\sin \omega = (\sqrt{8})\rho/(1 + 2\rho^2), \tag{28}$$

$$\cos \omega = (1 - 2\rho^2)/(1 + 2\rho^2). \tag{29}$$

Then calculate $f_0, \cdots, f_7$ according to (7):

$$\begin{aligned}
f_0(c, s, \rho, d, b) &= (24\pi/25)^{\frac{1}{2}}(b + d\{4 + \rho[3 + \rho(2 + \rho)]\}(1 - \rho)^2/8),\\
f_1(c, s, \rho, d, b) &= (1/8)^{\frac{1}{2}}\lambda \sin \omega,\\
f_2(c, s, \rho, d, b) &= \tfrac{1}{2}\lambda c\cdot(1 + \cos \omega),\\
f_3(c, s, \rho, d, b) &= \tfrac{1}{2}\lambda s\cdot(1 + \cos \omega),\\
f_4(c, s, \rho, d, b) &= \tfrac{1}{2}\lambda\cdot(c^2 - s^2) \sin \omega,\\
f_5(c, s, \rho, d, b) &= \lambda cs\cdot \sin \omega,\\
f_6(c, s, \rho, d, b) &= (5/16)^{\frac{1}{2}}\lambda\cdot c\cdot(1 - \cos \omega),\\
f_7(c, s, \rho, d, b) &= (5/16)^{\frac{1}{2}}\lambda\cdot s\cdot(1 - \cos \omega).
\end{aligned}$$

Substitute these into (8) to obtain

$$\bar{\Lambda}(c, s, \rho, d, b) = T_1 + T_2 + T_3 + T_4, \tag{30}$$

where

$$\begin{aligned}
T_1 &= \text{a constant which does not depend on } c, s, \rho, d, b,\\
T_2 &= (24\pi/25)(a_0 - b - d\,(4 + \rho[3 + \rho(2 + \rho)])\,(1 - \rho)^2/8)^2,\\
T_3 &= \lambda^2,\\
T_4 &= -2\lambda\,\{a_1 \sin \omega + a_2 c + a_3 s + (a_4 c + a_5 s) \cos \omega + [a_6(c^2 - s^2)\\
&\qquad + a_7 2cs] \sin \omega\}.
\end{aligned} \tag{31}$$

Notice that we have expressed $\bar{\Lambda}(c, s, \rho, d, b)$ in terms of the a_i instead of the α_i. The relation between α_i and a_i is the same as the relation between $\mathcal{H}_i$ and H_i. [Compare (6), (9)–(16), and (17).] Since T_1 has no bearing on the minima of $\bar{\Lambda}(c, s, \rho, d, b)$, assume that

$$T_1 = 0.$$

In (30) the only term which depends on b is T_2. Therefore, no matter what the values of c, s, ρ, and d are, if b is chosen to minimize T_2, then it turns out that

$$T_2 = 0. \tag{32}$$

Thus,

the locations of the minima of $\bar{\Lambda}(c, s, \rho, d, b)$ and $T_3 + T_4$ coincide and $\bar{\Lambda}(c, s, \rho, d, b) = T_3 + T_4$ at the minima. (33)

Define β to be the expression within braces in (31). It follows that

$$T_3 + T_4 = \lambda(\lambda - 2\beta).$$

No matter what the value of β is, by varying d appropriately, λ can be made to take on any given value. Compare with eq. (27). Thus, for any given β the expression $\lambda(\lambda - 2\beta)$ is minimal at

$$\lambda = \beta, \tag{34}$$

where it takes on the value $-\beta^2$. That is to say that

the locations of the minima of $T_3 + T_4$ and $-\beta^2$ coincide, and $T_3 + T_4 = -\beta^2$ at the minima. (35)

Using (18)–(20), the expression for β shortens to

$$\beta = e_0(c, s) + e_1(c, s) \cos \omega + e_2(c, s) \sin \omega. \tag{36}$$

Where β is extremal, ω must extremize the scalar product

$$(e_1(c, s), e_2(c, s)) \cdot (\cos \omega, \sin \omega) \tag{37}$$

because nothing else in β depends on ω. The sign of (37) will then have to be the same as the sign of $e_0(c, s)$.

The following equations for $\cos \omega$ and $\sin \omega$ result:

$$\cos \omega = e_1(c, s)/U(c, s), \qquad \sin \omega = e_2(c, s)/U(c, s). \tag{38}$$

[For $U(c, s)$ see definition (21).] Substituting the above into β at (36)produces $A(c, s)$ as defined at (22). Thus, we have shown that

the extrema of β and $A(c, s)$ coincide, and $\beta = A(c, s)$ at the extrema. (39)

Statements (33), (35), and (39) together prove the solution theorem, and (23), (24), and (25) of the theorem are the same relations as (38), (34), and (32).

LEMMA: $\bar{\Lambda}(c, s, \rho, d, b) = \bar{\Lambda}(-c, -s, -\rho, -d, b + d)$.

PROOF: This follows because the two argument quintuples specify the same edge.

6. *An Empirical Demonstration of the Operator's Performance*

Picture 1 was read into a computer by a vidicon camera whose output was digitized into a 297 × 231 integer array. The brightness in each array element was coded by

a grey scale of 16 values. This digitized array served as input to the picture processing system. In addition, the input disk size (see Section 2.1) was set at 69, and the "conf" and "diff" thresholds (see Section 4) were set at 0.88 and 2.50, respectively.

In order not to miss details, the program covers pictures exhaustively in a scanning manner with a hexagonal grid of overlapping operator applications. When an edge is found, scanning is interrupted and the edge is traced until lost. When that happens, a search-mode is entered, and depending on the outcome of the search, either the edge tracing or systematic scanning resumes. Areas which were investigated previously are skipped during further scanning or tracing.

Processing the entire array takes about 1.5 minutes on the PDP-10, but much computing time is wasted on empty (edge-free) areas. As the author sees it, this situation opens the only logical possibility for saving computing time in the picture processing program.

Picture 2 is the plotted output. However, it does not contain the brightness values on each side of the edges, information very valuable to other programs which carry the picture processing onward [2].

7. *Summary and Conclusion*

Edge recognition plays an important role in computer vision. This stimulated the author to formulate a set of requirements which should be met by an edge recognizing operator and then to find the optimum solution to those requirements. The most important consideration in choosing the requirements is the trade-off between speed and reliability in the presence of noise. The requirements are stated, and the formal mathematical definition of the resulting local nonlinear operator follows. The operator is then cast into a workable and practical form by proof of a theorem which shows that a certain sequence of calculations will lead to the correct output of the operator. A demonstration of the operator's performance is presented.

The way a biological organism and a computer best approach the problem of vision need not be similar. While nature can afford large amounts of information processing because of parallel processing techniques, programmers may take advantage of shortcuts provided by mathematical analysis.

ACKNOWLEDGMENTS. The author expresses his gratitude to Gary Feldman, Lester Earnest, and Raj Reddy, who contributed to the growth of the paper with valuable suggestions.

REFERENCES

[Note: Refs. 4 and 5 are not cited in text.]

1. HUBEL, D. H., AND WIESEL, T. N. Receptive fields, binocular interaction and functional architecture in the cat's visual cortex. *J. Physiol. 160* (1962), 106–154.
2. GRAPE, G. R. Computer vision through sequential abstractions. AI-MEMO, Artif. Intel. Proj., Stanford U., Stanford, Calif., June 1969.
3. ROSENFELD, A. Picture processing by computer. *Comput. Surveys 1*, 3 (Sept. 1969), 147–176.
4. WICHMAN, W. M. Use of optical feedback in the computer control of an arm. Eng. thesis, Dep. of Eng., Stanford U., Stanford, Calif., Aug. 1967.
5. ROBERTS, L. G. Machine perception of three-dimensional solids. In *Optical and Electro-Optical Processing of Information*, MIT Press, Cambridge, Mass., 1965, pp. 159–197.

RECEIVED JANUARY, 1970; REVISED JUNE, 1970

Graphics and Image Processing   W. Newman Editor

Use of the Hough Transformation To Detect Lines and Curves in Pictures

Richard O. Duda and Peter E. Hart
Stanford Research Institute,
Menlo Park, California

Hough has proposed an interesting and computationally efficient procedure for detecting lines in pictures. This paper points out that the use of angle-radius rather than slope-intercept parameters simplifies the computation further. It also shows how the method can be used for more general curve fitting, and gives alternative interpretations that explain the source of its efficiency.

Key Words and Phrases: picture processing, pattern recognition, line detection, curve detection, colinear points, point-line transformation, Hough transformation

CR Categories: 3.63

This work was supported by the Advanced Research Projects Agency and the National Aeronautics and Space Administration under Contract NAS12-2221.

1. Introduction

A recurring problem in computer picture processing is the detection of straight lines in digitized images. In the simplest case, the picture contains a number of discrete, black figure points lying on a white background. The problem is to detect the presence of groups of colinear or almost colinear figure points. It is clear that the problem can be solved to any desired degree of accuracy by testing the lines formed by all pairs of points. However, the computation required for n points is approximately proportional to n^2, and may be prohibitive for large n.

Rosenfeld [1] has described an ingenious method due to Hough [2] for replacing the original problem of finding colinear points by a mathematically equivalent problem of finding concurrent lines. This method involves transforming each of the figure points into a straight line in a *parameter space*. The parameter space is defined by the parametric representation used to describe lines in the picture plane. Hough chose to use the familiar slope-intercept parameters, and thus his parameter space was the two-dimensional slope-intercept plane. Unfortunately, both the slope and the intercept are unbounded, which complicates the application of the technique. In this note we suggest an alternative parameterization that eliminates this problem. We also give two alternative interpretations of Hough's method, one of which reveals plainly the source of its efficiency. Finally, we show how the method can be extended to find more general classes of curves in pictures.

2. Fundamentals

The set of all straight lines in the picture plane constitutes a two-parameter family. If we fix a parameteriza-

Reprinted with permission from *Commun. Ass. Comput. Mach.*, vol. 15, pp. 11–15, Jan. 1972.

tion for the family, then an arbitrary straight line can be represented by a single point in the parameter space. For reasons that become obvious, we prefer the so-called *normal parameterization*. As illustrated in Figure 1, this parameterization specifies a straight line by the angle θ of its normal and its algebraic distance ρ from the origin. The equation of a line corresponding to this geometry is

$$x \cos \theta + y \sin \theta = \rho.$$

If we restrict θ to the interval $[0, \pi]$, then the normal parameters for a line are unique. With this restriction, every line in the x-y plane corresponds to a unique point in the θ-ρ plane.

Suppose now, that we have some set $\{(x_1, y_1), \ldots, (x_n, y_n)\}$ of n figure points and we want to find a set of straight lines that fit them. We transform the points (x_i, y_i) into the sinusoidal curves in the θ-ρ plane defined by

$$\rho = x_i \cos \theta + y_i \sin \theta. \qquad (1)$$

It is easy to show that the curves corresponding to colinear figure points have a common point of intersection. This point in the θ-ρ plane, say (θ_0, ρ_0), defines the line passing through the colinear points. Thus the problem of detecting colinear points can be converted to the problem of finding concurrent curves.

A dual property of the point-to-curve transformation can also be established. Suppose we have a set $\{(\theta_1, \rho_1), \ldots, (\theta_k, \rho_k)\}$ of points in the θ-ρ plane, all lying on the curve

$$\rho = x_0 \cos \theta + y_0 \sin \theta.$$

Then it is easy to show that all these points correspond to lines in the x-y plane passing through the point (x_0, y_0). We can summarize these interesting properties of the point-to-curve transformation as follows:

Property 1. A point in the picture plane corresponds to a sinusoidal curve in the parameter plane.

Property 2. A point in the parameter plane corresponds to a straight line in the picture plane.

Property 3. Points lying on the same straight line in the picture plane correspond to curves through a common point in the parameter plane.

Property 4. Points lying on the same curve in the parameter plane correspond to lines through the same point in the picture plane.

In Section 3 we apply these results to the problem of detecting colinear points in the picture plane and show how significant computational economies can be realized in certain situations.

3. Applications and Alternative Interpretations

Suppose we map all of the points in the picture plane into their corresponding curves in the parameter plane. In general, these n curves will intersect in $n(n - 1)/2$ points corresponding to the lines between all pairs of

Fig. 1. The normal parameters for a line.

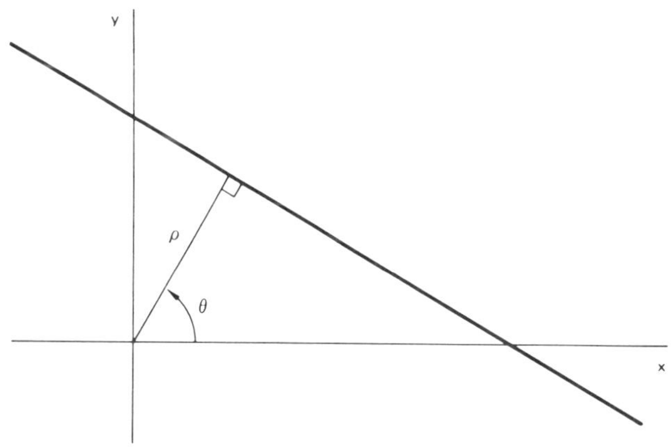

Fig. 2. Projection of colinear points onto a line.

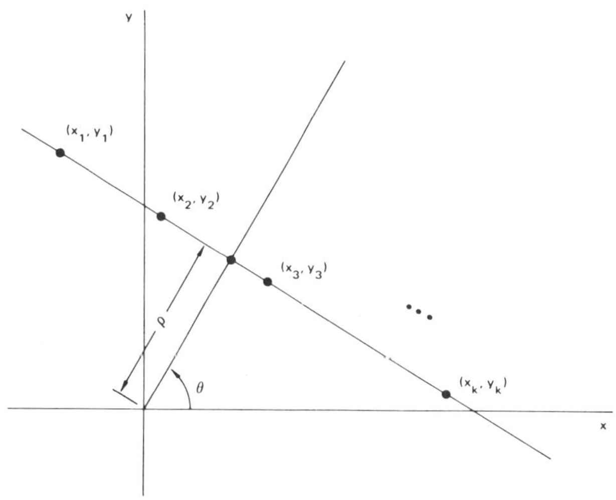

figure points. Exactly colinear subsets of figure points can be found, at least in principle, by finding coincident points of intersection in the parameter plane. Unfortunately, this approach is essentially exhaustive, and the computation required grows quadratically with the number of picture points.

When it is not necessary to determine the lines exactly, the computational burden can be reduced considerably. Following Hough's basic proposal, we specify the acceptable error in θ and ρ and quantize the θ-ρ plane into a quadruled grid. This quantization can be confined to the region $0 \leq \theta < \pi$, $-R \leq \rho \leq R$, where R is the size of the retina, since points outside this rectangle correspond to lines in the picture plane that do not cross the retina. The quantized region is treated as a two-dimensional array of accumulators. For each point (x_i, y_i) in the picture plane, the corresponding curve given by (1) is entered in the array by incrementing the count in each cell along the curve. Thus, a given cell in the two-dimensional accumulator eventually records the total number of curves passing through it. After all figure points have been treated, the array is inspected to find cells having high counts. If the count in a given cell (θ_i, ρ_j) is k, then precisely k figure points lie (to within quantization error) along the line whose normal parameters are (θ_i, ρ_j).

An alternative interpretation of the point-curve transformation can be obtained by recognizing that the ρ computed by (1) locates the projection of the point (x_i, y_i) onto a line through the origin with slope angle θ. Thus, if a number of figure points lie close to some line l, their projections onto the line normal to l are nearly coincident (see Figure 2). A given column in the θ-ρ accumulator array is just a histogram for these projections, so a high count in a given cell clearly corresponds to a nearly colinear subset of figure points. A variation of this approach was used by Griffith [3] to find long lines in a picture.

Let us investigate how the computation required by the accumulator implementation varies with the number of figure points. To be more specific about the quantization, suppose that we restrict our attention to d_1 values of θ uniformly spaced in the interval $[0, \pi)$. Suppose further that the ρ axis in the interval $[-R, R]$ is quantized into d_2 cells. For each figure point (x_i, y_i), we use (1) to compute the d_1 different values of ρ corresponding to the d_1 possible values of the independent variable θ. Since there are n figure points, we need to carry out this computation nd_1 times. When these computations are complete, the d_1d_2 cells of the two-dimensional accumulator are inspected to find high counts. Thus the computation required grows linearly with the number of figure points. Clearly, when n is large compared to d_1, this approach is preferable to an exhaustive procedure that requires considering the lines between all $n(n-1)/2$ pairs of figure points.

Another alternative interpretation exposes the source of this efficiency. Consider again Property 4 in Section 2: Points lying on the same curve in the θ-ρ plane correspond to lines through the same point in the picture plane. When the curve corresponding to figure point (x_i, y_i) is "added" to the accumulator, we are really computing and recording the parameters of the d_1 lines in the picture plane passing through (x_i, y_i), and because θ is quantized, these are "all the lines in the plane" passing through (x_i, y_i). Should a given parameter pair ever recur as a result of computing the d_1 lines through some other figure point, the recurrence will be reflected in an increased count in the appropriate accumulator cell. Roughly speaking, then, for each figure point the quantized transform method considers only the set of all d_1 lines through that point, whereas more exhaustive methods consider all $(n-1)$ lines between the given point and all other figure points.

4. Example

The following example illustrates some of the features of the transform approach. Figure 3(a) shows a television monitor view of a box, and Figure 3(b) shows a digitized version of that view. A simple differencing operation locates significant intensity changes and produces the binary picture shown in Figure 3(c). This 120×120 picture contains many nearly colinear figure points that can be fit well by a few straight lines.

Sampling θ at $d_1 = 9$ twenty-degree increments in θ and, quantizing ρ into $d_2 = 86$ two-element cells, we obtain the two-dimensional accumulator array shown in Table I. If the array entry at (θ_0, ρ_0) is k_0, then k_0 figure points lie on parallel lines for which $\theta = \theta_0$, and ρ lies between ρ_0 and $\rho_0 + 2$. When many points are nearly colinear, the entry for the line that fits them best is large. The largest entry in Table I occurs at $(0°, -5)$ and corresponds to the middle vertical edge of the box. The nine circled entries correspond to locally maximum values that exceed the arbitrary threshold of 35. The corresponding nine groups of nearly colinear figure points are shown in Figure 3(d). In this example, it happens that every group corresponds to some physically meaningful line in the picture. However, two significant lines on the top of the box were not found—one, because it contained very few points, and the other, because it fell between the lines at $\theta = 80°$ and $\theta = 100°$. The 20° angular quantization interval was chosen to keep the accumulator array small. Clearly, we were fortunate to have found as many lines as we did, and a smaller quantization interval would have to be used in practice.

A few remarks concerning some limitations of the transform approach are in order. First, the results are sensitive to the quantization of both θ and ρ. Finer quantization gives better resolution, but increases the computation time and exposes the problem of clustering entries corresponding to nearly colinear points. Second, the technique finds colinear points without regard to contiguity. Thus the position of a best-fit line can be distorted by the presence of unrelated figure points in another part of the picture. A related problem is that of meaningless groups of colinear points being detected. In our example, a false line would be detected if the threshold were reduced from 35 to 24, the value needed to detect the top left-hand edge of the box.

The transform approach does successfully find groups of colinear or nearly colinear figure points. If the minimum size of a significant group is known, all such groups can be detected. If additional properties such as contiguity are known, they can be used to reject meaningless results. In general, the transform approach should be viewed primarily as a computationally efficient way of accomplishing a conceptually simple step in scene analysis.

5. Extensions and Conclusions

The transform method can be generalized and specialized in several ways. We note immediately that any parametrization of the family of straight lines can be used. As we have mentioned, Hough used the slope-intercept parameterization. However, this parameterization has the disadvantage of being sensitive to the choice of coordinate axes in the picture plane. If several figure points lie on a nearly vertical line, for example, both the slope and the intercept may be arbitrarily large. Thus the entire two-dimensional parameter plane must be considered. As Rosenfeld [1] has pointed out, one could do the entire problem twice, interchanging the x- and y-axes, but this would introduce additional complications. The normal parametrization avoids these disadvantages, fundamentally for the same reasons that make it useful in integral geometry: It allows us to place an invariant measure on the set of all straight lines.

An important special use of the transform method is to detect the occurrence of figure points lying on a straight line and possessing some specified property. For example, suppose we want to find whether a significant number of figure points lie on a line through the point (x_0, y_0) in the picture plane. As we have seen from Property 4, the normal coordinates of any such line must lie on (or, in practice at least near) the curve $\rho = x_0 \cos\theta + y_0 \sin\theta$. Hence, the transform process can be carried out in the usual way, but attention can be restricted to the region of the θ-ρ plane near this curve. If we find a cell with count k near this curve, then we are assured that k figure points lie on a line passing (nearly) through the point (x_0, y_0). Similarly, suppose we are interested only in lines having a given direction, say θ_0. Again, we carry out the process in the usual way, but restrict our attention to a subset of the θ-ρ plane in the vicinity of $\theta = \theta_0$.

It is clear that the general transform approach can be

Fig. 3 An illustrative example.

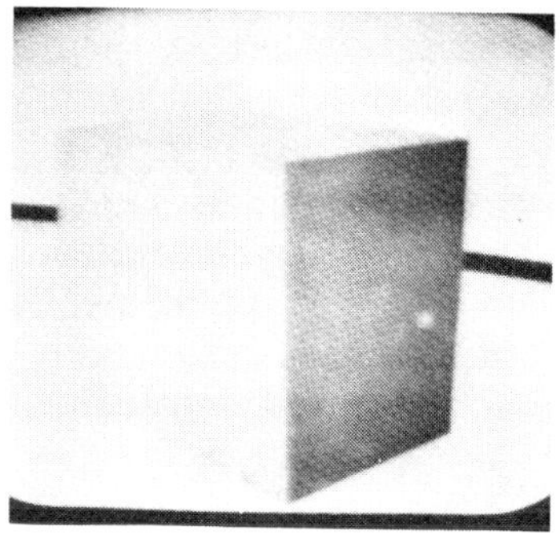

(a) Monitor

(b) Digitized

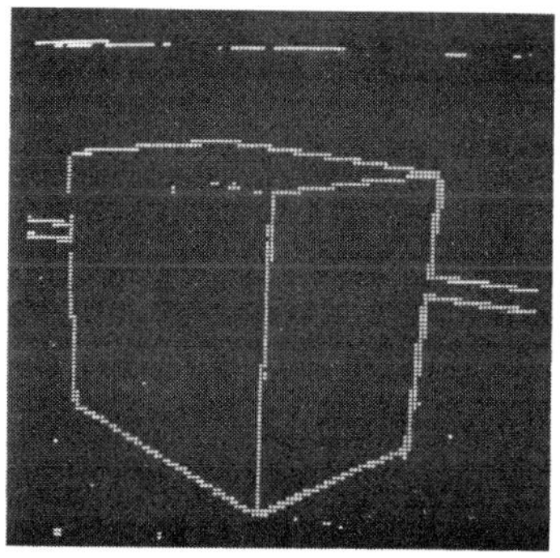

(c) Gradient

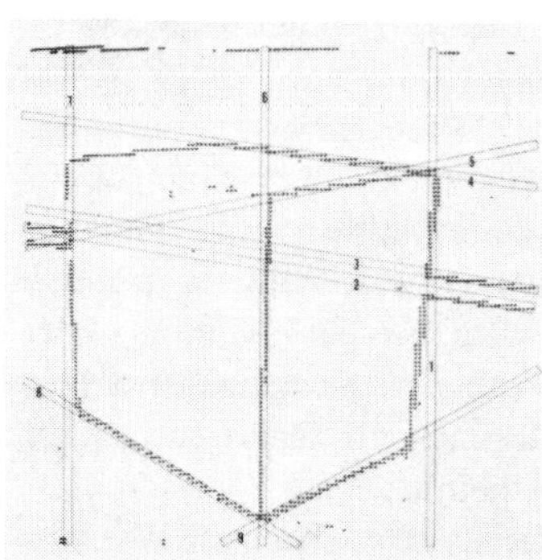

(d) Lines

Table I. Accumulator Array for Figure 3(c)

ρ \ θ	0°	20°	40°	60°	80°	100°	120°	140°	160°
85								2	
83							1	2	
81							4	4	
79			2				6	5	
77				2			8	4	2
75							6	6	2
73							4	3	3
71		2		1		1	4	2	3
69			1	4		12	4	3	5
67			3		2	14	2	3	4
65			1			11	1	2	4
63					5	2		2	4
61						1		3	9
59	4	1			11	9	1	8	12
57	4	3		3	10	12	3	10	15
55	9	5		4	5	4	5	11	12
53	6	6		4	10		11	9	14
51	4	9		4	20	2	11	10	8
49	5	6		2	10	3	11	13	8
47	8	4	4	4		2	13	10	10
45	4	7	14	3			11	6	8
43	4	18	21	5		1	12	10	8
41	9	17	21	15		25	18	7	8
39	8	20	21	13		22	11	11	7
37	12	17	22	17		9	10	9	10
35	(38)	14	17	17	38	8	7	9	6
33	37	16	22	21	(42)	10	5	9	9
31	35	11	21	23	23	8	11	9	10
29	13	18	18	23	20	14	13	9	9
27	7	16	12	30	20	20	7	9	6
25	7	18	12	32	19	27	8	7	8
23	8	12	11	20	17	(52)	11	6	7
21	7	17	12	23	8	11	15	11	10
19	9	14	12	16	7	7	14	6	7
17	9	12	12	16	6	9	16	12	7
15	8	13	13	11	7	10	16	14	10
13	10	9	15	11	7	10	16	13	6
11	12	11	13	14	(40)	10	16	13	13
9	10	10	16	14	8	9	14	21	22
7	10	8	22	12	(41)	6	7	12	21
5	11	12	15	11	23	6	11	14	14
3	13	15	15	8	18	7	11	16	15
1	10	14	17	11	7	8	9	10	12

ρ \ θ	0°	20°	40°	60°	80°	100°	120°	140°	160°
−85									
−83									
−81									
−79			3	1					
−77			1	3					
−75									
−73									
−71		2							
−69		2							
−67									
−65			1	2					
−63		1			2				
−61	5								
−59	7		2	1		1			
−57	6					2	1		2
−55	0	10	6			1	1		4
−53	16	13	12		18	4	1		6
−51	(40)	15	11		15	16			6
−49	32	18	11	(57)	15	23	1	1	5
−47	10	16	11	22	14	16	21	9	5
−45	7	17	11	11	16	18	(41)	21	6
−43	8	12	14	10	13	17	12	17	6
−41	6	7	14	11	14	14	7	19	12
−39	7	10	9	8	12	8	11	20	23
−37	7	7	14	8	17	9	12	18	24
−35	8	9	17	8	10	7	10	23	23
−33	6	12	15	8	12	9	11	22	26
−31	5	9	19	9	8	11	16	18	15
−29	9	10	12	9	8	9	18	18	15
−27	7	12	10	8	6	9	18	19	19
−25	5	10	8	8	7	7	22	9	14
−23	6	11	9	9	6	11	19	12	9
−21	7	15	9	7	10	10	16	10	11
−19	6	13	8	16	9	11	17	9	10
−17	7	17	9	15	7	11	16	14	13
−15	6	15	10	17	8	13	10	14	9
−13	10	15	9	15	9	17	11	13	12
−11	10	13	10	7	8	17	9	11	15
−9	7	14	8	7	8	23	8	12	15
−7	9	15	12	7	8	21	7	13	12
−5	(77)	13	15	9	7	14	10	12	15
−3	26	14	14	6	8	12	9	11	18
−1	10	13	18	9	8	8	11	12	15

extended to curves other than straight lines. For example, suppose we want a method to detect circular configurations of figure points. We can choose a parametric representation for the family of all circles (within a retina) and transform each figure point in the obvious way. If, as a parametric representation, we describe a circle in the picture plane by

$$(x - a)^2 + (y - b)^2 = c^2,$$

then an arbitrary figure point (x_i, y_i) will be transformed into a surface in the a-b-c parameter space defined by

$$(x_i - a)^2 + (y_i - b)^2 = c^2.$$

In this example, then, each figure point will be transformed into a right circular cone in a three-dimensional parameter space. If the cones corresponding to many figure points intersect at a single point, say the point (a_0, b_0, c_0), then all the figure points lie on the circle defined by those three parameters. As in the preceding case of straight lines, no saving is effected if the entire process is performed analytically. However, the process can be implemented efficiently by using a three-dimensional array of accumulators representing the three-dimensional parameter space.

In principle, then, the transform method extends to arbitrary curves. We need only pick a convenient parameterization for the family of curves of interest and then proceed in the obvious way. A parameterization having bounded parameters is obviously preferable, although this is not essential. It is much more important to have a small number of parameters since the accumulator implementation requires quantization of the entire parameter space and the computation grown exponentially with the number of parameters.

Received November 1970, revised April 1971

References

1. Rosenfeld, A. *Picture Processing by Computer*. Academic Press, New York, 1969.
2. Hough, P.V.C. Method and means for recognizing complex patterns. U.S. Patent 3,069,654, Dec. 18, 1962.
3. Griffith, A.K. Computer recognition of prismatic solids. Ph.D. Th., Dep. of Math., MIT, June 1970.

E. C. Greanias
P. F. Meagher
R. J. Norman
P. Essinger

The Recognition of Handwritten Numerals by Contour Analysis

Abstract: A character recognition system has been developed for the recognition of handwritten numerals. This system uses a logically controlled cathode ray tube scanner to generate basic measurements that characterize significant features of the numeral shapes. A contour-follower procedure is used to control the scanner. In addition, special scanner subroutines initiated by feedback from the recognition logic are utilized. Character shape data are generated in a sequential form, which can be analyzed for recognition with an easily realizable logic.

An experimental model has been built that recognized 99.3% of numerals written by 45 subjects after 30 minutes of training. The error rate for these characters was 0.11%. The rejected character rate was 0.59%.

Introduction

Over the past ten years, optical character recognition has developed as a method for entering printed characters into information processing systems. This paper describes a practical method for the recognition of numerals that have been recorded by hand with an ordinary pen or pencil.

Several authors[1,2] have noted that the general pattern recognition system can be described in two principal parts:

1. A first part (sometimes called the *receptor*) that produces a set of quantities which characterize the pattern in terms of a prescribed set of measurements. These measurements can be continuous or discrete, or both.

2. A second part (sometimes called the *categorizer*) that applies a decision criterion to the pattern measurements for a decision as to which, if any, of the allowable classes the pattern belongs.

The effectiveness of a recognition system will depend upon how well the significant differences between pattern classes are characterized by the prescribed set of measurements,[3] and the accuracy with which the decision criterion categorizes the pattern measurements. Pattern recognition literature contains a number of papers that discuss optimum decision functions[4] and practical methods for the selection of the decision criteria. The recognition system that is described in this paper employs a simple, direct method for generating significant measurements that characterize the fundamental shape differences among handwritten numerals. The categorizer that has been adopted employs detailed logical statements of the acceptable combinations of these measurements for each character to decide the identity of the unknown pattern. A weighted sum of probabilities decision function might have been used, but "detailed logical statements" were chosen instead to simplify the description of the handwritten numeral shapes that are acceptable for recognition. Futhermore, the implementation of the detailed logical statements for handwritten numerals was sufficiently economical to convince the authors that cost optimization of the categorizer is a secondary consideration for this system.

To provide a practical solution for the handwritten numeral recognition problem, it was required that preparing documents for the machine reader should not constrain the writer so severely as to inhibit any other functions that he had to perform at the same time. To accomplish this, considerable tolerance was provided for:

• *Character registration.* If necessary, characters can be located anywhere within an area several inches square.

• *Character size.* The height of characters can vary over a 4-to-1 range.

• *Shape.* The tolerance for shape is defined in terms of the shapes that occur in unconstrained handwritten numerals. The initial recognition criteria were selected to cope with more than 90% of the character shapes found in 3000 samples of unconstrained numbers.[5]

• *Line quality.* Lines from medium hard pencils and dark-ink ball point pens in reasonably good condition are acceptable.

• *Character rotation or slant.* Normal characters can be rotated ±20°.

This paper will outline the basic operation of the recognition system that was developed to solve this problem and some of the procedures that were used to define the recognition criterion.

Basic operation

A block diagram of the basic operation of this pattern recognition system is shown in Fig. 1. The characters are scanned on the stationary document by a flying-spot scanner. Signals from the scanner corresponding to the document reflectivity are usually the direct output of the receptor. In this system, they are not used directly, but feed a beam-control logic which in turn generates the signals that are analyzed. The beam-control logic is also controlled by feedback from the categorizer.

The characters are scanned optically along a variable path that depends upon the information obtained as scanning proceeds. The actual character on the document serves as the primary store of raw character information. The scanner is the means for retrieving the information as required from this "memory". Specific regions of the character are scanned selectively (and repetitively, when necessary) according to their significance in character identification. The scanner is also made to seek significant features in new areas of the character when prior scanning does not provide conclusive results.

Scanning starts with a search pattern that "looks" for a character on the document. This search pattern may take any form that is compatible with the document but would typically be a raster scan. When a character is detected by means of a "black" photo signal, the search pattern is stopped and signals are generated in the beam-control logic to cause the beam to follow the edge of the character with an exploratory circular motion, as shown by the photograph in Fig. 2. The signals that are passed on to the categorizer for analysis are essentially the beam deflection signals, with the circular pattern filtered out to provide a smoothed outline of the character[6] as illustrated by the photograph in Fig. 3. After recognition has been accomplished, the search operation is resumed in the region beyond the character.

Machine-reading capabilities

As a consequence of this mode of operation, several important machine-reading capabilities can be achieved in a straightforward and economical way.

• *Independence of character size and location*
To accomplish recognition, the contours of each character are traced two or more times. During the first tracing, the extreme positive and negative excursions of the character's outer contour in x and y are stored. On subsequent passes around the character, the beam is located relative to the character extremities and therefore is not dependent on the character location or the point of initial contact. The stored peaks of the character contours provide a measure of character size

Figure 1 **Block diagram of character recognition system utilizing contour analysis.**

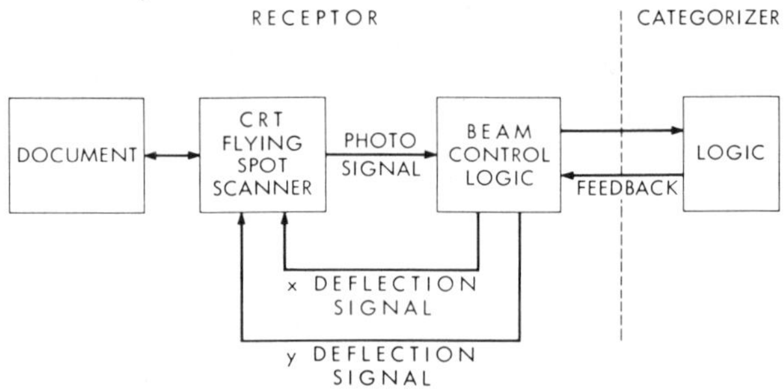

Figure 2 **Photograph of deflection signals, showing search and follow modes.**

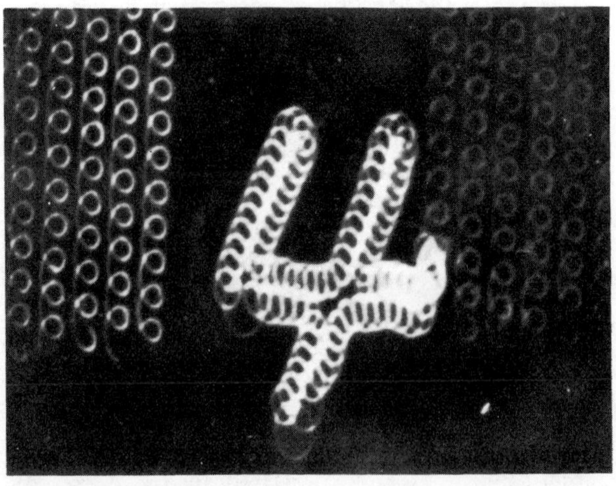

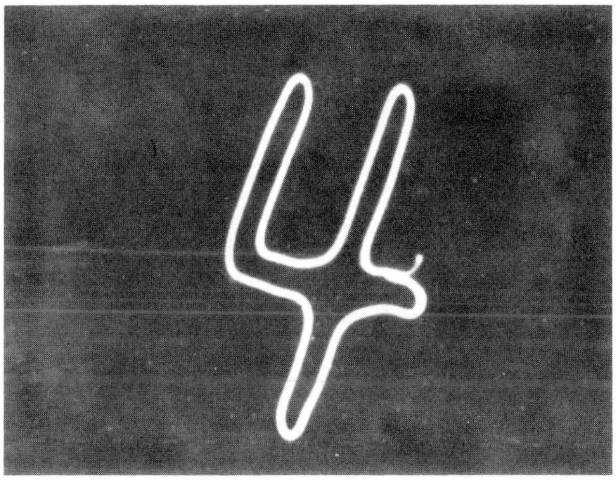

Figure 3 **Photograph of filtered deflection signals.**

which is used to adjust (i.e., to normalize) the vertical and horizontal distance measures of the system. The character size measurement is also used to adjust the scan resolution (circle size) on subsequent passes around the character. This ensures that all important shape features will be detected, and that the time to follow differently sized characters of the same shape will be approximately the same. Figure 4 contains photographs of actual scanning traces that occurred during the first and second pass around the character **m**. The improved resolution on the second pass makes the shape of the **m** apparent.[7]

• *Tolerance for line quality*

It is evident that successful recognition of a character in this system depends upon how well the line edge is followed. For this reason it is worth while to employ some fairly sophisticated methods to determine the location of lines having marginal quality. The follower method used causes the beam to move in and out of the line periodically. By using the recent histories of the "black" and "white" signals, the "local" black-white threshold that is maintained to discriminate the reflected light signals can be adjusted to automatically accommodate large variations in the character line and paper background reflectivities. Furthermore, the occurrence of "black" signals can be predicted on the basis of the direction of the line edge that is being followed. When a discontinuity occurs, a special scanning subroutine is used to check the validity of signals that do not occur as predicted. For example, in the case of weak lines, the failure of "black" to occur when predicted initiates a routine which causes the beam to explore an adjacent region of the line with a modified black-white threshold.

• *Tolerance for shape*

The analysis of the filtered beam deflection signals is based on two properties of the line edges that they represent:

a) The relative location of the line edge within the character.

b) The approximate direction of the line edge.

By considering both properties simultaneously, the significance of most line edge signals can be detected without resorting to precise measurements of location and direction. In some instances, such as certain **8**'s and **0**'s, identity cannot be established by coarse measurements on the outer contour of the character. For these cases, the scanner is redirected to obtain more information from the interior of the character, which also can be analyzed with simple, coarse measurements, to resolve the conflict. Since accurate recognition can be accomplished without resorting to precise measurements of line locations or directions, this system provides large tolerance for shape variations.

The experimental reader

The experimental model of this recognition system

Figure 4 **Photograph illustrating the effects of circle size adjustments following the character "m".**

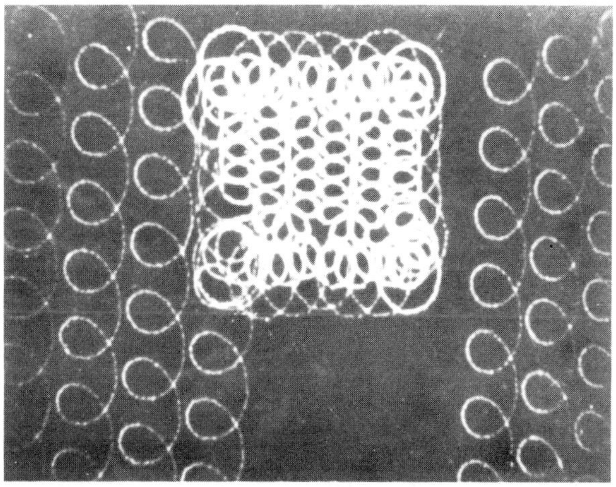

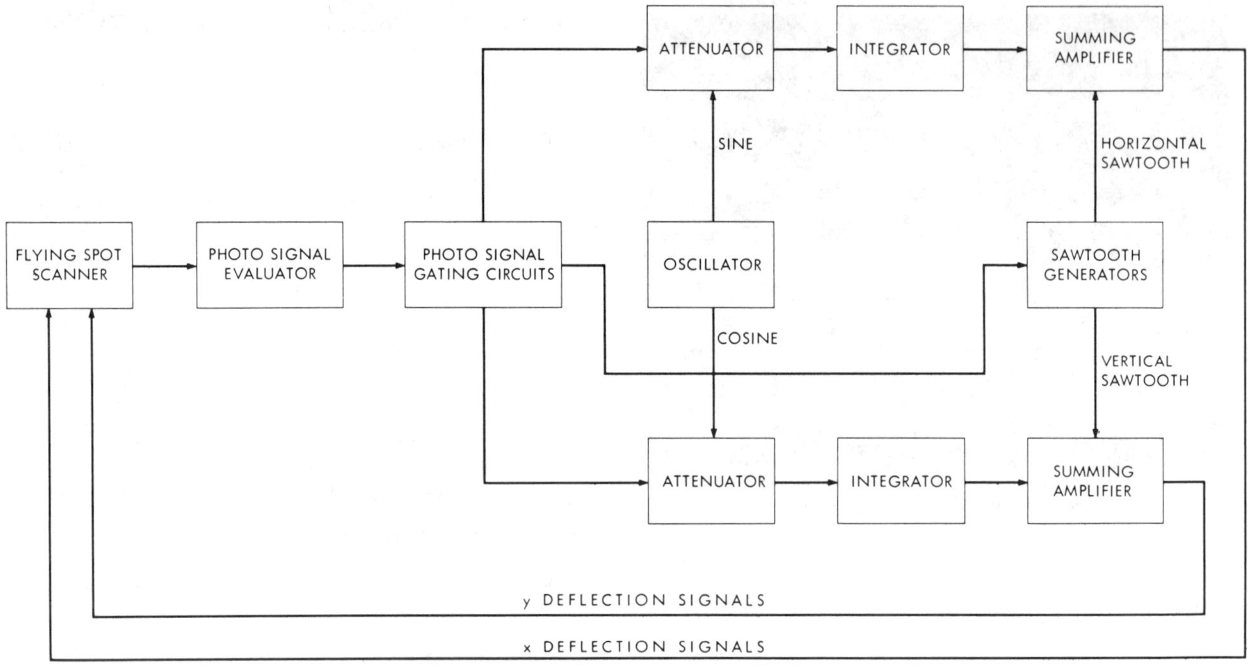

Figure 5 **Block diagram of beam control logic.**

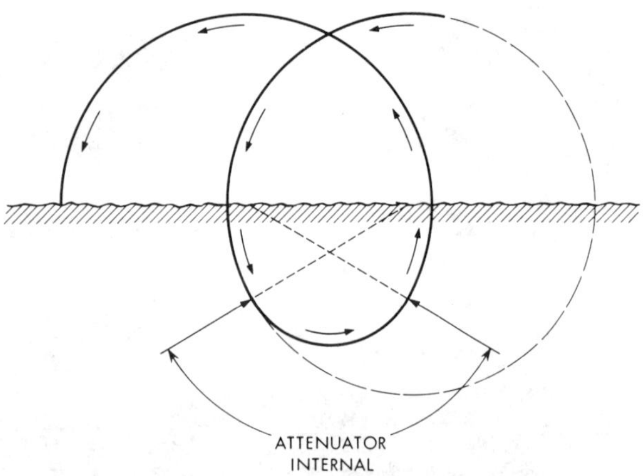

Figure 6 **Beam motion during line following.**

that has been constructed and tested employs a cathode ray tube flying-spot scanner designed to read numerals in a 2″ × 3″ area on an IBM card. The scanning procedure includes the search mode, which is used to find the characters, the follow mode, which is used to explore the outer contour of the characters, and an internal scan mode, which is used to explore the inner contours of selected characters. The basic principle of the beam-control logic is illustrated schematically in Fig. 5. The x and y deflection signals are generated by summing the exploratory circle component and sawtooth sweep signal for each axis. When a character is detected and the search mode is terminated, the outputs of the sawtooth generators are frozen. Circular deflection components are still fed to the "summers," but to achieve edge follower action the radius of the circular motion is attenuated for a fixed interval after the beam has entered the line. This attenuation interval is timed to occur when the beam is moving essentially parallel to the line edge but opposite to the direction of advance along the edge. Follower action is accomplished by integrating the x and y circle components that provide the circular motion. Because of the action of the integrators, the beam cannot move abruptly when the radius is suddenly attenuated. Instead, at the instant of attenuation the beam does not move, but the center of the circle is shifted in the direction of advance. When the attenuation interval is finished, the center of the circle is also shifted in the direction of advance. This in effect generates a cycloidal motion (see Fig. 6). The duration and magnitude of the attenuation determines the advance per circle.

To generate the waveforms for analysis, the outputs of the integrators are filtered by a low-pass network that cuts off at approximately one-third of the exploratory circle frequency. On the first pass around the character, the filtered signals are fed into capacitor peak-storage circuits. The outputs of the positive and negative peak storage circuits for each deflection axis are connected through a resistor divider network. In this model the vertical deflection range is divided into four equal increments and the horizontal deflection range into three equal increments by comparing the filtered deflection signals on subsequent passes with voltage thresholds defined by the resistor divider

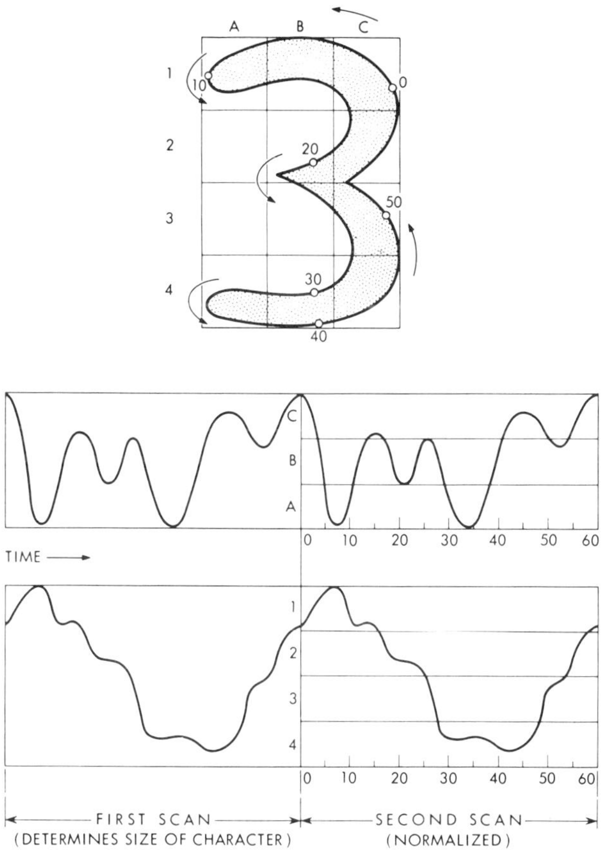

Figure 7 **Normalized scan matrix.**

networks. These thresholds define 12 possible zones in a 3 × 4 array that the deflection beam can enter as it travels around the character. Signals that identify the location of the beam within the 12 zones are continuously generated in the model. Figure 7 illustrates this operation.

Line directions are determined in this experimental model by first generating the time derivative of the filtered horizontal and vertical deflection signals. A comparison is then made of the instantaneous values of these derivatives to determine in which of eight 45° sectors (i.e., N, NE, E, SE, S, SW, W, NW) the beam is directed. This is accomplished by implementing the following relationships:

$A \qquad dx/dt < (\tan 22.5°)dy/dt$

$B \qquad -dx/dt > (\tan 22.5°)dy/dt$

$C \qquad dy/dt < -(\tan 22.5°)dx/dt$

$D \qquad -dy/dt > -(\tan 22.5°)dx/dt$

The conditions for the instantaneous line directions are:

$$\begin{aligned} \mathrm{N} &= A\cdot\bar{B} \\ \mathrm{NE} &= B\cdot\bar{C} \\ \mathrm{E} &= C\cdot\bar{D} \\ \mathrm{SE} &= D\cdot\bar{A} \\ \mathrm{S} &= \bar{A}\cdot B \\ \mathrm{SW} &= \bar{B}\cdot C \\ \mathrm{W} &= \bar{C}\cdot D \\ \mathrm{NW} &= \bar{D}\cdot A \end{aligned}$$

When the scanning beam passes the point of original contact with the character, a signal is generated by a circuit which monitors the condition of the follower integrators and the direction of beam motion. This pulse signifies the start of the second pass around the character and turns on all of the subsequent recognition operations.

Recognition is based upon selected sequences of line directions and locations that are detected after the first pass around the character.[10] To provide adequate shape tolerance, the filtered beam deflection signals from 3000 uncontrolled characters were recorded and studied by simulation on an IBM 7090 computer. The line direction and location signals of each character were sampled 60 times around the periphery to produce 120 code digits (one digit for direction and one digit for position, per sample) as the input for the simulation program.

First, significant features were identified for each character type. Then shape feature tests were defined, and recognition logics were assembled that satisfied 2752 of the characters in the sample. These recognition criteria were selected in accordance with the logic designers' subjective judgments regarding "normal" character shapes. The shape features that were emphasized for each character are summarized in Table 1.

Figure 8 illustrates a typical recognition criterion for the numeral **5**. The features that are required are identified by the small arrows adjacent to the character outline. For example, stroke labeled I is defined by the occurrence of an east (E) beam motion anywhere in the second (2) row of the array or east (E) or southeast (SE) motion in the third (3) row of the array. When this simple condition is satisfied for an adequate short interval, a latch is set to store the event and enable the detection of the second required stroke (II) by a corresponding logical circuit. Similar implementation is provided for strokes (III), (IV), (V) and (VI). The output that is obtained when stroke (III) is satisfied is used in the recognition statement of other characters and is defined as a shape feature test. Strokes (V) and (VI) call for extended occurrence of the indicated horizontal motion so longer durations of the signals are required to set the corresponding latches.

The characters **2**, **3**, **4** and **7** are also recognized by similar recognition logics but, when the possible occurrence of **0**, **1**, **6**, **8**, or **9** is detected, further scanning of the internal regions of the numeral is required to insure that errors do not occur.[12, 13] To accomplish this, the scanner is forced to break through the line and

Table 1 **Shape features identified for each character.**

Significant Feature	*Location*	*Use**
long N, NE or NW strokes	right side	0 1 7
long S, SW or SE strokes	left side	0 1 7
horizontal line end	top left	2 3 7
horizontal line end	top right	5 6
horizontal line end	bottom left	3 5
horizontal line end	bottom right	2
loop	bottom	2
west bay	bottom left	3 5
north bay	top center	4
northeast bay	top right	9 8
short arc	top left	9
NE stroke	top left	4
horizontal line end	right center	4
east bay	top right	6 5
notches	left	8
notches	right	8
southwest bay	bottom left	4 9
short arc	bottom right	6 5
large lake	center	0
small lake	top center	8 9
small lake	bottom center	8 6

* Note: Many of these features are also used as inhibit conditions in other characters.

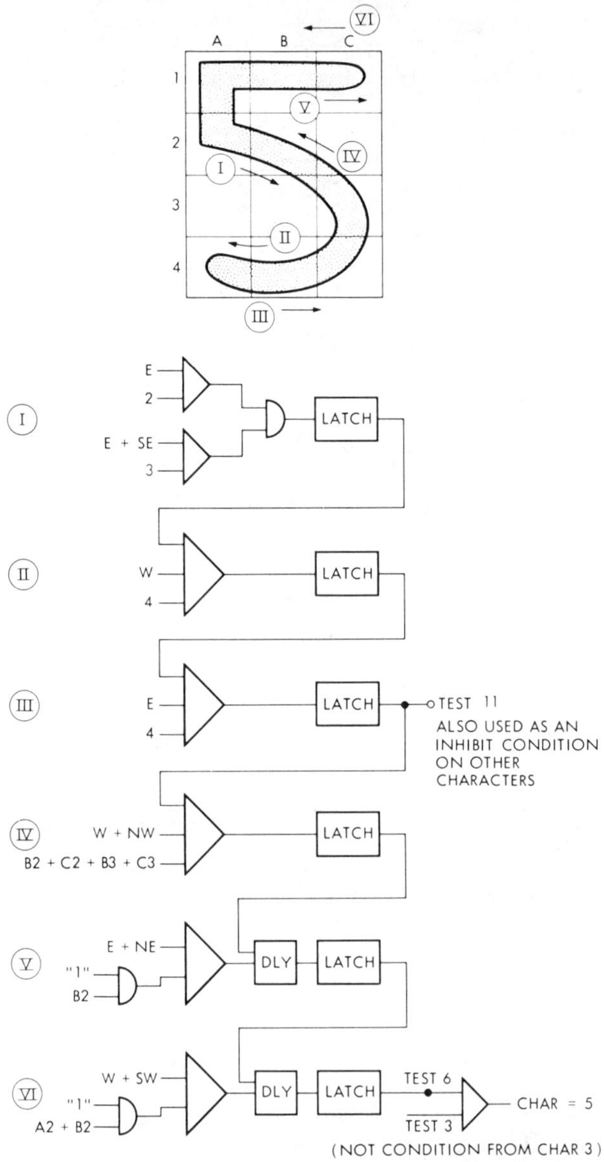

Figure 8 **Recognition criterion for the numeral 5.**

follow the internal contour. See Fig. 9. The information derived from the inner contour scan is used to define shape feature tests similar to those derived from the outer contours and final recognition takes place when appropriate inside shape features are detected.

Results

The recognition logics for this model were implemented with conventional circuits using approximately 800 transistors. An additional 200 transistors were required for the circuits that generated the line direction and location signals. The cathode ray tube deflection circuits and beam-control logic required approximately 400 transistors. The reading capabilities of this model were determined in an experimental study at Tufts University and are described in the Letter to the Editor by M. N. Crook and D. S. Kellogg.[14] Approximately 200 untrained subjects wrote 20,000 numerals, of which 92% were correctly recognized. This is comparable to the results of the original simulation. After 45 writers were given approximately 30 minutes of initial training to write "normal" characters, they wrote 23,500 numerals, of which 99.3% were recognized correctly. The error rate for these characters was 0.11%, and the reject rate was 0.59%.

The performance of this experimental reader cannot be described adequately in terms of error statistics alone. The variations in character shape and line quality with which the reader can cope are not apparent in these figures. Unfortunately, there are no suitable quantitative measures of character shape and quality for use in a precise description of the machine reader's tolerances. However, an indication can be obtained from a description of the most frequent causes of failure to read. These are:

- *Linked or touching characters.* When adjacent numerals touch or are linked together, at least one will be rejected.
- *Broken characters.* Large gaps (> 0.015″, beyond the limits of the subroutine that is provided to cope with

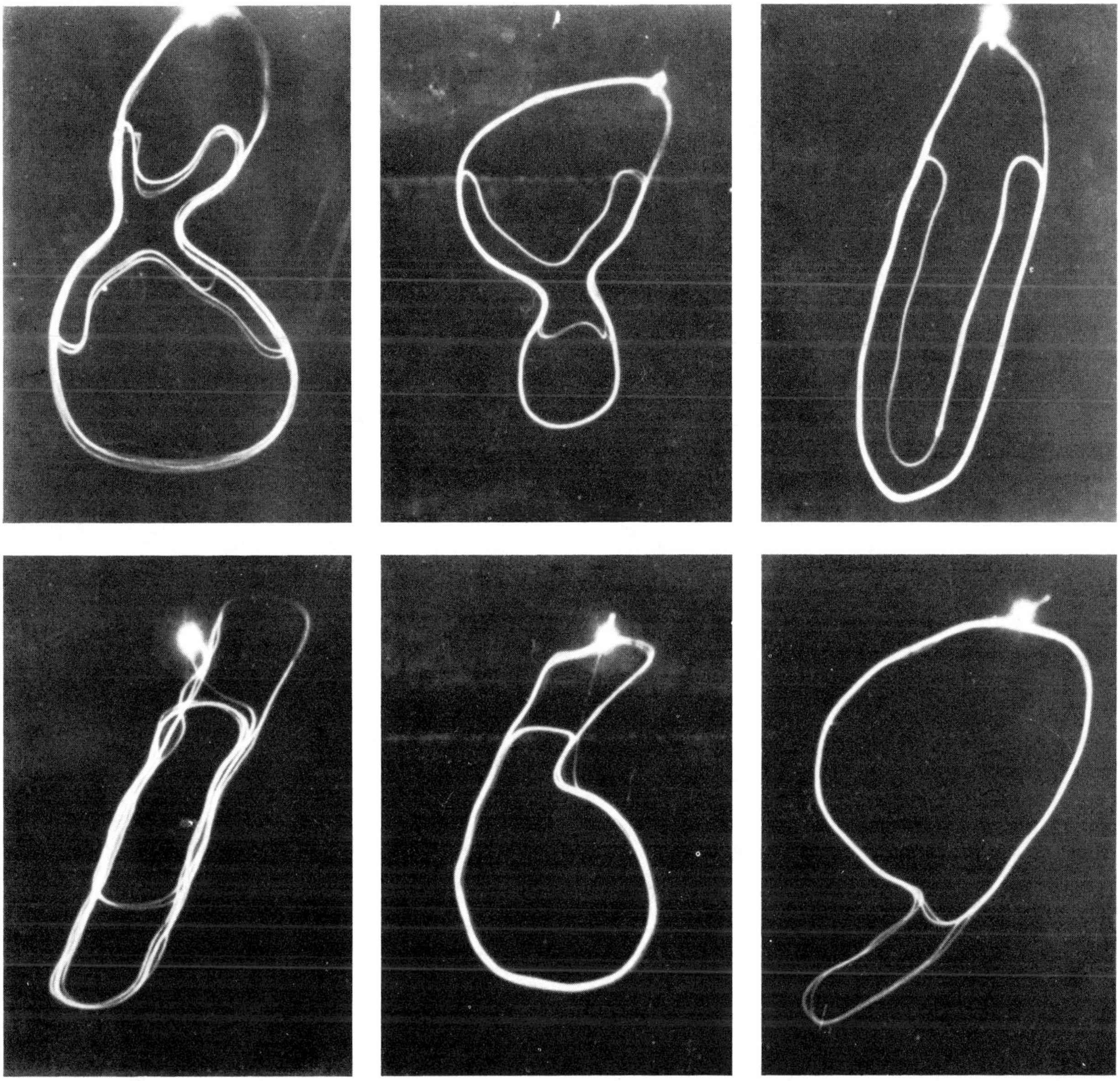

Figure 9 **Cathode ray tube display showing two internal contours of the numeral 8 and one contour each for numerals 0, 1, 6 and 9.**

discontinuities) due to lines that fail to join or to weak impressions can cause rejects or errors. This occurred most frequently in **5**'s with horizontal top bars that did not join the rest of the character.

• *Characters with extraneous line segments.* These lines occurred accidentally or as part of flourishes. Some examples were **7**'s with extraneous horizontal lines at the bottom that could cause ambiguities between **7**'s and **2**'s, and **1**'s with extraneous lines at the top that could cause ambiguities between **7**'s and **1**'s.

• *Characters with exaggerated proportions.* This can occur in any of the numerals. For example, extremely long stems on **9**'s or **7**'s would make the top appear insignificant and cause possible ambiguities with the **1**.

• *Characters similar to other character shapes.* The "closed top" type of **4** was frequently missed when the horizontal bar failed to cross the vertical stem because the resultant shape was very similar to the **9**. (When the subjects were trained to write for this reader, they were encouraged to use the open-top **4**.)

Conclusion

It should be noted that the recognition system described here does not employ a single omnipotent technique to read handwritten numerals. The indicated tolerances for character shape, size and quality of this reader were obtained by providing specific capabilities to handle specific problems. Although some features of this reader are similar to ideas published elsewhere[8, 9, 15, 16] the authors believe that by providing the combination of functions that have been described, a new high in pattern recognition flexibility has been achieved. Some noteworthy features of this reader are:

a) The special scanning subroutines that cause the beam to re-explore poor lines with a modified black-white threshold to accommodate weak regions of the line.

b) The special scanning subroutines that cause the beam to seek specific shape features in selected regions of the character for the resolution of ambiguities that still exist after normal scanning.

c) The simultaneous use of line-direction and position-relative-to-extremities information for recognition.

The development of more powerful readers might be achieved by improving the performance of these functions or adding new functions to the reader.

Acknowledgment

The authors wish to acknowledge the contributions of D. Alexander and P. Hurley in the design and construction of the experimental model, and of H. Penafiel in the programming and design by simulation of the recognition logic.

References and footnotes

1. D. M. Green and T. Marill, *IRE Trans. on Electronic Computers*, **EC-9**, 472-477 (1960).
2. W. H. Highleyman, *Proc. IRE*, **50**, 1501-1514 (1962).
3. E. C. Greanias and Y. M. Hill, *Proc. IRE*, **7**, 119-126 (1957).
4. C. K. Chow, *IRE Trans. on Electronic Computers*, **EC-6**, 247-254 (1957).
5. Subjective judgment was used in the selection of recognition criteria. In general, the numerals that were similar to most of the other samples in this population were considered "normal", and the recognition criteria were designed to accept them. The tolerance that was provided to accept all "normal" characters made it possible to recognize many numerals with shape combinations that did not occur in the "normal" group.
6. Recognition based on the analysis of analog waveforms that correspond to the shape of the character was also accomplished in the early work of W. Sprick (Ref. 8). Other references and suggestions regarding curve following recognition techniques are summarized in M. E. Stevens' N.B.S. report (Ref. 9).
7. Although this paper covers handwritten numerals, a lower case printed **m** is shown to provide a more effective illustration of the value of circle size control.
8. W. Sprick. U.S. Patents 2,738,499 and 2,838,602.
9. M. E. Stevens, National Bureau of Standards Technical Note 112, May 1961, pp. 53-55.
10. A similar sequential recognition procedure was described by Greanias et al. (Ref. 11).
11. E. C. Greanias, C. J. Hoppel, M. Kloomok, and J. S. Osborne, *IBM Journal*, **1**, 8-18 (1957).
12. The recognition of the closed-top **4** was based on the occurrence of a pointed top or an adequate crossing of vertical stem by the horizontal line.
13. Internal scanning is tried on the **1** because with a thick line it could be mistaken for a **0**. When internal scanning is tried on the **1**, the beam always finds an "outer" contour.
14. M. N. Crook and D. S. Kellogg, *IBM Journal*, this issue, p. 76.
15. *Communications Theory*, edited by Willis Jackson, Butterworth's Scientific Publications, 1953, pp. 323-325.
16. M. Nadler, "An Analog-Digital Character Recognition System," presented at *IFIP Congress 62*, Munich, September 1, 1962.

Received September 19, 1962

Graphics and Image Processing
W. Newman Editor

An Application of Heuristic Search Methods to Edge and Contour Detection

Alberto Martelli
Istituto di Elaborazione della Informazione del Consiglio Nazionale delle Ricerche

This paper presents a method for detecting edges and contours in noisy pictures. The properties of an edge are embedded in a figure of merit and the edge detection problem becomes the problem of minimizing the given figure of merit. This problem can be represented as a shortest path problem on a graph and can be solved using well-known graph search algorithms. The relations between this representation of the minimization problem and a dynamic programming approach are discussed, showing that the graph search method can lead to substantial improvements in computing time. Moreover, if heuristic search methods are used, the computing time will depend on the amount of noise in the picture. Some experimental results are given; these show how various information about the shape of the contour of an object can be embedded in the figure of merit, thus allowing the extraction of contours from noisy pictures and the separation of touching objects.

Key Words and Phrases: picture processing, pattern recognition, edge detection, contour detection, contour following, optimization problems, dynamic programming, shortest path, heuristic search methods, problem solving methods

CR Categories: 3.63, 3.64, 3.66, 5.42

A portion of this work was done while the author was with the Department of Electrical Engineering and Computer Science, New York University. Partial support was provided by the Air Force Office of Scientific Research under Grant AF-AFOSR-70-1854.
Author's address: Istituto di Elaborazione della Informazione del Consiglio Nazionale delle Ricerche, Via S. Maria 46, 56100 Pisa, Italy.

1. Introduction

Many edge detection methods have been proposed which apply local operations to every point of the input picture, in order to extract short edge elements. Some of these operations are very simple, like the gradient or the Laplacian [1], whereas other operations are more complex and allow the elimination of most of the local noise [2, 3]. When small edge elements have been obtained, they must be connected together to form edges: for this operation, edge following algorithms are used [4]. In order to detect meaningful edges in noisy pictures, edge following procedures need global information about the edges in the picture and the structure of the objects they represent, and usually this heuristic information is embedded into the whole program. However, this approach has the disadvantage that minor changes of the characteristics of the sought edges may require reprogramming of large sections of the program.

A general class of methods used successfully in picture processing are the optimization methods [5–12]. In particular, Montanari [6] has proposed a method for detecting curves in noisy pictures, where the properties of a curve are embedded in a figure of merit. A dynamic programming technique is used to determine the optimal curve with respect to the given figure of merit. A similar approach has been used by the author [9, 10] for detecting edges, but here the optimization problem has been posed as a shortest path problem on a graph, instead of solving it using dynamic programming.

In this paper we show how these concepts can be applied to the extraction of contours from noisy pictures. In the next section, we examine the relations between dynamic programming and graph search, and we show that any optimization problem which can be posed as a dynamic programming problem can also be stated as the problem of finding the shortest path on a graph. The latter representation of an optimization problem is usually advantageous with respect to the former, because there are many algorithms for finding the shortest path which use heuristic information to speed up the search.

In Section 3, we show how the properties of an edge or a contour can be embedded in a figure of merit and how the edge finding problem can be reduced to a shortest path problem on a graph. The best edge can then be obtained by applying a search procedure to the given graph. This search procedure can be seen as an edge following procedure which uses global information.

Some experimental results are given in Sections 4 and 5, where different amounts of noise have been added to various pictures. Contours of the objects in the pictures are detected even when the noise level is high, and the computing time for detecting these contours increases with the level of noise. Moreover, we show how information about the shape of the contour can be embedded in the figure of merit, thus allowing the detection of contours of known shape in very noisy pictures.

Reprinted with permission from *Commun. Ass. Comput. Mach.*, vol. 19, pp. 73–83, Feb. 1976.

Finally the technique described in this paper is compared with a standard edge detection technique. The experimental results show that when the level of the noise is high and the standard technique does not work any longer, our method is still able to detect the contour in the picture.

2. Dynamic Programming and Graph Search

The reduction of a dynamic programming problem to the problem of finding the shortest path in a weighted graph is well-known in the so-called "serial" case [13]. In this section, we review the concepts of a serial dynamic programming problem and its reduction to a shortest path problem and we present some methods which can be used for solving this problem. Furthermore, we show that a similar reduction can be used as well in the case of nonserial dynamic programming.

In general, an optimization problem can be formulated as follows. Given an *objective function* $f(x_1, x_2, \ldots, x_N)$, where variables x_i can assume only discrete values $(0 \leq x_i \leq l_i, i = 1, \ldots, N)$, find the minimum of the objective function f and a value $(\bar{x}_1, \bar{x}_2, \ldots, \bar{x}_N)$ for which the minimum is achieved. If the objective function f has the form

$$f(x_1, x_2, \ldots, x_N) = f_1(x_1, x_2) + f_2(x_2, x_3) + \ldots + f_{N-1}(x_{N-1}, x_N) \quad (1)$$

we have a so-called *serial optimization problem*, which can be solved with a multistage optimization process (*serial dynamic programming*), using the following recursion formula [14]:

$$h_1(x_1) = 0,$$
$$h_{k+1}(x_{k+1}) = \min_{0 \leq x_k \leq l_k} (f_k(x_k, x_{k+1}) + h_k(x_k))$$

where $k = 1, \ldots, N-1$ and $0 \leq x_{k+1} \leq l_{k+1}$.

At each stage we build two tables, one storing the values $h_{k+1}(x_{k+1})$, $0 \leq x_{k+1} \leq l_{k+1}$, and the other one storing the values $m_{k+1}(x_{k+1})$ of x_k for which the minimum is achieved. At the end of this process, the minimum of f is the minimum of $h_N(x_N)$ and the values for which this minimum is achieved are obtained by tracing back the tables m_i with the recursion formula: $\bar{x}_N$ = the value of x_N for which the minimum of $h_N(x_N)$ is achieved, and $\bar{x}_k = m_{k+1}(\bar{x}_{k+1})$, $k = N-1, \ldots, 1$.

For example, let us consider function f in Table I. Functions f_1 and f_2 are also given in Table I, together with functions h_i and m_i. The minimum of f, i.e. the minimum of $h_3(x_3)$, is 6 and $\bar{x}_3 = 1$. By tracing back the tables we get $\bar{x}_2 = m_3(\bar{x}_3) = m_3(1) = 1$ and $\bar{x}_1 = m_2(\bar{x}_2) = m_2(1) = 0$.

Let us show now how a problem of serial dynamic programming can be reduced to a shortest path problem in a weighted graph. The graph for the objective function (1) consists of N levels of nodes and is constructed as follows.

The first level has $l_1 + 1$ nodes corresponding to the $l_1 + 1$ values of variable x_1 and the second level has $l_2 + 1$ nodes corresponding to the $l_2 + 1$ values of variables x_2. There is a directed arc $(n_1\ n_2)$ from every node n_1 of the first level to every node n_2 of the second level. If node n_1 corresponds to the value $\bar{x}_1$ of x_1 and node n_2 corresponds to the value $\bar{x}_2$ of x_2, then a cost $f_1(\bar{x}_1, \bar{x}_2)$ is associated with the arc $(n_1\ n_2)$, and analogously for the other levels. Finally, we add a node s (called *start node*) connected with arcs of cost zero to the nodes of the first level. For example, the graph for the function in Table I is given in Figure 1.

Let us call the nodes of the last level *goal nodes*. It is easy to verify that each path from the start node to a goal node corresponds to a possible assignment of values to the variables $x_1, x_2, \ldots, x_N$ and its cost, i.e. the sum of the costs of the arcs belonging to it, is the value of f for this assignment. The problem then becomes one of finding the path between the start node and a goal node of minimal cost. For example, for the graph in Figure 1 the minimal cost path has cost 6 and corresponds to the values $\bar{x}_1 = 0$, $\bar{x}_2 = 1$, $\bar{x}_3 = 1$.

The problem of finding the shortest path in a weighted graph is common in artificial intelligence and operations research and many methods have been proposed for its solution (see [15]) for a thorough review of these methods). Here we give an algorithm, due to Hart, Nilsson and Raphael [16]), which uses information about the problem being represented by the graph, called *heuristic information*, to speed up the search.

Let $G(N, A)$ be a directed graph. If an arc is directed from node n_i to node n_j, then node n_j is said to be a *successor* of node n_i, and node n_i is said to be a *parent* of node n_j. A positive cost $c(n_i, n_j)$ is attached to every arc $(n_i\ n_j)$. We want to find a minimal cost path between the start node s and any node belonging to the set of goal nodes.

The algorithm starts with node s and generates step by step the graph G until a goal node is reached. At each step, a node of the subgraph of G already generated is selected, its successors are generated (this process is called *expansion* of the node), and pointers are set up

Table I.

$f(x_1, x_2, x_3) = f_1(x_1, x_2) + f_2(x_2, x_3)$

$l_1 = l_2 = l_3 = 1$

x_1	x_2	$f_1(x_1, x_2)$	x_2	x_3	$f_2(x_2, x_3)$
0	0	6	0	0	5
0	1	4	0	1	7
1	0	2	1	0	4
1	1	5	1	1	2

$h_1(x_1) = 0$

x_2	$h_2(x_2)$	$m_2(x_2)$	x_3	$h_3(x_3)$	$m_3(x_3)$
0	2	1	0	7	0
1	4	0	1	6	1

Fig. 1. Graph for the objective function of Table I.

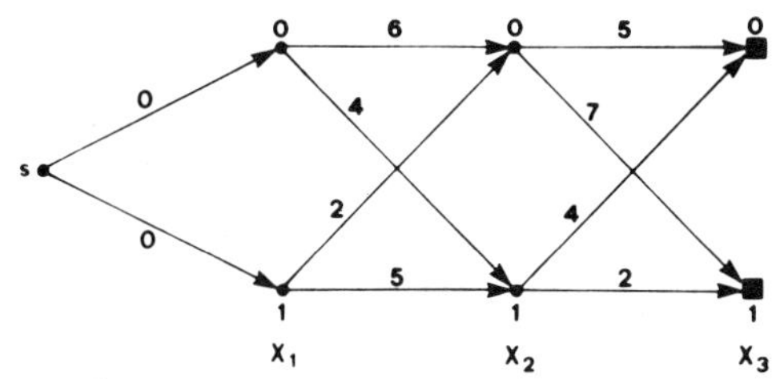

Fig. 2. An interaction graph (a) before and (b) after elimination of node V_1.

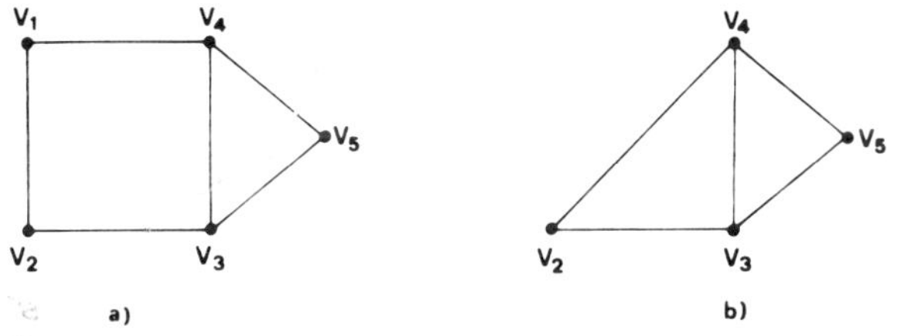

om each successor back to its parent node. The nodes are selected at each step according to an *evaluation function* $\hat{f}(n)$, which is an estimate of the cost of a minimal cost path from s to a goal node constrained to go through node n. This function $\hat{f}(n)$ can be expressed as

$$\hat{f}(n) = \hat{g}(n) + \hat{h}(n)$$

where $\hat{g}(n)$ is an estimate of the cost of a minimal cost path from s to n and $\hat{h}(n)$ is an estimate of the cost of a minimal cost path from n to a goal node. Here, $\hat{g}(n)$ can be chosen as the cost of the path from s to n given by summing the arc costs encountered while following the pointers from n to s (this path is the lowest cost path from s to n found so far) and $\hat{h}(n)$ can be obtained using the heuristic information available from the problem domain.

ALGORITHM A*

(1) Mark the start node s *open* and set $\hat{g}(s) = 0$.
(2) If no node is *open*, exit with failure; otherwise continue.
(3) Mark *closed* the *open* node n whose total estimate $\hat{f}(n) = \hat{g}(n) + \hat{h}(n)$ is smallest. (Resolve ties for minimal $\hat{f}$ values arbitrarily, but always in favor of any goal node.)
(4) If n is a goal node, exit with the solution path obtained by tracing back through the pointers; otherwise continue.
(5) Expand node n, generating all of its successors. If there are no successors go to (2).
(6) If a successor n_i is not marked, set $\hat{g}(n_i) = \hat{g}(n) + c(n, n_i)$, mark it *open* and direct a pointer from it back to n.
(7) If a successor n_i is marked *closed* or *open*, update its value

$$\hat{g}'(n_i) = \min(\hat{g}(n_i), \hat{g}(n) + c(n, n_i)).$$

Mark *open* those *closed* successors whose $\hat{g}$ values were thus lowered, and redirect to n the pointers from all nodes whose $\hat{g}$ values were lowered.
(8) Go to (2).

It is possible to prove that, if $\hat{h}(n)$ is a lower bound on the cost of a minimal cost path from node n to a goal node, then Algorithm A* will find an optimal path. In particular, if no heuristic estimate is available (i.e. $\hat{h} \equiv 0$), then Algorithm A* coincides with Dijkstra's algorithm [17].

The multistage optimization process can be considered, in terms of graph search, as a breadth-first method, i.e. a method which will expand first all nodes on the first level, then all nodes on the second level, and so on. By comparing the two methods, it is clear that Algorithm A* can lead to substantial improvements in computing time. In fact, while the first algorithm must expand all nodes of the search graph, Algorithm A* may expand only a small subset of nodes, if heuristic information is available from the problem to be solved. Furthermore, in many applications we are not concerned with finding the optimal solution and a "good" solution will be enough. In this case, more powerful heuristics can be used, thus obtaining a more efficient search of the graph.

These concepts can be extended to objective functions more general than (1), where some terms of f are functions of more than two variables. This is the so called *nonserial dynamic programming* case [18]. Such a problem can still be solved using a multistage optimization method, where the optimization is carried out separately for each variable x_i, but in this case, the optimization must be performed for all the values of the variables which "interact" with variable x_i. To help explain this concept we introduce an *interaction graph*. In this graph, nodes V_i correspond to variables x_i and two nodes V_i and V_j are connected with an undirected arc iff the variables x_i and x_j *interact*: i.e. if there exists at least one term of the objective function which depends on both x_i and x_j. For example, let us consider the objective function

$$f(x_1, x_2, x_3, x_4, x_5) = f_1(x_1, x_2) + f_2(x_2, x_3) + f_3(x_1, x_4) + f_4(x_3, x_4, x_5). \tag{2}$$

Its interaction graph is given in Figure 2(a).

Now we can carry out optimization with respect to one variable, for example x_1. We get

$$\begin{aligned} \bar{f}(x_2, \ldots, x_5) &= \min_{x_1} f(x_1, x_2, \ldots, x_5) \\ &= \min_{x_1} [f_1(x_1, x_2) + f_3(x_1, x_4)] \\ &\quad + f_2(x_2, x_3) + f_4(x_3, x_4, x_5) \\ &= h_1(x_2, x_4) + f_2(x_2, x_3) \\ &\quad + f_4(x_3, x_4, x_5) \end{aligned}$$

where h_1 is a function of the variables which interact with x_1.

After elimination of variable x_1, we have an optimization problem of the same type for the objective function $\bar{f}$ and the new interaction graph is obtained by deleting node V_1 and connecting with a complete subgraph all nodes which are adjacent to V_1, i.e. V_2 and V_4 (see Figure 2(b)). The process goes on by eliminating all variables in a certain order.

Usually, the order in which variables are eliminated

Table II.

$$f(x_1, x_2, \ldots, x_5) = f_1(x_1, x_2) + f_2(x_2, x_3) + f_3(x_1, x_4) + f_4(x_3, x_4, x_5)$$

x_1	x_2	$f_1(x_1, x_2)$
0	0	2
0	1	6
1	0	5
1	1	1

x_2	x_3	$f_2(x_2, x_3)$
0	0	3
0	1	5
1	0	4
1	1	2

x_1	x_4	$f_3(x_1, x_4)$
0	0	3
0	1	1
1	0	4
1	1	5

x_3	x_4	x_5	$f_4(x_3, x_4, x_5)$
0	0	0	4
0	0	1	2
0	1	0	6
0	1	1	7
1	0	0	4
1	0	1	4
1	1	0	5
1	1	1	3

is important because it determines the amount of computing time and storage required. Let the *dimension* d_i be the number of variables which interact with variable x_i when x_i is eliminated. The computational effort needed is exponentially dependent on the dimensions d_i of the various stages. Therefore, the problem of finding the order of variable elimination which minimizes the computing time is important in nonserial dynamic programming [18, 19].

Given an order of variable elimination, it is possible to reduce a nonserial dynamic programming problem to a shortest path problem on a graph. As an example, we will show the construction of such a graph for the objective function (2). Functions f_1, f_2, f_3, f_4 are given in Table II. We assume that variables are eliminated in the order x_1, x_2, x_3, x_4, x_5.

The graph consists of six levels of nodes. Node s is on level zero. The nodes on level 1 correspond to the values of variable x_1. The nodes on level 2 correspond to the values of the pair x_2x_4, since x_2 and x_4 are the variables which interact with x_1. Level 3 corresponds to the stage of the multistage optimization process where variable x_2 is eliminated and its nodes correspond to the values of the pair x_3x_4, since x_3 and x_4 interact with x_2 when it is eliminated. We proceed analogously for the other levels (see Figure 3).

Nodes on level i are connected to nodes on level $i + 1$ by arcs in such a way that equal variables are assigned equal values. For example, the node on level 2 corresponding to $x_2 = 0$ and $x_4 = 1$ is connected only to the nodes on level 3 with $x_4 = 1$. Finally, the costs associated with the arcs are the values of the terms f_i of f. To be precise, the values of $f_1 + f_3$ are associated with arcs between level 1 and level 2; the values of f_2 with arcs between level 2 and level 3 and the values of f_4 with the arcs between level 3 and level 4. The graph is shown in Figure 3.

We can easily see that, for the graph in Figure 3, each path from s to a goal node corresponds to an assignment of variables, and the cost of the path is the value of f for this assignment. Therefore, our problem becomes one of finding a minimal cost path and we can solve it by using one of several methods of graph searching, such as Algorithm A*.

The method of construction of the search graph described for our example can be generalized to any nonserial dynamic programming problem, and the reduction to a minimal path problem is therefore general for dynamic programming. There are, however, dynamic programming problems for which another reduction is more suitable, i.e. the reduction to the problem of finding a tree of minimal cost in an AND/OR graph. An example of such a reduction is given in [20].

3. Edge and Contour Detection

In this section we show how the properties of an edge or a contour can be embedded in an objective function and how the problem of extracting edges or contours from noisy pictures becomes the problem of minimizing this function.

A picture is given as a rectangular array $[g_{ij}]$, $i = 1, \ldots, r$, $j = 1, \ldots, c$, where the value g_{ij} gives the gray level of the point of coordinates i and j. Two points (i, j) and (h, k) are *neighbors* if

$$|i - h| + |j - k| = 1.$$

Let an *edge element* be an ordered pair of points (P, Q) such that P and Q are neighbors. By representing the picture on a square grid, an edge element (P, Q) defines the common boundary between the two adjacent squares P and Q. For example, the edge element $((i, j), (i, j + 1))$ in Figure 4 gives the boundary segment AB. The direction of this segment is obtained with the convention of moving clockwise around the first square. Let us call A the *tail* and B the *head* of the segment AB. An edge element (P', Q') is said to be a *successor* of the edge element (P, Q) if the head of the boundary segment of (P, Q) coincides with the tail of the boundary segment of (P', Q'). Therefore each edge element which is not in the first or last row or column has exactly three successors. For instance, the three successors of the edge element $((i, j), (i, j + 1))$ of Figure 4 are the edge elements $((i, j), (i + 1, j))$, $((i + 1, j), (i + 1, j + 1))$, $((i + 1, j + 1), (i, j + 1))$, corresponding respectively to the three boundary segments BC, BD and BE. Finally a *possible edge* is a sequence of edge elements (P_1, Q_1), $(P_2, Q_2), \ldots, (P_N Q_N)$ such that (P_i, Q_i) is a successor of (P_{i-1}, Q_{i-1}), $i = 2, \ldots, N$.

In general, global information is needed for detecting an edge, if a picture has many gray levels and is noisy. The method presented in this paper assigns a cost to every possible edge, taking into account the properties of the desired edge. The cost can be expressed as a sum of terms, each dependent on few variables only.

Fig. 3. Graph for the objective function of Table II.

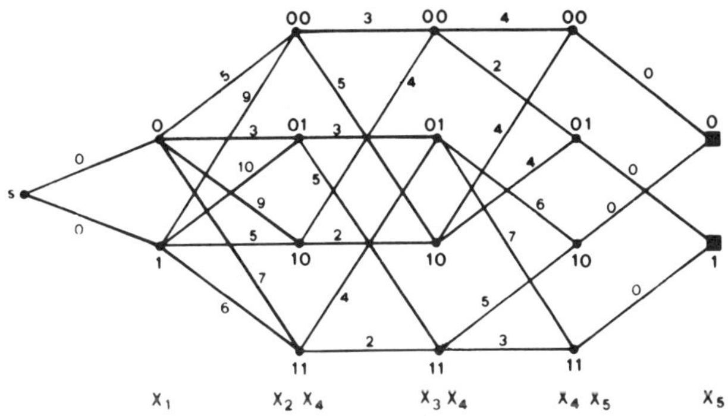

Fig. 4. Successors of an edge element.

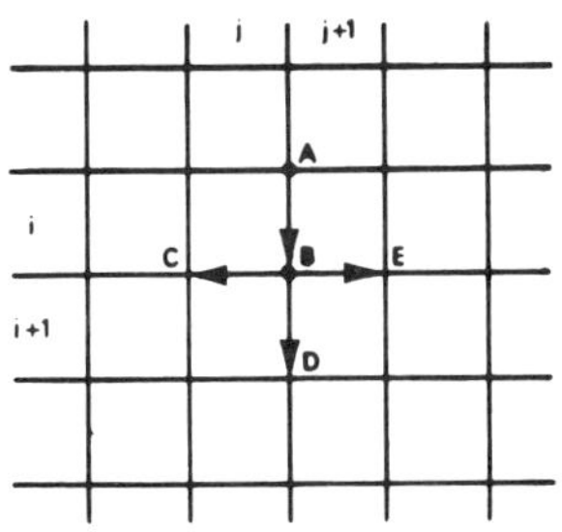

Fig. 5. Picture containing an edge.

	1	2	3
1	7	2	2
2	5	7	0
3	5	4	2

Fig. 6. The interaction graph of the objective function (4).

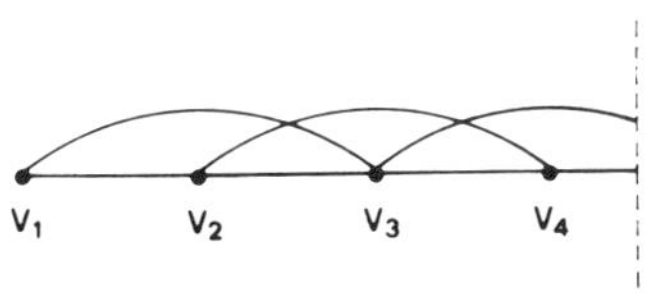

Fig. 7. Graph representing the possible edges in the picture of Fig. 5.

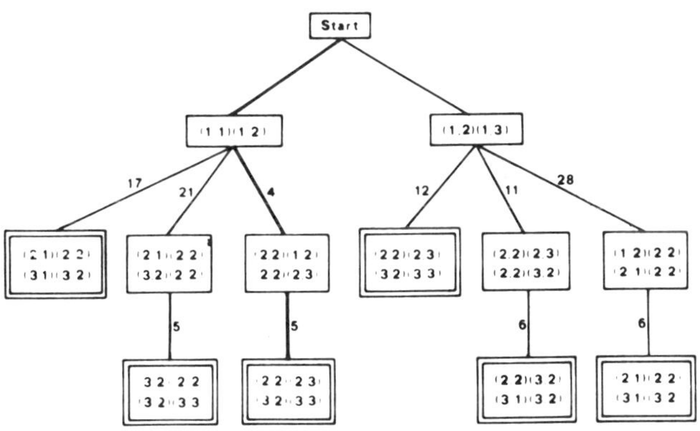

Let $x_i = (P_i, Q_i)$ be the ith element of the edge. The cost of the edge can be given as follows:

$$C(x_1, x_2, \ldots, x_i, \ldots) = c_1(x_1, x_2, \ldots, x_k) + c_2(x_2, x_3, \ldots, x_{k+1}) + c_i(x_i, x_{i+1}, \ldots, x_{i+k-1}) + \ldots \quad (3)$$

The functions c_i contain information about local properties of the edge such as gray levels and curvature, whereas their sum C conveys global information. The edge $(\bar{x}_1, \bar{x}_2, \ldots, \bar{x}_i, \ldots)$ whose cost C is minimal is the best edge according to our requirements.

For example, let us consider the 3×3 picture in Figure 5 and let us define an edge as the boundary between two regions of different gray levels. If we assume that the first point of an edge element is in the region with the highest gray level, then a good edge will be such that the sum of the differences of gray levels between the first and the second point of its elements is high. Therefore, the following cost is associated with each edge element $x_i = (P_i, Q_i)$:

$$c'(x_i) = M - (g(P_i) - g(Q_i))$$

where $g(P_i)$ and $g(Q_i)$ are the gray levels of the points P_i and Q_i and M is the maximum difference of gray levels between two adjacent points. In this case $M = 7$. This cost is low if $g(P_i)$ is high and $g(Q_i)$ is low, and it is always positive.

Moreover, let us assume that we are looking for a "smooth" edge, that is, we do not allow a sequence of three edge elements whose first or second points coincide. This constraint can be embedded in a function of three variables:

$$c''(x_i, x_{i+1}, x_{i+2})$$

which is zero if the values of x_i, x_{i+1}, x_{i+2} satisfy our constraint and very high otherwise.

The cost of an edge can be expressed as

$$C(x_1, x_2, \ldots) = c'(x_1) + c'(x_2) + \ldots + c''(x_1, x_2, x_3) + c''(x_2, x_3, x_4) + \ldots \quad (4)$$

Therefore the problem of finding an edge in the picture of Figure 5 has been reduced to an optimization problem whose objective function is (4). Our problem is now that of finding an assignment of variables which minimizes (4) and it can be solved by searching a graph, as explained in the preceding section.

For the sake of simplicity, let us introduce some constraints for the sought edge. We assume that this edge starts in the first row and ends in the last row. Furthermore, we assume that the region of highest gray level is on the left side of the picture. The interaction graph of (4) is shown in Figure 6. If variables are eliminated in the order $x_1, x_2, x_3, \ldots$, we see that each eliminated variable interacts with two variables, and thus the nodes of the search graph will correspond to a pair of variables.

The search graph for (4), with the above described constraints, is shown in Figure 7. The first element of

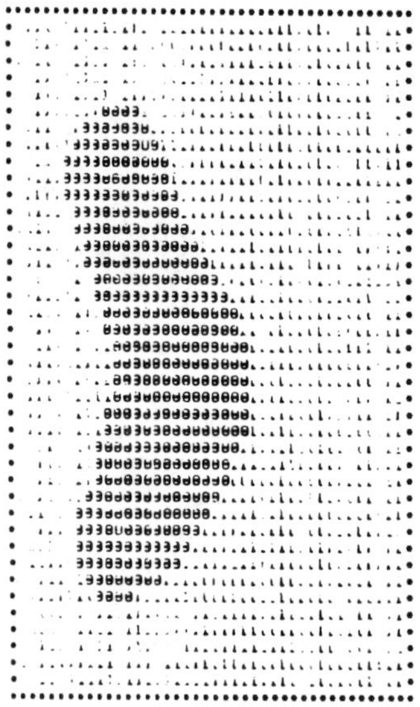

Fig. 8(a). An object with smooth contour.

Fig. 8(b). The object of Fig. 8(a) after addition of normal noise of mean value 0 and standard deviation 2.

Fig. 8(c). The object of Fig. 8(a) after addition of normal noise of mean value 0 and standard deviation 4.

the edge can only be either ((1, 1), (1, 2)) or ((1, 2), (1, 3)). Goal nodes are enclosed by a double rectangle and correspond to edge elements ending in the last row of the picture.

Each path between the start node and any of the goal nodes is a possible edge. The arcs whose cost c'' is high have not been displayed and only the nodes and the arcs leading to a goal node appear in the figure. The cost attached to an arc between a node on the first level $\bar{x}_1$ and a successor $\bar{x}_2$, $\bar{x}_3$ is $c'(\bar{x}_1) + c'(\bar{x}_2) + c'(\bar{x}_3)$.

The cost attached to an arc between two intermediate nodes $\bar{x}_i$, $\bar{x}_{i+1}$ and $\bar{x}_{i+1}$, $\bar{x}_{i+2}$ is $c'(\bar{x}_{i+2})$, since $c''(\bar{x}_i , \bar{x}_{i+1} , \bar{x}_{i+2})$ is zero. The dark path shows the minimal-cost solution, corresponding to the dark boundary in Figure 5.

Given a graph representing an edge finding problem, we can search it using, for example, Algorithm A*. Notice that this algorithm can be considered as an edge following algorithm. In fact, it starts with the first element of the edge and, at each step, extends by one element the most promising edge obtained so far. The number of expanded nodes depends on the noise and, as a limit case, we can consider a perfect picture without noise. For example, if we assume that the picture of Figure 5 consists of two regions of gray level 7 and 0, the best edge is uniquely determined as the edge separating these two regions. We see that the value of function c' for any edge element is zero and thus the total cost of the best edge is zero. If we apply Algorithm A* in this case, without any heuristic, we see that the only expanded nodes are those on the optimal edge and, therefore, this algorithm is similar to other edge following algorithms defined for pictures without noise [1].

In general, the objective function describing the characteristics of an edge will be of the type (3) where all terms of C are functions of k variables. If the variables are eliminated in the order x_1 , x_2 , x_3 , . . . , each eliminated variable interacts with $k - 1$ variables, i.e. each stage of the multistage optimization process has a dimension $k - 1$. This means that at every stage we have to build a table whose entries are all possible edges of length $k - 1$. Therefore we see that the complexity of the multistage optimization process will be greatly dependent on k, because the size of the tables is exponential with k.

If we consider now the search graph for (3), we see that each node corresponds to $k - 1$ consecutive edge elements, and the number of nodes on each level of the graph is equal to the number of entries of the tables built in the multistage process. But the number of nodes expanded by the search algorithm is usually only a very small fraction of the nodes of the graph. Moreover, the search algorithm goes on by expanding a node at each step and the number of successors of a node does not depend on k. In fact, if a node is associated with $(\bar{x}_i , \bar{x}_{i+1} , . . . , \bar{x}_{i+k-2})$, its successors will be associated with $(\bar{x}_{i+1} , . . . , \bar{x}_{i+k-2} , \bar{x}_{i+k-1})$, where $\bar{x}_{i+k-1}$ can assume only three values corresponding to the three edge elements which are successors of the edge element $\bar{x}_{i+k-2}$. Therefore we can increase k without substantially affecting the complexity of the search process, thus allowing the terms c_i to take into account more complex properties of edges such as their curvature.

Assume now that we are given a picture containing an object of uniform gray level on a background of a different gray level and that we want to extract from the picture the contour of this object. This contour can be obtained by applying a method similar to the one just described, if we consider a contour as an edge whose first and last element coincide.

We assume that we are given a region through which the contour must pass (for example a row of the picture), and that the properties of the contour (for example, difference of gray levels and smoothness) are embedded in an objective function. Then we start following the contour using Algorithm A*, taking as start edge elements the elements contained in the given region and expanding the search graph according to the objective function. Goal nodes are obtained dynamically during the search, in such a way that the resulting contour is closed. This can be achieved by recalling that for each node of the graph there is one optimal path from the start node to this node which can be obtained by tracing back the pointers, and therefore the first edge element of the corresponding edge is known.

4. Some Experimental Results

Algorithm A*, described in Section 2, was implemented in Fortran V on a UNIVAC 1108 computer and it was used to detect contours with different properties in noisy pictures. Information about the properties of the contour is given by specifying the start nodes, the characteristics of the goal nodes and the rules for expanding a node and for computing its cost. Thus this information can easily be changed by simply substituting some subroutines.

The program was tested by adding different amounts of noise to the image array of Figure 8(a). This image array consists of an object of constant gray level 10 against a background of gray level 5, and it has 40 rows and 40 columns. In Figures 8(b) and 8(c) an independent amount of noise has been added to every point of the image array of Fig. 8(a). The statistical distribution of the noise is normal with mean value zero and standard deviation 2 and 4, respectively. The image arrays are represented with 16 gray levels and are printed with overstruck characters to simulate the different gray levels.

In order to detect the contour of the object, we assume as start points for our search the edge elements on the eleventh row of the array between columns 2 and 10.

The edge element used in this and in the following examples consists of four adjacent points (P_i', P_i'', Q_i', Q_i'') which lie all in the same row or column and such that P_i' and P_i'' belong to the region of highest gray level. The cost takes into account both the fact that there must be a difference of gray level between the object and the background and that the contour must be smooth. The former requirement is taken into account by assigning to each edge element the cost

$$c'(x_i) = 10 - (g(P_i') + g(P_i'') - g(Q_i') - g(Q_i'')).$$

This cost becomes zero if $c'(x_i)$ is negative. With this definition, the cost of each edge element of the contour in the picture without noise is zero.

The smoothness is taken into account by assigning a cost to the curvature of the contour. We therefore consider 8 consecutive edge elements: we connect the tail of the first element to the head of the fourth element with a straight line and the tail of the fifth element to the head of the eighth element with another straight line, and we measure the angle between the two lines. This angle can be considered as proportional to the curvature. The cost c'' assigned to eight consecutive edge elements is given by this angle multiplied by a coefficient k. In this case, $k = 5$.

As mentioned above, we can introduce some heuristics into the search algorithm to speed up the search, even if we are no longer guaranteed to find the optimal solution.

In our case, nodes on low levels of the search graph will have usually a lower cost than nodes on high levels, because nodes on low levels correspond to shorter edges. If we search the graph using Algorithm A*, we will expand all the nodes whose cost is smaller than the cost of the minimal solution and these nodes will be mainly on the first levels of the graph. For this application, however, we can make the following simplification.

Suppose that, at a certain step, we have expanded a node n_1 of cost c_1 which corresponds to an edge of length l_1 and that there is a node n_2 not yet expanded of cost $c_2 > c_1$ corresponding to an edge of length $l_2 < l_1$. We assume that, if $l_2 \leq l_1 - t$, where t is a given threshold, then by expanding n_2 we cannot obtain an edge whose cost is less than the cost of an edge obtained by expanding n_1. Therefore, at each step of the algorithm, we keep track of the maximum level l_M of the graph reached so far and we do not expand any node whose level l is $l \leq l_M - t$. In our example we have $t = 3$.

Figures 9(a) and 9(b) (p. 80) give the result of the application of the program to the image arrays of Figures 8(b) and 8(c) respectively. The contour shown is the inner boundary of the object as detected by the program.

As pointed out before, the number of nodes expanded by this method, and therefore the computing time, increases with the noise. Table III shows the number of expanded nodes as a function of the standard deviation of the noise, for our example. The average time of expansion of a node is approximately 3 msec.

In this example we have assumed that the image array contains exactly one object and that we know a region through which the contour passes. However, the program can easily be modified in order to be able to detect any number of object contours.

Table III.

Standard deviation of noise	Number of expanded nodes
0	124
1	143
2	255
3	451
4	512

Fig. 9(a). Contour detected in the picture of Fig. 8(b).

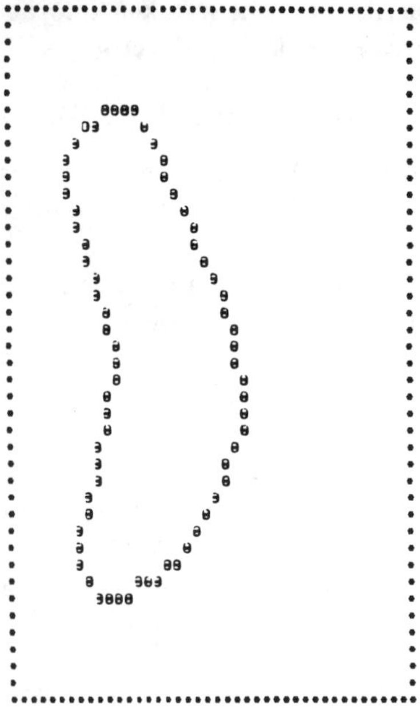

Fig. 9(b). Contour detected in the picture of Fig. 8(c).

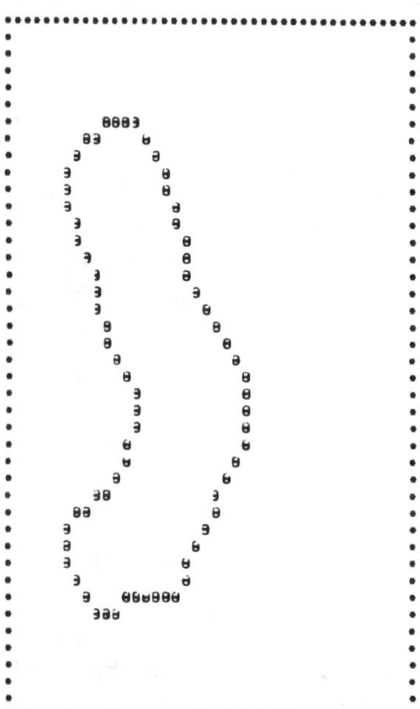

The image array is divided into contiguous rectangular windows. Starting with the window in the upper left corner, we scan the array from left to right and from top to bottom. For each window, we look for a contour passing through its lower edge, that is, the start nodes are associated with the edge elements belonging to the lower edge of the window. The search graph is then expanded according to the rules given by the algorithm. In order to detect a contour, for each node to be expanded we compute its relative cost, defined as the ratio between the cost of the optimal path to this node and the number of elements of the corresponding optimal edge. If the relative cost is higher than a given threshold, the node is not expanded. Therefore if there is no contour, the relative cost of the nodes will be higher than the threshold, and eventually the list of the nodes to be expanded will be empty and the algorithm will stop without encountering any goal node. On the other hand, if there is a contour, the relative cost of its nodes will be lower than the threshold and the algorithm will detect the contour. When a contour has been found, it is marked on the image array in such a way that it will not be followed in a subsequent step.

Let us consider the image array in Figure 10, containing two touching objects with a smooth contour with the addition of normal noise of mean value 0 and standard deviation 3. As we said before, the objective function used in the examples of this section takes into account the requirement that the contour must be smooth. Therefore we want our program to be able to separate the contours of the two objects.

The program, starting from a given region, will generate the sequence of edge elements of the contour of an object, say clockwise, until it finds the point of intersection with the contour of the other object. Here it could either follow the contour of the second object or follow an ideal smooth continuation of the contour of the first object. The choice of either path depends on their cost. If the former possibility is chosen, the term of the cost c' which takes into account the difference of levels will be low and the term of the cost c'' which takes into account the curvature will be high, because of the sharp change of direction of the contour. If the latter way is chosen, c' will be high, because in the common area between the two objects there is no difference of level, and c'' will be low. Therefore by suitably weighting the curvature the contours of the two objects can be separated.

The result of the application of the program to the image array in Figure 10 is shown in Figure 11. The program detects the two objects and interpolates their contours in the area where they are touching.

The technique for detecting edges and contours presented in this paper is more complex than other known edge following algorithms. In order to show that this complexity is necessary to improve the performance, a simple edge following algorithm has been implemented and tested on the examples in Figures 8(b) and 8(c). This algorithm is similar to the one described in [21] and is based on a simple gradient operator, which is used to determine the presence of an edge and its direction.

If we wish to apply the operator to a point i in the image, we extract the gray levels of the eight points adjacent to it. Let the nine elements of the 3×3 array centered on i be

$$\begin{matrix} a & b & c \\ h & i & d \\ g & f & e \end{matrix}$$

Fig. 10. Two touching objects with addition of normal noise of mean value 0 and standard deviation 3.

Fig. 11. Contours detected in the picture of Fig. 10.

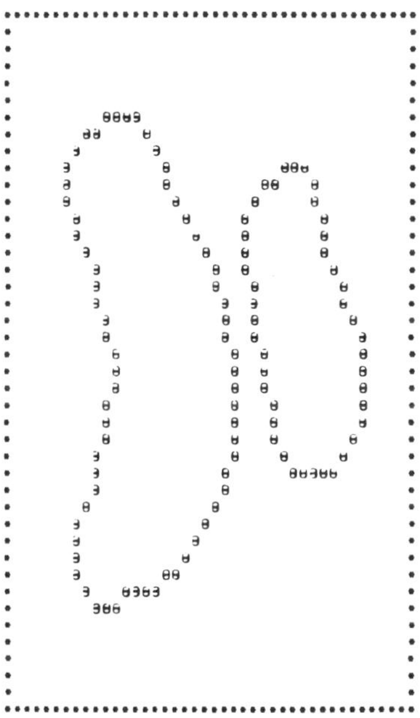

Then the x and y components of the gradient are

$$G_x = (c - g) + (e - a) + 2(d - h),$$
$$G_y = (c - g) + (a - e) + 2(b - f).$$

The operator returns the magnitude of the gradient

$$G = (G_x^2 + G_y^2)^{\frac{1}{2}}$$

and its direction given as an integer between 0 and 7.

The program does a raster scan of the image starting at the upper left corner and applies the operator to every point. If the magnitude of the gradient is smaller than a given threshold, the scan continues. Otherwise, the point is considered as the first point of an edge, and the program begins following this edge.

Assume that the program has detected an edge up to a point i. The program then moves to the point j adjacent to i in the direction perpendicular to the gradient of i. The gradient operator is applied to j and, if its magnitude is greater than the threshold, this point is added to the edge. Otherwise we compute the average gray level of the 3×3 array centered on j and compare it with a suitably chosen threshold, in order to determine whether j is inside or outside the object. The program then makes another attempt with a point k adjacent to i in the direction perpendicular to the gradient of i plus or minus 1,[1] according to the outcome of the previous test. If even the magnitude of the gradient of k is too small, the program abandons that edge and resumes the scan.

In order to simplify the detection of edges in noisy pictures, we assume that all edges are smooth; that is, if i, j, and k are three consecutive points of an edge, then the direction of the segment $i - j$ and the direction of the segment $j - k$ differ at most by 1. With this restriction, several noise generated edges with abrupt changes of direction will be rejected.

The program has been applied to the image arrays in Figures 8(b) and 8(c), and the results are shown in Figures 12(a) and 12(b) respectively. In these figures we have retained only the edges whose length is greater than 4. The ones denote the endpoints of the edges. The program took about 450 msec for the first image array and 750 msec for the second one.

In the first case, with standard deviation 2, the performance of this program is rather good. Although the edge detected by it is not closed, it would be quite easy to connect the end-points of the edges in order to obtain the closed contour of the object. However, in the case of standard deviation 4, this program detects only small parts of the contour, together with many wrong edges, and no closed contour can be recognized in Figure 12(b).

Therefore, when the level of the noise is low, contours can be detected with simple edge following algorithms which use only local criteria. However, above a certain level of noise, global methods are needed, such as the one described in this paper, in order to detect meaningful contours.

5. Detection of a Contour of Known Shape

As pointed out before, the characteristics of the sought contour can be embedded in the objective function. For example, we might insert in the objective function information about the shape of the object we want to detect. This new information should allow a more directional search and an easier recognition of the object in a noisy picture.

---

[1] Of course, these are modulo 8 operations.

Fig. 12(a). Edges detected in the picture of Fig. 8(b) by a simple edge following program.

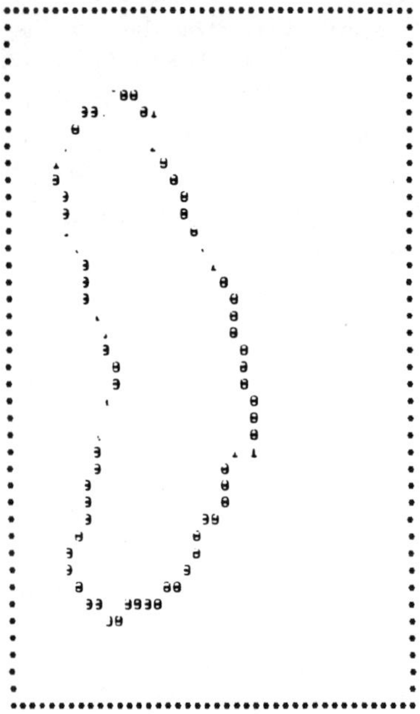

Fig. 12(b). Edges detected in the picture of Fig. 8(c) by a simple edge following program.

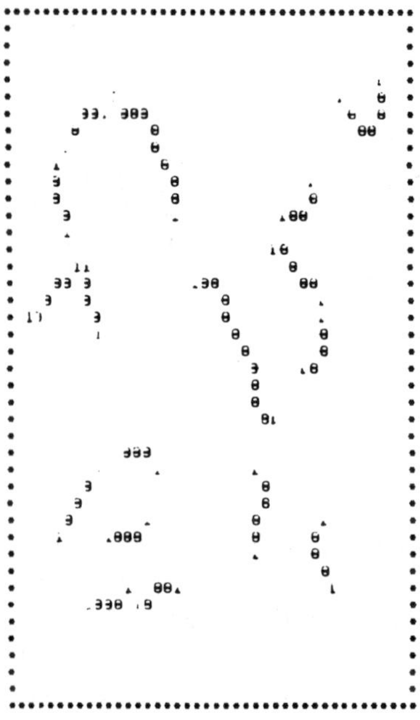

In this section we assume that the object whose contour we want to detect is a square. Therefore, we can use the information that its contour consists of four straight line segments of equal length and forming right angles. This information is taken into account as follows.

A node of the search graph is associated with nine consecutive edge elements which are assumed to belong to the same side of the square, and it can have two types of successors. The first type of successor is also associated with nine consecutive edge elements which are assumed to belong to the same side of the square. A cost proportional to the curvature is associated to an arc joining two nodes of this type, because the sides of the square must be straight. The second type of successor is used for passing from one side to the next one. It is considered as the start node of a new side and it is expanded as if it were a start node of the graph.

A cost taking into account the difference of gray level is computed for each edge element as in the examples of Section 4. Moreover, other costs are computed, taking into account the properties of the square, as soon as information is gathered during the search.

The program starts by following one side of the square. When a successor of the second type is generated, the program begins to generate edge elements of the second side of the square and so on. When the first side has been detected, we know that the direction of the second side must be perpendicular to it. Therefore the direction of each sequence of ten elements for the second side is computed and compared to the required direction and a cost proportional to the difference of the two directions is added. Analogously, when the first side has been detected, the length of the sides of the square is known. Therefore when a successor of the second type is generated, the length of the previously generated side is computed and compared to the required length. A cost proportional to the difference of the two lengths is then added.

The program was applied to the image array of Figure 13(a) where one corner of a square is covered by another object of the same gray level. The information about the shape of the contour enables the program to detect it, even when a high amount of noise is added to the image array. Figure 13(b) shows the image array of Figure 13(a) with the addition of normal noise of mean zero and standard deviation 5. The contour detected in this array is shown in Figure 14.

6. Conclusion

We have presented a global method for detecting edges and contours in noisy pictures. Properties of edges are embedded in an objective function and the edge which minimizes this function is sought. We have shown that, by representing the minimization problem as a shortest path problem on a graph, graph search algorithms can be used to find the solution. An important feature of our method is that the effort required for extracting edges and contours depends on the amount of noise in the picture and that the heuristics added to the search algorithm keep the computing time within acceptable limits even for very noisy pictures. Moreover, our examples show how rather complex properties can be embedded in the objective function without significantly affecting the complexity of the search process.

In the examples given in this paper we have purposely used a very simple edge element, in order to show that an edge can be recognized in a noisy picture using a global method, even if no edge element can be locally

Fig. 13(a). A square touching another object.

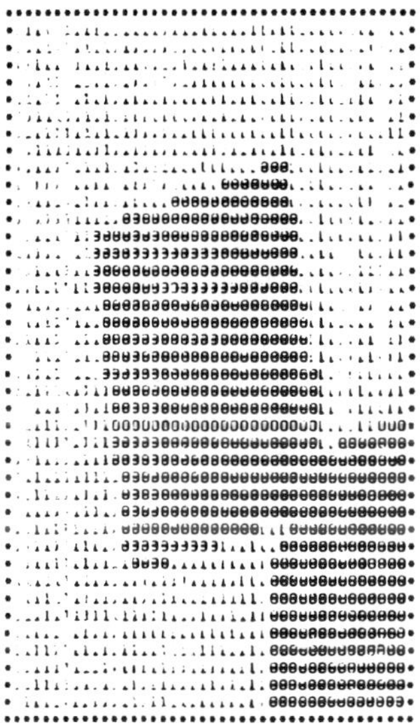

Fig. 13(b). The same as Fig. 13(a) with addition of normal noise of mean value 0 and standard deviation 5.

Fig. 14. Contour of the square detected in the picture of Fig. 13(b).

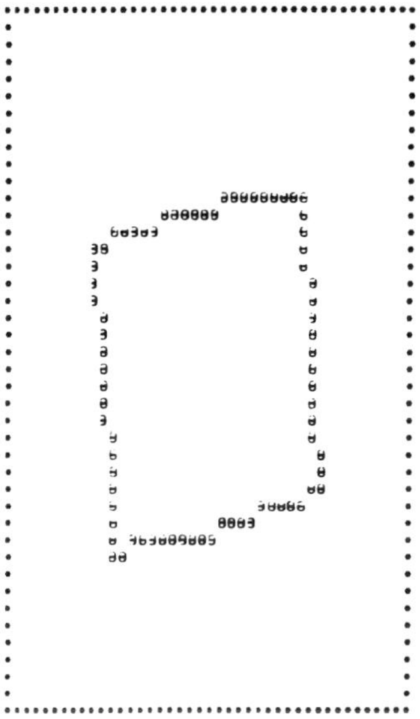

detected. However, if we want to apply this method to more complicated pictures, a better approach will probably be the one of choosing more complex edge elements, e.g. those detected by Hueckel's operator [2]. In this case, a great amount of noise is locally eliminated and therefore the search algorithm will be more directional and hence faster.

Received January 1974; revised December 1974

References

1. Rosenfeld, A. *Picture Processing by Computer*. Academic Press, New York, 1969.
2. Hueckel, M.H. A local visual operator which recognizes edges and lines. *J. ACM 20*, 4 (Oct. 1973), 634–647.
3. Rosenfeld, A., Thurston, M., and Lee, Y.-H. Edge and curve detection: further experiments. *IEEE Trans. on Computers C-21*, 7 (July 1972), 677–715.
4. Pingle, K.K., and Tenenbaum, J.M. An accommodating edge follower. In Proc. 2nd Int. Joint Conf. on Artificial Intelligence, London, Sept. 1971, pp. 1–7.
5. Kovalewsky, V.A. Sequential optimization in pattern recognition and pattern description. Proc. IFIP Congress, 68, North-Holland Pub. Co., Amsterdam, 1968 pp. 146–151.
6. Montanari, U. On the optimal detection of curves in noisy pictures. *Comm. ACM 14*, 5 (May 1971), 335–345.
7. Martelli, A., and Montanari, U. Optimal smoothing in picture processing: An application to fingerprints. Proc. IFIP Congress 71, North-Holland Pub. Co., Amsterdam, 1971 pp. 173–178.
8. Fischler, M.A., and Elschlager, R.A. The representation and matching of pictorial structures. *IEEE Trans. on Computers C-22*, 1 (Jan. 1973), 67–92.
9. Martelli, A. Edge detection using heuristic search methods. *Computer Graphics and Image Processing 1*, 2 (Aug. 1972), 169–182.
10. Martelli, A. Contour detection in noisy pictures using heuristic search methods. Proc. 1st Int. Joint Conf. on Pattern Recognition, Washington, 1973, pp. 375–388.
11. Ramer, U. Computer edge extraction from photographs of curved objects. Ph.D. dissertation, Tech. Rep. CRL-34, New York U., Dec. 1973.
12. Ballard, D.H., and Sklansky, J. Hierarchic recognition of tumors in chest radiographs. Proc. Second Int. Joint Conf. on Pattern Recognition, Copenhagen, August 1974, pp. 258–263.
13. Kaufmann, A., and Cruon, R. *Dynamic Programming*. Academic Press, New York, 1967.
14. Bellman, R., and Dreyfus, S. *Applied Dynamic Programming*. Princeton U. Press, Princeton, N.J., 1962.
15. Nilsson, N.J. *Problem-Solving Methods in Artificial Intelligence*. McGraw-Hill, New York, 1971.
16. Hart, P., Nilsson, N., and Raphael, B. A formal basis for the heuristic determination of minimum cost paths. *IEEE Trans. on System Science and Cybernetics SSC-4*, 2 (July 1968), 100–107.
17. Dijkstra, E. A note on two problems in connection with graphs. *Numer. Math. 1* (1959), 269–271.
18. Bertelè, U., and Brioschi, F. *Nonserial Dynamic Programming*. Academic Press, New York, 1973.
19. Martelli, A., and Montanari U. Nonserial Dynamic Programming: on the optimal strategy of variable elimination for the rectangular lattice. *J. Math. Analysis Appl. 40*, 1 (Oct. 1972), 226–242.
20. Martelli, A., and Montanari, U. Additive AND/OR graphs. Proc. 3rd Int. Joint Conf. on Artificial Intelligence, Stanford, Calif. 1973, pp. 1–11.
21. Pingle, K.K. Visual perception by a computer. In *Automatic Interpretation and Classification of Images*, A. Grasselli (Ed.), Academic Press, New York, 1969, pp. 277–284.

Edge Detection in Pictures by Computer Using Planning

M. D. Kelly

Department of Computer Science
Stanford University

Abstract

This paper describes a program for extracting an accurate outline of a man's head from a digital picture. The program accepts as input digital, grey scale pictures containing people standing in front of various backgrounds. The output of the program is an ordered list of the points which form the outline of the head. The edges of background objects and the interior details of the head have been suppressed.

The program is successful because of an improved method for edge detection which uses heuristic planning, a technique drawn from artificial intelligence research in problem solving. In brief, edge detection using planning consists of three steps. A new digital picture is prepared from the original; the new picture is smaller and has less detail. Edges of objects are located in the reduced picture. The edges found in the reduced picture are used as a plan for finding edges in the original picture.

This paper describes the computer solution of a problem in picture analysis and description. It is suggested that a very simple concept, here called planning, can be of general value in similar tasks. In basic terms, planning consists of tentative scene description in a picture of greatly-reduced size and detail.

I. PROBLEM DEFINITION

A program for extracting an accurate outline of a man's head from a digital picture has been developed. A typical input picture is shown in figure 1. When the methods described in this paper are applied to the input represented by figure 1, the result is the outline of the man's head as shown in figure 2.

The work described in this paper is a part of a larger project, the visual identification of people by computer. The main purpose of this research is to develop and improve techniques for picture processing by computer. The actual identification of a large group of individuals is only a secondary consideration. The method to be used for identification is the following: measurements such as height, width of head, and distance between eyes are obtained from the picture of the person; these measurements are then used in a simple pattern classification scheme. Measurements from the face and body should provide a good means of identifying people. The reason for this expectation is that physical measurements were the basis of the Bertillon system for identification of people, which had wide use in police work prior to the discovery of the usefulness of fingerprints (Thorwald 1965). In addition, Bledsoe has reported considerable success in identifying faces from their measurements in a man-machine system (Bledsoe 1966).

The emphasis in this work has not been on pattern classification, which forms so much of the pattern recognition literature. Rather, the main effort has focused on 'characterization' of the input picture, accurately locating features such as eyes, nose, or ears in digital pictures. 'It is generally recognized that characterization is really the principal problem in pattern recognition.' (Ho and Agrawala 1968 p. 2102).

The picture is represented inside the computer as a matrix of light intensity values. The pictures discussed in this paper contain the head of a man (or woman) against a background of normal room objects. All pictures contain the head in front view, looking at the TV camera.

From the picture of the head, it is necessary to find an accurate outline of the head so that dependable reference points can be found from which the width of the head can be measured. The location of the top of the head is also obtained from the outline. The position of the outline of the head is used in subsequent processing to determine where to search for smaller features, such as eyes.

This paper discusses a method for detecting edges (outlines of objects) in pictures. The method is an improvement on the most commonly used procedures because it considers global information more efficiently than traditional local techniques. The method is called 'planning' in this paper, a term drawn from artificial intelligence research, for the method fits precisely the description of planning given by Minsky (1963). It is believed that planning can be useful in many projects involving computer vision. In addition, it is helpful to relate the search for edges in pictures to the work on reduction of a given search space in the field of artificial intelligence.

Much work on edge detection in digital pictures has been reported. An 'edge' is the boundary between two objects or between an object and the background. It is contrasted with 'line' which denotes a thin stroke against a uniform background. (The line in figure 2 represents edges in figure 1.) Typically a gradient operation is applied to local areas of the picture to find the edges. The results of the gradient operation are improved with more global information in various routines which will follow an edge through a picture.

Specific desired edges remain elusive, in spite of many efforts at edge detection. Because of noise and the quantization of the picture, many false edges are found. Desired edges often vanish and reappear in grey scale pictures. Edge followers become confused on temporarily strong false paths. To reduce these problems, planning was adopted.

Edge detection by planning can be explained very simply. It is much like a defocused preliminary scan of a picture, an idea recommended by Kirsch *et al.* (1957). In spite of its simplicity, the idea has apparently never been used. Planning has three steps:

1. From the original picture extract a new small picture; figure 5 is an example, a small picture formed from figure 3.
2. Locate the desired edges in the new small picture; figure 7 shows the edges of the head found in figure 5.
3. Use the edges found in the small picture as a plan for finding edges in the original picture. See figure 2 for the results of this final operation.

The rest of this paper will discuss in more detail the ideas outlined above. Section 2 surveys previous work in finding edges. Section 3 introduces planning as used in artificial intelligence. Section 4 applies planning to edge detection. Section 5 gives the details of the program to find the outline of a head. The final section contains some results and a discussion of this and related methods.

2. COMPUTER EDGE DETECTION

Finding the edges in a picture is important, if the picture is to be analyzed and described by a computer. Much of the important information in a picture is contained in the edges. This may be seen in figure 4 which shows the edges present in figure 3. It is evident that most of the information of figure 3 is retained.

Edge detection has received considerable attention as a part of computer picture processing. Roberts (1965), in his pioneering work in machine perception, detected edges in the picture as the first step of processing. Narasimhan and Fornango (1964) emphasized the importance of edges in experiments with pictures of human faces. Guzman (1968), in his work on the analysis and description of scenes, assumed that all edges in the picture had been found accurately. Other research which illustrates the importance of edges in pictures includes the Stanford Artificial Intelligence Project (Feldman *et al.* 1969), the SRI Robot project (Forsen 1968), and Rosenfeld, Thomas and Lee (1969).

Edges in pictures are detected by the use of operators which examine and compare intensity values within a small region of the picture. Most often this operator is some variant of the gradient, the derivative in the direction of the maximum change of intensity. Roberts used the following discrete approximation to the gradient. If A is the matrix representing the picture, the magnitude of the gradient at point $A_{i,j}$ is given by

$$|\mathrm{grad}(A_{i,j})|=\sqrt{[(A_{i+1,j+1}-A_{i,j})^2+(A_{i+1,j}-A_{i,j+1})^2]}$$

Figure 4 is an example of the results of the application of a gradient operator to the picture of figure 3.

There are many other possible approximations to the gradient. Some of these are given by Pingle (1969) [describes an operator developed by Sobel and Feldman], Rosenfeld (1969), Hueckel (1969), and Sakai, Nagao and Fujibayashi (1969). The use of higher derivatives or the Laplacian operator is sometimes suggested for finding edges. These methods seem unsatisfactory because of their tendency to amplify noise.

Edge followers are used to combine the edge information produced by a local operator into a line that represents the edge. An edge follower is an algorithm that searches for and follows an edge. It uses the direction and curvature of edges to connect adjacent short segments. Edge following is discussed by Rosenfeld (1969, p. 136). Descriptions of working edge followers are given by Pingle (1966), Pingle, Singer, and Wichman (1968), and by Greenblatt, Holloway, and Sordillo (1966).

In spite of much effort, the accurate selection by computer of 'important' edges in fairly cluttered scenes has not been particularly successful. The methods given for edge detection are fine for presenting information to the human eye (as figure 4 shows), but what is desired is scene analysis and description within the computer. In reaching for this latter goal problems of noise, disappearing edges, and false edges intrude. Brief comments about the limitations of these methods encountered at the Stanford Artificial Intelligence Project have appeared (Pingle, Singer, and Wichman 1968, McCarthy *et al.* 1968, Paul, Falk, and Feldman 1969). Nilsson (1969, p. 513) states the problems with local methods of edge detection from experience at the SRI Robot project:

> 'The line drawing often contains flaws that seriously complicate its analysis. Some of these flaws could be corrected by more elaborate local processing. However, there is a limit to how well local processing can perform, and when significant edges cannot be told from insignificant edges on the basis of local criteria, the goal of producing a perfect line drawing in this way must be abandoned.'

You must know what you are looking for in order to make sense out of a complex picture. Global information about the edges in the picture and the structure of the objects they represent is needed. This is the idea behind the development of linguistic picture-processing techniques. A survey of these techniques can be found in Miller and Shaw (1968). Another approach to incorporating global structure into edge detection is the decision tree in use at SRI (Nilsson 1969).

The global methods for edge detection also encounter problems. Here the program looks for specific edges in specific relationships. The search space is very large, tens of thousands of points in a picture, and the combinatorics of the problem force unacceptably long search times. In the picture there are just too many lines and there is too much looking to do.

3. PLANNING

The previous section may be summarized as follows. Edge detection is necessary for picture analysis and description. For good results in edge detection, a global strategy is needed. When a global strategy is adopted, problems are encountered with a search space that is too large.

In the field of artificial intelligence, global search is one of the central problems. Much research has been done on developing heuristics for reducing search in such fields as problem solving and game playing. A generally useful concept which has emerged is planning, first used by Newell, Shaw, and Simon (1959) in their General Problem Solver.

Planning is discussed by Minsky (1963) in his paper 'Steps toward Artificial Intelligence'. Artificial intelligence is considered to be mechanization of the problem-solving process. Basic techniques for problem solving include search, pattern recognition, learning, planning, and induction. The following quotations from Minsky's paper define planning.

> Planning is the analysis of problem structure in the large.
>
> 'For really difficult problems, . . . step-by-step heuristics . . . will fail, and the machine must have resources for analyzing the problem structure in the large – in short, for "planning".' (p. 435)
>
> 'Perhaps the most straightforward concept of planning is that of using a *simplified* model of the problem situation. Suppose that there is available for a given problem, some other problem of essentially the same character with less detail and complexity. Then we could proceed first to solve the simpler problem. Suppose, also, that this is done using a second set of methods, which are also simpler, but in some correspondence with those for the original. *The solution of the simpler problem can then be used as a "plan" for the harder one.*' (p. 442)

The steps outlined in the preceding paragraph have been applied almost directly to the problem of finding useful edges in a picture. This represents an entirely new application for planning. It has not been used previously in picture processing.

4. EDGE DETECTION USING PLANNING

The difficult problem of locating important edges in pictures may be approached by using the three steps of planning.

(1) Extract a simplified problem from the given problem. To simplify the problem, a much smaller picture is prepared with the intensity at each point equal to the average intensity over an area of the original picture. The new problem: find the important edges in the new small picture.

The new problem is much easier than the original problem. Many of the small features in the original picture are no longer present since the picture has been greatly reduced in size. Only the important features of the picture remain. Less detail means fewer edges, a smaller search space, and much faster processing. Because of the smoothing done while averaging intensities,

DD

the small picture will contain little noise. Faint and blurred edges of large objects will be enhanced.

(2) Solve the simpler problem. The techniques for finding edges discussed in section 2 can be applied to find the desired edges in the small picture. Since only the principal objects from the original picture now remain, only major edges will be detected. As short edges are collected to form shapes, they may be tested as to their acceptability. This testing should incorporate knowledge of acceptable shapes for the objects to be recognized. Since there are few edges to consider, false paths are found relatively quickly. Backtracking is far less of a problem since the data structures which must be erased are much smaller.

(3) Use the solution to the simpler problem as a guide (a plan) for solving the actual problem. Within the small picture, certain edges have been found. Now it is a fairly simple matter to return to the larger picture and find the desired edges accurately. For example, a straight line may connect points P' and Q' in the small picture. P and Q are the corresponding points in the large picture. Therefore we know that in the large picture there is an approximately straight edge which runs from the vicinity of point P to the vicinity of point Q. Since the approximate location and direction of this edge is known it is possible to detect it quite easily and accurately. The search for this edge can be confined to a narrow band between P and Q. It will be easy to detect a false path for it will soon diverge from this narrow band.

5. PLANNING APPLIED TO A SPECIFIC PROBLEM

This section describes the program that was used to find the outline of the head.

Input

Picture input is provided by a vidicon television camera attached to a PDP-10 computer. The system permits rapid acquisition of a digital matrix representing light intensities in a scene. The size of the matrix is variable, the maximum size is 256×333 points. Each point of the matrix contains a four-bit number ranging from 0 (very dark) to 15 (very light). The picture presented in figure 3 will be used as an example throughout this section. The size of this picture is 226×325 points.

Reduce in size

A new small picture is produced from the original. Each point in the small picture is the average value of the 64 points in an 8×8 square in the original picture. Figure 5 is an example of a small picture. Only the shape of the head is clearly visible. Other details have been smoothed out. The size of figure 5 is 28×40 points. The decision to reduce in size by the factor eight was somewhat arbitrary. In smaller pictures the head tended to disappear; in larger pictures some unwanted details were still present.

Figure 1. Unprocessed input picture of a man's head.

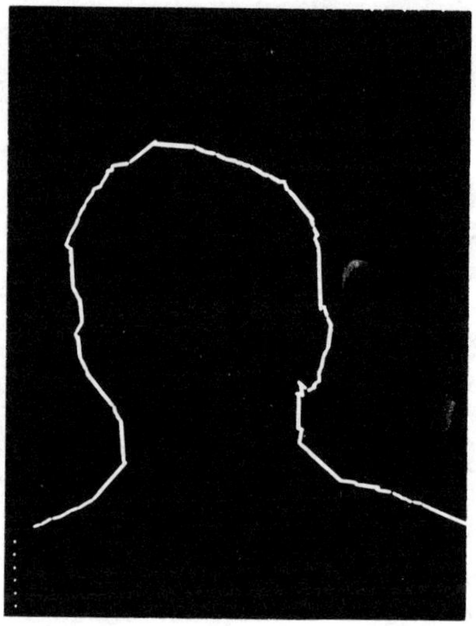

Figure 2. The outline of a head, the result of processing figure 1.

Figure 3. Unprocessed input picture used to obtain figures 4 through 8.

Figure 4. The results of applying a gradient operator to figure 3.

Figure 5 The results of reducing figure 3 in size by averaging.

Figure 6 The results of applying an edge-finding operator to figure 5.

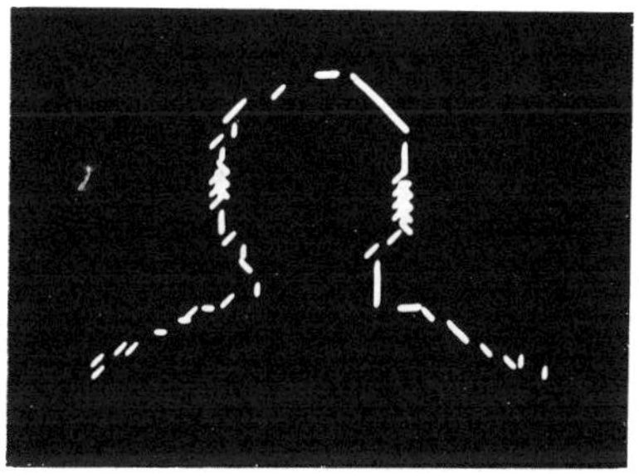

Figure 7. The 'plan' extracted from figure 6 which will be used to follow the head outline.

Figure 8. The outline of the head in figure 3, obtained using the plan of figure 7.

Find all edges in the reduced picture

The small picture obtained in the preceding step is processed to produce a new matrix which contains information on the edges in the small picture. This matrix will be called the *small edge matrix*. Figure 6 is an example of the small edge matrix derived from figure 5. Each element in the matrix contains four bits, representing the four directions (horizontal, vertical, diagonal right, diagonal left) in which edges may be detected. All bits are zero if no edge was found at a point. A one in any bit indicates that an edge in that direction was found.

The local operator which is used to detect edges is described below. Many operators were candidates for this purpose, including those mentioned in section 2. The choice of the particular operator used was made because of high confidence in its ability to reject false edges. Some edges may be missed but even this is preferable to reporting false edges. A more straightforward gradient operator was not used because they sometimes produce an edge indication in an uneven transition area of grey shades. In addition there is no natural cut-off threshold for the gradient operator. Hueckel's operator (1969), which can give very good results, was rejected because it was too large to be used effectively in the small picture. Most of the smoothing utilized by Hueckel's operator has already been done during averaging.

The edge operator used is applied to 3×3 squares of the picture. Let the points in this square be labeled as follows:

$A\ B\ C$
$D\ E\ F$
$G\ H\ I$

The algorithm for detecting a horizontal edge is:

if $B > H$ **then**
    if $[\min(A, B, C) - \max(G, H, I)] > 1$ **then** EDGE
        else NO-EDGE
else if $B < H$ **then**
    if $[\max(A, B, C) - \min(G, H, I)] > 1$ **then** EDGE
        else NO-EDGE
else NO-EDGE.

The EDGE or NO-EDGE results apply to point E. This algorithm is also applied in the vertical direction (using points A, D, G and C, F, I) and along both diagonals (points A, B, D and F, H, I for one diagonal; points B, C, F and D, G, H for the other). Although this algorithm may appear to be time consuming, it may be programmed efficiently.

The small edge matrix is searched as described in the next step to find the small head outline. Because of repeated search and backup during the search for the head, it is best to perform the local edge identification operation only once for each point. Since the picture is small, this does not consume excessive time.

Find head in reduced picture

The search for the outline of the head is done in the small edge matrix. Various heuristics are included in the search program which define an acceptable head shape. The final result of the search is a list structure containing the coordinates of the points which constitute the outline of the head. Each entry in the list structure designates a point where an edge is present in a direction which is reasonable for that part of the head. An example of an unreasonable edge direction would be a vertical edge among a row of horizontal edges which have been labeled the top of the head. Figure 7 shows the edges of the head found in this step. The representation of acceptable head shapes is incorporated into program statements. There are searches, branches, and possible back-up as the outline of the head is built up. This process is similar in some ways to the SRI decision tree search (Nilsson 1969). If no head was present in the original picture, its absence is detected at this stage.

The details of the search for the head are as follows. Three short-line segments are found which are candidates for being part of the top and sides of the head. The spatial relationship between these lines must be reasonable (for example, the top of the head must be above the sides). An attempt is made to connect these line segments to form a 'head' shape. The program searches the region between them for edges which are part of the somewhat semicircular top of the head. If this cannot be done, other possible short-line segments are tried as sides or tops. If the top half of the head is found, a search is made below it for the inward curves of the neck and then the curves outward toward the shoulders. When all of these requirements have been met, the head has been found. The sides of the head are then examined for indentations where the ears should be since the ears sometimes merge with the background. If indentations exist, these are filled in.

Use plan to find outline in original

This part of the program is a *plan follower*. Its input is the full size intensity picture and the list structure containing the small head outline. The output of the plan follower will be a new list structure containing the coordinates of an accurate outline of the head. Between successive points in the plan, the plan follower searches a narrow band for an edge which connects the points. This band is sixteen points wide, since the plan was reduced in size by a factor of eight. Although this band is narrow compared to the size of the picture, it is still wide enough to contain several edges. The proper edge is chosen primarily by direction. If there are two parallel edges running in the desired direction within the search band, that edge farthest from the center of the head is chosen. The reason for this choice is the assumption that edges within the head (in hair, ears, and so on) are more likely to follow the directions of the head outline than edges found on background objects.

The operator used to detect edges in the full-sized picture is almost the same as that used in the small picture. In the small picture the operator was applied to the nine points of a 3×3 square. In the large picture the same operator is applied to the nine points at the corners, the center of the sides, and the center of a 5×5 square, for example,

$$\begin{array}{c} A - B - C \\ - - - - - \\ D - E - F \\ - - - - - \\ G - H - I \end{array}$$

The reason for this change was to allow detection of faint edges in the large picture.

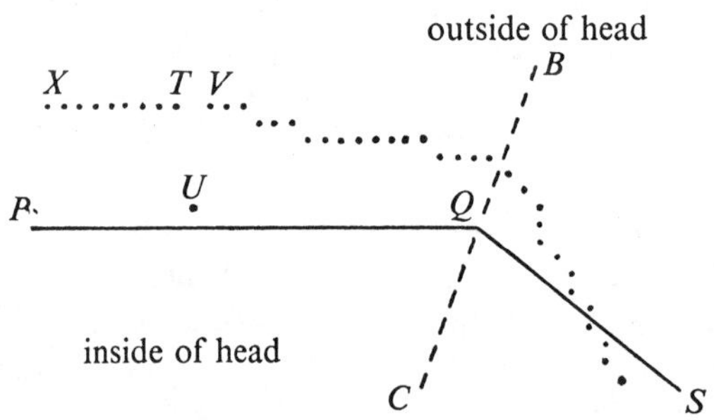

Figure 9. Operation of plan follower

The details of the plan follower are given below. Reference is made to figure 9 which is a schematic diagram of the operation of the plan follower in a small region. Two non-parallel straight lines are identified in the plan, lines *PQ* and *QS*. Point *X* is the last previous point that has been found on the edge. The direction of search is generally from *P* to *Q* to *S*. The line *BC* is found which bisects the angle *PQS*. Corresponding to the line *PQ* in the plan, the plan follower will search from point *X* until the edge crosses line *BC*. Within this region the edge will be accepted if its direction is roughly parallel to *PQ* or to *QS*. The edge search is done by moving one unit in the direction of *PQ* from the last point found and applying the edge detection operator along a line perpendicular to *PQ*. Normally the edge detection operator is applied only to five points immediately in front of the last point found. Under certain conditions, however, the edge detection operator is applied over all 16 points across the search band. These conditions are:

(a) The edge has just taken a sharp turn.

(b) The edge is lost.

(c) The intensity outside the head outline has changed abruptly.

(d) The edge zig-zags.

These measures are necessary to track correctly the edge when it is sharply curving. The three points most recently found are filtered to remove a single point which is far off the edge. For example, if U were the only edge point found between points T and V, point U would be rejected.

6. RESULTS AND DISCUSSION

The program which has been described does a good job of finding head outlines. It is difficult, however, to quantify what is meant by a good head outline. A qualitative measure has been used; the original picture is compared visually with the outline which is produced. In order to give an idea of the effectiveness of this program, three additional pictures with their resultant outlines are presented in figures 10, 11, and 12.

The remainder of this section will discuss briefly some observations about edge detection using planning.

Is 'planning' merely the application of a large edge detection operator? Certainly the search for edges made in the small picture could be a search of the large picture using an operator which examines a 24×24 point square. In fact, the two steps, large picture to small picture, and small picture to small edge matrix, could be one step, the application of a 24×24 point operator. Alternatively, the entire large picture could be heavily smoothed and the search for edges could be done in this smoothed picture. Planning, however, has two advantages over either of the approaches listed above. First, it is much easier to design and debug the program which searches for significant edges when using planning. This is because the reduced picture is so much smaller. The use of planning aids insight. While designing a program to find the edges in a picture of 625 points (for instance), the designer can comprehend easily both the overall structure and the detailed structure of the picture. It is very difficult to obtain the same level of understanding when the picture contains 40,000 points. The second advantage of planning is speed. This increase of speed is particularly important when designing search programs on an interactive computer system.

Some tests were made to determine the speed improvement obtained using planning. The results indicate that using planning is about 40 times faster than the same search without planning. The tests were performed as follows.

First the program just described, which finds the head outline using planning, was timed. The time to process a head picture from input to completion of the large outline was about 6 seconds. This time, as well as all others mentioned, was measured on the PDP-10 computer and is accurate to ± 20 per cent. Then the program was modified to perform without planning.

The first step was to eliminate the fine detail of the picture by smoothing in a fashion similar to the averaging used in planning. The entire picture was smoothed using averages over 8×8 squares. This time was about 64 times the speed without planning, namely, 160 seconds.

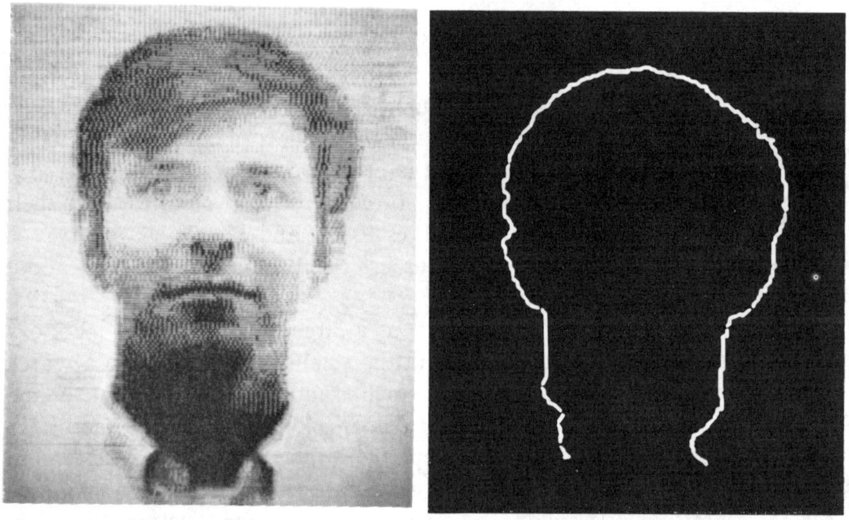

Figure 10

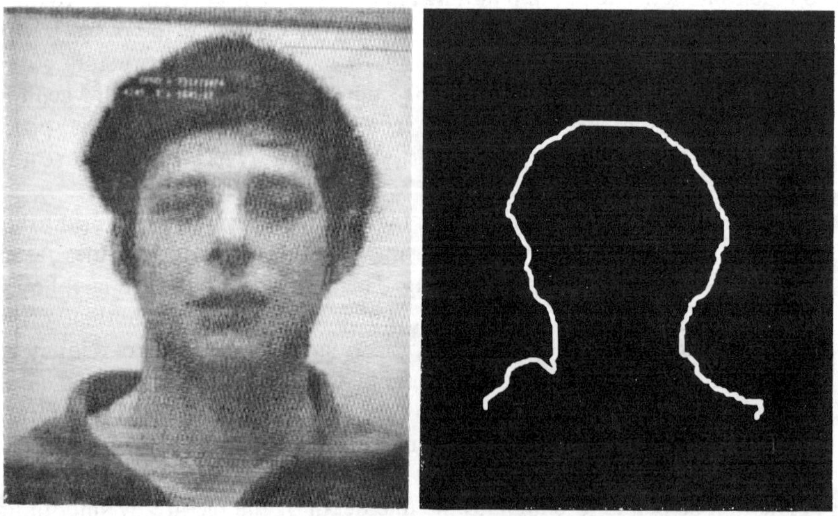

Figure 11

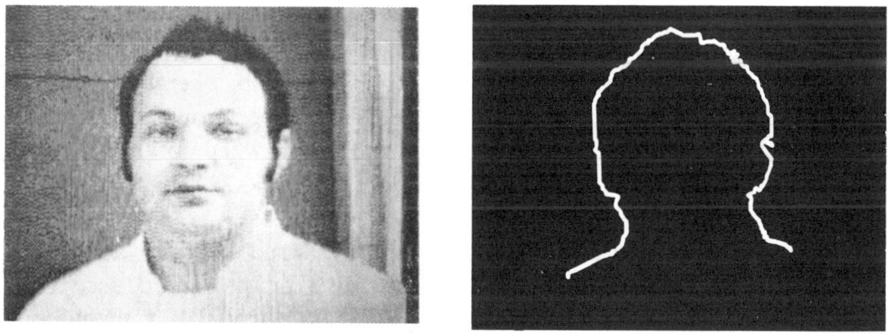

Figure 12

All edges were found in the smoothed picture. This took 34 seconds, again a factor of about 64.

Next the search program was used on the full-sized smoothed picture to search for the outline of the head. Here is where planning really showed its speed. This search, in a reduced size picture, takes about 0·3 seconds. In the large picture, the search took about 40 seconds. The reason for this is that quite a sizeable incorrect data structure is built up when following false paths and many points must be processed before the error is detected.

The plan-following stage has no counterpart time when not using planning.

In summary, the program without planning took 234 seconds, 40 times slower than the time using planning.

If a program incorporated more elaborate structural knowledge of the object to be recognized, could planning in a picture of reduced size be eliminated? No, such a program must be less effective for the following reasons. The combinatorics of the large search space will require excessive time. The effect of noise and irregularities in the picture will be even worse. This is emphasized by the elaborate provisions for search that had to be built into the plan follower. Even when the direction and approximate location of an edge are known, its detection can be hard. For such edges, planning is essential.

Sakai, Nagao, and Fujibayashi (1969) have recently reported their work on finding faces in photographs. They first produce a picture, the result of a gradient operation, which contains the edges of the input picture. Large templates corresponding to the head outline and facial features are then matched with the edge picture. All reasonable positions and sizes of the template are tried. This method appears to be time consuming when compared to the planning method, and the result is only an approximate location for the head in the picture.

An improvement on the planning technique presented in this paper would be recursive application of planning at varying size reductions. For instance, the original picture could be reduced in size twice, four times, and eight times. The plan found in the $\frac{1}{8}$ size picture could be used to find edges in the $\frac{1}{4}$ size picture. This could then be used as a plan to find edges in the $\frac{1}{2}$ size picture. Finally the accurate true edge would be found in the original picture. A hint of such a method was mentioned in the description of the plan follower. The local edge-finding operator examined only alternate points. It could be considered to be using a $\frac{1}{2}$ size picture. The program could decide itself how much size reduction and averaging to do. At each level, if there is too much detail, a smaller picture could be called for.

The greatest deficiency of the program described here is the lack of a high-level mechanism for describing picture structure. In order to find objects in cluttered pictures, a program must know what it is looking for. It would be desirable for that part of the program which finds the plan (the head outline) in the small picture to be general purpose. To such a general purpose program one could specify in some higher level 'language' that a head is desired where a head is:

round on top;
somewhat flat on the sides;
curving inward for the neck below the sides;
and so on.

The same program could then be used to identify cows or battleships (for instance) using a different structural description. Such high level descriptions are the goal of the picture-description languages discussed by Miller and Shaw (1968) which were mentioned earlier. Such a general-purpose program was not used for finding head outlines. Instead, the structure of the head was embedded implicitly in the program. The reason for this is that picture-description languages are not sufficiently well developed to permit the specification of the numerous heuristic tests necessary to find the head outline. Satisfactory formal description of picture structure remains a major obstacle to advances in picture processing.

Acknowledgement

The author would like to thank Professor D.R.Reddy for his ideas, encouragement, and criticism which played a major part in the development of the work reported here. In addition, discussions with J.M.Tenenbaum contributed many fresh viewpoints to this paper.

The research reported here was supported in part by the Advanced Research Projects Agency of the Office of the Secretary of Defense (SD–183).

The author is now at the School of Information and Computer Science, Georgia Institute of Technology.

REFERENCES

Bledsoe, W.W. (1966) Man-machine facial recognition. Panoramic Research Inc., Report PRI: 22.

Feldman, J.A., Feldman, G.M., Falk, G., Grape, G., Pearlman, J., Sobel, I. & Tenenbaum, J.M. (1969) The Stanford hand-eye project. *Proc. Int. Joint Conf. on Art. Int.* 1969, pp. 521–6.

Forsen, G.E. (1968) Processing visual data with an automaton eye. *Pictorial Pattern Recognition,* pp 471–501 (eds Cheng, G.C. *et al.*). Washington, DC: Thompson Book Co.

Greenblatt, R., Holloway, J. & Sordillo, D.A. (1966) Sides 21, *Art. Int. Memo No. 104-2 (MAC–M–320).* MIT, Cambridge, Mass.

Guzman, A. (1968) Decomposition of a visual scene into three-dimensional bodies. *Proc. AFIPS 1968 Fall Joint Comp. Conf.*, **33,** 291–304. Washington DC: Thompson Book Co.

Hueckel, M. (1969) An operator which locates edges in digitized pictures. *Art. Int. Memo No. 105.* Stanford University, California.

Ho, Y. & Agrawala, A.K. (1968) On pattern classification algorithms, introduction and survey. *Proc. IEEE,* **56,** 2101–14.

Kirsch, R.A., Cahn, L., Ray, C. & Urban, G.H. (1957) Experiments in processing pictorial information with a digital computer. *Proc. Eastern Joint Comp. Conf.*, pp. 221–9. New York: Spartan Books.

McCarthy, J., Earnest, L.D., Reddy, D.R. & Vicens, P.J. (1968) A computer with hands, eyes, and ears. *Proc. AFIPS 1968 Fall Joint Comput. Conf.*, **33**, 329–38. Washington DC: Thompson Book Co.

Miller, W.F. & Shaw, A. (1968) Linguistic methods in picture processing: a survey. *Proc. AFIPS 1968 Fall Joint Comp. Conf.*, **33**, 279–90. Washington DC: Thompson Book Co.

Minsky, M. (1963) Steps toward artificial intelligence. *Computers and Thought*, pp. 406–50 (eds. Feigenbaum, E.A. & Feldman, J.). New York: McGraw-Hill.

Narasimhan, R. & Forango, J.P. (1964) Some further experiments in the parallel processing of pictures. *IEEE Trans. EC–13*, 748–50.

Newell, A., Shaw, J.C. & Simon, H.A. (1959) Report on a general problem solving program. *Proc. Int. Conf. on Info. Proc.*, pp. 256–64. Paris: UNESCO House.

Nilsson, N.J. (1969) A mobile automaton: an application of artificial intelligence techniques. *Proc. int. Joint Conf. on Art. Int.*, pp. 509–15.

Paul, R., Falk, G. & Feldman, J.A. (1969) The Computer representation of simply described scenes. *Art. Int. Memo No. 101.* Stanford University, California.

Pingle, K.K. (1966) A program to find objects in a picture. *Art. Int. Memo No. 39.* Stanford University, California.

Pingle, K.K. (1969) Visual perception by a computer, *Automatic Interpretation and Classification of Images*, pp. 277–84 (ed. Grasselli, A.), New York: Academic Press.

Pingle, K.K., Singer, J.A. & Wichman, W.M. (1968) Computer control of a mechanical arm through visual input. *Proc. IFIP Conf.*, H pp. 140–6. Edinburgh.

Roberts, L.G. (1965) Machine perception of three-dimensional solids. *Optical and Electro-Optical Information Processing*, pp. 159–97 (eds. Tippett, J.T. *et al.*). Cambridge, Mass.: MIT Press.

Rosenfeld, A. (1969) *Picture Processing by Computer.* New York: Academic Press.

Rosenfeld, A., Thomas, R.B. & Lee, Y.H. (1969) Edge and curve enhancement in digital pictures. *University of Maryland Computer Science Technical Report 69–93.*

Sakai, T., Nagao, M. & Fujibayashi, S. (1969) Line extraction and pattern detection in a photograph. *Pattern Recognition*, **1**, 233–48.

Thorwald, J. (1965) *The Century of the Detective.* New York: Harcourt, Brace & World.

Part V
Boundary Representation and Description

Once a curve or boundary in a picture has been extracted, it can be represented and described in a variety of ways. As with region representations, some representations preserve all of the information about the contour, while others preserve only what is of interest for a particular task.

One basic way to represent a smooth curve mathematically is to give its slope angle (or its curvature) as a function of arc length. Such representations are known as intrinsic equations for the curve. Both the slope intrinsic equation and the curvature intrinsic equation are invariant to translation and are easily normalized to be independent of scale; in addition, the curvature intrinsic equation is invariant to rotation.

The first paper, by Freeman, develops the properties of a discrete version of the slope intrinsic equation, known as the chain code. A number of basic procedures are given for manipulating chain-encoded curves. See Freeman (1974) for a survey of subsequently developed procedures and applications based on chain encoding. For an alternative, basic coding procedure that is especially useful for computing area and intersection properties, see Merrill (1973).

Closed contours, such as complete region boundaries, have the important characteristic that their intrinsic functions either are or can easily be made periodic. This leads naturally to a mathematically complete Fourier series representation. As with moments, the low-frequency Fourier coefficients are insensitive to noise and quantization error, and provide an incomplete, but useful, gross representation of the contour.

This approach admits a number of variations, each having its own advantages and disadvantages. For example, the paper by Zahn and Roskies (1972) provides a detailed and elegant treatment of the properties of the Fourier expansion of the slope intrinsic function. Among other things, this function has the advantage of being directly related to the chain code, but has the disadvantage that a partial-sum approximation usually does not yield a closed contour. The second paper, by Granlund, avoids this defect by expanding the equations giving the *xy*-coordinates as a function of arc length. By treating the boundary points as points in the complex plane, Granlund obtains simple equations that show how the Fourier coefficients vary with translation, rotation, scaling, and the choice of a starting point. These equations suggest various ways to normalize the coefficients to obtain invariance to such transformations; this approach is reminiscent of, but somewhat simpler than, the methods described by Hu in Part III.

Perhaps the major disadvantages of Fourier descriptors are that interesting local characteristics of the boundary are obscured by being distributed over all of the coefficients, and that the required integration destroys their validity if any significant portion of the contour is obscured by occlusion. An alternative approach is to segment the contour and to characterize it through a sequence of local descriptions. Points of high curvature are popular candidates for breakpoints. The third paper, by Zahn, gives a procedure for detecting such curvature points and for smoothing and describing the resulting piecewise linear approximation. [See Rosenfeld and Johnston (1973) for a different procedure for detecting curvature points. An alternative way to smooth the contour is to find the so-called minimum-perimeter polygon; see Montanari (1970) and Sklansky, Chazin, and Hansen (1972).]

Piecewise-linear (polygonal) approximations have been used extensively in scene analysis. They are obviously appropriate for pictures of polyhedral objects, and thus have been the standard choice for the blocks world (see the papers by Roberts in Part VI and by Brice and Fennema in Part II). Since a single contour is in a sense one dimensional, many of the techniques for segmenting and describing waveforms are also applicable to contours; of particular interest here are a series of papers by Pavlidis and his students (1972–1975) using methods of functional approximation.

A more directly geometrical way to characterize shape is in terms of the presence of corners (convex and concave), notches, lakes, bays, peninsulas, etc. We have not included a paper that concerns this way to characterize shape, but feel obliged at least to mention some relevant references. Unger (1959) originally suggested the use of such features for character recognition, and they are certainly intuitively appealing shape measures (see Munson (1968) and Freeman (1974) for extensions of these ideas). Many of these features can be obtained relatively easily from the contour and its convex hull. This brings up an issue we have thus far ignored, namely, the definitions of such things as straight lines and convex regions on a digital image; these topics are treated by Rosenfeld (1974) and Sklansky, Chazin, and Hansen (1972), respectively.

The final paper in this part, by Aggarwal and Duda, concerns the analysis of pictures of moving polygonal objects. Motion adds another dimension to scene analysis, and introduces such important topics as change detection [Lillestrand (1972)], tracking [Smith and Phillips (1972; cited in Part I)] and motion stereo [Nevatia (1976)]. The paper by Aggarwal and Duda emphasizes the construction of models from occluded views, and thus leads us naturally into the blocks world of Part VI.

On the Encoding of Arbitrary Geometric Configurations*

HERBERT FREEMAN†, SENIOR MEMBER, IRE

Summary—**A method is described which permits the encoding of arbitrary geometric configurations so as to facilitate their analysis and manipulation by means of a digital computer. It is shown that one can determine through the use of relatively simple numerical techniques whether a given arbitrary plane curve is open or closed, whether it is singly or multiply connected, and what area it encloses. Further, one can cause a given figure to be expanded, contracted, elongated, or rotated by an arbitrary amount. It is shown that there are a number of ways of encoding arbitrary geometric curves to facilitate such manipulations, each having its own particular advantages and disadvantages. One method, the so-called rectangular-array type of encoding, is discussed in detail. In this method the slope function is quantized into a set of eight standard slopes. This particular representation is one of the simplest and one that is most readily utilized with present-day computing and display equipment.**

Introduction

CONCOMITANT with present-day advances in the fields of control and communications, a growing need is being felt for precise means of describing arbitrary geometric configurations. This need arises in two ways: 1) from the desire to communicate geometric information, and 2) from the requirement for analysis and synthesis of geometric configurations.

If one wishes to transmit the outline of a lake, the information relating to the contour must first be identified (a description process) and then encoded in a manner which permits efficient transmission. Decoding and reconstruction of the contour in terms of a pictorial representation is required at the receiving end of the transmission. The twin objectives of communication are efficient transmission and faithful reproduction. Any scheme for communicating geometric information must be able to withstand critical examination on these objectives.

The analysis and synthesis of arbitrary geometric configurations have, until recently, received only scant attention; the tedium of lengthy hand computation was too discouraging. With the availability of modern high-speed digital computers, however, labor of computation is no longer a deterrent. Thus it is now entirely feasible to use a computer to determine the surface area of a lake from its contour, or, given a set of past weather maps, to compute a most probable future weather map. One need merely develop the necessary algorithms for the computer and devise methods for encoding the geometric information in a way which will facilitate such analysis.

* Received by the PGEC, August 8, 1960; revised manuscript received, January 27, 1961.

† Dept. of Elec. Engrg., New York University, New York, N. Y.

An arbitrary geometric curve is a curve which is pictorially defined (*e.g.*, the outline of a knife), but for which no analytic description exists. It thus differs from a mathematical (or nonarbitrary) curve, such as a triangle, sine wave, or hypocycloid, which can be precisely formulated in mathematical terms. A circle is a mathematical curve perfectly described by a simple equation. In contrast, a wheel is an arbitrary geometric curve which, to a certain limited degree of precision, can be approximated (described) by a circle. If one desires descriptions which will take into account more and more of the wheel's minute irregularities, the descriptions must become progressively more complex and unwieldy.

Description Processes

Consider the problem of describing arbitrary geometric curves to permit their transmission over a telegraph communication link or their manipulation by a digital computer. The curves are assumed to be continuous plane curves, open or closed, and either singly or multiply connected. One approach to this problem is to cut the curve into arbitrarily fine segments and then to approximate each segment with a standardized mathematical curve.

The segmentation of curves raises some important questions. Should the segments be large and few in number, or should they be small and much greater in number? If the former, the mathematical representation of the segments will be relatively complex (*e.g.*, high-order polynomials may be required), while, if the latter, a straight-line approximation will suffice. Further, should the segments all be equal or should segmentation be dependent on the curve itself? Uniform segmentation has the advantage of simplicity; however, segmentation which is based on the properties of a particular curve may lead to more efficient encoding.

An encoding scheme for arbitrary geometric curves should be simple, highly standardized, and universally applicable to all continuous curves. Further, the encoding scheme should be such that it facilitates digital computer analysis of curve properties. For these reasons subsequent discussion will be limited to straight-line segmentation using either a single length or a small number of standard lengths. It is not intended to imply that other approaches are without merit; their consideration will, however, be outside the scope of this paper.

Reprinted from *IRE Trans. Electron. Comput.*, vol. EC-10, pp. 260–268, June 1961.

A question that arises at once concerns the fineness to which a curve is to be segmented. Consider the curve of Fig. 1, drawn in the x-y plane. One may introduce a parameter t which will be a measure of distance along the curve as the curve is traversed in the direction indicated, beginning at the starting point x_0y_0. This makes it possible to represent the original curve by two parametric curves, $x(t)$ and $y(t)$, as shown in Fig. 2.

Approximating a plane curve with straight-line segments is a sampling-type operation, and its validity can be tested by means of the sampling theorem. This can be done by obtaining the Fourier transforms of the two parametric curves and examining their frequency spectra. Generally, the frequency spectra will be infinite, with progressively increasing attenuation of the higher-frequency components. A qualitative interpretation of the sampling theorem states that the sampling interval T must be less than one half the reciprocal of the highest significant frequency component. If the precision of representation is to be such that over a square-meter x-y plane any errors in location of points on the curve will not exceed one millimeter, one may refer to the representation as having a precision of 60 db. In a loose sense this can be carried over into the frequency spectrum, where components which are attenuated 60 db or more may then be considered insignificant.

Once the desired segment-length T has been determined, one may replace the original arbitrary curve with a curve consisting of a chain of straight-line segments of length T. The curve can then be described simply by a sequence of angles and a standard distance. When reconstructing the curve from the given data, the scale can be arbitrarily selected by choosing the appropriate standard length. A coded representation of

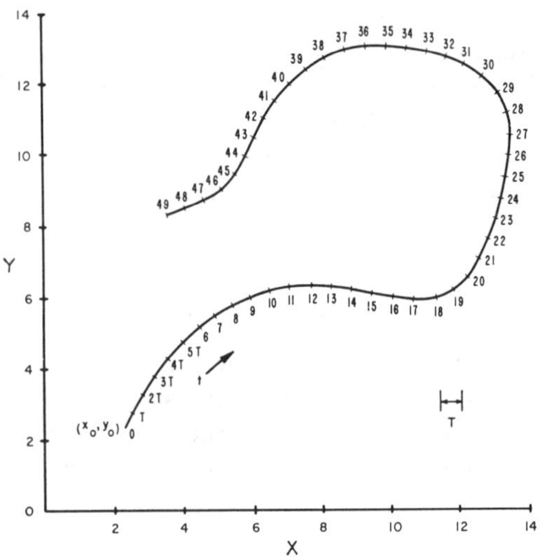

Fig. 1—Arbitrary plane curve.

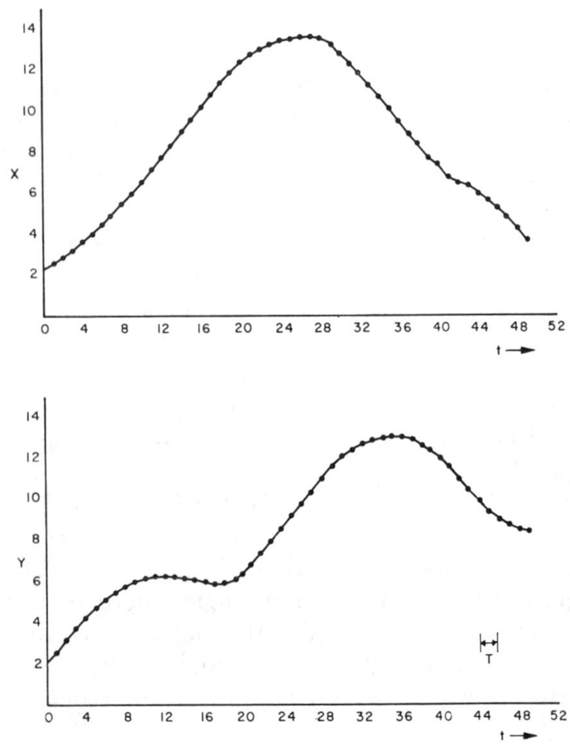

Fig. 2—Parametric curves.

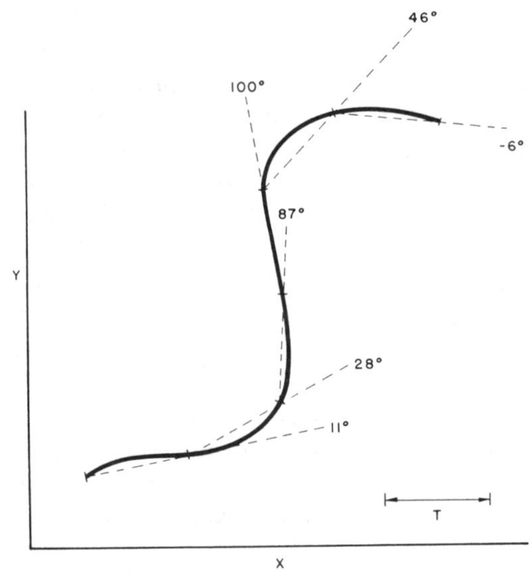

Fig. 3—Coarsely quantized arbitrary plane curve.

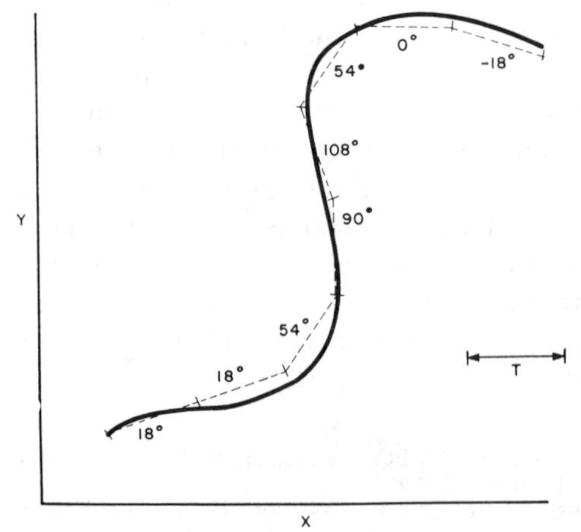

Fig. 4—Quantization based on a standard angle of 18°.

the curve shown in Fig. 3 would appear as follows:

$$11°, 28°, 87°, 100°, 46°, -6°.$$

The actual encoding procedure can be simplified by severely limiting the number of permissible angles. Thus if one were to limit the encoding to 20 standard angles (18° increments) and to number these consecutively from 0 to 19, the curve of Fig. 4 could be represented by the following number series (read from left to right):

$$1, 1, 3, 5, 6, 3, 0, 19.$$

It is evident that any desired precision (fineness of resolution) can be obtained with a relatively coarse angle quantization by making the standard distance arbitrarily small.

This method of describing an arbitrary geometric curve may be termed the unit-vector method. In essence, a curve is represented by a sequence of small vectors of unit length and a limited set of possible directions. For reasons of data handling, it is sometimes preferable to use a grid or array type of description. Thus, one may lay a rectangular grid over the curve and describe the curve by identifying the grid points which lie closest to it. Such a grid system has the advantage over the unit vector scheme of being much less susceptible to error build-up during encoding.

Rectangular-grid representations of arbitrary geometric curves are frequently encountered in connection with digital-computer output displays. Thus, a common type of display consists of a cathode-ray tube on the face of which the beam spot can be positioned to, say, one part in 1024 in both the x and the y direction. Two ten-place binary numbers (20 bits) are required to locate a particular point on the face of the tube. Although a typical curve display may require a very large number of points, the full 20-bit coordinates of each point are usually given. This is unnecessarily wasteful.

The rectangular grid method is somewhat similar to a unit-vector method based on eight angles. The significant difference lies in the use of two standard distances which are in the ratio of $\sqrt{2}$ to each other; *i.e.*, one is the side and the other is the diagonal of a small square in the grid.

The unit-vector representation for six angles has properties that merit special consideration; it can also be regarded as being a representation based on a hexagonal grid. This is the only grid utilizing a single standard distance. The use of a hexagonal grid for computer output display can be achieved with a cathode-ray tube having its pairs of deflection plates mounted at 60° to each other instead of the conventional 90°.

Coding Processes

The process of adopting a symbolism convention and then quantitatively identifying the describing properties in terms of the convention is called encoding. It is desired here to develop means for encoding arbitrary geometric curves so as to facilitate their handling in communications systems and digital computers. A general objective of all encoding is that the process be efficient, *i.e.*, that it use a minimum of bits to represent some given information.

Consider the rectangular-grid representation mentioned previously. The conventional manner of representing a curve consists of giving the full twenty-bit coordinates of each point of a curve (for a 1024×1024 array). Such coding would be necessary if all points were fully independent of each other. If the points describe a continuous curve, they are, of course, far from independent. An important restriction which holds for continuous curves (and which may be considered a definition of a continuous curve) is that successive points are adjacent to each other. Once one point on a continuous curve is given, the next point's location is very much restricted. This knowledge can be utilized to locate successive points with far fewer bits than the number required if each point were to be independently located.

For the rectangular grid it is clear that if a point on a continuous curve is known, the next point can assume only one of eight possible adjacent positions, as shown in Fig. 5(a). If one assigns the decimal digits zero through seven to these eight positions, starting with the one which is horizontally to the right and progressing in a counter-clockwise direction, the code of Fig. 5(b) is obtained. One notes that in this manner each point on a curve is located relative to the previous one, and only the first need be located in an absolute sense. To shift a curve vertically or horizontally, only the coordinates of the first point require changing. As an illustration, consider the curve of Fig. 6 which can be represented by the code sequence (read from left to right).

$$107643,$$

or in binary form,

$$001 \quad 000 \quad 111 \quad 110 \quad 100 \quad 011.$$

One notes that only three bits are required to specify one point on the curve. The required memory capacity for a continuous curve encoded in this way is then only 15 per cent of that required for a curve which has all its points independently specified in a 1024×1024 point array.

One property of the rectangular-array code is at once apparent. Adding "2" (modulo 8) to each of the digits in the decimal representation of a curve causes the curve to be rotated 90° in a counter-clockwise direction. A subtraction will result in clockwise rotation. The addition (or subtraction) of any even number n will cause a rotation equal to $n/2$ times 90°. If "1" is added to each digit, a rotation of 45° is achieved; however, this rotation is nearly always accompanied by distortion or at least a change of scale. This is true for the addition (or subtraction) of any odd number.

The encoding method discussed thus far is by no means unique. Nor is it necessarily the one which is most

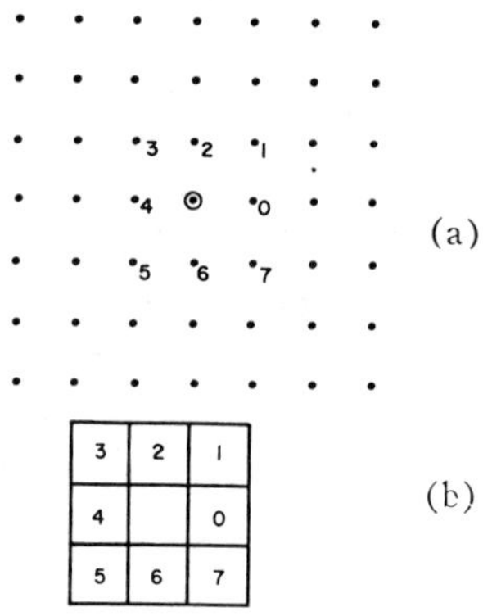

Fig. 5—Rectangular array.

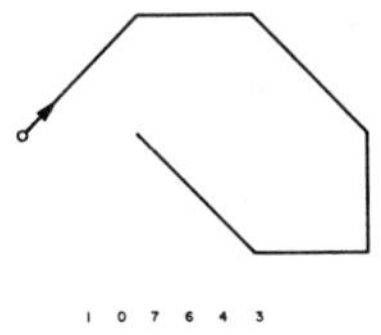

Fig. 6—Illustration of rectangular-array representation.

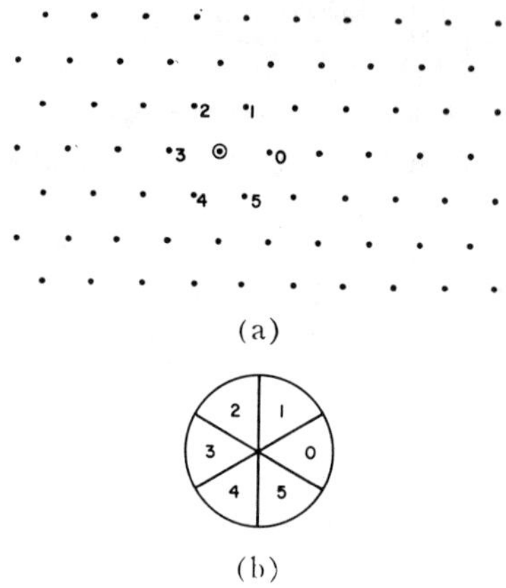

Fig. 7—Hexagonal array.

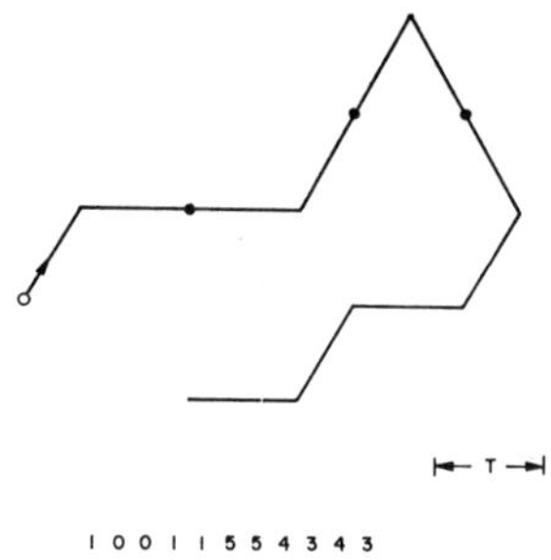

Fig. 8—Illustration of hexagonal-array representation.

economical in its use of memory capacity. In developing the method it was supposed that, given one point, the next point could assume one of eight neighboring positions with equal likelihood. If certain conditions are imposed on an arbitrary curve (in addition to continuity), a two-bit or even a one-bit code can be utilized. For example, if it is known that the slope of the arbitrary curve changes very slowly, *i.e.*, if the values of dx/dt and dy/dt at $t=n$ differ only slightly from those at $t=n+1$, then a one-bit code may suffice. Such a code would be associated with the second difference, and would be used as follows. Given the slope between the previous two points, the slope to the next point would be formed by increasing the last slope by 45° if the digit is "1," or decreasing the slope by 45° if the digit is "0." A straight line would be represented by a sequence of alternating ones and zeros.

The rectangular-array code of Fig. 5 has the disadvantage, as has been noted, that the addition of an odd number to each digit results in distorted rotation. This drawback can be avoided by using instead the hexagonal array shown in Fig. 7(a). For each given point, there are six possible locations for the next point, all of which are equidistant from the given point. The six next-point locations can be numbered from zero to five in a counter-clockwise direction, starting with the location horizontally to the right. Three bits per digit are required to represent the six next-point locations. The code is unsaturated since there are two unused combinations, and special techniques must be used to obtain modulo-six additions. An example of a hexagonal-array display is given in Fig. 8. The curve is represented by the sequence (read from left to right):

10011554343.

The hexagonal-array representation has the property that addition (modulo six) of unity to each digit causes a 60° distortion-free rotation of the curve. The method suffers from the disadvantage of being based on an uncommon, nonorthogonal coordinate system, of having its angular resolution limited to 60° (in contrast to 45° for the rectangular array), and of resulting in an unsaturated binary code representation.

The simplicity of the rectangular array, as well as the ease with which it can be adapted to standard display techniques, suggests that this array be given special attention. Accordingly, the remaining sections of this paper are devoted entirely to this encoding method. It is usually not difficult to see, however, that most techniques applicable to the rectangular array can be extended to the hexagonal and the unit-vector arrays.

Rectangular-Array Encoding

In the rectangular-array method, based on the code of Fig. 5(b), a number sequence utilizing only one kind of digit represents a straight line; *e.g.*, the sequence 22222222 · · · 2 describes a vertical straight line. A total of eight such straight lines can be represented in this manner with the digits "0" through "7." Each of these lines makes an angle with the horizontal which is equal to the digit multiplied by 45°. Straight lines making angles with the horizontal of other than multiples of 45° can be formed with digit pairs, triplets, or higher-order digit groups. Thus the sequence 121212121212 · · · 12 represents a "straight" line making an angle of +63.43°. The total number of straight lines which can be represented with digit pairs is 16. For triplets this increases to 32. One notes that these digit groups can contain only two kinds of digits, and that these cannot differ by more than unity, *e.g.*, 0 and 1, 1 and 2, etc.

Examples of a variety of straight-line representations are given in Fig. 9.

All first-quadrant straight-line digit groups up to five digits in length are listed in Table I. Angles in other quadrants can be obtained by adding either 2, 4 or 6 to each of the digits in a group.

The digit groups given in Table I are not necessarily unique. Thus, the digit groups 01010 and 00011 are equivalent; *i.e.*, they describe the same straight-line slope. However, the first digit group is to be preferred because of its greater smoothness. To assure minimum deviation from the true straight-line path, the two kinds of digits making up a digit group should alternate to the maximum extent.

The representation of an arbitrary curve of any significance will usually consist of a sequence of many hundreds of digits. The more precisely an arbitrary curve is to be represented, the more finely it must be quantized, and, therefore, the greater the number of digits required in its representation. If a curve is completely arbitrary (except for continuity), no condensation or abbreviated representation is possible. However, if there is some regularity in the curve, this *pattern* can be exploited to simplify the representation.

Consider, for example, the straight line which makes an angle of $+26.57°$ with the horizontal. It is represented by the sequence

$$0101010101010101 \cdots 01.$$

One may regard this number sequence as a function of a parameter n where $n=0, 1, 2, \cdots, N$. A more compact representation is then possible through the introduction of the concept of the generating function.[1]

Given a function $f(n)$, one defines another function $u(x)$ as follows:

$$u(x) = \sum_{n=0}^{N} f(n)x^n,$$

where $u(x)$ is the generating function of $f(n)$. For the straight line just mentioned, the generating function is given by the following:

$$(0) + (1)x + (0)x^2 + (1)x^3 + (0)x^4 + (1)x^5 + \cdots.$$

This infinite sequence has the closed form

$$\frac{x}{1 - x^2}.$$

In obtaining $f(n)$ from the corresponding generating function, one must recognize that a zero is a meaningful digit which represents a finite segment of the curve being described. Generating functions corresponding to the first eight entries of Table I are given in Table II.

[1] C. Jordan, "Calculus of Finite Differences," Chelsea Publishing Co., New York, N. Y., p. 20; 1950.

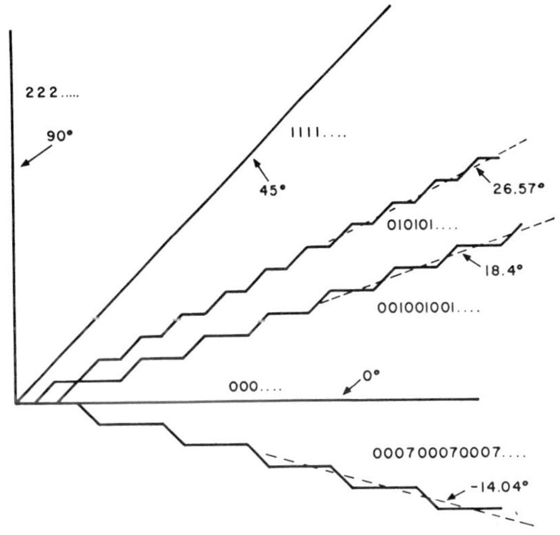

Fig. 9—Straight-line representations.

TABLE I
FIRST-QUADRANT, STRAIGHT-LINE DIGIT GROUPS

Digit Group	Tangent	Angle (Degrees)
0	0	0
00001	0.2	11.32
0001	0.25	14.04
001	0.33	18.4
01010	0.4	21.8
01	0.5	26.57
10101	0.6	30.09
110	0.67	33.7
1110	0.75	36.85
11110	0.8	38.65
1	1.0	45
11112	1.25	51.3
1112	1.33	53.1
112	1.5	56.3
12121	1.67	59.01
21	2	63.4
21212	2.5	68.2
221	3	71.6
2221	4	75.97
22221	5	78.7
2	∞	90

TABLE II

	Digit Group	u(X)
1	0	$\frac{0}{1 - X}$
2	00001	$\frac{X^4}{1 - X^5}$
3	0001	$\frac{X^3}{1 - X^4}$
4	001	$\frac{X^2}{1 - X^3}$
5	01010	$\frac{X + X^3}{1 - X^5}$
6	01	$\frac{X}{1 - X^2}$
7	10101	$\frac{1 + X^2 + X^4}{1 - X^5}$
8	110	$\frac{1 + X}{1 - X^3}$

The concept of generating function representation is not limited to straight lines. Consider the sequence 01234567. This sequence will be called the minimal circle and is shown in Fig. 10. It is the smallest figure which possesses the circle property of permitting arbitrary rotation without any visible effect. It is minimal in that there is no smaller figure for which this property holds.

The minimal circle is a closed curve and is represented by a short sequence of digits. Even when dealing with short sequences, it is sometimes desirable to employ the generating function representation. For the minimal circle the generating function is given by

$$\frac{x}{(1 - x)^2},$$

as is easily verified by performing the indicated long division. This generating function leads to an infinite series; however, since the numbers are all modulo eight, the series is periodic and simply causes the same path to be traversed over and over again.

Various minimal waveforms, such as square waves or triangular waves, can be represented in a straightforward manner. Minimal waveforms are waveforms which cannot be reduced without destroying the characteristics of the waveform. A set of common minimal waveforms, together with their digit groups and generating functions, is shown in Fig. 11. The numerators

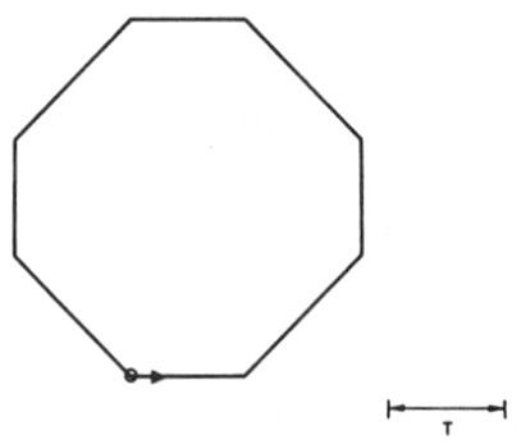

Fig. 10—Minimal circle 01234567.

NAME	WAVEFORM	DIGIT GROUP	GENERATING FUNCTION
SQUARE		0206	$\frac{2x+6x^3}{1-x^4}$
SAWTOOTH		16	$\frac{1+6x}{1-x^2}$
TRIANGULAR		17	$\frac{1+7x}{1-x^2}$
SINE	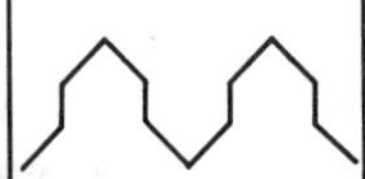	121767	$\frac{1+2x+x^2+7x^3+6x^4+7x^5}{1-x^6}$
RECTANGULAR	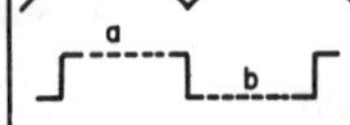	$2\underbrace{0\ldots0}_{a}6\underbrace{0\ldots0}_{b}$	$\frac{2+6x^{a+1}}{1-x^{a+b+2}}$
BACKWARD SAWTOOTH		27	$\frac{2+7x}{1-x^2}$
SPACED SAWTOOTH		160	$\frac{1+6x}{1-x^3}$

Fig. 11. Minimal waveforms.

of the generating functions bear a one-to-one correspondence to the digits of the respective digit groups, *i.e.*, to the digits in one period of the sequences. The denominators are determined by the period lengths of the sequences. Thus, if a sequence has a period of three digits, the denominator of the corresponding generating function becomes $1-x^3$.

ELEMENTARY MANIPULATIONS

One of the objectives of encoding an arbitrary geometric curve is to translate the significant information contained in the curve into a representation which will facilitate manipulation of the curve by a digital computer. The problem may be one of analysis, *i.e.*, to discern certain properties of a curve such as its area or axes of symmetry, or it may be one of synthesis, where a curve is to be constructed having specified properties. The following section will describe some of the simpler techniques for manipulating curves encoded in accordance with the three-bit rectangular array.

Expansion

Consider a curve stored in the memory of a digital computer. If displayed by the computer's graphical output unit (CRT or similar device), the curve will have an over-all size determined jointly by the coded representation of the curve and the scale factor of the display device. If it is desired to expand a portion of a curve relative to the rest of the curve, an operation must be performed on the coded representation of the curve. Changing the scale factor of the display device would have a uniform effect and could not be utilized for selective expansion.

The process of expansion is performed as illustrated by the following example. Given a curve represented by 012075, a curve exactly twice this size, but otherwise indistinguishable, is given by 001122007755. To expand a curve by a ratio n, each of the digits of the curve must be replaced by a set of n identical digits. One notes that n must be an integer.

If a curve is represented by a generating function, expansion can be performed analytically by replacing x by x^n and then multiplying by $(1-x^n)/(1-x)$. Thus the minimal circle 01234567 can be tripled in size by performing the appropriate operation on its generating function:

minimal form:

$$\frac{x}{(1 - x)^2},$$

expansion to triple size:

$$\frac{x^3}{(1 - x^3)^2}\,\frac{1 - x^3}{1 - x} = \frac{x^3}{1 - x - x^3 + x^4}.$$

Contraction

Over-all contraction is readily accomplished by an appropriate change of scale; however, selective contraction requires a special procedure and is almost always accompanied by distortion. This distortion is introduced because contraction implies (except for trivial cases) a reverting to a more primitive representation, and, consequently, a loss of information. Consider the contraction to one-half size of the curve 22101076. If the digits could be grouped into pairs of identical digits, the contracted curve would be given merely by selecting one digit from each pair. The first two digits are identical and meet this requirement. All the remaining digits, however, do not occur in pairs. One notices that the next four digits can be arranged to give two pairs, "11" and "00." The last two digits differ by unity and their mean does not favor any integer. In such a case, the previous digit is also considered and a mean of these three digits is obtained. The mean for 0 (*i.e.*, 8), 7, and 6 is the digit 7. The contraction is thus as follows:

22101076
22110076
22110077
2107.

Rotation

Rotation has already been discussed. Where the rotation is in multiples of 90°, one need merely add the appropriate even digit to each digit of the curve. Rotations of odd multiples of 45° are obtained by adding the appropriate odd digit. Such rotations are almost always accompanied by distortion. Rotations of arbitrary amounts involve digit-group substitutions and are beyond the scope of the present discussion.[2]

Curve Length

The length of a curve is given by the number of even digits plus the square root of two times the number of odd digits.

Inverse Digits

Inspection of the array of Fig. 5(b) shows that digits which are diametrically opposite in the array have a canceling effect on each other. Accordingly, such digits are referred to as inverses of each other. Thus the inverse of "1" is "5," etc. Numerically, the inverse of a digit is obtained simply by adding either plus or minus "4" to the digit—whichever gives a positive sum less than "8" (hence, adding "4" modulo 8).

Path Reversal

A curve is defined by selecting a starting point and giving a sequence of slope segments which trace out the path of the curve from this point. Path reversal is the process of tracing out a curve in reverse direction. It is achieved by replacing each digit of a curve sequence by its inverse and reversing the sequence end for end. As an example, path reversal changes the curve sequence 1232076 to 2346765.

Path Reduction

Path reduction is defined as the process of obtaining the shortest distance between the end points of an open curve and is found by reducing the given curve in the following manner. If a curve contains two digits which are inverses of each other, they may both be deleted without affecting the location of the end points. Further, there are many digit pairs which can be replaced by a single equivalent digit. Thus, the digit pair "42" can be replaced by the single digit "3." The two digits need not be adjacent for either a cancellation or a replacement to be possible. Table III lists all possible sum-replacement and inverse-digit pairs. There are some digit combinations which, although not permitting reduction, can be replaced by a pair of identical digits. Such digit combinations are presented by the double-digit entries in Table III.

TABLE III

	0	1	2	3	4	5	6	7
0			1	2	x	6	7	
1				22	2	x	0	00
2	1				3	4	x	0
3	2	22				44	4	x
4	x	2	3				5	6
5	6	x	4	44				66
6	7	0	x	4	5			
7		00	0	x	6	66		

x = Inverse

As an illustration, let it be desired to find the shortest curve connecting the end points of the curve 7107542. The length of this curve is $3+4\sqrt{2}$. The curve has two cancellation pairs, "15" and "04," which gives a partially reduced curve 772. Sum-replacement reduces this further to the two-digit curve 70. The minimum distance between the curve's end points is thus equal to $1+\sqrt{2}$.

Closure

To determine whether a given curve is open or closed, one performs a path reduction. If the curve vanishes, it is closed. If it does not vanish, it is open, and the residue (referred to as the closure) is the minimum distance between the end points.

As a corollary to the above, one notes that the digits of a given curve may be arbitrarily rearranged without changing the length of the curve or the minimum distance between its end points.

[2] For example, one notes from Table I that rotating the straight line which makes an angle of 26.57° with the horizontal so that its angle will be 30.09°, requires replacement of the digit group 01 by the digit group 10101.

Multiple-Loop Curves

To determine whether a given curve consists of multiple loops, one must examine the curve for subsets of adjacent digits for which the closure vanishes. Thus the curve 7012345667012345 has a total of seven subsets (including one for the curve itself), for which the closure vanishes, indicating the existence of seven possible ways of tracing out a closed path within the curve; however, only three of these loops are independent.

If a curve has more than one subset for which the closure vanishes, and if the subsets are nonoverlapping and separated by an ordered set of cancellation pairs, the curve is multiply connected except that the customary connecting branch has been provided (represented by the ordered set of cancellation pairs). The curve 011233455677232107656 is an example of a multiply connected curve. Both this and the previous curve are illustrated in Fig. 12.

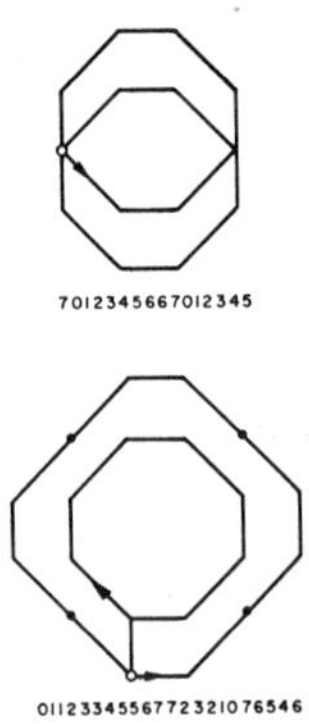

Fig. 12—Multiple-loop curves.

Intersection

To find the point of intersection of two curves, one must first specify a connecting branch which will place the two starting points into relative position. The connecting branch may be any simple curve tracing out a path from the starting point of one curve to the starting point of the other. It may not, however, touch the two curves except at their starting points. Tracing out first one curve and then the other (from the starting point of the connecting branch), a progressive path reduction is performed by considering one additional digit at a time. Each reduced curve is rearranged to place its digits in numerically ascending order. The corresponding lengths of the reduced curves are also computed. The two sets of path reductions are then compared. For each path-reduction sequence which is *common to both curves*, a nodal intersection exists (*i.e.*, the curves intersect at a grid point). If a comparison of the path-reduction sequences of the two curves shows them to differ by only one digit (either by a difference in value of one or the absence of one digit) twice in succession, and if simultaneously the corresponding lengths of the sequences interchange their roles as to which is the longer, a non-nodal intersection exists (*i.e.*, an intersection between grid points).

As an example, consider the two curves $A = 01066112$ and $B = 020117$, connected by a branch 66 from A to B. The successive path-reduction sequences and lengths are as follows:

		Sequence	*Length*
For A:	1)	0	1
	2)	01	$1+\sqrt{2}$
	3)	010=001	$2+\sqrt{2}$
	4)	0016=000	3
	5)	0006=007	$2+\sqrt{2}$
	6)	0071=0000	4
	7)	00001	$4+\sqrt{2}$
	8)	000012=00011	$3+2\sqrt{2}$
For B:	1)	6	1
	2)	66	2
	3)	660=67	$1+\sqrt{2}$
	4)	672=7	$\sqrt{2}$
	5)	70=07	$1+\sqrt{2}$
	6)	071=000	3
	7)	0001	$3+\sqrt{2}$
	8)	00017=00000	5

Inspection of the two sets of path reductions shows that $A(4)$ matches $B(6)$, indicating the presence there of a nodal intersection. Further, $A(6)$ and $B(7)$ differ by only one digit and the same is true for $A(7)$ and $B(8)$. Simultaneously, the length of A increases from 4 to 5.414 and that of B increases from 4.414 to 5, making A now longer than B. Hence the curves intersect non-nodally within the grid square defined by the end points of the four sequences $A(6)$, $A(7)$, $B(7)$, and $B(8)$.

Determination of Area

A fairly simple procedure exists for determining the area enclosed by an arbitrary curve. Consider the curve 10127 shown in Fig. 13(a). The area under this curve, *i.e.*, its integral with respect to x, can be obtained by adding the area covered by four vertical columns of unit width along the x axis:

$$2\tfrac{1}{2} + 3 + 3\tfrac{1}{2} + 4\tfrac{1}{2} = 13\tfrac{1}{2} \text{ square units.}$$

This suggests the following procedure for determining area. Define a modifier B to represent the instantaneous distance of the curve from the x axis. Thus in Fig. 13(a), the area A is initially zero and the modifier B is equal to "2." The initial slope of "1" (*i.e.*, 45°) causes the area to increase by the amount of the first vertical column to $2\frac{1}{2}$ and the modifier to change to 3. The next slope, "0," causes no change in the modifier since the curve is running parallel to the x axis, but the area is increased by the amount of the modifier, *i.e.*, to $5\frac{1}{2}$. It should be noted that the modifier gives the value of the leading edge of the particular vertical column. In the next step, the slope "1" increases the area by the modifier plus $\frac{1}{2}$ (to 9), and then increases the modifier to 4.

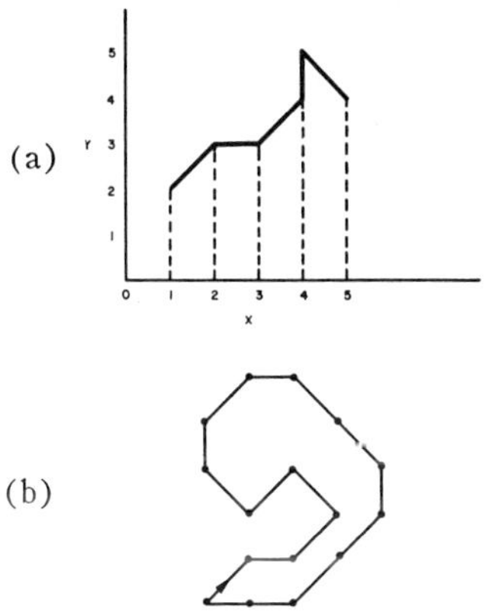

Fig. 13—Arbitrary curves to illustrate integration.

The slope "2" gives no change in area but increases the modifier to 5. The next slope, "7," increases the area by the modifier minus $\frac{1}{2}$ (to $13\frac{1}{2}$) and decreases the modifier to 4.

The foregoing procedure can be generalized to develop a set of rules for all eight slopes, as shown in Table IV. The table gives the slope (N), the change in area resulting from this slope (*i.e.*, the area in the vertical column), and the change to be made in the modifier. If the starting point coincides with the origin, the modifier is initially zero, if it does not, one must use the appropriate initial-condition value.

TABLE IV

Slope	Change in Area	Change in Modifier
0	$+B$	+0
1	$+B+\frac{1}{2}$	+1
2	+0	+1
3	$-B-\frac{1}{2}$	+1
4	$-B$	+0
5	$-B+\frac{1}{2}$	−1
6	+0	−1
7	$+B-\frac{1}{2}$	−1

As an example consider the curve 1013532107765544, shown in Fig. 13(b). This curve is traversed in a positive (clockwise) sense and has an area of 11 square units. The initial value of B is taken as zero.

N	A	B
1	$\frac{1}{2}$	1
0	$1\frac{1}{2}$	1
1	3	2
3	$\frac{1}{2}$	3
5	−2	2
3	$-4\frac{1}{2}$	3
2	$-4\frac{1}{2}$	4
1	0	5
0	5	5
7	$9\frac{1}{2}$	4
7	13	3
6	13	2
5	$11\frac{1}{2}$	1
5	11	0
4	11	0
4	11	0

Table IV can be expressed in the form of an algorithm and programmed for a digital computer. It is then a simple matter to determine the area enclosed by an arbitrary curve encoded according to the three-bit rectangular array. For open curves, single-valued in x, one can readily obtain their integral with respect to x. Integration with respect to y can be accomplished by first rotating the curve 90°.

Conclusion

A procedure has been developed for the description and encoding of arbitrary geometric curves which will facilitate the computer manipulation of such curves. Various approaches to the twin problems of describing and encoding have been discussed. Particular attention has been given to the rectangular-array type of representation which is one of the simplest and most natural, and one which lends itself easily to manipulation. Elementary manipulative techniques for the rectangular array have been described. Although consideration has been limited to plane (two-dimensional) curves, extension to three or more dimensions is readily visualized.

Fourier Preprocessing for Hand Print Character Recognition

G. H. GRANLUND, MEMBER, IEEE

Abstract—A pattern-recognition method, making use of Fourier transformations to extract features which are significant for a pattern, is described. The ordinary Fourier coefficients are difficult to use as input to categorizers because they contain factors dependent upon size and rotation as well as an arbitrary phase angle. From these Fourier coefficients, however, other more useful features can easily be derived.

By using these derived property constants, a distinction can be made between genuine shape constants and constants representing size, location, and orientation.

The usefulness of the method has been tested with a computer program that was used to classify 175 samples of handprinted letters, e.g., 7 sets of the 25 letters *A* to *Z*. In this test, 98 percent were correctly recognized when a simple nonoptimized decision method was used.

The last section contains some considerations of the technical realizability of a fast preprocessing system for reading printed text.

Index Terms—Character, contour, feature, Fourier, invariant, pattern, recognition, transform.

I. Introduction

Pattern recognition is a very broad field that includes various subjects such as character reading, medical diagnosis, weather prediction, and sound recognition. A general pattern-recognition system which can be taught to perform any arbitrary function is, so far, only a theoretical possibility. Present-day recognition systems, such as those for character recognition, can only deal with limited and well-defined classes of patterns.

A pattern-recognition system is usually considered to have two main parts. First, the pattern is treated by a feature extraction device which has to make measurements of parameters giving maximal information about the pattern. Secondly, this information is fed into a categorizer which makes a classification of the unknown samples. This latter problem is the better understood of the two and has been investigated in some detail. For example, considerable work has been carried out on the optimum decision problem [1], [2]. Feature extraction, however, has proved too difficult to allow a general approach. Nevertheless, the extraction of really significant features is considered more important for the final result than the optimality of the secondary decision process [3].

An important problem in pattern recognition is to find features that describe the processed patterns in an optimal way. Specifically when it concerns character recognition devices, it is important to find features that are invariant for different types of style, size, width of shape, slant of character, etc. These features are still expected to give enough information for an unambiguous classification.

Manuscript received March 3, 1970; revised August 20, 1971. This work was supported in part by the National Institutes of Health under Grant 3 P01 GM15006-03S2. This note describes the results of a work carried out in the Department of Network Theory, Chalmers University of Technology, Gothenburg, Sweden.

The author is with the Research Laboratory of Electronics, Massachusetts Institute of Technology, Cambridge, Mass. 02139.

Reprinted from *IEEE Trans. Comput.*, vol. C-21, pp. 195–201, Feb. 1972.

A great deal of work has been spent to find features which are invariant in different respects [4]–[6]. Different types of Fourier descriptors have been used because this is a representation that admits a separation of factors dependent upon size, angular orientation, and position [7], [8].

Some difficulty is involved if the Fourier coefficients as such are to be processed in a classifier. The Fourier coefficients contain information about size and orientation, and a phase factor dependent upon starting conditions. The phase factor is generally of no interest while information about size and orientation sometimes is useful.

In this note, features are described that are essentially shape parameters independent of size, angular orientation, position, and starting conditions. These features can be derived from the ordinary Fourier coefficients.

II. Mathematical Considerations

The pattern is, in this part, assumed to be a contour. We assume that the contour function, a closed-curve C, is included in a complex space according to Fig. 1. A point moving around the contour generates a complex-valued function u. Assume that the point is moving at a constant speed v along C. At every time t, there is a complex number u defined which is given by

$$u = u(t).$$

Now the parameter t does not necessarily mean time, but rather a parameter of length along the contour. The function u is periodic, which implies that there exists a value T such that

$$u(t + nT) = u(t).$$

It is now possible to express u as a complex Fourier series. The Fourier coefficients become

$$a_n = \frac{1}{T}\int_0^T u(t) \exp\,[-jn2\pi t/T]dt.$$

It can be shown that the Fourier series is convergent. This gives

$$u(t) = \sum_{-\infty}^{\infty} a_n \exp\,[jn2\pi t/T].$$

For simplicity, we now assume that the velocity v is chosen so that $T=2\pi$. The formulas are then simplified to

$$a_n = \frac{1}{2\pi}\int_0^{2\pi} u(t) \exp\,[-jnt]dt$$

and

$$u(t) = \sum_{-\infty}^{\infty} a_n \exp\,[jnt].$$

The Fourier coefficients of a specific contour are not unique. As one can easily realize, they are dependent upon the starting point of the contour tracking. The Fourier coefficients then differ from one another with respect to a parameter τ. This means that there is a set of Fourier coefficients for each τ of the function

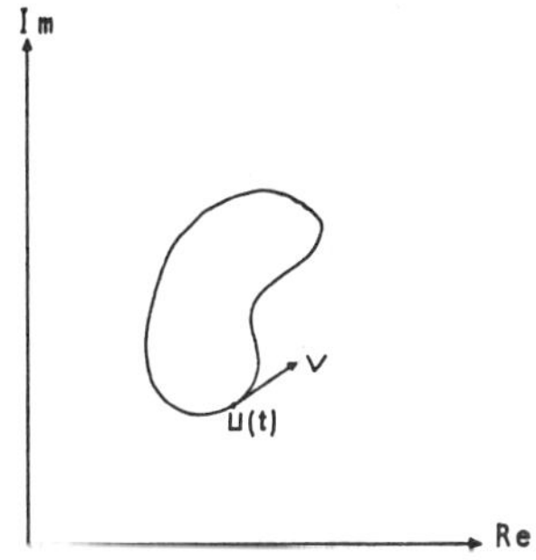

Fig. 1. Example of a contour function.

$$u = u(t + \tau).$$

Now we assume that there exists a certain function

$$u(t) = u^{(o)}(t)$$

and, subsequently, the other functions are given by

$$u(t) = u^{(o)}(t + \tau).$$

The index (o), both now and later, refers to a specific contour function. The resulting Fourier coefficients become

$$\begin{aligned} a_n &= \frac{1}{2\pi}\int_0^{2\pi} u^{(o)}(t+\tau) \exp\,[-jnt]dt \\ &= \exp\,[jn\tau]\cdot\frac{1}{2\pi}\int_0^{2\pi} u^{(o)}(t) \exp\,[-jnt]dt \\ &= \exp\,[jn\tau]a_n^{(o)}. \end{aligned}$$

It is now of great interest to consider the results of certain operations upon the contour such as translation, rotation, and dilation.

A. Translation

We assume $a_n^{(o)}$ to be a specific set of Fourier coefficients from a certain contour. Let us consider a translation of the contour with the complex vector Z. The resulting contour function, expressed in an inverse Fourier series, becomes

$$u(t) = u^{(o)}(t) + Z = \sum_{-\infty}^{\infty} a_n^{(o)} \exp\,[jnt] + Z.$$

We can now identify the Fourier coefficients of the translated contour as

$$a_n = \begin{cases} a_n^{(o)}, & \text{for } n \neq 0 \\ a_o^{(o)} + Z, & \text{for } n = 0. \end{cases}$$

All coefficients except a_o are invariant of translation. a_o is simply the complex vector indicating the position of the center of gravity of the contour.

B. Rotation

We assume here that the center of gravity is positioned at the origin. Rotation of the contour function $u^{(o)}(t)$ around the origin, with an angle ϕ in positive direction, gives another function $u(t)$ where

$$u(t) = \exp[j\phi]u^{(o)}(t).$$

The influence upon the Fourier coefficients is apparent if the function $u(t)$ is again expressed in an inverse Fourier series. The rotated contour then has the Fourier coefficients

$$a_n = \exp[j\phi]a_n^{(o)}.$$

C. Dilation

In a similar way, it can be shown that dilation of the contour with a factor R implies that the Fourier coefficients are multiplied with R.

$$a_n = Ra_n^{(o)}.$$

The general form for the Fourier coefficients of a contour generated by translation, rotation, and dilation of a specific contour is given by

$$a_n = \exp[jn\tau]R\exp[j\phi]a_n^{(o)}$$

where $a_n^{(o)}$ are the Fourier coefficients of the original contour.

These Fourier coefficients in their common form are not particularly useful because they contain information that, in some cases, may be insignificant. What is desired here is some operation that makes a distinction between shape properties and properties dependent upon translation, rotation, and dilation. However, other types of coefficients can be derived which include features more suitable for our purpose.

Let us consider the expression

$$\begin{aligned} b_n &= \frac{a_{1+n}a_{1-n}}{a_1^2} \\ &= \frac{a_{1+n}^{(o)}\exp[j(1+n)\tau]R\exp[j\phi]a_{1-n}^{(o)}\exp[j(1-n)\tau]R\exp[j\phi]}{[a_1^{(o)}\exp(j\tau)R\exp(j\phi)]^2} \\ &= \frac{a_{1+n}^{(o)}a_{1-n}^{(o)}}{[a_1^{(o)}]^2} = b_n^{(o)}. \end{aligned}$$

The last part of the expression does not contain either τ, R, or ϕ. As these coefficients do not contain a_o, they are also independent of translation. These b_n coefficients might thus be representative for the form of the contour.

In some cases, information of the orientation is necessary and the series of b_n coefficients can be completed by defining a coefficient

$$b_1 = \frac{a_2|a_1|}{a_1^2}.$$

The development of this expression becomes

$$\begin{aligned} b_1 &= \frac{a_2^{(o)}\exp[j2\tau]R\exp[j\phi]R|a_1^{(o)}|}{[a_1^{(o)}\exp[j\tau]R\exp[j\phi]]^2} \\ b_1 &= \frac{a_2^{(o)}|a_1^{(o)}|}{[a_1^{(o)}]^2}\exp[-j\phi] \\ &= b_1^{(o)}\exp[-j\phi] \end{aligned}$$

where b_1 is a parameter depedent upon rotation, but not on translation and dilation.

Another useful set of constants can also be defined. Consider the expression

$$d_{mn}' = \frac{a_{1+m}^{n}a_{1-n}^{m}}{a_1^{m+n}}, \qquad \begin{aligned} m &= 1, 2, 3, \cdots \\ n &= 2, 3, 4, \cdots. \end{aligned}$$

As in the case of the b_n coefficients, we assume that the pattern considered is derived by translating, rotating, and dilating a certain pattern in a normalized form. Then the coefficients can be written as

$$\begin{aligned} d_{mn}' &= \frac{[a_{1+m}^{(o)}\exp[j\tau(1+m)]R\exp[j\phi]]^n}{[a_1^{(o)}\exp[j\tau]R\exp[j\phi]]^n} \cdot \frac{[a_{1-n}^{(o)}\exp[j\tau(1-n)]R\exp[j\phi]]^m}{[a_1^{(o)}\exp[j\tau]R\exp[j\phi]]^m} \\ &= \frac{[a_{1+m}^{(o)}]^n[a_{1-n}^{(o)}]^m}{[a_1^{(o)}]^{m+n}} = d_{mn}'^{(o)}. \end{aligned}$$

We see that these d_{mn} coefficients are invariant in the same way as the earlier described b_n coefficients. Here, too, a coefficient containing angular information can be defined.

$$d_{m1}' = \frac{a_{1+m}|a_1|^m}{a_1^{m+1}}.$$

A development of the expression according to former examples gives

$$\begin{aligned} d_{m1}' &= \frac{a_{1+m}^{(o)}|a_1^{(o)}|^m}{[a_1^{(o)}]^{m+1}}\exp[-jm\phi] \\ &= d_{m1}'^{(o)}\exp[-jm\phi]. \end{aligned}$$

The coefficients d_{m1}' are then invariant of translation and dilation, but rotation with an angle ϕ causes the normalized term $d_{m1}^{(o)}$ to be multiplied with a factor $\exp[-jm\phi]$.

If d_{mn}' applies to numbers m and n that have a factor k in common, the expression d_{mn}' is a power with exponent k. We can thus define

$$d_{mn} = \frac{a_{1+m}^{n/k} a_{1-n}^{m/k}}{a_1^{(m+n)/k}}$$

which gives the relation

$$d_{mn}' = d_{mn}^k$$

to the earlier defined d_{mn}'. If m and n have no factor in common, ($k=1$), then $d_{mn}=d_{mn}'$, but in other cases, raising to a power causes an unnecessary loss of information which can be avoided by using d_{mn}. The properties of transformation of the d_{mn} constants are similar to those of the d_{mn}' constants.

An important notion is that $d_{nn}=b_n$ which means that the b_n coefficients are special cases of the d_{mn} coefficients. Later on, when b_n and d_{mn} coefficients are concerned, they are generally referred to as d_{mn} coefficients.

III. Qualitative Interpretation of Feature Constants

Let us make a qualitative interpretation of the various constants previously described and evaluate their significance for describing shape features.

The Fourier constants, and other expressions derived from them, indicate the degree of symmetry of the contour. A rotational symmetry of degree m means that a rotation with an angle $2\pi/m$ brings back the same pattern. For such patterns described by $u(t)$, the following expressions are valid.

$$\exp [j2\pi/m]u(t) = u(t + 2\pi/m)$$

$$\exp [j2\pi/m]a_n = \exp [j2\pi n/m]a_n$$

$$a_n[1 - \exp [j2\pi(n - 1)/m]] = 0.$$

If $a_n \neq 0$, then

$$1 - \exp [j2\pi(n - 1)/m] = 0$$

which means that $n-1$ has to be a multiple of m. For patterns with a rotational symmetry of degree m, significant values of the Fourier coefficients can be expected for indices of type $1 \pm pm$ where p is a positive integer.

If we now consider the previous expression for b_n,

$$b_n = \frac{a_{1+n}a_{1-n}}{a_1^2} \cdot$$

It is apparent that the expression contains Fourier coefficients that should be significant in case of rotational symmetry of degree n. The coefficient for normalization in the denominator is chosen to be a_1 because it is normally the biggest and works well in the invariance transformation.

An arbitrary contour can, to a certain extent, be divided into rotational symmetries of different degrees. However, some different contours can have the same set of coefficients b_n. Let us consider Fig. 2. It shows some coefficients of two triangular curves. They have rotational symmetries of degree two, as well as of degree three, and the computed values of b_2, respectively, b_3 (b_n with $n=2$ and 3) show very little difference, as can be expected. It is difficult to make a distinction between these two shapes based upon the constants b_2 and b_3 alone.

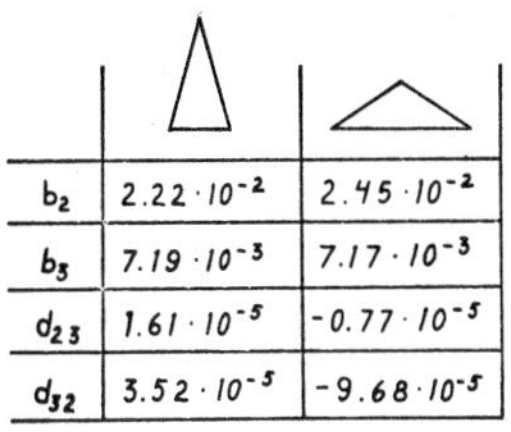

	(triangle 1)	(triangle 2)
b_2	$2.22 \cdot 10^{-2}$	$2.45 \cdot 10^{-2}$
b_3	$7.19 \cdot 10^{-3}$	$7.17 \cdot 10^{-3}$
d_{23}	$1.61 \cdot 10^{-5}$	$-0.77 \cdot 10^{-5}$
d_{32}	$3.52 \cdot 10^{-5}$	$-9.68 \cdot 10^{-5}$

Fig. 2. Triangles with rotational symmetries of different orientation

However, the d_{mn} coefficients can be used to give some helpful information. The d_{mn} coefficients indicated the relations between rotational symmetries of different degrees. This relation is expressed as a magnitude as well as an angle; the latter indicates primarily the angle between axes of different rotational symmetries. In Fig. 2 the coefficients d_{23} and d_{32} are given and a distinction based upon these coefficients is quite possible.

It should be noted that the d_{mn} and b_n coefficients are not complete descriptors of the shape and, therefore, the contour can not be reconstructed from the coefficients.

IV. Character-Recognition Experiments

In order to get an idea of the significance of the features previously described, an experimental test on handwritten letters was carried out.

The function of a character-recognition system using Fourier methods has been simulated with an IBM 360/65 computer. The test was performed upon a set of 175 letters. As the letters are fed to the computer via punched cards describing the pattern, it was rather time-consuming to produce an appropriate set. The same set of letters was therefore used as a learning set as well as a test set.

This fact certainly has an influence upon the final result and brings forth too optimistic results. It is difficult to estimate the discrepancy, and this performed test should merely be taken as a qualitative indication of the usefulness of the features. However, the aim has been to make a learning test set that is representative of what could be "normal" variations of carefully handprinted letters. No changes were made in the letter set after the start of the experiment. The different shapes appear in Fig. 3.

The processing starts with a character-acquisition phase, where each character, expressed as a 35×35 binary matrix, is fed into the computer. Each row of the matrix is represented by one card giving numbers of sequentially white and black elements. The computer then scans the matrix until it reaches the first black element. After that it tracks the contour, and, during the tracking, it evaluates the Fourier coefficients as defined previously. The d_{mn} coefficients are then computed.

In this test, the coefficients used were d_{11}, d_{22}, d_{21}, d_{12}, d_{31}, d_{13}, and d_{44}. It is not certain that these coefficients were the best choice. They were chosen after merely qualitative considerations. Five of the coefficients used contain information of the angular position of the letter, which may give too much dependence upon this orientational parameter. The character-classification scheme used is based on simple probability and distance measurements.

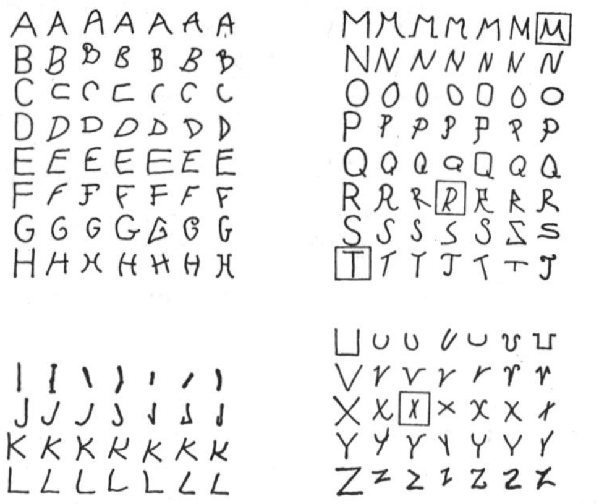

Fig. 3. Shapes of the processed letters.

First a training set was run. This set consisted of 7 handprinted samples of the 25 letters A to Z. W was unfortunately omitted because it is not included in the Swedish alphabet. In order to simplify the probability-function computations, it is assumed that the distribution of a certain d_{mn} constant of a letter class constitutes a two-dimensional Gaussian distribution. As the number of points defining it is as low as seven, the mathematical requirements are obviously not met.

The two-dimensional Gaussian distribution is defined by

$$f_{d_{mn}}(x, y) = \frac{1}{2\pi\sigma_1\sigma_2\sqrt{1-\rho^2}} \exp\left[-\frac{1}{2}Q(x, y)\right]$$

where

$$Q(x, y) = \frac{1}{1-\rho^2}\left[\frac{(x-x_o)^2}{\sigma_1{}^2} - \frac{2\rho(x-x_o)(y-y_o)}{\sigma_1\sigma_2} + \frac{(y-y_o)^2}{\sigma_2{}^2}\right].$$

In that expression, x and y mean real and imaginary components of a constant d_{mn}; x_o and y_o are means derived from the coefficients of the learning set; σ_1 and σ_2 are the variances of the variables in the learning set, and ρ is the correlation coefficient.

In the computer program, the correlation ellipse was computed. In Fig. 4, mean values and variance ellipses of coefficient d_{22} are plotted for the first few letters of the alphabet.

When the first set was analyzed, the d_{mn} coefficients of the unknown letters were computed. Any d_{mn} coefficient, together with the related variance ellipse, defines a one-dimensional distribution along the line joining the actual d_{mn} point and the mean. The new distribution function is of the form

$$f_{d_{mn}}(z) = \frac{1}{\sigma_z\sqrt{2\pi}} \exp\left[-\frac{1}{2\sigma_z{}^2}z^2\right]$$

where σ_z is the resulting variance derived from the operation earlier described and z is the distance between actual d_{mn} and the mean of d_{mn}. Assuming—with more or less validity—that the various d_{mn} coefficients are independent makes it possible to define a logarithmic relative probability of a letter belonging to a certain class c of letters.

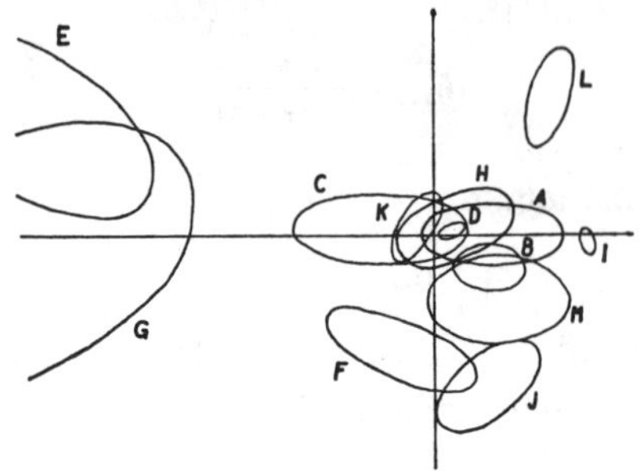

Fig. 4. Example of the distribution of coefficients d_{22} for the first 13 letters of the alphabet.

```
PROBABILITY
  OF
  A          -84865
  B          -172974
  C          1000
  D          -•2010347'+08
  E          -1501
  F          -3035
  G          991
  H          -944495
  I          -•1202279'+10
  J          -811
  K          -3977302
  L          -654
  M          930
  N          -552583
  O          -•1094149'+12
  P          -2069462
  Q          -•6672197'+08
  R          -4635
  S          -1245
  T          -271396
  U          901
  V          919
  X          -161182
  Y          -1346
  Z          -55214
CONCLUSION:
THE PATTERN REPRESENTS C
```

Fig. 5. Example of a printout from a recognition test.

$$P(c) = \sum_{m,n}\left[\left[-\frac{1}{2\sigma_z{}^2}z^2\right] - \ln \sigma_{z_{mn}}\right].$$

The index mn refers to a certain d_{mn} coefficient. The factor $\sqrt{\pi 2}$, common to all terms, is omitted in the computations. The decision is then based upon a comparison of $P(c)$ for different classes, and the class category c, assigned to the highest value of $P(c)$, is printed out. A typical printout is shown in Fig. 5. The numbers are normalized such that the highest value of $P(c)$ is set at 1000. The actual letter is C in series no. 2. It is apparent that we have a number of confusion pairs, such as C-G, C-M, C-U, and C-V, which have nearly the same distance measure. That seems reasonable because these letters have essentially a similar shape.

As is indicated in the printout, there is no "reject class" in the classification because an estimation of such a class would require a more careful analysis of the relationships and absolute values of the probabilities.

Of 175 test samples, all except four were recognized correctly. The incorrectly recognized characters were as follows. *M* in column seven was interpreted as an *R*; *R* in column four was interpreted as a *P*; *T* in column one was interpreted as an *I*; and *X* in column three was interpreted as an *R*. These letters are indicated with boxes in Fig, 3. The differences between the probabilities of the confusion pairs are, in most cases, small and could be used as rejection criteria.

The probability figures for some of the confusion pairs are given below.

*M*7	*R*4	*T*1	*X*3
H 998	*M* 989	*B* 987	*C* 986
M 999	*P* 1000	*I* 1000	*M* 983
R 1000	*R* 998	*P* 982	*R* 1000
X 1000	*T* 988	*T* 993	*X* 998

The reason for misrecognition has been that these letters have generated some d_{mn} coefficients that are far from the mean for that letter set. It is apparent from Fig. 3 that this does not necessarily mean that it is difficult to recognize the letter visually. It seems reasonable, however, that a more elaborate decision procedure using a weighting of the coefficients would give a better result.

V. Implementation Considerations

The classification experiments were carried out on a digital computer. However, the computation of Fourier coefficients on a general-purpose digital computer takes a relatively long time due to the great number of operations to be performed. There are also special-purpose computers available that make fast Fourier coefficient computations. They are, however, rather expensive.

In this section an analog system for the preprocessing is suggested that supposedly would be faster and less expensive than a digital system. A complete text-reading system consists of several parts. First, there has to be a scanning system and some system for administration of the material. These systems will not be considered here. It is assumed that the scanner can shift from scanning of the frame to following of a contour when the tracking point reaches the contour. Contour-tracking systems have been developed, for instance by IBM in their 1288 page reader [9]. We assume that the output from the scanner consists of two voltages x and y indicating the position of the tracking point. Referring again to Fig. 1, we have

$$\begin{cases} x = \operatorname{Re} u(t) \\ y = \operatorname{Im} u(t). \end{cases}$$

What is wanted is a system that makes a computation of the coefficients a_n where

$$\begin{aligned} a_n &= \frac{1}{T}\int_0^T u(t) \exp\left[-jn2\pi t/T\right]dt \\ &= \operatorname{Re} a_n + j\cdot\operatorname{Im} a_n \end{aligned}$$

and

$$\begin{aligned} \operatorname{Re} a_n &= \frac{1}{T}\int_0^T [\operatorname{Re} u(t) \cos n2\pi t/T \\ &\qquad + \operatorname{Im} u(t) \sin n2\pi t/T]\, dt \\ \operatorname{Im} a_n &= \frac{1}{T}\int_0^T [\operatorname{Im} u(t) \cos n2\pi t/T \\ &\qquad - \operatorname{Re} u(t) \sin n2\pi t/T]\, dt. \end{aligned}$$

The first values to be determined are T, cos nt, and sin nt. This can be done by using a system according to Fig. 6. The control logic causes the spot of the scanner to travel twice around the contour of the character. The first trip is to scale the system with respect to the processed letter. When the spot starts its first trip around the contour, the switch $s1$ is closed, which makes the following integrator start its action. At the time when the spot is back at the starting position, the switch $s1$ is opened and the integration stops. The voltage of the integrator is proportional to the time T to go around the contour. An analog-division unit D transfers the received value to $1/T$. This value can now be fed to the multipliers M controlling the feedback of an oscillator built around two integrators. The period time is given by

$$e\left(s^2 + \frac{n^2}{T^2}\right) = 0$$

which gives a period time T/n for the oscillator.

The initial value of the oscillators is set before the spot starts its second trip around the character. By closing the switches $s2$ and $s3$ the initial values are set to $1/T$ for the cosine integrator and zero for the sine integrator. During the time when the spot is making its second trip around the character, the oscillator is producing the signals $1/T$ cos nt and $1/T$ sin nt.

The remaining part of the integration can be performed by a system according to Fig. 7. As the system consists only of multipliers, adders, and integrators, the function should need no further explanation. The possibility of getting the elements of a_n, as well as of a_{-n}, is indicated.

One oscillator and one integrating system for every desired value n of a_n has to be included. The d_{mn} coefficients to be chosen might be the ones earlier treated, namely d_{11}, d_{22}, d_{21}, d_{12}, d_{31}, d_{13}, and d_{44}.

In order to make a computation of these d_{mn} coefficients, the following a_n coefficients are used: a_{-3}, a_{-2}, a_{-1}, a_1, a_2, a_3, a_4, and a_5. This means 5 oscillators; 2 "double" integrating systems (giving a_{-n} and a_n) and 3 single ones.

The computation of the d_{mn} coefficients may be best done by a digital computer connected on-line to the analog computer previously described. The digital computer might be

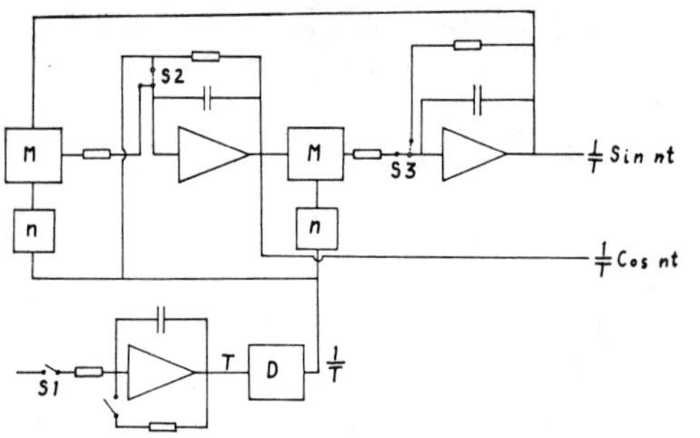

Fig. 6. System for generation of functions of weight.

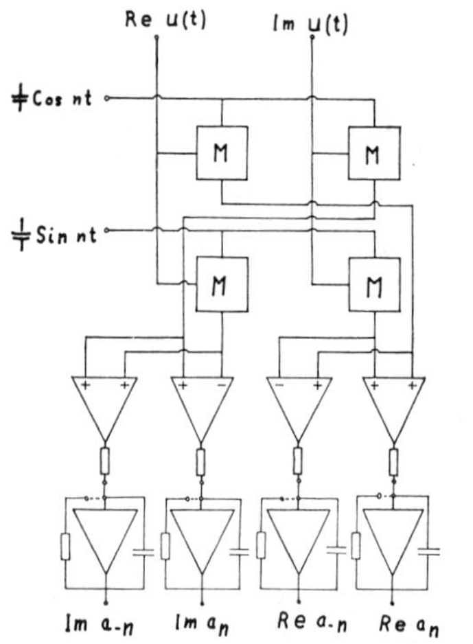

Fig. 7. System for computation of Fourier coefficients.

of a general-purpose type although a special-purpose computer of the same size would do faster work. As the tasks to be performed are quite simple and the amount of data per sample character is rather small, a computer, organized with parallel arithmetic units for each processing step and a small buffer memory between each unit, would be suitable.

The organization, in detail, is dependent upon the decision method used, but this part of the problem will be omitted here.

VI. Conclusions and Extensions

The outlines for a preprocessing system for character recognition has been described. A computer simulation of a recognition system working upon a specified set of unknown letters has given a recognition rate of 98 percent. The result has shown that a letter-recognition system based upon modified Fourier transform would be quite feasible. The system is inherently making a separation of shape properties and properties that in some cases may be insignificant, such as orientation and size. As is indicated in the test results, the features derived seem to be significant for shape properties. A better classification method would probably give better results. In the last part, the realization of a system for fast preprocessing of letters is described. Use of modern electronic components should allow a processing speed in the order of 5000 letters/s.

Acknowledgment

The author would like to thank Prof. E. F. Bolinder and Prof. B. Wedelin for their valuable suggestions and interest shown during the work, and also H. Wigstrom for his help with the computer programming and discussions concerning the interpretation of the various feature coefficients.

References

[1] W. H. Highleyman, "Linear decision functions, with application to pattern recognition," *Proc. IRE*, vol. 50, June 1962, pp. 1501–1514.
[2] F. W. Smith, "Pattern classifier design by linear programming," *IEEE Trans. Comput.*, vol. C-17, Apr. 1968, pp. 367–372.
[3] J. H. Munson, "Summary from the international conference on methodologies of pattern recognition," *IEEE Comput. Group News*, Mar. 1968, pp. 20–21.
[4] W. Doyle, "Operations useful for similarity-invariant pattern recognition," *J. Ass. Comput. Mach.*, vol. 9, Apr. 1962, pp. 259–267.
[5] J. R. Singer, "A self organizing recognition system," in *Proc. Western Joint Comput. Conf.* Washington, D. C.: Spartan, May 1961, pp. 545–554.
[6] R. F. Meyer *et al.*, "Analytic approximation and translational invariance in character recognition," *Optical Character Recognition*, pp. 181–195.
[7] J. K. Brousil and D. R. Smith, "A threshold logic network for shape invariance," *IEEE Trans. Electron. Comput.*, vol. EC-16, Dec. 1967, pp. 818–828.
[8] E. L. Brill, "Character recognition via Fourier descriptors," in *WESCON Tech. Papers*, Session 25, 1968.
[9] *IBM System 360 Component Description, IBM 1288 Optical Page Reader Model I*, Manual A21-9081.

A FORMAL DESCRIPTION FOR TWO-DIMENSIONAL PATTERNS*

C. T. Zahn

Stanford Linear Accelerator Center
Stanford University, Stanford, California

Summary

A "structural description" for two-dimensional black and white patterns is defined as the set of contour lines for an appropriate function which fits the binary pattern. These contour lines are mutually non-intersecting closed polygonal curves with edges in only eight different directions and they represent the boundaries between connected black and white areas of the pattern. Rigorous procedures are described to transform a "matrix pattern" into a "structural description" and vice versa. Advantages of this method for describing patterns previous to pattern recognition are discussed at some length.

Introduction

This paper presents a method for obtaining a "structural description" of any two-dimensional picture whose elements are a finite set of points in the plane each having value zero (white) or one (black) and such that the points are arranged in the form of a square lattice. Figure 1 shows one such picture. The "structural description" we employ consists of a set of simple polygonal closed curves whose vertices are ordered in either clockwise or counter-clockwise sense. These curves represent the continuous boundaries between connected sets of black points which we call "objects" and connected sets of white points which we call "holes." Fig. 2 depicts the structural description of the pattern in

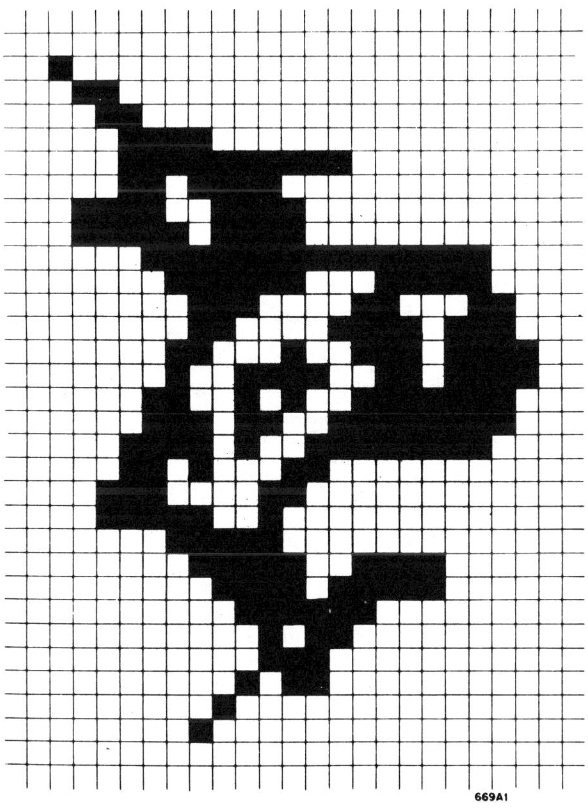

FIG. 1

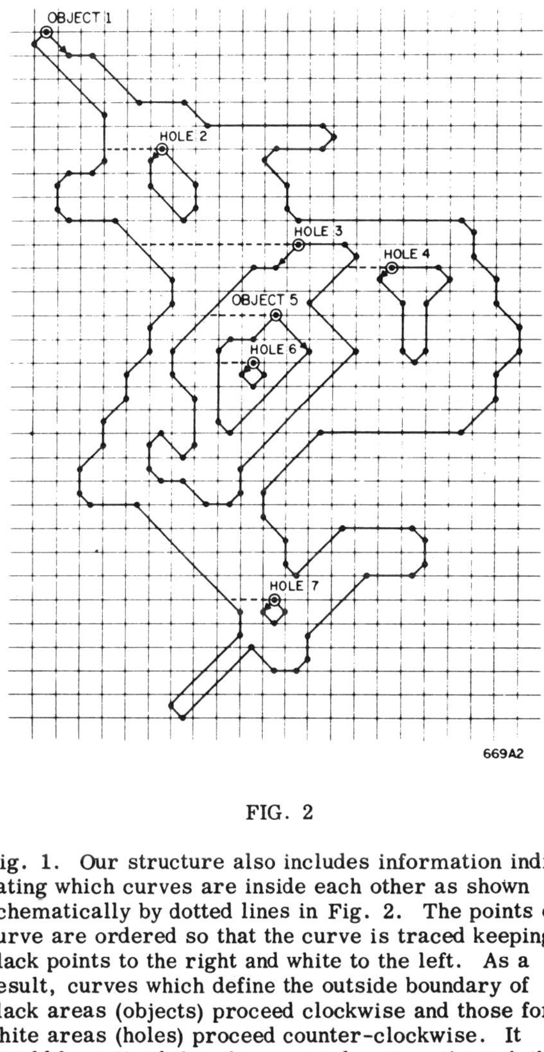

FIG. 2

Fig. 1. Our structure also includes information indicating which curves are inside each other as shown schematically by dotted lines in Fig. 2. The points of a curve are ordered so that the curve is traced keeping black points to the right and white to the left. As a result, curves which define the outside boundary of black areas (objects) proceed clockwise and those for white areas (holes) proceed counter-clockwise. It should be noticed that the curves do not go through the outermost picture points of an area but rather go between these outermost points and the adjacent points of opposite color.

Our selection of this particular format to describe the information content of a black/white pattern was motivated by a belief that the sequential trace of the boundary of an object contains the most useful data for

*Work supported by U.S. Atomic Energy Commission.

Reprinted with permission from *Proc. Int. Joint Conf. on Artificial Intelligence*, May 1969, pp. 621–628.

recognizing objects one from another in a great many applications. Experimental evidence from the psychology of human visual perception and the neurophysiology of animal visual perception support this belief to an amazing extent.

For a more detailed description of this "structural description" and an extended discussion of its uses, the reader is referred to Zahn.[1]

Structural Description as Contour

The closed curves of the "structural description" are, in fact, the contour lines at a height of 1/2 for a continuous function F(x, y) defined over the square area of the picture and such that F takes value 1 where the picture has a black point and 0 where it has a white point. The precise construction of F is as follows:

Consider the subdivision of the picture area into smaller squares defined by the square matrix of picture points by drawing horizontal and vertical lines through the points of the picture. It should be clear that a triangulation* of the picture area can now be effected by separately triangulating each smaller square, as shown in Fig. 3. We shall describe rules for determining which diagonal will be added to each square to complete the triangulation.

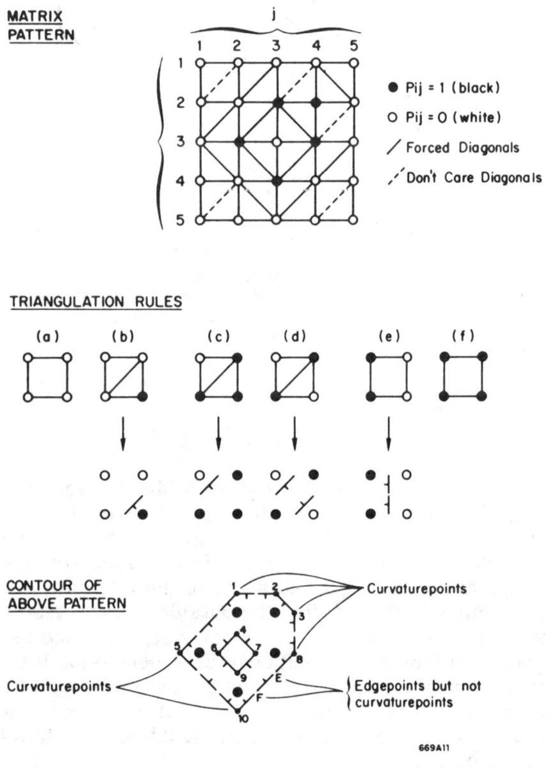

FIG. 3

In deciding how to triangulate the square mesh we follow two simple rules. If there is a choice of whether to connect two black points or two white points, choose the black and try to have as few curvaturepoints as possible. The triangulation rule embodied in (d) of Fig. 3 assures that black points have precedence, while rules (b) and (c) prevent a straight edge at 45° from being encoded as a saw-tooth contour line. The reader may verify these statements by choosing the other diagonal and then constructing the contour. The actual contour for the top pattern in Fig. 3 is presented at the bottom of Fig. 3 and hatch marks have been appended to show which side of a contour line is the high or black side.

There are only six essentially different combinations of values possible for the corners of a square. These are shown in Fig. 3 along with the proper diagonal. Where no diagonal is shown it means that either one may be chosen.

Now that the picture area has been triangulated, we define our function F separately for each triangle. Supposing some triangle to have vertices A, B, and C, we let P_A, P_B, P_C denote the picture values at points A, B, C which are indeed points of the "matrix picture." There is a unique linear function $F^*(x, y) = ax + by + c$ which takes on values P_A, P_B, P_C at points A, B, C. Each triangle of the triangulation determines uniquely an F^*_{ABC} and F is defined so that $F(x, y) = F^*_{ABC}(x, y)$ whenever (x, y) is inside or on triangle ABC.

The proofs that F is well-defined and continuous are easily derived from the exemplary behavior of linear functions. In fact, the only way discontinuity could possibly occur is by way of F being double defined for some point on a common edge $\overline{AB}$ between two triangles Δ_1 and Δ_2. The reason that this cannot occur is because F^*_1 and F^*_2 (the linear functions for Δ_1 and Δ_2) when restricted to the edge $\overline{AB}$ are both found to vary linearly between A and B and to have the same values at these two points; hence $F^*_1 = F^*_2$ on the edge $\overline{AB}$ and F is well-defined.

Having defined F over the picture area as a continuous piecewise-linear approximation to the original picture, we can simply repeat that the "structural description" of the picture is the set of contour lines of F drawn at a height of 1/2. This approach certainly corresponds well with any intuitive notion of edges in a binary picture.

Curvaturepoints

We use the term "curvaturepoints" to describe the points where the contour lines (called "edges") change direction. For example, in the bottom of Fig. 3 the numbered points 1 - 10 are the curvaturepoints whereas E and F are points on the contour but not at bends. This is because the two sections of contour line which meet at E are both in the same direction.† Referring to Fig. 4, we see that at E there is an incoming edge and an outgoing edge both in direction 5, whereas at point 8 there is an incoming edge in direction 6 and an outgoing edge in direction 5.

It turns out that curvaturepoints can only occur at points midway between two picture points. Furthermore, the question of whether or not a given point is a

*Subdivision of plane area into disjoint triangles.

†Contours are directed so that the hatch mark is on the right.

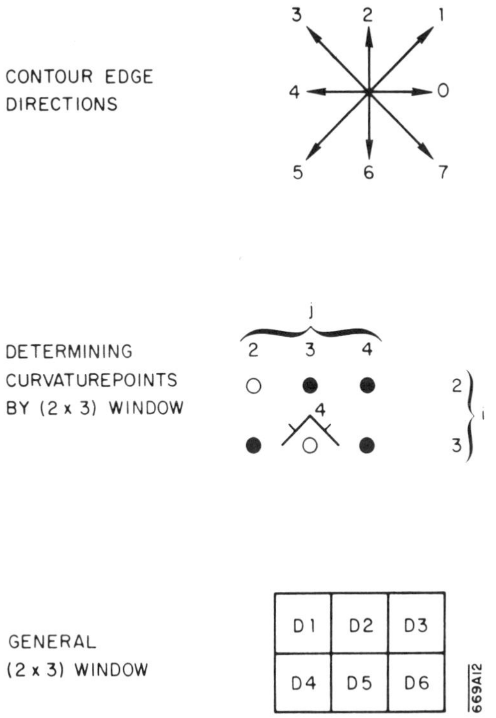

FIG. 4

curvaturepoint can be determined by referring to the values of just 6 picture points arranged in a (2×3) or (3×2) array. Figure 4 depicts how curvaturepoint 4 of Fig. 3 is determined by the (2×3) window shown. If P_{ij} denotes the value of picture point (i, j) then ($P_{33} = 0$, $P_{23} = 1$ and $P_{34} = 1$) implies that there is an incoming edge in direction 3; furthermore ($P_{33} = 0$, $P_{23} = 1$ and $P_{32} = 1$) implies an outgoing edge in direction 5. As a result, we record that the fourth curvaturepoint is at i = 2.5, j = 3 with EDGEIN = 3 and EDGEOUT = 5.

If D1 - D6 denote the picture point values as shown in Fig. 2, then the following hold for the point midway between D2 and D5:

First of all, the point is an edgepoint if and only if D2 and D5 are different in value. If not, it is useless to proceed further with this point. The values EDGEIN and EDGEOUT are determined by the following logic formulas, which are easily implemented by digital hardware:

D5 = 1, D1 = D2 = D4 = 0 → EDGEIN = 1

D2 = D6 = 1, D5 = 0 → EDGEIN = 3

D2 = D3 = 1, D5 = D6 = 0 → EDGEIN = 4

D2 = 1, D3 = D5 = D6 = 0 → EDGEIN = 5

D1 = D5 = 1, D2 = 0 → EDGEIN = 7

D4 = D5 = 1, D1 = D2 = 0 → EDGEIN = 0

D3 = D5 = 1, D2 = 0 → EDGEOUT = 1

D2 = 1, D1 = D4 = D5 = 0 → EDGEOUT = 3

D1 = D2 = 1, D4 = D5 = 0 → EDGEOUT = 4

D2 = D4 = 1, D5 = 0 → EDGEOUT = 5

D5 = 1, D2 = D3 = D6 = 0 → EDGEOUT = 7

D5 = D6 = 1, D2 = D3 = 0 → EDGEOUT = 0

The determination of which edgepoints are curvaturepoints is then simply a matter of whether or not EDGEIN ≠ EDGEOUT. It should be fairly clear from the simplicity of the above conditions that the digital logic involved is easily implemented.

Linkage of Curvaturepoints

So far we have described how an input "matrix picture" can be transformed into a set of "curvaturepoints" of its approximate contour lines. It is still not entirely obvious that these separate and sometimes widely scattered points can be linked together in the proper way to reconstruct the polygonal curves of its "structural description." The reader may have noticed that in an earlier section we spoke of the "structural description" as consisting of closed curves, whereas it is perfectly possible for contour lines to end at the edge of a map or in our case, a picture. To eliminate any chance of confusion, we shall assume that the picture points on the boundary of an input picture are all zero. This will force all contours to be closed curves but will not really hamper the validity of our approach. The only reason we make this assumption is to simplify the discussion below; in practice there is no need to border a picture with zeros this way, but the linkage of curvaturepoints is somewhat more complicated otherwise.

We shall assume that all the curvaturepoints of a given input picture are delivered to a general purpose digital computer in the order in which they are encountered during a normal left-to-right and top-to-bottom raster scan. The rows will be designated by increasing values of Y and the column positions within each row by increasing values of X. Also, the incoming and outgoing edge directions EDGEIN and EDGEOUT are considered an integral part of any curvaturepoint. For example, the simple pattern of Fig. 3 would be represented as:

POINT	Y	X	EDGEIN	EDGEOUT
(1)	1.5	3.0	1	0
(2)	1.5	4.0	0	7
(3)	2.0	4.5	7	6
(4)	2.5	3.0	3	5
(5)	3.0	1.5	3	1
(6)	3.0	2.5	5	7
(7)	3.0	3.5	1	3
(8)	3.0	4.5	6	5
(9)	3.5	3.0	7	1
(10)	4.5	3.0	5	3

A simple inspection of the above table shows the following interesting facts about the two POINTS 1 and 5. First of all, EDGEIN [1] = EDGEOUT [5] = 1 and X[1] + Y[1] = X[5] + Y[5] = 4.5; furthermore, there is no curvaturepoint P between* these two which satisfies the condition X[P] + Y[P] = 4.5. It should be clear that whenever two points P and Q (P < Q) have the same value of X + Y, then they lie on the same line drawn in direction 1 - 5 (see Fig. 4). If, in addition, there is no intermediate curvaturepoint and either EDGEIN[P] = EDGEOUT[Q] = 1 or EDGEOUT [P] = EDGEIN [Q] = 5, then the two points are adjacent curvaturepoints of the same contour curve and hence should be linked together. There are similar conditions for the linkage of points with the same (X - Y), X or Y values. Each case is associated with a pair of directions (respectively 3 - 7, 2 - 6 or 0 - 4) and the precise conditions are geometrically self-evident. These conditions are all that are needed to perform the linkage of the curvaturepoints into simple closed polygonal curves. The implementation of this procedure uses very strongly the fact that the points are ordered by increasing X within increasing Y since there is a great deal of geometric information implicit in this sorting order.

The data structure required to implement the linkage of curvaturepoints is essentially a cyclic list; this requirement reflects the natural structure of a closed polygonal curve as well as the fact that at various times during linkage we are required to join together two already created partial polygonal curves. This joining (concantenation) would be quite difficult if simple arrays were used.

Smoothing Curves

In Fig. 5, we have redrawn the outermost curve of the "structural description" in Fig. 2. Some of the curvaturepoints are not solid dots and these are to be deleted from the description. The criteria which control which points will remain are explained below, along with an intuitive justification for performing this smoothing operation.

An "inflection" will be defined as two adjacent curvaturepoints, one with a bend† of - 1, the other + 1, and such that the distance between the two points is 1 or $\sqrt{2}$. Referring to Fig. 5, the pair of points (3, 4) are at distance 1 from each other and have bends respectively + 1 and - 1 as shown in the insert. As a result, the incoming edge at 3 and the outgoing edge at 4 are in the same direction, namely 7. Similarly the pair (6, 8) are at distance $\sqrt{2}$ with bends - 1 and + 1; hence, we have another "inflection."

When points 3, 4, 6 and 8 have been deleted, the sequence of curvaturepoints becomes 1, 5, 9, ... and point 5 retains a bend value of + 1. Since we allow point deletions to occur only in pairs with bends that cancel (+ 1 and - 1), it is clear that the total bend around a

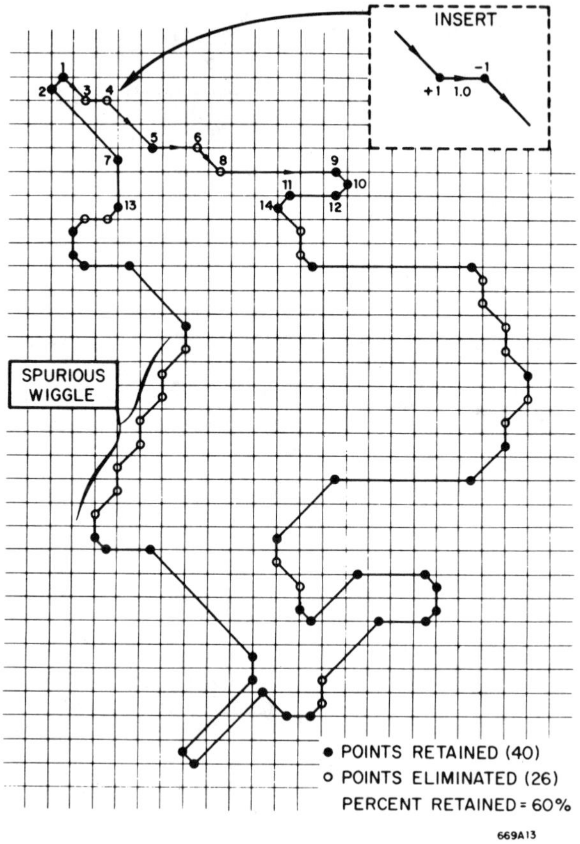

FIG. 5

curve is not disturbed at all. The bend at point 5 in Fig. 6 is not exactly + 1 as can readily be seen, but as an approximation it does not really go far wrong. Furthermore, if more precision is required, it can be calculated from the X and Y coordinate values of the 3 points 1, 5 and 9.

The main advantages of performing this smoothing of inflections are the reduction in the amount of data to be processed in subsequent discrimination algorithms and the elimination of some sequences of edge bends which we may call "spurious wiggles" since the corresponding edge in a typical original continuous pattern was probably straight. Figure 5 shows a "spurious wiggle" of 4 pairs of points. In the example depicted the reduction in data is substantial, amounting to 40%, although the picture pattern contains a great deal of real detail.

As a final point we want to make it perfectly clear that this smoothing procedure is not an absolutely necessary part of the overall method. Depending on the particular application, it may or may not be appropriate. We present it here in place of several other smoothing algorithms which we have experimented with because it has been found to accomplish a substantial reduction in data without eliminating many major details and while requiring only simple decision criteria.

*meaning a point with a number between 1 and 5.

†The bend at any curvaturepoint measures the change in direction of the edge at that point in units of 45°. Bends toward the right are negative and toward the left, positive.

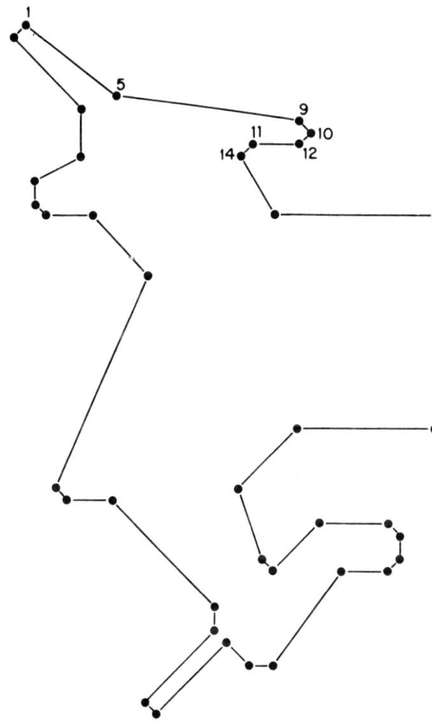

SMOOTHED EDGE CONTOUR

FIG. 6

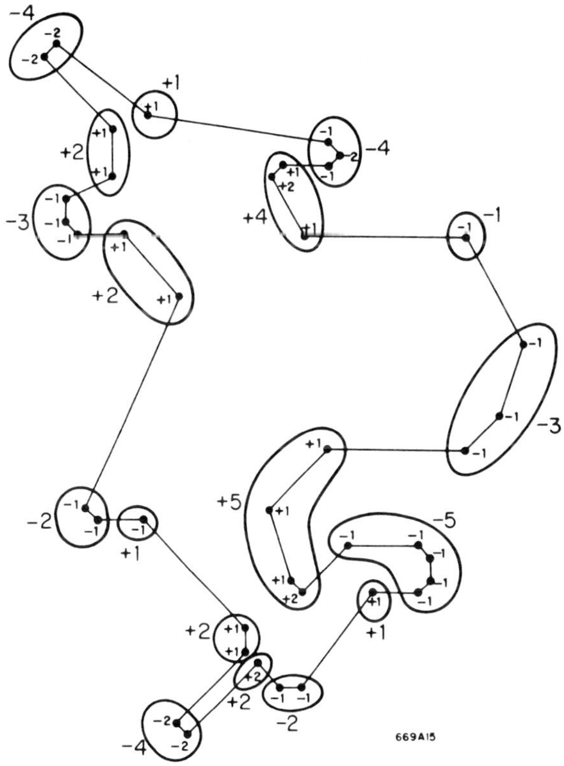

FIG. 7

Bend Groups

When the sequence of curvaturepoints describing a single closed contour has been constructed and the "inflections" have been deleted as described in the previous section, then a very compact but extremely characterizing signature can be computed by merging curvaturepoints according to the following rules:

Transform the closed curve into a sequence of integers representing amounts of bend by adding together the values of bend for two adjacent curvaturepoints of like sign whenever their distance apart is less than some threshold which is appropriate to a given application.

Figure 7 depicts how the grouping of like points would be made for our sample curve of the last section, assuming a threshold distance of approximately 4.0. The compact signature for this curve is (-4, +1, -4, +4, -1, -3, +5, -5, +1, -2, +2, -4, +2, +1, -2, +2, -3, +2).

Once again we must emphasize that special applications will generally suggest variations in the merging of bends in addition to guiding the determination of the distance threshold. Some applications may, of course, require the retention of more detailed edge length information in which case this simple scheme would not be appropriate.

Recovery Algorithm

An important property of "structural descriptions" is that an algorithm exists which will recover the original digitized binary matrix pattern given only the structural description. We shall give a brief outline of such an algorithm and discuss the theoretical foundation for its validity.

The famous mathematical theorem known as the "Jordan Curve Theorem" states that any simple closed curve in the plane divides the remainder of the plane into two distinct sets called the "inside" and "outside," such that any two points in the same set can always be connected with a continuous curve lying entirely in that same set; any continuous plane curve joining a point "inside" the curve to one "outside" must have at least one point in common with the closed curve. The proof of this theorem is much simpler for polygons than for general curves; Courant and Robbins[2] give a proof for the polygonal case which depends on the fact that if a point "outside" a closed curve Γ is connected by a straight line L to another point P, then P is also "outside" Γ if and only if L intersects Γ an even number of times. If L intersects Γ an odd number of times, then P is "inside" Γ.

The algorithm we propose is based on the latter fact, which is also proved in Ref. 2. Referring to Fig. 9, we shall show how it is possible to determine that the points in ranges β and δ (for $X = X_0$) are inside

Γ while ranges α, γ and ϵ are outside. The proper assignment of values then depends only on the status of Γ ("object" or "hole").

The determination that ranges β and δ are inside Γ proceeds as follows:

Curve Γ is traced from its top all the way around and back to the top again. Whenever the curve Γ crosses the vertical line $X = X_0$, we record the fact by making a mark in all points below the crossing. For example, Γ crosses first at e_1 and all points in ranges, β, γ, δ and ϵ are marked + as shown in the table. This means that each of these marked points can be connected by a straight line to the top of the picture area only by a line crossing Γ at e_1. The table is filled in for e_2, e_3, and e_4 in an entirely similar manner. After e_4, there are no more crossings and the curve returns to the starting point. We then determine which points have received an odd number of marks and find that all points of ranges β and δ are inside Γ.

When this procedure is carried out for all vertical lines X_0, then the "inside" points for Γ will have been determined. In any actual implementation, we assume that all X_0's would be handled simultaneously while Γ was being traced only once.

If the same process is repeated for all curves Γ in a "structural description" then the result is just as valid as for one curve because curve crossings always represent a color change.

No claim is made for the efficiency of this procedure; we merely wanted to show with some degree of mathematical rigor that the "structural description" contains all the information of the original digitized pattern and it can be mechanically recovered if necessary.

Advantages of the Curvaturepoint Method

The generality of the curvaturepoint method is one of its most important properties. The method applies to any "black and white" pattern whose significant content consists of connected sets of similar points. Hence, it also applies to input pictures which can be made to comply with this condition by suitable preprocessing; fairly simple local transformations on binary pictures have been found to be very successful at regularizing pictures in this way. Our method can be used in some cases to extract contrast information from grey-scale* pictures by converting the picture to binary several times using different thresholds. This idea is more fully elaborated in Zahn.[1] This applicability to multilevel pictures should come as no surprise when it is remembered that the method is simply a contouring of a two-dimensional distribution of numerical values.

Simplicity and mathematical rigor are properties of the method which we feel have been largely overlooked in most "edge-detection" schemes. The two properties are closely related, for the mathematical rigor with which "curvaturepoint extraction" and "linkage" are performed is a direct consequence of the simple and straightforward definition of the "structural description." Being contour lines of a simple function, the "structural description" is constrained to consist of non-intersecting closed polygonal curves whose edges are directed in only eight different ways and whose edge bends or "curvaturepoints" are also tightly constrained. It is precisely because of such constraints that the "linkage" of widely spaced "curvaturepoints" can be accomplished in a completely assured way. Not only is the transformation from binary pattern to structural description rigorous and uniquely defined; the reverse transformation (see Recovery Algorithm) exists as well, proving a unique one-to-one correspondence between a binary pattern and its structural description and also showing that the structural description contains total information from the binary pattern.

One of the most serious obstacles to the further development of digital picture processing is the volume of data implied by the two-dimensionality of pictures. When the resolution is doubled the data volume is quadrupled; a picture $1,000 \times 1,000$ has one million data points, an amount which is still prohibitively high for even the largest computers. The "structural description" on the other hand contains one-dimensional information (contour lines) and therefore the data volume increases only linearly with resolution. This means that if "curvaturepoint extraction" can be done in special digital hardware then the storage requirement for implementing the method on a general purpose computer will be greatly reduced and more richly detailed patterns can be handled. The "linkage" would be accomplished by programming. It turns out very happily that "curvaturepoint extraction" is defined by extremely simple Boolean logic (see Curvaturepoints) and its digital hardware implementation could be accomplished with the addition of two long shift registers. We consider this efficient implementation to be an extremely vital aspect of the method.

The intuitive character of our method should prove of considerable benefit in constructing recognition algorithms. In a recent article Uhr[3] claims that edges, angles and the interrelations between lengths and slopes are the important and meaningful properties for human pattern perception and recognition. If this is true then it is reasonable to expect that algorithms based on "structural descriptions" will be highly intuitive in nature and hence less mysterious than those based on information other than edges. In support of Uhr's claim we shall cite evidence from psychology and neurophysiology which tends to indicate the predominance of edge information in animal visual perception. Attneave[4] in experiments on human subjects found that the number of bends in polygonal shaped objects accounted for 80% of the difference between objects when rated by subjects according to judged complexity. This clearly suggests importance of curvaturepoints. Other researchers have found that an object's visibility is related to the length of its boundary, strongly suggesting that for purposes of perception the edge is the predominant information content.

The research of Lettvin et al.[5] on the optic nerve cells in the frog indicates that signals reaching the frog's brain from its eyes are highly contrast-oriented. Hubel and Wiesel[6] determined that nerve cells in the cat's brain are specific to the existence of edges in the visual field of given slope. This neurophysiological

*Picture values range over an ordered finite set such as (0, 1, 2, 3).

evidence if extrapolated to the human case lends further support to the claim that edge information is the raw data for visual perception mechanisms in man.

Although the method is clearly directed toward recognition via edge-bend-sequence, nevertheless familiar quantitative properties such as perimeter, area, moments, height and width are easily calculated. In addition, we can define and compute reasonably intuitive quantitative measures of oblongness, compactness, wigglyness and total absolute curvature.

For some pattern recognition applications (character recognition especially) it is useful to have a method which is position and size invariant. "Structural descriptions" contain position and size information but in such a way that it is quite easily disregarded. Rotation invariance is also easy to achieve by simply considering the sequence of curvaturepoints without paying attention to the initial edge direction. The information is always there when needed, however, so that recognition methods sensitive to position, size and rotation are not hampered in the least.

A pattern description method is greatly enhanced if it is compatible with some fairly elegant language for pictures. This is particularly true if the language has a formal phrase structure grammar which allows pattern recognition to be accomplished by the formal parsing procedures which have been developed for such grammars. The survey paper of Miller and Shaw[13] discusses this aspect of picture processing quite comprehensively. Ledley[7] for example, has shown that formal parsing techniques can be used to recognize different chromosome shapes by transforming their edge sequences into a string of primitives and then parsing this string. For example, in Fig. 8 O means no bend, E means sharp 180° convexity, Y means sharp 180° concavity, etc. These are the primitives of the language. The formal grammar would define "arm" as OEO, "double arm" as arm Y arm, etc.

An essentially one-dimensional data format seems to be advantageous for linguistic processing of pictures because phrase-structure grammars describe sets of "strings." When the data is not in a "string" format there are some subtleties involved in making the correspondence between the data and linguistic formalism. The recent work of Shaw[8] shows that automatic parsing recognizers can be quickly implemented when the structure of the picture can be represented by a suitable one-dimensional grammar. Syntax directed parsing methods are employed so that new recognition tasks require only a new syntax table and specially written primitive recognizers.

In addition to formal language parsing methods, our "structural description" can be used very readily in a decision tree approach to recognition. In fact, the cyclic list data format of the "structural description" lends itself naturally to sequential decisions made as the list is traversed. As with all pattern description schemes, it is possible to reduce to a vector of quantifiable properties and then use one of the many procedures based on the property vector representation of a pattern. The algorithms of Freeman[9, 10, 11] for "curve segment matching" are applicable with almost no change since "structural descriptions" are essentially "Freeman encodings" of the curves defining boundaries in a binary picture. For example, the Freeman chain-encoding of the closed curves of Fig. 3 would be (00766555333111) and (5713). Each digit represents a unit vector as shown in Fig. 4 and the curve is traced sequentially. Our "structural descriptions" vary from this format only to the extent of merging like-direction contiguous unit vectors into a single vector with length. The "structural description" for the outer curve of Fig. 3 is essentially (0, 2; 7, 1; 6, 2; 5, 3; 3, 3; 1, 3). These curve matching algorithms are capable[12] of putting together an "apictorial jigsaw puzzle" which attests to the subtlety of their shape discrimination.

Acknowledgements

The author would like to thank Alan Shaw of Cornell University* and Jeno Gazdag of IBM for many hours of very useful discussion on pattern recognition methods; Robert H. Penny of General Electric Co. for introducing us to and stimulating our interest in digital picture processing and William F. Miller of Stanford University for his continuing encouragement and support of this research. We also thank the referee for several constructive suggestions.

References

1. C. T. Zahn, "Two-dimensional pattern description and recognition via curvaturepoints," Report No. SLAC-70, Stanford Linear Accelerator Center, Stanford University, Stanford, California (1966).

2. R. Courant and H. Robbins, What is Mathematics? (Oxford University Press, London, 1941); pp. 267-269.

3. L.Uhr, "Pattern recognition," in Pattern Recognition, ed. Leonard Uhr (John Wiley & Sons, New York, 1966); pp. 365-381.

4. F. Attneave, "Physical determinants of the judged complexity of shapes," J. Exptl. Psychol. 53, 221 (1957).

5. J. Y. Lettvin, H. R. Maturana, W. S. McCulloch and W. H. Pitts, "What the frog's eye tells the frog's brain," Proc. IRE 47, (November 1959); pp. 1940-1951.

6. D. H. Hubel and T. N. Wiesel, "Receptive fields, binocular interaction, and functional architecture in the cat's visual cortex," J. Physiol. 160, 106 (1962).

7. R. S. Ledley, "High-speed automatic analysis of biomedical pictures," Science 146, 216 (1964).

8. A. C. Shaw, "The formal description and parsing of pictures, Report No. SLAC-84, Stanford Linear Accelerator Center, Stanford University, Stanford California (1968).

9. H. Freeman, "On the encoding of arbitrary geometric configurations," IRE Trans. on Electronic Computers EC-10, (1961); pp. 260-268.

*formerly Stanford University.

10. H. Freeman, "Techniques for the digital computer analysis of chain-encoded arbitrary plane curves," Proc. National Electronics Conference (1961); pp. 421-432.

11. H. Freeman, "On the digital computer classification of geometric line patterns," Proc. National Electronics Conference (1962); pp. 312-324.

12. H. Freeman and L. Garder, "Apictorial jigsaw puzzles: the computer solution of a problem in pattern recognition," IEEE Trans. on Electronic Computers (April 1964); pp. 118-127.

13. W. F. Miller and A. C. Shaw, "Linguistic methods in picture processing - a survey," Proc. Fall Joint Computer Conference 33, Part One (1968); pp. 279-290.

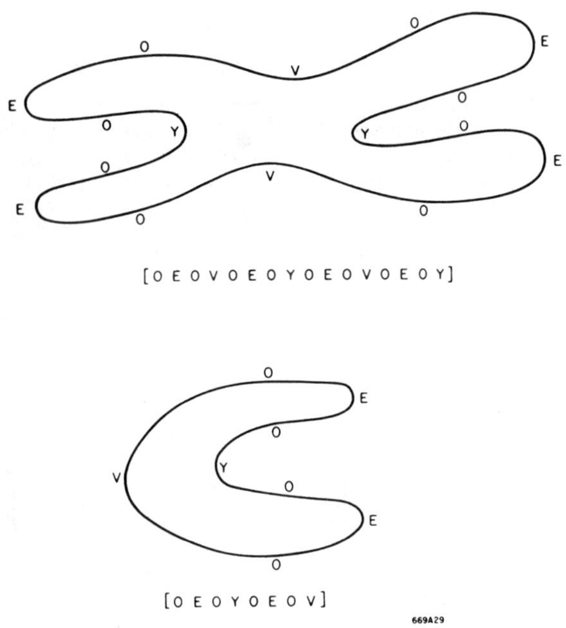

FIG. 8

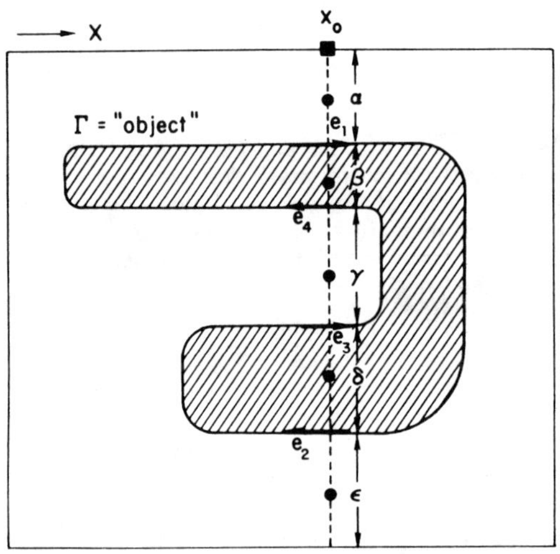

	α	β	γ	δ	ε
e_1		+	+	+	+
e_2					+
e_3				+	+
e_4			+	+	+
$+\Sigma$	0	1	0	1	0

669A53

FIG. 9

Computer Analysis of Moving Polygonal Images

J. K. AGGARWAL, MEMBER, IEEE, AND RICHARD O. DUDA, MEMBER, IEEE

***Abstract*—A general mathematical model is developed as an idealization of the problem of determining cloud motions from satellite pictures. The model consists of superimposed planes of rigid moving polygons. The problem is to determine from a sequence of scenes the linear and angular velocities of the figures, and to decompose the scene into its component figures. Study of the model reveals a number of fundamental relations that form the basis for an analysis program. In particular, a systematic anaylsis is given of the topological changes that can occur when overlapping figures move together or apart. A computer program based on these results is described, and experimental results are presented.**

***Index Terms*—Change detection, cloud motion, motion, moving images, occlusion, overlapping figures, pattern recognition, polygonal figures, scene analysis, topological changes.**

I. INTRODUCTION

THIS PAPER describes an investigation of a particular problem in the analysis of pictures of moving two-dimensional objects. The general problem of detecting significant changes in a sequence of pictures is a difficult one. Although Ulstad [1] traces its history back to 1920, most of the work in this area is relatively recent, depending as it does on digital image processing. In particular, considerable attention has been devoted to the determination of cloud motions from digitized satellite photographs [2]–[4]. Since that problem served to motivate the current investigation, some of its characteristics will be described briefly.

In a typical satellite photograph the regions of cloud cover are considerably brighter than the background, whether the clouds are low stratus layers or high cirrus layers. Although cloud velocities in any local region are relatively constant, nonuniform patterns of motion show up over larger areas. The cloud velocity fields are closely related to wind velocity fields, and major changes in wind speed and direction occur over large areas, such as the mid-Pacific region. The purpose of deriving cloud motions from satellite photographs is to obtain these wind velocity fields.

The basic technique used in programs that track cloud motions is cross correlation. In some programs the cross correlation is done directly, small local zones being matched from one picture to the next [2], [3]. With other programs the pictures are first clustered to find brightness centers, and these centers are matched from one picture to the next [4]. In either case, good results can be obtained provided that the pictures are properly registered, that clouds are not in a state of rapid formation or dissipation, and that there is only one layer of clouds in any given area.

Unfortunately, it is not at all uncommon for several cloud layers to be present at once, and each layer can be moving with its own speed and in its own direction. This invalidates the use of cross correlation, unless the various layers can somehow be separated before successive images are matched. If the photographs are viewed individually, it is surprisingly difficult to distinguish between the various layers. In particular, the brightness differences between different layers are quite small, and one must discriminate on the basis of rather subtle shape and texture information. Viewed in a time-lapse sequence, however, the various layers are much more clearly distinguishable, the joint motions of clouds in a given layer often being quite striking. This paper is primarily concerned with ways of extracting, from a sequence of pictures, the information implicit in these joint motions.

Both to avoid the many complications introduced by the use of real imagery and to obtain somewhat greater generality, we have investigated an idealized model of this problem. The model, defined precisely in the next section, can be understood intuitively by imagining a black background, overlaid by a number of transparent sheets, upon which opaque white polygons have been drawn. As the sheets are translated (and perhaps rotated), the exposed and hidden parts of the different polygons shift and change, but the figures on any given sheet maintain their shapes and relative positions. The basic problem is to derive a description of the figures on each sheet and the motion of the sheets from a sequence of views of the superimposed sheets.

Clearly, this model is closely related to the way in which animated cartoons are made. Viewed in a time-lapse sequence, the pictures would form animated silhouettes. Thus, the model is relevant to many other problems involving motion in scene analysis besides the problem of tracking cloud motions. Similar models have been used before in some of the scene analysis literature concerning line drawings. Specifically, Potter [5] used a two-layer version of this model in investigating the use of joint motions to segment a scene of moving, nonoccluding objects, and Badler [6] used a spherical projection model in developing a system that produces natural language descriptions of pictures of known, moving three-dimensional objects. Of course, all such models are highly idealized, and only a first approximation to actual imagery. However, even our simplified model poses problems too difficult to solve by

Manuscript received October 20, 1974; revised April 10, 1975. This work was supported in part under NSF Grant GK42790.

J. K. Aggarwal is with the Department of Electrical Engineering, University of Texas, Austin, Tex. 78712.

R. O. Duda is with the Artificial Intelligence Center, Stanford Research Institute, Menlo Park, Calif. 94025.

Reprinted from *IEEE Trans. Comput.*, vol. C-24, pp. 966–976, Oct. 1975.

mathematical analysis. Our purpose is to understand better how these problems can be solved before additional complications are introduced.

II. MATHEMATICAL MODEL

The following mathematical model is a slight generalization of the transparent sheet model just described. It is based upon a collection of planar figures moving in several parallel planes. For simplicity, the shapes of the figures are restricted to polygons, and the polygons are assumed to be rigid. Thus, each figure retains its shape and size during the course of its motion. The polygons are otherwise unrestricted. In particular, they may have one or more holes and need not be convex. The motions of the polygons in a given plane may be correlated or may be independent. That is, the figures may share a joint motion, or each figure may have its own linear and angular velocities. It is envisaged that there will be several planes and several figures moving in each plane. Although the present paper concentrates on independent motions, the programs generated are equally applicable to correlated motions. Fig. 1 shows a simple example in which Plane A contains a single triangle with a rectangular hole and Plane B contains a nonconvex pentagon.

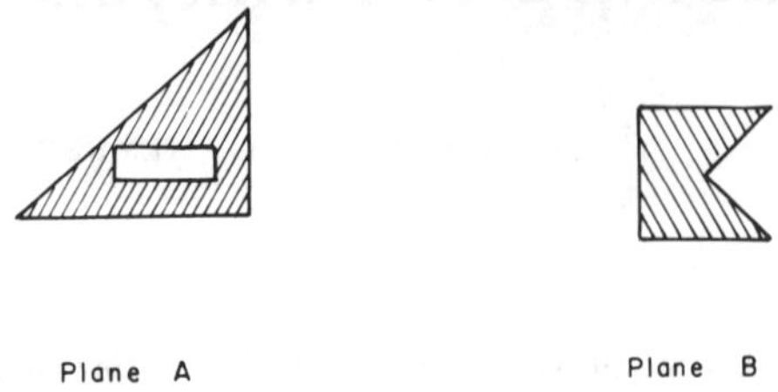

Fig. 1. The individual planes.

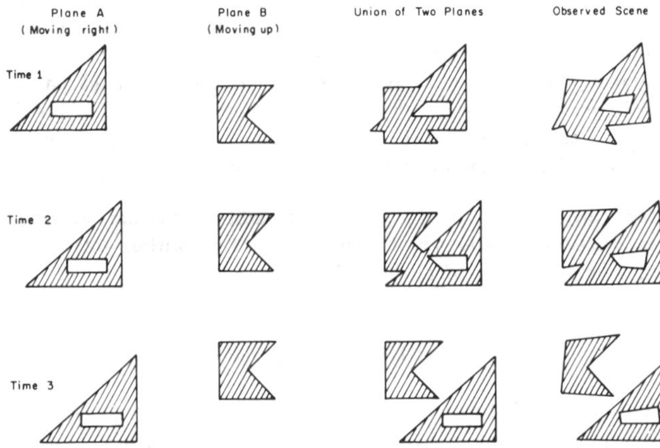

Fig. 2. Motion as seen by observer.

It is assumed that the observer sees a perturbed, parallel projection of the planes, and is unable to distinguish the different planes from a single view. To be more specific, the input picture at any instant in time is taken as the union of the polygons in the various planes, with the coordinates of the vertices perturbed by the addition of random noise. It is further assumed that the time interval between successive pictures is sufficiently small that only "small" changes occur from picture to picture. Fig. 2 illustrates the observer's view of the polygons shown in Fig. 1.

The objective of the present investigation is to determine the linear and angular velocities of the figures, and to decompose the scene into its component figures from a time sequence of scenes as described above. In the general case, both the number of planes and the number of polygons in each plane are unknown to the observer.

The difficulty of this problem in the general case will be apparent from the following sections. Here, some additional comments on the model are presented to show some of the considerations involved in its selection. The restriction of the figures to polygonal shape is motivated by computational convenience. The nature of the study would not be changed significantly by assuming curvilinear boundaries for the figures, but the computational burden would be severely increased. It is believed that the introduction of curved boundaries would bring in extraneous issues without adding significantly to the generality of the problem.

The assumption that the polygons are rigid is also made for simplicity. In a more realistic situation, one will encounter cases where figures appear and disappear, or change in size or shape, or both. By adding random noise to the coordinates, we are in fact allowing small changes in size and shape to occur. In addition, this noise precludes the trivial solutions that can be obtained if the rigid polygons can be observed with unlimited precision. If one assumes that the polygons are obtained by fitting straight lines to gray scale data, it would be more realistic to model uncertainty by randomly inserting or deleting certain vertices. We hope to be able to investigate a more complicated model of this type in the future.

In the present formulation there is no way to tell whether one plane is occluding or is occluded by another. This idealization, which is a good first approximation to the satellite cloud photograph problem, means that no depth clues are available to aid in resolving the figures. Inherent in the model is the assumption that there are no distinguishing features for the various planes. Such features do exist in real applications, and their introduction will certainly make the solution to the problem easier and not more difficult.

III. THE ANALYSIS PROBLEM

The analysis problem is as follows. Given a sequence of scenes generated according to the model, find the linear and angular velocities of the figures and decompose the scene into its component figures. Although the problem as posed is essentially mathematical, our approach to its solution is heuristic and computational. The general problem can be arbitrarily complicated, and it is not sensible to ask for a general solution. However, a reasonably general and interesting computational solution must take into account the problems posed by time sampling, overlapping figures, topological changes, and uncertainty, whether due to quantization or noise, and must solve them

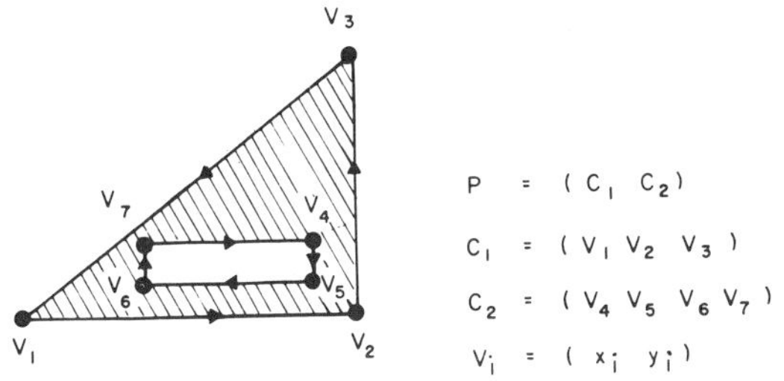

Fig. 3. List representation of a polygon.

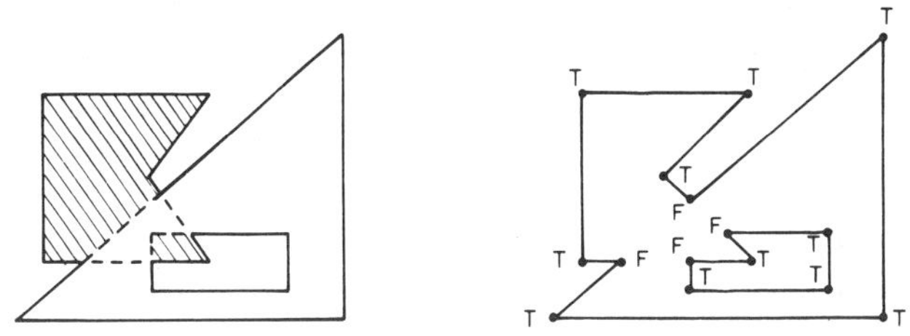

Fig. 4. True and false vertices.

under the usual requirements for economy in the use of time and memory. Some of the various kinds of information that can be exploited by the analysis program are described in this section, starting, for simplicity, with the noise-free case.

A. The Noise-Free Case

In our computer programs a polygon P is represented by a list of boundary components—the external boundary and, if present, the internal or hole boundaries. Each boundary component is a list of vertices appearing in counterclockwise order for the external component and in clockwise order for the internal components (see Fig. 3). With this convention, if one follows the edges joining successive vertices of a boundary component, then the interior of the polygon always lies to the left. Finally, a vertex is described by a pair of coordinates (x,y). From this description, the lengths of the edges, the angles at the vertices, and other properties of the polygon are readily computed.

Let us temporarily assume that the input scene is free of noise and that none of the polygons overlap. Then, since the polygons are assumed to be rigid, the lengths of the edges and the angles at the vertices will remain constant as the polygons move. Given two successive scenes of this type, the only real problem is the identification of corresponding vertices. Once any vertex in the first scene is paired with a vertex in the second scene, all other vertices for that polygon can pair in only one way, and the two polygons either will or will not match. While the matching problem is essentially rather trivial, it raises some problems that also appear in more complex cases. For example, multiple solutions can be encountered when there is symmetry or several identical figures. More important, when there are many polygons to match, the straightforward exhaustive procedures become very inefficient, and the need for good search strategies becomes apparent.

If two polygons overlap, the resulting polygon will contain two kinds of vertices, vertices from the original polygons and vertices where the polygons intersect (see Fig. 4). For brevity, we call the former *true vertices* and the latter *false vertices*. In the absence of noise, true and false vertices can be easily distinguished, since the angles at true vertices stay constant as the polygons move, while the angles and the lengths of the edges incident on false vertices generally change. Of course, if an angle does not change, the vertex is not necessarily true, but if an angle changes then the vertex must be false, and if the length of an edge changes then at least one of the vertices at the ends of the edge must be false.

This suggests a straightforward way of analyzing noise-free pictures of moving rigid polygons, provided that the changes from picture to picture are sufficiently small that no old vertices disappear and that no new vertices are formed. One merely pairs vertices as before, demanding a perfect match for one angle but requiring, say, only a topological match for the rest of the polygon (same number of boundary components and the same number of vertices for each component). If a polygon in the first scene matches more than one polygon in the second scene, a criterion function that measures the differences between corresponding lengths and angles can be used to determine a best-match solution. A variation of this procedure plays an important role in the analysis program described in the next section.

The efficiency and even the accuracy of the search procedure can be increased significantly by exploiting additional facts about the problem. For example, if a polygon has been successfully tracked from one scene to the next, one can easily estimate its linear and its angular velocities, and thus predict its location and orientation in the next scene. Efficiency can be increased by searching for matches only in the vicinity of the expected vertex locations; accuracy can be increased by including positional errors in computation of the matching criterion function. For another example, if it is expected that several figures will share a joint motion, then the accuracy of the velocity estimates can be increased by clustering. In particular, when a number of vertices have been paired, angular velocities can be computed and clustered, using any of a number of simple one-dimensional clustering procedures. All vertices belonging to the same polygon must yield the same angular velocity, of course, as must vertices for polygons sharing the same motion. If the angular velocity is zero, then all of these vertices must have the same linear velocity. More generally, if the angular velocity vector is $\boldsymbol{\omega}$ and if the velocity of a vertex located at $\boldsymbol{r}_i$ is $\boldsymbol{v}_i$, then the velocity of a vertex at $\boldsymbol{r}_j$ must be $\boldsymbol{v}_i + \boldsymbol{\omega} \times (\boldsymbol{r}_j - \boldsymbol{r}_i)$. These relations express the information provided by joint motions, and should be particularly useful when noise introduces uncertainty.

B. Topological Changes and Noise

When enough time elapses so that old vertices get hidden and new vertices appear, a general matching strategy becomes much more complicated. Here having good velocity estimates becomes especially important. If the positions of all of the polygons in the next scene can be predicted, then future overlap can be anticipated and properly taken into account. It is more difficult to anticipate how or even when two overlapping polygons will separate, but at least the possibility of matching problems can be anticipated if some of the vertices are known to be false.

The problems arising because of topological changes are extemely interesting, but difficult to solve in a way that is both efficient and general. The nature of the problem changes with the sampling rate. If the interval between successive pictures is sufficiently short, topological changes will occur infrequently, and, although they may have global consequences, they can be analyzed essentially locally. However, there is a tradeoff between the computational burden of local processing of many pictures taken at short time intervals and that of more global processing of a few pictures taken at long time intervals. At present that tradeoff is not well understood. The program described in the next section takes an essentially local approach, based on the assumption that topological changes can be treated as rarely occurring special cases to be considered when the standard matching procedure fails.

When the vertex locations are perturbed by noise, all of the angles and edge lengths change at least slightly from scene to scene. Thus, as long as no old vertices disappear and no new vertices appear, one can only state the probability that a vertex is true or false. Of course, with enough independent observations these probabilities will tend to certainty, but this is a rather uninteresting result. There is little need to try to distinguish between a noisy true vertex and a false vertex that is changing extremely slowly. For most practical purposes the two figures involved might just as well be treated as one. The main reason for wanting to detect false vertices is to avoid the serious errors in velocity estimates that can arise when two or more overlapping figures moving in different directions are treated as one rigid figure.

Actually, a single scene provides considerable information about which vertices are true and which are false. For example, true vertices are often acute, having an interior angle of less than 180°, while false vertices are often obtuse. As illustrated in Fig. 4, this is true whether the angle is on the external boundary or on an internal boundary. In fact, while true vertices need not be acute, false vertices, with one exception, must be obtuse. The exception[1] arises when a false vertex is formed where two true vertices coincide, as illustrated in Fig. 5. Excluding this as a probability zero event, we can automatically label all acute vertices as true vertices. It is also possible in principle to make more subtle inferences from other geometrical relations. For example, by extrapolating lines such as the dotted lines in Fig. 4, one can make plausible conjectures about the composition of the scene. These conjectures might be rejected after subsequent scenes are analyzed, but if they are confirmed they might speed the analysis greatly. However, we have had no experience with such procedures, and their true usefulness remains to be seen.

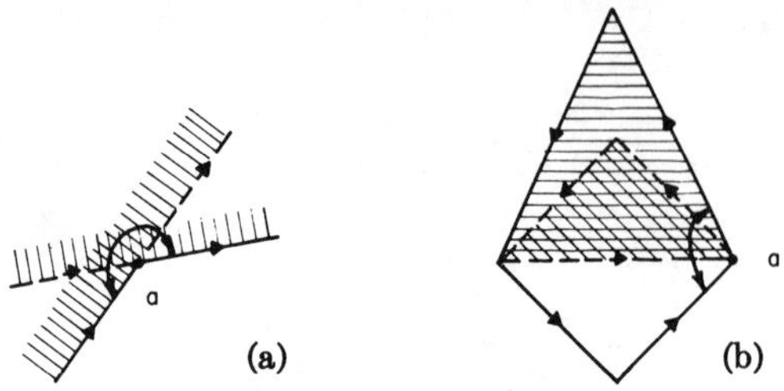

Fig. 5. (a) Normal obtuse false vertex. (b) Acute false vertex formed by coincidence of two true vertices.

[1] First pointed out to us by R. Petermann.

IV. THE ANALYSIS PROGRAM

A. Overview

In this section we describe the major parts of a Lisp program—implemented by Petermann [7]—that analyzes pictures of moving polygons. The input to the program is a sequence of scenes $S_0, S_1, \cdots, S_k, \cdots$ where each scene is a collection of polygons. As the program runs it constructs and continually refines a model of the component polygons, including their estimated linear and angular velocities. This information is the principal output of the program. In analyzing a new scene and updating the model, the program has access to the new scene, the immediately preceding scene, and the current model. Mathematically, if M_k is the model at time k, then the program computes M_{k+1} as a function of M_k, S_k, and S_{k+1}.

Although the details of the program occasionally get rather involved, the overall strategy is straightforward. The basic operations performed are summarized in Fig. 6. Because the program begins with no past history of the motions, the initial procedure for matching the first two scenes is somewhat different from the strategy for matching subsequent scenes. This latter strategy is the more significant of the two. It is based on the assumption that the time interval between successive scenes is so short that most of the time there will be little trouble in matching a polygon in scenes S_k with a polygon in S_{k+1}, there being only one match that is at all reasonable. On occasion, however, the matching procedure will fail, primarily because the way that the figures overlap one another has changed, resulting in a change in topology. When this happens a special routine is invoked to identify the type of topological change that has occurred, and to complete the matching process.

This is clearly a heuristic strategy. No attempt is made to guarantee that matches made between successive scenes are in any sense globally optimal. No provisions for backtracking are included to allow the program to recover from an erroneous match. The primary purpose of the program is to demonstrate that a fairly simple, computationally feasible strategy for tracking overlapping figures

Initialization

1. Read first two scenes, S_0 and S_1.
2. Mark acute angles as "True," obtuse angles as "Potentially False."
3. Under the assumption of no topological changes, match each polygon in S_0 with a polygon in S_1. (Failure to find a match causes program termination.)
4. Compute the linear velocities of the vertices and the angular velocities of the edges.
5. Compare the angular velocities of edges incident on each "Potentially False" vertex. If the difference exceeds a preassigned threshold, mark the vertex as "False".
6. Generate the initial model, splitting off components bounded by "False" vertices.

Main Loop

7. Read the next scene.
8. For each polygon in the previous scene, attempt a match.
9. If the match succeeds and if the numbers of vertices in the two matched polygons are the same, go to Step 13.

Topological Analysis (See Section IV-D)

10. From vertex counts, determine the type of topological change. (If there was no topological change, the program fails at this point.)
11. Locate the vertex which changed its visible/occluded status.
12. Update the model to account for topological changes.

Model Updating

13. Compute and update the linear velocities of vertices and the angular velocities of edges.
14. Compare the angular velocities of edges incident on each new "Potentially False" vertex. If the difference exceeds a preassigned threshold, mark the vertex as "False."
15. Update the model, splitting off components bounded by "False" vertices.
16. Go to Step 7.

Fig. 6. Basic steps in the tracking program.

can handle a nontrivial class of problems that frequently arise.

B. Polygon Matching

The following procedure for matching a polygon P in S_k with a corresponding polygon P' in S_{k+1} is a basic component of the tracking program. It begins by trying to pair a vertex V on the external boundary of P with external vertices of polygons in S_{k+1}. The vertex V selected for matching is not crucially important, but two criteria are used to screen out poor choices. To avoid troubles with false vertices that might appear, disappear, or change greatly in angle, a vertex is automatically excluded if its angle is obtuse. To avoid troubles with multiple matches, a vertex is excluded if its angle is within δ_1 of some other angle on the external boundary of P. If no vertex meets this latter criterion, as would happen with a square, then a repeated angle will be selected for matching, but this is assumed not to be a typical situation. If more than one vertex passes these tests, the first one encountered is used.

Associated with the vertex V is an estimated velocity vector $\boldsymbol{v}$, which initially is zero. The routine uses this estimated velocity to predict the location of V in S_{k+1}. In addition, it uses a radius of uncertainty Δr chosen with due consideration of velocities, time interval between successive scenes and the dimensions of the scene. The list of vertices of the external components of polygons in S_{k+1} is then searched to find a list of candidate vertices satisfying two conditions.

1) The location of a candidate must be within the radius of uncertainty of the predicted position of V.
2) The angle of a candidate must be within $\delta_2 (= \delta_1/2)$ of the angle at V. The parameters δ_1 and δ_2 are chosen with due consideration of noise.

If no vertex in S_{k+1} passes these tests, another vertex on P is chosen and the process is repeated until either a match is made or P is declared to be unmatchable. In this latter case a topological change is presumed to have occurred, and a special procedure for analyzing topological changes, described in Section IV-D, is invoked.

If one or more candidates are found, the best matching one is tentatively selected and a criterion function is computed that measures the angular and distance errors encountered in matching all of P with P'. If the resulting mismatch is below some arbitrary threshold, P is declared to be matched to P', P is removed from S_k, P' is removed from S_{k+1}, the model M_k is updated as described below, and the next polygon in S_k can be matched against the remaining polygons in S_{k+1}. However, if the mismatch is too great, another candidate is selected and the process is repeated until either a match is made or P is declared to be unmatchable. This latter case usually results in an uncorrectable error, although it sometimes can be handled by the section of the program that handles topological changes.

The success of this suboptimal, sequential matching approach depends to a large extent on the fact that if fairly good velocity estimates can be obtained then it is quite unlikely that an incorrect match with a nearby polygon will yield small errors. In other words, it is implicitly assumed that the scenes contain a fairly heterogeneous collection of polygons rather than a field of virtually identical figures. If the polygons are quite similar and if they move appreciably between successive scenes, then this approach places quite a burden on obtaining good initial velocity estimates. Such estimates might be obtained by using a more exhaustive or perhaps an "optimal" matching procedure whenever new figures appear and good initial velocity estimates must be obtained.[2] Another approach is to eliminate the various thresholds and to use a tree searching procedure rather than a sequential procedure during matching, allowing backtracking and rematching whenever large mismatches are encountered. We have chosen this faster and simpler approach because we were more concerned with the problem of topological changes than the problem of optimal matching.

C. Updating the Model

A polygon P in scene S_k may, of course, be the union

[2] This approach was used in the radar echo tracking program of Blackmer, Duda, and Reboh [8].

of two or more component polygons. After P has been matched with P' in S_{k+1}, changes in angles and edge lengths provide evidence about the existence of these components. The program stores these components, to the extent that they are known, in a model. In our present program, each component of P can itself be split into further components whenever angular changes at pairs of obtuse vertices are significantly greater than the changes expected due to noise (see Fig. 7). No provisions are currently made for rejoining components that might previously have been erroneously separated. The rationale here is that our chief concern is to avoid the meaningless results that are obtained when a velocity estimate is computed by averaging the different velocities of different components of a polygon. If a single polygon is occasionally represented by several components in our model, the resulting loss in accuracy in the velocity estimates is not considered to be as serious a problem.

After a polygon has been separated into components, the estimates of the velocity for each vertex are updated. In addition, the angular velocities for the edges of each component are averaged and used to update the angular velocity estimate for the whole component. This information is then available for clustering to determine the number of layers, and possibly to reunite incorrectly separated components, although neither of these operations has yet been programmed.

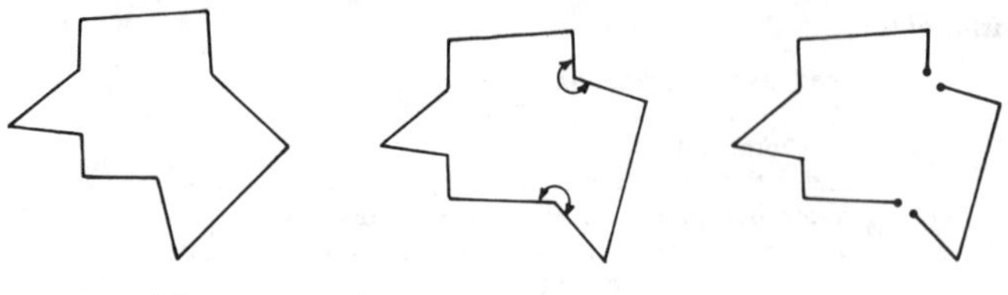

Fig. 7. Detecting components for the model. (a) Polygon at time k. (b) Obtuse angles that change significantly. (c) The two model components.

D. Topological Changes

When old vertices disappear or new vertices appear, the matching procedure described above fails and leaves one or more polygons in S_k unmatched. As was mentioned earlier, our program tacitly assumes that this is a fairly rare situation. To be more specific, we assume that the time interval between successive scenes is sufficiently short that no more than two components are involved in any topological change.

Even with this restriction there are a surprisingly large number of situations that can arise. Consider, for example, the situation in which a vertex, each of whose incident edges are visible, passes under the edge of another component and becomes hidden. The case that first comes to mind is when the two components are initially separated and then come together, as shown in Fig. 8(a). This case is obviously characterized by the fact that the number of figures in the scene is different at the two times. However, there are several other possibilities. Fig. 8(b) and (c) show that if the components overlap initially then the penetration of the vertex will create a hole. In Fig. 8(b) the vertex is initially on an external boundary component, while in Fig. 8(c) it is initially on an internal boundary component. Note, however, that we have tacitly assumed that the vertex is acute. If it is obtuse, then the components *must* overlap initially, and although no new holes are ever formed, a small old hole can disappear. In Fig. 8(d) the vertex is a true vertex on an external boundary component. In Fig. 8(e) it is a false

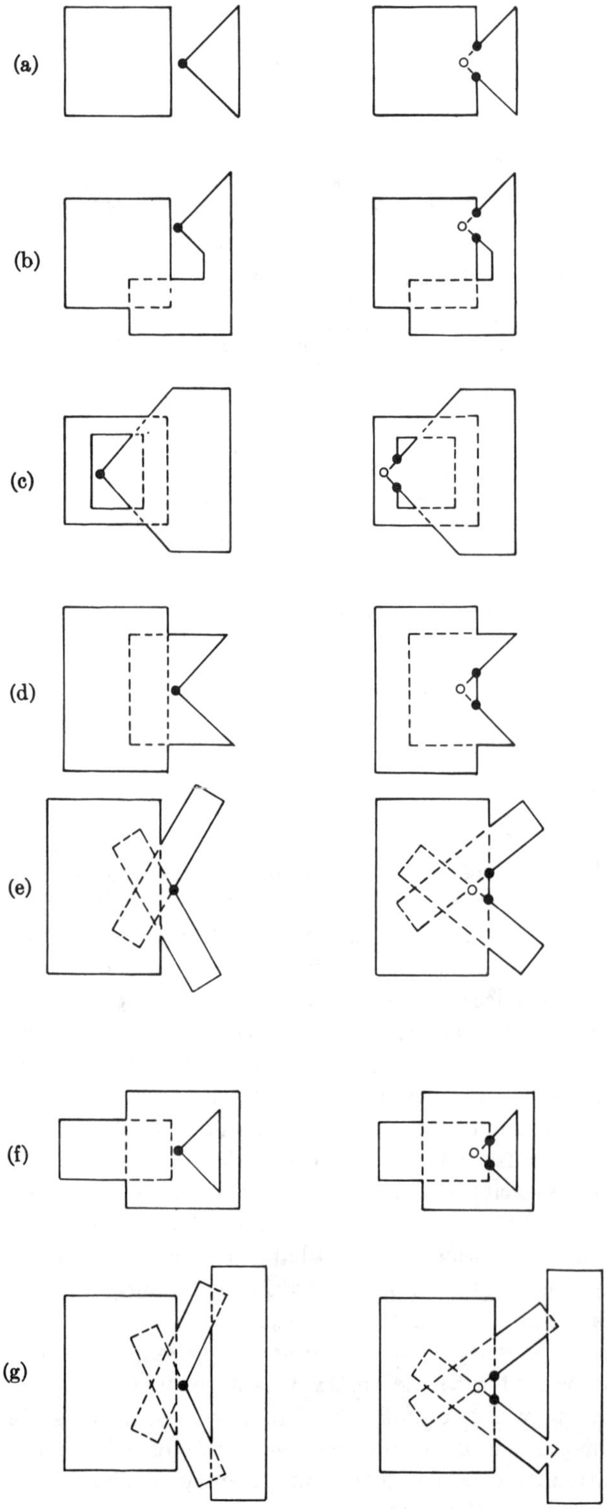

Fig. 8. Cases in which a vertex, both of whose edges are visible, becomes hidden.

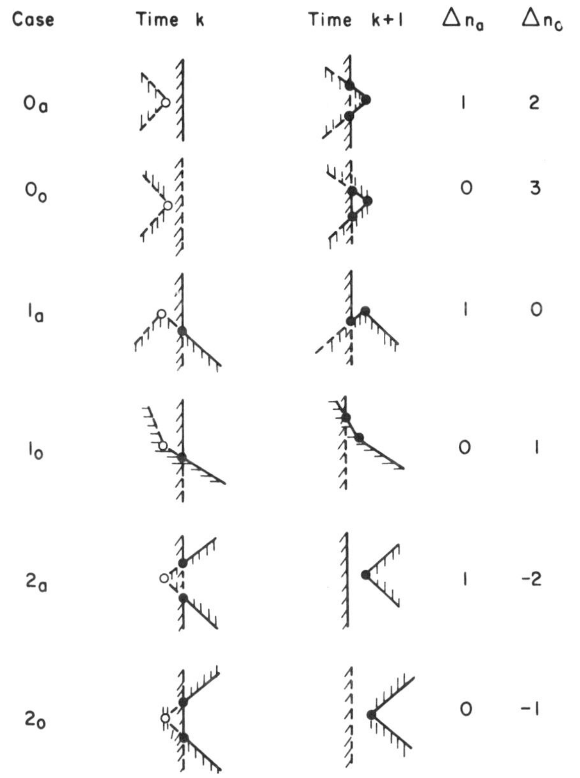

Fig. 9. Case analysis for topological changes.

vertex on an external boundary component. The two remaining situations arise when the vertex is on an internal boundary component, as illustrated in Fig. 8(f) and (g).

The various cases that can arise can be systematically classified using the following criteria.

1) Whether the number of visible edges incident on the hidden vertex is 0, 1, or 2.
2) Whether the vertex angle is acute or obtuse.
3) Whether the vertex was initially hidden and becomes visible, or initially visible and becomes hidden.
4) Whether the vertex is true or false.
5) Whether the incident edges are on an external or an internal boundary component.

The first criterion—the number of visible incident edges—provides a simple basic classification. Let case i ($i = 0,1,2$) denote the situation in which i incident edges are visible. These three cases can be further subdivided according to the second criterion—whether the vertex angle is acute or obtuse. We denote these subcases by case ia and case io, respectively. These six cases are illustrated in Fig. 9. Note that case $0o$ is rather special, since a convex vertex with no visible incident edges occurs only just before a hole is exposed.

Additional subcases can be distinguished on the basis of the remaining three criteria. Some of these distinctions are more important than others. For example, the time sequence is certainly important. In Fig. 9 the vertex is shown as initially hidden, but Fig. 9 obviously illustrates the opposite sequence equally well, and in effect shows twelve cases. Knowledge of whether the vertex is true or false can be quite helpful at times, but this information is not always available, and whenever possible we have tried to avoid depending upon it. Finally, whether the incident edges are on external or internal boundary component is primarily helpful in keeping the bookkeeping straight in case 2.

Fig. 9 also shows the change in the number of acute and obtuse angles, indicated by Δn_a and Δn_o, respectively. The numbers shown assume that the vertex in question was initially hidden; for the opposite sequence the algebraic signs should be reversed. Examination of Fig. 9 discloses that all but two of the twelve resulting cases can be distinguished by these gross vertex counts. The two that cannot be distinguished are case $1o$, in the sequence shown, and case $2o$, in the opposite sequence, since $\Delta n_a = 0$ and $\Delta n_o = 1$ in each instance. These two cases in fact cannot be distinguished without knowing which vertices are true and which are false, and to establish this may require waiting several additional time periods, if it can be done at all.

There are two reasons for wanting to distinguish between these various cases. One is to prevent the matching program from becoming confused by the topological changes in matching the remaining vertices. The other is to allow the model updating program to make the proper adjustments to the model. Fortunately, the inability to distinguish the two cases mentioned above does not effect the matching procedure. However, it does effect the updating procedure, forcing a postponement of the decision as to which model component receives the new vertex.

E. Matching and Updating When Topological Changes Occur

In this section we describe briefly the procedure used to match vertices when a topological change is known to have occurred. The first step is to count the changes Δn_a and Δn_o in the number of acute and obtuse vertices to determine which of the ten distinguishable cases has occurred.

Consider first case $0a$ (see Fig. 9). This case is characterized by the appearance of a new acute vertex having two short incident edges. The two other new vertices terminating those edges must be obtuse and close together, and the two other edges incident on those vertices must be collinear. These properties are used by a routine that searches the boundary components to locate these three new vertices. Once they are located, they are tagged so that the matching program will skip over them as it matches corresponding vertices at the two times.

This general strategy is followed in all of the other cases. The vertices that change are characterized, and the figures are searched to locate the new vertices. Once located, the vertices are tagged so that they will not interfere with the matching program, which can then proceed essentially as if no topological changes had occurred. Case 2 does present some additional difficulties, since corresponding vertices at times k and $k + 1$ can sometimes lie on different kinds of boundary components, external at one time and internal at another. Fig. 8 illustrates the various ways that this can occur. However, the difficulties essentially involve care in bookkeeping, and present no fundamentally new problems.

V. EXPERIMENTAL RESULTS

The tracking program has been run on a large number of different problems to verify that the many different cases involving topological changes are properly treated. The following two examples are relatively simple cases that are intended to illustrate the kinds of problems that the tracking program can handle, rather than to evaluate the effectiveness of the heuristic parts of the procedure.

The first example is shown in Fig. 10. The objects in the scene are a large triangle that initially completely hides a smaller triangle containing a triangular hole. Thus, this second triangle is not visible in scenes S_0 or S_1, and the initial model, M_1, is just the larger triangle. In S_3 a vertex of the obscured triangle appears (case 0*a*) and the model M_3 shows that this triangle is treated as a separate component of the scene. Note that when a new vertex appears in this way, the associated obtuse vertices are known to be false without the need for computing angular velocities.

At time 5 a vertex of the hole appears (case 0*o*). Although in this instance one can logically conclude that the two edges incident on that vertex must be associated with the partly exposed triangle, this association is not made in the model. This is one of many special instances in which logical inferences could be made to achieve better performance, but at a cost of greater complexity.

At time 8 a second external vertex of the hidden triangle appears (case 1*a*), at time 9 a second internal vertex appears (case 1*o*), at time 11 the third internal vertex appears (case 2*o*), and at time 17 the third external vertex appears (case 2*a*). Thus, this simple example involves all six of the cases listed in Fig. 9. Every one of these events resulted in a matching failure that had to be corrected by topological analysis. The final model, M_{17}, looks just like the corresponding scene, but of course the existence of two distinct components in the scene was known from time 3. It is this ability of the tracking program to decompose the scene that is its most important attribute.

The second example is shown in Fig. 11. It consists of three figures, two as shown in scene S_0, and a triangle with a triangular hole hidden completely by the polygon on the right hand side. In the following sequence of scenes, the triangle moves from its initial position to a position where it is completely hidden by the polygon on the left hand side. The triangle is never visible in its entirety. However, the program correctly constructs the description of the triangle together with that of the hole. In scene S_3 a vertex of the hidden triangle becomes visible, in scene S_6 the vertex of the hole becomes visible, in scene S_8 one of the vertices of the triangle gets hidden by the left hand polygon. These facts are correctly displayed in M_8. Eventually, in scene S_{19}, the triangle is completely hidden by the left hand polygon, and the program has generated the correct description of the triangle, as shown in model M_{19}. The ability of the program to generate the complete description of objects, from its partial descriptions, is another important feature of this program.

The examples above are relatively simple in terms of the number of moving figures. However, the program is applicable to scenes containing an arbitrary number of figures and has been successfully run on fairly complex scenes.

VI. DISCUSSION

We have presented an idealized model of the problem of analyzing pictures of overlapping layers of essentially two-dimensional objects. This model is a natural one for investigating the problem of tracking the motions of several layers of clouds, and, as was indicated in the Introduction, undoubtedly has other applications as well. However, like any model of real phenomena, this one focuses on certain aspects of the situation and ignores others. Here the emphasis was on the behavior of joint motions, shape changes, and false vertex motions that accompany images of moving occluding objects.

With our model, this was virtually the only kind of information available for decomposing the scene and determining the velocities. In most real applications, of course, much additional information would be available. For example, in the cloud tracking problem there are measurable differences in brightness, boundary shape, and texture between clouds in different layers. In principle, at least, such additional information should simplify the problem of decomposing the scene. In the future we hope to address the question of how such different types of information can in fact be jointly exploited.

Even within the limitations of the current model a great variety of problems arise. The model allows both very simple and arbitrarily complex scenes to be generated. This makes it difficult to provide an objective evaluation of any analysis program, for one can always generate scenes that are within the ability of the program, or that can overwhelm the program in a variety of ways.

In designing the program described in Section IV we chose to address some questions and to ignore others. Specifically, we were primarily concerned with seeing what was required to match a relatively small number of overlapping figures when only "small" changes occurred between successive scenes. The case analysis given in Section IV shows that even this restricted problem can involve a surprising variety of situations, but that it does yield to a systematic approach.

Some of the restrictions on the present program could be relaxed with relatively little effort. In particular, the restriction that no more than two components be involved in any topological change is not essential to our approach. This restriction was imposed merely to allow the case analysis to be decided by gross vertex counts. If several topological changes were allowed between successive

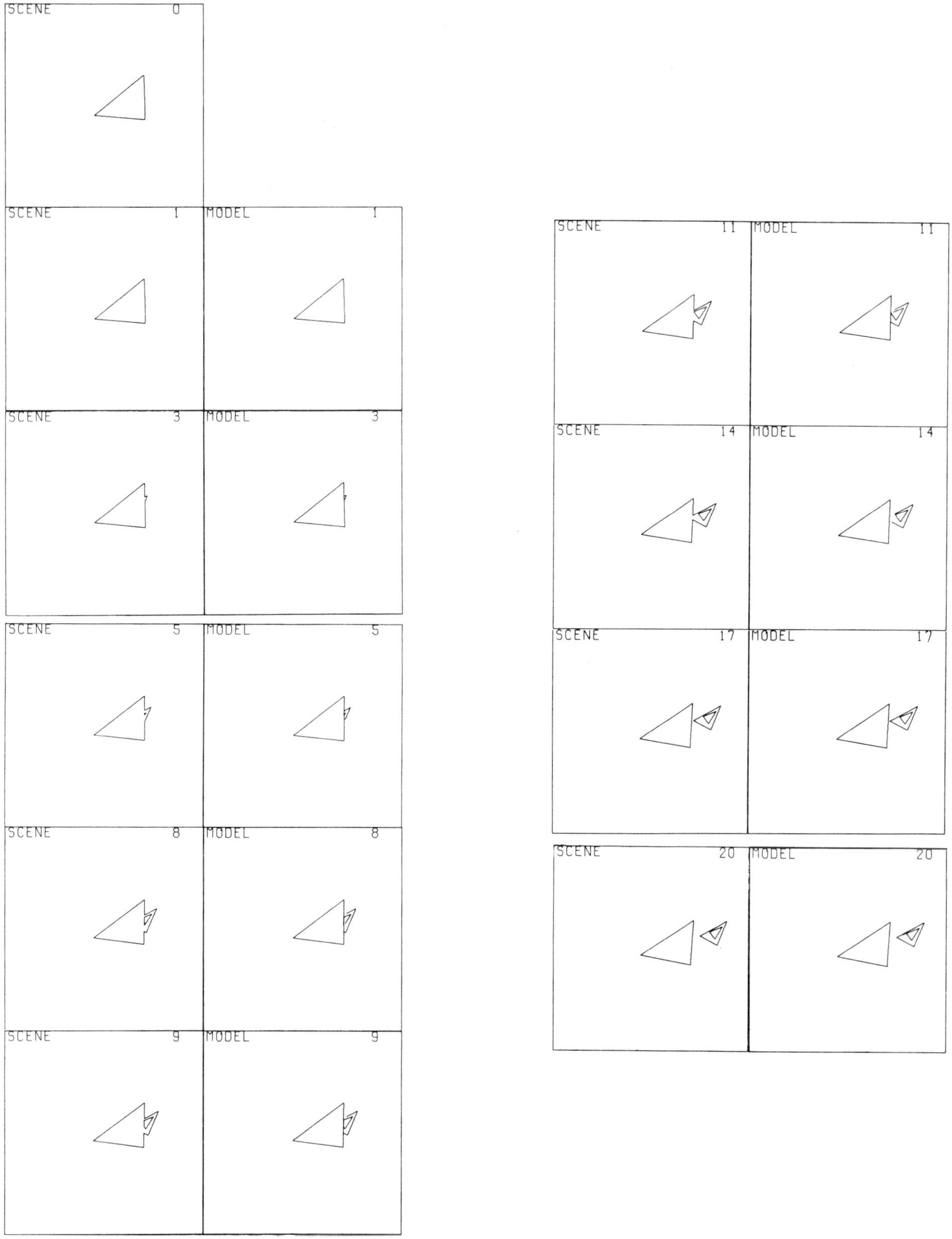

Fig. 10. Tracking two moving, overlapping triangles.

scenes, more elaborate methods would have to be used to localize the troublesome areas, but the procedure would be otherwise essentially unchanged. It would be more difficult to change the program to cope with large numbers of polygons, nonrigid polygons, or polygons whose number of sides and/or components could change from time to time. All of these problems present interesting and challenging topics for further study.

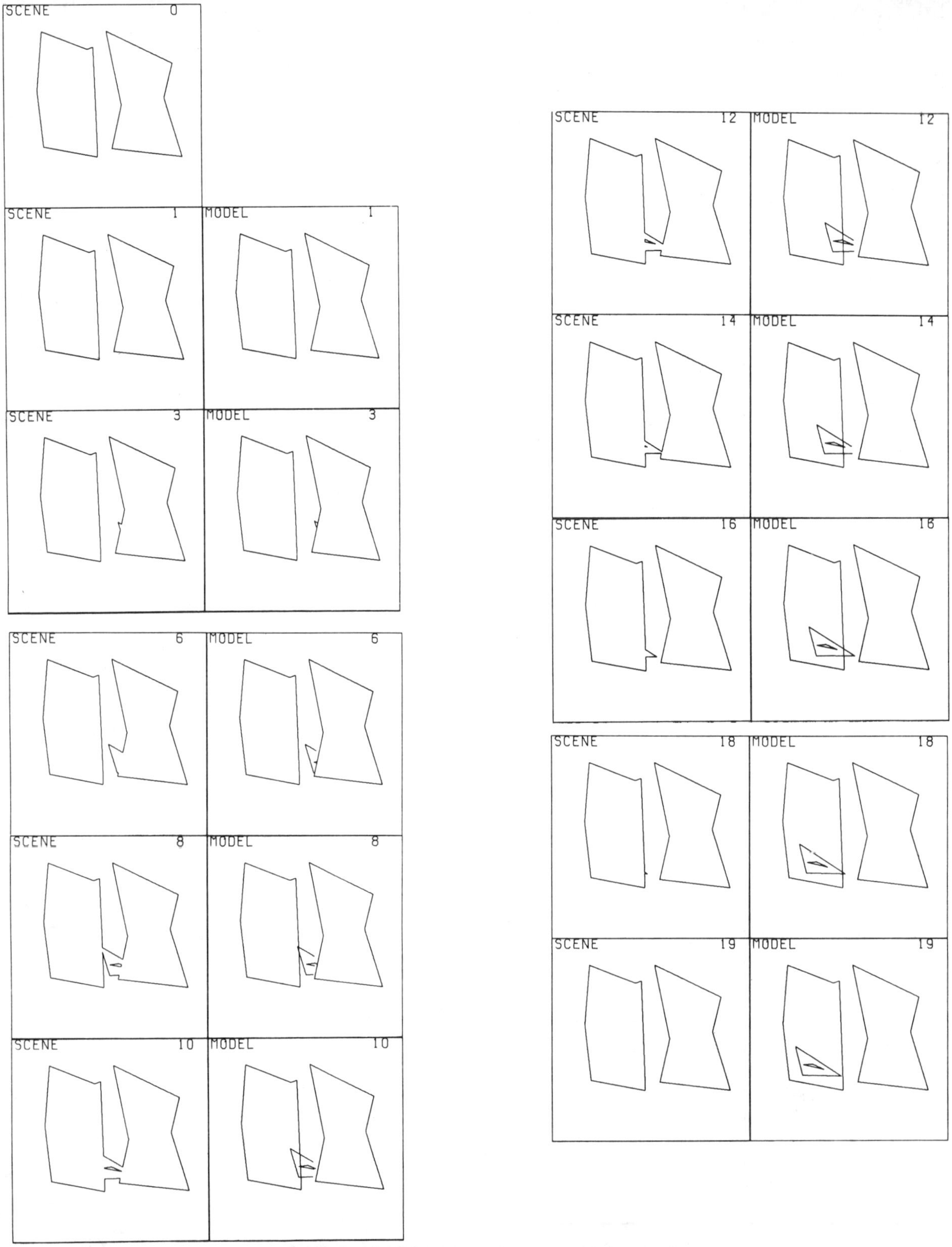

Fig. 11. Tracking three moving figures.

ACKNOWLEDGMENT

The authors would like to acknowledge the valuable assistance of R. Petermann, who wrote the analysis program described in Section IV and who contributed significantly to its final design. We would also like to thank C.-C. Lin, who devised and implemented the algorithm for generating the data used to test the analysis program, and Dr. S. A. Underwood for his help during this research.

REFERENCES

[1] M. S. Ulstad, "An algorithm for estimating small scale differences between two digital images," *Pattern Recognition*, vol. 5, pp. 323–333, Dec. 1973.

[2] J. A. Leese, C. S. Novak, and V. R. Taylor, "The determination of cloud motion patterns from geosynchronous satellite image data," *Pattern Recognition*, vol. 2, pp. 272–292, Dec. 1970.

[3] E. A. Smith and D. R. Phillips, "Automated cloud tracking using precisely aligned digital ATS pictures," *IEEE Trans. Comput.*, vol. C-21, pp. 715–729, July 1972.

[4] D. J. Hall, R. M. Endlich, D. E. Wolf, and A. E. Brain, "Objective methods for registering landmarks and determining cloud motions from satellite data," *IEEE Trans. Comput.* (Short Notes), vol. C-21, pp. 768–776, July 1972.

[5] J. L. Potter, "Motion as a cue to segmentation," in *Proc. Milwaukee Symp. Automatic Control*, Milwaukee, Wis., Mar. 1974, pp. 100–104.

[6] N. Badler, "Three-dimensional motion from two-dimensional picture sequences," in *Proc. 2nd Int. Joint Conf. Pattern Recognition*, Aug. 1974, pp. 157–161.

[7] R. A. Petermann, "Computer analysis of planar motion of polygons," M.S. thesis, Dep. Elec. Eng., Univ. Texas, Austin, Tex., Jan. 1975.

[8] R. H. Blackmer, Jr., R. O. Duda, and R. Reboh, "Application of pattern recognition techniques to digitized weather radar data," Stanford Res. Inst., Menlo Park, Calif., Final Rep., Contract 1-36072, SRI Project 1287, May 1973.

Part VI
Scene Analysis: The Blocks World

In general terms, scene analysis is concerned with extracting from an image a description of a scene, including the objects present and the relations among them. The goal of much scene analysis research has been machine perception—the interpretation of a two-dimensional image in terms of a configuration of three-dimensional objects. Since the projection from a three-dimensional world to a two-dimensional image is many-to-one, such an interpretation is an extrapolation that either implicitly or explicitly makes use of additional information about the world. The key questions concern the kind and extent of world knowledge that is needed, how this knowledge is represented, and how it is used.

A great many factors influence the visual appearance of three-dimensional objects, such as illumination, shadows, surface reflectance, geometrical shape, viewing angles, and occlusion by intervening objects or different parts of the same object. To simplify the problem, initial investigations were done in a world of diffusely illuminated polyhedral blocks having uniform surface reflectance—the blocks world. Images of piles of blocks are naturally represented as line drawings, and much of the work in blocks world scene analysis has concerned the interpretation of line drawings.

The first paper, by Roberts, was one of the first and most influential papers on machine perception. That paper describes a computer program whose input was a digital picture and whose output was a composition of three-dimensional models that "explained" that picture. The basic models were a cube, a wedge, and a hexagonal prism. These models could be scaled and freely combined to fit a wide variety of polyhedra.

To "fit" these models to the picture, the program executed a sequence of edge detection and line-fitting procedures that had to produce a topologically flawless line drawing. Heuristics were used to establish a correspondence between nodes in this line drawing and vertices in the model. In general, the mapping between a model and its image is a projective transformation involving parameters for scaling, translation and rotation. [For an alternative and illuminating approach to interpreting projected views of polyhedra, see Mackworth (1973).] By employing homogeneous coordinates, Roberts was able to find choices for these parameters that gave a minimum-squared-error fit. By repeatedly fitting and removing models, the program produced the desired three-dimensional interpretation.

One of the steps in this process that is the most difficult to accomplish is the production of the topologically perfect line drawing. Several methods for using high-level knowledge to aid this process have been investigated. Assuming that the predominant problem is caused by missing lines, Falk (1972) used a set of eight fixed-size models and an extension of the heuristics described by Guzmán in the second paper to create a system that could function with imperfect line drawings. Griffith (1973) used heuristics for hypothesizing possible missing lines, whose presence or absence could be verified by a more sophisticated line detector. Grape (1973) used perspectively consistent two-dimensional models of edge structures for convex objects to organize edge information and to complete missing lines, including occluded edges. Like Griffith, Shirai (1973) used heuristics to hypothesize lines. However, Shirai intermixed the edge-finding and line-fitting procedures constantly, using previous results to guide the search for more edges. [This kind of interaction between high-level and low-level procedures is often referred to as heterarchical, as opposed to fixed-sequence, or hierarchical, control. See also Shirai (1975).]

Even with the aid of homogeneous coordinates, the successive fitting and removal of models is a complex and time-consuming process. In the second paper, Guzmán showed that quite complicated blocks world line drawings could be analyzed without the explicit use of any models. That paper describes a program whose input was a perfect line drawing and whose output was a list of all of the objects found and the regions that correspond to faces of those objects. This is a less complete analysis than that of Roberts, but it provides a three-dimensional interpretation by simple heuristic methods. The procedure is based on scanning the input for local clues (*T* junctions usually mean occlusion; *Y* junctions usually mean the intersection of three touching surfaces) and then using a clustering procedure to link these clues together.

The simplicity of Guzmán's procedure made it attractive as a component of larger systems. (For example, Winston (1975) used it in his program for learning scene descriptions from examples; see Underwood and Coates (1975) for a program that learned convex object descriptions from multiple views.) However, Guzmán's heuristics did not always work, and attempts to correct them with more elaborate rules weakened the virtue of simplicity.

By coincidence, Clowes (1971) and Huffman, in the last paper in this part, simultaneously arrived at virtually the same approach to clarifying the situation. Their approach was to view each line in the drawing as representing one of three kinds of edges of trihedral solids—a convex edge, a concave edge, or an occluding edge. They then systematically investigated the possible ways that lines could form junctions, producing a catalog of allowed line labelings. With this catalog of local constraints, one can use a tree-searching procedure to examine the entire drawing and find all of the possible edge interpretations. (Huffman was particularly interested in those cases in which there was no consistent three-dimensional interpretation, as happens with drawings of "impossible" objects.)

This systematic approach has since been extended significantly by Waltz (1975), who relaxed the restriction to trihedral solids and introduced the complication of shadow lines. One result was an enormous expansion of the catalog of legal line configurations. Another result was the discovery that the constraints between adjacent vertices were so powerful that in many cases a preliminary constraint satisfaction filtering eliminated any need for subsequent tree searching. These results have had consequences beyond the blocks world, some of which are reflected in the last two papers in Part VII.

MACHINE PERCEPTION OF THREE-DIMENSIONAL SOLIDS *

L. G. Roberts

Massachusetts Institute of Technology, Lincoln Laboratory, Lexington, Massachusetts

I. Introduction

The problem of machine recognition of pictorial data has long been a challenging goal, but has seldom been attempted with anything more complex than alphabetic characters. Many people have felt that research on character recognition would be a first step, leading the way to a more general pattern recognition system. However, the multitudinous attempts at character recognition, including my own, have not led very far. The reason, I feel, is that the study of abstract, two-dimensional forms leads us away from, not toward, the techniques necessary for the recognition of three-dimensional objects. The perception of solid objects is a process which can be based on the properties of three-dimensional transformations and the laws of nature. By carefully utilizing these properties, a procedure has been developed which not only identifies objects, but also determines their orientation and position in space.

Three main processes have been developed and programed in this report. The input process produces a line drawing from a photograph. Then the three-dimensional construction program produces a three-dimensional object list from the line drawing. When this is completed, the three-dimensional display program can produce a two-dimensional projection of the objects from any point of view. Of these processes, the input program is the most restrictive, whereas the two-dimensional to three-dimensional and three-dimensional to two-dimensional programs are capable of handling almost any array of planar-surfaced objects.

* This report is based on a thesis of the same title submitted to the Department of Electrical Engineering at the Massachusetts Institute of Technology on 10 May 1963, in partial fulfillment of the requirements for the degree of Doctor of Philosophy.

In order to implement the three-dimensional processing of pictures, perspective effects must be considered. For this reason, a four-dimensional, homogeneous system of coordinates will be used. In this system a single 4×4 matrix can modify a position vector by a linear transform, a translation, and a perspective transformation. Although many books discuss this homogeneous system of coordinates, their presentations are either incomplete or too involved for our purposes.[1] Therefore, the system is explained in Appendix A. Without the notational simplicity provided by using homogeneous transformations, most of the following analysis would not have been accomplished.

II. Background

A. *Pattern Recognition by Computer*

There have been numerous attempts to recognize simple patterns by machine. There is the work with neuronlike nets of threshold elements which divide the set of all input patterns into a number of classes by correlating a set of adaptive weights with some functions of multiple input cells. For this type of system there may not be many output classes and the transformations of the patterns must be minimal or nonexistent. Because of these restrictions, the patterns worked on so far have been those which, although complex, are not subject to much transformation such as characters and spoken digits. My paper on character recognition is typical and gives the other references.[2] This type of system would be of no value for multiple object recognition, except perhaps for finding the lines originally.

In the work by Selfridge and Neisser *et al.*,[3] a more useful set of tests is made on the input, whereas the output processing is similar to that of the system described earlier. That is, computation routines are developed to extract the useful information from the input, and their outputs are weighted to determine the most likely output class. Here again a small set of outputs is expected and characters were the patterns tested. One problem with both these methods is that they were intended for specific groups of abstract patterns, such as characters, and not for the well-defined geometry of photographs. They are better suited for looking at my resultant list structure of objects and deciding whether a group of objects is a chair or a table.

The closest that any researcher has come to the problem I propose is Leo Hodes in his work on processing line drawings.[4] His main result was to produce a list of lines and vertices by following out lines. Then he suggested a few simple tests which might be made on this list to find triangles, etc. Although his main purpose was to study abstract line patterns, he did describe a working line follower which was of value to me in this part of my effort.

The only work I know of on machine depth perception is that on binocular images. Julesz has reported a procedure which shifts the binocular pictures to find the areas at different depths.[5] This procedure uses only texture, not edges, to develop the depth information and shows that the binocular information alone is sufficient for depth perception. This work is similar in goal but completely different in procedure from mine. Other work in machine photograph processing has mainly been in the field of information reduction for bandwidth compression and my paper in this area summarizes this work.[6]

B. Psychophysical Theory

There has been a large volume of psychophysical research on human depth perception and shape recognition. From all this I have tried to isolate the ideas and theories which are used to explain our monocular perception of a three-dimensional world. It will be apparent, however, that the work of Gibson is dominant in my mind, since his book is both clear and complete.[7]

Of all the monocular depth cues perhaps the most written about is that of known object size. Ittelson reports experiments in which only one object could be seen, thus eliminating other depth cues.[8] Although the assumed size of objects such as playing cards tended to vary, the subjects would judge the depth reasonably well for normal-sized cards and proportionately shorter for jumbo cards. Thus he, for one, showed that the size of familiar objects is a good relative depth cue and fair for absolute depth. Gibson points out, however, that this type of distance perception is rarely used in the everyday world, since we look at arrays of objects rather than at single objects and can use more general depth cues.

Gibson's favorite cue is that of texture gradient. This is the effect of perspective on the grain or fine structure of large surfaces. As these surfaces recede, the apparent grain becomes finer. Another gradient cue is the illumination variation which puts curved surfaces in relief. This shows us the surface depth variations. The final depth cues are those of aerial perspective or blur with depth, and the angular upward position of objects toward the horizon which is a depth measure in most outdoor scenes.

Recognition of forms, shapes, and objects is often discussed from the Gestalt point of view, where shadowy forms and plane geometry figures are the forms to be recognized. Attneave and Arnoult spend many pages explaining random shape generation and the useful procedures for analyzing them.[9] They discuss contour following, differentiating pictures, and some of the simple measures of shape complexity. If they were discussing character recognition, it might be reasonable to use these tools; however, they say they are investigating "natural forms." This preoccupation with the abstract projected form is strongly attacked by Gibson. He feels that the visual world of objects and surfaces should be studied rather than the visual field on the retina. A perspective transformation does not reduce a solid object to a shadowy form. Rather, it defines the set of shapes which go with a single perception.

Perspective variations in a cube were tested by Langdon in an experiment on three-dimensional solids.[10] He found that perspective plays only a minor part in the perception of the size and depth and that the subjects always saw a cube, even when it was badly distorted by the perspective transform. The continual perception of a cube, even when transformed, is consistent with Gibson's idea that shape perception is and must be invariant under perspective transformation. My idea of models also follows from this, since each model represents an invariant percept, and can be identified with any projection of itself.

III. Depth Perception

The perception of depth in a monocular picture is based completely upon the assumptions of the observer. Some of the assumptions are about the nature of the real world and some are based on the observer's familiarity with the objects. Without these assumptions the picture is just another two-dimensional image, whereas with them the human is rarely confused about the depth relationships represented in the picture. Since humans agree so closely on their depth impressions, it is fair to assume that their major assumptions are the same, and are therefore subject to identification and analysis. The following is an attempt to set down some of the likely assumptions and derive what depth information can be obtained if they were used.

A. Transformation of Real World

The first assumption is that the picture is a view of the real world recorded by a camera or comparable device and therefore that the image is a perspective transformation of a three-dimensional field. This transformation is a projection of each point in the viewing space, toward a focal point, onto a plane. The transformation will be represented with a homogeneous, 4×4 transformation matrix P such that the points in the real world are transformed into points on the photograph (see Appendix A for an explanation of homogeneous coordinates). The transformation depends on the camera used, the enlargement printing process, and, of course, the coordinate system the real world is referred to. Let us fix the real world coordinates by assuming that the focal plane is the $x = 0$ plane and that the focal point is at $x = f$, $y = 0$, $z = 0$. In order that the picture not be a reflection, we choose the focal plane in front of the camera. Then the objects seen will be in the x half-space. Thus the focal plane is really the plane of the print, not of the negative. Figure 1 shows this arrangement.

A particular camera will have some focal distance f. We shall consider the square on the focal plane which was enlarged to create the print. The center of this square will be at some coordinates y_0, z_0, and the size of the square from the center to an edge will be some distance S. The actual size to which the square is enlarged is unimportant, since we shall measure the print with a normalized system which has the origin at the center and the

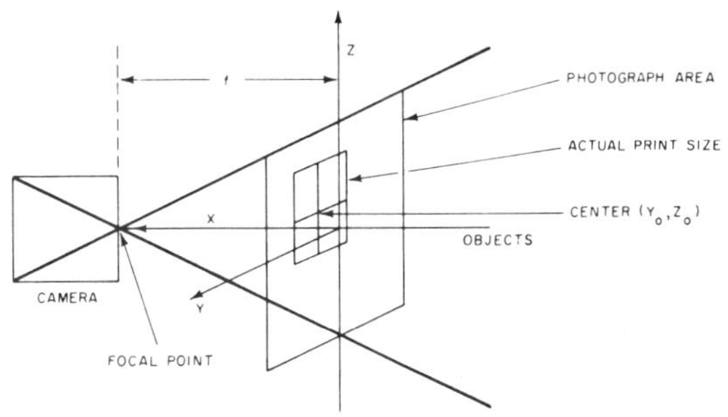

FIG. 1. Camera transformation.

edges at $y, z = \pm 1$. Normally, the whole film area is printed, and in this case $y_0 = z_0 = 0$ and S is equal to half the film size. For standard cameras without special lenses, the ratio of focal distance to film size is usually the same since this corresponds to a fixed viewing angle. Thus the ratio S/f is fairly constant at about $\frac{1}{4}$. In my case, the ratio is known if the camera is known and could be supplied with the photo.

It is not necessary to know the variables y_0, z_0, f, and S since they can be computed from the picture, given other assumptions later on. However, for the sake of simplicity we shall assume that S/f is known and that $y_0 = z_0 = 0$. The numerical values of S and f alone are not necessary, since this just affects the scale of the real world. Thus we can assume that $S = 1$ and with $r = S/f$ obtain a simple transformation P,

$$P = \begin{bmatrix} 1 & & & -r \\ & 1 & & \\ & & 1 & \\ & & & 1 \end{bmatrix}$$

If $\bar{v}$ is a point in real space, then $\bar{v}P$ is a point in a perspective space such that its Y and Z coordinates are the original point's projection in the picture plane. The X coordinate of $\bar{v}P$ is also obtained and will be useful for hidden line computation during display of three-dimensional objects.

Thus a transformation from the real world to a picture has been described, and to go the other way simply requires the inverse transformation P^{-1}. Of course, the x component of the real-world points will not be known in this case.

B. Objects Observed

We shall further assume that the three-dimensional field observed consists of a set of solid objects which occupy a definite region of space. Since we realize that it is usually possible to pick out the lines which define the boundaries of the objects and their surfaces, we shall assume that this has been accomplished and that the picture has been reduced to a line drawing. Because the objects are solid, we do not expect to see the boundaries which are hidden from the focal point by another solid.

Second, we shall assume that the objects seen could be constructed out of parts with which we are familiar. That is, either the whole object is a transformation of a preconceived model, or else it can be broken into parts that are. The models could be anything from a cube to a human body; the only requirement is that we have a complete description of the three-dimensional structure of each model.

The transformation from the model to the real world object will be a suitably restricted homogeneous transformation matrix R. We must allow an arbitrary rotation and translation of the model in order to position it properly in space. We should also like to allow three degrees of freedom for size change of the model so that a cube model can represent any parallelepiped. So far we have allowed nine degrees of freedom. The 4×4 matrix R can allow fifteen degrees of freedom since it has 16 elements and the total scale of the matrix is arbitrary in the homogeneous coordinate system. The last six degrees of freedom represent skew and perspective deformations. Skew deformations are size changes in the x, y, and z directions after the model has been rotated and will change the sides of a cube to parallelograms. A perspective deformation is most easily visualized as a compression of one end of the model. Objects that have been deformed in either way are not usually considered to be simple instances of the model. Furthermore, objects deformed in these ways could be constructed from smaller parts, so it is not necessary to allow skew and perspective deformations.

We cannot allow perspective deformation and still obtain a unique transform R from the picture; therefore, we require the top three elements in the last column of R to be zero. Skew variations can be allowed if we maintain very high accuracy in our computations, so our derivation will allow them, but later on they will be eliminated.

Now R transforms a model into an object and P transforms the object onto the picture so that if

$$H = RP$$

H transforms the model points into picture points. Therefore, in order to identify a group of points and lines in the picture with a particular model, we must find out if there is any transformation H which will take the model's points and lines into those of the picture. If such a model and transform are found, it can be said that the object represented in the picture could be that model under the transformation $R = HP^{-1}$.

C. Model Identification

Let us say that we are given a picture of a parallelepiped, and it has been reduced to a line drawing. We can then find the interior polygons which correspond to the surfaces of the object. There will normally be three quadrilaterals visible. These polygons all come together at one point which can be used for a reference point. If we look through our list of models, we find that a cube and perhaps other models have three quadrilaterals about one point. Therefore, we can pick a point in the cube model which has the proper polygons around it, pick a polygon from both the cube and the picture as starting points, and proceed to list topologically equivalent point pairs. When we have finished, we have a list of seven three-dimensional points from the model and a corresponding list of seven two-dimensional points from the picture. By adding a homogeneous coordinate $w = 1$ to each point vector, we obtain a 4×7 matrix of model points A and a 3×7 matrix of picture points B.

Now by means of the similarity test derived in Appendix B, we obtain the best transform H which will take A into B. We also obtain a mean-square error which indicates whether or not the model chosen really can fit the picture. We can then choose the model that causes the least error. For the parallelepiped, the cube model should fit with very little error. The transform obtained is a 3×4, since no depth data accompanied the picture points. Since we know P and thus P^{-1}, we can start to obtain R, the real space transform of the model. Since we have required that the three perspective components of R should be zero, we can specify the top three components of R in the first column as $(-1/r)$ times the corresponding elements of the last column of H.

$$R = HP^{-1}$$

$$H = \begin{bmatrix} y_1 & z_1 & w_1 \\ y_2 & z_2 & w_2 \\ y_3 & z_3 & w_3 \\ y_4 & z_4 & w_4 \end{bmatrix} \qquad P^{-1} = \begin{bmatrix} 1 & & & r \\ & 1 & & \\ & & 1 & \\ & & & 1 \end{bmatrix}$$

$$R_0 = \begin{bmatrix} -w_1/r & y_1 & z_1 & 0 \\ -w_2/r & y_2 & z_2 & 0 \\ -w_3/r & y_3 & z_3 & 0 \\ 0 & y_4 & z_4 & w_4 \end{bmatrix}$$

The lower left element of R still is not known since it is the x position of the whole object, and if an object grows in size as it moves away from us, it can maintain the same projection on the focal plane. Thus this depth value must be found some other way. For the present, we can call it zero and label the matrix R_0.

At this point, by assuming that an object in a picture is a transformation of a known model and by utilizing our knowledge of perspective geometry, we have been able to find the model and transformation which best represent the object. We know the precise orientation and position of the object except for one depth variable. We also know all the dimensions of the object relative to its total size. We also should know the skew deformation of the object, since we have obtained eleven variables. However, compression or expansion in the x direction produces only a slight change in the picture due to the perspective, since we are looking along the x axis. Thus the determination of the x skew can easily be in error. This problem of skew error in derived transformations was not realized until the computer program implementing these ideas began to produce distorted transformations of models. Even though the program could match every point in the picture with model points within one part in 4000, the x skew of the transformation might be off by a factor of 2. Thus it is clear that the x skew is not really obtainable from a picture, even though it can be derived mathematically.

If x skew deformations of the model must be restricted, it is logical to eliminate all skew variations for consistency. To restrict skew it is necessary to ask that the top three rows of R be orthogonal. Since the introduction of this requirement into the similarity test requires solving second-order equations, it is simplest to modify R after it has been computed. Thus the top three rows of R are forced to be orthogonal by modifying the first column, which is the one with the worst error. This means that the focal ratio r and w_1, w_2, and w_3 are not needed for the computation of the model transformation. Now in fact, r may be computed from the ratio of the new first column to the w_j.

Thus, by assuming that the objects seen in a picture are nondeformed transformations of known models, we can find the model and transformation without knowing the camera characteristics. Even if the picture is an orthogonal projection, as is almost the case with long telephoto lenses, we can compute the proper transformation. This would be impossible if we did not eliminate skew since r, w_1, w_2, and w_3 would all be zero. Thus the process accounts for, but does not depend on, perspective information.

The information required to obtain a transformation is obtained from the points in the line drawing. These points have two dimensions each and we need to determine eight degrees of freedom in the transformation. Therefore, at least four points from the picture must be used. These points cannot all lie in the same plane of the object or the equations will be degenerate. If more than four points are available, the mean-square error will indicate whether they are consistent with the model, and therefore help in the selection of the proper model.

D. Depth Information

After the matrix R_0 is obtained for an object or object part, it is still necessary to obtain the x translation or depth. Here we must resort to another assumption. There are several possibilities which I have chosen to ignore. If we were to assume that we know the models so well that we knew their size, this would fix their depth also; or, if we wished to interpret shadows, this might determine the depth if shadows existed. The fact that one object is partly in front of another supplies depth information, but only in the form of inequalities. Lastly, the various gradients—intensity, blur, and texture—might be useful for determining the depth gradients of each surface, but this information has already been found through the use of models. All these cues may be useful to humans, but each one is restricted in its generality and only useful in special cases.

The one depth perception concept which is suitably general, and sufficiently accurate to position all objects properly, is the use of a support theorem. We assume that each object must be supported somehow, either by another object or by a ground plane. This assumption allows us to project each object back in the x direction until it hits the ground plane or another object. While it is being projected back, it must be expanded so as to maintain the same image on the focal plane. The slope of the ground plane can be determined by examining each object for parallel planes, choosing a plane which goes under the focal point and is as parallel to the $z = 0$ plane as possible. When such a plane is found, we know only its slope, not its distance, from the origin. However, this single variable can be set arbitrarily, since it affects only the total scale of the picture. Actually it can be guessed rather accurately for the majority of pictures just by assuming or knowing the distance the camera was held from the ground and the focal ratio r. Since r may be computed, we could assume that the camera was held five feet above the floor, and now we can state the dimensions of each object in feet.

For compound objects, we know the pieces should fit together, so their relative depths are determined and the compound object can then be treated as one object and projected onto the ground plane. The whole procedure is relatively simple so long as the ground plane is really planar. If the ground curves, this could be taken care of by computing the curvature from the slopes of several objects. If there are breaks in the floor's slope such as walls, the breaks will be seen as lines and walls treated as objects. Thus the support assumption enables us to properly place all the objects in space.

To review the depth perception assumptions and results, we assumed that the picture was, in fact, a perspective view of the real world, that the objects shown in the picture could be described by means of one or more transformations of known models, and that all objects were supported by others or by a ground plane. The transformations allowed were restricted to rotation, translation, and size changes. Then, from a single picture, each object which has four or more points showing can be described in terms of the models and positioned in a three-dimensional space. The scale of this space in feet can even be determined if the distance of the camera from the floor can be supplied. The whole representation in three dimensions should be accurate, except for a simplification of hidden details and occasional problems due to the breakdown of the assumptions. However, humans have the same problems.

IV. Picture Input and Reduction to Lines

Pictures are presently being entered into the computer by means of a facsimile scanner, although many types of optical scanners would be suitable. The facsimile scanner, however, was already connected to the computer for some of my previous experiments.[6] A 4 × 5 photographic print is placed on the drum of the scanner and the computer made ready. Then, during each rotation of the drum, a photomultiplier output scans a line of the picture. An analog-to-digital converter samples the photomultiplier output at about 600 cps and sends the computer ten-bit digital intensity values. Thus, in about three minutes, a 256 × 256 raster of intensity samples can be read into the computer. Each sample is compressed to eight bits in the computer, so the storage of one picture requires about half a million bits of memory. Thus four pictures can be stored in the TX-2 memory. Figure 2*a* and *b* shows a picture before and after computer sampling.

When the scanning is completed, the picture is processed with a local differential operator to produce a new raster which has the appearance of a line drawing. The choice of a differential operator is very critical and many variations were tried. Three main criteria can be used to judge such an operation. The edges produced should be as sharp as possible, the background should produce as little noise as possible, and the intensity of the lines produced should correspond closely to a human's ability to perceive the edge in the original picture. Edge sharpness depends upon the number of samples used by the differential operator. Background noise seems to be reduced by using operators symmetric in x and y. In order to make equally apparent edges have equal derivatives, the intensity values of the picture can be subjected to a gamma change so as to make intensity differences proportional to a human's ability to perceive them. According to psychophysical theory, the square root of the intensities should be used in order to achieve the desired effect.[11]

Therefore, after a picture is read in, a differential picture is created according to the functions

$$y_{i,j} = \sqrt{x_{i,j}}$$
$$z_{i,j} = \sqrt{(y_{i,j} - y_{i+1,j+1})^2 + (y_{i+1,j} - y_{i,j+1})^2}$$

where $x_{i,j}$ is the initial intensity value, $z_{i,j}$ is the computed derivative value, and i and j are the coordinates in the two dimensions. The resulting z values indicate the probability of a line through that cell. Even though the square-

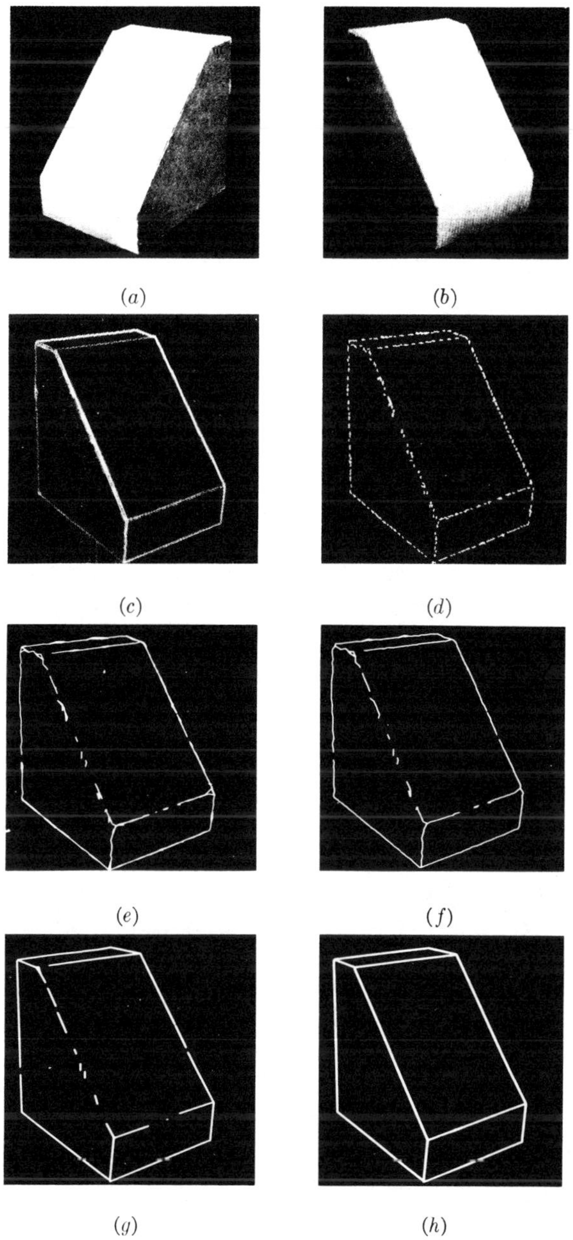

FIG. 2. Picture to line drawing. (*a*) Original picture. (*b*) Computer display of picture (reflected by mistake). (*c*) Differentiated picture. (*d*) Feature points selected. (*e*) Connected feature points. (*f*) After complexity reduction. (*g*) After initial line fitting. (*h*) Final line drawing.

root operation involved takes more time than several simpler operations which were tried, the extra line sharpness and background noise reduction obtained are well worth the additional time. Figure 2*c* is an example of the result of this differentiation.

After obtaining the differential picture, the problem is to determine a set of lines and end points which correlate well with the raster. It is no simple chore to obtain a list of lines and their end points from a half-million-bit array of data. A simple procedure might be to choose a clip level and start tracing out lines which correspond to a string of adjacent cells in the raster, all of whose values are above the clip level. The hopelessness of this procedure is easily seen when one looks at typical pictures and considers them as a three-dimensional surface where the z values are used as the height. Even a very clean input picture when viewed in this way looks like a bumpy, hilly landscape, with a broken-down stone wall representing the lines, and where some hills are higher than the top of other stone walls. If we imagine the clip level as a flood over this landscape, there is no water level which covers all the hills and yet does not submerge some stone walls. In fact, even by adjusting the water level to be optimal for a particular area, a line will look like stepping stones in a rock-strewn brook rather than a smooth dam. Thus it can be seen that the problem of mapping the walls is not a simple one.

The procedure I have developed for finding the lines from the differential picture makes mistakes in complex pictures and is a complex special-purpose program demonstrating very few general concepts. However, it does manage to produce accurate line drawings from a sufficient group of pictures so that the transformation techniques can be tested on data from real photographs. The description of this procedure will be rather general in nature.

The over-all concept was to look at local features first and build up to the determination of long lines in a series of steps. There are two reasons for this procedure. First, at each step the complexity of the processing goes up since more data are being considered, but at the same time the number of features to process is being reduced at each step. Thus the total time to process the picture is much less than if a one-step process were used. Such a one-step process was tried by Leo Hodes in an attempt to find the lines in black and white line drawings.[4] For each possible line a correlation with the raster was made, and then the line position was corrected and recorrelated. This type of procedure becomes more and more time consuming as the raster size, line length, and accuracy required increase, whereas a multistep process need not consume additional time.

The second reason for a multistep process is that the initial local feature extraction can take into consideration the local noise level and thus detect lines which an over-all clip level would miss.

The first step of the process looks for points in the differential picture which seem to be on a line. The maximum sample in each 4×4 square of raster points is considered as a possibility if its value exceeds a small threshold. This threshold is low enough to include all the lines and much of the noise but it does eliminate further processing of smooth areas of the picture. Its value has been determined by experiment to be about one-tenth of the maximum intensity in the original picture. After a point has been chosen in this way, four correlations are performed to find the direction of a line through the point which best fits the data. Four lines, having a length of five samples with slopes of 0, 1, ∞, -1, are correlated with the data around the point, and the ratio of the best to the worst fit is taken. If the ratio is greater than a second threshold (about 3), the point and the best direction are recorded. In this way a set of feature points is obtained, along with approximate line directions through them. The procedure can be considered to be a ridge detector which locates points along each ridge with limits on the height and width of the ridge. The number of feature points obtained is usually between 100 and 1000. Figure 2*d* shows a short line at each feature point, pointing in the recorded direction.

After the feature points are found, the next step is to connect lines between neighboring points. The line directions of the points are used to limit cross connections between adjacent lines. Specifically, a pair of points are connected if they are in touching 4×4 squares and if the line's direction will be within $\pm 23°$ of the direction recorded for either point. Any points left unconnected are eliminated, thus filtering out most of those created by noise. The result is a preliminary line drawing composed of many short lines. There are two major problems with this line structure: sections of a long line may be missing, and sections may be complicated by multiple interconnections. The interconnections are most obvious at corners but also appear along a line as extra width. Figure 2*e* gives a more graphic illustration. To reduce these small networks of lines to a single line or neat corner, two reduction techniques are used. Their important property is that they do not change the over-all connectivity and topology of the line structure. First, the longest side of each triangle is deleted. A triangle is defined here as any three lines in a loop. This cleans out most of the unwanted lines; however, there are a few quadrilaterals left. Therefore, each group of four lines connected in a loop is compressed along its shortest diagonal; that is, the two closest, nonadjacent points are merged. There may still be a few pentagons left but most of the networks have been reduced. The last step in smoothing out the line structure is to remove all small tails or spurs which are unconnected at one end and connect to more than one line at the other end. These smoothing operations are all on short lines so no major features are changed. They do, however, limit the resolution of the input system to about four to eight samples out of 256. This restriction is on resolving short, close lines, not on the accuracy of longer lines. The result of smoothing Fig. 2*e* appears in Fig. 2*f*.

Now that a structural outline of the lines in the picture has been obtained, longer lines can be fitted to sections, missing segments filled in, and extra segments removed. The segments along the path of a true straight line may weave in and out but each point was obtained by the ridge detection technique and is accurate to about one sample width if it was caused by the edge. A sequence of singly connected points with no intersections will probably be caused by the same edge or sequence of edges. By a least-mean-square error-line-fitting routine, a straight line can be put through the points and result in a very high accuracy line. Curve segments could also be fitted to the points if this were desired; the main additional problem would be the choice of the type of curve to try—circle, parabolic, cubic, etc. It becomes more and more obvious, as one considers fitting curves to the picture edges, that it is advantageous to have a set of points already determined, through which to put a curve, instead of having to correlate various curves with the picture data.

To fit a straight line to a sequence of points, a sequential least-mean-square error-fitting routine is used. The problem is to find the best coefficients (a, b, c) for the line equation

$$ax + by = c$$

The data are supplied sequentially in the form of points (x, y) and it is desirable to recompute as little as possible each time a point is added. However, upon each addition of a point, the coefficients (a, b, c) and the new mean-square error E should be available. It is sufficient to keep a history of five numbers, the cumulative sums

$$\sum x, \qquad \sum y, \qquad \sum xy, \qquad \sum x^2, \qquad \sum y^2$$

and the number of entries made n. Then, after each addition of a new point (x, y), the coefficients are computed as

$$\begin{aligned} a &= \sum x \sum y^2 - \sum y \sum xy \\ b &= \sum y \sum x^2 - \sum x \sum xy \\ c &= \sum x^2 \sum y^2 - \sum xy \sum xy \\ E &= c(nc - a \sum x - b \sum y)/n(a^2 + b^2) \end{aligned}$$

Since these equations represent the mean-square best reduction of the unnormalized error $E(a^2 + b^2)$, they do not always produce the least-mean-square distance error as represented by E. However, the solution is much easier than the complete form and just as good in almost all cases.

The procedure for fitting a line to a series of connected points starts by choosing any small line segment as a starting place and moving in one direction until a point is reached with other than two line segments attached, or until the mean-square error exceeds a threshold. When the error threshold is reached, a bend in the true lines has probably been passed, so the procedure is to back up until the angle between the little line segments and the computed line has decreased by a factor of 2. Usually this condition will occur at the bend sought for, since on the other side of the bend the angles must be negative. This procedure works very well and needs no threshold adjustment because of the backup procedure.

When the line has been finished in one direction from the starting point, the other direction is investigated in the same manner, still modifying the same computed line. The method of using cumulative sums as the only history for the line computation allows points to be subtracted out during backup, eliminating the need for large tables of past history. Also, the procedure is fairly independent of the starting point since, during the second direction's backup, the starting point can be passed, thus creating a line totally to one side of the starting point. This will occur when a break point appears just after the starting point, and as the break point is passed in the forward direction, the error does not build up to threshold.

Thus the line-fitting procedure replaces groups of small line segments with longer, more accurate lines. The ends of these new lines are at the intersections of several lines, at break points as detected by the error criteria, or are free and unconnected. Each time a long line is computed, the points at its ends are moved onto the line since the line is more accurate than the points. If several long lines meet at a point, the point's coordinates are computed to be the intersection point of the two longest nonparallel lines. Thus the points become as accurate as the lines connected to them. A special case may come up due to the incomplete removal of a network of small lines: Two lines may be constructed between the same end points. These must be merged and the lines connected to the false intersections reprocessed into one line, if possible.

When all the lines have been fitted to the small segments, the representation of the picture consists of a set of lines and end points mapping the edges in the picture. There still may be sections of lines missing and extra segments near intersections. Figure 1*g* shows the result at this point. Now line filling and merging are done to complete the line drawing. Each line is considered for modification. If the line is of significant length, the nearest points to both ends of the line, which are within about three raster units of the extended line, are considered as possibilities for new end points. A new line is correlated with the differential picture between an end point and the new point and if the average value along the line is greater than a threshold, the line is put in. When a line is very short, it is not extended but is considered for merging or elimination. If the line's end points both connect to one other line, then the end points are merged, otherwise the line is deleted. After extending, merging, or deleting the proper lines, the whole line structure is again processed with the mean-square line-fitting program in order to eliminate extra joints which may have been created. The resulting line drawing is the finished version as shown in Fig. 1*h*.

The entire picture-to-line-drawing process is not optimal but works for simple pictures. It has several useful parts; the differentiation, the feature point extraction, and the mean-square line fitting are the best parts. In the future, I hope to recombine these sections to produce a more general system.

V. Construction of Three-Dimensional Objects from Line Drawing

The program described in this section starts with a planar line drawing and produces a three-dimensional description of the objects shown in the drawing in terms of models and their transformations. The line drawing may be one generated by the picture input process or some other computer program such as the three-dimensional display program (Sec. VI). The main restriction on the lines is that they should be a perspective projection of the surface boundaries of a set of three-dimensional objects with planar surfaces. Any line drawing produced by the three-dimensional display program is acceptable as an input to this program, and since this program's output is an input to the display program, the two programs can be used to check each other. The models used for construction can be any set of three-dimensional building blocks which seem useful so long as all their surfaces are planar. Since the models can be put together so that their joint lines disappear, almost any complex object can be constructed with a very few models. There are only three models presently used in the program: a cube, a wedge, and a hexagonal prism (Fig. 6, in Sec. VI illustrates these models). Section III describes the general procedure used for this two-dimensional to three-dimensional transformation, whereas we now wish to develop the specific mathematics and techniques used in the program.

A. Polygon Recognition

The line drawing which is produced by the three-dimensional display program is just a list of end-point pairs, one pair for each line. This type of input is specially processed to put it in the form wanted. Each line is assigned to a line block in a line list and each point to a point block. Each point indicates which lines are connected to it and each line block points to its end points. Thus, for each end-point pair in the input list, a line block is created and the point list searched for point values close to the end points. If the points already exist, the line is just tied into them; otherwise, a new point block is created. Upon completion of this phase, each point is checked against all lines to see if it lies on a line but is not connected to the line. If this occurs, the line is broken in two and both new ends are tied to the point. The list format produced is a good form for topology processing and is the same format as that produced by the picture input program.

The first problem is to find the polygons described by the lines. In order to trace the polygons easily, the lines tied to each point are ordered by their angle of exit. This allows us to start with any line, choose an adjacent line at one end point and continue around the polygon to the first line, without ever getting off the polygon. This procedure can be made to go clockwise or counterclockwise around polygons, and thus record two polygons for each line. A list of polygon blocks is prepared, each tied to the lines that compose it. The lines also point to their two polygons. As each polygon is produced, the exterior angle at each vertex is computed and the sum of these angles is kept for the polygon. These angles are computed in semicircles, so they are between $+1$ and -1. The sum will therefore be $+2$, if the polygon was hollow on its inside, but if the polygon was really an exterior boundary of an object, the "hollow" part is outside and the sum will be -2. Figure 3 shows the polygons of a cube projection and their exterior angles.

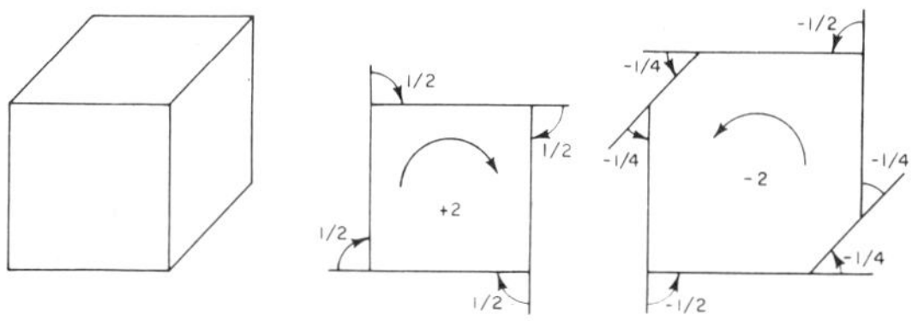

FIG. 3. Exterior angles. The exterior boundary of an object is separated from the normal polygons by the sum of the exterior angles, taking the unconnected direction as the inside. The direction of travel and the sum of the exterior angles appear in the center of (*b*) and (*c*). Angles are in semicircles. (*a*) Complete cube. (*b*) Interior polygon. (*c*) Exterior boundary.

It may not be apparent that exterior boundaries are difficult to separate from the real polygons, but as the computer traces a polygon, it has no concept of the inside; it just traces a closed path with all connecting lines on one side of the boundary. It must expect some negative angles because the polygon may be concave. Thus the sum of the exterior angles is necessary, if the computer is going to separate exterior boundaries from real polygons.

Some further information is obtained as the polygon is traced out: the number of sides of the polygon, the number of negative angles encountered, and the number of near-zero exterior angles. One or more negative angles indicate that the polygon is concave, whereas the zero angles indicate collinear joints which most likely were produced by another object partially hidden behind this one. Thus a first guess at the number of sides the surface really has is the number of lines minus the number of zero angles. The polygon is then marked as complete and convex if there are no negative angles, it is not an exterior boundary, and it does not include a point where a zero angle was observed on another polygon. The last condition eliminates from initial consideration polygons that are most likely partially hidden by another object. Figure 4 shows an example where each complete and convex polygon is labeled with its first-guess number of sides.

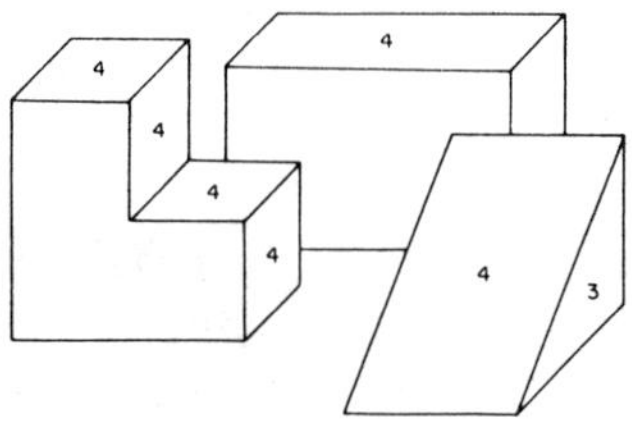

FIG. 4. Complete convex polygons. The polygon selection procedure would select the numbered polygons as complete and convex. The number indicates the probable number of sides. A polygon is incomplete if one of its points is a collinear joint of another polygon.

The reason concave polygons are not considered at first is that the models presently being used in the program are all convex. However, if a model with concave polygons is included, the appropriate concave polygons from the line drawing should be considered. In this connection, the rule is that both the number of positive angles and the number of negative angles of a polygon are invariant under any perspective transformation. Thus the only polygons which need be considered at this point are those which have the same number of plus and minus angles as some model surface. With the models included in the present program, the acceptable polygons are further restricted to have 3, 4, or 6 sides.

A comment on incomplete polygons may be useful at this point. A perfect projection of solid objects with visible width and no two-dimensional markings is being assumed for the input. Since such projections are completely composed of surface boundary projections, they will never contain any points connected to only one line, unless the points are on the boundaries of the picture. If a special external boundary square is included, there will be no incomplete polygons as is assumed. The three-dimensional display program will always generate acceptable input, but the picture input process could miss lines, generate false lines, or include two-dimensional markings. These cases could be taken care of by further checking the picture and other techniques, but at present this task was ignored. The program eliminates such problems by deleting all lines not connected in a closed polygon. Also any isolated polygon is considered to be a two-dimensional marking and is deleted. Thus all letters except *B* are deleted as well as most other simple two-dimensional markings. When the process is completed, these markings could be transferred to the appropriate surface, if desired, but this was not done.

One further computation is performed on each polygon even though it is used only to order the investigation of the polygons. The area of each polygon is found as the program moves clockwise around the points (x_j, y_j),

$$A = \tfrac{1}{2} \sum (x_{j+1}y_j - x_jy_{j+1})$$

This procedure, which is really summing the signed areas of the triangles formed by each line and the origin, is very simple and works for any shape polygon. If the motion is counterclockwise instead of clockwise, as is the case for external boundaries, the area will come out negative. This formula for the area of a polygon was derived by the author since no suitable formulation could be found easily.

B. Model Matching

The first tool used to match the polygon structure to the models is topology. Basically, we wish to find points in the line drawing which fit a transformation of some model. The polygon structure is used to find a suitable model with a set of topologically equivalent points. Then the mean-square error technique is used to find out whether the point positions are related by a simple transformation or not. Topology matching proceeds in four steps. First, each point is examined to see if it is completely surrounded by approved polygons. When such a point is found, the number of polygons of each type is counted. At present, since the only approved polygons are those with 3, 4, or 6 sides, three counts are obtained. A list of triads corresponding to distinct points on the models is maintained so that a quick search will indicate which model points are surrounded by the same polygon structure as the picture point. For example, if a point has three quadrilaterals around it, the list will specify a particular point on the cube model. The other points on the cube need not be listed because they are all similar to each other. Once a model point has been chosen, the program cycles around the picture and model points to line up the order of the polygons. If the orders cannot be matched, other listed model points are tested; however, if they are matched, a list of equivalent point pairs is constructed.

The computation of the optimal transformation matrix from the point pairs is presented in Appendix B. Besides producing the transformation, the procedure generates the mean-square error. A threshold is placed on this error to eliminate models which fit the picture topologically, but do not fit exactly without being deformed. Models having acceptably small error can now be transformed to produce the lines and points which were not part of the fitted area. The points are checked against the picture to make sure they do not fall outside the object's external boundary. If a point does exceed this boundary, the model must be discarded since it would produce new lines not in the picture. Models that pass this test, however, could represent at least part of the object, and are accepted. If a transformed model completely accounts for a group of connected lines, the transform and model are used to represent that object; however, if some lines are left, the procedure described under object construction must be used. Figure 5*a* through *d* shows the processing of a photograph in which a single cube model was used to describe the three-dimensional object, whereas Fig. 6*a* through *d* illustrates a situation in which two models were needed.

The examination of all points surrounded by polygons is only the first step of the topology matching. When all the points are tested, lines are examined for approved polygons on both sides. A second list of model information is searched for any models with such a pair of adjoining polygons. When a line and model are found, the polygons are aligned and a list of point pairs produced. From here on the transformation procedure is the same as before.

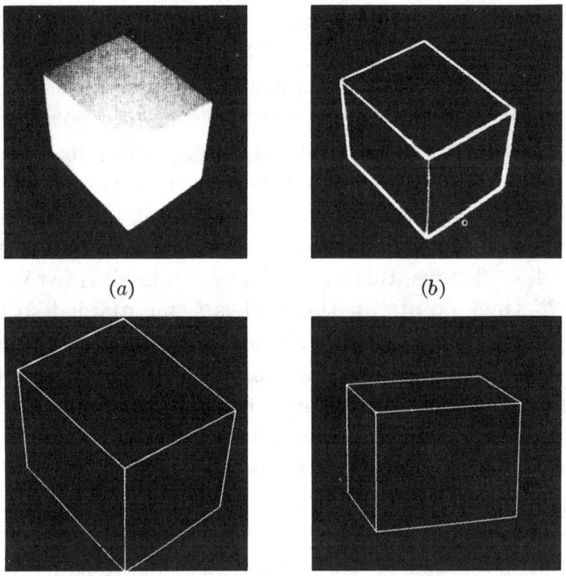

(a) (b) (c) (d)

FIG. 5. Single model: reduction of photograph to line drawing and display of three-dimensional construction from another viewpoint. (*a*) Original picture. (*b*) Differentiated picture. (*c*) Line drawing. (*d*) Rotated view.

The third step, after all lines have been examined, is to test each remaining approved polygon. The polygon must have a line attached to one vertex. The model information lists each distinct model, polygon, and vertex for each type of polygon, and the point pair list is easily prepared. The fourth step, if necessary, is to take each point with three lines out of it and test these four points against each distinct vertex and line orientation in each model. This test is a last resort because there are so many model possibilities and in every case a transformation can be found which makes the mean-square error zero. Only finding a point outside the boundary can eliminate these cases.

The four steps are ideal for the cube model since the number of points found decrease by one with each step. Almost any visible piece of a cube will be processed with the maximum number of points possible. Even though this is not true for all models, the four steps are a good approximation to a comprehensive topology search. Topology tests and matching are very difficult to implement on a computer, even with the aid of a good list structure. Computer languages seem to be far superior at numerical and symbolic manipulation than the decision-loaded searching associated with topology. The inherent limitations of the serial processing of a line structure, without some global picture of what one is looking at, make specific tests such as the four steps far easier to achieve than any general procedure. Even to accomplish these steps without continual searching and backtracking, the list structure must be organized in a very special manner. At each vertex the connecting lines are ordered by angle of exit in a counterclockwise ring. Thus any time the program arrives on one line, it can easily exit on an adjacent line. Further, the lines in each polygon are ordered in a clockwise ring, with the end points of each line and the two polygons of the line specifically related. This is important because, upon arrival at an end point of a line, the program can immediately identify the right and left polygons and proceed around either one. Both the model structure and the line drawing are organized in this way. The definition of clockwise and

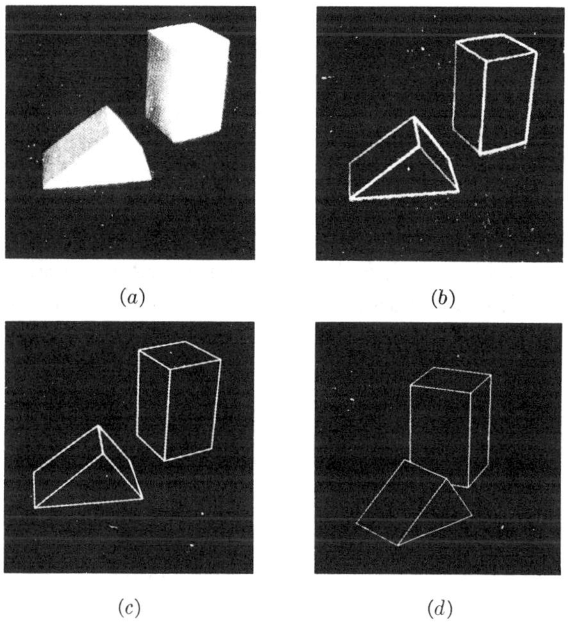

FIG. 6. Multiple objects: reduction of photograph to line drawing and three-dimensional construction, involving ground plane determination of depth. (*a*) Original picture. (*b*) Differentiated picture. (*c*) Line drawing. (*d*) Rotated view.

counterclockwise in a three-dimensional model structure is not obvious but, for compatibility with the projections of a solid model, all such orderings are made while looking in from the outside.

C. Compound Object Construction

A compound object is a single solid object which is not a transformation of a single model, but must be formed by piecing together several models. Whenever two models are fitted together such that they have a frontal plane in common, the three-dimensional display program will eliminate any piece of line which touches both models and is in that plane. Since these joint lines do not reflect any surface discontinuity, it is expected that they will also be missing in the line drawing input. If joint lines happen to appear in the input, the resultant structure will probably be the same as if they were missing; therefore, the display will not include them. As models are found which fit a part of an object, their lines are projected back onto the drawing and the joint lines found. Thus the complex unapproved polygons are cut up by these joint lines into smaller polygons until each piece is approved. The following rules are for deleting an accepted model from an object and reforming the picture to make ready for the next model identification. The expression "*T* joint" refers to a vertex at which two collinear lines and one other line, the "stem," meet. During the process, parts of the drawing are deleted, but these changes are never allowed to modify the external boundary polygon since it must be used to test new models. The concept of a "visible" model line or point refers to the points and lines on the frontal surfaces of the model.

1. Each visible point of the transformed model is projected on the drawing.

   a. Any new point pairs disclosed by the proximity of model and picture points are used to recompute the transformation for better accuracy.

   b. If a model point falls on a picture line, the line is cut in two and the point inserted.

2. All the model lines and points are added to the picture if they are not already there. Any picture point which lies on a model line, but not on either end, is separated from the model line structure. Thus a picture line that ends in a T-joint stem may be extended to its proper end point, and the collinear T-joint lines unified into one line. Joint lines are those visible model lines which were not in the picture and which divide a picture polygon.

3. Each visible model point in the picture which does not connect to any nonmodel lines is now marked "used." Also, all points on polygons with more than two "used" points are marked "used." The joining polygon between parts is the polygon which includes the joint line but is not divided

by it, and all the points on such a polygon must be unmarked. Now all "used" points are deleted along with their attached lines and polygons. Also, any line in a joining polygon which is the stem of a T joint at both ends should be deleted.

4. All remaining model lines should be marked as unnecessary. If all lines left connected to them are unnecessary or if no lines are left connected, the object has been finished.

Each time a model is stripped from an object in this way, its transform and model name are saved as part of an object block. For each model, a point which was connected to the remainder of the object is remembered, so that the depth relations between the parts may be computed. When an object is finished, there will be a string of points connecting its parts. An object transformation is set up to position the depth of the whole object; hence, the first model can be assumed to have the correct depth. Then each model whose point connects to the first has its transformation modified so that the points have the same depth. Then the models connected to those are updated and so on.

To correct a transformation R_0 when a known point $\bar{v}$ should be equal to a point $\bar{p} = \bar{q}R_0$, the new lower-left element x of R_0 is computed as

$$x = \frac{q_4 v_1 w - p_1 v_4}{q_4(v_4 - r v_1)}$$

Here w is the present lower-right element of R_0 and the new $w' = w + rx$. Note that the focal ratio r for the picture must be known at this point. However, the accuracy of r will not affect the accuracy of positioning the parts with respect to each other.

Eclipsed objects, or objects partially hidden from view by other objects, are automatically taken care of by the construction rules. One case, however, needs further attention. When an object is so well hidden that a dimension cannot be determined, this dimension must be estimated. An example of this case would be when only the top of a building is visible over another. The first assumption we make is that the object is supported by the ground plane. But a second assumption is needed to place the object, and the program assumes that the hidden object just touches the object in front. This is not a very good assumption, but there are no good assumptions.

Figure 7 illustrates the construction of a compound object. The original line drawing appears in A1 and includes a compound object and a partially hidden object. Since there are no points surrounded by acceptable polygons, we must look for a line with good polygons on both sides. There is only one such line to which a model can be fitted and this is in the upper object. Both the cube and wedge models fit this object; however, the cube is always tested first to avoid splitting cubes into wedges. The lines of the cube model are then projected onto the line drawing as in A2, and the transformed model is entered into the three-dimensional structure as displayed in A3. After a model has been identified, the "used" points and lines are deleted, thus producing the line drawing in B1. Now a new search for fitting models is made. The lower-right quadrilateral and the bottom line adjoining it are found to fit a cube model, resulting in B2 and B3. Next, a cube is fitted at

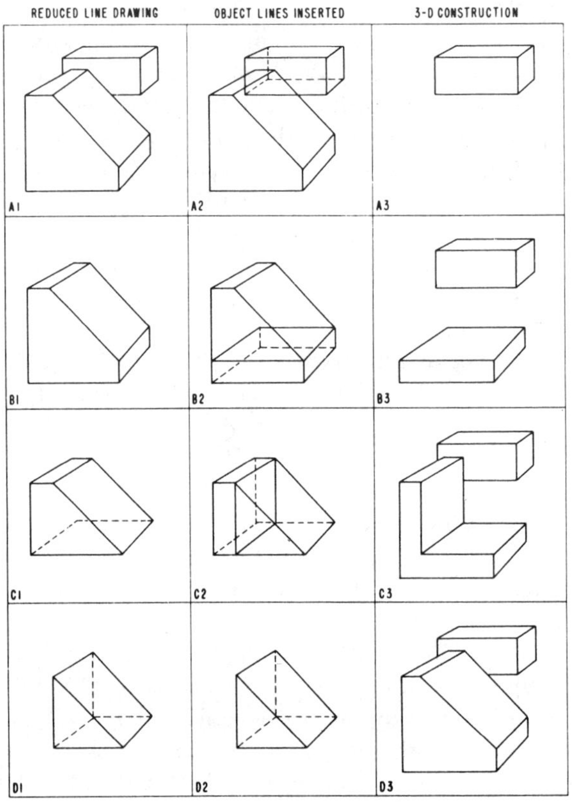

FIG. 7. Compound object construction: original line drawing in A1 is processed to obtain three dimensions in D3 by sequential recognition and deletion of four models in steps A, B, C, and D.

the left, in C1, producing C2 and C3. Finally, just a wedge is left in D1 and since all the back lines have already been determined, D2 appears the same. When this model is added to the three-dimensional structure, the result is a complete description of the objects and can be displayed as in D3 or from any other point of view. Figure 8*a* through *d* shows the computer processing of a similar compound object from a photograph. The collection of models describing the three-dimensional object can then be rotated as in Fig. 8*e* through *h*.

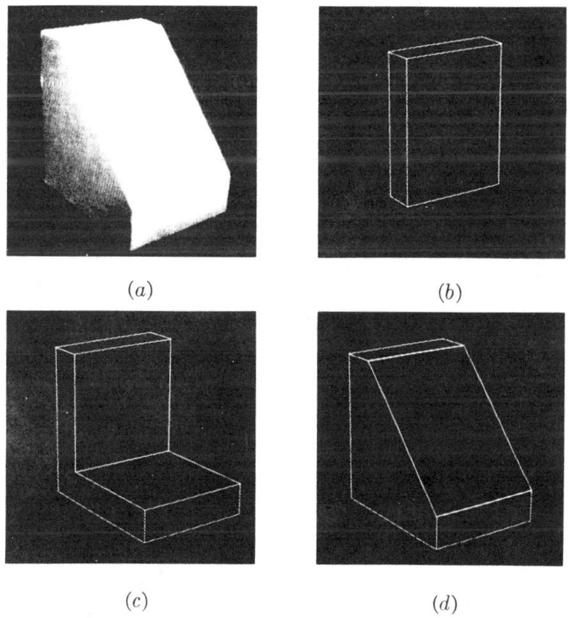

FIG. 8. (*a–d*) Compound object construction: processing of Fig. 2 to obtain three-dimensional description of compound object. (*a*) Original picture. (*b*) First construction model. (*c*) Two construction models. (*d*) Complete three-dimensional object.

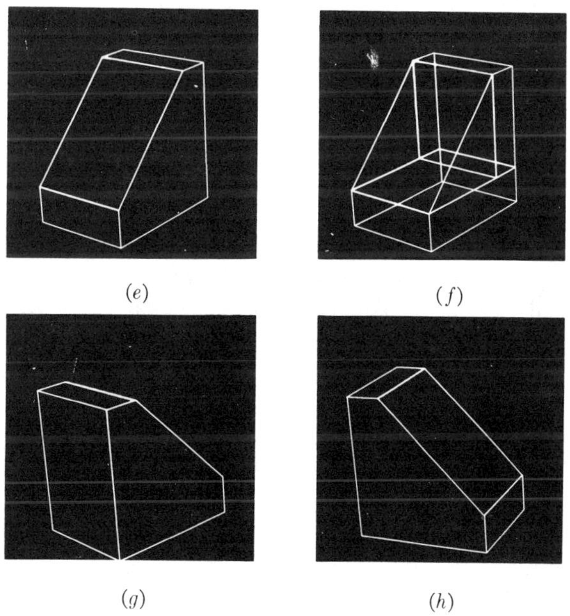

FIG. 8. (*e–h*) Compound object construction: rotated views of object obtained in Fig. 8(*d*). (*e*) Rotated view. (*f*) All lines of Fig. 8(*a*). (*g*) Second rotated view. (*h*) Third rotated view.

D. Ground Plane Assumption

When all the objects in the picture have been constructed, they still need their object transformations adjusted for depth. If an object consisted of a single model, the transformation listed is R_0. If the object was compound, an identity matrix was set down as the object transform and the model transforms all relate to it. In both cases each object has one free variable related to depth and size. The support theory would place each object on top of another or on a ground plane. In order to simplify the present program, the ground plane is the only support assumed. However, the addition of object support will not be very difficult.

Finding the ground plane is the most difficult part of the depth computation. Each object could be examined for possible support planes and all objects compared for a common plane. An object could be supported by anything from three points to a full plane in contact with the ground. To simplify this chore, it was assumed that the picture was upright and each object had a full plane in contact with the ground. A simple test of the slope of each plane of an object is used to determine if that plane is the bottom. This test merely asks that the bottom is not visible, that it faces the downward z direction, and that the tilts in the y and x directions are moderate. The best such plane is chosen from each object, and all are expected to agree. The slope of this plane is the only information available from the incomplete transformations, so the distance of the plane below the origin or focal point must be assumed. This distance just sets the numerical scale of all distances, so it might as well be unity. Thus a ground plane is determined and all the object bottoms are now made to lie in this plane.

With each model surface is stored its plane equation vector for the use of a three-dimensional display program. The dot product of this vector with any position vector is zero if the position is on the surface, and positive if the position is inside the solid. These plane vectors should be transformed by R^{-1} to become plane vectors of the transformed model. If the plane vector of an object plane is transformed by R_0^{-1}, it has the correct slope but not necessarily the correct length. In other words, its first three components are correct. These are used to find the ground plane. Now with the ground plane distance equal to one, we must find the depth x for each R. The lower-right component of R_0 we shall call w, and the same component in the final R, w'. We first set $x_1 = -w/2r$ and $w_1 = w/2$ to obtain R_1. If we transform the plane vector by R_0^{-1} and R_1^{-1}, we obtain $\bar{v}_0$ and $\bar{v}_1$. Normalizing these vectors so that the sum of the squares of the first three components is 1, we obtain normalized fourth components p_0 and p_1. The plane vector for R has a fourth component which is a linear combination of p_0 and p_1. Setting this combination equal to one and solving for the depth x, we obtain

$$x = \frac{w(p_0 - 1)}{r(1 + p_1 - 2p_0)}$$

$$w' = w + rx$$

Here again the focal ratio r is needed, and this time if r is zero, the computation of x blows up. This just reflects the fact that an infinite projection has infinite depths; however, it restricts r to nonzero for this procedure.

VI. Three-Dimensional Display

After a list of three-dimensional objects has been obtained in some manner, it should be possible to display them from any point of view. The sections of objects behind other objects should not be seen, nor should the back lines and construction lines of individual objects. The three-dimensional display program will do all this and more. It allows macrolike instances of objects so that a single object construction can be used many times with different transformations. It allows structures of models to be built up by the use of the knobs, push buttons, and light pen. Any object can be duplicated, deleted, or transformed. These extras make possible the construction of test cases for the two-dimensional to three-dimensional program to process. However, the most significant feature of this program is the mathematical technique which makes possible the hidden line removal.

A. Storage Structure

A good method of storing three-dimensional data is extremely important. The structure used is the basis for both the display program and the three-dimensional construction process. Therefore, the data necessary for hidden line removal must be quickly available and at the same time the topological structure must be suitable for model matching.

The list structure used is a list of tied blocks connected in rings. Ring list structures were developed for the TX-2 computer by Sutherland for his Sketchpad system.[12] Sketchpad allows a user to draw two-dimensional line drawings on the computer display with the aid of the light pen, knobs, and push buttons. An extension of this work to three dimensions is currently being completed by Johnson.[13] These two systems use ring list structure and, in order to be compatible with them, the same format is used in the three-dimensional display program. However, the exact block form used is different because of the different data requirements. In the ring structure, a block of registers is used for each item and contains pairs of ties to other blocks. Each pair of ties is part of a ring which allows the program to move from block to block around the ring in either direction.

The basis of all three-dimensional forms is the set of models. Each model block is tied to lists of its points, lines, and surfaces. The point blocks are tied to the lines connected to them and include a four-component position vector. Since a homogeneous vector can be normalized without changing the point, each component of the vector can be represented by a fixed-point, 36-bit number. The line blocks are tied to two point blocks and two surface blocks, since two planes determine a line. The surface blocks are tied to a ring of lines which represent the surface polygon and also include a plane vector. This detailed structure is needed for models only, since the objects are to be composed out of transformations of models. The models are always in the list structure and must be referred to by instances in order to be displayed. An instance is an intermediate block between a picture and either a model or another picture. Each instance includes a 4×4 transformation matrix and also the inverse of this matrix. The picture blocks may be referred to by any number of instances and have as their parts any number of instances. Thus each picture represents an object or a collection of objects and is composed of transformations of other pictures or models by means of instances. One picture is the current picture being displayed, and it has only one instance containing the picture transformation. Figure 9 shows a possible structure leading from the models to the current picture. The instances have been compressed to ties with matrices on them for simplicity.

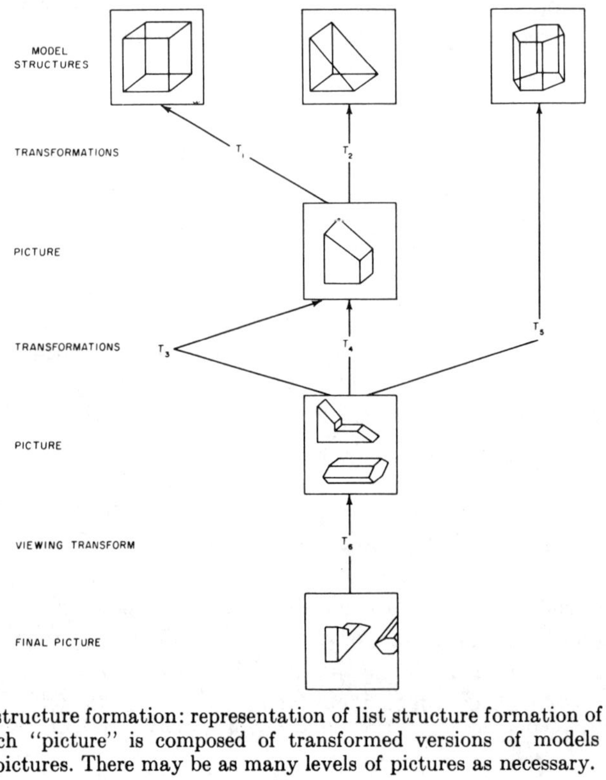

FIG. 9. List structure formation: representation of list structure formation of compound pictures. Each "picture" is composed of transformed versions of models and other pictures. There may be as many levels of pictures as necessary.

B. Display Generation

In order to display the picture represented by the current storage pattern, a recursive procedure is used, starting with the current picture. Each picture block has a temporary storage area to store the effective transformation and its inverse at that picture level. One recursion involves taking the transformation at one picture level, premultiplying by the matrix of one of the instances the picture refers to, and placing that transformation in the picture block referred to by the instance. The inverse matrices are carried along also, but postmultiplication must be used for them. If the instance refers to a model, the matrices obtained are put in a transform block along with the model name. Each time a model is reached, the process backs up to the last picture block and proceeds to the next instance, or if all instances have been processed, it backs up to the preceding picture block. In this way a list of transform blocks is obtained. Each transform block is now processed by tying to it a list of point-pair blocks obtained by transforming the end points of each line in the selected model. Each position vector is postmultiplied by the transformation in the transform block to obtain the new position vector. Also, the plane vectors of the model are transformed by the inverse matrix and collected as the columns of a volume matrix.

Thus a list of transform blocks is obtained, each block of which has a volume matrix and a list of point pairs. The point pairs represent all the lines in the display and can now be processed to eliminate the hidden lines. The display coordinates of a point are obtained from the four-component position vector $\bar{v}$ as

$$x = v_2/v_4$$
$$y = v_3/v_4$$

C. Hidden Line Elimination

Three steps are required to prepare a line for display. First, it is trimmed off at the edges of the display. Next, the back lines of each model are deleted. Third, the sections of each line which are hidden by other models are removed. It is the third part which is difficult and time-consuming. As far as I know, no one has ever devised a procedure for determining hidden line segments. One can imagine brute force methods such as calculating all the line intersections on the focal plane and then computing which lines were in which polygons and tracing out the frontal lines. But procedures such as this are hard to make complete for all cases, and the processing time could be fantastic. Therefore, a new mathematical method was conceived which utilizes volume inequality matrices to find out whether a point is inside or outside a volume. This test can then be extended by linear inequality solutions to tell which segment of a line is behind a volume. This is why the inverse transformations, plane vectors, and volume matrices are needed.

Since they are available, they can be used to advantage in the first two steps.

The volume matrix V of a model is a $4 \times n$ matrix with the n-plane vectors as columns. In the homogeneous coordinate system, a dot product of a position vector and a plane vector produces a measure of the distance of the point from the plane. The plane vectors of the models have had their signs adjusted so that the dot products will be negative if the point is outside the solid and positive if the point is inside. Thus, when a position vector $\bar{v}$ is postmultiplied by a volume matrix, the resulting vector $\bar{v}V$, since it is a collection of dot products, will have at least one negative term if the point is outside the volume. If any terms are zero, the point is on the model's surface, and if all are positive, the point is inside. In order that this test may work, the models must be convex. This is a small restriction, since the objects constructed with the models need not be convex,

$$(\bar{v}V \geq 0) \Rightarrow \bar{v} \text{ inside volume } V$$

In the expression above, the inequality sign means that all components of the vector should be nonnegative. Expressions and formulations of inequality matrices have been used in the field of linear programing to express the interior region of convex polyhedral cones where the general problem was the optimization of a function in such a region. However, literature in this field does not seem to cover the geometrical type of problem we are concerned with.

A volume matrix can be used for the first step of the line reduction: the elimination of lines from the display. As well as cutting off lines at the display edge, we wish to cut off lines in front of the focal plane. Thus a volume matrix can be designed which has planes at $y = \pm 1$, $z = \pm 1$, and $x = 0$. These planes form a semi-infinite box, and we wish to find the section of each line inside the box. First, we shall define the volume matrix V_0,

$$V_0 = \begin{bmatrix} 0 & 0 & 0 & 0 & -1 \\ 1 & -1 & 0 & 0 & 0 \\ 0 & 0 & 1 & -1 & 0 \\ 1 & 1 & 1 & 1 & 0 \end{bmatrix}$$

The set of all lines is the list of point-pair blocks. We shall want to use a line equation of the form

$$\bar{v} = \bar{s} + t\bar{d}, \qquad 0 \leq t \leq 1$$

where t is a running variable, moving the position $\bar{v}$ from $\bar{s}$ to $\bar{s} + \bar{d}$. A point-pair block has two points, $\bar{s}$ and $\bar{r}$. If the fourth component of these vectors is made to agree by normalizing one vector, the difference vector $\bar{d} = \bar{r} - \bar{s}$ can be formed. With this line representation we can proceed to find the values of the variable t for which the following inequalities are true:

$$\bar{s}V_0 + t\bar{d}V_0 \geq 0$$

This set of inequalities requires that the line be inside V_0 and we can solve for the minimum and maximum t. If we define

$$\bar{p} = \bar{s}V_0, \qquad \bar{q} = \bar{d}V_0$$

then for all j,

$$p_j + tq_j \geq 0$$

For $q_j > 0$, $t \geq -p_j/q_j$. For $q_j < 0$, $t \leq -p_j/q_j$. From these relations it is fairly obvious that we can obtain a minimum and maximum value for t and thus define the segment of line inside the volume. This segment then replaces the old point pair.

The second step of the line elimination is the removal of all the lines hidden by their own volume. Although this would be done automatically if we went right on to the third step and included the line's own volume among all the others, it is faster to eliminate these back lines ahead of time. The volume matrix associated with a line contains the plane vectors of the model, and two of these planes intersect to form the line. In each point-pair block we keep track of which columns of V are the line's plane vectors. The first component of a plane vector is the x component and will be negative if the plane faces toward the observer. Thus the second step is very simple: a line should be deleted if both the x components of its plane vectors are positive.

The third method of eliminating a line is to see if it is hidden by some other volume. For this test, each remaining line is tested against every volume matrix except its own. The procedure is similar to that of step one except that a two-variable inequality solution must be found this time. The segment of line found is the part to be deleted, and this may be an end, the center, or the complete line.

A volume matrix produces inequalities which tell if a point is in the volume. In step one we used a point with one degree of freedom t along a line. Now the line may be behind the volume so a second degree of freedom must be used to move the point forward through the volume. If a point on the line is moved forward in x in this transformed space, it will have to move through the volume, if it is hidden. The variable point will thus be represented as before with a variable t along the line, but also with a variable α, in the $\bar{x} = [1, 0, 0, 0]$ direction.

Variable point: $\bar{v} = \bar{s} + t\bar{d} + \alpha\bar{x}$
Inequalities: $\bar{v}V \geq 0$
Thus: $\bar{s}V + t\bar{d}V + \alpha\bar{x}V \geq 0$

Now $\bar{x}V$ is just the top row of V, which we can call $\bar{w}$, and $\bar{s}V$ and $\bar{d}V$ can be computed as before.

Define: $\bar{p} = \bar{s}V, \qquad \bar{q} = \bar{d}V, \qquad \bar{w} = \bar{x}V$
Thus for all j: $\cdot\, p_j + tq_j + \alpha w_j \geq 0$
Where: $0 \geq \alpha \geq 1, \qquad 0 \geq t \geq 1$

These inequalities must be solved for the minimum and maximum t for any nonnegative α. A few simple tests allow the removal of equations which are always satisfied or the termination of the test if an equation will never be satisfied. Ignore equation:

$$(q_j > 0) \wedge (p_j \geq 0) \wedge (p_j + w_j \geq 0) \quad (1)$$

$$(q_j < 0) \wedge (p_j > 0) \wedge (p_j + q_j \geq 0) \wedge (p_j + q_j + w_j \geq 0) \quad (2)$$

Quit, leave line:

$$(q_j > 0) \wedge (p_j < 0) \wedge (p_j + q_j \leq 0) \wedge (p_j + q_j + w_j \leq 0) \quad (3)$$

$$(q_j < 0) \wedge (p_j \leq 0) \wedge (p_j + w_j \leq 0) \quad (4)$$

These tests speed up the process considerably since only a few planes of each volume are really involved in the determination of t. To solve the remaining equations, many methods are possible and a very simple one was chosen. Each inequality is considered as an equality and all intersections between these equalities are found. Also, the intersections of these equalities with the boundaries, $\alpha = 0, 1$ and $t = 0, 1$, are found. Each intersection results in a pair (α, t), and these pairs are tested in the inequalities. Any pair satisfying all inequalities and the boundary conditions is used to compute a minimum and maximum t. Because the actual minimum and maximum t must occur at intersections, this is a complete solution. It is also fairly fast since there are usually very few equations. If no pair will satisfy all the inequalities, then the line is not hidden by the volume and is left intact. Otherwise, the section of line corresponding to the solution area, between minimum and maximum t, is deleted.

One case of interest can be modified during this process. We want to eliminate joint lines between touching, tangent objects. A joint line, when processed with the touching solid, is detectable since the inequality caused by the plane through the line will have $q_j = p_j = 0$. If this occurs, the solution space is limited to $\alpha = 0$. Thus the joint line can have a proper solution and will be eliminated. However, this also occurs for joints between objects where the planes are not tangent. We want these lines and must act to preserve them. Therefore, whenever $q_j = p_j = 0$, we make a special test to find out if either of the planes through the line from its own volume is parallel to the plane being tested. We also ask that the tangent planes face the same direction. If parallel planes facing the same way are not found, the line under consideration is kept intact and the rest of the test skipped.

This completes the testing of one line with one volume. The line is then tested against the other volumes. The final result of the complete hidden line removal is a modified list of point pairs. These lines are then displayed. The complete process tends to be time consuming for complicated displays, but the time does not seem to go up as the square of the number of objects, as might be expected. This is because objects far away from the line being processed are quickly disposed of by conditions (3) and (4). Thus the time consumed is on the order of one second per object for displays of up to thirty objects. It is not necessary to do the complete hidden line removal, though, if a faster moving display is required. The complete process can be reserved until the desired view is obtained.

D. Display Construction

Besides being able to display the list structure, the display program has provisions for modifying the list. The picture transformation in the first instance can always be changed by a rotation about each of three axes, a translation in three directions, or a size change. All transformation changes are obtained through the use of four shaft encoder knobs on the computer console. The function of these knobs is selected by means of push buttons. In addition to modifying the picture transformation, the light pen can be used to point out any instance transformation for modification. The pen is pointed at the object to be modified, and a level register indicates which instance level of the object to modify. This method is somewhat crude but does allow any instance to be modified. The transformation changes allowable for objects include rotation, translation, three size components, three skew components, and an over-all size factor. Beyond these transformation controls, any object can be deleted or duplicated. New instances of models can be generated and instances made of present pictures. These controls allow the construction of any list structure possible or the modification of any existing structure. Thus test pictures may be generated to facilitate the testing of this program and the three-dimensional construction program. Some sample pictures generated on the computer appear in Figs. 10*a* through *h*. These photographs illustrate the complexity of three-dimensional arrays which can be constructed on the computer in a few minutes.

VII. Conclusion

In the past, research in the pattern recognition field has been limited to the identification of two-dimensional shapes, mainly because it was thought that any three-dimensional analysis would be more difficult. The idea seems to have been that the two-dimensional work would pave the road for future three-dimensional work. However, progress has been slow and it may well be that the study of three-dimensional projections is an easier step. The human visual field is the result of a projective transformation, and the shapes perceived are independent of this transform. Thus it makes sense to utilize this transformation, since our goal is to recognize the same similarity classes which humans do.

The mathematics necessary to go from a photograph to an object list have been described. A set of transforms is found which takes a set of models

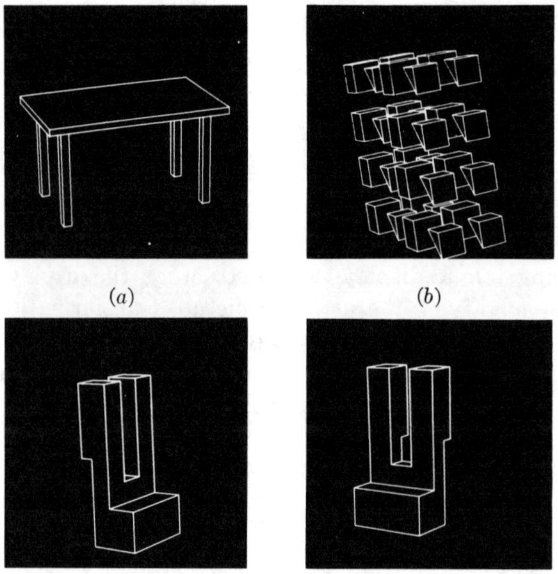

FIG. 10. (*a–d*) Three-dimensional displays: pictures constructed with three-dimensional display program. (*a*) Table. (*b*) Array made with instances. (*c*) Compound object. (*d*) Rotated view of object.

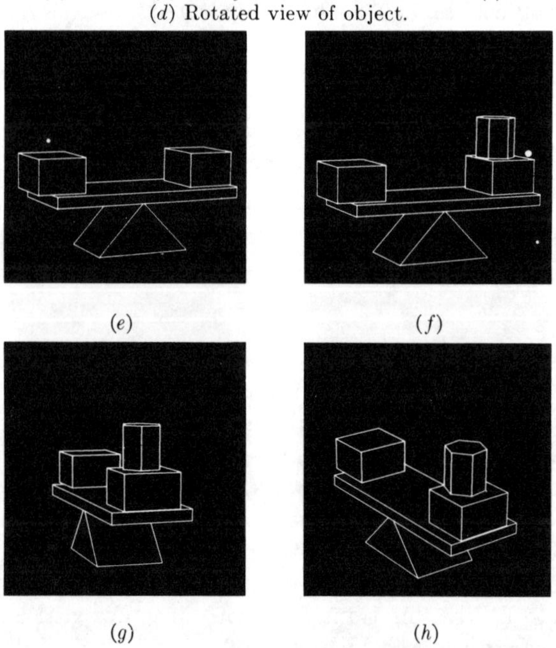

FIG. 10. (*e–h*) Three-dimensional displays: pictures constructed with three-dimensional display program showing hidden line elimination. (*e*) Seesaw. (*f*) With hexagonal prism. (*g*) Second view. (*h*) Balanced.

into the shapes observed. The models are then the invariant shapes which we perceive as the same object from any point of view. Individually or in specific groups, they could be given names.

The process of creating an object list from a picture is mainly mathematical, based on the natural laws of the space around us. It is based on the assumptions of model construction and object support and shows the theoretical implications of these concepts. Four or more points on an object must be seen in order to find the correct model and its transformation. Further, the depth relationships between objects depend on the focal ratio which can either be assumed as a camera constant, or calculated from the picture if the data are accurate enough. If the focal ratio used is wrong, the effect will be a contraction or expansion of the depth dimension. When both the focal ratio and the distance from the camera to the ground are known, the exact size of each object can be calculated.

The two-dimensional to three-dimensional and three-dimensional to two-dimensional programs are completely general, as long as the assumptions of model construction and support are not violated. The display program can handle any structure made up of transformations of models. The construction program will always produce a three-dimensional structure which projects to the given line drawing except that it will eliminate any two-dimensional markings, isolated polygons, and superfluous joint lines. Any drawing produced by the three-dimensional display program will be correctly reconstructed into the three-dimensional structure if the objects are properly supported and sufficiently visible.

The programs are all written for the TX-2 computer at Lincoln Laboratory. The TX-2 is ideal for this type of work because of its large memory for picture storage and its special input-output equipment. During the course of developing these programs, I designed a vector-drawing display for the computer which can draw line and circle segments. This display enables the computer to display a line drawing continuously and still spend most of its time computing new data. Thus it is possible to display rotating objects and have them move fairly smoothly.

The input program has about 5000 instructions and uses over 40,000 registers of data storage for its pictures and lists. It takes about one minute to process a picture into a line drawing of which half is for differentiation. The three-dimensional construction and display programs are each about 3000 instructions and use from 5000 to 40,000 registers of data storage depending upon the number of objects. Both construction and display take about one second per object. All told, a rotated view of the objects in a photograph might be obtained in two minutes.

I foresee at least two uses for this type of picture handling system. First, it could be used for an information reduction system to aid in the transmission of pictorial information. However, the necessity of an ultrahigh-speed computer will probably limit this use. Second, the computer programs will be useful input-output tools for future investigations of three-dimensional processes. The biggest benefit of this investigation, however, is an increased understanding of the possible processes of visual perception.

APPENDIX A
Homogeneous Coordinates

In a homogeneous coordinate system, a fourth coordinate, or scale factor, is used in such a way that the total scale of a vector is unimportant. That is,

$$k\bar{v} \equiv \bar{v}$$

I am using $\equiv$ to indicate that the same point is represented even though the individual components may not be equal. The above form is achieved by defining the point's coordinates, X, Y, Z, in terms of the homogeneous components x, y, z, w as below,

$$X = x/w, \qquad Y = y/w, \qquad Z = z/w$$

When new points are introduced into the system, w may be assigned to any nonzero value and these equations used to find x, y, and z. When points are to be displayed, the same equations are used to find X, Y, and Z. An added advantage for a fixed point computer is gained by using a homogeneous system: w may always be chosen so as to keep the numbers normalized.

A plane is represented by l, m, n, p such that on the plane

$$lx + my + nz + pw = 0$$

I have chosen to represent points by row vectors; therefore, to transform a set of points $\bar{v}_i$, each is postmultiplied by a 4 × 4 matrix H,

$$\bar{v}_i' = \bar{v}_i H$$

The advantage of homogeneous coordinates is that a single transform H can accomplish a full projective transformation. Normally, it is convenient to separate the various functions provided by a transform until they are needed and then multiply them. Below is a breakdown of a transform H, consisting of a rotation by a standard 3 × 3 R matrix, a translation by a vector $\bar{v} = (x, y, z, w)$, a perspective transform from a focal point at f on the x axis, a translation after perspective to a center (y_0, z_0), and a total picture scale factor S. This sequence might represent the transform made by taking a picture with an arbitrary camera orientation and making an enlargement of a section. The transform would take the real space points X, Y, Z into a Y', Z' on the print. An X' will be formed which can be used for eliminating object overlap.

$$H = \underset{\text{Rotation}}{\begin{bmatrix} & [R] & & 0 \\ & & & 0 \\ & & & 0 \\ 0 & 0 & 0 & 1 \end{bmatrix}} \underset{\text{Translation}}{\begin{bmatrix} w & & & \\ & w & & \\ & & w & \\ x & y & z & w \end{bmatrix}} \underset{\text{Perspective}}{\begin{bmatrix} f & & & -1 \\ & f & & \\ & & f & \\ & & & f \end{bmatrix}} \underset{\text{Translation}}{\begin{bmatrix} 1 & & & \\ & 1 & & \\ & & 1 & \\ & y_0 & z_0 & 1 \end{bmatrix}} \underset{\text{Scale}}{\begin{bmatrix} 1 & & & \\ & 1 & & \\ & & 1 & \\ & & & S \end{bmatrix}}$$

It should be noted that a set of points $\bar{v}_i$ can be written as the successive rows of a matrix V and transformed by one matrix equation, $V' = VH$.

Also, a plane equation can be expressed as a scalar product in terms of $\bar{n} = (l, m, n, p)$,

$$\bar{v} \cdot \bar{n} = 0$$

A plane normal n can be transformed along with the points of a space,

$$\bar{v}' = \bar{v}H, \qquad \bar{n}' = \bar{n}(H^{-1})^T, \qquad \bar{v}' \cdot \bar{n}' = 0$$

Of course, if the transform is orthogonal, $(H^{-1})^T = H$.

APPENDIX B
Similarity Test

We are given a matrix A of n points (x, y, z, w) from a model and want to find a transform H that will most nearly fit n points (y, z, w) in a matrix B. Thus we hope for

$$AH \cong B$$

However, we cannot write an equality sign above without introducing a diagonal scale matrix D which will allow the w_i values of AH to differ from the w_i of B,

$$AH = DB$$

We now have 12 unknowns in H (3×4) and n unknowns in D $(n \times n)$. Matrix A is $4 \times n$ and matrix B is $3 \times n$, creating $3n$ equations. Therefore, $n \geq 6$ should produce a complete solution. We shall use a minimum square-error technique to solve the equations. Thus we wish to minimize the squared error in each equation. We shall use A' to indicate the transpose of A,

$$\frac{\partial}{\partial h_{lm}} \left[\sum_i^n \sum_k^3 \left(\sum_j^4 a_{ij} h_{jk} - d_i b_{ik} \right)^2 \right] = 0$$

$$\sum_j^4 h_{jm} \sum_i^n a_{il} a_{ij} = \sum_i^n d_i a_{il} b_{im}$$

or $A'AH = A'DB$. Thus

$$H = (A'A)^{-1}A'DB$$

Now we must find D,

$$\frac{\partial}{\partial d_l} \left[\sum_i^n \sum_k^3 \left(\sum_j^4 a_{ij} h_{jk} - d_i b_{ik} \right)^2 \right] = 0$$

$$\sum_j^4 a_{lj} \sum_k^3 h_{jk} b_{lk} = d_l \sum_k^3 b_{lk}{}^2$$

Thus the diagonal terms of AHB' equal those of DBB'. Substituting for H and making the definition

$$G = A(A'A)^{-1}A' - I$$

we obtain a matrix $GDBB'$, which has zero diagonal terms.

$$\text{Define: } Q = BB'$$

If we now multiply the terms of G by those of Q', term by term, we get a new $n \times n$ matrix S,

$$S_{ij} = g_{ij}q_{ji}$$

$$\text{Define: } \vec{d} = d_1 \cdots d_n$$

Now the vector $\vec{d}$ or diagonal terms of D can be found by solving

$$Sd = 0$$

This equation requires S to be singular, with degeneracy at least one. If the degeneracy is one, the problem is solved since the common scale factor of D and H is unimportant. A degeneracy more than one means that too few equations were used ($n < 6$) or that the picture had no perspective. However, by assuming a value of unity for each undefined d_j, an accurate, but not complete solution, can be found.

When D is found, H can be found,

$$H = (A'A)^{-1}A'DB$$

It should be noted that for $n = 4$, A^{-1} will probably exist and in this case the best solution obtainable is one with no perspective,

$$H = A^{-1}B$$

If solutions without perspective are expected, the matrix D is unnecessary and the ordinary minimum square-error solution holds,

$$H = (A'A)^{-1}A'B$$

An error criteria can be found to indicate the mismatch of model and picture. An error matrix E is found,

$$E = AH - DB \quad \text{or} \quad E = GDB$$

Now if the sum of the squares of the components of E is taken, this number can be used to indicate the error magnitude. If one row of E contributes the main error, this point of B probably should not be mapped to the model A.

Acknowledgments

I am indebted to Professor P. Elias for his valuable suggestions and criticisms in his capacity of Thesis Supervisor. His advice, as well as that of Professors Claude Shannon, Murray Eden, and Thomas Stockham, has been very helpful in encouraging and guiding this work. I also wish to thank Ivan Sutherland for his help and list structure subroutines; Leonard Hantman, who programed the major part of the three-dimensional display process; and the M.I.T. Lincoln Laboratory staff for their cooperation in the use of the TX-2 computer.

References

1. Somerville, D. M. Y., *Analytical Geometry of Three Dimensions,* Cambridge University Press, 1959.
2. Roberts, L. G., "Pattern Recognition with an Adaptive Network," *IRE Intern. Conv. Record,* Pt. 2, 66–70 (1960).
3. Selfridge, O. G., and U. Neisser, "Pattern Recognition by Machine," *Sci. Am., 203,* 60–68 (August 1960).
4. Hodes, L., *Machine Processing of Line Drawings,* Report 54G-0028[U,] Lincoln Laboratory, Massachusetts Institute of Technology (March 1961).
5. Julesz, B., "Toward the Automation of Binocular Depth Perception," *Proceedings of the I.F.I.P. Congress,* Munich (1962).
6. Roberts, L. G., "Picture Coding Using Pseudo-Random Noise," *IRE Trans., IT-8,* 145–154 (1962).
7. Gibson, J. J., *The Perception of the Visual World,* Houghton Mifflin Co., Boston, 1950.
8. Ittelson, W. H., "Size As a Cue to Distance," *Am. J. Psychol., 64,* 54–67 (1951).
9. Attneave, F., and M. D. Arnoult, "The Quantitative Study of Shape and Pattern Perception," *Psychol. Bull., 53,* 452 (1956).
10. Langdon, J., "The Perception of 3-D Solids," *Quart. J. Exptl. Psychol., 7,* 19–36 (1955).
11. Stevens, S. S., "The Psychophysiology of Vision," *Sensory Communication,* W. Rosenblith, Ed., M.I.T. Press, Cambridge, and John Wiley & Sons, Inc., New York, 1961, p. 13.
12. Sutherland, I. E., *Sketchpad: A Man-Machine Graphical Communication System,* Technical Report No. 296[U], Lincoln Laboratory, Massachusetts Institute of Technology, 30 January 1963.
13. Johnson, T., *Sketchpad III, 3-D, Graphical, Communication with a Digital Computer,* S.M. Thesis, Department of Mechanical Engineering, Massachusetts Institute of Technology, Cambridge, Massachusetts, June 1963.

Decomposition of a visual scene into three-dimensional bodies

by ADOLFO GUZMÁN

Massachusetts Institute of Technology
Cambridge, Massachusetts

INTRODUCTION

We consider visual scenes composed by the optical image of a group of bodies. When such a scene is "seen" by a computer through a film spot scanner, image dissector, or similar device, it can be treated as a two-dimensional array of numbers, or as a function of two variables.

At a higher level, a scene could be meaningfully described as a conglomerate of points, lines and surfaces, with properties (coordinates, slopes, ...) attached to them.

Still a more sophisticated description could use terms concerning the bodies or objects which compose such a scene, indicating their positions, inter-relations, etc.

This paper describes a program which finds bodies in a scene, presumably formed by three-dimensional objects. Some of them may not be completely visible. The picture is presented as a line drawing.

When SEE—the pretentious name of the program—analyzes the scene **TRIAL** (see Figure 1 'TRIAL'), the results are:

(BODY 1. IS :6 :2 :1)

(BODY 2. IS :11 :12 :10)

(BODY 3. IS :4 :9 :5 :7 :3 :8 :13)

SEE looks for three-dimensional objects in a two-dimensional scene. The scene itself is not obtained from a visual input device, or from an array of intensities or brightness. Rather, it is assumed that a preprocessing of some sort has taken place, and the scene to be analyzed is available in a symbolic format (to be described in a later Section), in terms of points (vertices), lines (edges), and surface (regions).

SEE does not have a pre-concieved idea of the form or model of the objects which could appear in a given scene. The only supposition is that the bodies are solid objects formed by plane surfaces; in this way, it can not find "cubes" or "houses" in a scene, since it does not know what a "house" is. Once SEE has partitioned a scene into bodies, some other program will work on them and decide which of those bodies are "houses."

Thus, SEE is intended to serve as a link between a pre-processor [1,2] which transforms intensity pictures into point or line pictures, [5] and a recognizer (such as TD [3] or DT [4]), which handles this line picture and finds bodies, objects or zones matching with certain patterns or models. Instead of searching through the whole scene looking for parts to match its models, the work of the recognizer becomes simpler after SEE has partitioned the scene into bodies, because the data to be searched (matched) are smaller and better organized.

The analysis which SEE makes of the different scenes generally agrees with human opinion, although in some ambiguous cases it behaves rather conservatively. Distributed over these pages, the reader will find examples of scenes analyzed by SEE, and the peculiarities and behavior of the program will become clear.

The program SEE, written in LISP, has been tested in the PDP–6 machine of the Artificial Intelligence Group, Project MAC, at Massachusetts Institute of Technology. A preliminary version, written in CONVERT, [6] was used extensively for a quick test of ideas which shaped the program to its actual form. The analysis of a scene takes from 30 to 90 seconds, with the program running interpreted under the interpreter of the LISP programming system.

A more technical description of SEE can be found in an unpublished memorandum. [7]

Related work

Rudd H. Canaday [8] in 1962 analyzed scenes composed of two-dimensional overlapping objects, "straight-

Reprinted with permission from *Proc. AFIPS Fall Joint Comput. Conf.*, vol. 33, Dec. 1968, pp. 291–304.

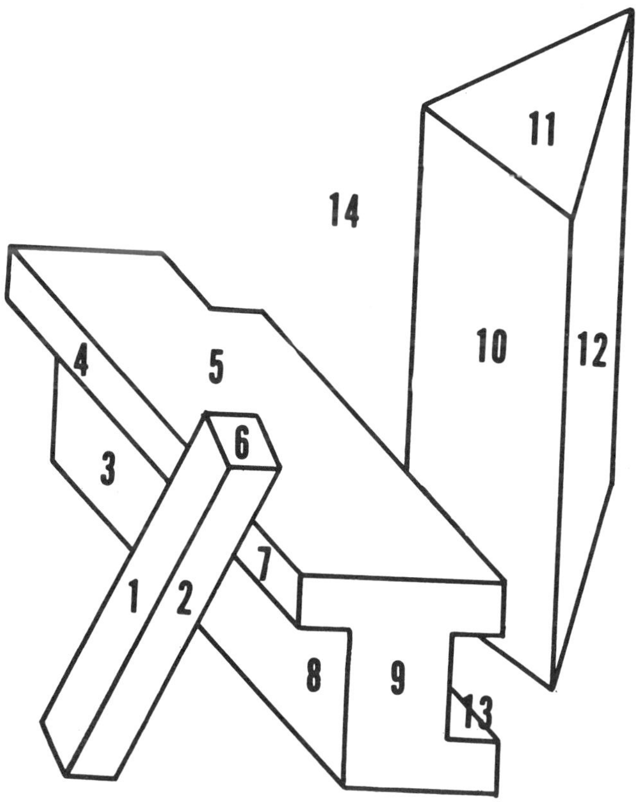

FIGURE 1—TRIAL

The program analyzes this scene and finds 3 bodies:

(BODY 1 IS :6 :2 :1)
(BODY 2 IS :11 :12 :10)
(BODY 3 IS :4 :9 :5 :7 :3 :8 :13)

sided pieces of cardboard." His program breaks the image into its component parts (the pieces of cardboard), describes each one, gives the depth of each part in the image (or scene), and states which parts cover which.

Roberts [9] in 1963 described programs that (1) convert a picture (a scene) into a line drawing and (2) produce a three-dimensional description of the objects shown in the drawing in terms of models and their transformations. The main restriction on the lines is that they should be a perspective projection of the surface boundaries of a set of three-dimensional objects with planar surfaces. He relies on perspective and numerical computations, while SEE uses a heuristic and symbolic (i.e., non-numerical) approach. Also, SEE does not need models to isolate bodies. Roberts' work is probably the most important and closest to ours.

Actually, several research groups (at Massachusetts Institute of Technology, [10] at Stanford University, [11] at Stanford Research Institute [12]) work actively towards the realization of a mechanical manipulator, i.e., an intelligent automata who could visually perceive and successfully interact with its enviornment, under the control of a computer. Naturally, the mechanization of visual perception forms part of their research, and important work begins to emerge from them in this area.

Organization of the paper

It is formed by the following headings:

- Introduction and related previous work.
- Input Format. The representation of the scene as it is entered to the computer.
- Format of a Scene. The representation of the scene as SEE expects.
- Type of Vertices. Classification of vertices according to their topology.
- The program. Analysis of the algorithm, description of heuristics.
- Interesting examples. Discussion. Future work.

Input format

For testing purposes, the scenes are entered by hand in a simplified format (called input format), and then some routines convert this to the form required by SEE. Eventually, the data will come from a visual input device, through a preprocessor.[2,6]

Examples of a scene

Suppose we want to describe the scene 'CUBE.' We begin by giving (in LISP) a value to 'CUBE.' (See Figure 2 'Cube')

```
(SETQ CUBE (QUOTE (A 1.0 1.0 (:1 B :4 G)
                   B 1.0 5.0 (:1 E :2 C :4 A)
                   C 3.0 7.0 (:2 D :4 B)
                   D 8.0 7.0 (:2 E :3 F :4 C)
                   E 6.0 5.0 (:2 B :1 G :3 D)
                   F 8.0 3.0 (:3 G :4 D)
                   G 6.0 1.0 (:1 A :4 F :3 E)
                   )))
```

Thus we associate with each vertex its coordinates and a list, in counterclockwise order, of regions and vertices radiating from that vertex.

The conversion of the scene, as just given, into the form which SEE expects, is made by the function LLENA; thus, (LLENA CUBE) will put in the prop-

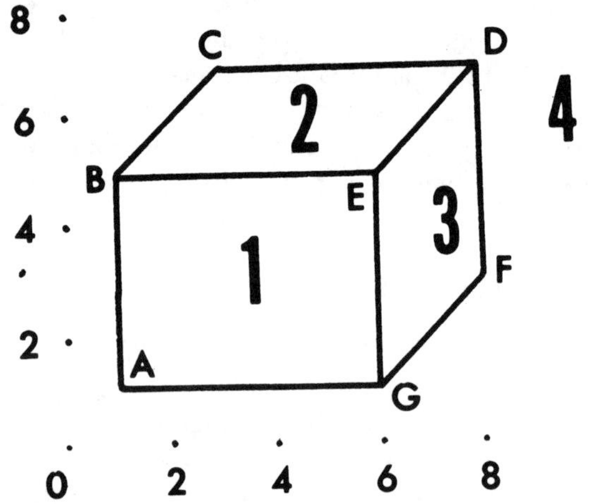

FIGURE 2—'CUBE'—A scene.

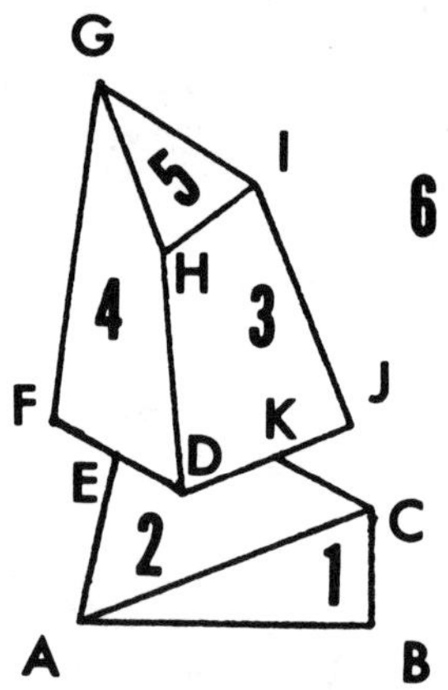

FIGURE 3—'ONE.' A scene. Vertices and surfaces (regions) are the main components of a scene.

erty list of CUBE the properties REGIONS and VERTICES; in the property list of each vertex, the properties XCOR, YCOR, NREGIONS, NVERTICES and KIND are placed; and in the property list of each region, it places the properties NEIGHBORS and KVERTICES. It also marks region :4 as part of the background.

In other words, LLENA converts a scene from the 'Input Fonnat' to the 'Format of a Scene' described in the next section.

Format of a scene

A scene is presented to our program as a scene in a special symbolic format, which we now describe. Essentially, it is an arrangement of relations between vertices and regions.

A scene has a name which identifies it; this name is an atom whose property list contains the properties 'REGIONS,' 'VERTICES' and 'BACKGROUND.' For example, the scene ONE (see Figure 3 'ONE') has the name 'ONE.' In the poperty list of 'ONE' we find

REGIONS —(:1 :2 :3 :4 :5 :6)
Unordered list of regions composing scene ONE.

VERTICES —(A B C D E F G H I J K)
Unordered list of vertices composing the scene ONE.

BACKGROUND— (:6)
Unordered list of regions composing the background of the scene ONE.

Region

A region corresponds to a surface limited by a simple connected curve. For instance, in ONE, the surface delimited by the vertices A B C is a region, called :1, but G I J K D H is not.

Each region has as name an atom which possesses additional properties describing different attributes of the region in question. These are 'NEIGHBORS,' 'KVERTICES,' and 'FOOP'. For example, the region in scene ONE formed by the lines AE, ED, DK, KC, CA, has ':2' at its name. In the property list of :2 we find:

NEIGHBORS — (:3 :4 :6 :1 :6)
Counterclockwise ordered list of all regions which are neighbors to :2. For each region, this list is unique up to cyclic permutation.

KVERTICES — (D E A C K)
Counterclockwise ordered list of all vertices which belong to the region :2. This list is unique up to cyclic permutation.

FOOP — (:3 D :4 E :6 A :1 C :6 K)
Counterclockwise ordered list of alternating neighbors and kvertices of :2. This list is unique up to cyclic permutation.

The FOOP property of a region is formed by a man who walks on its boundary always having this region to his left, and takes note of the regions to his right and of the vertices which he finds in his way.

Vertex

A vertex is the point where two or more lines of the scene meet; for instance, A, G, and K are vertices of the scene ONE.

Each vertex has as name an atom which posseses additional properties describing different attributes of the vertex in question. These are 'XCOR,' 'YCOR,' 'NVERTICES,' 'NREGIONS,' and 'KIND.' For example, the vertex H (see scene ONE) has in its property list:

XCOR	—	3.0 x-coordinate
YCOR	—	15.0 y-coordinate
NVERTICES	—	(I G D) Counterclockwise ordered list of vertices to which H is connected. This list is unique up to cyclic permutation.
NREGIONS	—	(:3 :5 :4) Counterclockwise ordered list of regions to which H belongs. This list is unique up to cyclic permutation.
KIND	—	(:3 I :5G :4 D) Counterclockwise ordered list of alternating nregions and nvertices of H. This list is unique up to cyclic permutation.

The KIND property of a vertex is formed by a man who stands at the vertex and, while rotating counterclockwise, takes note of the regions and vertices which he sees.

NREGIONS and NVERTICES are then easily derived from KIND: take the odd positioned elements of KIND, and its even-positioned elements, respectively.

Types of vertices

Vertices are classified according to the slope, disposition and number of lines which form them. The function (TYPEGENERATOR L), where L is a list of vertices, performs this classification, putting in the property list of each one of the elements of L, the type to which it belongs.

The TYPE of a vertex is always a list of two elements; the first is the *type-name*: one of 'L,' 'FORK,' 'ARROW,' 'T,' 'K,' 'X,' 'PEAK,' 'MULTI'; the second element is the *datum*, which generally is a list, whose form varies with the *type-name* and contains information in a determined order about the vertex in question. (See Table I 'VERTICES').

TABLE I—'VERTICES' Classification of vertices of a scene.

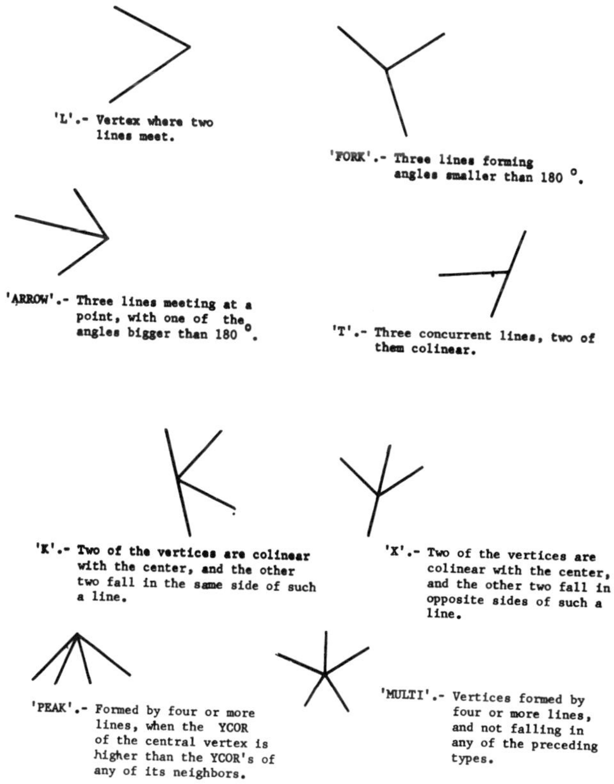

The program

The program named SEE accepts a scene expressed in the notation described above and produces outputs lists identifying and describing the bodies present in the scene.

In this section we describe the program, and how it achieves its goals, by discussing the procedures, heuristics etc., employed and the way they work. We begin with several examples.

Example 1. Scene **'STACK'** This scene (see Figure 4 'STACK') is analyzed by SEE, with the following results:

```
(LOCAL ASSUMES (:5) (:13 :14) SAME BODY)
    (BODY 1. IS :1 :3 :2 :18 :17)
    (BODY 2. IS :4 :16 :15)
    (BODY 3. IS :7 :6 :11 :12)
    (BODY 4. IS :9 :8 :10)
    (BODY 5. IS :13 :14 :5)
    (BODY 6. IS :20 :19)
```

Results for scene STACK

Example 2. Scene **'BRIDGE.'** With this example,

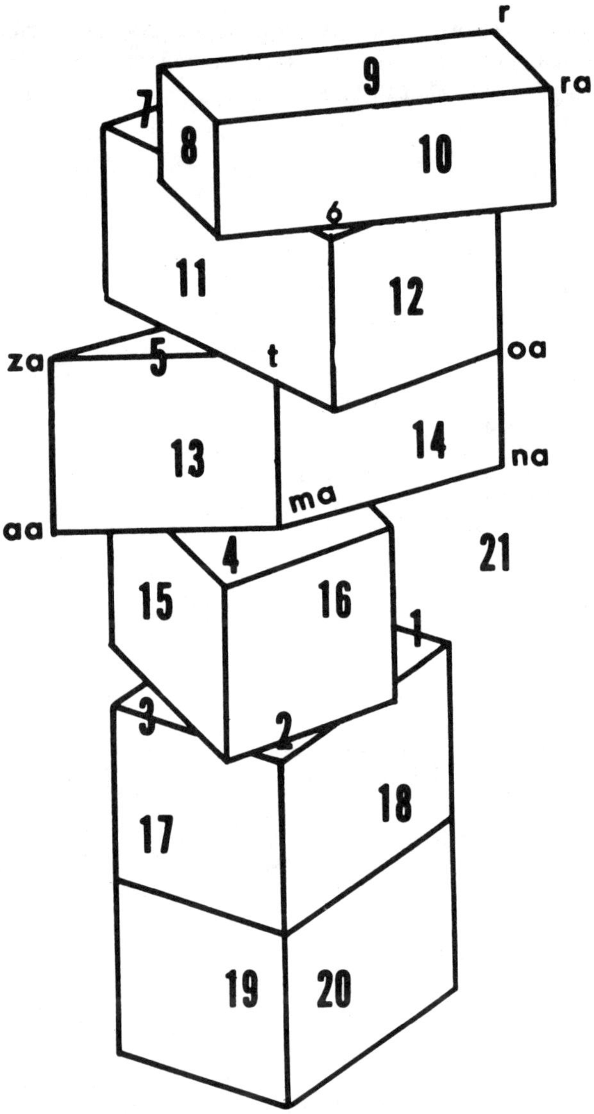

FIGURE 4—'STACK.' This scene is analyzed by SEE, with results detailed in Example 1. All bodies are correctly found. Some of the vertices appear in the drawing with their names; in other drawings we will omit the names; we are also omiting the coordinate axes.

we will give details of the program's operation. We start by loading into LISP the programs for SEE and its auxiliary functions.
Then, we evaluate:

(UREAD BRIDGE SCEN3) ↑Q
This causes scene BRIDGE (see Figure 5 'BRIDGE'), to be read (in the format described previously) and transformed to the proper form which SEE expects (also described before).

Finally, we evaluate

(SEE (QUOTE BRIDGE))
This calls SEE to work on BRIDGE.

Results appear in Table II, 'RESULTS FOR BRIDGE.'

(LOCAL ASSUMES (:18) (:19) SAME BODY)
(LOCAL ASSUMES (:28) (:29) SAME BODY)
(LOCAL ASSUMES (:10) (:8 :11 :5 :6 :4) SAME BODY)
(LOCAL ASSUMES (:7) (:8 :11 :5 :6 :4 :10) SAME BODY)

(SINGLEBODY ASSUMES (:19 :18) (:16) SAME BODY)

RESULTS
(BODY 1. IS :24 :9 :21 :27 :12 :25)
(BODY 2. IS :22 :26 :23)
(BODY 3. IS :17 :3 :20)
(BODY 4. IS :1 :2) Results for scene 'BRIDGE'
(BODY 5. IS :14 :15 :13)
(BODY 6. IS :19 :18 :16)
(BODY 7. IS :29 :28)
(BODY 8. IS :8 :11 :5 :6 :4 :10 :7)

TABLE II—RESULTS FOR BRIDGE

SEE finds eight bodies in scene 'BRIDGE' (See Figure 5 'BRIDGE'). Unnecessary detail has been removed from the drawings; the reader must bear in mind that the complete scene contains vertices (whose names do not appear in the drawings) and their coordinates (which are also not shown), as well as edges and regions.

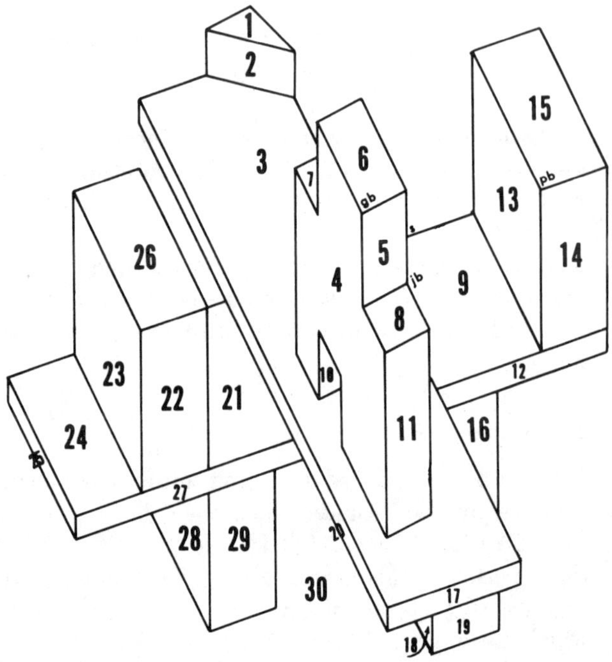

FIGURE 5—'BRIDGE.' The long body :25 :24 :27 :21 :9 :12 is correctly identified. (See Table II)

SEE and its parts

The operation of SEE is quite straightforward; the program is not recursive and does not do any tree search. Its main parts, which are executed one after another, unless otherwise indicated, are:

FILLSCENE. The properties SLOP and TYPE are generated for each vertex, if they were not already present.

CLEAN. Old properties used internally by SEE are removed from vertices and regions.

EVIDENCE. An analysis is made of vertices, regions and associated information, in search of clues that indicate that two regions form part of the same body. If evidence exists that two regions in fact belong to the same body, they are linked or marked with a "gensym" (both receive the same new label). There are two kinds of links, called strong (global) or weak (local).

LOCALEVIDENCE. Some features of the scene will weakly suggest that a group of regions should be considered together, as part of the same body. This part of the program is that which produces the 'local' links or evidence.

GLOBAL. The 'strong' links gathered so far are analyzed; regions are grouped into "nuclei" of bodies, which grow until some conditions fail to be satisfied (a detailed explanation follows later).

LOCAL. Weak evidence is taken into account for deciding which of the unsatisfactory global links should be considered satisfactory, and the corresponding nuclei of bodies are then joined to form a single and bigger nucleus. Assumptions made by LOCAL are printed (see output of SEE). LOCAL may call GLOBAL again, or go on.

SINGLEBODY. If a single region does not belong to a larger nucleus, but is linked by one strong evidence to another region, it is incorporated into the nucleus of that other region. This part of the program may call GLOBAL or LOCAL again, if necessary, or continue. SINGLEBODY also prints its assumptions.

RESULTS. The regions belonging to the background are screened out, and the results are printed.

Operation

Here is explained in considerable detail each of the parts of SEE that have been sketched above. This will help the reader understand the behavior of the program, its strength and deficiencies.

Example. Scene **'TOWER.'** First, we begin by showing a typical analysis of SEE with a somewhat complicated scene (see Figure 6 'TOWER'). Most of the scenes contain several "nasty" coincidences: a vertex of an object lies precisely on the edge of another object; two nearly parallel lines are merged into a single one, etc. This has been done on purpose, since a non-sophisticated preprocessor will tend to make this kind of error.

The output is given in Table III, 'RESULTS FOR TOWER.'

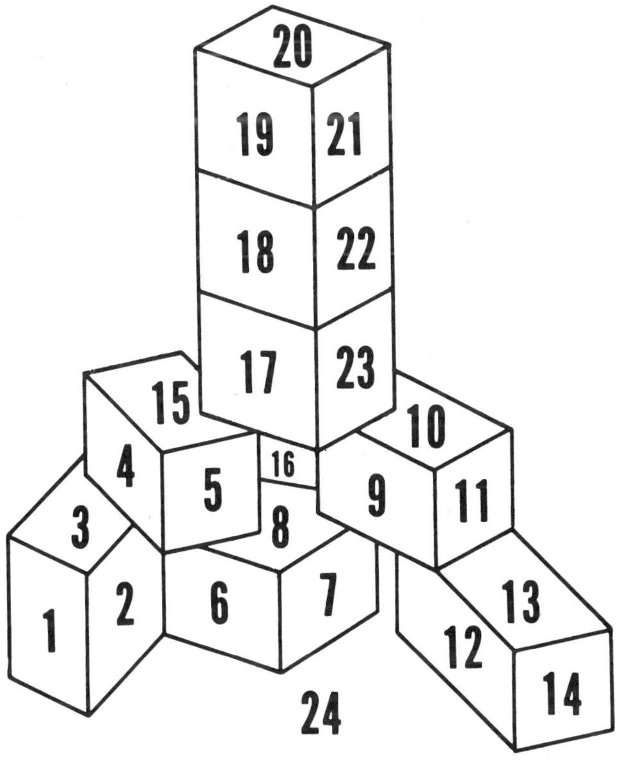

FIGURE 6—'TOWER.' Neither LOCAL nor SINGLEBODY are necessary to correctly parse this scene into the bodies which form it. There are many 'strong' links in this scene and SEE makes good use of them. (See Table III)

```
(BODY 1. IS :3 :2 :1)
(BODY 2. IS :5 :15 :4)
(BODY 3. IS :23 :17)
(BODY 4. IS :6 :7 :8)
(BODY 5. IS :10 :11 :9)          Results for TOWER
(BODY 6. IS :13 :14 :12)
(BODY 7. IS :18 :22)
(BODY 8. IS :20 :19 :21)
```

TABLE III—Results for Figure 6 'TOWER'

FILLSCENE, CLEAN. These two parts of SEE are simple; if necessary, FILLSCENE calls SLOPGENERATOR and TYPEGENERATOR; CLEAN removes some unwanted properties.

EVIDENCE. Strong or global links are placed by this part of the program. Two functions are used: EVERTICES and TJOINTS.

EVERTICES. This function considers each vertex of the scene under the following rules:

L. Vertices of the type 'L' are not considered.

FORK. If three regions meet at a vertex of the FORK type (Table IV (b)), and none of them is the background, links between them will be formed. For instance in Figure 6 'TOWER,' we establish the links :19— :20, :19— :21, and :20— :21. Nevertheless, some of these links may not be produced: in Figure 7 'FORK,' the link between :3 and :2 is *not* produced because Q is a "passing T"; the link between :1 and :3 is *not* generated because R is an 'L.' The link between :1 and :2 *is* generated.

This heuristic is powerful, and (see Figure 5 'BRIDGE') allows us, for instance, while working on vertex jb, to omit the link between regions :5 and :9 in scene 'BRIDGE.' Nevertheless, this same heuristic puts a link between regions :8 and :9 of the same scene. As we shall see later, this situation is not too bad.

ARROW. This type of vertex (Table IV (c)) causes two of its regions (the 'left' and the 'right' one) to be joined; in Table IV (c), we will put a link between :1 and :2, which counts as evidence that :1 and :2 belong to the same body. Nothing is said about :3.

For instance, this type of vertex joints 1: and :2 in Figure 5 'BRIDGE.'

X. Two cases are distinguished:
- (a) The X is formed by the intersection of two lines. No evidence is formed.
- (b) Otherwise (Table IV (f)), links :1— :2 and :3— :4 are formed. This type of X occurs when we have piles of objects; for instance, in Figure 6 'TOWER,' :18 and :22 are considered as belonging to the same body, due to this type of vertex.

PEAK. All its regions (table IV (h)), except the one containing the obtuse angle, are linked to each other. See also Figure 14 'CORN.'

T. A search is made for another T to match the vertex currently analyzed; two T's match if they are colinear and "facing each other;" if there are several pairs, the closest is chosen. For instance (Table IV (d)), A and B are paired. An indicator 'NEXTE' is placed in such vertices.
- (a) Once the NEXTE is determined, EVERTICES establishes a link between :1 and :2 (Table IV (d)), and another between :3 and :4. These links will not be produced if the results of them is to associate the background with something that is not the background.
- (b) The following test is done (Figure 8 'PARALLEL'): If neither :1 or :2 is the background, but both have it as neighbor, and in some part of the boundary of :1 (and the same holds for :2) with the background, the segment them (M — K in Figure 8 'PARALLEL') is parallel to the central segment (O—P) of the T, then :1 and :2 are linked. For instance, in Figure 4 'STACK,' this analysis applied to the vertex T will produce evidence that :13 and :14 belong to the same body (since ZA—AA, T—MA and OA—NA are parallel). This is a rather global heuristic although only useful for bodies with parallel faces. Also, EVERTICES classifies T's according to the slope of their central segments.

A summary of the strong links put by EVERTICES is found in Table IV 'GLOBAL EVIDENCE.'

TJOINTS. This function actuates on T's and established global evidence as described in part (a) of T (Table IV (d)).

LOCALEVIDENCE. An arrow is a leg if one of its left or right vertices is a corner (if necessary, through a chain of matched T's) which has a side parallel to the central segment of the arrow. The function LEGS finds legs in the scene, and stores this information in a list of 'weak' links (see Figure 9 'LEGS').

GLOBAL. Strong evidence is analyzed in this part of the program. When this section is entered, several links (strong and weak) exist among the different regions of the scene. These links are shown pictorially in Eigure 10 'LINKS,' for the scene 'BRIDGE' (see both). All the links to the background (:30) are deleted: the background cannot be part of any body.

Definition: a *nucleus* (of a body) is either a region or a set of nuclei which has been formed by the following rule.

TABLE IV—'GLOBAL EVIDENCE' The strong links are represented by dotted lines

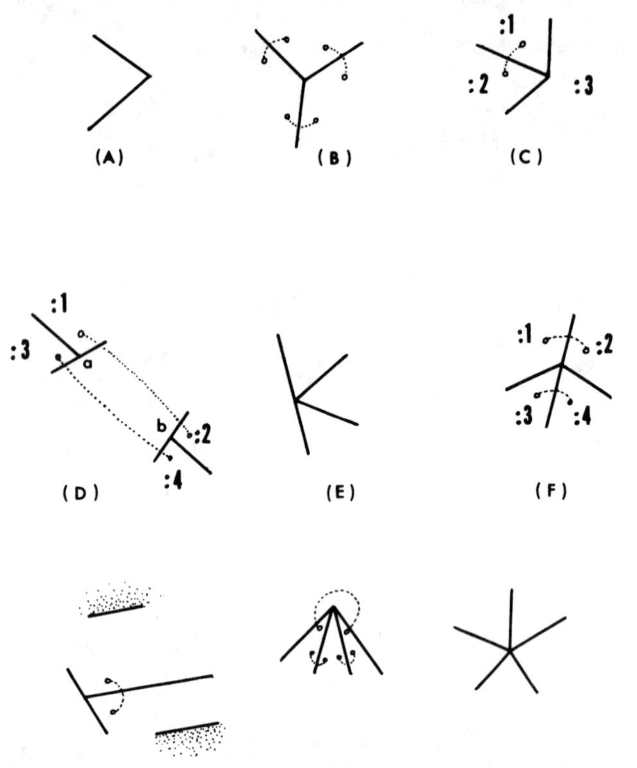

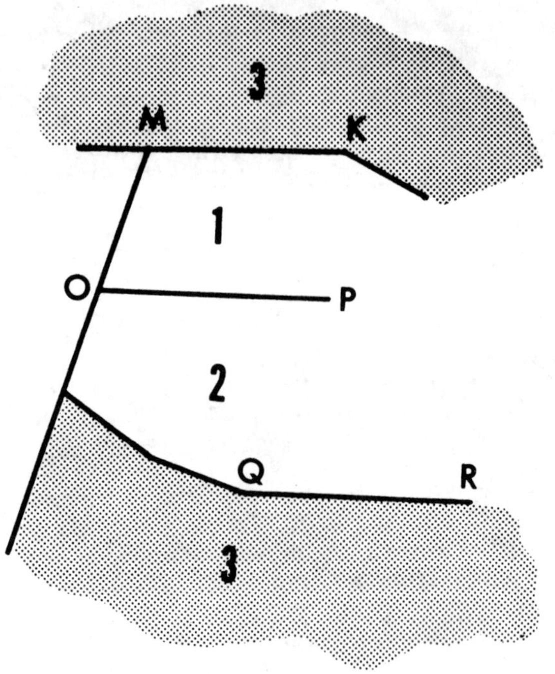

FIGURE 8—'PARALLEL.' A link is established between :1 and :2 because they do not belong to the background, but the background is a neighbor of both of them, and the segment that separates :1 from the bacgkround (and the same is true for :2) is parallel to OP, the central segment of the T in question (:3 is the background).

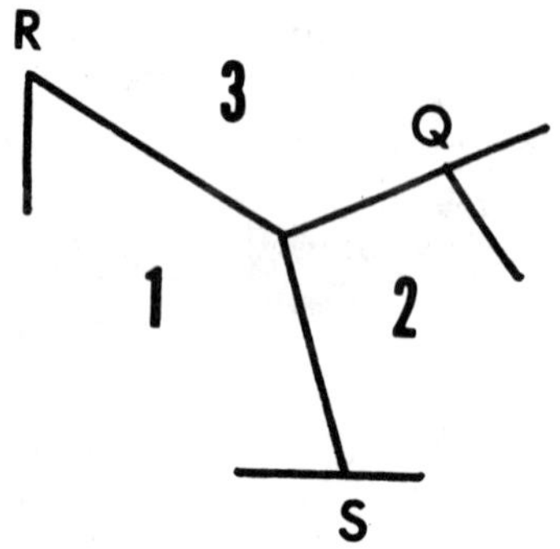

FIGURE 7—'FORK.'

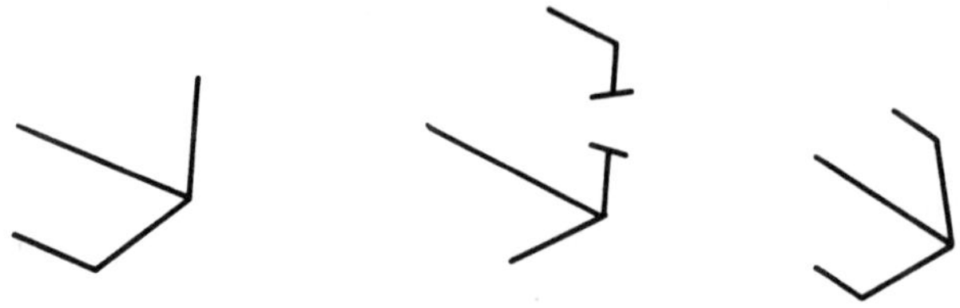

FIGURE 9—'LEGS.'
Three different types of legs.

Rule: If two nuclei are connected by two or more links, they are merged into a larger nucleus by concatenation.

(Two nuclei A and B are linked if regions a and b are linked where a ϵ A and b ϵ B). For instance, regions :8 and :11 are put together, because there exists two links among them, to form the nucleus :8—11. Now, we see that region :4 (see Figure 10 'LINKS') has two links with this nucleus :8—11, and therefore the new nucleus :8—11 = 4 is formed.

We let the nuclei grow and merge under the former rule, until no new nuclei can be formed. When this is the case, the scene has been partitioned into several "maximal" nuclei; between any two of these there are zero or, at most, one link. For example, figure 10 'LINKS' will be transformed into Figure 11 'NUCLEI.'

LOCAL. If some strong link joining two "maximal" nuclei is also reinforced by a weak link, these nuclei are merged.

For example, in scene BRIDGE (see Figure 5 'BRIDGE') the following local links exist (among others): :9 to :4, :10 to :4, :28 to :29, :18 to :19. Therefore, the corresponding nyclei are merged and now figure NUCLEI is transformed into figure 'NEW NUCLEI.'

A weak link does not cause the regions which it links to be merged or considered as belonging to the same body unless there is, in addition, one strong evidence between such regions. LOCAL may call GLOBAL again, if necessary.

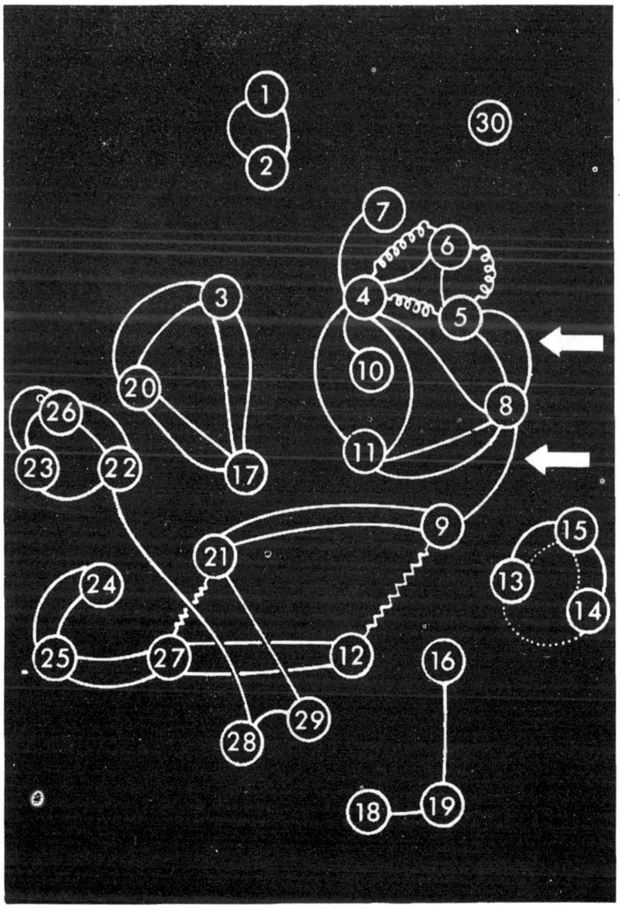

FIGURE 10—'LINKS.' This figure represents scene 'BRIDGE,' with the strong links between its regions (represented here by circles) shown. The dotted links represent the evidence generated by the vertex PB (see Figure 5 'BRIDGE'). The short arrows show the links put by vertex JB; note that a link between :5 and :9 is not put, because S (see Figure 5 'BRIDGE') is a passing t-joint. Zig-zag links are produced by the mechanism described in part (b) of T. (See Figure 8 'PARALLEL'). Curled links are produced by vertex GB; even if this vertex were occluded, and the links were missing, there is still enough evidence to identify regions :4 :5 and :6 as belonging to the same body. Weak links are not shown.

SINGLEBODY. A strong link joining a nucleus and another nucleus composed by a single region is considered enough evidence and the nuclei in question merged, if there is no other link emanating from the single region. For instance, in Figure 12 'NEW NUCLEI,' nucleus :16 is merged with nucleus :18—:19 (see Figure 13 'FINAL'). Nucleus :28 - 29 is not joined with :26—22—23 or with :24—25—27—12—21—9. Even if nucleus :28—29 were composed by a single region, it still will not be merged, since two links emerged from it: two nuclei claim its possession.

RESULTS. After having screened out the regions that belong to the background, the nuclei are printed as "bodies."

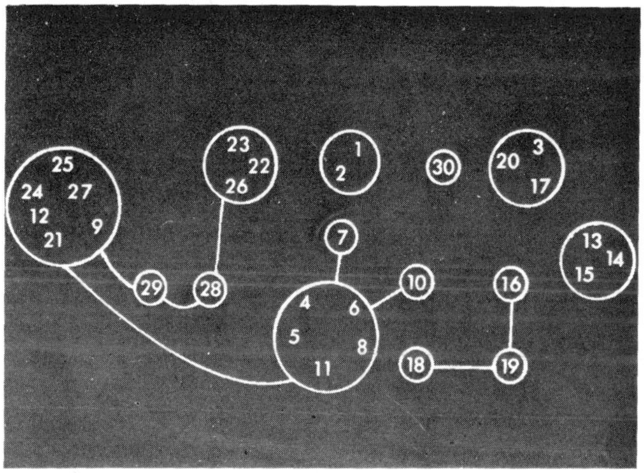

FIGURE 11—'NUCLEI.' After joining all nuclei having two or more links in common, the representation for the scene 'BRIDGE' changes from that shown in figure 10 'LINKS' to the one shown here.

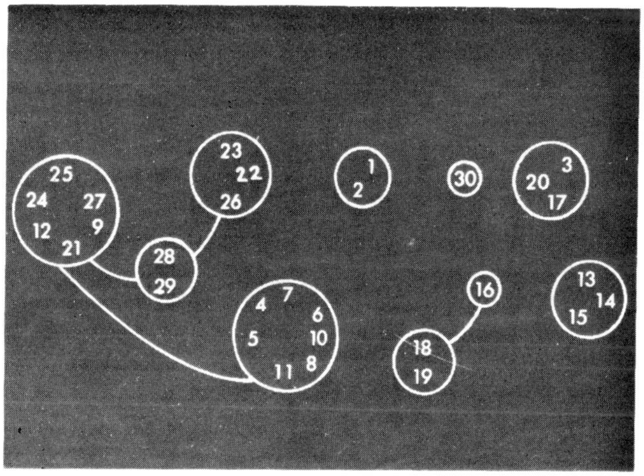

FIGURE 12—'NEW NUCLEI.' The figure shows the scene 'BRIDGE,' after LOCAL transforms it from the representation in Figure 11 'NUCLEI' to the one shown here.

In this process, the links which may be joining some of the nuclei are ignored: RESULTS considers the links of Figure 13 'FINAL', for instance, as non-existent.

Summary

SEE uses a variety of kinds of evidence to "link" together regions of a scene. The links in SEE are supposed to be general enough to make SEE an object-analysis system. Each link is a piece of evidence that suggests that two or more regions come from the same object, and regions that get tied together by enough evidence are considered as "nuclei" of possible objects.

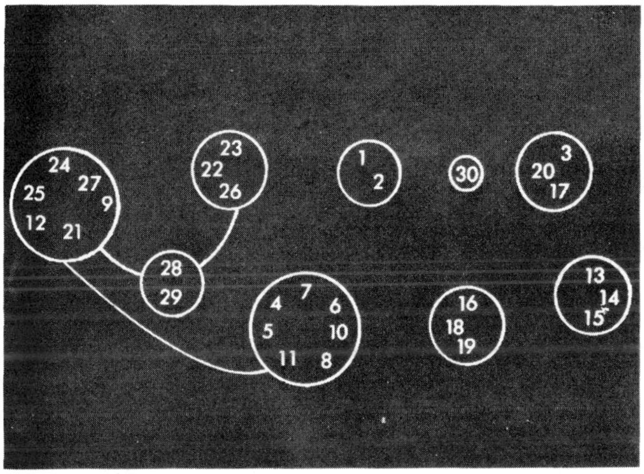

FIGURE 13—'FINAL.' SINGLEBODY joints the lonely region :16 with the nucleus :18–19.

Some interesting examples

We present in this section several scenes with the results obtained by SEE.

Example. **'CORN.'** The program analyzes the scene CORN (see figure 14 'CORN') obtaining the following identification of bodies:

(SINGLEBODY ASSUMES (:2 :3 :1) (:4) SAME BODY)
(BODY 1. IS :2 :3 :1 :4)
(BODY 2. IS :7 :6 :5)
(BODY 3. IS :13 :11 :12)
(BODY 4. IS :15 :16 :14) Results for 'CORN'
(BODY 5. IS :19 :18 :17)
(BODY 6. IS :21 :20)
(BODY 7. IS :8 :9 :10)

The region :4 got a single link with :3, and SINGLEBODY had to join it with the nucleus :1 :2 :3. Note that :4 and :12 did not get joined. The pyramid at the top was easily identified because its peak produces many links.

Example. **'MOMO'**. (See Figure 15 'MOMO'). The results are as follows:

(LOCAL ASSUMES (:17) (:9) SAME BODY)
(LOCAL ASSUMES (:9 :17) (:18) SAME BODY)
(BODY 1. IS :3 :2 :1)
(BODY 2. IS :32 :33 :27 :26)
(BODY 3. IS :28 :31)
(BODY 4. IS :19 :20 :34 :30 :29)
(BODY 5. IS :36 :35)
(BODY 6. IS :24 :5 :21 :4)
(BODY 7. IS :25 :23 :22) Results for 'MOMO'
(BODY 8. IS :14 :13 :15)
(BODY 9. IS :10 :16 :11 :12)
(BODY 10. IS :18 :9 :17)
(BODY 11 IS :7 :8)
(BODY 12. IS :38 :37 :39)

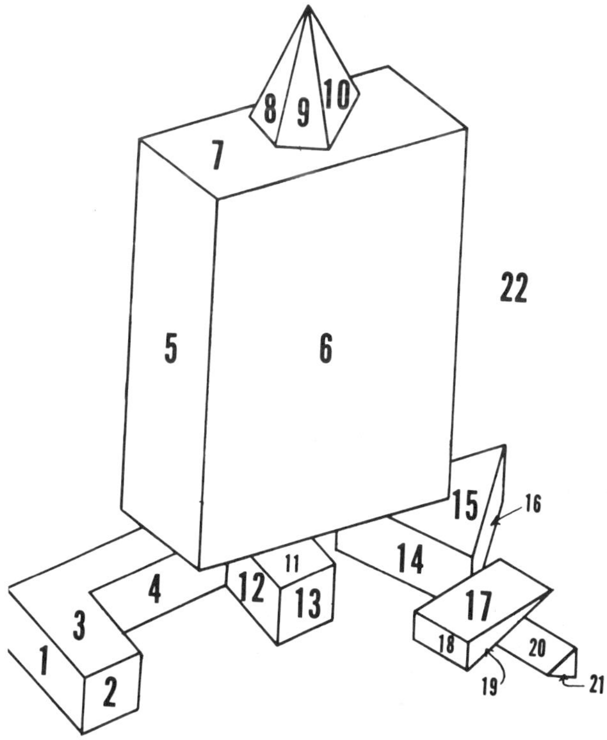

FIGURE 14—'CORN.' Since a link between :4 and :12 is not established, the bodies found in that part of the scene are :1 :2 :3 :4 and :11 :12 :13.

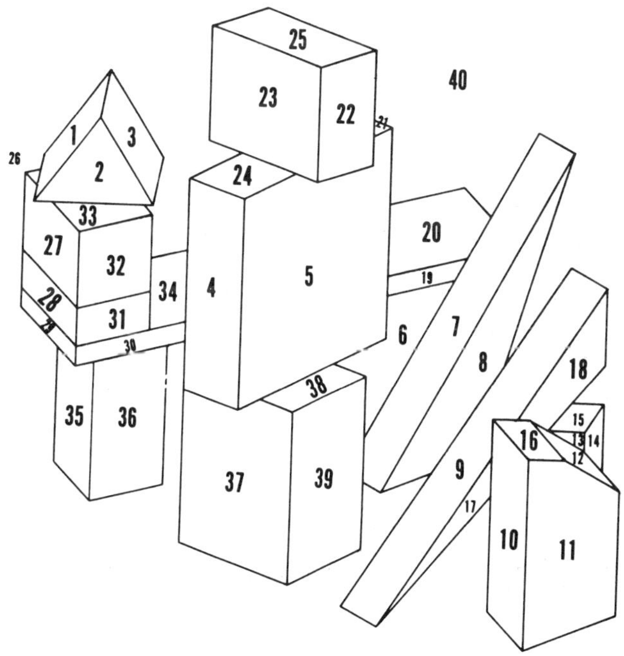

FIGURE 15—'MOMO.'

Comments on the solution for 'MOMO': The central cubes are easily isolated. :21 gets a link with :5 and another with :24, so there is no need to use LOCAL in it. The same is true of :26 with :27 and :33.

There is enough strong evidence to joint :28 and :31.

The links for :29 :30 :34 :20 :19 and :9 :17 :18 are indicated in Figure 16 'MOMO-LINKS.'

The dotted links between regions :30 and :34, and between :20 and :19 (see Figure 15 'MOMO') are due to the heuristic of parallel lines (of Figure 8 'PARALLEL'). These links made it possible to join the otherwise disconnected nuclei :29- :30- :19 and :34- :20. In particular, if :34 or :20 did not have parallel sides, SEE would have failed to merge these nuclei, and would have reported two bodies. The disposition of regions in MOMO is such that no link is present between :29 and :34, since :28 and :31 fall "exactly" parallel to :29 and :30. In a less extreme scene, some of these links would be present, allowing correct identification even if :29-:30-:34-:20-:19 were not a prism. Anyway, SEE did not make any mistakes.

The triangle of links formed by :9, :18 and :17 normally would have produced 3 separate bodies (since we need at least one double link to merge two regions); nevertheless, in this case, LOCAL makes some assumptions and saves the situation. Again, if :9 were not a parallelogram, there would be no weak links between :9 and :18 or between :9 and :17, and 3 bodies would have been reported instead of :9-:17-:18. But we should notice that, in general, two links instead of one would be between :17 and :18.

Links were extablished between :12 and :13, without serious consequences, because we need two mistakes, two wrong strong links, to fool the program. But that could happen.

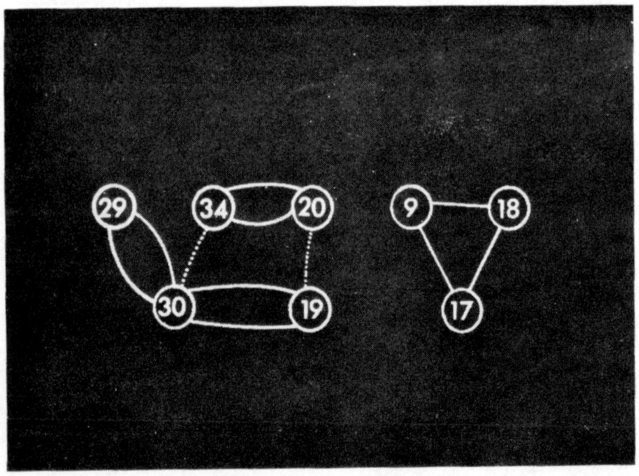

FIGURE 16—MOMO—LINKS'. Some links of figure 15 'MOMO'

Example. **'HOME'**.—(see scene HOME). The results are in Table V 'RESULTS FOR HOME.'

(SINGLEBODY ASSUMES (:38) (:39) SAME BODY)
(SINGLEBODY ASSUMES (:34) (:33) SAME BODY)
(SINGLEBODY ASSUMES (:25 :24 :22) (:23) SAME BODY)
(SINGLEBODY ASSUMES (:20) (:21) SAME BODY)
RESULTS
(BODY 1. IS :3 :6 :10 :2)
(BODY 2. IS :14 :13 :11 :12)
(BODY 3. IS :9 :8)
(BODY 4. IS :18)
(BODY 5. IS :15 :5)
(BODY 6. IS :38 :39)
(BODY 7. IS :7 :26 :27)
(BODY 8. IS :20 :21) Results for HOME
(BODY 9. IS :29 :28)
(BODY 10. IS :34 :33)
(BODY 11. IS :31 :32 :30)
(BODY 12. IS :25 :24 :22 :23)
(BODY 13. IS :16)
(BODY 14. IS :19)
(BODY 15. IS :17)
(BODY 16. IS :36 :37 :35)

TABLE V—Results for HOME

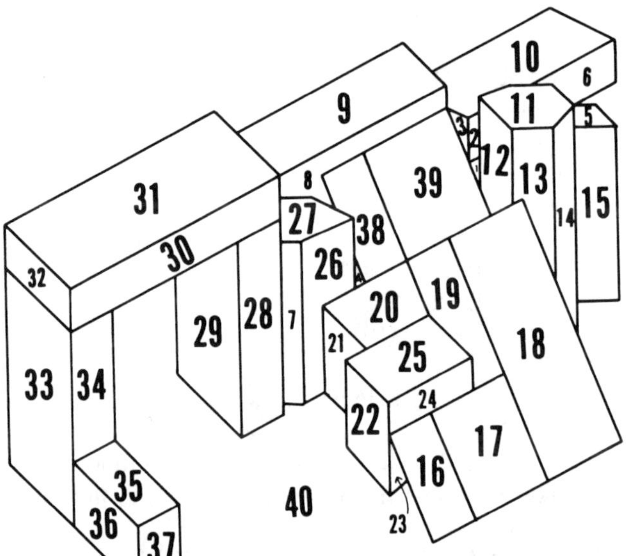

FIGURE 17—'HOME.' The hexagonal prisms did not create difficulties; SINGLEBODY was needed to group :34 with :33. The body :38–39 was identified as a single one, but :16–17 was reported as two. Note that there does not exist local link between :21 and :20; nevertheless, SINGLEBODY makes the correct identification. (See Table V).

Comments on the solution to HOME: There is a certain degree of ambiguity in this scene which human beings tend to resolve by assuming parallelepipeds. :16 and :17

are reported as two separate bodies, instead of one, which is probably a more natural answer.

In the picture from which 'HOME' was drawn, :19 and :18 were the two faces of a single parallelepiped leaning on :38-:39; SEE reports :19 as one body and :18 as another.

Nevertheless, SEE reports :38 and :39 as forming part of the same body (which was in fact the case in the picture in question), due to the fact that :4 and :1 are background.

Example. **'SPREAD.'**—(see Figure 18 'SPREAD'). The results are in Table VI 'RESULTS FOR SPREAD'.

(LOCAL ASSUMES (:36) (:4) SAME BODY)
(LOCAL ASSUMES (:30) (:31 :32 :29) SAME BODY
(LOCAL ASSUMES (:16) (:17) SAME BODY)
(LOCAL ASSUMES (:5) (:20) SAME BODY)
(LOCAL ASSUMES (:3 :11 :9) (:12 :10) SAME BODY)
(SINGLEBODY ASSUMES (:23) (:22) SAME BODY)
(SINGLEBODY ASSUMES (:4 :36) (:37) SAME BODY)
(SINGLEBODY ASSUMES (28 :26 :27) (:25) SAME BODY)
RESULTS
(BODY 1. IS :39 :40 :38)
(BODY 2. IS :23 :22)
(BODY 3. IS :42 :41)
(BODY 4. IS :4 :36 :37)
(BODY 5. IS :24)
(BODY 6. IS :28 :26 :27 :25) Results for SPREAD
(BODY 7. IS :31 :32 :29 :30)
(BODY 8. IS :20 :5)
(BODY 9. IS :12 :10 :3 :11 :9)
(BODY 10. IS :13 :7 :1)
(BODY 11. IS :21 :6)
(BODY 12. IS :8 :18)
(BODY 13. IS :17 :16)
(BODY 14. IS :45 :43 :44)
(BODY 15. IS :19)
(BODY 16. IS :15 :14)

TABLE VI—Results for SPREAD

Comments on the solution to SPREAD: The body :22-23-24, due to insufficiency of links, was split in two: :22-23 and :24.

Since there is only one link between :6 and :5, this body gets split into two: :6-21 and :5-20. Note that :21 is *not* the same face as :20, and there is where SEE gets confused and refuses to see evidence toward linking :21 with :20.

The long body :9-10-11-12-3 gets properly identified.

Example. **'HARD.'**—This scene is analyzed with the following results:

(LOCAL ASSUMES (:11) (:12) SAME BODY)
(LOCAL ASSUMES (:15) (:16) SAME BODY)
RESULTS
(BODY 1. IS :12 :11)
(BODY 2. IS :16 :15)
(BODY 3. IS :32 :31 :30)
(BODY 4. IS :9 :10 :8)
(BODY 5. IS :18 :19 :20)
(BODY 6. IS :13 :17 :14) Results for 'HARD'
(BODY 7. IS :5 :4)
(BODY 8. IS :1 :2 :33)
(BODY 9. IS :24 :23 :22 :3 :21 :28 :29)
(BODY 10. IS :25 :26 :27)
(BODY 11. IS :7)
(BODY 12. IS :6)

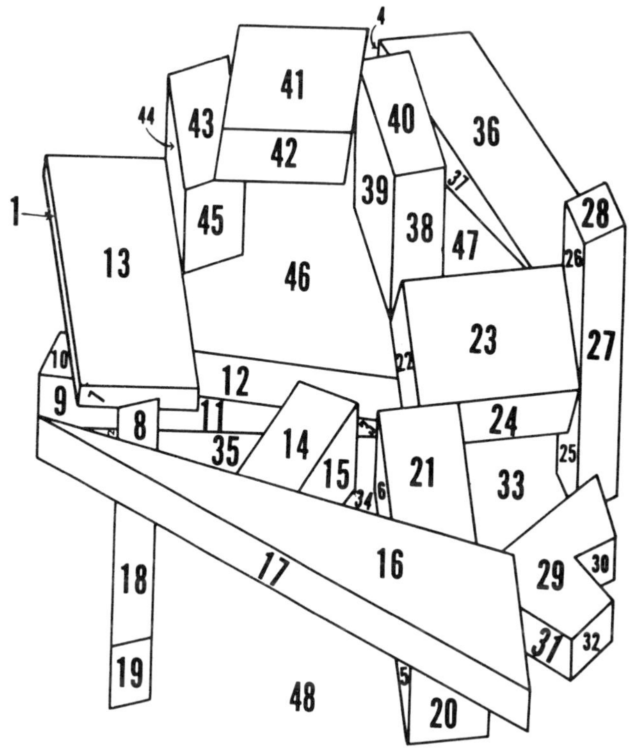

FIGURE 18—'SPREAD.' :41 and :42 are identified as a single body. Nevertheless, :8–18–19 gests broken into :8–18 and :19. :28–27–26–25 gets correctly identified, as well as the funny looking :29–30–31–32. (See Table VI).

Comments on the solution to HARD: :15 and :16 have to make use of weak evidence to get joined and recognized as forming part of the same body. Nevertheless, tois is not necessary with :28 and :29, because, through a chain of Ts, there is enough evidence in :3, :21 and :22 to join successfully all that long and twice occluded body.

There is one serious bug in this identification: regions :7 and :6 get identified as two separate bodies, and not as a single one, as one would normally do. This is caused by the fact that neither :7 nor :6 have visible 'useful' vertices, and there are not enough parallel lines in them to use the heuristic of Figure 8 'PARALLEL.'

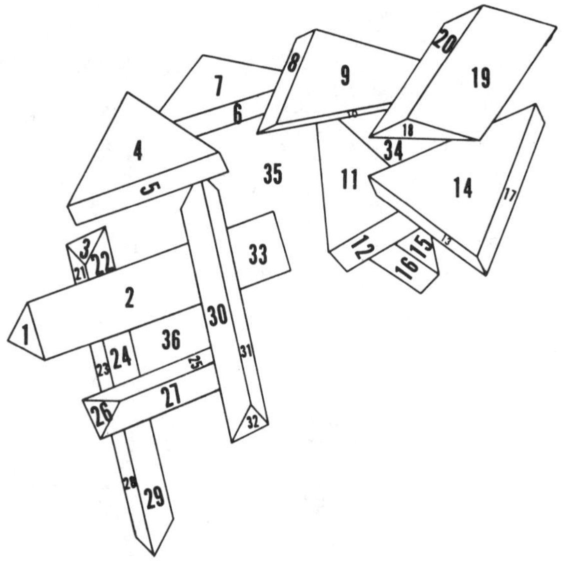

FIGURE 19—'HARD.' Body :21–22–3–23–24–28–29 is reported as a single object, which is correct. Nevertheless, regions :7 and :6 get reported as two bodies.

:33 was recognized as part of :1-2, as it should be.

DISCUSSION

We have described a program that analyzes a three-dimensional scene (presented in the form of a line drawing) and splits it into "objects" on the basis of pure form. If we consider a scene as a set of regions (surfaces), then SEE partitions the set into appropriate subsets, each subset forming a three-dimensional body or object.

The performance of SEE shows to us that *it is possible to separate a scene into the objects forming it, without needing to know in detail these objects;* SEE does not need to know the 'definitions' or descriptions of a pyramid, or a pentagonal prism, in order to isolate these objects in a scene containing them, even in the case where they are partially occluded.

The basic idea behind SEE is to make global use of information collected locally at each vertex: this information is noisy and SEE has ways to combine many different kinds of unreliable evidence to make fairly reliable global judgments.

The essentials are:

(1) Representation as vertices (with coordinates), lines and regions
(2) Types of vertices.
(3) Concepts of links (strong and weak), nuclei and rules for forming them.

The current version of SEE is restricted to scenes presented in symbolic form. It does not have resources for dealing with curves, shadows, noise, missing lines, etc. So it represents a "geometric theory of object identity" at present.

Since SEE requires two strong evidences to join two nuclei, it appears that its judgments will lie in the 'safe' side, that is, SEE will almost never join two regions that belong to different bodies. From the analysis of scenes shown above, its errors are almost always of the same type: regions that should be joined are left separated. We could say that SEE behaves "conservatively," especially in the presence of ambiguities.

Divisions of the evidence into two types, strong and weak, results in a good compromise. The weak evidence is considered to favor linking the regions, but this evidence is used only to reinforce evidence from more reliable clues. Indeed, the weak links that give extra weight to nearly parallel lines are a concession to object-recognition, in the sense of letting the analysis system exploit the fact that rectangular objects are common enough in the real world to warrant special attention.

Most of the ideas in SEE will work on curves too. However, we do not yet have a good idea of how sensitive the system will be to "symbolic noise," i.e., missing misplaced, and false region boundaries. As indicated in the scenes above, the system seems to have good resistance to "accidents" in which objects happen to to "line up" unusually well. This feature may be necessary if the preprocessor that (eventually) will feed data SEE decides to report two nearly colinear lines as one alone, or if it lumps several vertices into one, because they lie close to each other.

Extensions. (None of this incorporated in the actual program.) More heuristics could be added to increase the number of links; in particular, given a good number of "link proposers," parameters set outside (by the user) would tell SEE which set of heuristics to use; for instance, if we knew that the scene is formed by prisms, we could use the heuristics that ask for parallel lines having a given configuration, we could check the length of certain edges, etc.

Different kinds of links could also be established; in this way, 'contradictory' links (such as the three links of Figure 13 'FINAL' which SEE just ignores) could be studied further, in order to discover ambiguities. In particular, a "conditional link" would be useful: regions :2 and :3 belong to the same body if region :4 does not.

SINGLEBODY could be improved so as to analyze in more detail the regions that by themselves form bodies (that is, the bodies formed by only one region); in this way, we could hope to joint regions :6 and :7 of scene 'HARD.'

ACKNOWLEDGMENTS

The author is grateful to Professors Marvin Minsky and Seymour Papert for several helpful discussions, and to the referees of this paper for their constructive recommendations.

Michael Speciner made valuable suggestions; Cornelia Sullivan helped to codify most of the scenes.

"The author was partially supported by a scholarship from the Instituto Nacional de la Investigación Científica (México)."

The work reported here could not have been produced had the author not had access to the computational facilities of Project MAC, an M.I.T. research program sponsored by the Advanced Research Projects Agency, Department of Defense, under Office of Naval Research Contract Number Nonr-4102(01). This paper, having had this support from MAC, may be reproduced in whole or in part for any purpose of the United States Government.

REFERENCES

1 R GREENBLATT J HOLLOWAY
Sides 21
Memorandum MAC–M–320 A I Memo 101 Project MAC MIT August 1966 A program that finds lines using gradient measurements

2 K K PINGLE
A program to find objects in a picture
Memorandum No 39 Artificial Intelligence Project Stanford University January 1966 It traces around objects in a picture and fits curves to the edges

3 A GUZMÁN
Scene analysis using the concept of model
Technical Report No AFCRL–67–0133 Computer Corporation of America Cambridge Massachusetts January 1967

4 A GUZMÁN
A primitive recognizer of figures in a scene
Memorandum MAC–M–342 AI Memo 119 Project MAC MIT January 1967

5 A GUZMÁN
Some aspects of pattern recognition by computer
MS Thesis Electrical Engineering Department MIT February 1967 Also available as a Project MAC Technical Report MAC–TR–37 This memorandum discusses many methods of identifying objects of known forms.

6 A GUZMÁN H V McINTOSH
CONVERT
Communications of the ACM
9,8 August 1966 pp 604–615 Also available as Project MAC Memorandum MAC–M–316 AI Memo 99 June 1966

7 A GUZMÁN
Decomposition of a visual scene into bodies
Memorandum MAC–M–357 AI Memo 139 Project MAC MIT September 1967 unpublished This memorandum is a more technical description of SEE

8 R H CANADAY
The description of overlapping figures
MS Thesis Electrical Engineering Department MIT January 1962

9 L G ROBERTS
Machine perception of three-dimensional solids
Optical and electrooptical information processing
pp 159–197 J T Tippett et al eds MIT Press 1965

10 M L MINSKY S A PAPERT
Research on intelligent automata
Status report II
Project MAC MIT September 1967

11 J McCARTHY
Plans for the Stanford artificial intelligence project
Stanford University April 1965

12 C A ROSEN N J NILSSON eds
Application of intelligent automata to reconnaissance
Third Interim Report Stanford Research Institute December 1967

Impossible Objects as Nonsense Sentences

D. A. Huffman

Board of Studies in Information and Computer Science
University of California at Santa Cruz

INTRODUCTION

To every 3-dimensional scene there correspond as many 2-dimensional pictures as there are possible vantage points for the camera. It is, however, possible to construct pictures for which there is no corresponding scene containing physically-realizable objects. Pictures of such 'impossible objects' can be useful in giving insight into the constraints or grammatical rules associated with the 'language' of pictures, just as nonsense sentences can be useful in illustrating the rules of other languages. Impossible objects have been used by psychologists (Penrose and Penrose 1958) to create visual illusions which successfully challenge the ability of our perceptual systems to synthesize a 3-dimensional world from 2-dimensional information. The incompatibilities among the various portions of pictures of these objects are a novel way of testing our picture analysis procedures. The purpose of this paper is to demonstrate some possible decision procedures and to test them on pictures of both possible and impossible objects.

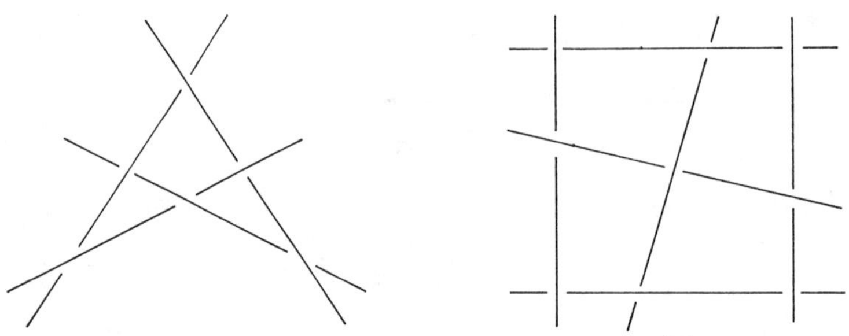

(a) (b)

Figure 1. Two pictures of sets of line-segments

Examples of pictures of objects for study are shown in figures 1, 2 and 3. The first of these are from a paper by the author (Huffman 1968) which dealt with objects consisting of straight line segments only. It was assumed that straight lines in the pictures were representative of straight-line segments in the 3-dimensional scene; the sense of 'passing' at the various intersections was also indicated. With these assumptions the object shown in figure 1(a) is not possible; the one in figure 1(b) is.

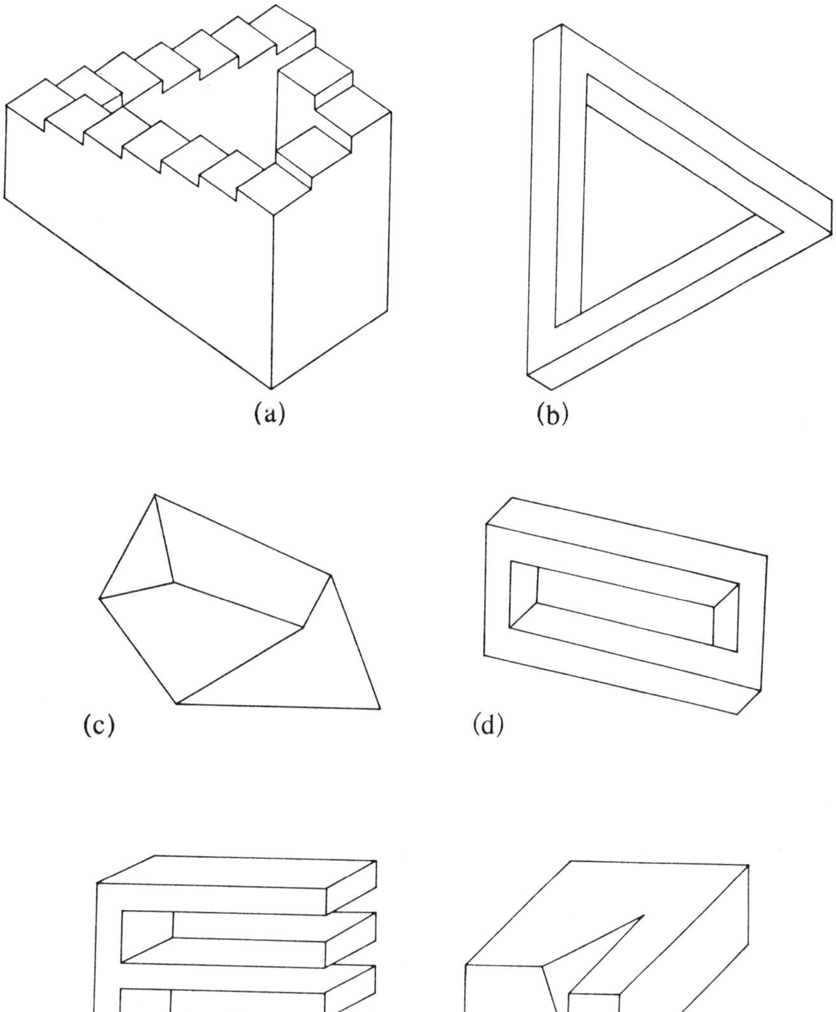

Figure 2. Examples of impossible polyhedra

The pictures in figure 2 purport to be of plane-bounded polyhedra. [The first two of these are adapted from Penrose and Penrose (1958)]. It is primarily with such polyhedral objects that this paper will deal. Under certain natural assumptions none of the objects in figure 2 is possible.

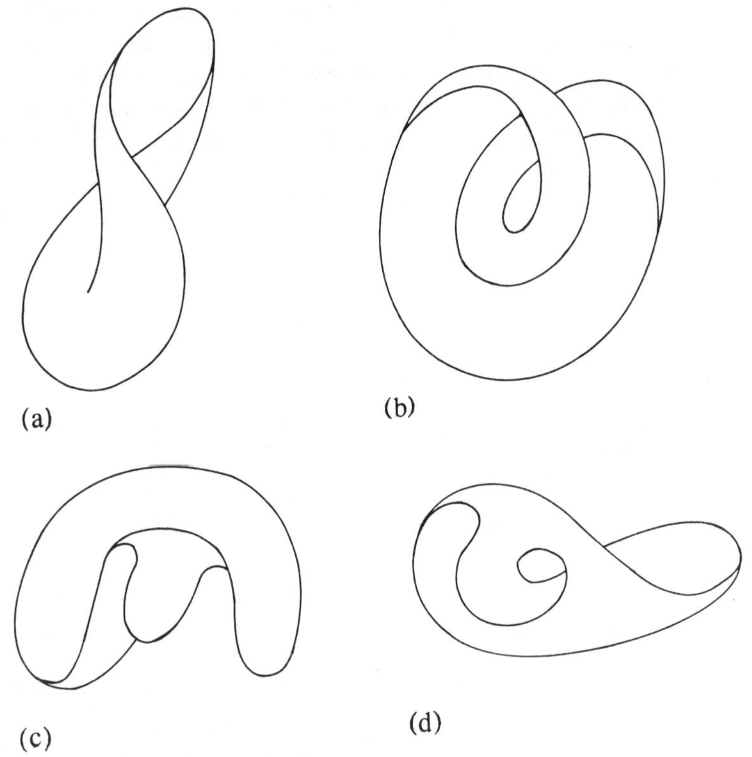

Figure 3. Examples of 'smooth' objects

In figure 3 are shown pictures of four different 'smooth' objects. One of these is impossible; the other three are possible. The constraints which must be taken into account in making decisions about such objects will be the topic of the final sections of this paper.

Of course the possibility or impossibility of the objects in any picture depends upon what assumptions we make about the physical nature of the 3-dimensional elements which we can, in the picture, view in only two dimensions. Typical assumptions we *could* make are: that what appear to be straight lines are actually straight, that lines which appear to be parallel are parallel, that objects are either 'thin' or have appreciable thickness, that all lines which appear to meet at a point actually meet at that point, that all edges which are on the side of the object toward the camera actually are represented in the picture, and so forth.

APPROACHES FOR PICTURE ANALYSIS

One assumption we shall make throughout this paper is that all pictures are taken from a 'general position'; that is, that a slight change of the position from which the picture is taken would not change the number of lines in the picture or the configurations in which they come together. In the case of pictures of polyhedra this eliminates the possibility of pictures in which two vertices of the objects in the scene are, by coincidence, represented at the same point in the picture, or two edges in the scene are seen as a single line in the picture, or a vertex is seen exactly in line with an unrelated edge.

The assumption of a general position for the camera is a practical one since it reduces the number of types of local configurations of lines which we must deal with in the picture. Furthermore, if this assumption leads us to judge as impossible an object or set of objects which we know to exist (and therefore by definition 'possible') we can conclude that the camera was probably not in a general position (or that some other assumption was unjustified). In that case we can either move the camera slightly and retake the picture, or go to an augmented list of local configurations which are possible and reanalyze the picture accordingly.

The assumption of a general position for the camera, strictly interpreted, would prevent us from considering pictures having several line-segments which, if extended, met at a common point unless the corresponding edges in the scene would also meet at a common vertex. A special case of this issue arises if several line-segments in a picture are mutually parallel (and would therefore meet at a point-at-infinity if extended). Because it is annoying to have to concern ourselves about this issue in drawing pictures to be analyzed we shall make whatever special commentary is necessary if several line-segments in a picture would meet if extended.

DERIVATION OF LOCAL CONSTRAINTS FOR PICTURES OF DEGREE-3 POLYHEDRA

Introductory Remarks

Pictures of scenes which contain only plane-bounded solid objects (that is, assortments of polyhedra) are especially attractive for purposes of logical analysis. On the one hand there are arbitrarily many objects of that type since there can be for each object as many vertices as desired, and its planar surfaces can meet each other so that in the region of a given vertex the object can be locally convex or concave in a variety of ways. Thus there is a rich and interesting set of objects which can be viewed. On the other hand the number of ways in which individual edges can be viewed and can obscure each other is limited. For these reasons an environment containing polyhedra is a sensible choice as an initial one into which to put a robot which looks at the scene with its television camera 'eye' and which moves about the environment, deciding what it has seen and how best to look for what it has not seen. Other authors [for example, Guzman (1968)] have also considered the problem of the analysis of scenes containing polyhedra.

In the work reported here no account has been taken of the effects of gravity or of the possible mutual interaction effects among various members of a set of polyhedra. If we did not make this assumption we could soon find ourselves immersed in such (for us, secondary) problems as determining angles of repose for piles of tetrahedra, or of deriving the result that a stack of identical unit cubes is just barely stable if the amount by which the nth cube down in the stack extends beyond the one below it is equal to $1/n$.

We shall assume that exactly three plane surfaces come together at each vertex of the polyhedra. (The techniques for less-restrictive assumptions can be shown to be relatively simple extensions of this special case.) An exhaustive listing of all (only four) inherently different vertex types is made, together with an exhaustive listing of the essentially different ways they can be viewed. (Certain views of different vertex types will look the same to the camera even though they 'mean' different things.) Similarly, straight lines in a picture can have any of four possible interpretations. Picture analysis will progress as an alternation between decisions about the correct interpretations of lines and of the points where sets of lines are incident in the picture. Correct interpretations of these lines and points yield correspondingly correct deductions about the true nature of the associated edges and vertices of the objects in the actual scene.

The single most important insight which makes this analysis possible is the realization that there are exactly four possible interpretations of a line in a picture. It can represent either a 'convex' or a 'concave' edge with both associated planes in view, or it can represent a (convex) 'hiding' edge which can obscure more distant parts of the scene either to one side of the edge or the other. A given line segment cannot have two different 'meanings' in two different parts of the picture and it is this constancy of interpretation of what a line must mean in a picture which is the key to the method of analysis presented here.

Definition of terms

The environment is assumed to contain an assortment of solid polyhedra which have exactly three planar *surfaces* at each of the *vertices* and, of course, two surfaces associated with each *edge*. We shall call the actual collection of polyhedra the (3-dimensional) *scene*; the projected (2-dimensional) view which the camera sees will be called the *picture*. The visible edges and vertices of the scene are associated with the straight *lines* (actually, line-segments) and *nodes*, respectively, of the picture. These lines and nodes determine a picture which is actually a graph, which we may refer to, from time to time, as the *picture-graph*. On each side of each line in a picture is an area which may or may not be associated with one of the surfaces of the same polyhedron associated with the line.

We shall use the term 'picture' to refer not only to projected views of possible 3-dimensional scenes but also to 2-dimensional line drawings which purport to be views of scenes but for which there may be no corresponding physically-realizable set of polyhedra. Such pictures of 'impossible objects' may be constructed with a view to exercising our analysis procedures in a special way. They may also come about naturally if, for example, an edge in the scene generates no corresponding line in the picture because of the lighting conditions under which the picture was taken.

Types of vertices and edges in the scene

There are four basic ways in which three plane surfaces can come together at a vertex. All four can be illustrated in the picture of a fireplace and raised hearth shown in figure 4(a). In figure 4(b), pictures of these four vertex

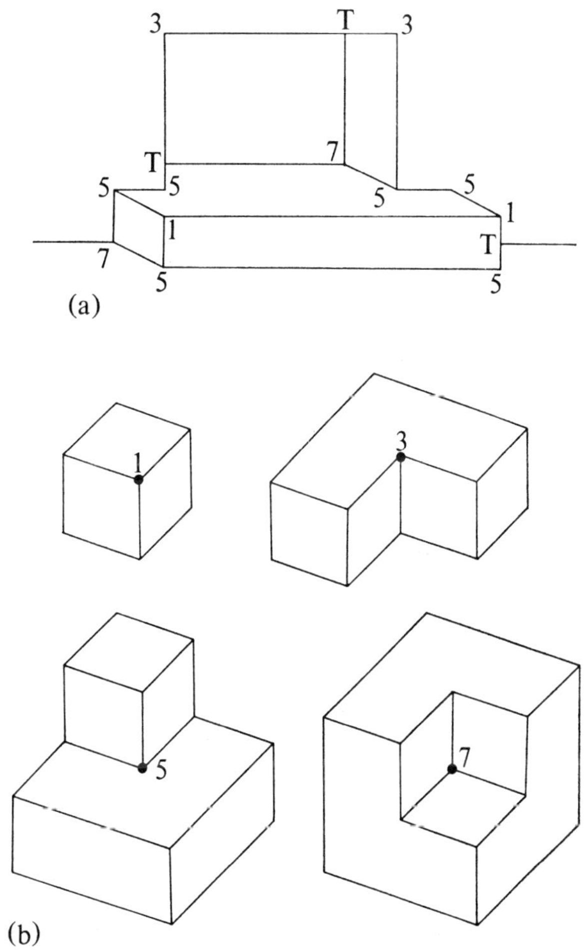

Figure 4. Illustrating the four types of vertices: (a) a fireplace and hearth; (b) the four vertex types

types are shown associated with pictures of four polyhedra. The identifying numbers are determined in accordance with the following reasoning. The three planes which meet at a vertex partition the surrounding space into eight octants. (Even if the three planes and associated three edges are not mutually orthogonal the term 'octant' will be used and the general comments offered here will nevertheless apply.) The number of octants which are occupied by solid material at the vertex is chosen as the type-number for the vertex. (The 'T-nodes' are not associated with physical vertices and will be commented on separately later.)

There are only two types of edges possible for our plane-bounded poly-hedra: concave and convex. Each of the four vertex types we have allowed is uniquely associated with a trio of edge-types. The edges associated with type-1 vertices are all convex; those associated with type-7 vertices are all concave. Two of the edges incident at a type-3 vertex are convex and the other is concave. Two of the edges incident at a type-5 vertex are concave and the other is convex.

Types of nodes and lines in the picture

A vertex can be viewed from any one of the octants which is not occupied by solid material and all views from a given octant give essentially the same 'configuration'. (The exact meaning of this comment will be apparent later.) For instance, a type-3 vertex can be viewed from the complementary five octants in which the eye or camera may be placed. Except in the case of a type-1 vertex, where rotational symmetries reduce the number of possibilities, the view from each of the octants gives an essentially different way in which the edges can meet.

The possible views of each of the four types of vertices are summarized in figure 5. Note that although there are always exactly three edges incident at a vertex, only two may actually be visible as lines in the picture; the third line is, in these cases, associated with an edge which is hidden (from the camera position). Such lines are shown dotted in the figures.

The (undotted) lines in each of the pictures of the vertices are labelled as follows:

(a) a '+' line represents a convex edge which has both of its corresponding planes visible from the camera;

(b) a '−' line represents a concave edge which has both of its corresponding planes visible from the camera;

The other two types of label for picture lines correspond to those convex edges in the scene which have both of their associated planes on the same side of the edge as viewed from the camera, one hiding the other. An arrow is used as the label for such lines with the convention that, as one moves in the direction indicated by the arrow, the pair of associated planes is *to the right*. Since an arrow may lead toward or away from a given node of the picture we conclude that, with respect to that node, we can label an incident line with either

(c) an 'in' arrow, or

(d) an 'out' arrow,

as well as with the '+' or '−' label.

It is clear that a line in a picture of a 'possible' scene must have a *single* one of the four labels associated with it along its entire length. (Otherwise the pair of planes associated with it would have different orientations in different parts of the scene.)

Our convention for labelling dotted lines (representing edges hidden from the camera) is consistent with the one for labelling lines representing visible edges. In order to determine the label for a line corresponding to a hidden edge, imagine that the surfaces between that edge and the camera are removed, and note the orientations of the pair of planes associated with the edge. The label which is to be assigned to the picture line is the same as if the intermediate surfaces had not been present. Examples of all four types of dotted lines are shown in figure 5.

Note that a hidden *line* may be labelled '+', even though the corresponding *edge* is concave, and that a hidden line may be labelled '−', even though the corresponding edge is convex. Similarly, hidden lines which are labelled with arrows may be associated with either concave or convex edges. (In each of these cases it is the parity of the number of surfaces between the camera and the edge which is the controlling factor.)

All visible edges of a scene correspond in a picture to lines which can be thought of as having zero *depth*. The dotted lines in figure 5 have unit depth. In general the *depth-index* for a line is the number of surfaces which would have to be removed to expose the corresponding edge in the scene. Each line in a picture has not only one of four *types*; it also has a non-negative depth index. We shall, in general, in this paper deal only with exposed, or zero-depth, lines. Thus we are giving here only a 'surface' analysis. A 'deeper' analysis will not be necessary for our present purpose.

From figure 5 it can be seen that the arrowed lines (both dotted and undotted) in each of the twelve possible vertex views satisfy a 'continuity' or 'conservation' rule: the number of arrows into and out of a node are equal. (This result is also true of nodes which represent vertices having more than three associated surfaces and it applies as well to pictures representing non-planar surfaces.)

In a picture an additional kind of node can occur which is associated with *no* corresponding vertex in the scene. These are the 'T-nodes' noted earlier. Each gives direct evidence that the bar of the 'T' is an arrowed line in which the direction of the arrow is from right to left (when the 'T' is in the standard upright position). The line type for the obscured line may be '+' or '−' or arrowed (in either direction). A listing of all of the other labelled line

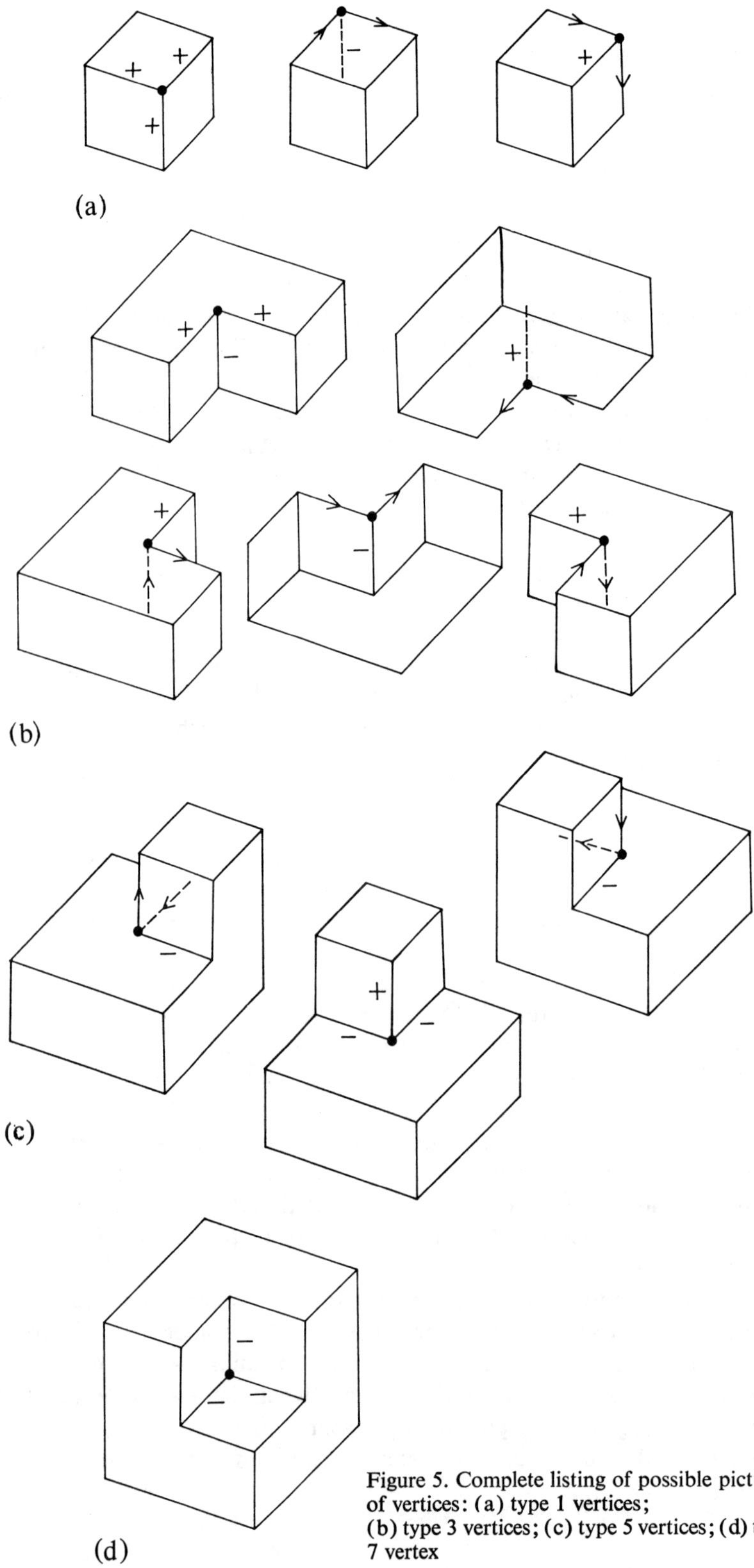

Figure 5. Complete listing of possible pictures of vertices: (a) type 1 vertices; (b) type 3 vertices; (c) type 5 vertices; (d) type 7 vertex

configurations possible in the vicinity of a node in a picture-graph is given in figure 6. It is convenient to call each either a 'V', 'W', or 'Y'. (Each of these configurations may, of course, appear with an arbitrary rotation in a given picture.) The integer shown is the associated vertex-type. Note that each of these three types of configuration around a node of the picture-graph can have several different 'meanings' in terms of the (3-dimensional) vertex-types it can represent. The proper interpretation of the configuration of lines incident at a given node cannot be determined unless the surrounding 'context' is taken into consideration.

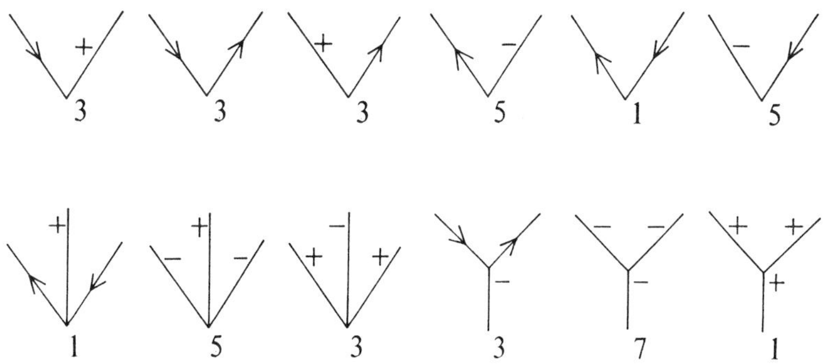

Figure 6. Possible labelled-line configurations around a picture node

THE PROBLEM OF PICTURE ANALYSIS

The mapping from scene to picture

Imagine a (3-dimensional) scene containing one or more polyhedra. In the process of producing a corresponding (2-dimensional) picture-graph we can imagine several different stages, at each of which some information is lost. For example, consider a rectangular parallelopiped through which a hole with a cross-section which is a parallelogram is completely drilled. A close approximation to a true representation of this object is given in the *labelled 'X-ray' picture* shown in figure 7(a). (For completeness a depth-index for all hidden lines and labels for the nodes could also be given but they would only unnecessarily clutter the picture here.) Some information about the actual dimensions of the object has already been lost, unavoidably, particularly information about the distances in the direction of the rays which pass through the camera position. This lost information is associated with the now-missing 'third' dimension.

If no indication is given of hidden lines we obtain the *labelled picture-graph* in figure 7(b). (The labels for the nodes of a picture-graph of a polyhedron with degree-3 vertices are uniquely determined by the line-labels; *see* figure 6. Thus these labels would be redundant.) Here we have irretrievably lost even more information about the real object since there is an infinite variety of hidden protrusions, cavities and obscured objects which would not be apparent in this type of picture. At this stage there is an inherent ambiguity about the hidden parts of the object. For example, in our example we have no evidence about how far the hole penetrates into the object. Finally, by dropping the line-labels we obtain the *unlabelled picture-graph* itself [figure (7c)].

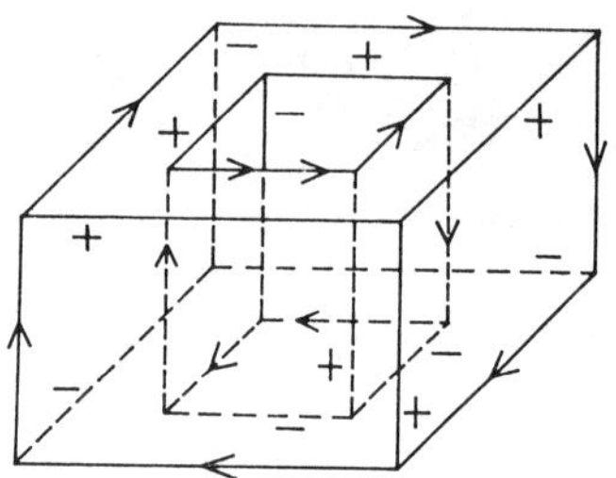
(a)

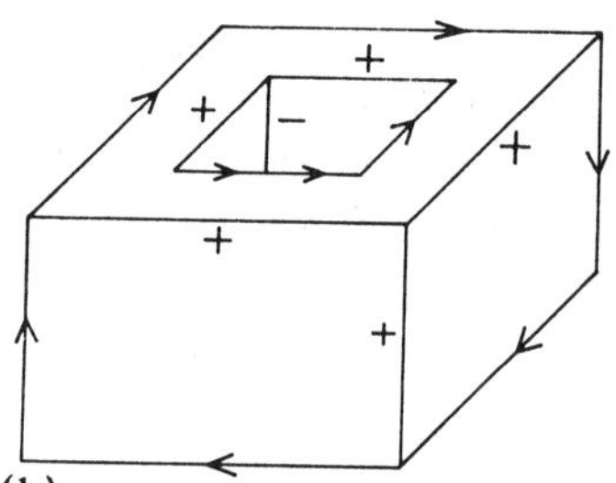
(b)

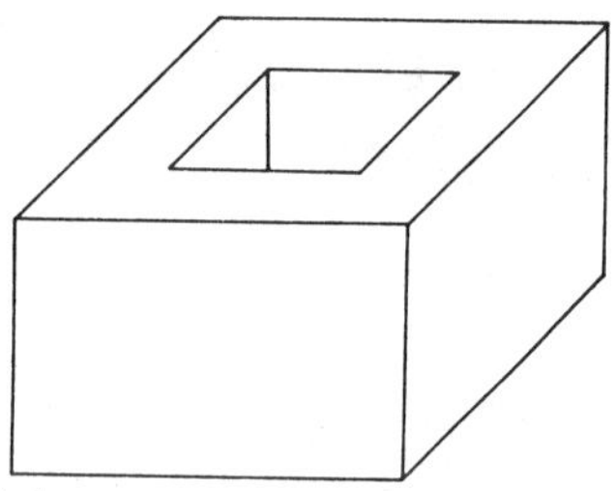
(c)

Figure 7. Illustrating three kinds of pictures: (a) labelled 'X-ray' picture-graph; (b) labelled picture-graph; (c) unlabelled picture-graph

Reconstruction of a scene from a picture

Certainly a necessary condition for the realizability of an unlabelled picture-graph as a scene containing polyhedra is that there exists some labelling which places exactly one of the four possible line-labels on each line so that around each node of the graph there exists just one of the allowable configurations given in figure 6. (The 'T-nodes' are exceptions which have already been noted.) The successful labelling of the lines around each node in a picture uniquely corresponds to an identification of the types of vertices and edges in the associated scene. It is also clear that the two picture areas on either side of a line which has been labelled with a '+' or a '−' must belong to the same object.

The labelling process for the lines and nodes of an unlabelled picture is analogous to the parsing of sentences in other languages. In our language the 'alphabet' consists of four types ('T', 'V', 'W', and 'Y') of nodes. In order to be 'syntactically' correct a picture containing these symbols must have them interconnected only in certain allowed ways. As we have noted above, even a properly-labelled picture can have a wide variety of interpretations. A totally-satisfactory analysis procedure would allow us to specify in detail at least one labelled 'X-ray' picture from which a corresponding object or set of objects could be constructed.

Forbidden picture-subgraphs

The syntactic rules for any language impose constraints on the manner in which the symbols of its alphabet may be related to each other. This is certainly true of our pictures of polyhedra. Some forbidden combinations are shown in figure 8. The lines of none of them can be consistently labelled. (The reader should verify this fact for himself.) The configuration consisting of two 'T-nodes' [figure 8(a)] is worthy of special notice since it is forbidden

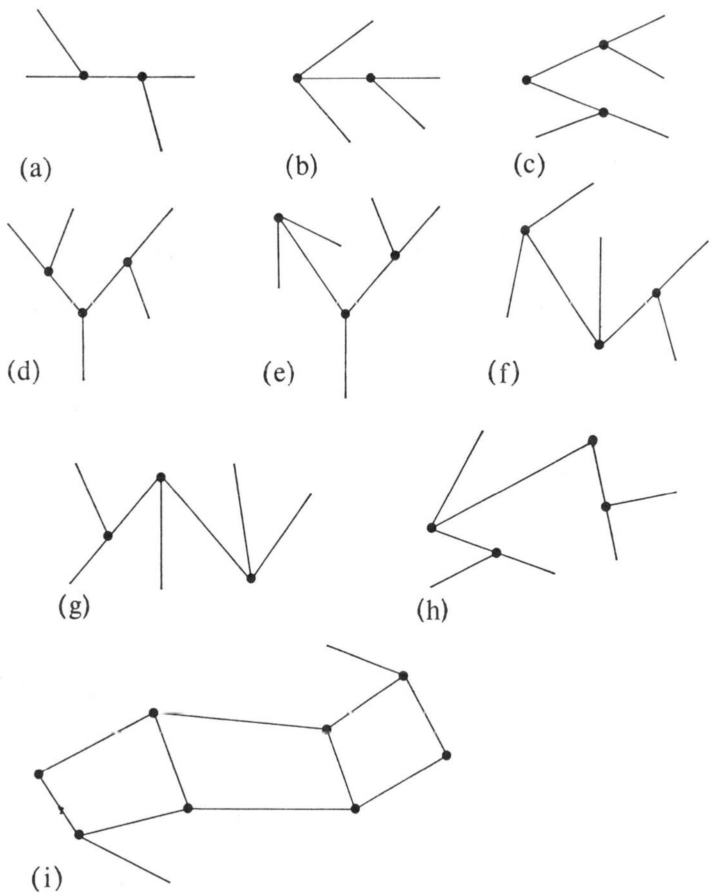

Figure 8. Examples of forbidden subpictures

even in pictures of polyhedra which may have an arbitrary number of edges associated with a vertex.

Pictures which cannot be labelled

In figure 9 eight examples of pictures of objects are given which are impossible to realize as degree-3 polyhedra. In these examples each of the lines can be labelled so as to satisfy the necessary constraints at all but one node. The identity of the node (in general, nodes) at which the labelling is not an allowed one may depend upon the order in which the labels are derived. Our primary concern is with whether or not a labelling is possible for the picture as a whole rather than with where the labelling fails. Often, however, the

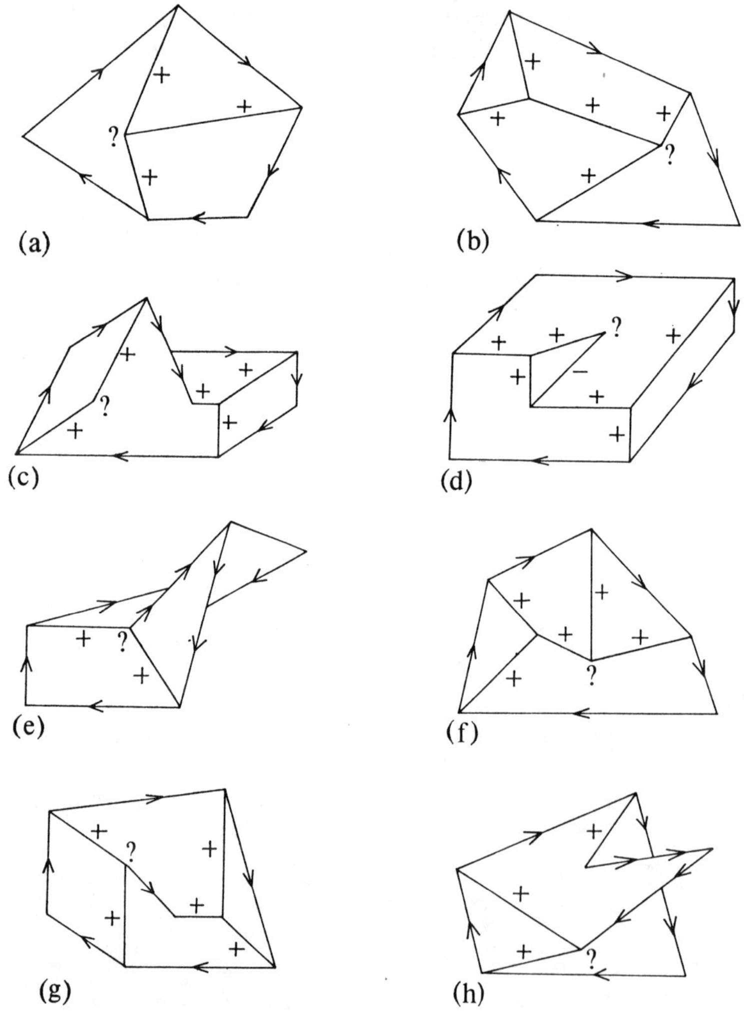

Figure 9. Pictures of objects for which there is no labelling possible

location of a picture region where labelling is impossible will be where a human subject reports that the picture 'doesn't look right' to him.

We have assumed that each of the eight pictures in figure 9 is of a complete object rather than, for example, a recess in the surface of a larger object. This assumption allows us to label each of the lines on the periphery of the picture with an arrow pointing clockwise. Other lines incident on nodes on the periphery can be labelled '+' or '−' depending on whether the arrows on the periphery take a right or left turn at those nodes (see the appropriate W and Y configurations in figure 6). Additional lines which can be labelled immediately with an arrow are those associated with a T-node. The remaining lines are labelled so that, if possible, an allowable configuration exists at each node. In this labelling procedure it may be necessary to assume a given label for a line in order to make further progress and to revise this assumption if it leads to a stage at which further labelling is impossible.

Pictures of 'unlikely' objects

The pictures in figure 10 can all be labelled consistently (in a unique way) and each describes an object which would probably be 'unlikely' to be seen by most of us. The objects in figures 10(a) and (b) can be visualized more easily if one imagines that a cavity in the form of a skew parallelopiped is removed from the more familiar part of each object pictured. In these two pictures we have violated the condition of a general position for the camera. If the camera position were moved somewhat certain edges of the cavities would no longer appear parallel to other edges of the object.

The object in figure 10(c) is in the general position. However, if the camera were moved a significant amount to the right of the object it would record the hidden construction on the back side of the object which was necessary to make the object 'possible'. The object in figure 10(d) can be realized by combining two tetrahedra.

The 'skew cube' of figure 10(e) looks impossible to many human subjects but it is realizable regardless of the exact orientation of the lines in the picture. In order to prove this we can imagine the construction of the object itself. First, orient three plane surfaces in any of the (infinite number of) ways in which they meet so that the edges associated with their pairwise intersections appear to the camera to be the three lines at the middle of the picture. Next, locate two more edges in each of these three surfaces so that the set of six appear to the camera to be in the positions of the six peripheral lines of the picture. Each of the vertices corresponding to the three 'W' configurations has associated with it a pair of edges determining uniquely the location of one of the three hidden surfaces of the object and the point common to these three plane surfaces is also uniquely determined.

If this point would appear to the camera to be inside the periphery of the object in the picture the object can be constructed with the six surfaces referred to above. If this point would appear to the camera to be outside

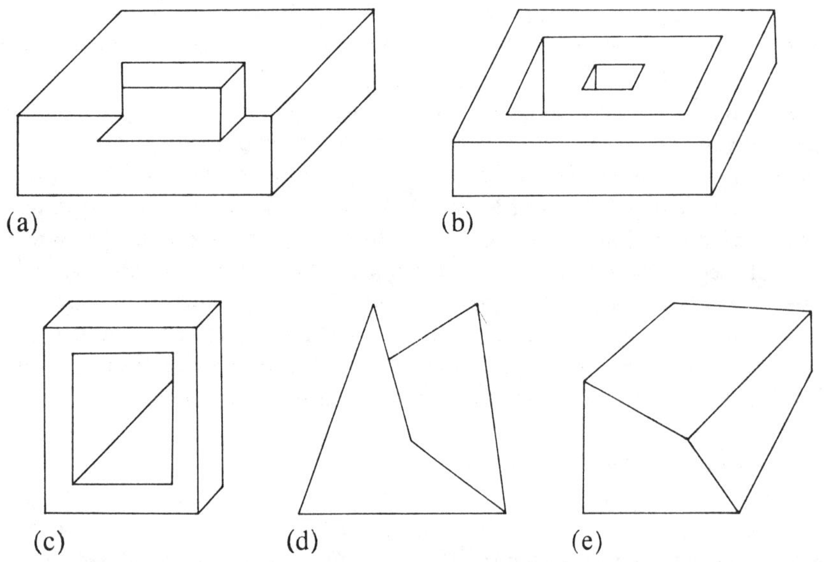

Figure 10. Examples of 'unlikely' objects

that periphery (and in front of one of the three visible surfaces) a seventh surface would be necessary in order to realize the object pictured. The orientation of the picture lines in the example given would require this construction.

In many applications the assumption that 'unlikely' objects do not exist in the scene can be as important a constraint as some of the others mentioned in this paper. The fewer the assumed number of types of objects the easier the picture analysis is.

ADDITIONAL PICTURE CONSTRAINTS

Introductory remarks

It will be seen in this section that even if a picture can be labelled consistently it may not be of a realizable object or set of objects. Additional constraints corresponding to more refined geometric tests must also be satisfied. Examples of pictures which can be labelled but which are not possible are given in figure 11. Figure 11(a) illustrates that the same object surface cannot be associated with two different sides of the same line (independent of which of the four labels the line has). The objects in figures 11(b), (c), and (d) are not possible since the picture gives evidence that two planes intersect along two different edges. In figures 11(e) and (f) the pairwise intersection of three planes gives three lines in the picture which do not intersect (when extended) at a common point, as they should. The same argument eliminates the picture in figure 11(g) as possible; this time the third plane is hidden from

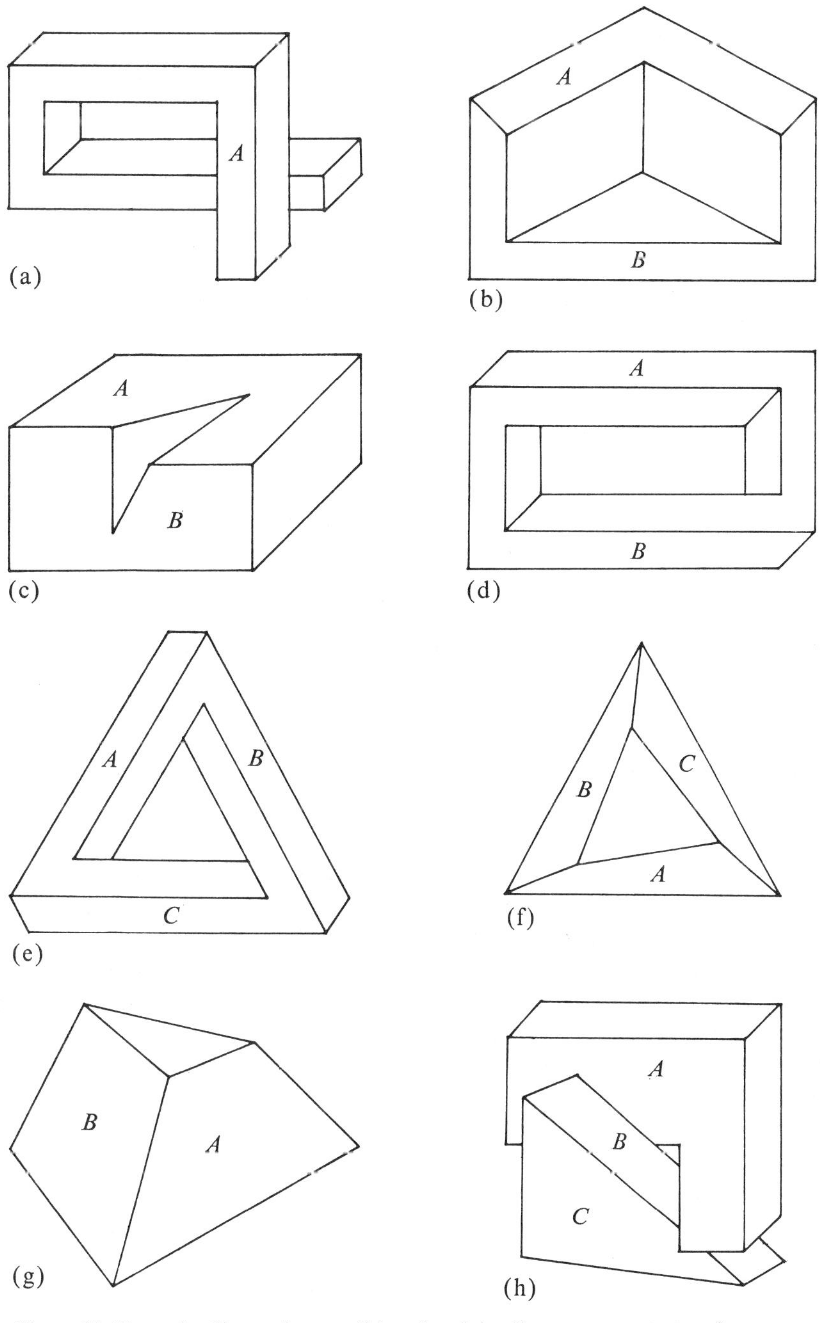

Figure 11. Examples illustrating conditions involving lines common to two faces

view even though the edges at which it intersects the other two planes is visible to the camera. The objects (or object: the two subobjects could be joined by a hidden coupling) in figure 11(h) are possible. This example shows pairs of areas in the picture which are associated with two different lines, but in this case the lines do not represent edges common to the two associated surfaces.

The geometric constraints which were violated in the examples above are simple illustrations of a more general set of necessary constraints. Let us define a *cyclically-ordered edge set* (or 'cyclic-set' for short) to be a cyclically ordered sequence of edges purporting in the picture to have the property that each pair of edges which are consecutive in the sequence determine a distinct plane. We know from the examples above that a cyclic-set with one member is not physically possible. A cyclic-set with two members is not possible unless the two edges are identical.

A cyclic-set with three members is not possible unless the three edges (extended if necessary) would meet at a common vertex. Since we have assumed that our pictures were taken from a general position this implies that the associated triple of lines in the picture would also have to meet at a common node. The three lines incident at a picture node representing a vertex automatically meet the realizability condition. When a cyclic-set has four or more edges the realizability condition is more complicated. For example, it is not necessary that either they or the corresponding picture lines meet at a common point. [See, for example, the four lines extending from the 'base' of the object in figure 13(a).]

The 'gain' concept

Consider the drawing of figure 12(a). The heavy lines (a, b, c, d and e) are intended to represent the edges of a single plane and the lines L_i to represent the directions taken by the 'third' edges at the points P_i. These lines (L_i) represent the cyclic set in question. The lines L_1 and L_2 (and a) are in one plane, the lines L_2 and L_3 (and b) in the next, and so forth.

There is no loss of generality in assuming that the given plane is parallel to the plane of the picture itself. Now imagine that a second plane, parallel to the first, is constructed with edges a', b', c', d', and e'. The lines representing these edges would have to be parallel to the corresponding lines constituting the boundary of the first plane. If the lines L_i have orientations corresponding to a physically-realizable scene the lines representing the boundary of the second plane would have to intersect on the lines L_i as shown in the example.

If the lines L_i have arbitrary orientations the construction illustrated in figure 12(a) will, in general, not be possible. Assume, for instance, that the orientation of L_1 is changed and that those of the other lines associated with the cyclic-set are not. Choose a point X on L_1 and construct a' parallel to a. From the intersection of a with L_2 construct b' parallel to b, and so forth. In general the procedure will determine a point Y (the intersection of e' and

L_1) which is different from X, as has been illustrated in figures 12(b), 12(c), and (12d). The ratio between the distance $\overline{P_1 Y}$ and the distance $\overline{P_1 X}$ will be defined as thc (clockwisc) *gain* associated with the cyclic-set. We could have defined a counterclockwise gain; it would be reciprocally related to the clockwise gain. The gain is independent of how far the initial point X chosen for the construction procedure is from P_1.

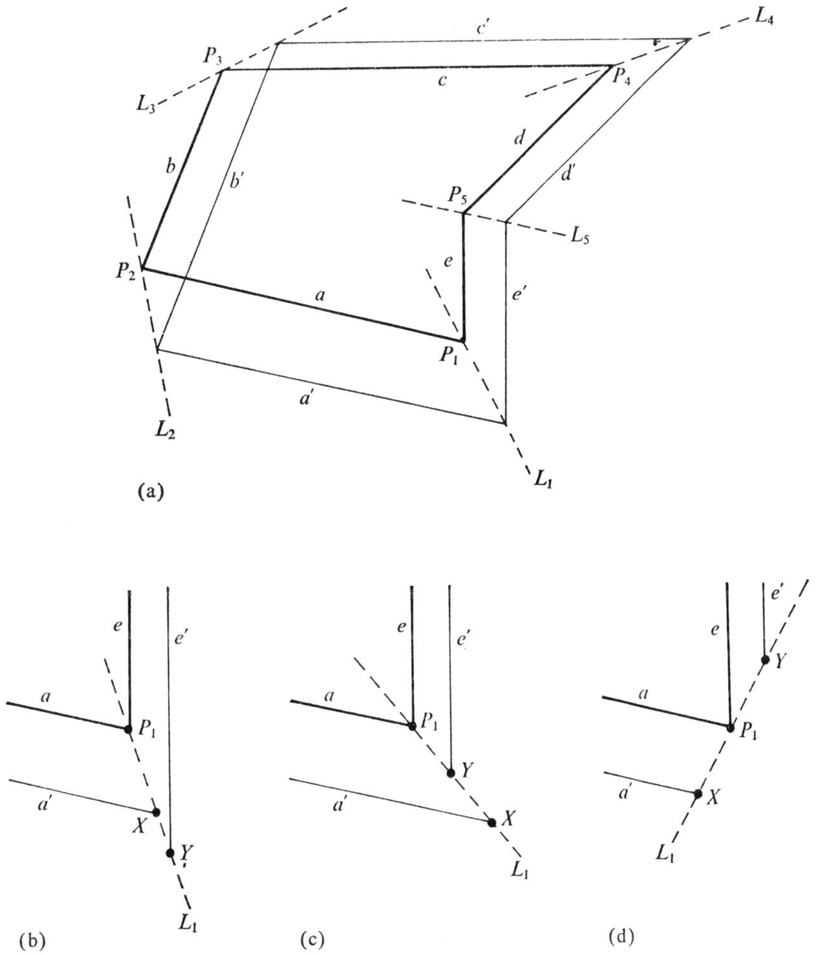

Figure 12. Illustrating the 'gain' concept: (a) illustrating unity gain condition; (b) illustrating gain$\simeq$3/2; (c) illustrating gain$\simeq$1/2; (d) illustrating gain$\simeq -1$

It is clear that a necessary condition for the realizability of an object depicted by a picture is that there be a *gain of unity associated with each cyclically-ordered edge-set*. When these gains differ from unity by only small fractions a picture will generally, but not always, look 'right' (*see* figure 13). Those in which the gains are appreciably different from unity will often look 'warped' (*see* figure 11(f), for example). In a sense the gain (or its reciprocal) is a measure of how 'ungrammatical' a picture is. It would seem possible to draw sets of pictures of objects in which the gain differed more and more from unity so as to test the sensitivity of human subjects to this type of picture distortion. In this application the advantage of having a precise numerical value for the distortion is obvious.

It is fairly easy to show that when the gain associated with a cyclic-set is negative the picture lines cannot even be labelled properly. This is beyond the scope of this paper and is left as an exercise to the reader.

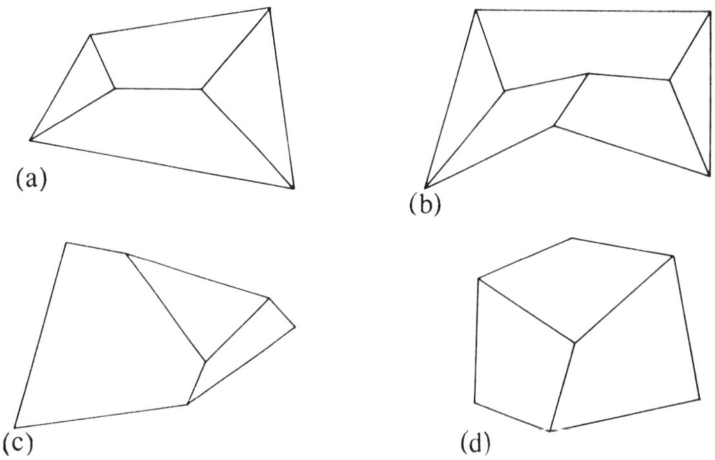

Figure 13. Pictures satisfying the unity-gain condition

The dual representation

A representation which is dual to the picture-graph will be summarized here briefly. Such a representation can be useful in visualizing the orientation of the surfaces of an object and in ascertaining whether or not they can be made mutually compatible.

A *dual picture-graph* is a graph in which there is a node corresponding to every surface represented in the picture, a (triangular) area corresponding to every (degree-3) node represented in the picture and a line corresponding to every line in the picture (*see* figure 14). If the picture is of a possible object it will be possible to construct the dual-graph and orient it so that every line in the dual-graph is at right angles to the corresponding line in the picture itself. The lines a, b, c and d in figure 14(a) represent the boundary of a plane surface in a picture. The directions of the lines e, f, g and h have been chosen so that the unit-gain requirement has been met. Therefore it is possible to construct surfaces of an associated object so that the surfaces A (defined by lines a, e and h), B (defined by the lines b, e and f), C (defined by the lines c, f and g) and D (defined by the lines d, g and h) are plane surfaces.

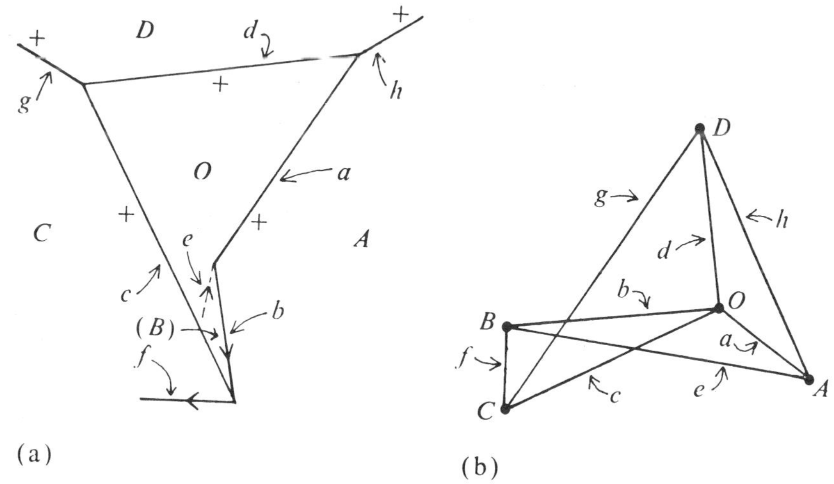

Figure 14. Illustrating duality: (a) picture-graph; (b) dual picture-graph.

We note that the directions of the three lines associated with a node in the picture uniquely determine the proportions of a corresponding triangle drawn in the dual-graph. By progressing around the boundary of a plane represented in the picture we generate in the dual a sequence of triangles the locations of which are uniquely determined once the first triangle in the sequence has been drawn. (Note, for example, in figure 14 the sequence of triangles *abe*, *bcf*, *cdg* and *dah*.) In this process a sequence of lines corresponding to a cyclic-set is generated in the dual graph (for example, the sequence *e*, *f*, *g*, *h*). This sequence will close on itself if and only if the unit-gain requirement is met.

With the dual-graph constructed in this manner its directed line-segments correspond to *gradients* and differences between gradients of planes in the picture. By 'gradient' in this context we shall mean a (two-dimensional) vector associated with a plane surface represented in the picture which, by its direction and magnitude, shows the direction and magnitude of the greatest positive rate of change of distance from the camera with respect to motion in the picture.

Assume in figure 14(a) that the object surface bounded by *a*, *b*, *c* and *d* is parallel to the picture plane. (The gradient associated with that plane is the zero vector.) The gradient associated with *D* would have the direction of the directed line segment $\overrightarrow{OD}$ in the dual-graph. The gradient associated with *B* would have the direction of $\overrightarrow{OB}$ in the dual-graph. (Note that the hidden surface *B* in the picture is under the reference plane and is tilted down toward the left.) The lines *a* and *c* have similar interpretations.

The vectors associated with the other lines can be interpreted as (vector) differences between pairs of the other gradients. With the reference plane assumed to have a zero gradient the node 0 in the dual can be considered to be the origin. Moving the origin away from this point corresponds to transforming the object pictured in such a way that its projection on the picture plane remains constant. Reducing the lengths of all lines in the dual-graph by a given factor corresponds to 'flattening' the object pictured by the same factor.

The (perpendicular) distance from the origin to an arbitrary line in the dual-graph is a measure of the slope of the corresponding object edge with respect to the picture-plane. Thus lines passing through the origin correspond to those edges which are parallel to the picture-plane. Pairs of surfaces which are parallel in the scene would be associated with the same point in the dual-graph.

It is possible to consider the dual-graph itself as a picture of a dual-scene in which information about the 'third dimension' has been lost. In that case the directed line segments in the original picture correspond to gradients of the surface in a 3-dimensional dual-scene. The other obvious dual relationships also hold. Special comment should, nevertheless, be made about the way pairs of lines are oriented with respect to each other in the scene and the dual-scene. If at the apparent point of intersection of two lines, L_i and L_j, in the picture L_i is in the scene actually closer to the camera than L_j is, then in the dual-scene L_j will be closer to the camera than L_i is. It follows that lines which intersect in the scene will also intersect in the dual-scene.

Summary remarks

We considered first a comparatively simple set of labelling constraints which we later found were not a sufficient test for the realizability of an object represented in a picture-graph. A more refined set of necessary conditions resulted from the unit-gain requirement and the alternate viewpoint available from the dual picture-graph. These constitute a productive way of demonstrating whether or not all of the surfaces in the scene can be realized as plane surfaces and whether or not they can have compatible orientations. In spite of this there are still examples of pictures of objects which cannot be analyzed by using only these tests in a straightforward way.

Consider, for example, the object(s) in figure 15. The edges can all be labelled consistently. The unit-gain requirement is met for all cyclic-sets of edges. And yet it is obvious that the scene is impossible to realize. One way of demonstrating this is to note that a properly-oriented piercing line would intersect the A-plane in two different points, p and q. Alternately, one could conclude from the picture that there must be at least two different lines in which the surfaces A and B must intersect: one in the left half of the picture and one in the right half. Other relatively simple tests give equivalent results.

It is not difficult to imagine other more complex sets of objects which can be proved to be possible or impossible *if* an appropriate set of auxiliary points, lines, or planes are added to the picture. Once these are added a proof of impossibility may be quite short and based on easily-understood principles of geometry. But to know *where* to apply these principles will probably require, in general, an analysis, the details of which are specific to the given picture.

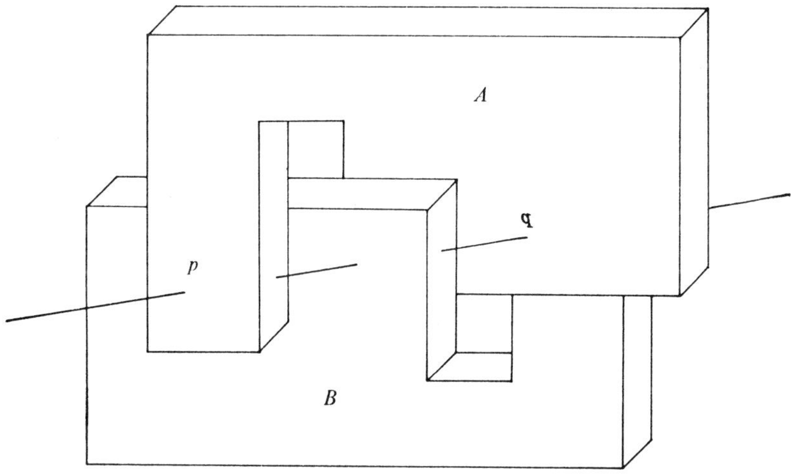

Figure 15. An impossible pair of objects

We have assumed for the purposes of this paper that the polyhedra which are pictured have only three edges associated with each vertex. This assumption is not so restrictive as it might at first seem. Vertices having four or more associated edges can be converted to sets of vertices having only three by a procedure of 'grinding' and/or 'filling' in the vicinity of the vertices. The vertex of the original polyhedron can then be considered to be the limit reached by the inverse procedure (in which the amount of ground or filled material is reduced to zero) and the configuration of lines at the corresponding picture node to be the conjunction of several of the basic configurations shown in figure 6. If we may assume that the polyhedra in our pictures can have arbitrarily large numbers of edges associated with their vertices the primary condition for the realizability of the objects pictured is that no two 'T' configurations exist which would imply arrow labels in different directions on the same line-segment [*see*, for example, figure 8(a)]. The results relating to the unit-gain requirement and the dual formulation apply without significant modification to pictures of these more general polyhedra.

A LANGUAGE FOR PICTURES OF SMOOTH OBJECTS

Types of picture lines and nodes

In this section I will present a brief summary report of work involving pictures of 'smooth' objects. A more detailed paper will be forthcoming at a later date.

'Smooth' here will be given an informal definition. By a *smooth surface* we shall mean one having, for each surface point, a well-defined directed line orthogonal to, and pointed away from the surface such that the orientation of this line is a continuous function of the position of the associated point on the surface. For the sake of simplicity we shall allow in our pictures only smooth surfaces bound by *smooth edges* (defined in a manner analogous to the one in which smooth surfaces were defined) and solid objects bound by smooth surfaces having no edges. Smooth objects may, of course, be quite complex. For instance, it is possible to associate either orientable ('two-sided') or nonorientable (' one-sided') smooth surfaces with arbitrarily large sets of edges which are arbitrarily knotted together. Consequently, smooth objects furnish another rich source of pictures for analysis.

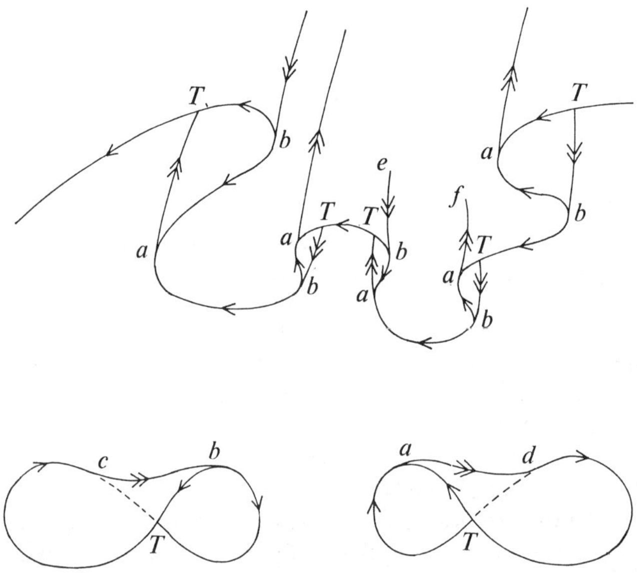

Figure 16. Cloth and saddles

In figure 16 there are some examples of pictures of smooth objects. Note that picture lines can represent either *edges* of a surface or *folds* in the surface. We distinguish between these two types of lines by labelling the first type with *single arrows* and the second type with *double arrows*. In each case we orient the arrow so that, as we travel in the direction of the arrow, the associated surface (one for a line representing an edge) or surfaces (two for a line representing a fold) are to the right in the picture. Thus the number of arrows in the label for a line is also a measure of the difference between the numbers of surfaces represented in the two areas on the two sides of the line.

It will be noted that at each indicated node in the picture (except the T-nodes) two of the lines meet at a *cusp*. This is true even for the nodes identified by '*c*', '*d*', '*e*' and '*f*' (at these nodes one of the lines represents a hidden edge or fold). This is a consequence of the fact that the surfaces and edges represented are smooth. For nodes of types '*e*' and '*f*' only one visible line is incident and this line always represents a fold. For pictures of *solid* smooth objects only these two latter types of nodes (and T-nodes) are necessary.

Derived constraints

The possible configurations of lines which can exist at picture nodes is summarized in figure 17. The last four configurations are possible in pictures of *creases* and are added to show how this style of picture language could be expanded. Surfaces which are creased are, however, not smooth in the sense that that term has been defined here and consequently in this paper we shall not consider further these last four types of picture nodes.

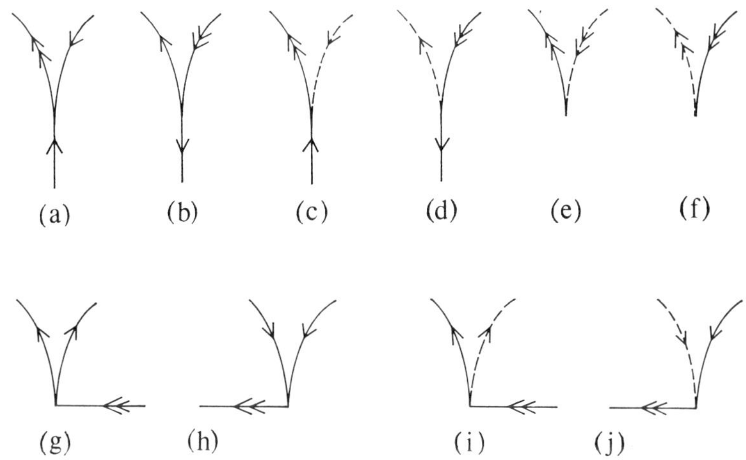

Figure 17. Configurations possible in a picture of smooth objects

By examining the first six configurations of figure 17 we can conclude that they occur in pairs which are related by being mirror images of each other. The mirror image of a configuration has its arrows pointing in directions opposite to the corresponding arrows in the original.

The hidden lines in figures 17(c), 17(d), 17(e), and 17(f) are all hidden under exactly one surface. When each of these configurations is placed under one or more surfaces none of the lines will be visible but the one will have a depth-index which is one more than those of the other lines.

In each of these configurations the number of arrows into a node equals the number of arrows out of that node (when hidden lines are taken into account). This is a consequence of the fact that the number of surfaces associated with a given area of the picture (bounded by lines which may be visible or hidden) is a constant. We can conclude that the net number of arrows into *each* closed picture region is also zero.

The surfaces, edges, and folds in a scene map into three basic unlabelled configurations (other than the 'T') in a picture: the 'gamma' node [figures 17(a), 17(b)], the 'lineal' node [figures 17(c), 17(d)] and the 'terminal' node [figures 17(e), 17 (f)]. In making decisions about the possibility or impossibility of smooth objects the lineal nodes cause special concern since one must consider the possibility that one or more lineal nodes may be concealed along any curved line segment. Nevertheless, even though the *number* of arrows associated with a line label changes as a lineal node is passed, the *direction* of the arrow(s) is unchanged. Consequently we know that the direction of the arrows appropriate for labelling a given segment of line is fixed over that line segment (until a gamma node or a terminal node is reached).

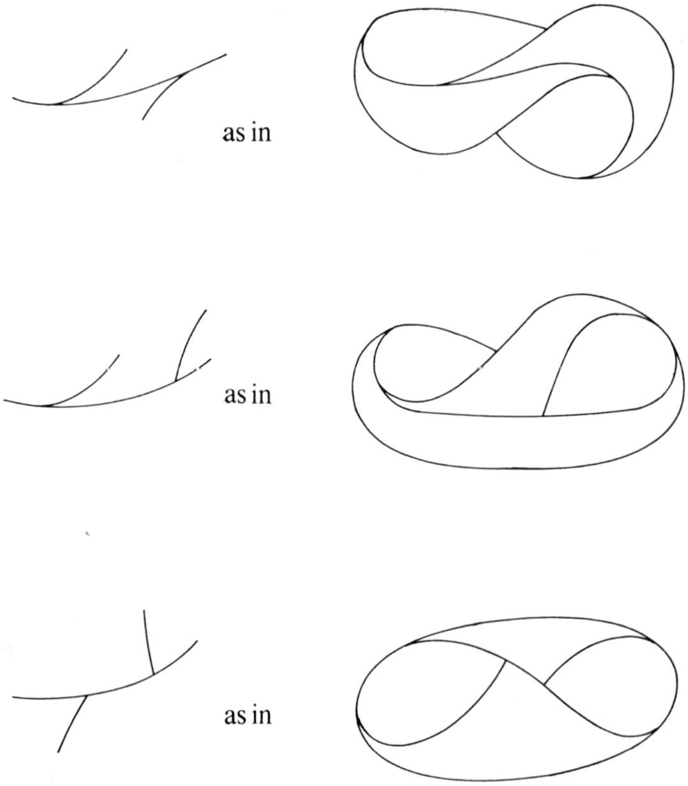

Figure 18. Forbidden configurations in pictures of smooth objects

An equally important conclusion follows from the observation of the direction of the arrows in the 'arms' of a gamma configuration: the arrow on the right arm is always into the node and the arrow on the left arm is always out of the node. This fact and the constancy of arrow direction in the vicinity of lineal nodes allow us to conclude that the three configurations in figure 18 (and their mirror images) cannot occur in pictures of possible objects. A slightly subtle version of the configuration of figure 18(b) is imbedded in the picture of figure 3(d) and prevents that object from being possible.

An example for study

The pictures in figure 19 show four different ways of completing the hidden line structure for the example of figure 3(a). The first of these shows the simplest and perhaps the most obvious way of continuing the visible lines as hidden lines. The labellings given on the hidden lines are compatible with those on the visible lines. Nevertheless the object is *not* possible.

In order to prove the impossibility of this object we examine the line which goes from p past q to r. This line indicates an edge which disappears between p and q and reappears between q and r. The line which *crosses* the path between p and q has a double arrow; this gives evidence that the edge represented at q is under *two* surfaces. On the other hand the line which crosses between q and r has a single arrow and this gives evidence that the edge represented at q is under only *one* surface. Since the line at q can have only one depth-index the object with the hidden line structure given in figure 19(a) is clearly impossible.

Other ways of completing the hidden lines are given in figures 19(b), (c), and (d). For each of these there is a way not only of labelling the lines in a consistent way but also of assigning a depth-index (shown by integers in the figure) to each line segment. Each configuration at a hidden picture node is basically the same as one of those in figure 17 but with the depth-index of all lines of the configuration increased uniformly by a number equal to the number of surfaces existing between the basic configuration and the camera. The picture in figure 3(b) is of a realizable object. One lineal node and one hidden gamma node suffice. The picture in figure 3(c) is also of a realizable object. Two lineal nodes and one hidden gamma node suffice. These examples are left as exercises for the reader.

Conditions for realizability

Any one of the three successful ways of adding hidden lines to the picture example above demonstrates the possibility of the object. The question of which is 'best' cannot, of course, be answered on an absolute basis. A reasonable preference is for the one (or ones) which require the fewest additional picture nodes. A reasonable constraint to impose is a minimum allowable radius of curvature for the hidden lines. Such a constraint would be somewhat analogous to an upper frequency limit or bandwidth restriction for continuous signals.

The author conjectures that a necessary and sufficient condition for the realizability of a picture of a smooth object (or objects) is that there be no conflict among the arrow *directions* imposed on a picture line by its association with an arm of a gamma node or the bar of a T-node (recall figure 18) or because of its identification as a part of the boundary of an object. The

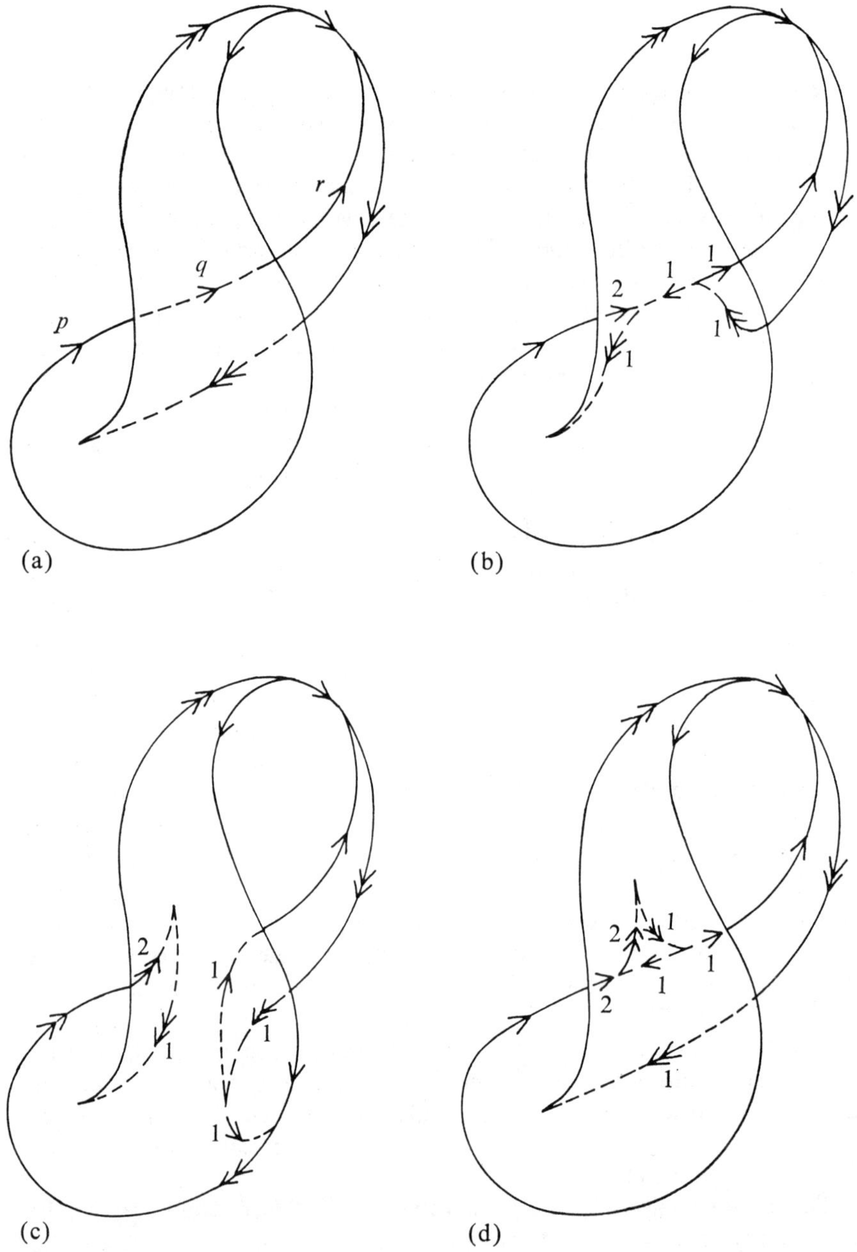

Figure 19. Possible hidden line configurations for a smooth object.

necessity of the condition is obvious. Its sufficiency would follow from the existence of a procedure for bringing together in a compatible way the hidden lines associated with T-nodes, lineal nodes, and terminal nodes. A key construction in this procedure was illustrated in figure 19(d). It allowed a single-arrowed line to go from one depth to another in the picture. Such techniques allow us to bring together hidden lines at the appropriate depth-indices.

In the proposed procedure (or in others which achieve the desired result) we must take into account the following issues. When two hidden lines cross in the picture the depth-index of the lower line must be changed by the number of arrows on the upper line. (This is a generalization of the rules appropriate at a T-node.) The depth-index of a line cannot exceed the number of surfaces associated with the areas on either side of the line, nor can the number of surfaces associated with a picture area be negative.

SUMMARY REMARKS

Very few real-world scenes contain only degree-3 polyhedra or simple smooth objects. Nevertheless the study of the picture languages associated with such simple scenes serves its purpose by giving us some perspective about what we might and might not expect to accomplish from the study of more complicated picture languages. As a minimum we should at least be better able to decide when sentences from a 'polyhedral language' or 'smooth language' are or are not present in a picture containing a babble of several languages. Pictures of objects with combinations of smooth and plane surfaces and having additional types of 'vertices' (for example, cones and cylinders) may be analyzed using combinations of the techniques illustrated in this paper.

As we have mentioned earlier additional picture configurations are possible if we allow creases in the surfaces. Other special types of surface features correspond to their own element types. For instance, the configurations around picture points which are associated with 'crumpled paper' surfaces are constrained by the fact that in the scene the angles which are represented at the picture point must sum to 360°. Wrinkles, puckers, scoring, and indentations all cause their own distinctive configurations in a picture and add to the list of symbols possible in a picture language.

Stereo pairs of pictures, sequences of pictures taken at slightly different times, and pictures containing shadow information (they are all closely related) offer other examples worthy of study.

Acknowledgement

I plead guilty to the charge that I deal with pictures of impossible objects because it is fun. It is, and that is reason enough. However, in addition to this I believe that much can be learned in the study of any language by asking '*Is* that a nonsense sentence?' and '*Why* is that a nonsense sentence?'.

Much of the material presented here was stimulated by contact with and tested on the members of the Artificial Intelligence Group at the Stanford Research Institute. I am especially grateful to the group manager, Charles A. Rosen, for his encouragement, enthusiasm, confidence, and patience.

REFERENCES

Guzman, A. (1968) Decomposition of a visual scene into three-dimensional bodies. *Proc. of the Fall Joint Computer Conference*, pp. 291–304.

Huffman, D.A. (1968) Decision criteria for a class of 'impossible objects'. *Proceedings of the First Hawaii International Conference on System Sciences.* Honolulu.

Penrose, L.S. & Penrose, R. (1958) Impossible objects: a special type of illusion. *Brit. J. Psych.*, **49**, 31.

Part VII
Scene Analysis: The Natural World

Without the regularities of the blocks world, it becomes more difficult to find viable, general approaches to the problems of three-dimensional scene analysis. Models of nonpolyhedral three-dimensional objects are much more complicated, and are much more difficult to exploit without explicit depth data. [For the use of a three-dimensional generalization of Blum's skeletal representation, see Agin and Binford (1976) and Nevatia and Binford (1973).] Huffman, in his paper in the preceding part, began an analysis of line drawings of smooth objects, and this work has been significantly extended by Chien and Chang (1974) and Turner (1974). Investigations have been made of such local depth clues as shading [Horn (1975)] and texture gradient [Bajcsy and Lieberman (1976)]. Nevertheless, most work on scene analysis in the natural world has avoided a direct confrontation with the problems caused by projecting a three-dimensional world onto a two-dimensional image plane.

Perhaps the best explored general formalism has been the syntactic or, more generally, the linguistic approach. Inspired by the successful development of formal techniques for programming languages, Eden (1962), Narasimhan (1964, 1966), Kirsch (1964), and many others investigated ways to extend and to apply these ideas to image analysis. By now this work has produced a rather large literature of its own, including conference proceedings edited by Kaneff (1970) and Nake and Rosenfeld (1972), and a textbook by Fu (1974).

The first paper, by Fu and Swain, is both a tutorial paper and a survey of work on syntactic pattern recognition. (Another useful survey by Miller and Shaw (1968) provides additional information, particularly about the earlier investigations. See also Evans (1969) for general approaches to descriptive analysis.) One of the major strong points of the syntactic approach is the conceptual clarity that comes from separating knowledge about the domain, procedures for detecting primitive elements, and strategies for performing the scene analysis. In particular, with the syntactic approach, knowledge about how the primitive elements can be interconnected to form larger structures is expressed explicitly in the form of a grammar. Subroutines for extracting low-level information from the image are isolated in the primitive detectors. The problem of deciding how to use this general knowledge together with the evidence supplied by the primitive detectors to describe the scene is delegated to the parser.

This separation of functions is also maintained in the second paper, by Barrow, Ambler, and Burstall. This paper assumes the existence of a syntactic system that can produce an essentially flawless but relatively low-level description in terms of primitive objects and simple relations among them. The resulting relational structure is then matched against models to provide a high-level description. Much of the paper is devoted to examining and comparing different strategies for doing this matching. The most successful method, which they call hierarchical synthesis, makes the greatest use of the structure of the objects in the scene. (See Hayes-Roth (1976) for a complementary approach to matching.)

Needless to say, the problem of obtaining error-free primitive detection or flawless low-level descriptions is, if anything, even more difficult with natural scenes than it is in the blocks world. Thus, methods that recognize the inevitability of errors and that can use high-level knowledge to cope with them are ultimately mandatory. The best solutions to these problems obtained to date have concerned scene verification, in which the general scene organization is known *a priori*. For example, Harlow (1973) shows how a graph representation of high-level knowledge can be used to guide the top-down analysis of images such as chest X-ray photographs and pictures of printed-circuit boards. Another example is given by work on face recognition, which relies heavily on prior knowledge of facial structure. (See the paper by Fischler and Elschlager in Part I, and the papers by Sakai, Nagao, and Kanade (1972) and Kaufman and Breeding (1976) for three different ways to use this knowledge.)

The third paper, by Yakimovsky and Feldman, shows how Bayesian decision theory can be used to provide robustness in a general-purpose, bottom-up segmentation procedure. The problem addressed is that of partitioning simple outdoor scenes into meaningful regions. Here knowledge about natural constraints between adjacent regions is expressed in probabilistic terms, and is used to guide the behavior of a region-growing procedure. The ideal segmentation is the one that maximizes the probability of a correct interpretation, given both the image and the prior contextual knowledge. Since the computation needed to find this optimal solution is prohibitive, the authors develop an approximate, sequential procedure that appears to give good, suboptimal solutions. [For further developments of this approach, see Feldman and Yakimovsky (1974).]

The fourth paper, by Tenenbaum and Barrow, takes a

related but different approach to the problem of using general knowledge to guide the segmentation of a scene. Rather than viewing the process as one of optimizing a probabilistic score, the authors view the problem in terms of constraint satisfaction. Regions are assigned one or more possible interpretations (class labels), and merging of regions having mutually incompatible interpretations is forbidden. Moreover, when regions are merged, the allowed interpretations can change. This can eliminate possible interpretations for adjacent regions, and these eliminations can propagate throughout the image.

It is possible to extend this approach to allow degrees of uncertainty in assigning class labels to regions, and to allow probabilistic or fuzzy relations between regions. The problem then becomes more one of constraint optimization than constraint satisfaction. The methods of Yakimovsky and Feldman provide one approach to the optimization problem. Another approach is through iterative, relaxation procedures [see Tenenbaum and Barrow (1975) and Rosenfeld, Hummel and Zucker (1976)]. The last paper, by Zucker, shows how relaxation can be used to reduce or to eliminate ambiguity in scene analysis. While the properties of the relaxation approach to constraint optimization are not yet thoroughly understood, it is particularly attractive because of its conceptual simplicity, versatility, and amenability to possible implementation in parallel hardware.

Many of the references cited in this and the preceding parts are quite recent. This reflects the fact that the field of digital image and scene analysis is still advancing rapidly. Enough time has passed to provide a degree of perspective on the field, making it possible to choose the papers in this collection with some assurance that they will remain standard references for many years to come. At the same time, new developments continue at a fast pace, so that in a few more years, further collections of papers in this field will undoubtedly appear.

On Syntactic Pattern Recognition

K.-S. Fu and P. H. Swain
PURDUE UNIVERSITY
LAFAYETTE, INDIANA

I. Introduction

This chapter is primarily a review and evaluation of the state of the art of syntactic pattern recognition, a subject of considerable current interest due to the increasing need for powerful new approaches to pattern recognition problems of great complexity. Aspects of the syntactic approach that seem to be in urgent need of attention are also discussed at some length. It is hoped that this chapter will provide an enlightening introduction and unifying background for the other chapters and at the same time will provide a contribution in its own right to the further development of syntactic pattern recognition.

Early attempts at machine generation and analysis of handwriting (Eden, 1961, 1963) demonstrated that a set of primitive elements together with a set of rules for combining the primitives could be used to generate script which was virtually indistinguishable from human handwriting. These studies had important implications for handwriting recognition since they suggested a method for (1) describing complicated two-dimensional patterns in a one-dimensional format well suited for input to communication and data processing systems and (2) distilling the essential information from complex patterns so as to reduce to practical levels the amount of information to be transmitted or processed. A serious obstacle to the recognition of handwriting remained in the problem of ambiguity, i.e., the fact that handwritten words apparently could not be partitioned into unique sequences of the primitives considered, particularly when the handwriting was far from "textbook" script. But it was postulated that contextual considerations could be helpful in solving this problem.

The concept of describing patterns in terms of primitive elements, subpatterns, and their relationships can be applied to other forms of pattern data for which concise numerical descriptions are often not easily obtainable, (e.g., nuclear bubble-chamber tracks, human speech, geometric figures, games, digital display data, etc.). With respect to such data, one can often formulate

Reprinted with permission from *Software Engineering*, J. T. Tou, Ed., vol. 2, 1969, pp. 155–182.

a hierarchical structure of subpatterns analogous to the syntactic structure of languages. Syntactic pattern recognition is an attempt to adapt the techniques of formal language theory, which provide both a notation (grammars) and an analysis mechanism (parsing) for such structures, to the problem of representing and analyzing patterns containing a significant syntactic content. As this apparently includes many kinds of patterns of interest, syntactic pattern analysis has recently become the focus of an increasing amount of pattern recognition research.

Regardless of the precise form of the data, syntactic analysis can proceed only if a grammatical model for the data generation and/or analysis process can be formulated. (This does not necessarily imply that the data is actually generated by a mechanism which operates in the same way as the model.) A related practical problem is the development of efficient analysis procedures based on the grammatical model. The available literature concerning syntactic pattern recognition deals almost exclusively with the formulation of grammatical models, although the development of analysis algorithms is currently receiving more attention. The following sections discuss the problem of selecting pattern primitives which can be used to describe patterns of interest, methods for assembling the primitives and their relationships into pattern grammars, and techniques for actually performing pattern analysis in terms of the grammars. Finally some problems involved in dealing with "real life" patterns containing the inevitable distortion and noise are noted and an approach to such problems is proposed.

It might be worth mentioning at the outset that the terms "linguistic" and "syntactic" are used almost interchangeably herein, as has become common practice in the literature, although the former is a somewhat broader term. Strictly speaking, "syntactic" refers only to the structural aspects of languages; analytical techniques have been considered, both for natural languages (Shank and Tesler, 1969) and for patterns (Fischler, 1969) which may be construed as linguistic but not syntactic.

II. Grammars, Syntax, Parsing, and Concatenation

Since some of the terminology of formal language theory will be used extensively throughout this paper, a necessarily brief introduction to the terminology is in order. Some additional material in this respect will be provided as needed.

The central idea of formal language theory is the generation and/or analysis of the strings (sentences) of languages in terms of grammars, typically *phrase-structure grammars* as defined informally in Figure 1. The structural description of the language (note that a language is just a set of strings over the set of terminals—often referred to herein as "primitives"—of the grammar) in

A <u>Phrase-Structure Grammar</u> is a four-tuple

$$G = (V_N, V_T, P, S)$$

V_N: finite set of "nonterminals" or variables

V_T: finite set of "terminals"

$S \in V_N$: "start" symbol or "sentence" symbol

P: finite set of "productions" or rewriting rules

Productions have the general form

$$\alpha \rightarrow \beta$$

where α and β are strings over

$$V_N \cup V_T,$$

and "$\rightarrow$" is read "is replaced by."

L(G) denotes the language (set of strings) generated by G.

Example: $V_N = \{S, A\}$ $\quad V_T = \{0, 1\}$

P: $S \rightarrow 0A \quad A \rightarrow 1A$

$A \rightarrow 0A \quad A \rightarrow 11$

$L(G) = \{x \mid x \in V_T^* \text{ and } x \text{ begins with 0, ends with 11}\}$

Sample generation: $S \Longrightarrow 0A \Longrightarrow 00A \Longrightarrow 001A \Longrightarrow 0010A \Longrightarrow 001011$

FIGURE 1. A phrase-structure grammar.

terms of a grammar is called a *syntax* of the language. Analysis in terms of this structure is called *syntactic analysis*, or often simply *parsing*.

In formal language theory, the only relation between the elements in a string, either a terminal string or an intermediate string containing nonterminals, is concatenation, i.e., the juxtaposition of adjacent elements. It will be seen throughout this chapter that a crucial point involved in adapting the techniques of formal language theory to pattern analysis is the generalization of this simple notion to include other (almost arbitrary) relationships and mechanization of the resulting, considerably more general, formalism in as systematic a way as has been done for formal "string" languages and grammars.

III. Selection of Pattern Primitives

The first step in formulating a syntactic model for pattern analysis is the determination of a set of primitives in terms of which the data of interest may be described. This will be largely influenced by the nature of the data, the specific application in question, and the technology available for implementing the analysis. The primitives should provide an adequate description of the data, i.e., the primitive descriptions of patterns from different classes should be distinguishable by whatever method is to be applied to analyze the descriptions. In addition to this obvious requirement, the primitive descriptions should be readily obtainable from the raw data and should be compact enough so as not to overtax the memory and processing capacities of the analysis system. In many instances these may be conflicting requirements, and it may be necessary to find an acceptable trade-off among them.

One of the earliest papers describing the decomposition of patterns into primitives (Grimsdale *et al.*, 1959) presented a conceptually appealing method which allowed the recognition system to determine the primitive elements by inspection of training samples. Each primitive consisted of an observed point set (i.e., part of a pattern) the horizontal extent of which displayed no sudden changes on successively scanned lines (quantitatively, "sudden" meant within a user-specified tolerance set to account for noise in the pattern). Such primitives are easily detected by a simple raster scan of the pattern. The analysis scheme developed was apparently quite successful, allowing for such annoyances as varied pattern size, pattern rotation, etc., but required an impractical amount of processing time and a large dictionary to remember the valid pattern structures. Simpler pattern descriptions and analysis procedures may have been achieved (possibly at the expense of increased hardware complexity) if explicit consideration could have been given to formalizing the pattern syntax.

More recent efforts concerned with the primitive selection problem may be grouped roughly as follows: general methods emphasizing boundaries, general methods emphasizing regions, and special methods which take advantage of the peculiarities of specific applications. A set of primitives commonly used to describe boundaries is the chain code due to Freeman (1961). Under this scheme, a rectangular grid is overlaid on the two-dimensional pattern and straight line segments are used to connect the grid points falling closest to the pattern (see Figure 2). Each line segment is assigned an octal digit according

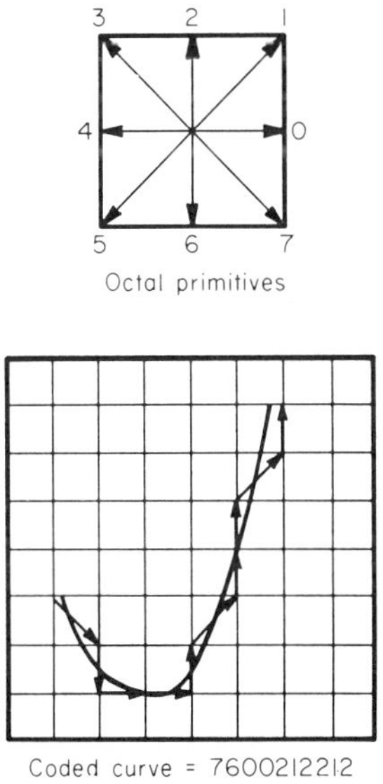

FIGURE 2. Freeman's chain code.

to its slope. The pattern is thus represented by a (possibly multiply connected) chain or chains of octal digits. This coding scheme has some useful properties. For example, patterns coded in this way can be rotated through multiples of 45; simply by adding an octal digit (modulo 8) to every digit in the chain (although only rotations by multiples of 90° can be accomplished without some distortion of the pattern). Other simple manipulations such as expansion, measurement of curve length, and determination of pattern self-intersections are easily carried out. Also, the code is a "saturated" three-bit code (i.e., there are 8 possibilities for each digit) which makes it both economical and easy to manipulate. Finally, any desired degree of resolution can be obtained by adjusting the fineness of the grid imposed on the patterns. Of course, this method is not limited to closed boundaries; it can be used for coding arbitrary two-dimensional figures composed of straight or curved lines and line segments.

Notable works using Freeman's chain code include efforts by Knoke and Wiley (1967) and by Feder (1967). Knoke and Wiley attempted to demonstrate that linguistic principles can usefully be applied to describe structural relationships within patterns (hand-printed letters). After preprocessing of a CRT-input pattern to obtain a Freeman-code description of the pattern, they use a "transformational grammar" to reduce irregularities (i.e., to smooth the pattern). The result of this step is a description of the pattern in terms of a list of vertex interconnections and the shapes of the corresponding connecting arcs, which constitutes a sentence in the pattern language (see Figure 3). The

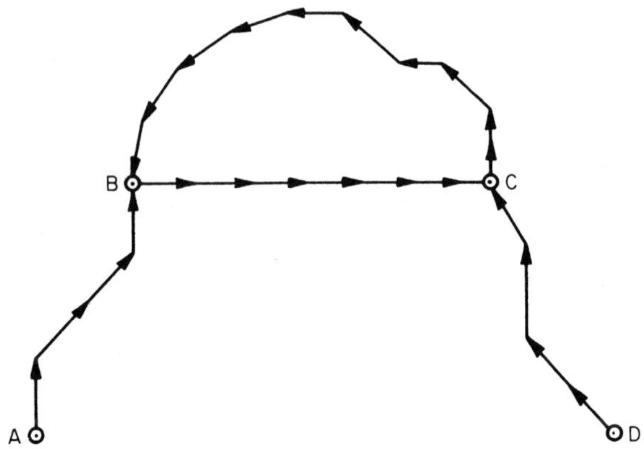

Preprocessed input

	Arc-shape	Relative direction	Initial vertex	Terminal vertex
Phrase 1	LIN	0	A	B
Phrase 2	LIN	2	B	C
Phrase 3	CURL	6	C	B
Phrase 4	LIN	7	D	C

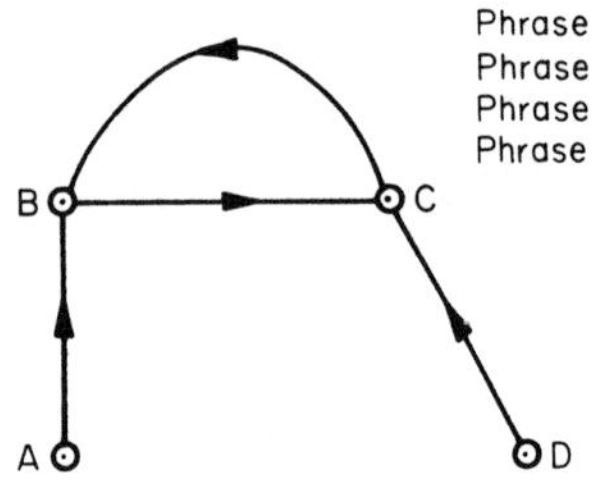

"Parsed" input

FIGURE 3. Parse of a chain-encoded pattern. (From Knoke and Wiley, 1967.)

sentence is then checked against a dictionary of prototypes and "recognized" if a match is found. Although some limited success has been achieved in applying this method to character recognition, the authors readily admit some serious difficulties, namely, (1) it has not been possible to find a pattern grammar which generates all and only valid patterns; (2) it is necessary to restrict the allowable length of pattern sentences; (3) it has not been possible to utilize context; (4) the syntactic analysis is slow and inefficient; and, (5) only connected patterns are considered.

Feder's work considers only patterns which can be encoded as strings of primitives (the Knoke and Wiley vertex–arc shape–vertex scheme can deal with multiply connected chains of primitives). Several bases for developing pattern languages are discussed, including equations in two variables (straight lines, circles and circular arcs, etc.), pattern properties (self-intersection, convexity, etc.) and various measures of curve similarity. Much of the work is taken up with determining the computational power (automaton complexity) required to detect the elements of these languages (which are mostly context-sensitive but not context-free), a problem considerably complicated by the facts that (1) the chain code yields only a piecewise linear approximation of the original pattern, and (2) the coding of a typical curve is not unique, depending to a degree on its location and orientation with respect to the coding grid. Unfortunately, little attention is given to how one would actually synthesize a pattern recognizer based on the proposed pattern languages.

The efforts of Knoke and Wiley and Feder provide confirmation for the hypothesis that the linguistic approach may be applied to pattern recognition, although some serious problems need further attention. On the negative side, however, it appears that although Freeman's chain-encoding method is one of the simplest ways to present patterns to data processing equipment, patterns so encoded usually must undergo considerable preprocessing (essentially data reduction, smoothing, and a sort of feature extraction), which helps to emphasize structural relationships (syntax), before syntactic analysis can be effectively applied. In this sense the Freeman encoding may be "too primitive" in much the same way as the binary elements in the familiar "photocell array" encoding used in some character recognition devices. The required preprocessing may not necessarily be best achieved by linguistically oriented techniques [for example, see Zahn (1969)].

A set of primitives for encoding geometric patterns in terms of regions has been proposed by Pavlidis (1968). In this case, the basic primitives are halfplanes in the pattern space (this could be generalized to halfspaces of the pattern space). If can be shown that any arbitrary polygon may be interpreted as the union of a finite number of convex polygons. Each convex polygon can in turn be interpreted as the intersection of a finite number of halfplanes. By defining a suitable ordering on the convex polygons composing the arbitrary polygon, it is possible to determine a unique minimal set of maximal (in an appropriate sense) polygons, called *primary subsets*, the union of which is the given polygon. Pavlidis' approach provides a formalism for describing the syntax of polygonal figures and more general figures which can be approximated reasonably well by polygonal figures. However, his analysis procedures require the definition of suitable measures of similarity between polygons. The similarity measures he has considered are quite sensitive to noise in the patterns and/or are very difficult to implement practically on a digital computer.

Another approach to the analysis of geometric patterns is discussed by Guzmán (1967, 1968). This approach assumes that a picture can be reduced by preprocessing to a list of vertices, lines, and surfaces. Various heuristics, based primarily on the analysis of types of intersections of lines and surfaces, are applied to this list to compose its elements into two- or three-dimensional regions. In effect, this amounts to a data compression process after which further processing for the purpose of, say, pattern recognition can be greatly simplified. Some candidate pattern recognition schemes are investigated (Guzmán, 1967), all of which involve methods for matching the reduced pattern descriptions against a prototype dictionary.

The chapter by Rosenfield and Strong proposes another selection of pattern primitives based on regions and apparently of a somewhat more general nature than those previously discussed. Interesting examples of rather specialized selections of pattern primitives include Eden (1963), Ledley *et al.* (1965), and the chapter by Pavlidis.

One aspect which most of the foregoing approaches share is a concern with the reduction of the pattern to be analyzed into some sort of string representation. This is partly due to the relative ease with which string representations can be handled, but it may also be attributed to a desire to take advantage of the existing results in formal language theory, since to use these results it is necessary either to develop effective methods to accomplish the reduction of patterns to strings of concatenated primitives or to extend the existing theory to include syntactic relations more general than concatenation. The former has, to date, been the most common approach. One effort to formulate generalized syntactic relations has recently been undertaken by Anderson (1968) for the syntactic analysis of handwritten mathematical expressions. Anderson has proposed an entity called a *syntactic unit* as the multidimensional analog of the one-dimensional syntactic variable (he defines both terminal and nonterminal syntactic units). A syntactic unit is a subscripted variable, say A_p, where A is the name of the unit and p is an ordered set of coordinates specifying the location of the subpattern named A in the pattern space. A pattern described as a list of syntactic units differs in a very fundamental way from the conventional string: Since each syntactic unit contains its own location information, the position of the unit in the list no longer has any particular significance. This gives considerably more flexibility and power to the corresponding grammars than one might expect. It also complicates the analysis. More will be said about this work in Section IV,B.

There is a group of syntactic pattern recognition researchers which is not particularly concerned about string-encoding of patterns because they are primarily interested in developing equipment capable of processing all elements of a digital picture array in parallel (Narasimhan, 1964, 1966; McCormick, 1963; Narasimhan and Mayoh, 1963). Under their picture-processing formalism, entire picture arrays are treated as operands of picture-processing operations which produce a new picture array as a resultant. Pictures can thus serve as syntactic primitives and higher-order syntactic constructs as well. Near the top of the hierarchy, however, their syntactic elements are very much like Anderson's syntactic units, each construct consisting of a name and an attribute list [see especially Narasimhan (1966)]. Interestingly, Rosenfeld and Pfaltz (1966) have noted that many (perhaps most) useful picture transformations which have been defined under parallel processing formalisms involve local operations (i.e., functions which define a value for each element in the new picture in terms of the values of the corresponding element and a small set of its neighbors in the original picture), and they have proved that parallel processing of this sort is computationally equivalent to sequential processing. In other words, there are no local parallel processing functions which cannot be accomplished by local sequential processing, and vice versa. What is perhaps more surprising is that sequential processors can be significantly more efficient in terms of the total number of operations involved (the number of points times the number of operations per point) than parallel processors in performing many useful types of picture processing. In fact, parallel processors are sometimes forced to simulate sequential processing and can do so only very inefficiently. But one can still expect that, in general, parallel processing will yield a significant time advantage over sequential processing in spite of the relative computational efficiency of the latter.

IV. Pattern Grammars and Syntax-Directed Analysis

Now if one assumes that a satisfactory solution of the "primitive selection" problem is available for a given application, the next question is: How does one carry out the analysis of the primitive pattern description? (Of course, this is not to imply that the primitive selection and pattern analysis aspects can be separated in practice; but the fact that they cannot will not prohibit a dichotomous survey of the work that has been attempted.) Before plunging too far into this part of the problem, it would be well to recall a bit more terminology commonly associated with methods of syntactic analysis.

A convenient way to exhibit syntactic structure is by means of a *parsing tree*. The root of the tree, conventionally displayed at the top, is labeled with the start symbol; the terminating branches, at the bottom, are labelled with terminal symbols the union of which (concatenation in the case of string languages) is a sentence of the language. The structure leading from root to terminating branches describes how the sentence is derived from the start symbol by successive applications of the productions in the grammar.

Some parsing techniques develop the parse by proceeding downward from the root of the tree, i.e., by attempting to synthesize the pattern through use of the syntactic productions. These are called *top-down* methods. Other methods work upward from the bottom of the tree by attempting to decompose the pattern by applying the productions in reverse. These are called *bottom-up* methods. There has been considerable debate among formal linguists as to the relative merits of the two approaches for automatic parsing, but from a practical point of view the best approach generally depends on the particular form of the grammar involved (Griffiths and Petrick, 1965).

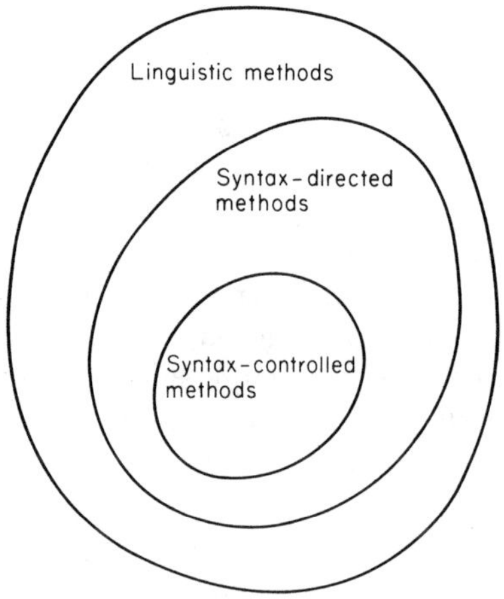

FIGURE 4. Hierarchy of linguistic analysis techniques.

One can describe a "hierarchy" of linguistic or syntactic analysis procedures (Figure 4). Any procedure which is at least incidentally concerned with decomposing a pattern into primitive components (alternatively, synthesizing it from primitive components) will be called a *linguistic* procedure. A procedure which utilizes in any way the syntactic description of an entity (sentence, pattern) is a *syntax-directed* procedure. If the grammar on which such a description is based is intimately associated with the procedure (e.g., the grammar may in some manner direct the analysis), then the procedure is said to be *syntax-controlled.* [This terminology is analogous to that used with respect to syntax directed compilers (Floyd, 1964a, b).] For example, a pattern categorization procedure might consist of matching the parse of an unknown against various prototype parses. This is syntax-directed but not syntax-controlled pattern recognition. A top-down or bottom-up procedure might have the category names as higher-order syntactic elements so that a successful parse necessarily accomplishes the categorization. This procedure would constitute syntax-controlled pattern recognition.

Surprisingly, some of the approaches which have been taken to syntactic pattern analysis are hardly syntax-directed, much less syntax-controlled. For example, in Pavlidis' work mentioned earlier, cognizance is taken of the fact that a syntactic model may be formulated for the composition of polygonal figures from halfplanes; but although higher-level constructs are defined in terms of the intersections of the primitive halfplanes, the syntactic analysis stops at that point, with no consideration given to the relationships at the higher level. This is like terminating the analysis of a FORTRAN statement once it has been ascertained to consist of additive terms and divisor terms without determining specifically how the terms are related. Such a scheme may be sufficient for a class of problems involving fairly elementary patterns and distinct pattern classes, but it is probably not an adequate model for more complicated situations.

On the other hand, a successful application of syntax-controlled pattern recognition is the FIDAC system (Film Input to Digital Automatic Computer) developed by Ledley *et al.* (1965) for the detection and classification of chromosome types in photomicrographs. In this case a set of primitives was determined (line segments of various shapes) which was at once adequate for the characterization of the patterns of interest and readily detectable by a suitable combination of hardware and computer software. Furthermore, the primitive-encoded patterns amounted to strings which could be generated by a fairly simple context-free phrase-structure grammar (Figure 5). The syntax-controlled analysis system permits rapid and dependable processing (according to its developers) of sufficient numbers of chromosome photomicrographs for statistical studies of chromosome variability.

Most attempts at syntactic pattern analysis fall somewhere between the two extremes cited above in terms of the degree to which they exploit the "syntactics" of the problem. Sometimes the syntactic structure is more readily definable than for Pavlidis' polygon figures, but only occasionally can it be defined as directly as appears to have been the case for FIDAC. General solutions for the primitive selection, grammar formulation, and analyzer synthesis problems have not yet appeared. The primitive selection problem has been discussed in the previous section. Research pertaining to the formulation of pattern grammars and some approaches to the synthesis of syntax-directed pattern analyzers will now be considered. It is in this area that one hopes to take advantage of the body of results available from research in formal language theory and syntax-directed compilers and perhaps some results from automata theory. Some references which may be of particular interest in this respect are included in the bibliography.

Much of the work yet to be mentioned can be conveniently described in terms of the following generalized syntactic formalism (GSF).

A *generalized syntactic element* (GSE) is a construct consisting of two parts; a *name*, which is actually a convenient way of referring abstractly to sets of invariant properties and property relationships; and an *attribute list*, which is a set of modifiers or variable properties which take on values associated with specific occurrences of the GSE.

A *generalized syntactic production* (GSP) is a rule which specifies how a syntactic element is composed from or generates other (usually more primitive) syntactic elements. A GSP also consists of two parts; the first part indicates the relationship of names; the second specifies the necessary relationships of attributes.†

To illustrate, consider three GSE's defined as follows. CEE(RADIUS, DIRECTION) represents the set of semicircular arcs (invariant property)

†In fact, a GSP may express a relationship of predicates in the sense of Evans (1967), Banerji (1968), or Sherman and Ernst (1969), thereby providing possibly the most general type of syntactic formalism.

V_T = {a,b,c,d,e}

V_N = {S, T, Bottom, Side, Armpair, Rightpart, Leftpart, Arm}

P: S → Armpair · Armpair
T → Bottom · Armpair
Armpair → Side · Armpair
Armpair → Armpair · Side
Armpair → Arm · Rightpart
Armpair → Leftpart · Arm
Leftpart → Arm · c
Rightpart → c · Arm
Bottom → b · Bottom
Bottom → Bottom · b
Bottom → c
Side → b · Side
Side → Side · b
Side → b · d
Arm → b · Arm
Arm → Arm · b
Arm → a

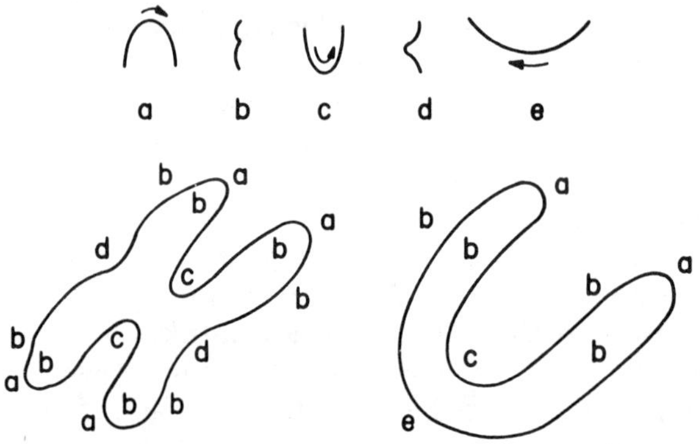

FIGURE 5. The grammar for FIDAC. From Ledley *et al.*, 1965. Used with the permission of Dr. Robert H. Ledley, National Biochemical Research Foundation, Silver Spring, Maryland.

of arbitrary radius and orientation (variable properties). BAR(LENGTH, DIRECTION) represents the set of straight line segments of arbitrary length and orientation. And DEE(HEIGHT, DIRECTION) represents the set of D-shaped figures of arbitrary size and orientation. The GSP

$$\text{DEE}(a, b) \rightarrow \text{BAR}(a, b)\ \text{CEE}(a/2, b)$$

describes how a D-shaped figure is composed of (or generates) a semicircular arc and straight line segment of suitable size and orientation.

This example may be far too simple to be realistic; for example, no connectivity specifications are given. But any such additional properties could be added to the attribute lists as desired. In fact, deciding how complex to make the attribute list and how much structure to subsume as part of the invariant property set associated with the GSE name constitutes a critical aspect of the selection of suitable syntactic elements and relations. This decision affects at once the form of the productions, the adequacy of the model, and the nature of the analysis algorithm. In practice the selection must be such that the required analysis is achievable within any specified cost and efficiency (processing speed) limitations.

A. Pattern Grammars

The Picture Description Language (PDL) developed by Shaw (1968, 1969) is actually a pattern grammar which appears to be a reasonably general and flexible formalism for the description and analysis of picture data (and possibly other forms of multidimensional data with syntactic content). The PDL is a generalized syntactic formalism and also a string grammar, so that there is reason to hope not only that it can be applied to a large class of pattern data, but also that some of the results and techniques of formal linguistics and/or automata theory can be brought to bear through it to assist in the realization of efficient pattern analysis techniques.

By means of the PDL formalism, one-, two-, and three-dimensional data can be put into string form. This is accomplished by requiring that all syntactic elements (pictures, subpictures, and primitives as well) have a *head* and a *tail*; concatenation can occur only at the head and tail of syntactic elements (head to tail). All pictures are assumed connected, which is made to hold even for pictures with disjoint elements by the definition of suitable *blank* primitives. A *null point* primitive is also defined which consists only of coincident head and tail. Four binary operators are defined which, together with the null point primitive, are sufficient to describe all possible concatenations of syntactic elements (Figure 6). Two unary operators are defined, one of which acts as a *reverser* (head-to-tail interchange), the other specifying superposition of primitives. The superposition operator in conjunction with the definition of blank primitives and *label operators* provide the PDL the ability to describe multiply connected figures. Thirteen context-free productions generate all valid PDL sentences. A PDL grammar for a particular application results when a set of suitable primitives and applicable productions is specified (Figure 7). Shaw gives several examples.

Possibly the most severe limitations of the PDL is its restriction to context-free languages. There is a strong feeling in some quarters that context-free

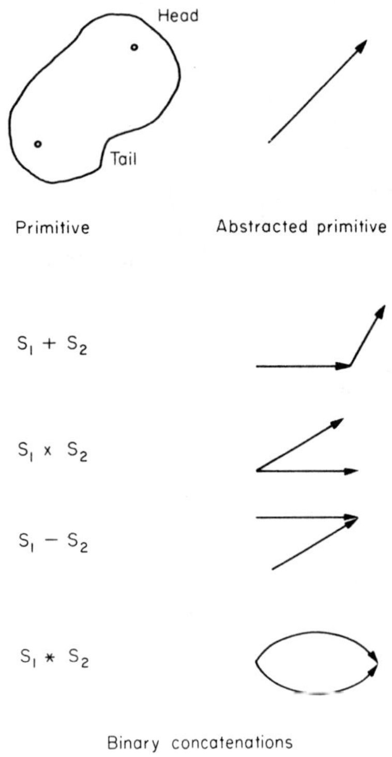

FIGURE 6. A partial definition of the PDL. (From Shaw, 1969.)

languages cannot be expected to provide adequate models for a very general class of complex pattern analysis problems. But this is an open question. The full potential of the PDL deserves to be investigated in this respect. The PDL has some other relatively minor limitations which are discussed by Shaw but do not seem insurmountable. It is significant that the PDL has proved useful in at least one practical application (Shaw, 1968) for which a picture analyzer based on the PDL has been implemented. The nature of this implementation is interesting and will be discussed in Section IV,B.

The *web grammars* due to Rosenfeld and Pfaltz (1969) appear to offer a natural two-dimensional generalization of string grammars. A *web* is a directed graph with nodes bearing labels from a vocabulary of symbols. Since a string is a special case of a graph, a string of symbols is a special case of a web. A web grammar consists of a vocabulary with terminal and nonterminal elements, and a set of rewriting rules (productions)—much like the familiar phrase-structure string grammar. However, the productions of the web grammar are more complicated: Each rule specifies the replacement of a subweb in the original web by another subweb to form the rewritten web. The

V_T = { dp, dm, h, vm }

V_N = {Triangle, House, S, A}

P: S → A
S → House
House → ((vm + (h + (∼ vm))) * Triangle)
A → (dp + (Triangle + dm))
Triangle → ((dp + dm) * h)

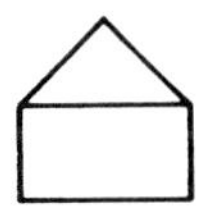

S ⟹ ((vm + (h + (∼ vm))) * ((dp + dm) * h))

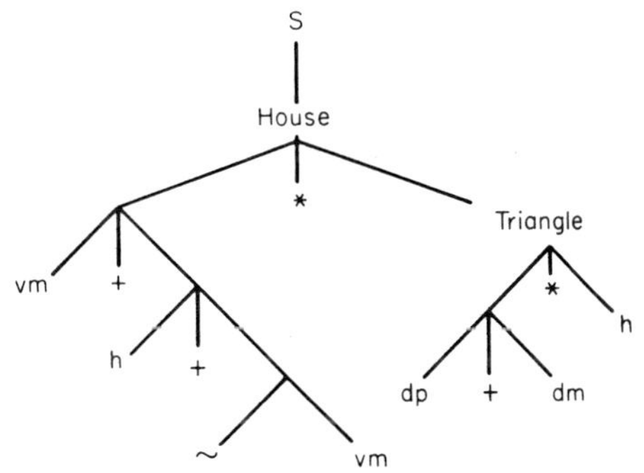

FIGURE 7. A PDL grammar and example. (From Shaw, 1969.)

rule must specify precisely how the new subweb is to be "embedded" in the "host" web. This can be done in any number of ways as long as the embedding rule does not depend on the host web, a requirement which assures that the rule can be applied to any occurrence of the variable (subweb) being rewritten (see Figure 8).

Context-sensitive and context-free web grammars are defined. A web grammar is a generalized syntactic formalism, the embedding rules constituting at least part of the attribute relationship. Shaw's PDL is a special case of a web grammar, but the unrestricted web grammar formalism seems to provide a more natural way of representing multiply connected graphs. Although the practical applicability of web grammars remains to be demonstrated, there would seem to be no reason why this formalism should not be at least as useful as the PDL. Thus, web grammars represent another step toward the

$G = \{V_N, V_T; S, P\}$

where

$V_N = \{A\}$ nonterminal vocabulary

$V_T = \{a, b, c\}$ terminal vocabulary

$S = \{\dot{A}\}$ initial web

P: productions

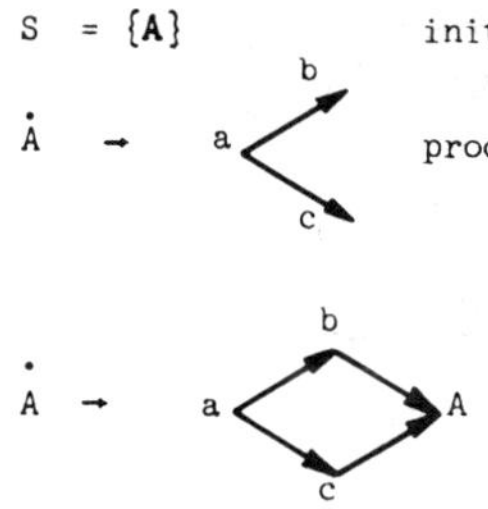

where the embedding rule for both productions is

$E = \{(p,a) \mid (p,A) \text{ is an edge of the host web}\}$

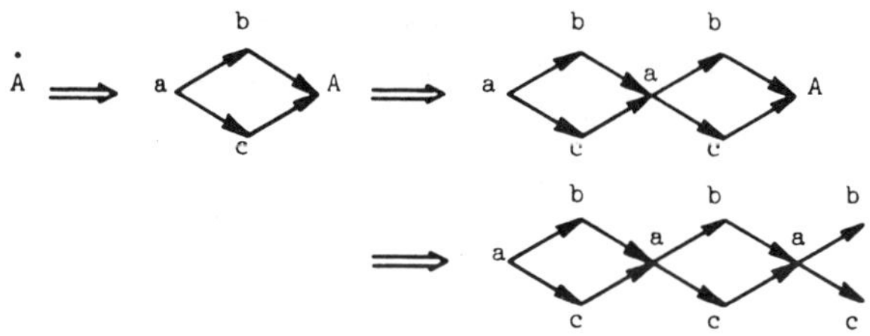

FIGURE 8. A simple web grammar. (From Rosenfeld and Pfaltz, 1969.)

generalization of phrase-structure grammatical concepts from strings to more general syntactic entities.

B. Pattern Analysis Mechanisms

Evans (1968a, b) has developed a "grammar-controlled pattern analyzer" (a LISP program) which he describes as a very useful tool for testing pattern grammars. Utilizing a GSF, the analyzer accepts as input both the pattern grammar (syntactic elements, productions, and predicates which define the syntactic relationships) and patterns to be analyzed (expressed in terms of the primitive syntactic elements). The output of the system consists of all hierarchical descriptions possible under the specified grammar, which amounts to an accounting of all possible subpatterns (GSE's) detected and their relationships. The grammar could be constructed so that part of this accounting would be the name(s) of any pattern class(es) to which the pattern might belong.

There is virtually no restriction on the form of the grammars acceptable by Evans' analyzer. They may be finite-state, context-free, or context-sensitive in form, i.e., in the same sense as is usually defined for string grammars. However, a string language generated by a generalized syntactic formalism grammar which is, say, context-free in this sense may *not* be a context-free string language, i.e., capable of being generated by a context-free string grammar (see Anderson, 1968, Chapter 10). Unfortunately, as Evans remarks, "the price paid for this generality may preclude any practical application of the system so that its principal use may be as an experimental tool in the choice of the primitives to be included in particular special purpose systems." This is not surprising considering the inefficiency of the general parsing techniques which have been developed for string languages and grammars. Again what appears to be needed is a generalization of some of the special string grammars for which there exist very efficient analysis procedures (Floyd, 1964; Wirth and Weber, 1966; Crespi-Reghizzi and Presser, 1969).

Anderson (1968) uses a top-down analysis procedure to parse handwritten mathematical expressions (see the discussion of his formalism in Section II). His grammars are GSF's which have been carefully formulated to enhance parsing efficiency, which is essential since top-down syntactic analysis procedures often tend to be very inefficient, particularly in rejecting "ungrammatical" sentences. He takes advantage of the fact that mathematical expressions tend to be written in a left-to-right manner and also utilizes to a limited extent some of the techniques of precedence analysis (Wirth and Weber, 1966). The idea is to arrange the grammar and the analysis procedure so that it becomes obvious as quickly as possible that the analysis is proceeding down an invalid branch of the parsing tree. Although Anderson provides no precise timing information, he indicates that his analysis prodecure is fast enough for the application even though it has not been rigorously optimized. This provides an indication that sufficiently specialized syntactic analysis procedures can be designed which are practical to implement.

Another interesting aspect of Anderson's work is an evaluation of the possible benefits of parallel processing. He shows that even two processors working in parallel on different branches of the parsing tree would yield a dramatic overall efficiency advantage over strictly serial processing. The efficiency gains tend to level off quickly as the number of parallel processors increases. Note that this "procedure-parallel" processing is considerably different from the "picture-parallel" processing concept of Narasimhan discussed earlier.

To implement an analysis system based on PDL, Shaw (1968) uses a syntax-controlled, top-down, goal-oriented picture parsing algorithm. As a top-down algorithm, the analyzer uses a PDL grammar to try to synthesize a PDL description of the input pattern. (A bottom-up algorithm would try to reduce groupings of syntactic variables and terminals into successively higher-level variables by reverse application of the productions in the grammar.) The algorithm is goal oriented because at each stage of the analysis it tries to attain a specific goal, namely, to determine whether a specific syntactic construct—the right-hand side of the production being applied—describes part of the input pattern. A successful generation of a PDL description of the input pattern indicates that the pattern belongs to the pattern class of interest. The analysis may also yield additional useful information about the pattern depending on any semantic analysis carried out in conjunction with the syntactic analysis.

As pointed out by Shaw, a goal-oriented procedure of this nature has some important advantages over other approaches. The logic of the syntactic analysis is straightforward, consisting primarily of stepping through the pattern grammar, with backtracking as necessary in case false goals are generated at higher levels in the analysis. [There has been some skepticism in the past regarding the use of top-down analysis for practical analysis schemes because of the inefficiency that results from a large amount of backtracking [see, for example, Griffiths and Petrick (1965)]. But Shaw's experience indicates that, assuming due care in defining the PDL grammar, the total time consumed by the syntactic analysis tends to be negligible compared to that required for the detection of primitives in the raw data.] Also, it is helpful to use a goal-oriented procedure in dealing directly with unpreprocessed pattern data because the syntax specifically prescribes which primitives must be located in the pattern and what relationship these primitives must have to the rest of the pattern. This is, in effect, one way of utilizing context to aid in the recognition process which, as a result, may not need to be as precise as methods which must search out arbitrary primitives (the job of some preprocessors, for instance, is to compile a list of possible primitives in the pattern without reference to the syntax). It should be clear that this may also reduce to some degree the effects of noise in the data.

A few restrictions must be imposed on the precise form of the PDL grammar to be implemented. For example, left-recursive productions are prohibited because top-down parsers can get into infinite loops when processing grammars with left-recursive productions. However, most of the restrictions are automatically satisfied if the grammar is transformed into Shaw's PDL *standard form*. Shaw believes, although he has not been able to prove, that any PDL grammar can be so transformed.

Shaw's work is the most convincing demonstration yet to appear that syntactic pattern analysis is a feasible approach to problems involving complexly structured data. It will be interesting to see if equally or more effective analysis algorithms can be developed employing the web grammars discussed earlier.

V. "Learning" Pattern Grammars by Grammatical Inference

The researcher's intuition and familiarity with the data are relied on heavily at present in formulating grammars for specific applications. The grammar synthesis procedure is now largely trial and error, sometimes aided by user-interactive computer graphics (Knoke and Wiley, 1967; Evans, 1968b). Clearly a facility is needed for automatically inferring grammars from representative samples of the languages in question, much as the training procedures of "conventional" pattern recognition technology are used to synthesize pattern classifiers from "training samples" of known classification. This is apparently a rather difficult problem which has only recently begun to receive attention in the literature (Gold, 1967; Sherman and Ernst, 1969; Rosenfeld *et al.*, 1969; Sauvain and Uhr, 1969). A more comprehensive consideration of the problem is the subject of the chapter by Evans.

VI. The Roles of Probability in Syntactic Pattern Analysis

The syntactic pattern analyzers found in the literature are generally well shielded from the "real world" by extensive preprocessing of the raw patterns (smoothing, gap-filling, thinning, etc.) intended to extract pure patterns and eliminate noise [for example, see works by Ledley *et al.* (1965), Narasimhan (1964), Knoke and Wiley (1967), and Evans (1968b).] The preprocessing techniques commonly used make little or no use of the pattern syntax even though the syntax could be of great assistance in distinguishing noise and detecting distortions of pure patterns. Unfortunately, however, it is precisely distortion and noise which make quite difficult the description of "real life" patterns by means of familiar nonprobabilistic grammatical models. Recalling that probabilistic approaches have often yielded some success in dealing with distortion and noise, one is led to consider the formulation of a probabilistic model for syntactic analysis.

In attempting to build a universal theoretical foundation for "grammatical analysis of patterns," Grenander (1967a) has proposed a model called a *probabilistic deformation grammar*. The model is somewhat vaguely specified in order to maintain generality, but it essentially involves a set of deformation transformations applied to the syntactic variables. A probability measure is defined over the set of deformation transformations. This model appears to be applicable only to the variety of syntactic analysis procedures which do not depend on a parse of the patterns but only on the description of the patterns in terms of primitives and deformations. These procedures sometimes turn out to be variations on the familiar minimum distance recognition rule.

In another work Grenander (1967b) has suggested the possibility of defining a probability distribution over the productions of a context-free pattern grammar. This in turn impresses a probability distribution on the set of parsing trees which may be generated by the grammar and thence on the language generated by the grammar. The model turns out to be a multitype branching process which has been studied in some detail (Harris, 1963). Grenander has investigated the constraints on the probability measure over the set of terminating parsing trees generated by context-free grammars [see also Booth (1969)]. This model comes much closer in spirit to the concept of syntax-controlled pattern analysis than does the probabilistic deformation grammar (although only empirical studies can be expected to demonstrate which model is truly most appropriate for a given problem). In addition to the work by Grenander and Booth, Kherts (1968) has defined and studied briefly the entropy of stochastic context-free grammars; and Fu and Li (1969) have described the realization of a stochastic finite-state automaton which recognizes a stochastic finite-state language, and have shown how such an automaton can be synthesized if a grammar generating the language is known. The remainder of this section will summarize the limited progress which has been made toward developing stochastic grammars and mention some problems which must be faced in order to develop an adequate model for stochastic syntactic analysis.

Following the formulation of Kherts (1968), the simplest type of stochastic grammar is defined as follows.

A stochastic finite-state grammar with independent production probabilities is a five-tuple (V_N, V_T, P, D, S), where V_N is a finite nonterminal vocabulary; V_T is a finite terminal vocabulary; P is a finite set of productions, each having one of the following forms:

$$A_i \to a_j A_k \qquad \text{or} \qquad A_i \to a_j,$$

where $A_i, A_j \in V_N$; $S \in V_N$ is the distinguished sentence symbol; D is a probability measure over P. The probability associated with $A_i \to a_j A_k$ is denoted by p^i_{jk}; the probability associated with $A_i \to a_j$ is denoted by p^i_{j0}.

Clearly it is necessary that $0 \le p^i_{jk} \le 1$. It is also required that

$$\sum_j \sum_k p^i_{jk} = 1$$

for each i such that $A_i \in V_N$. Moreover, it is desirable that the probability distribution over the productions impress a valid probability distribution over the language generated by the grammar. The probabilities of the terminal strings generated by the grammar should sum to unity. This is not automatically true. Grenander (1967b) has shown that a valid distribution over the language is obtained if each nonterminal of the finite-state grammar is contained in at least one string derivable with nonzero probability from the sentence symbol and each nonterminal generates with nonzero probability at least one string containing only terminals.

Unfortunately, this delightfully simple model is of little practical utility, failing by far to describe a sufficiently broad class of languages of interest. One might argue that for practical problems the maximum number of possible patterns which could be observed is certainly bounded (in most cases) and that finite-state grammars are capable of describing such situations. Indeed this is a theoretically valid observation, but the finite-state grammars which are necessary whenever the number of possible patterns is large and/or their structure is at all nontrivial are unwieldly and by no means revealing of the significant pattern structure. The reader may prove this to his satisfaction by trying to write a finite-state grammar for the language

$$L = \{a^n b^n c^n d^n \mid 0 < n \leq m\}$$

for m equal to, say, ten. This language might represent a family of squares with sides of length 1, 2, . . . , 10. Also, refer to the example at the end of this section.

The stochastic grammatical model can be generalized in either or both of two ways: by extension to more powerful grammar forms (context-free, context-sensitive), or by the use of more complex probabilistic mechanisms, e.g., various forms of conditional probabilities [see Booth (1969)]. The use of more powerful grammar forms seems to be the most direct approach [see Swain (1970) for a more detailed discussion of this point].

A stochastic context-free grammar can be defined in much the same manner as in the finite-state case. Grenander (1967b) and Booth (1969) have specified the conditions for a valid set of context-free production probabilities, which again involve guaranteeing that every derivation (or generation) from the grammar terminates with unit probability. Some successful applications of (nonprobabilistic) context-free grammars have been described earlier in this chapter, which attests to their greater practical utility as compared with finite-state grammars for pattern analysis purposes. But they still lack the descriptive power available to context-sensitive grammars which are generally most capable of producing the types of structural descriptions most useful for pattern analysis purposes.

The increasing descriptive power of the context-free and context-sensitive grammars is paid for in terms of increased complexity of the analysis system (Hopcroft and Ullman, 1969). Whereas deterministic finite-state automata are capable of recognizing finite-state languages, nonfinite nondeterministic devices are required, in general, to recognize the languages generated by context-free and context-sensitive grammars. This is at worst a theoretical difficulty, however, which is unlikely to cause serious problems in practice as long as some care is exercised in developing the required grammars. This is especially true when the languages of interest are actually finite-state even though the form of the grammars may be context-sensitive (possibly the most likely situation), or when the languages may be approximated by finite-state languages (Booth, 1969). An important step yet to be carried out is the development of a characterization of stochastic context-sensitive grammars; for example, the conditions required to guarantee a valid probability distribution over the productions and terminal strings have yet to be determined. Another practical problem is the complicated and altogether unintuitive manner with which context-sensitive phrase-structure grammars must perform routine operations such as maintaining counters which are essential to the derivations. Such "roundabout" operation tends to obscure the structural relationships within the derived sentences, relationships which are of paramount importance. Fortunately, this problem can be alleviated to a considerable extent by using special types of grammars which have only recently been defined and investigated. Examples include programmed grammars (Rosenkrantz, 1967) and indexed grammars (Aho, 1968). Stochastic versions of such specialized syntactic formalisms, once characterized, should constitute a step in the direction of the sort of compact and efficient grammatical models needed for practical analysis of real data. The authors are currently investigating the properties and potential applications of stochastic programmed grammars and expect to publish some results in the near future.

The following simple nonprobabilistic example illustrates some of the points discussed above, particularly the increased power of the productions of the more general classes of grammars.

EXAMPLE. It is desired to construct a grammar to generate the finite-state language $L = \{a^n b^n c^n \mid 1 \leq n \leq 3\}$. This might be the language describing, say, the set of equilateral triangles of side-length one, two, or three units. So that the grammar will be compatible with a top-down goal-oriented analysis procedure, the grammar must produce terminals in a strictly left-to-right order and at most one terminal may be produced by a single application of any production. Nonterminals may not appear to the left of terminal symbols, but the generation of nonterminals is otherwise unrestricted.

1. A finite-state grammar: $G_1 = (V_N, V_T, P, S)$, where

$$V_N = \{S, A_1, A_2, B_{10}, B_{20}, B_{30}, B_{21}, B_{31}, B_{32}, C_1, C_2, C_3\}$$

is the nonterminal vocabulary and $V_T = \{a, b, c\}$ the terminal vocabulary.

$$\begin{array}{lll} P: & S \to aA_1 & B_{21} \to bC_2 \\ & S \to aB_{10} & B_{30} \to bB_{31} \\ & A_1 \to aA_2 & B_{31} \to bB_{32} \\ & A_1 \to aB_{20} & B_{32} \to bC_3 \\ & A_2 \to aB_{30} & C_1 \to c \\ & B_{10} \to bC_1 & C_2 \to cC_1 \\ & B_{20} \to bB_{21} & C_3 \to cC_2 \end{array}$$

2. A context-free grammar (in Greibach normal form): $G_2 = (V_N, V_T, P, S)$, where

$$V_N = \{S, A_1, A_2, B_1, B_2, B_3, C\}, \qquad V_T = \{a, b, c\}$$

$$\begin{array}{llll} P: & S \to aA_1C & & B_3 \to bB_2 \\ & A_1 \to b & & B_2 \to bB_1 \\ & A_1 \to aB_2C & & B_1 \to b \\ & A_1 \to aA_2C & & C \to c \\ & A_2 \to aB_3C & & \end{array}$$

3. A programmed grammar†: $G_3 = (V_N, V_T, J, P, S)$ where $V_N = \{S, B, C\}$ is the nonterminal vocabulary, $V_T = \{a, b, c\}$ the terminal vocabulary, and $J = \{1, 2, 3, 4, 5\}$ the production labels.

P: Production label	Production core	Success branch	Failure branch
1	$S \to aB$	2,3	$\in$
2	$B \to aBB$	2,3	$\in$
3	$B \to C$	4	5
4	$C \to bC$	3	$\in$
5	$C \to c$	5	$\in$

† The programmed grammar operates as follows: Production 1 is applied first; in general, if one tries to apply production k to rewrite a nonterminal N and the current string contains at least one N, then the leftmost N is rewritten by k and the next production is selected from the "success branch" field; if the current string does not contain N, then k is not used and the next production is selected from the "failure branch" field (of k); if the applicable branch field contains $\in$, the derivation halts.

DISCUSSION. Even for this simple case, the context-free grammar is considerably more compact than the finite-state grammar. For this example, a context-sensitive grammar would not be much different from the context-free grammar and hence has not been given. However, the programmed grammar is still more compact than the context-free grammar.

VII. Summary and Further Remarks

Syntactic pattern analysis and recognition have been found to offer an approach to dealing with pattern data which cannot be conveniently described numerically or are otherwise so complex as to defy analysis by conventional techniques. It has often been observed that such data are highly structured and that the structural descriptions contain the essential information for recognition while being perhaps orders of magnitude more convenient to process than the raw patterns. Research in syntactic pattern recognition appears to have demonstrated the feasibility of this approach, at least for some nontrivial applications. However, some formidable hurdles remain to be cleared before syntactic pattern recognition can be widely applied. Some problems and areas of promise include:

1. A systematic approach to the determination of appropriate syntactic elements. Researchers presently depend largely on intuition to select reasonable sets of pattern primitives and higher-order syntactic elements (a situation analogous to that which has existed with respect to feature extraction in conventional pattern recognition techniques). The selection of syntactic elements affects the form of the pattern grammar and hence both the efficiency and the computational complexity of the analyzer, not to mention the pattern recognition accuracy. Certainly some sort of interaction between the primitive selection procedure and the evaluation of recognition performance is needed, both to optimize the performance and to minimize the analyzer complexity.

2. Grammar synthesis based on samples of pattern data (grammatical inference). Here again intuition is relied on heavily at present. A pattern grammar must be hypothesized and then iteratively tested by simulation and improved by the designer. User-interactive computer graphics aid considerably in this respect, but it is not difficult to imagine pattern recognition problems for which computer graphics could not be conveniently employed. Grammatical inference techniques are needed by which analyzers could learn grammars from sets of training patterns.

3. Generalization of concatenation to multiple dimensions and more complex syntactic relationships. A better generalization seems almost essential if efforts in formal linguistics and automata theory are to be brought to bear on the problems of syntactic pattern recognition. Shaw's PDL and the web grammars due to Pfaltz and Rosenfeld appear promising in this respect.

4. Development of special pattern languages and grammars which lend themselves to efficient parsing. Presently attainable processing speeds need to be improved by as much as several orders of magnitude.

5. Detailed formulation of a stochastic syntactic pattern analysis model capable of processing patterns with distortion and noise. Given future development of training methods for determining the necessary stochastic grammars (including vocabularies, productions, and production probabilities), one can imagine a syntax-controlled pattern recognition scheme such as the following. A stochastic grammar is found for each individual pattern class; for each pattern to be classified, a parse and its associated probability are obtained

(whenever possible) according to each class grammar; based on the parse probabilities (and any *a priori* information available), the classification is then made according to a criterion such as minimum risk (Bayes). Such a scheme has the advantage of applying syntactic analysis directly to the raw data rather than going through an initial "noise cleaning" stage which makes little or no use of the wealth of prior information stored in the pattern grammar.

6. Further contributions of automata theory to the syntactic analysis problem. Work is continuing on characterization of the classes of languages recognizable by various types of automata [see especially Jones (1968)]. As mentioned in Section VI, stochastic finite-state automata can be synthesized based on stochastic finite-state grammars and can recognize the languages generated by such grammars. It has recently been shown that stochastic automata can recognize some languages that are context-free but not finite-state (Salomaa, 1968); it has been conjectured that stochastic automata also may recognize nonfinite-state languages which are not context free (Fu and Li, 1969). There is room for speculation that stochastic automata theory may eventually provide both a theoretical foundation for stochastic syntactic analysis and an alternative approach to the realization of analytical mechanisms based on stochastic grammars.

The range of important problems to which syntactic pattern analysis could be applied and which otherwise appears to be beyond the scope of presently known techniques will continue to stimulate research in this new direction.

ACKNOWLEDGMENTS

This work was supported by NSF Grant GK-1970, AFOSR 69-1776, USDA Contract No. 12-14-100-9549(50), and NASA Contract No. NGR 15-005-112.

REFERENCES

Aho, A. V. (1968). "Indexed Grammars—An Extension of Context-Free Grammars." *J. Assoc. Comput. Mach.* **15**, 647–671.

Aho, A. V. and Ullman, J. D. (1968). "The Theory Languages." *Math. Systems Theory* **2**, 97–125.

Anderson, R. H. (1968). "Syntax-Directed Recognition of Hand-Printed Two-Dimensional Mathematics." Ph.D. Dissertation, Harvard Univ., Cambridge, Massachusetts.

Banerji, R. B. (1963). "Phrase Structure Languages, Finite Machines, and Channel Capacity." *Information and Control* **6**, 153–162.

Banjeri, R. B. (1966). "Some Studies in Syntax-Directed Parsing." *In* "Computation in Linguistics: A Case Book." (P. L. Garvin and B. Spolsky, eds.), Chapter 4. Indiana Univ. Press, Bloomington, Indiana.

Banjeri, R. B. (1968). "A Language for Pattern Recognition." *Pattern Recognition* **1**, 63–74.

Booth, T. L. (1967). "Sequential Machines and Automata Theory." Wiley, New York.

Booth, T. L. (1968). "Discrimination between Discrete Random Processes Using Sequential Testing Techniques." *Proc. Ann. Conf. Information Sci. and Systems, 2nd, Princeton, 1968*, pp. 214–218.

Booth, T. L. (1969). "Probabilistic Representation of Formal Languages." *IEEE Conf. Record, 9th Ann. Symp. Switching and Automata Theory, Univ. of Waterloo, Ontario, Canada, 1969.*

Chartres, B. A., and Florentin, J. J. (1968). "A Universal Syntax-Directed Top-Down Analyzer." *J. Assoc. Comp. Mach.* **15**, 447–464.

Chomsky, N. (1956). "Three Models for the Description of Languages." *IRE Trans. Information Theory*, **2**, 113–124.

Crespi-Reghizzi, S., and Presser, L. (1969). "Extensions to Precedence Techniques for Syntactic Analysis." *Proc. Purdue Centennial Year Symp. on Information Processing, Purdue Univ., 1969.* Purdue Univ., Lafayette, Indiana.

Eden, M. (1961). "On the Formalization of Handwriting." *Proc. Symp. Appl. Math.* **12**, 83–88. Am. Math. Soc., Providence, Rhode Island.

Eden, M. (1963). "Handwriting and Pattern Recognition." *IRE Trans. Information Theory* **8**, 160–166.

Evans, T. G. (1967). "A Formalism for the Description of Complex Objects and Its Implementation." *Proc. Internat. Congr. Cybernetics, 5th Namur, Belgium, 1967.*

Evans, T. G. (1968). "A Grammar Controlled Pattern Analyzer," *In* "Information Processing 68." North-Holland Publ., Amsterdam.

Evans, T. G. (1969a). "Descriptive Pattern Analysis Techniques: Potentialities and Problems." in "*Methodologies of Pattern Recognition*" (S. Watanabe, ed.). Academic Press, New York, 1969.

Evans, T. G. (1969b). "Grammatical Inference Techniques in Pattern Analysis." *Internat. Symp. Comput. Information Sci., 3rd, Bal Harbour, Florida, 1969.* Academic Press, New York. See also his chapter in this book.

Feder, J. (1967). "Languages, Automata, and Classes of Chain-Encoded Patterns." Tech. Rep. 400–165. N.Y.U. Lab. for Electroscience Res.

Fischler, M. A. (1969). "Machine Perception and Description of Pictorial Data." *Proc. Joint Internat. Conf. Artificial Intelligence, Washington, D.C., 1969.*

Floyd, R. W. (1964a). "Bounded Context Syntactic Analysis." *Comm. ACM* **7**, 62–67.

Floyd, R. W. (1964b). "The Syntax of Programming Languages—A Survey." *IEEE Trans. Electronic Computers* **13**, 346–353.

Friedman, J. (1969). "Directed Random Generation of Sentences." *Comm. ACM* **12**, 40–46.

Freeman, H. (1961). "On the Encoding of Arbitrary Geometric Configurations." *IRE Trans. Electronic Computers* **EC10**, 260–268.

Fu, K. S., and Li, T. J. (1969). "On Stochastic Automata and Languages." *Information Sci.* **1**, 237–256.

Gilbert, P. (1966). "On the Syntax of Algorithmic Languages." *J. Assoc. Comput. Mach.* **13**, 90–107.

Gold, E. M. (1967). "Language Identification in the Limit." *Information and Control* **10**, 447–474.

Grenander, U. (1967a). "Foundations of Pattern Analysis." Preliminary Version, Tech. Rep. No. 2 (Project Graphics). Div. of Appl. Math., Brown Univ., Providence, Rhode Island.

Grenander, U. (1967b). "Syntax-Controlled Probabilities." Div. of Appl. Math., Brown Univ., Providence, Rhode Island.

Griffiths, T. V., and Petrick, S. R. (1965). "On the Relative Efficiencies of Context-Free Grammar Recognizers." *Comm. ACM* **8**, 289–300.

Grimsdale, R. L., Summer, F. H., Tunis, C. J., and Kilburn, T. (1959). "A System for the Automatic Recognition of Patterns." *Proc. IEE (London) Pt. B* **106**, 210–221.

Guzmán, A. (1967). "Some Aspects of Pattern Recognition by Computer." Project MAC Rep. No. MAC-TR-37. M.I.T., Cambridge, Massachusetts.

Guzmán, A. (1968). "Decomposition of a Visual Scene into Three-Dimensional Bodies." *Proc. Fall Joint Comput. Conf., 1968*, Thompson, Washington, D.C.

Harris, T. E. (1963). "The Theory of Branching Processes." Springer, Berlin.

Harrison, M. A. (1965). "Introduction to Switching and Automata Theory." McGraw-Hill, New York.

Hopcroft, J. E., and Ullman, J. D. (1969). "Formal Languages and Their Relation to Automata." Addison-Wesley, Reading, Massachusetts.

Irons, E. T. (1963). "An Error-Correcting Parse Algorithm." *Comm. ACM* **6**, 669–673.

Jones, N. D. (1968). "Classes of Automata and Transitive Closure." *Information and Control* **13**, 207–229.

Kherts, M. M. (1968). "Entropy of Languages Generated by Automated or Context-Free Grammars with a Single-Valued Deduction." *Nauch. Tekh. Inform. Ser. 2*, pp. 29–34.

Kirsch, R. A. (1964). "Computer Interpretation of English Text and Picture Patterns." *IEEE Trans. Electronic Computers* **EC13**, 363–376.

Knoke, P. J., and Wiley, R. G. (1967). "A Linguistic Approach to Mechanical Pattern Recognition." Appl. Res. Lab., Curry, McLaughlin and Len, Inc., Syracuse, New York; also *Proc. IEEE Comput. Conf. 1st Chicago, 1967*, pp. 142–144.

Ledley, R. S., Rotolo, L. S., Golab, T. J., Jacobsen, J. D., Ginsburg, M. D., and Wilson, J. B. (1965). "FIDAC: Film Input to Digital Automatic Computer and Associated Syntax-Directed Pattern Recognition Programming System." *In* "Optical and Electro-Optical Information Processing Systems." (J. Tippet, D. Beckowitz, L. Clapp, C. Koester, and A. Vanderburgh, Jr., eds.), Chapter 33. M.I.T. Press, Cambridge, Massachusetts.

McCormick, B. H. (1963). "The Illinois Pattern Recognition Computer (Illiac III)." Digital Comput. Lab. Rep. No. 148. Univ. of Illinois, Urbana, Illinois.

McLaren, R. W. (1968). "A Technique for Generating the Structure of an Automaton to Satisfy a Given Input/Output Relationship." *Proc. Ann. Conf. Information Sci. and Systems, 2nd, Princeton, 1968*, pp. 209–213.

McMurty, G. J., and Fu, K. S. (1964). "On Learning Behavior of Finite-State Systems in Random Environments." *Proc. 2nd Ann. Allerton Conf. on Circuits and System Theory, Univ. of Illinois, 1964.*

Mattews, G. H. (1963). "Discontinuity and Asymmetry in Phrase Structure Grammars." *Information and Control* **6**, 137–146.

Miller, W. F., and Shaw, A. C. (1968). "Linguistic Methods in Picture Processing—A Survey." *Proc. Fall Joint Comput. Conf., 1968.*

Narasimhan, R. (1964). "Labeling Schemata and Syntactic Descriptions of Pictures." *Information and Control* **7**, 151–179.

Narasimhan, R. (1966). "Syntax-Directed Interpretation of Classes of Pictures." *Comm. ACM* **9**, 166–173.

Narasimhan, R., and Mayoh, B. H. (1963). "The Structure of a Program for Scanning Bubble Chamber Negatives." Digital Computer Lab. File No. 507. Univ. of Illinois, Urbana, Illinois.

Nasu, M., and Honda, N. (1968). "Fuzzy Events Realized by Finite Probabilistic Automata." *Information and Control* **12**, 284–303.

Pavlidis, T. (1968). "Analysis of Set Patterns," Tech. Rep. No. 25. Dept. of EE, Information Science and Systems Lab., Princeton Univ., Princeton, New Jersey. See also *Pattern Recognition* **1**, 165–178.

Pavlidis, T. (1969). "Linguistic Analysis of Waveforms." *Internat. Symp. Computer and Information Sci., 3rd, Bal Harbour, Florida, 1969.* Academic Press, New York. See also his chapter in this book.

Paz, A. (1966). "Some Aspects of Probabilistic Automata." *Information and Control* **9**, 26–60.

Rabin, M. O. (1963). "Probabilistic Automata." *Information and Control* **6**, 230–245.

Rosenfeld, A. (1969). "Picture Processing by Computer." *Computing Surveys* **1**, 147–175.

Rosenfeld, A., and Pfaltz, J. L. (1966). "Sequential Operations in Digital Picture Processing." *J. Assoc. Comput. Mach.* **13**, 471–494.

Rosenfeld, A., and Pfaltz, J. L. (1969). "Web Grammars." *Proc. Joint Internat. Conf. Artificial Intelligence, Washington, D.C. 1969.*

Rosenfeld, A., and Strong, J. P. (1969). "A Grammar for Maps." *Internat. Symp. Computer and Information Sci., 3rd, Bal Harbour, Florida, 1969.* Academic Press, New York. See also his chapter in this book.

Rosenfeld, A., Huang, H. K., and Schneider, V. B. (1969). "An Application of Cluster Detection to Text and Picture Processing." *IEEE Trans. Information Theory* **15**, 672–681.

Rosenkrantz, D. J. (1967). "Programmed Grammars—A New Device for Generating Formal Languages." Ph.D. Dissertation, Columbia Univ., New York. Also *IEEE Conf. Record Ann. Symp. Switching and Automata Theory, 8th Austin, Texas, 1967* (summary of thesis).

Salomaa, A. (1968). "On Languages Accepted by Probabilistic and Time-Variant Automata." *Proc. Ann. Conf. Information Sci. and Systems, 2nd, Princeton, March 1968.* Princeton Univ., Princeton, New Jersey.

Sauvain, R. W., and Uhr, L. (1969). "A Teachable Pattern Describing and Recognizing Program." *Pattern Recognition* **1**, 219–232.

Shank, R. C., and Tesler, L. G. (1969). "A Conceptual Parser for Natural Languages." *Proc. Joint Internat. Conf. on Artificial Intelligence, Washington, D.C., 1969.*

Shaw, A. C. (1968). "The Formal Description and Parsing of Pictures." SLAC. Rep. No. 84. Stanford Linear Accelerator Center, Stanford Univ., Stanford, California.

Shaw, A. (1969). "A Formal Picture Description Scheme as a Basis for Picture Processing Systems." *Information and Control* **14**, 9–52.

Sherman, R., and Ernst, G. W. (1969). "Learning Patterns in Terms of Other Patterns." *Pattern Recognition* **1**, 301–313.

Swain, P. H. (1970). "On Nonparametric and Linguistic Approaches to Pattern Recognition." Ph.D. Dissertation, Purdue Univ., Lafayette, Indiana, June, 1970.

Tou, J. T. (1969) "On Feature Encoding in Picture Processing by Computer." *Proc. Ann. Allerton Conf. on Circuits and System Theory, 7th, Univ. of Illinois, 1969.*

Wirth, N., and Weber, H. (1966). "Euler: A Generalization of Algol, and Its Formal Definition." *Comm. ACM* **9**, 13–25.

Zahn, C. T. (1969). "A Formal Description for Two-Dimensional Patterns." *Proc. Joint Internat. Conf. Artificial Intelligence, Washington, D.C., 1969.*

SOME TECHNIQUES FOR RECOGNISING STRUCTURES IN PICTURES

H. G. Barrow, A. P. Ambler and R. M. Burstall

DEPARTMENT OF MACHINE INTELLIGENCE AND PERCEPTION
UNIVERSITY OF EDINBURGH

Introduction

Scope

The task to which this paper is addressed is that of scene analysis, that is, of deducing from a single two-dimensional image the organisation of the scene which it depicts, in terms of objects and their interrelationships.

The aim of our research is to develop techniques whereby a machine may observe its surroundings and then use its observations to achieve goals in an effective and efficient manner. To fulfill such requirements the machine will inevitably use 'knowledge' gained from past experience and observation to plan its activities, and also to interpret its sensory data.

In the present early stages of development of picture analysis and Artificial Intelligence (the two are inevitably and inextricably linked) we are not yet close to achieving an analysis system of reasonable power and versatility. The rest of this paper describes some steps in what we think may be the right direction.

We shall treat the task of scene analysis as primarily that of describing the picture and matching the description against models.

The purpose of the paper is to discuss the idea of a finite relational structure, that is a set of elements with given properties and relations between them, as a useful mathematical tool for describing pictures, and to describe general techniques for matching such structures against each other. These may be used to recognise known objects in a picture, without dependence upon special properties of objects or the environment.

A number of experiments are described in which these ideas and techniques are applied to picture interpretation. In particular, a program is described which is used in conjunction with an on-line T.V. camera to recognise such simple, but irregular objects as cups and spectacles (Barrow and Popplestone 1971). The T.V. picture is first converted into a set of regions of uniform brightness and properties of those regions (compactness, triangularity) and relations between them (adjacency, leftof) are determined. This structure of regions with given properties and relations is then matched against standard structures previously stored. Other programs have been written embodying more sophisticated techniques which match relational structures with as many as 40 elements, but these have not yet been used on structures derived from actual T.V. pictures.

The methods of this paper might also have application in the fields of inferential systems, problem solving and question answering.

Representation of Pictures

The starting point in our investigations is the description of the picture and object models in relational terms. We do not here advocate any particular set of elements or properties or relations. We have experimented with descriptions couched in terms of straight lines, of areas and of boundary curves, using a variety of properties and relations. For example Figure 1 shows a picture *'Bear on rug with ball'*, and Figure 2 shows the relational structure which corresponds to it using regions as elements, the property *circular* and two-place relations *adjacent, immediately-inside* and *left-of* (abbreviated to *circ, adj, ins* and *left*). This relational structure leaves a great deal unspecified, for example the arms could be just above the legs and the mouth could be above the eyes. Introducing extra relations such as *above* and *larger-than* would make the description more precise. Figure 3 shows the relational structure defining the known object *'Bear'*, a substructure of the one for *'Bear on rug with ball'*. By using only a limited number of relations and perhaps not specifying all possible instances of these we enable a wide class of pictures to be recognised as bears; evidently we must try to be neither too strict nor too lax in our definition.

Clowes (1971) has endeavoured to focus attention on the distinction between a 2-dimensional picture and a 3-dimensional scene and he and Huffman (1971) improve on ideas of Guzman (1968) to give effective methods for deducing the 3-dimensional structure of polyhedra from pictures, using local clues (they do not deal with irregular shapes). This gives us two alternatives:

A. (i) Analyse picture into regions, lines points or whatever.
(ii) Convert into a relational structure.
(iii) Match this relational structure with stored structures describing views of known objects.
(iv) Use knowledge of 3-dimensional structure of the objects

found to deduce the 3-dimensional structure of the scene.

B. (i) Analyse the picture into regions, lines, points or whatever.
(ii) Use local clues or stereo information to determine 3-dimensional analysis of scene.
(iii) Express this 3-dimensional analysis as a relational structure (now using relations like *behind*).
(iv) Match this relational structure with stored structures describing known objects.

Indeed we can think of the picture and its subpictures as relational structures using one set of properties and relations and one kind of element, and the 3-dimensional objects and their parts as relational structures using another set of properties and relations and another kind of element.

The Barrow and Popplestone program proceeds by (A), but both alternatives are worth pursuing, and both involve a matching step (Aiii, Biv), to find what *known* objects are depicted.

Why do we use relational structures as representations? The object recognition process is essentially one of abstraction, where we say that a number of pictures all represent the same object, or possibly objects forming a class such as cups or chairs. Thus we must discard most of the information in the picture, and it is plausible that we should discard the *metric* information and keep only properties and interrelationships of parts of the picture. Thus the distance between two lines may not be important but the fact that they are parallel may be. This approach has been taken by people who have worked with 'logical' descriptions of pictures (Evans 1964, Burstall 1968, Clowes 1971). In the 2-dimensional recognition case this means storing relational descriptions of a small number of characteristic views of each object, since small rotations leave the relationships intact.

The element set and relations are usually carefully selected so that the 'useful' information is retained while the 'noise' is discarded. The notion of noise is here extended to mean not only an extra spurious quantity mixed with the original correct signal, but also variations in the picture which are not useful in analysing and interpreting.

In practice throwing away *all* the numerical information may be too rash. As a compromise we may extend the notion of relational structure slightly so that instead of a relation which takes n-tuples of elements to truth values we use a function which takes n-tuples of elements to real numbers. The Barrow and Popplestone program does this. A relational structure is not, however, an efficient representation for all mathematical structures, e.g. representation of the spatial organisation of an array as a relational structure requires many statements of positional information, perhaps one for each pair of points.

The matching process for relational structures attempts to find whether one structure "occurs in" or is a substructure of another structure. More precisely we need a function which assigns to each element of the first structure a distinct element of the second structure in such a way as to preserve the properties and relations which subsist in the first structure. (If a 'kluge' is a round region on top of a square region, then if the picture shows a kluge it must have a round region on top of a square region.) Such a function is called a *monomorphism* from one relational structure to the other, that is a homomorphism which is one-to-one. In the next section we will give more precise definitions.

Relational Structures As Picture Descriptions

In this section we will define the notions of relational structure and monomorphism, giving examples of their use for describing pictures and solid objects.

Definition. A *predicate set* is a set of predicates each having a given number of arguments.

Definition. A *relational structure* over a given predicate set is a set X, called the *carrier*, and an assignment to each n-argument predicate of an n-ary relation, i.e. a subset of X^n, the set of all n-tuples of elements in X. (We will call a 1-ary relation a property.)

Definition. Given two relational structures R and S over the same predicate set and carriers X and Y, we say that a function $f: X \longrightarrow Y$ is *homomorphism* if for any predicate p, $p(f(x_1), \ldots, f(x_n))$ holds in S whenever $p(x_1, \ldots, x_n)$ holds in R. We write $f: R \longrightarrow S$ if f is a homomorphism.

Definition. A *monomorphism* is a homomorphism which is one to one (or "injective") i.e. $f(x_1) = f(x_2)$ implies $x_1 = x_2$.

Definition. The homomorphism (or monomorphism) f is said to be *full* if, for any p, $p(f(x_1), \ldots, f(x_n))$ holds in S if and only if $p(x_1, \ldots, x_n)$ holds in R.

Definition. A relational structure R is said to be a *substructure* of a relational structure S if there is a monomorphism $f: R \longrightarrow S$.

We may sometimes allow a set of *operators* each denoting a function in $X^n \longrightarrow X$ and extend the definition of homomorphism to require that $g(f(x_1), \ldots, f(x_n)) = f(g(x_1, \ldots, x_n))$ for any operator g.

Relational structures are used to give the semantics of first order logic, see, for example, Schoenfield (1967) or Cohn (1965) for details. A sentence of first order logic *denotes* a class of relational structures, namely all those which satisfy it.

If R is a relational structure describing a view of a known object and S is a relational structure describing a picture which is presented for analysis we translate the question "Is the object in the picture?" to the question "Is there a monomorphism f: R⟶S?". Similarly given a repertoire of relational structures $R_1, \ldots, R_n$ representing views of known objects, we wish to find all monomorphisms from the R_i to S.

The choice of elements of the carrier of the relational structure and of the predicates is left free. The elements may be points, lines, regions in the two dimensional case, or vertices, edges, surfaces and solid regions in three dimensions.

For example in Figure 2, 'Bear on rug with ball' we had a relational structure which used predicates *circular* (one place) and *adjacent, immediately-inside* and *left-of* (two place). The same predicate set is used in Figure 3 'Bear', but the two relational structures are different, with different carriers and different relations. Thus the relations for 'Bear on rug with ball' have the following sets of n-tuples.

circ: (1), (5), (6), (7), (8), (9), (10)

adj: (2,11),(2,14),(2,15),(3,5),(4,5),(5,11),(11,12),(11,13),(11,14), (11,15),(11,2) etc. for symmetric pairs

ins: (1,2), (6,5), (7,6), (8,5), (9,5), (10,9)

left: (3,4),(6,9),(6,10),(7,9),(7,10),(12,13),(14,15)

The relations for 'Bear' have

circ: (3), (4), (5), (6), (7), (8)

adj: (1,3),(2,3),(3,9),(9,10),(9,11),(9,12),(9,13),(3,1) etc. for symmetric pairs

ins: (4,3), (5,4), (6,3), (7,3), (8,7)

left: (1,2),(4,7),(4,8),(5,7),(5,8),10,11),(12,13)

'Bear' is a substructure of 'Bear on rug with ball' since there is a monomorphism f:

$f(1) = 3,\ f(2) = 4,\ f(3) = 5,\ f(4) = 6,\ f(5) = 7,\ f(6) = 8,\ f(7) = 9,$

$f(8) = 10,\ f(9) = 11,\ f(10) = 12,\ f(11) = 13,\ f(12) = 14,\ f(13) = 15.$

(Note: Because we have omitted the relation *above*, and hence have not specified that arms must be above legs, there is a second monomorphism, f_2:

$f_2(1) = 3,\ f_2(2) = 4,\ f_2(3) = 5,\ f_2(4) = 6,\ f_2(5) = 7,\ f_2(6) = 8,$

$f_2(7) = 9,\ f_2(8) = 10,\ f_2(9) = 11,\ f_2(10) = 14,\ f_2(11) = 15,$

$f_2(12) = 12,\ f_2(13) = 13.$)

An alternative way of looking at a relational structure, at least so long as only unary or binary predicates are involved is to think of it as a graph, with labels or 'colours' associated with the nodes and arcs. This point of view connects the problem of finding monomorphisms between relational structures with work on Graph Isomorphism (Unger 1964, Sussenguth 1965, Corneil and Gottlieb 1970).

We would like to point out one way in which the notion of relational structure might be improved on as a picture description tool. If we take points, lines and regions as the elements the distinction between these classes is more fundamental than predicates such as vertical or straight. Similarly the "boundary operator" which takes a line onto its endpoints and a region onto its bounding lines is a fundamental relation. We may think of a relational structure whose carrier is not a set but a complex, in the sense of algebraic topology. One may speculate that taking account of this particular form of the structures might lead to more efficient algorithms, and we have some tentative ideas along these lines.

Methods of Matching Relational Structures

The methods considered

We wish to match a known relational structure R against another structure S derived from the picture, that is we wish to know whether there is any monomorphism from R to S so showing that R is a substructure of S. We will first outline each method briefly and later give a fuller description.

A naive strategy would be to generate all possible one-to-one functions from R to S and check each of them to see whether it is a monomorphism. One improvement on this has been used in the Barrow and Popplestone object recognition program. The function is built up pair by pair by taking an element from R and one from S. As each pair is added it is checked to see that any relations holding in R for its first element hold for the corresponding element in S. A mismatch causes consideration of all functions which could be constructed by extending the current partial function to be deferred, and another pair of corresponding elements is tried, thus generating a search tree. By always developing the tree in the most promising direction the best possible match can be found.

Another strategy is to progress systematically through the structures from some initial pair of nodes, adding new pairs which have elements related to some already specified pair. This is analogous to the notion of parsing a string of characters from left to right and this analogy has been exploited in the QUAC program (Ambler and Burstall 1969) which uses relational search techniques for question-answering and for some matching of picture relational structures.

For complex structures, such generative approaches usually lead to consideration of an unnecessarily large number of candidates. It is more efficient to guide the search process in some way. Unger (1964) proposed a technique for matching graphs which has been extended by Rastall (1969) following some suggestions by Burstall. This technique uses the given properties and relations to classify the elements of R and S. For example, a class of elements of R may consist of all elements with a certain property which are related by a certain relation to two other elements. The corresponding class of S elements will consist of all elements with that property which are related by that relation to *at least* two other elements. The only functions which could possibly be monomorphisms are those which carry elements of an R class into elements of the corresponding S class. Alternating classification and enumerative choices are used to produce all monomorphisms.

Another approach, suggested by Barrow and programmed by D. Milner (1971) is that of hierarchical synthesis. In this method we not only specify a set of known structures representing objects which might occur in the picture but also specify a hierarchy of substructures of these. The recognition proceeds by first finding the smaller substructures and then checking combinations of them to recognise larger known substructures in the picture. This process may be extended over several stages. This process has been described and programmed in more algebraic terms by Burstall and Barrow.

We will first describe these methods in more detail and then make some comments about their respective merits and demerits. In the following section we will point out some connections with logical inference systems.

An object recognition program using a simple matching algorithm

When we had successfully connected a T.V. camera as a peripheral to our I. C. L. 4130 computer, we started to tackle the problem of object recognition with everyday objects including curved and irregular objects, rather than rely too heavily upon the neat projective properties of polyhedra. A program was accordingly written to recognise single irregular objects in pictures (cups, spectacles). It is reported in full elsewhere (Barrow and Popplestone 1971), so only an outline is given here.

The work of Brice and Fennema (1970) upon region analysis was shown to us, and Popplestone suggested describing a picture in terms of properties of and relations between the significant regions, matching the resulting relational structure against models.

The first stage of the program is based upon the concepts of region analysis, but is intended to be much faster and more economical than Brice and Fennema's algorithm, sacrificing the exhaustiveness to do so. Starting from 256 points regularly spaced over the picture, the program tries to find an area of approximately constant brightness round each. Since several points may lie within one region, regions are recognised by their centroid co-ordinates, area, perimeter and average contrast across their boundaries. Thus only one copy of each region is retained; the regions may at this stage over-lap. The next step is to coalesce adjacent regions if the average contrast a-cross their common boundary is less than a threshold value. When this pro-cess can no longer be performed, the picture has been partitioned into areas of significant size and with strong contrast at their boundaries.

The description phase is then entered. In this program the notion of a relational structure has been extended. Relations (and properties) are not simply true or false, but may take a real value, that is, they are functions to real numbers. For each region several Fourier coefficients are calculated as a measure of the shape of the region. Thus seven property measures for each region are determined. Relational measures computed include the adjacency of two regions, which contains topological information, a measure of the (normalised) distance between the centroids of two regions, which contains geometrical information about relative position, and a measure of convexity of common boundary. The properties and relations are computed exhaus-tively to form the description, but only as a matter of experimental conven-ience.

Models of views of objects are stored as relational structures with the same relation set. The extension to real values results in storage of a mean and standard deviation for each relation measure. Picture and model rela-tions are taken to agree if the picture value lies within 3 standard deviations of the mean.

The final phase of the program is the matching. Matching is performed in an incremental fashion by building up correspondences, one pair at a time. A partial match is evaluated by determining how many of the relations in the model structure agree with their corresponding picture relations. Only relations involving elements in the new pair need be considered.

At any stage of the process many partial correspondences will exist and the program selects for incrementing that which is most promising (accord-ing to a function computed from number of agreements etc.). In fact the partial matches for all the models are mixed together, so the program at-tempts to extend the most promising match of all at each step.

The result of the process is the best match, according to the evaluation function, that can be found between the picture and any object model. Such a match need not necessarily be complete i.e. the best result may be the match of part of the picture with part of the model. Thus the program can be used even if not all the object is visible, or the picture is 'dirty', producing errors in the description.

The program has a repertoire of nine objects; ball, pencil, doughnut, wedge, cylinder, hammer, tube, cup and specta ¹ ³s. It can correctly recognise these objects about 85% of the time taking .om one to ten minutes, depending upon the complexity of the picture.

One limitation of the existing program is that the description in terms of regions is too global in nature, and is easily affected by partial occlusion. More local information, and redundancy are required. However, this would increase the number of elements upon which the description is based and would exacerbate the second limitation, which is that the matching process, while being more efficient than an exhaustive, naive strategy, takes a time which increases rapidly with the number of elements. It is after all a tree search. Much time is taken if there are more than about six regions.

Relational composition search

This technique grow out of a relational question answering project along the lines of Raphael's SIR (1964) but with means of defining new relations as compositions, unions, intersections, inverses, etc. of old ones, based on an analogy with context-free grammars. The program 'QUAC' is described in Ambler and Burstall (1969), and it was used to recognise some objects from relational structure descriptions of them, as well as to answer questions in a small subset of English using a relational data base.

Consider again our friend the bear (Figure 3). If we use . to denote composition of relations, $\cup$ and $\cap$ to denote union and intersection (considered as operations on their pairs) and $^{-1}$ to denote inversion we can say that the defined relation

$$\text{adj} \cap (\text{left} . \text{adj})$$

holds between 1 and 3. Also

$$(\text{ins} . (\text{ins} \cap (\text{left} . \text{ins})) \cap \text{left} . \text{ins} . \text{ins}))\ \text{ins}^{-1}$$

holds between 5 and 6.

Given any element and such a definition of a relation it is easy to find all elements so related to it, using a recursive top to bottom 'parsing' technique. Starting from such a given element the existence of such a related element defines the intermediate structure, so that we have a representation of the relational structure, or at least some paths through it, basically in terms of relational composition. The definitions are a little awkward and unnatural but the program succeeded in recognising a 'cube' (9 elements in 9) in about 60 seconds and the 'bear' in 8 seconds.

The technique does not seem appropriate for large structures but may be useful for detecting local structure in large pictures, when used in combination with another method such as the hierarchical one described below. It does seem to use the adjacency of nodes in a natural way.

Relational structure matching using extensions of graph isomorphism techniques

The problem of finding monomorphisms from one relational structure to another is a generalisation of the better known problem of finding isomorphisms between graphs. In this context, by "graph" we mean a set of nodes N (the carrier) together with a binary relation $R \subseteq N \times N$ defined over them. In the relational structure case we have many relations, not necessarily binary, and for monomorphisms the structures will not necessarily have the same number of nodes. What is more we are not just looking for *full* monomorphisms. Relations of 3 or more arguments can be converted into sets of binary relations, and unary relations (properties) present no problems.

Techniques for finding (full) isomorphisms of graphs have been described by Unger (1964), by Sussenguth (1965), and Corneil and Gottlieb (1970). Sussenguth also deals with monomorphisms. It is possible to generalise Unger's techniques to find monomorphisms of relational structures (Rastall 1969). In essence Unger uses the structure of a graph to derive additional properties of the nodes of the graph and hence to classify the nodes. For example, the number of arcs leaving each node may be used as an extra property for classification. We want a function from the nodes of the first graph onto nodes of the second. Unger classifies the nodes of each graph and so restricts the number of possible functions since a node in a certain class of the first graph must correspond to some node in the same class of the second graph. For example, a node with two arcs leaving it can only be mapped onto a node with two arcs leaving it. He uses the resulting classes to define further properties and thus refine the classification, and so on until either it has been shown that no isomorphism is possible, or until one has been found, or until no further refinement is possible, at which stage an enumerative choice is made and the classification phase is repeated.

To refine a classification Unger determines for each node how many neighbours it has in each of the classes. Two nodes in the same class may differ in the distribution of their neighbours among the other classes and this will yield a refinement of the classification. Schematically:

procedure examine (classification);

begin

```
refine:  try to refine classification;
         if inconsistent then exit;
         if each class has just one node then check for isomorphism
            and exit;
         if no refinement was possible then
            enumerate: for new classification := choice_1,...,choice_n do
                              examine (new classification)
         else goto refine
end;
examine (initial classification);
```

Example

The non-directed graphs in Figure 4 can be initially classified by number of neighbours thus:

Graph 1. A: (1,3), B:(5,6,7) C:(2,8) D:(4)

Graph 2. A: (5,7) B:(1,4,3) C:(6,8) D:(2)

This reduces the initial 8! functions to 24 functions consistent with the classification. We now note the classes of the neighbours of each node.

Graph 1	Graph 2
1 CC 3 CD	5 CC 7 CD
5 BBD 6 BBD 7 BBC	1 BBC 4 BBD 3 BBD
2 AACD 8 ABCD	6 AACD 8 ABCD
4 ABBCC	2 ABBCC

This yields a refined classification with only two possible functions

Graph 1 E:(1) F:(3) G:(7) H:(5,6) I:(2) J:(8) K:(4)

Graph 2 E:(5) F:(7) G:(1) H:(4,3) I:(6) J:(8) K:(2)

Repeated examination of neighbours fails to distinguish the nodes in H (in fact there is an automorphism carrying 5 into 6 and 4 into 3, so they are indistinguishable). We enumerate, putting 5 into a class of its own, say L, and *either* 4 *or* 3 into L in the second graph. Both alternatives give isomorphisms without further processing.

Corneil and Gottlieb have a means of further refining the classification before enumerating, but this does not seem extendable to the monomorphism case.

Turning now to relational structure monomorphisms the main difficulty in extending the graph isomorphism techniques is not in dealing with a number of relations. It is that the image element in the picture structure may have relations to a larger number of elements than does the source element in the structure describing the known object. Thus element 9 in 'Bear' (Figure 3) has only 5 elements adjacent to it whilst its image 11 in 'Bear on rug with ball' (Figure 2) has 6. This occurs even if we seek only *full* monomorphisms since an element in the substructure may be related to the elements outside it. Thus if we classify the source ('Bear') elements and make the classes of target ('Bear on rug with ball') elements which might correspond to each of these, the latter classes will be overlapping rather than disjoint. Rastall's method is to follow the same general scheme of computation as in the *'procedure'* given about but to allow overlapping classes of target elements.

Consider finding monomorphisms from R, 'Bear', to S, 'Bear on rug with ball'. Let us use the property *circ* to get an initial classification

Structure R. A: (1,2,9,10,11,12,13) B: (3,4,5,6,7,8)

Structure S. A: (1,2,3,4,5,6,7,8,9,10,11,12,13,14,15) B: (1,5,6,7,8,9,10)

Class A in R does not have *circ*, class B has. The corresponding class A in S contains both non-*circ* elements and *circ* elements, whilst B contains only *circ* elements, since if an element is *circ* its image must be *circ*.

We now refine the classification using the relations *adj, ins* and *left* successively to get classes

Structure R. C:(1) D:(2) E:(9) F:(10,12) G:(11,13) H:(4) I:(5)

Structure S. C:(3,4) D:(4) E:(11) F:(3,12,14) G:(4,13,15) H:(6) I:(6,7)

Structure R continued J:(6) K:(8) L:(7) M:(3)

Structure S continued J:(6,7,8,9,10) K:(9,10) L:(9) M:(5)

For example elements 1, 2 and 9 of R have a circular node adjacent to them and so can only go into nodes 3, 4 or 11 of S. Using the other relations eliminates some of these possibilities to give the classification shown above. At this stage we notice that 4 of S is committed to element 2 of R, so that we can remove it from classes C and G of S. Such observations lead to

Structure R. C:(1) D:(2) E:(9) F:(10,12) G:(11,13) H:(4) I:(5)

Structure S. C:(3) D:(4) E:(11) F:(12,14) G:(13,15) H:(6) I:(7)

Structure R continued J:(6) K:(8) L:(7) M:(3)

Structure S continued J:(8) K:(10) L:(9) M:(5)

Further attempts to refine this are fruitless so we make a pair of enumerative choices, say 10 goes to 12 *or* 10 goes to 14. Each of these on refinement immediately leads to an unambiguous classification which is a monomorphism.

In practice, Rastall's program starts off with an initial partition based on a property-relation-property classification (circles adjacent to circles, etc.) or a still stricter one and reaches a solution even more quickly. Going from isomorphisms to monomorphisms reduces the power of the classification method somewhat, but the program has recognised such things as a description of a cube in a collection of objects (9 elements in 40, 32 seconds) and a 'cat' in a 'cat on table' (10 elements in 18, 257 seconds). Times are rather variable depending both on the nature of the structures and on certain parameter settings which control the effort expended in various parts of the search. The general impression is that refined combinatorial analysis, in the spirit of say Corneil and Gottlieb (1970), is less important than avoidance of tedious attempts at classification which do not pay off. The program needs more flexibility in its choice between refinement and enumeration. It is restricted to dealing with properties and 2-place relations, but n-ary relations can be translated into sets of 2-place ones with some loss of efficiency.

In general the method is quite fast. Unfortunately we see no simple way of extending it to deal with 'dirty' or incomplete pictures with missing nodes or instances of relations, nor will it easily cope with 'real-valued' properties and relations (that is, functions).

Recognition by hierarchical synthesis

Suppose a relational structure, S, has N elements, and we wish to match it against another structure, R, with n elements. A simple combination generating-checking approach will involve testing of the order of N^n combinations (assuming $N >> n$). It will be clear that in practice such approaches become unworkable if n is larger than about 10.

We now make the following observation. In general, for substructures of non-trivial complexity, the number of occurrences of a particular substructure in S will be of the order of N. (This is not a law; but it provides a useful heuristic technique.) Suppose we break R into k substructures of r_i elements ($i = 1, \ldots k, k < n$). We now first identify all occurrences of each substructure in S: there will be of the order of N for each of these. Then we generate and test combinations of the substructures, searching for combinations which match R. The number of combinations will be of the order of N^k. Thus the total number of checking operations will be of the order of

$$N^k + \sum_{i=1}^{k} N^{r_i}$$

which will be much less than $N^n = N^{\sum_i r_i} = \prod_i N^{r_i}$.

Thus, if we can break up the model structure into suitable substructures, we can reduce considerably the amount of work required in matching. In fact, we can apply this procedure recursively to make very great economies of time and effort.

This principle has been used in practice, in a program written by D. Milner which endeavours to find complex objects in pictures. The data in this case is provided by hand, and is, arbitrarily, a description in terms of regions: we have not yet connected the program to the real picture analysis routines.

Each object that the program 'knows about' is composed of several sub-objects, and these may be composed of smaller sub-objects, in a hierarchy with picture regions at the bottom-most level. See for example Figure 5 which shows a hierarchical structure for 'Bear' (many other hierarchical decompositions would be possible).

The program is written in a modular fashion, for each object there is a module, and the interrelations of the modules form a network. It is the job of each module to produce a set of all the objects of a certain type in the picture. It does this by accepting similar sets from the modules associated with its sub-objects, generating combinations of sub-objects and checking them. At the lowest level, modules accept as their input the set of all regions of the picture, and produce sets of regions of particular shapes. (At present, it seems that the lowest level requires the bulk of the picture processing time, as every member module checks all the picture regions. A more subtle classification technique could easily be used here.)

Regions are classified into about 40 categories but there are only 3 relations *adjacent, above, left-of*. Once more, however, *adjacency* is a real-valued measure, and not a strictly binary predicate. About a dozen objects are defined at present, the most complex possessing 12 regions and 5 levels of sub-object.

The program performs encouragingly well. It takes only seconds to recognise objects compared with minutes for even simple objects with the previous single-level matching program. A 10-region man-figure is recognised in 10 seconds, while for a 37-region picture (Figure 6) containing a man, a car and a house, only 26 seconds are required to identify all three objects. The program is in a high-level language and could be speeded up by machine coding.

Having described the organisation of the network, it is perhaps now worthwhile to describe the organisation of the operations. A module may be activated, or remain dormant. If it is activated and all the modules upon which it depends have produced their results, it performs its task of generation and testing. When it has produced its set of results, it then makes a request for activation of all modules which accept its results as data. If the module is activated and it is unable to function because it lacks data, it requests that the offending lower level modules be activated. At every stage, the most popular module is activated, and since some modules never become activated at all operating efficiency is secured.

The program can be run 'bottom-up' or 'top-down'. If a particular object module is activated, it will call only those modules upon whose results it depends. Thus a search for a particular object in the picture may be carried out. Alternatively, if all the lowest level modules are activated, activity will spread up the hierarchy to modules which might be able to produce results. Thus all objects in the picture may be recognised.

The remaining problem of this approach is that of setting up the modules and network automatically in a 'learning phase'. At present we have not attempted to tackle this problem. The ideas of Winston (1970) about learning descriptions in the form of relational structures may be relevant.

A mathematical discussion of the hierarchical matching method may be of interest both to understand it better and make it easier to program; we give such a discussion in the Appendix. We separate the idea into two parts, one of which is dependent on the notion of relational structures and their monomorphisms, and the other is a more general algebraic notion of a hierarchical descriptive system couched in category theoretic terminology, and hence allowing other interpretations of the notion of morphism and structures other than relational structures, (see Cohn 1965 or Maclane and Birkhoff 1967 for the terminology used). This extra freedom might have practical significance in cases where metric or topological structure turned out to be more significant than relational structures. It can be used to cope with the case where the relations are replaced by real-valued functions and approximate matches are desired.

The method as described in the Appendix has been programmed to check that there are no snags. It ran more slowly than Milner's version, but it did recognise a small structure 'House' (see Appendix) in 17 seconds and "Bear" in 68 seconds and should run faster if more tightly coded.
The program is simple and is made up almost entirely of general functions which could be used in other programs about algebras and relational structures.

At first sight, hierarchical synthesis may appear to be a picture grammar by another name. In fact it has some similarities but is in reality a different approach. The idea that it does take over from grammars is the notion of *constituency:* that a symbol (substructure) consists of a collection of lower-order symbols, i.e. that a substructure may itself possess substructures. However, in the formulation in terms of relational structures, the constituent substructures need not necessarily be disjoint.

Comparison of methods

The methods described above have been applied to a large variety of problems. Table 1 gives results for some that were definable in a form which enabled all the methods to be used. There are types of problem to which not all methods are applicable (e.g. those involving numerical information or incomplete data). We wish to emphasize that the times given in Table 1 should not be taken too seriously, since a different representation of the data was required for each method. (All times in this paper are for programs written in the POP-2 language on an ICL 4130, a machine roughly similar to the IBM 7090.)

The tree-search method of building up a monomorphism one pair at a time is robust, enabling one to handle 'real-valued' properties and relations, imperfect matches, and simultaneous matching against many models. It was used successfully with actual T.V. pictures. But it seems to be too combinatorially inefficient to handle structures of any great size.

The relational composition search is quite fast, but the construction of definitions of objects is unnatural. It uses definitions of paths through a relational structure, rather than the relational structure itself. This means that the definition of "parallelepiped" is very cumbersome. It can only look for one object at a time, but since it can memorize sub-paths (i.e. sub-objects) that it finds in the course of its search, the labour involved in looking for a second object that shares some structure with the first, is reduced. The relational composition search method cannot cope with incomplete data. We feel that, used in conjunction with other methods, something along along the lines of relational composition search should be useful for finding local structures in larger pictures.

The method based on Unger's graph isomorphism technique suffers from its generality. There is at present no way of giving it extra information about the type of picture to be processed (in the way that the sub-objects of hierarchical synthesis method represent such special information). As demonstrated by the Picture example, it is badly affected by irrelevant data. In this case it has spent most of the time in processing completely useless information (e.g. that region 35 is to the left of region 20, or that region 6 is

TABLE 1

	Tree Search	Relational Composition	Extended Graph Isomorphism	Hierarchical Synthesis
Flower (5 elements in 5)	680 secs	2 secs	3 secs	0.85 secs
Maid (8 elements in 11)	10^6 secs (by extrapolation)	10 secs	36 secs	2 secs
Parallelepiped (9 elements in 9)	10^{10} secs (by extrapolation)	65 secs	408 secs	22.5 secs
Bear (13 elements in 15)	10^{10} secs (by extrapolation)	8 secs	17 secs	5 secs
Picture (6, 11, 12 elements in 37)	—	—	4000 secs (estimate)	26 secs

Description of problems

Flower. A 5 element object to be recognised in a 5 element picture, with properties "petal-shaped", "leaf-shaped", "stalk-shaped" and "circular", and with relations "'left", "adjacent" and "above".

Maid. An 8 element object to be recognised in an 11 element picture, with properties "circular", "elliptical", "toppart", "line" and "skirt", and with binary relations "ontop", "onleft", "onright", "onrightbottom", "onleftbottom" and the tertiary relation "aligned".

Parallelepiped. A 9 element object to be recognised in a 9 element picture. This is a line picture of a cube as described in Burstall (1968) but using only the binary relation "parallel" and the tertiary "between" and not using "equidistant". Recognition of this object was expected to be difficult for all of the methods, because the definition is highly symmetrical. There is a six-fold automorphism (three rotations and a reflection) and hence six parallelepipeds must be found.

Bear. A 13 element object to be found in a 15 element picture, properties and relations as described elsewhere in this paper.

Picture. 12 objects to be looked for in a 37 element picture containing three objects: Man (11 elements), Car (12 elements) and House (6 elements). This is the Picture of Figure 6, with 42 shape properties and the binary relations "adjacent", "above", "below", "left" and "right".

below region 37). Furthermore, it can only look for one object at a time. Since twelve were to be sought, this irrelevant work had to be done twelve times. The Parallelepiped example took a long time because of the tertiary relation "between". The program has recognised such a line drawing of a parallelepiped, but using the binary relations "meets" and "parallel" in 17 secs.

The method of hierarchical synthesis is quite fast. Moreover, since objects have sub-objects in common and basic elements may be sorted by property, it is possible to search for a number of objects in parallel with much greater efficiency than searching for them in sequence. Real-valued relations may be used in descriptions and they may be evaluated when required, rather than in advance. Since it builds up structures via their parts it offers hope of guissing at partly occluded objects. It can be used either to look for a given object (top-to-bottom) or to say which of a repertoire of known objects occur in a picture without checking each one separately (bottom-to-top).

Hierarchical synthesis thus seems to be the most promising method of those investigated.

Connection with Logical Inference

Given a description of a picture using certain predicates and definitions of objects which might occur in it using the same predicates, instead of talking in terms of relational structures we can set up the recognition problem as a proof problem in first-order logic in the obvious way. We will need constants to denote the elements (points, lines, etc.) in the picture. The picture description will be a set of sentences, each of which states that a predicate applies to an n-tuple of these constants. The object definitions will be clauses involving universally quantified variables. For example, with a similar decomposition to Figure 5, we may define a bear in first-order logic using the basic predicates *circular, adjacent, inside, leftof* and auxiliary predicates *head, eyes, body* and *triple* (we omit universal quantifiers by convention):

$$\mathrm{head}(x_1, \ldots, x_8) \;\&\; \mathrm{body}(x_9, \ldots, x_{13}) \;\&\; \mathrm{adj}(x_3, x_9) \Rightarrow \mathrm{bear}(x_1, \ldots, x_{13})$$

$$\mathrm{triple}(x_1, x_3, x_2) \;\&\; \mathrm{circ}(x_3) \;\&\; \mathrm{eyes}(x_4, x_5, x_7, x_8) \;\&\; \mathrm{circ}(x_6) \;\&\; \mathrm{ins}(x_4, x_3)$$
$$\;\&\; \mathrm{ins}(x_6, x_3) \;\&\; \mathrm{ins}(x_7, x_3) \Rightarrow \mathrm{head}(x_1, \ldots, x_8)$$

$$\mathrm{circ}(x_1) \;\&\; \mathrm{circ}(x_2) \;\&\; \mathrm{circ}(x_3) \;\&\; \mathrm{circ}(x_4) \;\&\; \mathrm{ins}(x_2, x_1) \;\&\; \mathrm{ins}(x_4, x_3)$$
$$\;\&\; \mathrm{left}(x_3, x_1) \Rightarrow \mathrm{eyes}(x_1, x_2, x_3, x_4)$$

$$\mathrm{triple}(x_2, x_1, x_3) \;\&\; \mathrm{triple}(x_4, x_1, x_5) \Rightarrow \mathrm{body}(x_1, \ldots, x_5)$$

$$\mathrm{adj}(x_2, x_1) \;\&\; \mathrm{adj}(x_3, x_1) \;\&\; \mathrm{left}(x_2, x_3) \Rightarrow \mathrm{triple}(x_2, x_1, x_3)$$

The picture 'bear on rug with ball' to be recognised would be represented by variable-free sentences, in terms of constants $a_1, \ldots, a_{15}$ denoting regions of the picture.

$circ(a_1)$. $circ(a_5)$. $circ(a_6)$. $circ(a_7)$. $circ(a_8)$. $circ(a_9)$. $circ(a_{10})$.

$adj(a_2, a_{11})$ . . . . $adj(a_{11}, a_{15})$.

$ins(a_1, a_2)$ . . . . $ins(a_{10}, a_9)$

$left(a_3, a_4)$ . . . . $left(a_{14}, a_{15})$

Now we can prove there is a bear in the picture i.e. $bear(a_3, \ldots, a_{15})$ using the usual logical inference rules of instantiation and modus ponens. Thus from the eyes definition and the sentences

$circ(a_6)$. $circ(a_7)$. $ins(a_7, a_6)$. $circ(a_9)$. $circ(a_{10})$. $ins(a_{10}, a_9)$

we infer

$eyes(a_6, a_7, a_9, a_{10})$.

similarly we obtain $triple(a_3, a_5, a_4)$ from $adj(a_3, a_5)$. $adj(a_4, a_5)$ and $left(a_3, a_5)$ and use this with some of the original sentences to obtain $head(a_3, \ldots, a_{10})$, and so on to get $bear(a_3, \ldots, a_{15})$.

If we think of the resolution method of mechanical theorem proving (Robinson, 1965a) we see that this proof, easy though it is, involves a large number of resolution steps. However the inferences using a definition each correspond to a single *hyper-resolution* inference (Robinson, 1965b), and only 5 such 'macro' inference steps are required. But we still need a more efficient method of finding a set of unit ground clauses than the very tedious one of comparing the literals on the left of the definition one by one with the data clauses in a tree-search manner until a match is found. This is impracticable if the definition has more than half a dozen or so literals on its left hand side.

To deal with larger definition clauses we may use one of the techniques for finding monomorphisms between relational structures. We are dealing with the restricted kind of hyper-resolution inference where the definition clause (the 'nucleus' or negative clause) contains only predicates applied to variables and where picture clauses (the 'electrons' or positive clauses) are unit clauses consisting of a predicate applied to some constants.

Suppose we wish to hyper-resolve the 'head' definition clause with the original picture clauses together with $eyes(a_4, a_5, a_7, a_8)$ and $triple(a_3, a_5, a_4)$. Then we must find a unifying substitution σ, that is a function from the variables $x_1, \ldots, x_8$ in the 'head' clause to the constants $a_1, \ldots, a_n$, such that if $p(x_{i_1}, \ldots, x_{i_n})$ occurs in the definition then $p(\sigma x_{i_1}, \ldots, \sigma x_{i_n})$ occurs as a ground unit clause. Clearly we need a monomorphism $\sigma: R \longrightarrow S$, where (i) R is the relational structure whose carrier is the variable symbols $x_1, \ldots, x_8$ and whose predicates *circular, adjacent, inside, leftof* and *eye* hold for some n-tuple of x's just if this n-tuple and predicate occurs as a literal on the left hand side of the definition clause and (ii) S is the relational structure whose carrier is the constant symbols $a_1, \ldots, a_{15}$ and whose predicates hold for some n-tuple of a's just if this n-tuple and predicate occurs as a unit clause in the picture description.

A hyper-resolution theorem prover was written by Ambler, and the Rastall relational structure monomorphism program was incorporated to find the unifying substitutions for the hyper-resolution inferences. This was indeed much faster for long clauses than the naive method of matching literals one at a time, which was only used for short clauses. The matching was generalised a little to allow ground terms in the picture clauses as well as constants and terms in the definition clauses. In one example "Maid", with 8 nodes and 18 literals on the left hand side of the definition, the naive hyper-resolution method took 50 minutes, whilst the hyper-resolution using the monomorphism program took 20 seconds (the hyper-resolution was not very efficiently programmed but even so the difference is quite startling). The Rastall monomorphism program was only written to cope with unary and binary relations, so that some pains had to be taken at the interface between the theorem prover and the monomorphism tester to convert n-ary relations $(n \geq 3)$ to a number of binary ones; this slowed up the program somewhat.

One pleasing feature of this approach is that it connects the relational structure matching techniques which are quite efficient with logical inference techniques which are more general in their application and might be necessary for making deductions about non-pictorial aspects of the situation. The connection between relational structure monomorphism finding and hyper-resolution is not confined to picture processing problems, although it is only of any practical interest where long clauses are encountered.

Appendix

Mathematical discussion of the hierarchical method

Our task is to compare a hierarchy of structures with a given picture structure and find all monomorphisms from the structures in the hierarchy to the picture structure. Such monomorphisms 'describe' the picture by saying that some known object occurs in it. We think of the hierarchy with its monomorphisms as a *category*, i.e. a collection of 'objects' (structures) and, morphisms,(monomorphisms from one structure to another) with an associative composition operation for morphisms and an identity morphism for each object (Cohn 1965, Maclane and Birkhoff 1967). Indeed the hierarchy is a subcategory of a larger category containing also all possible picture structures. For generality we will not at first restrict the category to be one of relational structures and their monomorphisms, only introducing this specialisation when necessary. We think of an object in the hierarchy being constructed out of lower objects in the hierarchy and for each such construction give a rule for trying to find the higher object in the picture when the lower objects have been found. Such a step forms an operation of a general algebra whose elements are monomorphisms from objects of the hierarchy to the picture. We start off with monomorphisms from the lowest objects to the picture and try to find monomorphisms from the higher objects be repeatedly applying these operations, that is by closing the initial subset of this algebra.

Let $\mathcal{C}$ be a category and $\mathcal{V}$ be a subcategory of $\mathcal{C}$ (the vocabulary of objects to be used in descriptions). Let V_o be a subset of the objects of $\mathcal{V}$. By a description of an object c of $\mathcal{C}$ we will mean a morphism $h: b \longrightarrow c$ where be is an object of $\mathcal{V}$.

By a construction of b from a_i, where b and the a_i are all objects of $\mathcal{V}$, $i = 1, \ldots, n$, we mean an n-tuple of morphisms, f, where $f_i: a_i \longrightarrow b$ for $i = 1, \ldots, n$. Let F be a set of such constructions. (The number n may vary from one construction in F to another.)

Definition. We will call $(\mathcal{V}, V_o, F)$ a hierarchical descriptive system for $\mathcal{C}$ if the following conditions hold.

(i) For any construction f in F with $f_i: a_i \longrightarrow b$, $i = 1, \ldots, n$, with a_i and b in $\mathcal{V}$ and any n-tuple g with $g_i: a_i \longrightarrow c$, $i = 1, \ldots, n$, with c in $\mathcal{C}$ there exists at most one $h: b \longrightarrow c$ such that the diagram

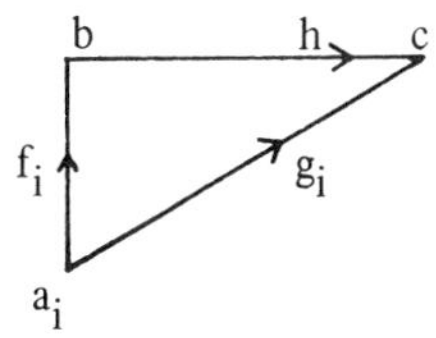

commutes for all i.

We denote this h by $\alpha_f(g_1, \ldots, g_n)$, considering, α_f as a partial function in $\mathrm{Hom}(a_1,c) \times \ldots \times \mathrm{Hom}(a_n,c) \longrightarrow \mathrm{Hom}(b,c)$.

(ii) For any b in $\mathcal{V}$ either b in V_o or else there exists a construction f in F with $f_i: a_i \longrightarrow b$, for some a_i.

(iii) There is a partial ordering $\leq$ over the objects V such that if f is in F with f_i: $a_i \longrightarrow b$ then $a_i < b$ for each i.

(iv) Hom(b,c) is finite for any b in $\underline{V}$ and c in $\underline{C}$.

(v) F is finite (and hence so is the set of objects of $\underline{V}$, by condition ii above).

If (V, V_o, F) is a hierarchical descriptive system for C we may associate with it a partial algebra F whose carrier is the set of all morphisms g: b→c, for b in V, and which has an operation α_f associated with each f in F.

By the closure of a subset of the carrier of F we mean the least subalgebra containing that set.

Our algorithm derives from the fact that, under the above assumptions, we can describe an object c in forms of the vocabulary V by starting with descriptions of c in terms of the basic objects in V_o and then applying the operations of F repeatedly to get descriptions in terms of higher objects. i.e. compute the closure of the set of initial descriptions. Any description, h, of c can be obtained in this way. Formally we have:

Theorem. For any h: b c where b is in V, h is in G_c where

$$G_c = \text{Closure}_F(\bigcup_{a \in V_o} \text{Hom}(a,c))$$

Proof. By induction on $\leq$ we have immediately

(i) If b is in V_o then h is in Hom(b,c), which is a subset of G_c

(ii) Otherwise, suppose the theorem holds for all $b' < b$. There exists f in F with f_i: $a_i \to b$. Put $g_i = f_i \cdot h$. Then since $a_i < b$ we have g_i is in G_c for each i, and since $h = \alpha_f(g_1, \ldots, g_n)$ h is in gG_c.

This theorem gives us 'completeness' of the closure process (it com putes all descriptions of c). Each element of the closure is clearly a morphism to c from some b, by the definition of F, so the process is 'sound'. Since $\underline{V}$ has finitely many objects, each with finitely many morphisms to c, the closure is finite and the process of computing it must terminate.

Hierarchical descriptive systems for relational structures

Let us take as the category C the category of finite relational structures over a particular predicate set and their monomorphisms. We take as V the subcategory of structures representing meaningful objects and substructures of these which are likely to be useful for recognising them.

If c is the relational structure obtained from the picture to be described then G_c is the set of descriptions, with the maximal elements under $\leq$ being the "most comprehensive" descriptions, the largest identifiable substructures. We take V_o as some easily obtainable initial structures e.g. all single nodes of c, or perhaps some or all of the n-tuples occurring in the relations of c.

For the constructions in F we take n-tuples of monomorphisms f_i: $a_i \rightarrow b$ such that the union of the images of the a_i is the whole carrier of b.

To define $h = \alpha_f(g_1, \ldots, g_n)$, such that $f_i \,.\, h = g_i$, $i = 1, \ldots, n$ we note that since f_i is a monomorphism it has an inverse, say f_i^{-1}, which is a partial injection, composing this with g_i, considered as an injection, we get the partial injection $f_i^{-1} . g_i$. Taking the union of these we get $\bigcup_i (f_i^{-1} . g_i)$: $b \rightarrow c$ which is total, but may or may not be single valued and injective. If it is a total injection it still may not be a homomorphism. If it is a homomorphism $\bigcup_i (f_i^{-1} . g_i)$ is the required unique monomorphism h. If it fails to be single valued, injective or a homomorphism then $\alpha_f(g_1, \ldots, g_n)$ is undefined.

To save some of the labour of checking that $\bigcup_i (f_i^{-1} . g_i)$ is a homomorphism from b to c, we may express b as the 'superimposition' of a pair of structures b_1 and b_2 on the same carrier as b, such that any relation holding in b holds either in b_1 or in b_2 but not in both; also a relation holds for $x_1, \ldots, x_k$ in b_1 iff it holds for $f_i^{-1}(x_1), \ldots, f_i^{-1}(x_k)$ in a_i for some i. Thus b_1 is the image of the f_i, and b_2 represents the extra structure in b, not contained in the a_i. Now to ensure that $\bigcup_i (f_i^{-1} . g_i)$ is a homomorphism we need only check the relations in b_2 since those in b_1 are automatically satisfied.

Example. Suppose V consists of the following, using one binary directed relation.

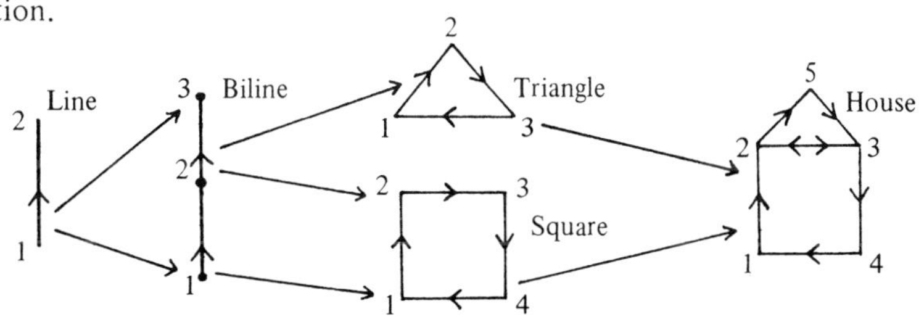

Then denoting the function f: $\{1,..,m\} \longrightarrow \{1,..,n\}$
$f(1) = i_1, ..., f(m) = i_m$

by $(i_1,..,i_m)$ we have constructions

Line, Line ⟶ Biline $f_1 = (1,2),\ f_2 = (2,3)$
Biline ⟶ Triangle $f_1 = (1,2,3)$
Biline, Biline ⟶ Square $f_1 = (1,2,3),\ f_2 = (3,4,1)$
Triangle, Square ⟶ House $f_1 = (2,5,3),\ f_2 = (1,2,3,4)$

If we take

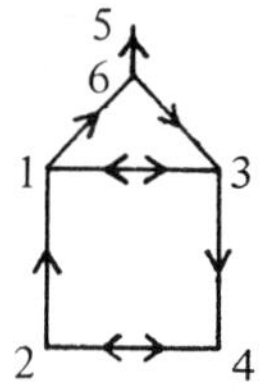

Picture

The required h: House ⟶ Picture is $h = (2,1,3,4,6)$ which is obtained from g_1: Triangle ⟶ House, $g_1 = (1,6,3)$ and g_2: Square ⟶ House, $g_2 = (2,1,3,4)$, which are obtained eventually from g_{10}: Line ⟶ House, $g_{10} = (2,1)$, g_{11}: Line ⟶ House, $g_{11} = (1,6)$ etc.

Computing closures for an algebra

Having shown how to compute the individual operations α_f it remains to explain how the closure of the starting set of monomorphisms with respect to these operations is to be computed. There are some obvious techniques. We can start with the initial elements and apply each operation to all possible n-tuples of them repeating the process with the new elements added to the set. This process can be much improved by noticing that the operation α_f with $f_i: a_i \longrightarrow b$ only applies to $g_i: a_i \longrightarrow b$, so that the object a_i may be thought of as the 'sort' or 'type' of the ith argument. By collecting elements of the same sort together it is no longer necessary to generate all n-tuples, but only those which have the correct sorts. Better still if F is large is to keep a table showing which constructions in F require arguments of a given type a, and only apply a construction in F when a suitable argument for it has been generated.

Alternatively if we are looking for a particular object b in the picture c we may proceed from "top to bottom" and use V and F to explore only the possible constructions of b.

A further improvement might be gained by looking for an object d before using f such that

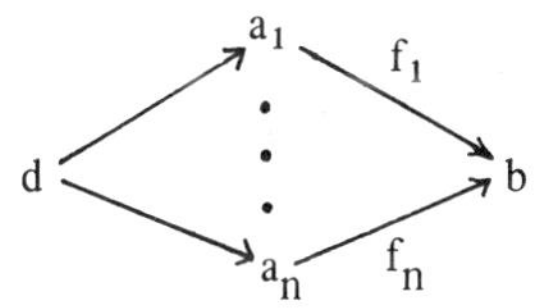

Application of f will not succeed unless $g: d \longrightarrow c$ has already been found and $d < a_i$ for each i. This again cuts down the number of n-tuples $g_i: a_i \longrightarrow c$ which need be tried. In the above example we only need try injections of triangles and squares which have a line in common.

Thus the distinction between computing the results of the operations α_f and closing the algebra breaks up the computation into two parts each capable of refinements for efficiency.

Acknowledgements

This work has been carried out at Edinburgh University with the help of a grant from the Science Research Council. We would like to thank Professor D. Michie for his encouragement of the vision work, Mr. Steve Salter for building the equipment and Miss Eleanor Kerse for her speedy and accurate typing of the paper. We are grateful to Mr. John Rastall and Mr. David Milner for allowing us to describe their contributions to this research which were carried out as student projects.

REFERENCES

1. Ambler, A. P. and Burstall, R. M. (1969) Question-Answering and Syntax Analysis, *Experimental Programming Reports: No.18,* Department of Machine Intelligence and Perception, University of Edinburgh.
2. Barrow, H. G. and Milner D. (1971) A Hierarchical Picture Interpretation System with Directed Search. *MIP - R - 85,* Department of Machine Intelligence and Perception, University of Edinburgh.
3. Barrow, H. G. and Popplestone, R. J. (1971) Relational Descriptions in Picture Processing. *Machine Intelligence 6* (eds. B. Meltzer and D. Michie) Edinburgh: University Press, pp. 377-396.
4. Brice, C. R. and Fennema, C. L. (1970) Scene Analysis Using Regions, *Artificial Intelligence 1,* 205-226.
5. Burstall, R. M. (1968) Computer Recognition of Rectilinear Solids from an Imperfect T.V. Picture. *MIP-R-40,* Department of Machine Intelligence and Perception, University of Edinburgh.
6. Clowes, M. B. (1971) On Seeing Things. *Artificial Intelligence 2,* 1, 79-116.

7. Cohn, P. M. (1965) *Universal Algebra.* New York: Harper and Row.
8. Corneil, D. G. and Gottlieb, C. C. (1970) An Efficient Algorothm for Graph Isomorphism. *J. Assoc. Comput. Mach.,* 17, 51-64.
9. Evans, T. G. (1964) A Heuristic Program to Solve Geometric-Analogy Problems. *AFIPS Spring Joint Comput. Conf., 25,* 327-338.
10. Green, C. (1969) The Application of Theorem-Proving to Question-Answering Systems. *A. I. Technical Memo. No. 8, S.R.I. Project 7494,* Stanford Research Institute.
11. Guzman. A. (1968) Decomposition of a Visual Scene into Three-Dimensional Bodies. *AFIS Proc. Fall Joint Comput. Conf.,* 33, 291-304.
12. Huffman, D. A. (1971) Impossible Objects as Nonsense Sentences. *Machine Intelligence 6* (eds. B. Meltzer and D. Michie) Edinburgh: University Press, pp. 295-324.
13. MacLane, S. and Birkhoff, G. (1967) *Algebra.* New York: Macmillan.
14. Raphael, B. (1964) SIR: A Computer Program for Semantic Information Retrieval. *MAC-TR2, Project MAC,* Massachusetts Institute of Technology. Cambridge: Massachusetts.
15. Rastall, J. (1969) Graph-Family Matching. *MIP-R-62,* Department of Machine Intelligence and Perception, University of Edinburgh.
16. Robinson, J. A. (1965a) A Machine-Orientated Logic Based on the Resolution Principle. *J. Assoc. Comput. Mach., 12,* 23-41.
17. Robinson, J. A. (1965b) Automatic Deduction with Hyper-Resolution. *Internat. J. Assoc. Comput. Math., 1,* 227-234.
18. Schoenfield, J. R. (1967) *Mathematical Logic.* Reading, Massachusetts: Addison -Wesley.
19. Sussenguth, E. H. (1965) A Graph-Theoretic Algorithm for Matching Chemical Structures, *J. of Chem. Documentation, 5,* 36-43.
20. Unger, S. H. (1964) GIT - A Heuristic Program for Testing Pairs of Directed Line Graphs for Isomorphism. *Comm. Assoc. Comput. Mach., 7,* 26-34.
21. Winston, P. H. (1970) Learning Structural Descriptions From Examples. *MAC-TR-76, Project MAC,* Massachusetts Institute of Technology. Cambridge: Massachusetts.

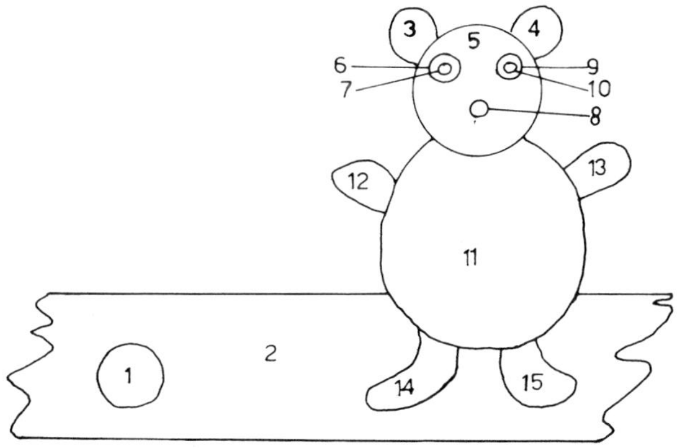

Figure 1. 'Bear on rug with ball'

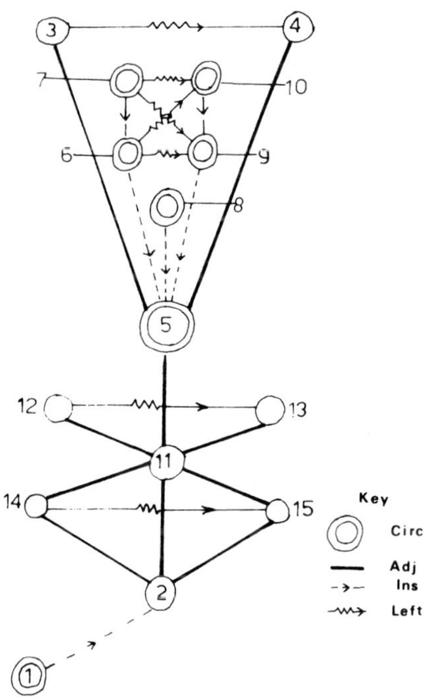

Figure 2. 'Bear on rug with ball' relational structure

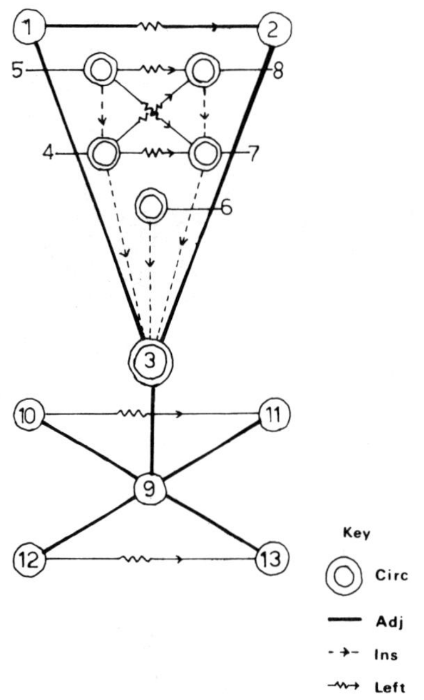

Figure 3. 'Bear' relational structure

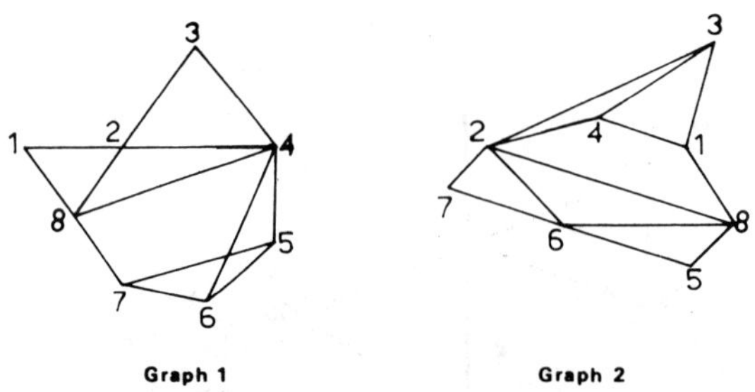

Figure 4

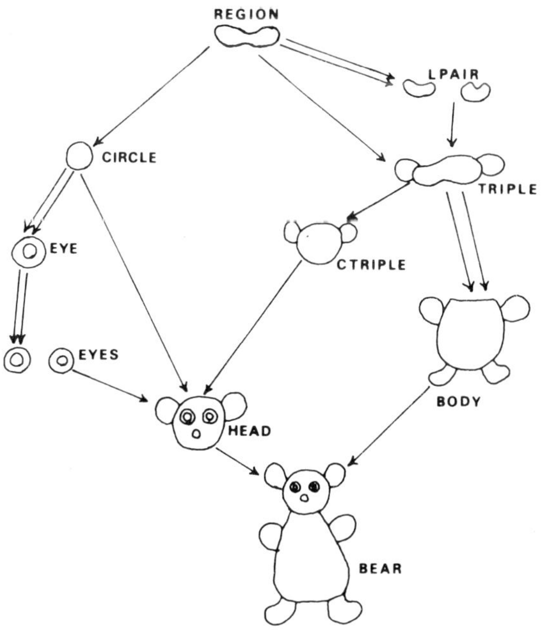

Figure 5. Hierarchical synthesis of 'Bear'

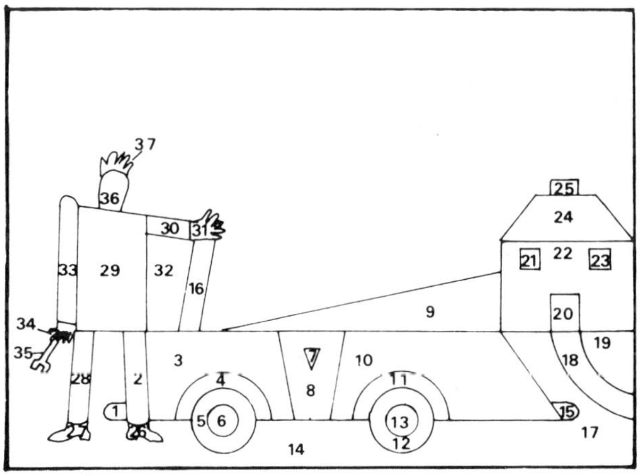

Figure 6. Picture

A SEMANTICS-BASED DECISION THEORY REGION ANALYZER

by

Yoram Yakimovsky and Jerome A. Feldman
Artificial Intelligence Laboratory
Stanford University
Stanford, California

ABSTRACT

The problem of breaking an image into meaningful regions is considered. Bayesian decision theory is seen to provide a mechanism for including problem dependent (semantic) information in a general system. Some results are presented which make the computation feasible. A programming system based on these ideas and their application to road scenes is described.

INTRODUCTION

The problem of segmentation, breaking a complex image into sections, is a central problem in machine perception. The analogous problem arises in the analysis of speech and, for that matter, in any problem of overwhelming size.[26] We will concentrate on the image segmentation problem, but most of the ideas are of wider applicability. The main idea is to apply (Bayesian) decision theory techniques and the use of problem-dependent information (semantics) to attack the image segmentation problem.

The segmentation problem for T.V. images is as follows: given a picture of some scene, we have a rectangle composed of some 200 x 300 points and for each such point some information on the light intensity and perhaps color. For any further processing 60000 points is far too many, so, depending on the perception task that we have in mind, the image should be segmented into regions. Each of these regions should be meaningful in the problem domain and the relevant information needed for the specific task should be easily obtainable.

There has been a great deal of work on segmenting images and a certain limited success has been achieved.[21] Some biological and meteorological images can be effectively segmented using known techniques. However, for images like those arising in road scenes or presented to assembly-line robots, the existing algorithms do not suffice. A major problem is that the existing algorithms use absolute criteria such as intensity difference, boundary strength, etc. to form regions.[2,3] But the criteria for what is a "region" will surely vary with context. Certain shades of green, yellow and brown might be merged into a single region of grass in a scene, yet distinguishing the same set of colors might be crucial for region separation in another scene or even in another part of the same scene (assume for instance that a yellow car occludes part of the grass). Another critical consideration is the goal of the perception. For some problems separating the green grass from yellow grass will be essential, in others completely confusing.

The importance of goal direction and context-dependence for effective problem solving is now well understood in artificial intelligence and scene analysis is just another example. One can certainly write a special purpose region analyzer for a fixed class of images and it will work better than any general algorithm. This, in fact, has been done successfully by Brice and Fennema[3] and Harlow and Eisenbeis[13] and is sometimes just the right thing to do. The obvious difficulty with this ad-hoc approach is that it requires a lot of work to build or modify each individual program.

In this paper we present a theoretical framework for a general system incorporating context dependence in a region analyzer. The theory has been developed into a computer program for region analysis and a number of experiments on real pictures have been performed. There is an enormous amount of work remaining to be done, but we feel that a promising start has been made.

Before describing the system in more detail, we must make one additional point of clarification. It is a tenent of artificial intelligence research that any information that can be brought to bear will be helpful in a given task. This is especially true in machine perception, but our current efforts do not attempt to exploit it fully. Region analysis is assumed to be a preliminary (relatively fast) partitioning of an image before further processing. For this reason, we have made no attempt to include semantic features like three-dimensional shape analysis in the region analyzer. We are still studying the capabilities of our semantic structure. As more experience becomes available we will be able to determine which information should be used in the segmentation process and which should be left for higher level processing.

1. THEORY

The underlying theory of our system is Bayesian decision theory. The ideas are quite beautiful and powerful, but have not received as much attention as they should from artificial intelligence workers. Even a brief description would be beyond the scope of this paper: "Elementary Decision Theory" and "Decision Analysis - Introductory lecures on choice under uncertainty" are good introductions and "Optimal Statistical Decisions" and "Mathematical Statistics a Decision Analysis Approach" are advanced texts.[4,20,5,8] The two central ideas are the use of a utility function to measure the value of various alternatives and an optimality theorem. This theorem shows that any adequate (admissible) strategy is equivalent to a strategy of maximizing expected utility for some choices of utility function and probabilities. The theory provides a complete world view (like, e.g. logic) and has been applied to many management problems. The difficulty in practice is that it is difficult to select the utility and probability functions and to actually carry out the computations. We describe below how we attack these problems.

For region analysis, we can define the utility to be the probability that the analysis is correct, contingent upon two factors: the (a priori) context knowledge about the picture domain, and the values of measurements on this particular image, i.e. we are to maximize:

P(global-interpretation | context,values of measurements)

An interpretation will divide the image into regions and attach a type label (meaning) to each region. One choice of evaluating the overall interpretation would be obtained by considering each region independently. If for a given partition of the image into regions we have {R(i) i=1,n} regions then the interpretation assigns label INT(i) to region R(i). The values of INT(i) will be sky, grass, road etc. depending on context and goals. Then assuming independence we will want to maximize the expression:

$$\prod_i P\{R(i) \text{ is } INT(i) \mid \text{context,values of measurements on } R(i)\}$$

over all partitions of the image into regions and assignments of labels to regions. This is quite conventional so far and is, in fact, too simple for our purposes. We want to account for two additional considerations. First we must use the model to get a good segmentation of the image into regions. For example, we might want to merge green, yellow and brown patches to create the whole area that we call grass. Secondly, we want to use additional semantic constraints e.g. the grass is below the sky, to influence the total probability of an analysis of the scene.

In an attempt to enrich the semantic structure to support more of the problem knowledge and to provide for control mechanism on the region growing algorithm, the semantic structure was allowed to have also "first order structure". In addition to the properties of each individual region we have, for each pair of adjacent regions of some interpretations, expected relative properties and some expected features of their common boundary line. For instance if we have two adjacent regions, one which is named "sky" and the other "hill" then we expect that the sky is above the hill, is a brighter blue than the hills, and that the boundary is usually a more or less horizontal smooth line. The relative properties are usually more significant than the absolute properties since they are less sensitive to variation between pictures. This semantic model is too limited to describe all that is known of a scene, but many classes of scenes can be segmented properly with first order methods.

Recall that we want to get a partition of the input and interpretation for the segments (regions) and boundaries so as to maximize the likelihood of having the right interpretation. Let R(i) be the i-th region, B(i,j) the boundary between region R(i) and R(j) (if it exists) and the label of R(i) be INT(i). Then with our first order assumption the expression that we want to maximize is:

$$P\{\text{global interpretation} \mid \text{values of measurements}\} = \prod_i P\{R(i) \text{ is } INT(i) \mid \text{values of measurements on } R(i)\} \times \prod_{B(i,j)} P\{B(i,j) \text{ is between } INT(i) \text{ and } INT(j) \mid B(i,j)\text{'s measurements}\} \qquad [1]$$

The use of [1] represents more than just our belief that properties of individual regions and boundaries will suffice for our semantics. It also entails an assumption that the probability can be factored into the product above. This amounts to assuming that the probabilities of interpretations of each region (boundary) are dependent on the local properties of the individual region (boundary) and are independent of all other measurements. Thus boundary B(i,j) which is the boundary between R(i) and R(j) is evaluated as boundary between INT(i) and INT(j), where R(i) is labeled INT(i) and R(j) is labeled INT(j).

For example, if INT(i) is "sky" and INT(j) is "hill", the evaluation will include factors involving the expected direction, smoothness, etc. of a boundary between sky and hill. These factors are assumed to be independent of the particular color etc. of the sky and hill. If the independence assumption seems to be unreasonable, consider the following argument:

$$P\{\text{interpretation} \mid \text{values of measurement, context}\} = \frac{P\{\text{values of measurements} \mid \text{interpretation,context}\} \times P(\text{interpretation} \mid \text{context})}{P(\text{values of measurements} \mid \text{context})} \qquad [2]$$

Now

P{values of measurements on R(i) | R(i) is INT(i), context}

and

P{values of measurements on B(i,j) | R(i) is INT(i) and R(j) is INT(j),context}

are plausibly considered independent of each other. A similar argument can be used for the factorization of the other two terms in the expression on the right of [1]. Putting these terms together give us back what we have in [1]. From all picture models which were described in the literature the model described in reference 17 is most similar to ours in the basic methodology aspects.

For a given utility function (like [1]) there are standard techniques in decision theory for finding the maximum utility. Unfortunately, the general techniques are too slow and much of our effort has gone into developing algorithms for efficiently computing an approximately optimal partition. The region growing algorithm starts with many small regions and on each iteration merges two adjacent regions (regions with a common boundary). The two basic decisions are which pair of regions to merge on each iteration and when to stop the algorithm. These two decisions can be controlled directly by the limited probabilistic semantic world model that we have.

In general, on each iteration of region growing the pair of regions whose common boundary is the weakest in the current image partition will be merged. Hence the control of the region growing algorithm is by evaluation of the boundary strength. We will show how our semantic representation can be used directly to compute the goundary strenth.

The second task of the semantics is to produce the stopping criterion for the region grower. In our case since we want to maximize

P{interpretation | measurements' values, context}

the optimal partition will be the one with that interpretation which maximizes this likelihood estimate over all partitions and all possible interpretations of partitions which do not allow false boundaries (boundaries between two regions which are interpreted as parts of the same final object).

In order to have an effective way to determine that probability we need a relatively fast way to obtain or approximate for a given partition the optimal interpretation and its value.

In the next section, we will describe relatively fast methods for computing upper and lower bounds on the optimal value of the probability of a given partition.

The bounds on the value of the global interpretation will be used as follows: The algorithm will collapse regions, and generate a sequence of image partitions. For each partition generated, the bounds on the possible value of the best interpretation will be evaluated. Then when the collapsing has been carried too far (as observed by strong decline of the possible interpretation value) the system will back-up to the most promising partitions observed while growing the regions (as indicated by the lower and upper bounds estimates of the quality of the partition observed). Next we will search for the best interpretation for the partitions observed whose bounds were high enough to make it possible that they are the best partition observed. The current algorithm will simply choose the best of these, but more sophisticated procedures can be used if necessary.

Althernatively we can use the procedure which assigns meaning to all regions directly as the stopping criterion. That is, if the best global assignment found for the given image partition does not interpret any boundary as an erroneous boundary (e.g. a boundary between regions of the same interpretation) we stop merging (see discussion below on lower bound estimates for the image partition for more details).

2. PROGRAM ORGANIZATION

The program has four basic sections: 1) the initialization, 2) initial assignment and boundary strength evaluation, 3) heuristics to evaluate a partition by approximating its optimal assignment, and 4) a limited interactive learning system.

The first step is to take a vector of values of measurements at a set of sample points (in our system usually 500-1000). The local measurements currently indicate only the dominant gray level around the sample point or if the image is observed through several color filters the dominant gray level through each of the color filters.

We employ a preliminary region-merging for initialization. The idea is to use a very crude algorithm on the reduced problem. The simple merging algorithms considered were:

1) Take an image point and grow a region around it consisting of all image points which can be connected to the starting point by a path of points which satisfies the following conditions: a) each adjacent pair of points along the path are adjacent geometrically, b) the jump in the value of the measurements vector between two adjacent points along the path is less than some threshold.[2,3] Note that some modifications are needed to treat gradual but strong changes.

2) This initialization is an extension of 1. It initializes as in 1 and then collapses, independent of order, all boundaries with strength less than some threshold. The advantage here over method 1 is the option to use more sophisticated boundary strength evaluation.[3]

Our initialization method utilizes a sampling technique initialized as in 1 (taking the connecting path to be a path of sample points) and then merges regions iteratively by eliminating the globally weakest boundary first. That is, on each iteration the pair of regions whose common boundary is weakest in the current image partition is merged into one region.

The boundary strenth in this stage is evaluated directly from the differences across the boundary and its geometrical structure.[28] The stopping crietrion in this case can be a threshold on the weakest boundary, that is the merger is stopped when the weakest boundary is stronger than a given threshold. This threshold is chosen very conservatively so as to stop this simple minded region grower before it produces false merges. In our experiments it turned out that the simple initialization algorithm had to be stopped quite early so that the semantic control was called with about 100 regions present, see reference 28 for more details.

The main algorithm first computes additional properties (like shape) of the regions and boundaries resulting from the initialization. It then assigns probabilities to the tentative interpretations of the regions, i.e. computes

$$P\{R(i) \text{ is } X \mid \text{values of measurement on } R(i)\}$$

The boundary strength may be evaluated by two related methods: 1) the probability that the boundary is a real boundary (a boundary between different objects in our semantic world model) and 2) the change in the value of the interpretation as a result of eliminating the boundary. We will describe here the first of these which is the one currently used. The second method has some advantages and will be discussed below.

We approximate the probability of the boundary to be real as follows:

$$\frac{\displaystyle\sum_{INT1 \neq INT2} \left\{ \begin{array}{l} P \text{ is a boundary between INT1 and INT2} \mid \\ \quad B\text{'s features}) \\ \times\, P(R1 \text{ is INT1} \mid R1 \text{ features}) \\ \times\, P(R2 \text{ is INT2} \mid R2 \text{ features}) \end{array} \right\}}{\displaystyle\sum_{INT1, INT2} \left\{ \begin{array}{l} P(B \text{ is a boundary between INT1 and INT2} \mid \\ \quad B\text{'s features}) \\ \times\, P(R1 \text{ is INT1} \mid R1 \text{ features}) \\ \times\, P(R2 \text{ is INT2} \mid R2 \text{ features}) \end{array} \right\}} \qquad [3]$$

The next step is to search for a partition which yields a good value of [1]. This involves both forming partitions and computing the value attainable from a labelling of regions in each partition.

A lower bound on the value of an image partition is computed by actually finding a good global interpretation using a simple fast algorithm. Briefly what we are doing is to take the region of highest confidence interpretation and assign to it its most probable interpretation. This assignment allows the program to update the probabilities of adjacent regions of the newly interpreted region by considering the boundary features of the newly assigned region. Then the region of highest confidence from all un-interpreted regions is assigned, etc. This is essentially starting a depth first search of the tree of interpretations and yields a value for the partition which is the desired lower bound. Extending this search to full tree search would yield the optimal interpretation. More details on the sequential assignment process are given below.

Recall that we want to approximate the maximum possible value of the expression in [3] over all possible values of INT(i) for a given picture partition.

The assignment procedure we use to estimate the best possible assignement of INT(i) for all R(i) for a given image partition is as follows:

1) Compute for each region the ratio (based just on local measurements of the region) between the most likely interpretation and the next most likely interpretation. This ratio will be called the CONFI(REG). Let x1 be such that P(R is x1 values of measurements on R) is maximized for R and let x2 be such that P{R(i) is x2 values of measurements on R(i)} is the next highest. Then

$$\text{CONFI}(R(i)) = \frac{P(R(i) \text{ is x1 } \text{ measurements of } R(i)}{P(R(i) \text{ is x7 } \text{ measurements of } R(i)}$$

2) Sort the regions by their confidence ratio.

3) Assign the region with highest confidence (the one with highest ratio) its most likely interpretation.

4) Update probabilities of different assignment to regions that were not assigned already, assuming that the last assignment is true. Let the region assigned most recently be R(1) and its interpretation be INT(1). Now if R(i) has boundary B(1,i) with R(1), then for any interpretation x of R(i) in evaluating equation 1 above, there will be a term of the form

P{R(i) is x values of measurement on R(i)}

from the first product and one of the form

P{B(1,i) is boundary between INT(1) and x
B(1,i)'s measurements' values}

from the second product. Therefore a better approximation of the probability of R(i) being x, assuming that R(1) is INT(1), is

$$\text{Pnew}(R(i) \text{ is } x) = \text{Pold}(R(i) \text{ is } x) \times P(b(1,i) \text{ is between INT}(1) \text{ and } x \;\; B(i,j)\text{'s features})$$

Thus we use the new information to find a more accurate probability for the different possible assignments for R(i), by counting the newly interpreted region R(1).

We do that updating to all possible interpretations for all adjacent regions of R(1).

5) Compute the new confidence ratio and resort the regions by the new confidence ratios.

6) If any region is still unassigned go to step 3 else exit.

This process of assigning interpretations iteratively provides a good guess on the possible best interpretation, but it does not guarantee the total maximization of our product. We can extend the current algorithm into a full tree search (undoing some assignments and trying alternative ones) to get the best interpretation. This will be a depth first search in the tree of all possible assignments, where each node will stand for the assignment of a meaning to a region. For efficiency purposes we can use various pruning techniques to reduce the search effort required to secure optimality.[16] Our current algorithm is the portion of the tree search up to the point where we get to the first terminal node (first global assignment). One should also note that the same sequential assignment and extension into tree search can be applied to an extended first order world model, where we allow relations between any two regions (not necessarily adjacent) if we continue to assume independence. The only difference is that we will have to update the probabilities and confidence of all regions not just those regions adjacent to the newly interpreted region.

We have the option to use this assignment procedure as a region grower by taking all pairs of adjacent regions that were assigned the same meaning and merging them. To avoid false merging we consider all regions which were assigned meaning with a low confidence level not mergable into other regions. This approach may be extended by adding it to the meaning assignment algorithm as a new step 3.5. If any adjacent region of the newly interpreted region is already assigned a meaning and it is identical with the meaning of the newly interpreted region, then merge the two together. From that point on the unified region will be considered in updating probabilities of other, not yet interpreted regions.

We can use the two extensions (merging on the run, and full tree search) together. This will generate a very reliable meaning assignment concurrent with a region growing procedure which has backup capabilities. It will be, however, relatively slow.

Upper Bound

The upper bound is computed by relaxing the consistency constraint. This condition means that a boundary between two regions of known interpretation has to be counted in as a boundary between those two interpretations. We relax this condition by breaking the product (1) into local sub-products and finding the best local interpretation for the terms involved in this subproduct. We take the best possible value for each sub-product separately, and, multiplying them, obtain in an upper bound on the value of the best global interpretation. One such relaxation is to consider all regions and boundaries independently and to assign for each the best possible interpretation considering only its own properties. The product of all these probabilities is an upper bound on the value of equation 1. It is this sort of estimate which would be used to approximate the single step improvement in the second method of boundary evaluation mentioned above. An exact computation of the change in interpretation value would be too time consuming. We do not yet know which boundary strength computation will be better.

Given the lower and upper bounds of the value of the best possible interpretation for a given picture partition, a variety of graph searching techniques can be applied to find a suitable interpretation and to pick out the best partitions observed for more detailed investigation using the full semantic knowledge.

3. LEARNING

One of the basic problems with any recognition system is the development of sharp classification capabilities for objects (in our case interpreting regions and boundaries). Our case is especially complicated since we need to recognize portions of objects, spurious boundaries and to overcome partial occlusions. To make the game of developing these capabilities easier we developed an interactive learning system. Its main task is to carry out bookkeeping jobs, to estimate probabilities, and to point out pitfalls and options for improvement in the classification scheme.

The non-parametric approximation of the probability density function works as follows: Given a set of measure functions on some class of objects, we break the space of all possible combinations of values of those measurements into cells (not necessarily cartesian), trying to get an effective classification. That is, given that the values of the measurements of an object fall into some cell, we want that often the probability estimate of the real meaning of the object to be high. Given a fixed partition of the measurement vector space into cells we want to learn the probabilities of different interpretations of objects whose measurements' values fall into a cell. This is done by keeping, for each cell and for each possible interpretation, the count of how many times in the past the value of the measurements of objects of that interpretation fell into this cell. The probability of that interpretation is just the number of times the measurements of an object of that interpretation fell into the cell, divided by the total number of objects which fell into that cell.

This brings us to the second learning system which would try to create a cell structure with as few cells as possible while attaining a good classification among the possible interpretations. For this purpose we sould utilize an augmented classification tree whose leaves correspond to the cells. The augmented tree also allows representation of the fact that two measurements are independent.

Currently this tree is generated interactively. To generate a sub-optimal classification tree automatically the system would keep a whole history list containing objects observed in the past, their properties and their real meaning. Based on this history the system could try to order the application of measurements so as to get good and cheap classification, creating as few as possible cells (leaves), while still keeping the good classification probability high. It also has the ability to point out cells that are not sufficiently discriminating so that they may be worked on interactively or automatically (primarily by breaking each such cell into finer subcells, so that for each subcell the classification is more reliable). A detailed description of this learning and classification system is given in reference 28. Techniques for organizing the classification tree so as to get near optimal sequential classification is described in Slagle and Lee, where game (α-b) type tree search is utilized in creating the decision tree.[23]

Such learning techniques are common to many pattern recognition and sequential decision problems. A vast amount of research, both theoretical and experimental, has been done in this area. Reference 10 is a good description of the theory and Reference 6 is a good introduction to various applicable techniques. It is interesting to comment that a learning scheme similar to the first (emphasizing correlations) was developed by Arthur Samuel.[22] Attempts are now being made to apply this learning scheme to speech segmentation and recognition.[25]

It is interesting to compare our technique with the nearest neighbor classification which is investigated in various papers.[29] This principle is to take, for a new unknown occurence of an object, the interpretation of the object observed in the past whose features of the new object. There are two deficiencies in this approach. First, only rarely is there an obvious metric on the space of values of measurements, and hence only rarely is it clear exactly how to measure distance in the features of two objects. Secondly, it is very hard to search for the nearest object observed in the past (unless we are in one dimension) since we have to compute the distance from many examples observed in the past to get the minimal distance. An effective way of reducing the search time will call for breaking the space into cells the way we do. That is, locating first the cell into which the measurements of the new object fall and then searching only among known objects whose measurements fall into that cell for the nearest one, ignoring objects which fall into other cells. Thirdly, the answer returned is just one possible interpretation and not a list of different possible interpretations with various probabilities. Extending the nearest neighbor principle to find the n- nearest objects and computing the probabilities of different interpretations based on them will make the computation even less efficient because of search time and will force even more reliance on space partitioning than the method we currently use.

4. RESULTS AND CONCLUSIONS

As a first experiment, the program was applied to the problem of segmenting images which might be seen driving on a road in the vicinity of the laboratory. The analysis was simplified by assuming the camera was in an upright position. There were six possible labels for a region: sky, road, roadside vegetation, car, shadow of car and tree. The non-terminal nodes in the classification tree are calls on integer valued functions. Some properties and the number of possible values for each are: light intensity (4), color hue(8), color saturation (4), size (5, logarithmic), vertical position (4), horizontal position (4), position on edge of image (4) and some crude shape descriptors.

With six types of region, we get 18 (6x6/2) types of boundary. Some boundary properties used, with the number of values for each are: relative size (6 logarithmic), relative intensity (6, logarithmic), relative color (3 green relation x 3 red x 3 blue), boundary shape and orientation (21 classes), relative position (4 right extremes x 4 left extremes x 4 above extremes x 4 below extremes), boundary length (5) and position of boundary in frame (5x5). The algorithms for computing the various properties and the discrimination in each were chosen intuitively.

Initial values for the counts in cells were also set intuitively, and the learning routine was used interactively to refine them (approximating the probabilities and breaking cells to finer cells if desired for better classification).

The program was able to segment the scenes correctly using only region properties except that it had difficulty isolating the image on the car on the road. Since a car can be of any color, the program either needs to make use of boundary relations, (e.g. the car is on the road) or perhaps shape discrimination should be made more sophisticated.

A region growing algorithm based on absolute properties would not work in these scenes mainly for the following reasons: 1) The trees and sky generated very many regions that were more varied in the values of their measurements than any other thing in the picture. 2) The sides of the road are patches of brown, green and yellow. 3) Strong shadows appear frequently on the road.

The second domain to which the system was applied was left ventricular angiograms (x-ray images of the left ventricular made visible by injection of a radio-opaque dye). These angiograms are useful for various cardiologic applications since they allow observation of myocardial movement. The semantics used for this application described the heart interior, chest cavity background and the dark frame border. No color was available here, and as a result light intensity, position and shape was the major recognition tools. In addition the non-semantic region grower had to stop at a relatively early stage because of noise and lack of high contrast border. The number of regions on termination of the non-semantic region grower was two hundred. It is encouraging that the adjustment to the second domain was very easy. We hope that in the future a general and rich library of feature extracting routines with the capability of working on many models will be achieved.

Shown below are illustrations of the results of experiments. All pictures are taken from a graphics terminal with gray level capabilities. There are six bits available per image point. Five are used for displaying the original picture, while the high order bit is used for the overlay of displaying the boundary lines.

This is a preliminary version of a general system for utilizing decision theory in scene analysis. There are a number of ideas from both areas that have yet to be tried and many experiments yet to be run. However, there are already some additional considerations which should be mentioned.

The most restrictive assumption in the current program is assumption that the interpretation of a region depends only upon adjacent regions. There are ways of selectively relaxing this rule so that occluded objects can be understood without having each region depend on all others. In fact, the entire approach will stand or fall on the question of whether there is sufficient independence to allow for good performance without prohibitive calculation cost.

The choice of local measurements around each point is, of course, another crucial factor. The idea of relatively coarse sampling allow us to apply more operators, including ones like Hueckel's, or texture finders which inherently involve many points.[14] There is the additional important potential for variable density sampling, possibly using planning in the manner of Kelly.[15]

A more difficult task would be to effectively incorporate 3-D constraints, as done so successfully for blocks by, i.e., Falk and Waltz (this would call for the addition of vertex properties).[7,27] There are many possible refinements to the learning procedure, especially on the question of what measurements are important.

BIBLIOGRAPHY

1. Barrow, H.C., Ambler, A.P. and Burstal, R.M. "Some techniques for recognition structures in pictures" Department of Machine Intelligence and Perception, University of Edinburgh, 1972.

2. Barrow, H.G. and Popplestone, R.J. "Relational Description in Picture Processing" Machine Intelligence, 6, 377-396, 1971.

3. Brice, C.R. and Fennema, C.L. "Scene Analysis Using Regions" Journal of Artificial Intelligence, 1, 205-226, 1970.

4. Chernoff, H. and Moses, L. "Elementary Decision Theory" Wiley Publishing Company, N.Y., 1959.

5. De-Groot, M.H. "Optimal Statistical Decisions" McGraw-Hill, N.Y., 1970.

6. Duda, R.D. and Hart, P.E. "Pattern Classification and Scene Analysis" Wiley, 1973.

7. Falk, G. "Computer Interpretation of Imperfect Line Data as a Three Dimensional Scene" Artificial Intelligence Memo 139, Stanford University, 1970.

8. Ferguson, T.S. "Mathematical Statistics, A Decision Analysis Approach" Academic Press, 1967.

9. Feldman, J.A. "A Way of Looking at Things" Dept. of Computer Science, Hebrew University, Jerusalem, January 1971.

10. Fu, K.S. "Sequential Methods in Pattern Recognition and Machine Learning" Academic Press, N.Y., 1968.

11. Griffith, A.K. "Mathematical Models for Automatic Line Detection" JACM, 62-80, 1973.

12. Guzman, A. "Computer Recognition of Visual Object in a Three-Dimensional Scene" M.I.T., 1969.

13. Harlow, C.A. and Eisenbeis, S.E. "The Analysis of Radiographic Images" Technical Report, Dept. of Electrical Engineering, University of Missouri-Columbia.

14. Hueckel, M.H. "An Operator Which Locate Edges in Digitize Pictures" Artificial Intelligence Memo 105, Stanford University, 1969.

15. Kelly, M.D. "Visual Identification of People by Computer" Artificial Intelligence Memo 30, Stanford University, 1969.

16. Nilsson, N.J. "Problem Solving Methods in Artificial Intelligence", McGraw-Hill.

17. Preparata, Franco P. and Ray, S.R. "An Approach to Artificial Nonsymoblic Cognition" Information Science, 4, 65-86, 1972.

18. Pingle, K.K. and Tenenbaum, J.M. "An Accomodation Edge Follower", IJCAI-2.

19. Quam, L.H. "Computer Comparison of Pictures" Stanford Artificial Intelligence Memo 44, Stanford University, 1968.

20. Raiffa, H. "Dicision Analysis - Introductory Lectures on Choice Under Uncertainty" Addison Wesley, July 1970.

21. Rosenfeld, A. "Picture Processing by Computer" Academic Press, N.Y. 1969.

22. Samuel, A.L. "Some Studies in Machine Learning Using the Game of Checkers II - Recent Progress" IBM Journal of Research and Development, 11, 601-617, 1967.

23. Slagle, J.R. and Lee, R.T.C. "Application of Game Tree Searching Techniques to Sequential Pattern Recognition" CACM, 103-110, Feb. 1971.

24. Thomas, A. "Color Perception" Stanford University 1973.

25. Thosar, R.B. and Samuel, A.L. "Some Preliminary Experiments in Speech Recognition Using Signa-

ture Table Learning" Artificial Intelligence Memo, Stanford University, (to appear).

26. Vicens, P. "Aspects of Speech Recognition by Computer" Artificial Intelligence Memo 85, Stanford University.

27. Waltz, D.L. "Generating Semantic Description from Drawing of Scenes with Shadows" M.I.T. TR-271.

28. Yakimovsky, Y. "Picture Processing, Use of Probabilistic Semantics for Region Growing and Recognition," Stanford University. (Forthcoming dissertation).

29. Cover, T.M. and Hart, P.E. "Newest Neighbor Pattern Classification" IEEE Trans. Information Theory, Vol. II-13, 21-27, 1967.

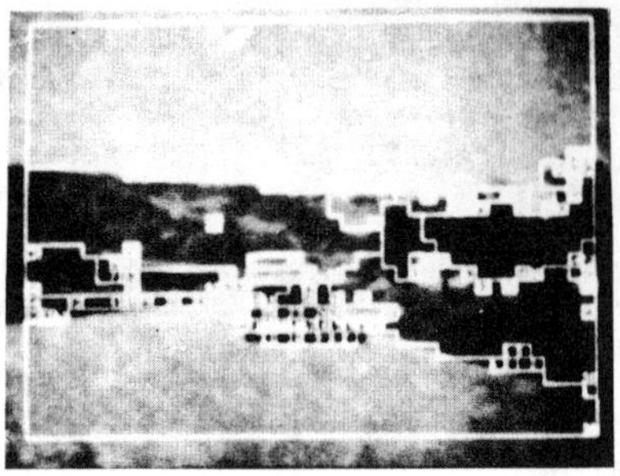

(A-3) The output of the region grower which melts weakest boundary first with non-semantics boundary strength evaluation. This is the result of stopping with the default stopping criterion.

(A-1) The output of the segmentation base on path connectivity when it is stopped by the default stopping criterion. The result is image segmented into a few hundred regions.

(A-4) Result of merging regions down to 30 regions using weakest boundary first algorithm and non-semantics boundary strength evaluation. Note that the top of the car is melted into the roadside vegetation.

(A-2) The effect of reducing the number of regions to 40 using path connectivity algorithm (using more liberal threshold than our current stopping threshold).

(A-5) Result of attempt to reduce the number of regions to 20 without using semantics (melting weakest boundary first non-semantic boundary strength evaluation).

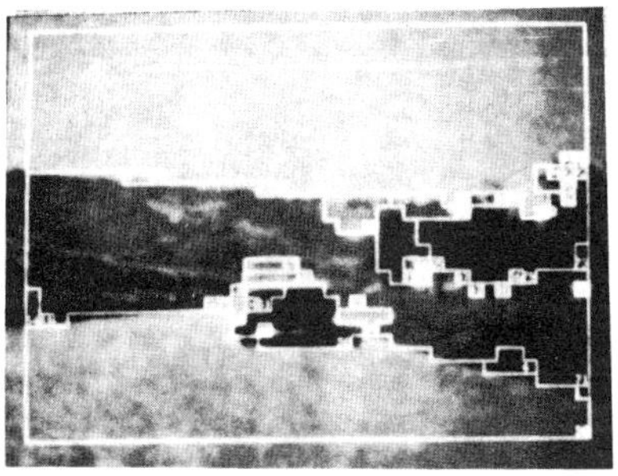

(A-6) Output of region grower based on semantics. (Melting weakest boundary first where boundary strength is computed using the semantic world model).

(A-7) Final grouping of regions based on the interpretation assigned to them by the world model. Regions whose meaning was assigned with confidence less than 10 are not mergable. They occur usually on the real boundary between two regions.

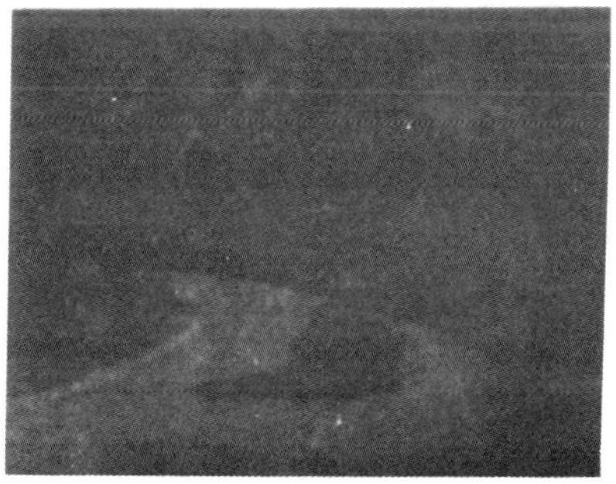

(B-1) Original picture.

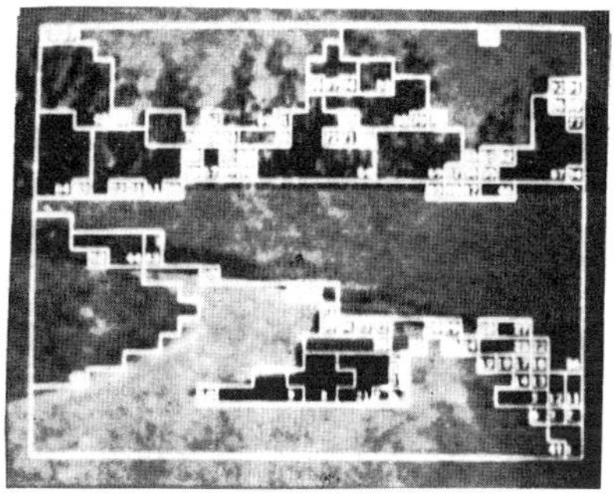

(B-2) Output of the non-semantic weakest boundary melted first region grower.

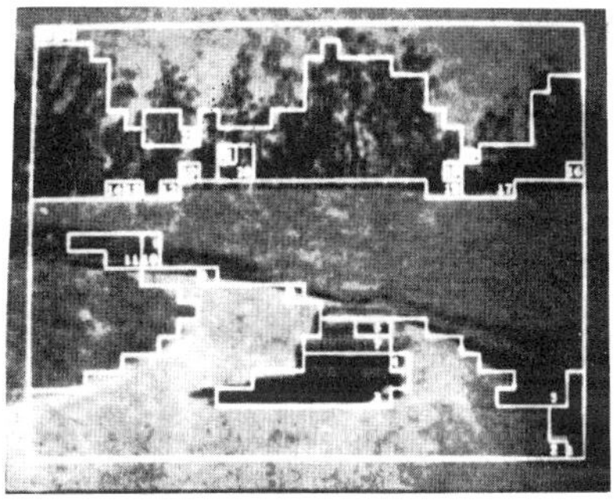

(B-3) Output of the semantic based region grower

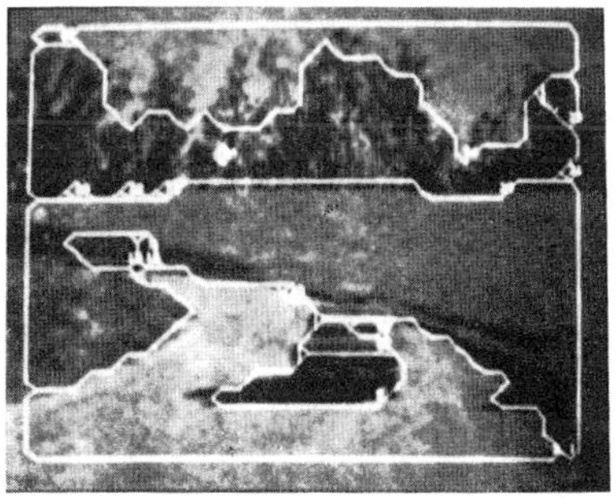

(B-4) Result of grouping regions by their assigned meaning. Taking only regions which were assigned meaning with confidence over 10 to be mergable.

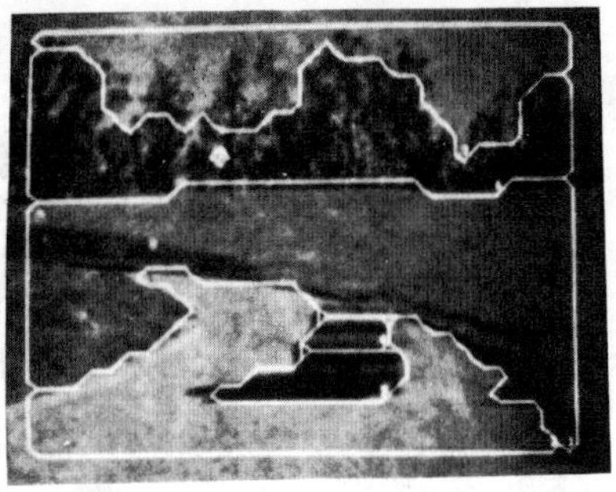

(B-5) Grouping regions by their assigned meaning, all regions considered mergable.

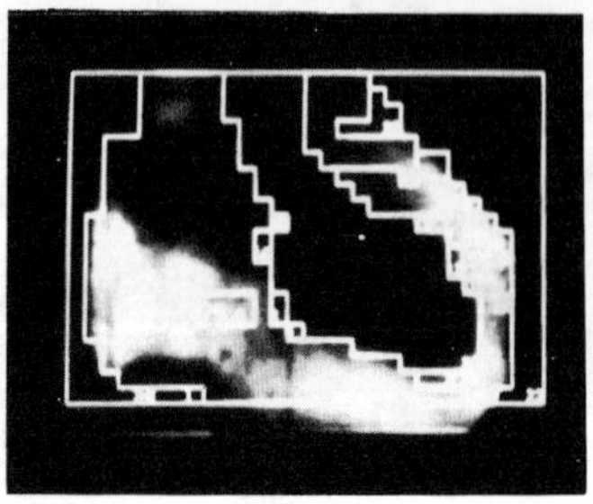

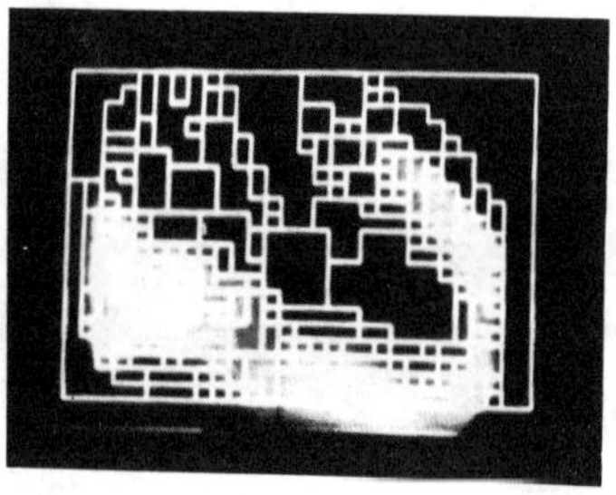

(F-1) Left ventricular angiogram. Output of the non-semantics weakest boundary first region grower. The stopping criterion is to stop when the merger gets down to two hundred regions.

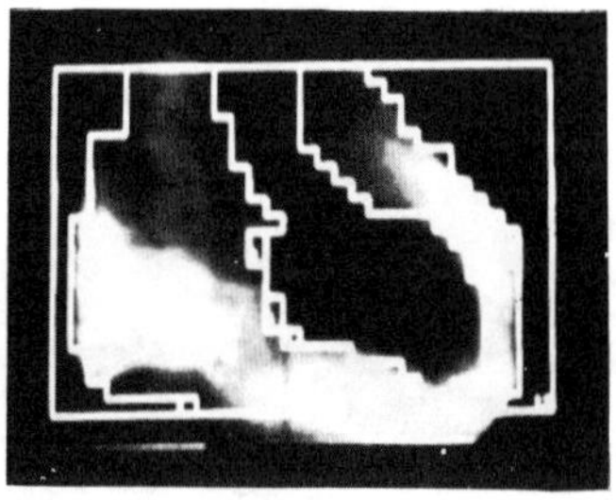

(F-2-3-4) Iterations of semantic region grower. The region grower used is grouping of all adjacent regions which are assigned the same meaning by the sequential assignment procedure, before the first assignment with low confidence level occurs. On each iteration the confidence threshold is lowered.

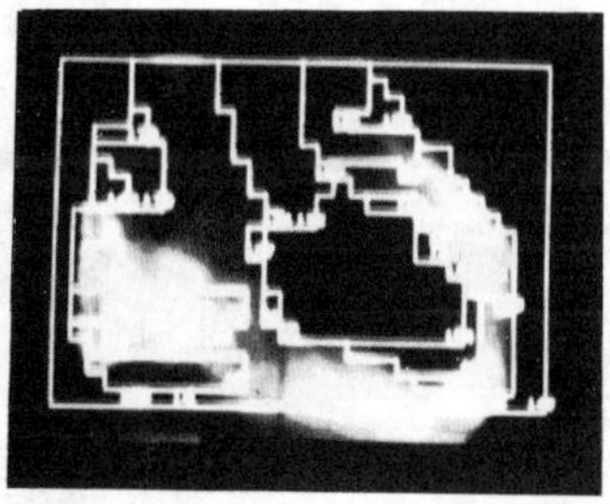

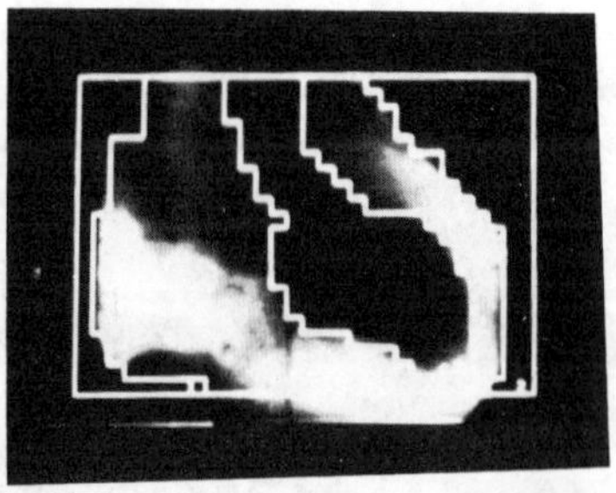

(F-5) Final output . The heart interior is the dark center, around it is the chest cavity and on the two sides there is the dark frame border.

IGS: A PARADIGM FOR INTEGRATING IMAGE SEGMENTATION AND INTERPRETATION*

by

J. M. Tenenbaum
H. G. Barrow
Artificial Intelligence Center
Stanford Research Institute
Menlo Park, California 94025

Summary

This paper presents a new approach for integrating the segmentation and interpretation phases of scene analysis. Knowledge from a variety of sources is used to make inferences about the interpretations of regions, and regions are merged in accordance with their possible interpretations.

The deduction of region interpretations is performed using a generalization of Waltz's filtering algorithm. Deduction proceeds by eliminating possible region interpretations that are not consistent with any possible interpretation of an adjacent region. Different sources of knowledge are expressed uniformly as constraints on the possible interpretations of regions. Multiple sources of knowledge can thus be combined in a straightforward way such that incremental additions of knowledge (or equivalently, human guidance) will effect incremental improvements in performance.

Experimental results are reported for a room scene using a human collaborator and a set of relational constraints as sources of knowledge. The experiment demonstrates that segmentation is much improved when integrated with interpretation. Moreover, the integrated approach incurs only a small computational overhead over unguided segmentation.

Introduction

A major goal of scene analysis has been the ability to partition images into regions that could then be interpreted as semantically important entities such as objects or surfaces. The conventional approach of performing segmentation followed serially by interpretation is frequently ineffective because semantically important pictorial features can be visually indistinct while visually prominent features can be artifacts in the context of the current task. Consider, for example, partitioning an image of a black and white cat on a black and white rug.

This paper presents a new approach, called Interpretation-Guided Segmentation (IGS), for integrating segmentation and interpretation in the analysis of natural scenes. Knowledge from a variety of sources is used to make inferences about the interpretations of regions, and regions are merged in accordance with their possible interpretations. General knowledge, domain-specific knowledge, and picture-specific knowledge (contributed interactively by a human collaborator) are used uniformly. Thus, the quality of the analysis can be progressively improved by incremental additions of knowledge.

Background

The key concept underlying IGS is that picture elements (pixels) can be clustered into regions more reliably if clustering is based on semantic interpretations rather than raw pictorial data. The development of this idea can be traced through almost a decade of region analysis research. The earliest algorithms, such as those of Brice and Fennema[1] and Barrow and Popplestone,[2] clustered raw brightness data and were consequently limited to blocks-world scenes where surfaces had uniform brightness and high contrast with adjacent surfaces. Additional pictorial features such as color and texture[3-8] extend the applicability of region growing but do not alter the fact that pictorial similarity is not always a reliable criterion for segmenting a scene into regions that correspond to objects.

Some of the first explicit attempts at clustering pixels on a semantic basis occurred in the analysis of satellite imagery. Individual pixels first were classified into crop categories by comparing their multispectral signatures with training sets and then were grouped with contiguous pixels of the same category.[9-10] Unsupervised classifiers were alternatively used to invent categories empirically.[10-11] Pixel classification was subsequently used for segmenting landscape scenes viewed from ground level.[12-13] Here, pixel attributes were inadequate for unique semantic interpretation; the context in which the attributes occur proved crucial.

Recent research has emphasized semantic interpretation at the region and interregion levels. Yakimovsky and Feldman,[14] for example, classified regions on the basis of brightness, hue, saturation, size, position in image, and overall shape as well as the length, shape, orientation, and contrast of their boundaries with adjacent regions. Harlow and Eisenbeis[15] used brightness, position in image, and the interpretations of adjacent regions to label areas in chest X rays.

Our work can be viewed as an extension of these efforts in two directions: The use of additional sources of knowledge for constraining region interpretations and the use of more effective deductive mechanisms for integrating knowledge. The central concern

*The research reported herein was supported by the National Aeronautics and Space Administration under Contract NASW-2086. Additional support was furnished by the Advanced Research Projects Agency under Contract DAHC04-75-C-0005.

Reprinted from *Proc. 3rd Int. Joint Conf. on Pattern Recognition*, Nov. 8-11, 1976, pp. 504-513.

in this paper is to promote the concept of IGS which is latent or subjugated in the works summarized above.

Interpretation-Guided Segmentation: An Overview

If interpretations could be assigned to every pixel, then segmentation would be reduced to the trivial process of collecting adjacent pixels with the same labels. There are two difficulties in automating interpretation at the pixel level; namely, the excessive volume of data and the absence of global attributes (e.g., shape, texture, and boundary relations). These attributes emerge only after a region structure has been imposed on the pixels, but without them, interpretation is usually ambiguous.

Following Yakimovsky,[14] the integration of segmentation and interpretation is accomplished in our system by proceeding incrementally. A scene is partitioned first into elementary regions consisting of individual pixels or perhaps groups of adjacent pixels with identical attributes. Beginning at this level, the system performs the most complete interpretation possible in the current partition. On the basis of this interpretation and previous knowledge about the domain, the system next merges the pair of adjacent regions that are least likely to represent different objects. The process then iterates by revising the interpretation to fit the current partition and performing another merge (see Figure 1).

The risk associated with a merge is evaluated by assessing the likelihood that the adjacent regions (or pixels) are fragments of the same object. Merges are never allowed between regions whose sets of possible interpretations are known to be disjoint because they must correspond to distinct objects. The safest merge combines regions that already have been assigned the same unique interpretation. It is also safe to merge adjacent regions whose interpretations, while not yet unique, have been narrowed to the point where previous knowledge constrains both regions to take the same interpretation. For example, suppose that Pictures were constrained to hang only on Walls and thus could never appear adjacent to Doors in an image. Two adjacent regions with Door and Picture as possible interpretations, could thus be safely merged since both regions must be interpreted either as parts of a Door or as parts of a Picture. If there are no safe merges, as defined by the above criteria, then the regions separated by the lowest contrast boundary are merged, provided that their possible interpretations are not disjoint. When the possible interpretations of all adjacent regions in the current partition are disjoint, the analysis terminates.

After each merge, the resulting partition is reinterpreted. This process assigns to each region a set of possible interpretations that are consistent with the region's attributes and also with at least one possible interpretation of each adjacent region. The reinterpretation is performed incrementally as follows: When regions merge, the resultant region initially inherits the possible interpretations shared in common by its parent regions. (These are obtained by intersecting the interpretation sets of the parent regions.) Some of these common interpretations may be incompatible with the expanded range of attribute values found in the enlarged region and can therefore be immediately ruled out. A small region, for example, will admit interpretation as either a small object or part of a large object, but a large region can correspond only to a large object.

Interpretations eliminated in the course of region merging may, in turn, allow semantically related interpretations to be dropped as possibilities of other regions. For example, if a newly merged region becomes too large to be a Chairseat, the possibility Chairback can be dropped for the region above it. These secondary eliminations may themselves propagate additional eliminations extending throughout the image. The deduction of region interpretations is explained more fully in the next section.

The interpretation process requires knowledge of how given region interpretations can appear in legal scenes. In the absence of such knowledge each region is assumed to have all possible interpretations. Hence, any adjacent regions can be merged, but no merge is guaranteed to be safe. The system will thus function as a conventional region-grower, merging regions in order of boundary contrast. Knowledge acts by constraining the possible interpretations of regions, and thereby prevents merges between semantically incompatible regions. Ideally, with sufficient knowledge, all permitted merges will be safe and therefore correct.

Deducing Region Interpretations with Relational Constraints

The process of deducing region interpretations using constraints is a generalization of Waltz's[16] filtering algorithm similar to that developed independently by Rosenfeld.[17] Every region initially is assigned a set of possible interpretations that are consistent with its own attributes. Deduction proceeds by eliminating those locally possible interpretations that are not consistent with any possible interpretation of an adjacent region. After a region interpretation is eliminated, the interpretations of all adjacent regions must be reexamined to determine whether they are still compatible with the surviving interpretations. Deductions may thus propagate, as in Waltz filtering.

The consistency of an interpretation is determined by relational constraints that specify, for each interpretation, the allowed interpretations for adjacent regions in a specified relationship. A region labeled Lake, for example, cannot appear above an adjacent region labeled Sky. For each pair of adjacent regions, a set of applicable relations is determined on the basis of properties of their common boundary. For example, one region may be spatially above and to the right of the other, with strong brightness contrast along the boundary. An interpretation is allowed if at least one interpretation of the other region simultaneously satisfies all the applicable relational constraints in conjunction with the interpretation being filtered.

An Example

The deduction of region interpretations by filtering is illustrated in Figure 2. The example involves an image of an empty room that has been correctly partitioned into six regions corresponding to the objects Floor, Wall, Door, Baseboard, Picture, and Doorknob. The problem is to determine an assignment of interpretations to regions that satisfies the constraints defined at the bottom left of Figure 2. Initially, every region is assigned all six possible interpretations, but immediately Picture and Knob are dropped from regions 1, 3, and 6 because their size violates constraint 4. This stage of labeling is shown in Figure 2(a). Regions are now filtered as pairs in order of region number, beginning with the pair R1, R2. Region 2 is located within region 1 and thus constraint 1 applies. This constraint can be satisfied in two ways: by instantiating region 2 to Picture and region 1 to Wall or region 2 to Knob and region 1 to Door. Consequently, Wall and Door are retained as possible interpretations for region 1 while Picture and Knob are kept as possibilities for region 2. Next, regions 1 and 3 are filtered by constraint 2. This constraint is satisfied when regions 1 and 3 are instantiated jointly as either Door and Wall, Wall and Door, or Door and Baseboard. The interpretation Floor, cannot appear beside any interpretation and is thus eliminated as a possibility for region 3. Finally, regions 1 and 5 are filtered by constraint 3, which is satisfied only when region 5 is instantiated as Floor or Baseboard.

The state of interpretation after filtering region 1 with all its neighbors appears in Figure 2(b). Region 2 is now due for filtering against its neighbor region 1, but because no interpretations have been eliminated from either region since the last time the pair was filtered, we proceed to region 3 and its neighbors. Applying constraint 1 (within) to regions 3 and 4, constraint 2 (beside) to regions 3 and 5, and constraint 3 (above) to regions 3 and 6, leaves the interpretation possibilities shown in Figure 2(c). The few remaining ambiguities are quickly resolved: Constraint 1 applied to regions 4 and 3 eliminates the interpretation Picture from region 4 and Constraint 3 applied to regions 5 and 1 eliminates the interpretation Door from region 1. At this point all pairs of regions have been filtered once. The process now reconsiders those pairs of regions whose interpretation sets have changed since the last time they were filtered. In particular, constraint 1 is reapplied to regions 1 and 2, and eliminates the interpretation Knob from region 2. The final (correct) interpretation of the scene is shown in Figure 2(d).

Experiments

Introduction

An experimental test of IGS was performed using a modified version of a previously described region analysis program.[12] The existing implementation in Fortran imposed three restrictions on the general paradigm:

(1) Each region was allowed a maximum of 18 possible interpretations, represented as a pattern of bits.

(2) Constraints were limited to binary valued relations between a pair of adjacent regions.

(3) The initial partition could contain a maximum of 1,000 elementary regions.

The experiment was performed on an elementary room scene, Figure 3, which resembles the idealized scene used in Figure 2. In this experiment, the six possible region interpretations--Wall, Door, Picture, Floor, Baseboard, and Doorknob--were constrained by eight relations defined by the boxes in Table 1. Each box gives for each interpretation of a region R1, the permissible alternative interpretations for a related region R2. For example, if region R1 is above R2, then R1 can be Floor only if R2 can also be Floor. On the other hand, if R1 is below R2, then R1 can be Floor provided R2 is either Floor, Door, or Baseboard. These constraints were compiled into the filtering program in the form of bit tables so that bits representing required interpretations could be rapidly matched with logical operations against those representing possible region interpretations. When several constraints apply between a pair of regions, the interpretation sets allowed by each constraint are intersected to obtain the set of interpretations that simultaneously satisfy all constraints.

The applicable relations between a pair of regions were determined in this experiment by factors that could be most easily extracted from an existing region data structure. Applicability of the relations above, below, and beside was based on the relative image coordinates of the regions' centers of mass and vertices of their bounding rectangles (derived from X, Y boundary extrema). A region R1, for example, is defined to be above a region R2 provided that its highest boundary point is higher in the image than the highest point of R2 and that its lowest point is higher than R2's centroid. It was also required that the horizontal extents of R1 and R2 overlap and that the size of both regions exceed 5 pixels. The two last requirements minimize the possibility that a relation will be prematurely applied at an early stage of partitioning. (Refer to the Shortcomings portion of the Discussion section.) Below is defined as the converse of above. Beside is a symmetric relation that applies when regions with vertical overlap are sufficiently displaced in a horizontal direction.

Adjacency is a universal relation that applies between any regions with a common boundary. Inside and outside refer to regions that are holes within other regions. These three relations are topological properties of the region data structure and not subject to the geometrical artifacts of merging. They are therefore applied regardless of region size.

The relation contrast applies whenever the average difference in brightness of two regions across their common boundary exceeds a conservatively large threshold, T1. The relation no-contrast applies when the difference is less than a second conservatively small threshold (T2). For the current scene, these thresholds were empirically set at T1 = 42 and T2 = 15 (assuming 256 brightness levels). The contrast constraint ensures that two adjacent and strongly contrasting regions will not receive the same interpretation if a surface with that interpretation is known to be approximately uniform in brightness. (It is assumed that all objects in the room domain except Picture have uniform brightness.) The relation no-contrast ensures that two regions with similar brightnesses will not

Table 1

RELATIONS GOVERNING INTERPRETATIONS OF ADJACENT REGIONS IN ROOM SCENE DOMAIN

R1 Above R2*	
R1	R2
Baseboard	Floor, Baseboard
Door	Knob, Floor, Door
Floor	Floor
Wall	Picture, Wall, Baseboard
Picture	Picture, Wall
Knob	Knob, Door

R1 Below R2	
R1	R2
Baseboard	Wall, Baseboard
Door	Knob, Door
Floor	Floor, Door, Baseboard
Wall	Picture, Wall
Picture	Picture, Wall
Knob	Knob, Door

R1 Beside R2	
R1	R2
Baseboard	Door, Baseboard
Door	Knob, Wall, Door, Baseboard
Floor	Floor
Wall	Picture, Wall, Door
Picture	Picture, Wall
Knob	Knob, Door

R1 Adjacent to R2	
R1	R2
Baseboard	Wall, Floor, Door, Baseboard
Door	Knob, Wall, Floor, Door, Baseboard
Floor	Floor, Door, Baseboard
Wall	Picture, Wall, Door, Baseboard
Picture	Picture, Wall
Knob	Knob, Door

R1 Contrasts with R2	
R1	R2
Baseboard	Knob, Picture, Wall, Floor, Door
Door	Knob, Picture, Wall, Floor, Baseboard
Floor	Knob, Picture, Wall, Door, Baseboard
Wall	Knob, Picture, Floor, Door, Baseboard
Picture	Knob, Picture, Wall, Floor, Door, Baseboard
Knob	Picture, Wall, Floor, Door, Baseboard

R1 No Contrast with R2	
R1	R2
Baseboard	Knob, Picture, Door, Baseboard
Door	Picture, Door, Baseboard
Floor	Knob, Picture, Wall, Floor
Wall	Knob, Wall, Floor
Picture	Knob, Picture, Floor, Door, Baseboard
Knob	Knob, Picture, Wall, Floor, Baseboard

R1 Inside R2	
R1	R2
Baseboard	Baseboard
Door	Door
Floor	Floor
Wall	Wall
Picture	Picture, Wall
Knob	Knob, Door

R1 Outside R2	
R1	R2
Baseboard	Baseboard
Door	Knob, Door
Floor	Floor
Wall	Picture, Wall
Picture	Picture
Knob	Knob

*Box lists the interpretations of Region R2 that are compatible with each interpretation of Region R1, given that R1 is above R2. Other relations are analogously defined.

receive different interpretations whose brightnesses are known to be significantly different--for example, Wall and Door.

Methodology

The image in Figure 3 was digitized to 8 bits at 60 X 60 resolution. All pixels were initially assigned the set of all possible interpretations. An arbitrarily chosen pixel within the baseboard area of the image was then manually assigned the unique interpretation Baseboard. This assignment was made to exclude explicitly the case where every region in the image receives the interpretation Picture (i.e., the image portrays a picture of a room scene rather than a room scene). Also, a thin vertically elongated rectangular region was drawn to isolate the bright specular reflection at the top of the image between the Door and Wall. This was done to avoid complicating the initial experimentation by introducing semantics to handle specularity. All pixels within the isolated region were assigned a special universal interpretation that both supports and is supported by any adjacent interpretation. With this interpretation, the anomalous region was effectively removed from the analysis because it could not participate in filtering or safe merges and could merge unsafely only with another region that has the universal interpretation.

The interpreted image array was partitioned into elementary regions composed of adjacent pixels with identical brightness (to 4 significant bits) and identical interpretations. The resulting partition, shown in Figure 4, contained 264 regions. At this stage, all regions admitted all possible interpretations, except regions 15 (baseboard) and 16 (universal).

The region interpretations were filtered using relational constraints applicable in the initial partition. The results of filtering are shown for selected regions in the caption of Figure 4. Note that many parts of the scene already have acquired unique interpretations. These parts include large areas of the Wall (regions 7-10) and Door (11), as well as the Baseboard (14), the Doorknob (12), and the lower (bright) half of the Picture (6). Many of the smaller regions contained within these areas are also labeled with the same interpretation as the containing region.

During filtering, eliminations propagated from the manually assigned Baseboard interpretation. The possibilities for region 14, adjacent to and noncontrasting with region 15 (known to be Baseboard) were immediately reduced to Door or Baseboard. Regions 10 and 14 could then be filtered by the relations above and contrast, thus leaving those regions with the unique interpretations Wall and Baseboard, respectively. The interpretation of region 13, beside and noncontrasting with Baseboard region 14, was then narrowed to the alternatives Door and Baseboard. The interpretation Wall propagated upward from region 10 to region 9 through the relations above and no-contrast, and subsequently to regions 5, 7, and 8. This, in turn, allowed region 6 to be interpreted as Picture since it is above and contrasting with region 9, now known to be Wall. Meanwhile, region 11, which is beside and contrasting with region 9 (Wall) and adjacent and noncontrasting with region 13 (Door or Baseboard), is uniquely constrained to be Door.

The initial stage of filtering leaves two main areas of the image with uncertain interpretations. Region 13 and its interior regions still admit the possibilities Door or Baseboard, while regions 1-6 in the upper left part of the scene can each be interpreted as either Wall or Picture. The Door/Baseboard ambiguity persists because regions 11 and 13 do not satisfy the formal conditions defining the relation above. The second ambiguity arises because of a brightness gradient across the Wall such that regions 5 and 8 do not fulfill the conditions for either contrast or no-contrast. Consequently, the interpretation Picture cannot be eliminated from region 5 and the resulting Wall/Picture ambiguity then propagates to the other regions in the area. A third and relatively minor area of ambiguity exists among the small regions on the border between Wall and Door. These regions, adjacent to both Door and Wall, are classified as either Door, Wall, or Baseboard.

Based on the interpretations surviving the initial filtering, approximately 200 safe merges are performed. The resulting partition containing about 68 regions, after filtering, yielded the results shown in Figure 5. The safe merges primarily involved adjacent regions already having the same unique interpretations. Regions in the upper part of the Wall with possible interpretations Wall and Picture also could be safely merged where the contrast constraint did not apply (because a Picture/Wall boundary is required to have contrast). Although the resulting partition appears much cleaner, the basic ambiguities described above persist. These ambiguities must now be resolved by postulating unsafe merges, based on the region brightnesses included in the caption of Figure 5.

The first unsafe merge significantly affecting interpretation occurred with approximately 43 regions remaining. Regions 6 and 9 (in Figure 5), with a contrast of 4, were merged into a single region with the unique interpretation Door (the intersection of the interpretation possibilities for regions 6 and 9) and an average brightness of 15. Next, with approximately 25 regions remaining, regions 1 and 4 (contrast 37) were merged to form one large region of Wall with brightness 87. As a result of this merge, the contrast relation could now be applied to eliminate the interpretation Wall from region 3. Finally, with about 20 regions left, the small regions, such as region 10 between Door and Wall, were merged unsafely into Wall. At this point, after a total of 43 unsafe merges and 214 safe merges, the analysis terminated with 11 regions remaining, all having unique and disjoint interpretations.

Results

The final partition and associated region interpretations are shown in Figure 6. Based on the limited semantics used in the experiment, the analysis is correct. A wall-mounted thermostat was fragmented into three regions (5-7), which were then interpreted as Pictures. A noisy pixel in the center area of the wall area was also assigned the interpretation Picture. These interpretations occurred because Picture was the only legal possibility for a contrasting region contained within a region labeled Wall. The interpretation errors could have been avoided by introducing explicit interpretations for Thermostat and Noise (which would be distinguished from Picture by additional constraints on region size). Finally, the so-called picture, actually a calendar, was split into two regions--a landscape and numeric data--because the spiral binding connecting them was invisible in the digitized image.

A point worthy of note is that no intrinsic classification of regions was made. With two minor exceptions, interpretations were derived entirely from context.

Acquisition and Debugging of Knowledge

The constraints used in this experiment were developed empirically. Incorrect or inadequate constraints manifest themselves ultimately by the elimination of all interpretations from some region. When this situation occurred, the analysis was immediately halted so that the constraints could be interactively refined. Unfortunately, because of the way eliminations propagate, the source of the error was often not the uninterpretable region. Hence, debugging was difficult.

There are two basic types of error: the elimination of a correct region interpretation and the unsafe merging of regions that should represent distinct objects. The first type of error occurs when a valid supporting interpretation is omitted from the definition of a relation or when a relation is inappropriately invoked. An erroneous unsafe merge occurs because an incorrect region interpretation was not eliminated early enough in the analysis and is therefore an indication that additional constraints are needed.

Discussion

Features

The scene analysis paradigm described in this paper has three main features. First, segmentation and interpretation are completely and effectively integrated; second, many diverse sources of knowledge can be used to guide the analysis; and third, the process of interpretation and its integration with segmentation can be made tolerably efficient.

Segmenting on the basis of interpretations rather than raw attributes allows retention of semantically important, but visually indistinct, boundaries (such as the Door/Baseboard boundary in Figure 6) and at the same time, deletion of high-contrast edges that are insignificant to the current task (such as interior details of the picture on the wall). A related advantage is the existence of a well-defined criterion for terminating the analysis--namely, when all adjacent regions have disjoint interpretations.

The more knowledge sources available, the quicker the filtering process will converge to unique region interpretations. Determining region interpretations at an early stage of partitioning reduces the need to resort to unsafe merges, which are a principal source of error. Two types of knowledge were used in the Experiments section to deduce region interpretations: manual interaction and relational constraints. Additional experiments have been performed in which interpretations were deduced from region attributes such as color and texture[12] and from a geometric model.[18] All these knowledge sources can be expressed uniformly as constraints on the possible interpretations of regions. Thus, they can be combined in a straightforward way. Many other sources of knowledge are expressible as constraints on the possible interpretations for given areas of a scene and can, therefore, also be integrated. In particular, this provides a possible mechanism for integrating partial descriptions of the scene produced by several independent modules, or even other scene-analysis programs.

In the current implementation, the overheads associated with region merging completely dominate the computation expended in filtering. The room scene experiment took 5 minutes on a KA-10 processor, of which only about 1 minute was spent deducing region interpretations. Consistent region interpretations are conventionally found by performing a case analysis or backtrack search.[19-21] Such search methods are ordinarily impractical on the scale with which we are dealing (hundreds of regions). Filtering, however, drastically reduces the number of cases that must be considered by eliminating inconsistent interpretations, once and for all, the first time the inconsistency is discovered. This also forestalls the characteristic thrashing of a backtrack search.

Shortcomings

The paradigm has some unresolved problems. The first concerns relations that apparently were applied appropriately, but that may later prove to have been inappropriate--after they have already been used to eliminate "correct" interpretations. (See Figure 7.) Again, because of the way eliminations propagate, this type of error is extremely difficult to trace.

A second fundamental issue concerns whether the IGS approach can be extended to less constrained domains. The existing system has been demonstrated in a variety of domains where each contain fewer than 10 objects and 10 relations. The real world contains millions of objects and millions of relations. Clearly, a more concise initial symbolic description at a general level is required, perhaps in terms of surface characteristics such as curvature (planar, convex, concave), orientation (vertical, horizontal), texture and material (e.g., metal, plastic, wood), and relations such as occlusion, joining, and support. Bajcsy's[22] work on symbolizing texture and that of Marr[23] and Paton[24] on symbolizing shading are relevant. Once the scene has been analyzed in general terms, that analysis can be used to guide a more detailed specific analysis.

In addition to the above conceptual problems, some implementational compromises were adopted for efficiency. First, constraints currently are restricted to be binary-valued (all or nothing) relations between pairs of adjacent regions. Clearly, the filtering process could be extended to handle not only relations among more than two entities that are not necessarily pictorially adjacent, but also relations that involve degrees of uncertainty. We have experimented with a LISP program called MSYS,[25] which performs region interpretations based on real-valued, procedurally represented constraints. However, MSYS has not yet been integrated with a segmentation program to perform a complete scene analysis. Second, there is no provision for backup in the event of an erroneous unsafe merge. Elsewhere, we have conjectured that a vision system with <u>enough</u> knowledge would not need backup,[26] but this ideal has yet to be realized in practice. Finally, candidate regions for unsafe merges are currently selected on the basis of average color contrast. Comparing the textures and brightness gradients of regions, in addition to their average colors, should significantly improve the basic decision regarding whether two regions belong to the same surface and thereby reduce the risk of an erroneous merge.

In spite of the shortcomings, the apparent facility with which multiple sources of knowledge may be integrated and the resulting improvements observed in segmentation make Interpretation-Guided Segmentation a promising paradigm for future research and application.

Acknowledgments

The authors wish to acknowledge the help of Stephen Weyl, who programmed most of the FORTRAN region-finding system, and Helen Wolf, who provided valuable additional programming support.

References

1. C. Brice and C. Fennema, "Scene Analysis Using Regions," Artificial Intelligence, 1, pp. 205-226 (1970).

2. H. Barrow and R. Popplestone, "Relational Descriptions in Picture Processing," in Machine Intelligence, 6 (Edinburgh University Press, Edinburgh, Scotland, 1971).

3. M. Yachida and S. Tsuji, "Application of Color Information to Visual Perception," Pattern Recognition, 3, p. 307 (1971).

4. J. L. Muerle and D. C. Allen, "Experimental Evaluation of Techniques for Automatic Segmentation of Objects in a Complex Scene," in Pictorial Pattern Recognition, Cheng et al. (Eds.), pp. 3-14 (Thompson Book Co., Washington, D.C., 1968).

5. A. Rosenfeld and M. Thurston, "Edge and Curve Detection for Visual Scene Analysis," IEEE Trans. Computers, C-21, p. 582 (1971).

6. A. Rosenfeld, M. Thurston, and Y. Lee, "Edge and Curve Detection: Further Experiments," IEEE Trans. Computers, C-21, p. 677 (1972).

7. F. Tomita, M. Yachida, and S. Tsuji, "Detection of Homogeneous Regions by Structural Analysis," Proc. 3IJCAI, pp. 564-571 (August 1973).

8. R. Ohlander, "Analysis of Natural Images," Ph.D. Thesis, Department of Computer Science, Carnegie-Mellon University, Pittsburgh, Pennsylvania (April 1975).

9. J. Gupta and P. Wintz, "Multi-Image Modeling," TR-EE 74-24 School of Electrical Engineering, Purdue University, Lafayette, Indiana (September 1964).

10. T. V. Robertson, K. S. Fu, and P. H. Swain, "Multispectral Image Partitioning," Information Note 071373, Laboratory for Application of Remote Sensing, Purdue University, Lafayette, Indiana (1973).

11. A. G. Wacker and D. A. Landgrebe, "Boundaries in Multispectral Imagery by Clustering," 1970 IEEE Symposium on Adaptive Processes, University of Texas, Austin, Texas (December 1970).

12. J. M. Tenenbaum and S. A. Weyl, "A Region Analysis Subsystem for Interactive Scene Analysis," Proceedings Fourth International Joint Conference on Artificial Intelligence, Tblisi, USSR (September 1975).

13. A. Hanson, E. Riseman, and P. Nagin, "Region Growing in Textured Outdoor Scenes," Technical Report 75C-3, Department of Computer and Information Science, University of Massachusetts, Amherst, Massachusetts (February 1975).

14. Y. Yakimovsky and J. Feldman, "A Semantics-Based Decision Theoretic Region Analyzer," Proc. 3IJCAI, pp. 580-588 (1973).

15. C. Harlow and S. Eisenbeis, "The Analysis of Radiographic Images," IEEE Trans. Computers, C-22, pp. 678-688 (1973).

16. D. G. Waltz, "Generating Semantic Descriptions from Drawings of Scenes with Shadows," AI TR-271, Artificial Intelligence Laboratory, Massachusetts Institute of Technology, Cambridge, Massachusetts (August 1972).

17. A. Rosenfeld, R. Hummel, and S. Zucker, "Scene Labelling by Relaxation Operations," TR-379, Computer Science Department, University of Maryland, College Park, Maryland (1975).

18. J. M. Tenenbaum and H. G. Barrow, "Experiments in Interpretation Guided Segmentation," Technical Note 123, Artificial Intelligence Center, Stanford Research Institute, Menlo Park, California (March 1976).

19. R. O. Duda, "Some Current Techniques for Scene Analysis," Technical Note 46, Artificial Intelligence Center, Stanford Research Institute, Menlo Park, California (October 1970).

20. A Guzmán, "Analysis of Curved Line Drawings Using Context and Global Information," Machine Intelligence, 6, pp. 325-376, B. Meltzer and D. Michie (eds.) (Edinburgh University Press, Edinburgh, Scotland, 1971).

21. Y. Y. Yakimovsky, "On the Recognition of Complex Structures: Computer Software Using AI Applied to Pattern Recognition," Proc. International Joint Conference on Pattern Recognition, pp. 345-353, Copenhagen, Denmark (August 1974).

22. R. Bajcsy, "Computer Description of Textured Surfaces," Proc. 3IJCAI, pp. 572-579 (August 1973).

23. D. Marr, "Early Processing of Visual Information," AI Memo No. 340, AI Laboratory, Massachusetts Institute of Technology, Cambridge, Massachusetts (December 1975).

24. K. Paton, "Picture Description Using Legendre Polynomials," Computer Graphics and Image Processing, 4, No. 1, pp. 40-54 (March 1975).

25. H. Barrow and J. M. Tenenbaum, "MSYS: A System for Reasoning About Scenes," Tech. Note 121, Artificial Intelligence Center, Stanford Research Institute, Menlo Park, California (March 1976).

26. H. Barrow and J. M. Tenenbaum, "Representation and Use of Knowledge in Vision," Tech. Note 108, Artificial Intelligence Center, SRI, Menlo Park, California (July 1975). (Also published in SIGART, June 1975).

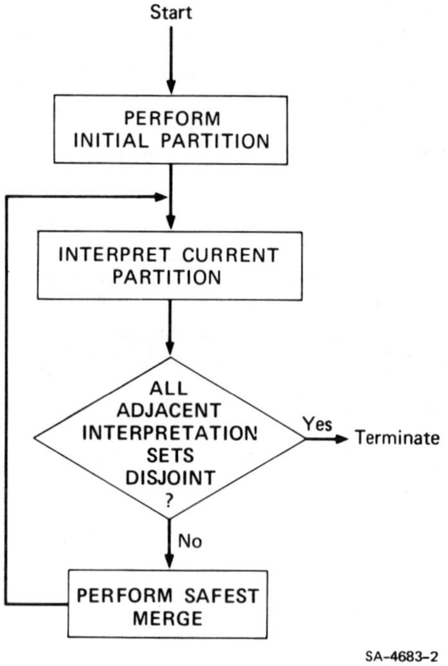

FIGURE 1 OVERVIEW OF INTERPRETATION-GUIDED SEGMENTATION PARADIGM

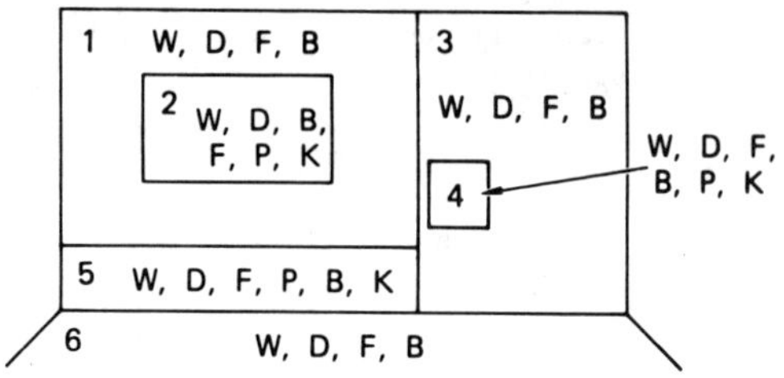

(a) POSSIBLE REGION INTERPRETATIONS* BASED ON REGION SIZE CONSTRAINTS

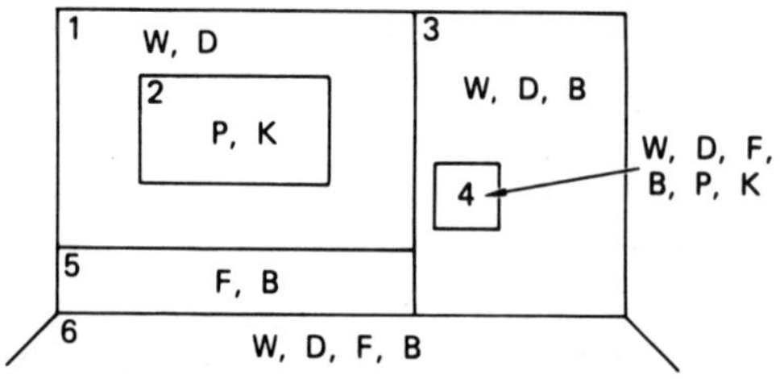

(b) INTERPRETATIONS SURVIVING AFTER REGION 1 HAS BEEN FILTERED AGAINST ALL ADJACENT REGIONS

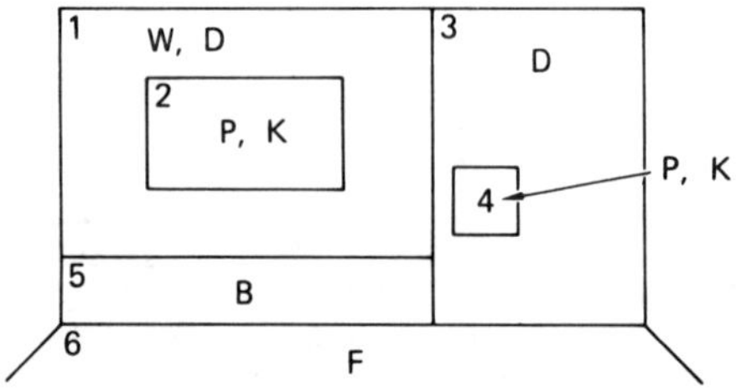

(c) INTERPRETATIONS REMAINING AFTER FILTERING REGION 3

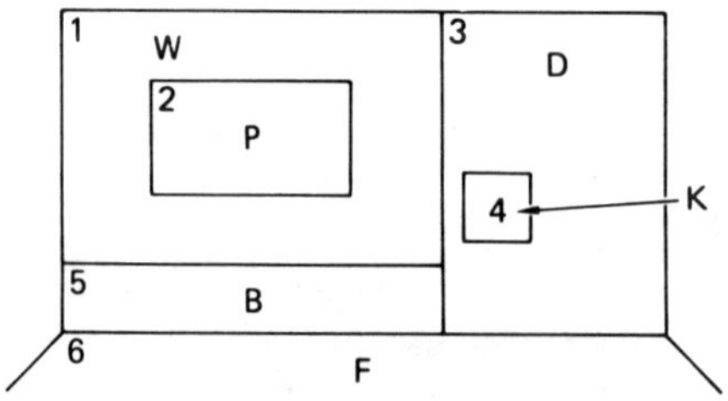

(d) FINAL (CORRECT) REGION INTERPRETATIONS

*Legend: D-Door, W-Wall, F-Floor, P-Picture, B-Baseboard, K-Doorknob

Constraints:
1. Within (Picture within wall) (Doorknob within door)
2. Beside (Wall beside door) (Baseboard beside door)
3. Above (Wall above baseboard) (Baseboard above floor) (Door above floor)
4. Size (Small doorknob) (Small picture)

APPLICABLE CONSTRAINTS BETWEEN ADJACENT REGIONS

REGION	1	2	3	4	5	6
1	X	W	B	X	A	X
2	W	X	X	X	X	X
3	B	X	X	W	B	A
4	X	X	W	X	X	X
5	A	X	B	X	X	A
6	X	X	A	X	A	X

W - Within/Contains
A - Above/Below
B - Beside

SA-4683-5

FIGURE 2 DEDUCING REGION INTERPRETATIONS USING RELATIONAL CONSTRAINTS

SA-4683-24

FIGURE 3 DIGITIZED IMAGE OF ROOM SCENE (8 BITS AT 256 x 256 RESOLUTION)

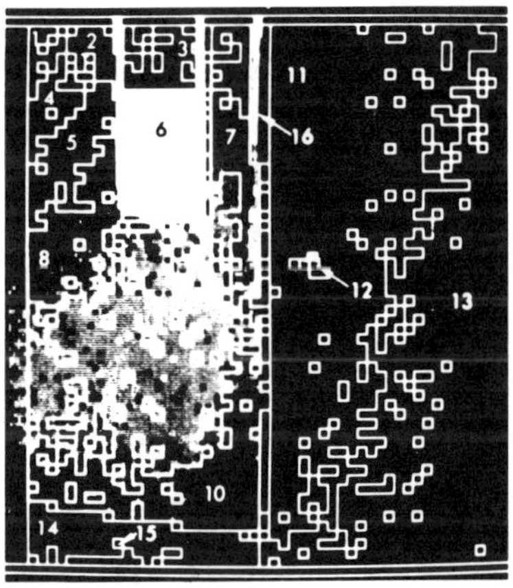

INTERPRETATION POSSIBILITIES
FOR SELECTED REGIONS
FOLLOWING INITIAL FILTERING

Region	Possible Interpretations
1-5	Picture Wall
6	Picture
7-10	Wall
11	Door
12	Knob
13	Door, Baseboard
14-15*	Baseboard
16*	Universal

*Manually assigned interpretation.

SA-4683-25

FIGURE 4 INITIAL PARTITION OF ROOM SCENE (264 REGIONS BASED ON 4 SIGNIFICANT BITS OF BRIGHTNESS AT 60 x 60 RESOLUTION)

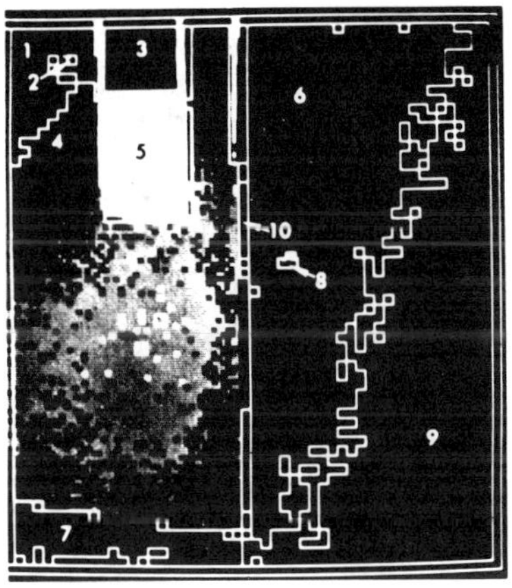

INTERPRETATION POSSIBILITIES
FOR SELECTED REGIONS
FOLLOWING REFILTERING

Regions	Possible Interpretations	Average Brightness
1	Picture, Wall	52
2	Picture	188
3	Picture, Wall	16
4	Wall	89
5	Picture	182
6	Door	13
7	Baseboard	27
8	Knob	64
9	Door, Baseboard	17
10	Wall, Door	48

SA-4683-26

FIGURE 5 ROOM SCENE PARTITION AFTER 200 SAFE MERGES

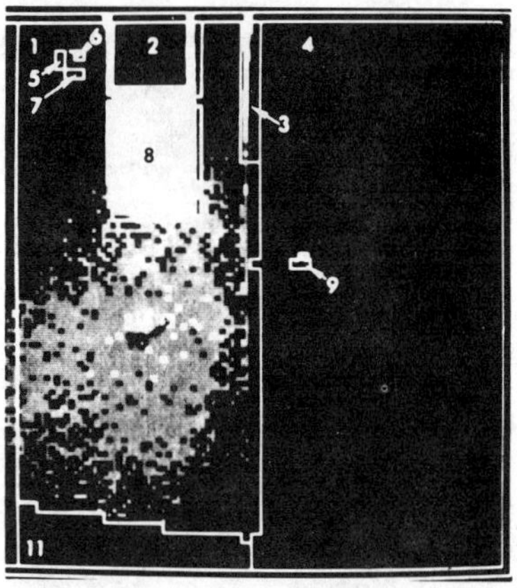

FINAL REGION INTERPRETATIONS

Region	Interpretation
1	Wall
2	Picture
3	Universal
4	Door
5	Picture
6	Picture
7	Picture
8	Picture
9	Knob
10	Picture
11	Baseboard

SA-4683-27

FIGURE 6 FINAL PARTITION OF ROOM SCENE

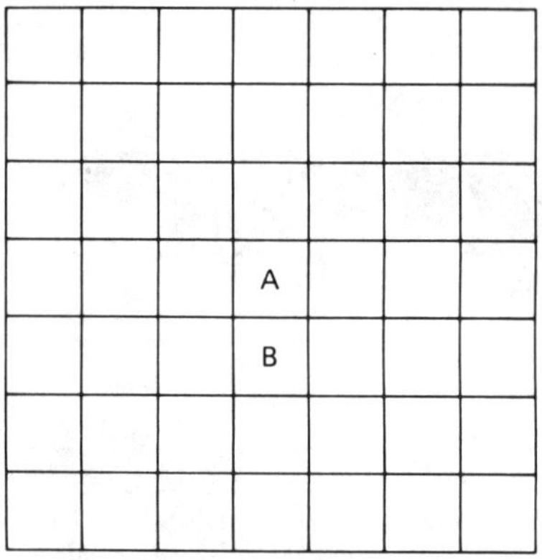

(a) SECTION OF IMAGE AT AN EARLY STAGE OF PARTITIONING (REGION A ABOVE REGION B)

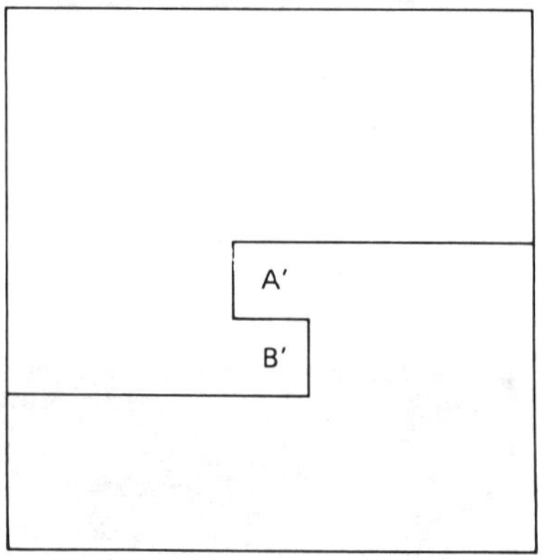

(b) SAME SECTION OF IMAGE: REGIONS A AND B HAVE BEEN MERGED INTO REGIONS A′ AND B′ RESPECTIVELY, WITH (REGION B′ ABOVE REGION A′)

SA-4683-23

FIGURE 7 INAPPROPRIATE USE OF THE RELATION "ABOVE" AT AN EARLY STAGE OF PARTITIONING

RELAXATION LABELLING AND THE REDUCTION OF LOCAL AMBIGUITIES

Steven W. Zucker*
Computer Science Center
University of Maryland
College Park, MD 20742

Abstract

The use of knowledge in the form of constraint expressions arises very naturally in many problem solving systems and in many machine perception systems. Knowledge in such a form is useful at various levels in vision systems, particularly in the resolution of local ambiguities or uncertainties. For example, ambiguities may arise when an original intensity image is interpreted into a low-level, symbolic description. To reduce this uncertainty, it is often sufficient to use processes which are restricted to a purely local perspective. Relaxation labelling is one such class of processes. The knowledge that it employs is in the form of compatibility relationships between neighboring objects. Abstractly, relaxation labelling is a class of computational processes which manipulate labels on graphs. This paper describes relaxation labelling processes and several of their applications in vision systems. It is shown that they can provide consistent sets of local symbols to serve as enhanced input for more active, global processes.

1. Introduction

Computing correct interpretations for local symbolic variables is an integral part of many problem-solving and understanding systems. Ambiguities in these variables arise when each may assume one of a large set of possible values. In order to restrict this set of possibilities, that is, to reduce local ambiguities, additional sources of knowledge must be used[33]. One very natural source arises when considering interactions between the variables. Such knowledge can be represented in the form of constraint or compatibility expressions. This paper describes the use of knowledge in this form by a class of computational processes, called relaxation labelling processes, to reduce or remove ambiguities in local descriptions. Many different kinds of problems fit nicely into this general format; see, e.g.,[7,10,12,14,19,26].

The main thrust of this paper is that there are many situations which arise within vision systems, particularly at the lower levels, which also fit into this general paradigm. It must be noted at the outset, however, that there is an important difference between the relaxation labelling algorithms to be developed here and the traditional approaches to constraint satisfaction. Most of the computational mechanisms developed previously (such as generate and test, backtrack programming, and sequential filtering) were sequential in nature*. However, in vision systems there are many places in which sequential algorithms create difficulties. For one, there is the problem of selecting a starting point for the algorithm and of determining whether the final result is independent of it[40]. There is also the problem of handling the very large amounts of data which must be processed, particularly in the lower levels of vision, and of maintaining the resultant large numbers of potential hypotheses and their alternatives. These difficulties, together with the prevalence of parallelism in human low-level vision[5], provide sufficient incentive to develop parallel relaxation labelling processes.

2. Relaxation Labelling Processes

Many different kinds of symbolic objects, at various levels of abstraction, arise in the course of understanding an intensity array. For our present purposes it is convenient to think of these objects as being assigned to positions in the visual array. This induces a neighbor relation over the set of objects and suggests a graph representation. In such a graph the nodes correspond to the objects and the edges indicate neighboring pairs of objects. Attached to each object-node is a set of labels indicating possible semantic interpretations for that object. The relaxation labelling process attempts to decide, on the basis of local evidence, which of these possible interpretations is the correct one.

Selection of the label set for an object determines the space of possible interpretations for (or relevant assertions about) that object. For example, we will consider the interpretation of picture points into a space of oriented line segments, and the interpretation of lines into a space of physical edge configurations. Both of these interpretive spaces consist of symbols which provide a basis for describing (or computing) compound objects within a larger interpretive universe (see also Marr's remarks about the principle of explicit naming[22]). Namely, long curves can be constructed from oriented unit line segments and solid objects can be described by their edges. More generally, relaxation labelling processes should be useful whenever labels are being assigned to data (i.e., assertions made about the data) and whenever knowledge is available about how these labels

---

*Current address: Dept. of Electrical Engineering, McGill University, Montreal, Quebec H3C 3G1, Canada

*Mackworth[21] gives an excellent review of these techniques, with important insights into their inherent computational efficiency.

Reprinted from *Proc. 3rd Int. Joint Conf. on Pattern Recognition*, Nov. 8–11, 1976, pp. 852-861.

can interact. These labels need not fit as neatly into a descriptive hierarchy as the above examples indicate, even in low-level vision. Another, more abstract possibility is suggested by stereo imaging problems, in which it may be useful to define second order labels such as positional difference or relative error symbols.

Once the label set and semantic universe are determined, it is necessary to evaluate a priori the ways in which the labels can interact. This knowledge is represented in the constraint or compatibility functions.

Given an input in this form, i.e., a graph with a label set attached to each node, relaxation labelling processes use the constraint expressions to eliminate node labels which are inconsistent with the labels still residing on neighboring nodes. This is done in parallel on every node in the graph and then re-iterated until there is no change in the label set. The final labelling ordinarily contains a smaller number of interpretations for each object than was originally possible. With fewer local interpretations, the combinatorial number of possible global interpretations which can be synthesized from them is reduced. The reduction of local ambiguities in this manner simplifies the design of the algorithms which subsequently build the global interpretations. It also improves their operational efficiency, because some of the (incorrect) data which they would have had to process have already been eliminated. These general remarks will become more clear in the next section, in which we begin to discuss relaxation labelling processes in some detail.

2.1 Discrete Relaxation Labelling

Semantic labels can be assigned to objects in an all-or-none fashion in which the label is either appropriate or not, or they can be assigned with some level of certainty (or appropriateness) attached to them. We will first consider the all-or-none case. This will lead to the discrete relaxation labelling process, a parallel, iterative version of the Waltz algorithm[38].

Discrete relaxation labelling will be developed in the context of simplified line drawing interpretation, in which the problem is to build the interpretation of the entire drawing from the interpretations of its component lines. This simplified domain is used because it offers a very clear picture of what is going on in the abstract symbol-manipulating processes. The particular universe of interpretation is illustrated in Figure 1a (after [17, 4]). In this universe a line can denote (i) a convex edge, (ii) a concave edge, (iii-iv) an occluding edge with the forward object below or above*. It is now straightforward to determine all pairs of labels ($\Lambda_{ij}(\lambda,\lambda')$) which can actually occur on neighboring line segments (i.e., line segments which meet at a vertex); see Figure 2.

*Note that while (iii) and (iv) are rotationally equivalent, for clarity in this example it is convenient to label them distinctly.

Suppose that, as an example, the particular input to the discrete relaxation process is the one shown in Figure 1b. It involves the set of objects a_i ($i = 1,2,3$), the neighbor relationships, and the initial labelling. In principle, these initial label sets were determined a posteriori. That is, measurements were made on the original image to determine which labels were appropriate.* The label set for each object a_i will be denoted Λ_i.

The measurements for sides a_2 and a_3 were ambiguous, resulting in the labels shown. Side a_1 was impossible to observe, leaving only the default alternative of assigning it the complete label set Λ.

With this as input, the relaxation process discards labels according to the following rule: Keep the label λ at node a_i if and only if for every neighbor a_j there is a label $\lambda' \in \Lambda_j$ which is compatible with λ (i.e., $\Lambda_{ij}(\lambda,\lambda') \neq \Phi$). Beginning with the initial label set at each node, this rule is applied to every node in parallel until no additional labels are discarded. (For a proof that this procedure always terminates, see[30].) If this sequence of operations is performed on the input picture in Figure 1b using the constraints in Figure 2, then the final labelling in Figure 3a is obtained. Note that while sides a_1 and a_2 still remain uncertain of their interpretation (but less so), side a_3 is now completely certain. It has been entirely determined by information contained in the distribution of labels in its neighborhood.

The size of the neighborhood is a function of the iteration number. It grows radially outward as the iteractions proceed. This implies that the number of iterations required for termination is determined by the maximum distance that information must travel in order to disambiguate the worst (i.e., the least locally determined) node.

Finding the legitimate global interpretations for Figure 3a should now be easier because there are fewer possible interpretations for each side. One approach is to use the labelling network (Figure 3b), a graph in which the nodes correspond to object-label pairs and in which two nodes, (a_i,λ) and (a_j,λ'), are joined by an edge if and only if $(\lambda,\lambda') \in \Lambda_{ij}$. Closed cycles in this graph which span the nodes then correspond to consistent global interpretations -- these are drawn in Figure 3c.

*For an attempt to measure the information contained in local intensity configurations suggestive of interpretations such as these, see B.K.P. Horn, Image Intensity Understanding, A.I. Memo 335, M.I.T., August, 1975.

There is, however, a limitation inherent in the relaxation labelling process which arises because of its restricted local perspective. This can be demonstrated by attempting to label a Penrose impossible figure[28] (Figure 4a). Note that while this figure is well-formed locally, in that each corner could actually occur in a real triangle, it is ill-formed globally. There is no single point of view from which a real triangle would give this projection.

Since the irregular aspect of this figure concerns the depth orientations of its sides, the label set describes these orientations as either into (λ_I) or out of (λ_O) the plane of the paper. To obtain the constraint set, it is necessary to examine the ways in which these labels can be combined at a vertex. It is easy to see (by examining a real triangle) that all pairwise combinations are possible. Thus the constraint set is

$$\Lambda_{ij}(\lambda,\lambda') = \{(\lambda_O,\lambda_O),(\lambda_I,\lambda_I),(\lambda_O,\lambda_I),(\lambda_I,\lambda_O)\}$$

The process begins with both labels on every side; each label is then consistent with one of the labels on each of its neighbors, so no labels are discarded. The labelling network, however, immediately reveals the unusual side configuration for the Penrose figure (Figure 4c) -- namely, that the object-spanning cycles over this network are of length six. This is clearly impossible for a figure with only three sides.

2.2 Probabilistic Relaxation Labelling

Often, during the understanding process, interpretations are not just possible or impossible, but are possible with some degree of certainty. Such is the case in low-level vision, where the problem is to interpret the intensity array into a vocabulary of low-level symbols. The typical approach to the problem has been to evaluate a local feature detector on the intensity array, and then, on the basis of this measurement, to assert the existence of an actual feature in the original picture. However, most feature detectors are not uniquely invertible[22]. They respond not only to the presence of the feature, but also to various noise configurations. This often makes it difficult to confidently assert the existence of pictorial features solely on the basis of single detector measurements. The probabilistic relaxation labelling process introduced in this section is extremely useful at this level in vision systems, because it can be used to disambiguate feature detector responses on the basis of neighboring feature detector responses. At the same time it aids the translation of the original array of intensity values into another kind of data structure. In this new data structure, symbolic assertions are attached to spatial locations derived from the original array.

The probabilistic relaxation process can be thought of as a generalization of the discrete relaxation process. The discrete model was developed on the basis of label sets attached to nodes in a graph. Labels inconsistent with their neighbors were discarded from these label sets. This discarding process can be modeled by attaching an additional binary variable to each label. When the label is present, the variable is set to one; when the label is discarded it is changed to zero. The probabilistic relaxation process can now be described as an extension of the discrete relaxation process in which the range of this variable is extended to the interval [0,1]. This variable, denoted $p_i(\lambda)$, can now be interpreted as an estimate of the probability that label λ is correct for object a_i. In other words, it serves as a measure of our confidence in assigning label λ to object a_i.

Now, instead of discarding the labels, the probabilities on the labels are updated. This is done using heuristic knowledge embedded in what are termed compatibility functions. Compatibility functions are the probabilistic counterparts to the discrete constraint relations. More specifically, the compatibility function $r_{ij}(\lambda,\lambda')$ controls the contribution that the probability attached to label λ' on object a_j makes to the probability attached to label λ on object a_i during the updating process. In a loose sense, the compatibility function resembles a correlation function, because it assumes values according to the following benchmarks:

$$r_{ij}(\lambda,\lambda') = \begin{cases} -1 & \text{if } \lambda \text{ at } a_i \text{ is incompatible with } \lambda' \text{ at } a_j \\ 0 & \text{if } \lambda \text{ at } a_i \text{ is independent of } \lambda' \text{ at } a_j \\ +1 & \text{if } \lambda \text{ at } a_i \text{ is perfectly compatible with } \lambda' \text{ at } a_j \end{cases}$$

Since the compatibility can assume both positive and negative values, both positive and negative influences can propagate around the graph. This often has the effect of organizing the different interpretations into sets of cooperating and competing[24] subprocesses.

With the compatibility functions specified (often in a procedural form), the probabilities are updated according to an appropriate rule, e.g.,

$$p_i^{k+1}(\lambda) = \frac{p_i^k(\lambda)\ [1+q_i^k(\lambda')]}{\sum_\lambda p_i^k(\lambda)[1+q_i^k(\lambda')]} \qquad (1)$$

$$\text{where } q_i^k = \sum_j c_{ij} \sum_{\lambda'} r_{ij}(\lambda,\lambda')\ p_j^k(\lambda') \qquad (2)$$

This rule states that the probability for label λ on object a_i at the (k+1)st iteration is a function of both the previous estimate for that probability, $p_i^k(\lambda)$, and the contribution from the probability distributions on the neighboring label sets, $q_i^k(\cdot)$.

The most straightforward way in which the neighboring probabilities can contribute to the updating process is shown in Eq. 2. For each neighbor j, the probability on each label λ' is weighted by the compatibility function, and the result is then summed. The total neighborhood contribution is then a weighted (c_{ij}) sum of all these neighbor contributions. Thus the more certain a label is, the more influence it has. More complex schemes could consider pairwise (or higher order) combinations of the probabilities by incorporating conditional expressions into the above rule[41].

Once the neighborhood contribution has been determined, it enters the updating rule in a nonlinear fashion (i.e., the term $[1+q_i^i]$). A nonlinearity is essential to the process because it can be proven that linear schemes iterate to terminal values which are independent of the initial probabilities[30]. Thus linear processes are independent of any a posteriori information. The denominator in Eq. 1 guarantees that the updating rule maps probabilities into probabilities.

In this probabilistic model, as in the discrete model, the process is limited to a local perspective on the neighborhood of each node. The sphere of influence for each node does, however, grow with the iteration number.

3. Applications of Probabilistic Relaxation Labelling

3.1 Line and Curve Enhancement

The first applicaton of the probabilistic model will be the enhancement of lines and curves. Consider interpreting an intensity array into the extremely simple symbolic universe of oriented line segments. A picture point can either correspond to a unit line segment in any of eight orientations, or it can correspond to a background point (the no-line label)[41]. See Figure 5.

The underlying object graph for this application is composed of the individual picture points, with each point joined to its eight nearest neighbors. The label probabilities are initialized by evaluating a local line detector[32] in eight orientations at every point. The results are scaled so that a single strong response gives a large value to the appropriate line-label probability; otherwise, the weight is distributed proportionally over all the line labels. The no-line probability is one minus the sum of the line probabilities.

The line-line compatibilities are shown in Figure 5b. Line segments oriented in the same direction support one another, while perpendicular line segments detract support. Since it is possible for curves to make right angle turns, however, the negative coefficient is smaller than the maximum permited. The no-line label is supported by neighboring no-line labels and line segments not directed toward (or away from) it, and it is weakened by line segments directed toward it.

The result of applying this process to a satellite photograph of a winding river is shown in Figure 6. While the local line detector responds to some extent everywhere in the noisy background (Figure 6b), the relaxation process enhances the river quite dramatically (Figure 6f). It is now a simple matter to design a tracking routine to actually follow the river.

The next figure (Figure 7) shows a satellite picture of two crossing roads. The composition of one of these roads changes from concrete to asphalt, and, because of this change in reflectivity, it is extremely difficult to detect solely from the line detector outputs. The relaxation process, however, enhances the asphalt portions so that they are directly comparable to the more distinct lengths of road. Once again these enhanced roads would be easy to track.

The curve enhancement application makes apparent two additional and more general aspects of probabilistic relaxation processes. The more detailed paper[41] demonstrated the insensitivity of the process to proportional changes in the compatibility coefficients. In other words, the final result is not particularly sensitive to specific numerical values, but rather is sensitive to their proportional relationships. If this were not the case, the usefulness of these processes would be severely limited. For then, it would be insufficient to determine proportional working values from an understanding of the problem domain. Instead, specific numbers would have to be determined, and, because these numbers would be picture dependent, the process would have no generality.

The second additional aspect of the probabilistic model is the empirical observation that the line enhancement process can be conceptualized as two competing subprocesses. One subprocess attempts to establish a line label at each point, and the other subprocess attempts to establish a no-line label. More generally it has been found that while relaxation labelling processes are designed with only one label set in mind, they often act as if there were several label subsets, each defining its own subprocess. It seems as if competition takes place between the subprocesses and cooperation takes place within each subprocess. This conceptualization of the internal organization of these processes is often a useful aid in their design. Similar labels can be lumped together and treated as a single module in relation to the others.

3.2 Labelling Dot Clusters

The second application of probabilistic relaxation labelling extends the low-level symbolic universe to include shapes with interiors, not just thin lines and curves. More specifically, the problem is to label all of the dots in a planar distribution with one of the following interpretations (i) interior point, (ii) edge point (with some orientation, and (iii) noise point. This dot labelling problem is relevant to two different areas, clustering algorithms and scene segmentation. Most clustering algorithms are based on a metric notion of proximity[2,15]. Under some conditions this produces final clusters which are rather different from those which humans would perceive[39]. The above labels can provide some indication of the role each point plays in determining the overall shape of the dot cluster. This, in turn, should lead to more visually satisfying results. The labelling of dot clusters can also be considered as a first step toward the general scene segmentation problem, i.e., the problem of segmenting a graylevel picture into regions of homogeneous intensity separated by well-marked lines and edges[31].

The underlying graph for this dot labelling application is constructed by considering each dot as an object and joining it with its k nearest neighbors. For the examples to be presented in this paper, k = 4 and k = 8 produced effectively the same results. The initial probabilities are estimated loosely in terms of the disparity between a point and the center of gravity of its k nearest neighbors. If a point is surrounded uniformly, the interior label is given a strong probability. If most of a point's neighbors are skewed to one side, the appropriate edge-value is given a strong initial probability. If a point is very far away from its neighbors (with respect to most interpoint distances), the noise label is given a strong initial weight. In general, the initial probabilities are a weighted combination of these three possibilities.

The compatibility functions are shown in Figure 8b. Basically the edge-to-edge compatibilities behave like the line-to-line compatibilities in the line enhancement example, except that the mutual influence decreases as the distance separating them increases. Interior points are strengthened by nearby edge points that are on the proper side of the edge, by other nearby interior points, and by noise points which are far away. They are weakened by distant interior points, by improperly oriented edge points, and by nearby noise points. The analogous compatibilities for edge and noise points are straightforward.

The results of applying this labelling process to a figure eight shaped dot cluster are shown in Figure 9 (after[39]). Note that most points are labelled unambiguously, except for the two points in the neck of the figure. The process was unable to decide whether this figure consisted of two clusters just touching or whether it consisted of just one connected cluster. Rather than forcing a decision one way or another on the basis of purely local information, the labelling process will pass the two possibilities to a more global process for a decision. Thus the different interpretations are made explicit with respect to the shape of the dot cluster.

A second example of dot labelling consists of the two intersecting C-shapes in Figure 10. Note that both C's are properly enclosed inside edges and that they remain perfectly distinct.

3.3 Additional Applications

Relaxation labelling is an approach to the reduction of local ambiguities which only requires knowledge about how local symbols can interact. These techniques should be applicable to problems in any domain which are specifiable in this fashion.

That many aspects of the visual world are highly determined locally has been known at least since the pioneering work of Helmholtz. One example of this local determinism in vision, rather different from those already presented, is that of perceptual transparency. Matelli[23] has recently shown that degrees of transparency can be computed on the basis of certain lightness ratios that are evaluated at the borders between adjacent regions. Another example is the scene analysis problem of region growing, in which other kinds of regional interaction have been found to be useful. The system of Feldman and Yakimovsky[9], in particular, has many ideas in common with the present work on relaxation processes. Within a region, knowledge about the distribution of gray levels can be used for noise cleaning[20]. More generally, because arbitrary scene segmentation is a problem involving both regional interactions and the kinds of edge and line interactions discussed in Section 3.1 and 3.2, it too should be appropriate for a relaxation labelling formulation (e.g.,[34]).

Another kind of local interaction amenable to relaxation labelling takes place in structured pattern matching. For example, a template can be thought of not as a unit whole, but rather as a relational structure of subtemplates. Instead of trying to match the entire template to every possible data point, it is more efficient to match the subtemplates[37] and then combine them together. Fischler and Elschlager[11] proposed linking the subtemplates with elastic springs, and then using dynamic programming to find the optimal solution. Recently there have been attempts to solve this problem by using a relaxation labelling formulation, in which the elastic relations between the subtemplates are incorporated into the compatibilities. Initial experiments have been encouraging[6] and hopefully the results will extend to more general pattern matching problems. Working on other, related approaches to structural pattern matching, Ullmann[35] has recently proposed an algorithm for subgraph isomorphism which is very close to the discrete relaxation process, and Hinton[16] has been using relaxation techniques to find globally best figures in ambiguous drawings.

4. Perspectives and Extensions on Probabilistic Relaxation Labelling

Probabilistic relaxation labelling introduced an explicit notation for the likelihood that a label is correct. This took the form of a continuous variable (in the interval [0,1]) attached to each label. We called this variable the probability that the label is correct because it serves, in the sense of Carnap[3], as a measure of the evidence supporting such an hypothesis. As the process iterates, and more evidence is examined, this probability value is updated. The process terminates either when all of the evidence has been examined or when sufficient evidence has been examined for each probability to become sufficiently stable. Thus the number of iterations required depends on the distance that information must travel in order to disambiguate the most uncertain assignments. The more that a label is determined by its immediate neighbors, the smaller the number of iterations needed.

The probability estimates are updated by the rule which defines how neighborhood information is used. One of the simplest ways to do this was to form a weighted sum of neighboring label probabilities which act through the compatibility functions. Sometimes, as in the line enhancement application[41], more complex schemes are required. In that example, in order to prevent lines from thickening during updating, it was necessary to incorporate some information about the distribution of probabilities over the neighborhood rather than just information about their algebraic sum. This new information was obtained from a difference relation across two lables. More generally, it might be necessary to include arbitrary conditionals in the updating rule so that structural characteristics of the neighborhood can enter the updating computation. Along the same lines, rather than just considering binary constraints between variables, n-ary constraints could be used[25].

For applications in which many labels have to be updated at each node, it may be advisable to parametrize some of the variable label dimensions. In the line enhancement application, for example, instead of having a separate label for each possible line orientation, a single label could be used to denote a unit line segment. The orientation of this line segment would then be indicated by a parameter. With this formulation, only two labels would have to be updated -- the no-line label and the line label. However, the line orientation parameter would then have to be updated on the basis of the line orientations around it[36]. This last point reveals the price that has to be paid for increased computational efficiency: less accuracy in the orientation variable. That is, instead of updating orientation on the basis of a distribution of probabilities over all possible orientations, only a single value is available from each distribution. However, the more accurately this value represents the distribution, the closer it comes to the non-parameterized formulation.

The foundations for this remark about parameterization were already laid in the section on line and curve enhancement. There it was described how the process behaved as if there where two subprocesses -- one for lines and one for no-lines. Each of these subprocesses had to remain distinct, hence at least two labels were needed. However, the internal conflicts within each subprocess could have been lumped into a single parameterized variable just as orientation was in the above reformulation.

In addition to parameterization, knowing the type of an object or some of its characteristic features, can often be used to partially restrict the label set which may legitimately apply to it. To take advantage of such partial evidence, a program could be attached to each node. This program could then either generate the label set (cf.[38]) or refine the node label set on the basis of partial evidence accumulated by the relaxation process. This, in turn, would improve the operational efficiency of the process, because only subsets of the labels would have to be updated on each iteration. A more extreme possibility is to allow these node programs to treat a portion of the object graph (or perhaps the entire graph) as data, thereby giving the local nodes more of a global perspective. In this way, many diverse sources of knowledge could be used simultaneously (cf.[8]).

Another possibility for generalization is the actual construction of the object graph. For example, in the line enhancement application, the graph was completely determined by the matrix underlying the intensity array. In the clustering application, on the other hand, the graph was built by a program which searched for the k nearest neighbors of a given point. In other situations, the design of this graph can have important effects on the final results. For instance, in the sub-template matching application, a basic criterion had to be met (i.e., a significant match) before a node was established. Varying this criterion affected the number and the distribution of the nodes that were established[6].

In addition to these remarks about how relaxation labelling processes can be extended, it is interesting to note that entirely different approaches to them are possible. For example, one can conceive of the label probabilities as inequalities over a set of variables which take values in the range [0,1]. More traditional, mathematical relaxation techniques can then be applied[1]. Or, from another point of view, these processes can be conceived of as networks of automata. The discrete model would then correspond to a network of finite automata[29], and the probabilistic model to a network of probabilistic automata[27]. Perhaps these alternate points of view will reveal more general results about the convergence of relaxation labelling processes.

5. Conclusions

Relaxation labelling processes, as described in this paper, are techniques for deciding among alternative symbolic interpretations for local variables. They use a specific kind of knowledge about the ways in which these variables can interact. When the variables are domain-specific, e.g., when their interpretations are only meaningful within a domain such as the blocks world, then the knowledge about how they can interact will also be specific to the domain. However, when the variables are general purpose[42], i.e., when they are common to many different semantic domains, then the knowledge of how they can interact is also general purpose.

General purpose knowledge, at least in terms of low-level vision, is knowledge about the structure of the intensity array and how it relates to symbolic constructs such as edge and line segments*. It is useful in a variety of semantically-specific domains, including bubble chamber physics, satellite imagery, and the blocks world. Furthermore, it is useful even before domain-specific questions or expectations are formulated. For example, in Figure 8 it was possible to use very general knowledge about how line segments can interact to enhance lines. It was not necessary to know that the lines specifically represented roads. Moreover, once these long lines were labelled, they might be able to serve as highly salient features important for associating the image with more specific and appropriate semantic domains.

Thus, relaxation labelling processes begin to provide us with a mechanism through which general-purpose and domain-specific processes can communicate. One possibility is to let the compatibility functions be sensitive to the requirements of domain-specific experts. In terms of the line enhancement application, for example, bubble chamber experts might bias the compatibilities in favor of very straight lines. In the absence of this specialization, values derived from more general-purpose considerations could be used, and the processing would proceed to accumulate evidence useful in the search for more relevant and specific domains. While this discussion does not suggest a mechanism for directly inferring the correct semantic domain, it does offer a means for obtaining reliable low-level data to be used by the mechanism.

This paper stems from earlier, more mathematical work done with A. Rosenfeld and R. Hummel. I would especially like to thank A. Rosenfeld for his continued interest and for our many fruitful discussions.

The support of the National Science Foundation under Grant MCS-72-03610 is gratefully acknowledged, as is the help of Shelly Rowe in preparing this paper.

---

*For a discussion of a more complete vocabulary for low-level vision, see Marr[22].

References

1. Agmon, S. The relaxation method for linear inequalities, Can. J. Math., Vol. 6, No. 3, pp. 382-392.

2. Anderberg, M. R. Cluster Analysis for Applications, Academic Press, New York, 1973.

3. Carnap, R. Logical Foundations of Probability (2nd Ed.), The University of Chicago Press, 1962.

4. Clowes, M. B. On seeing things, Artificial Intelligence, Vol. 2, 1971, pp. 79-116.

5. Cornsweet, T. Visual Perception, Academic Press, New York, 1970.

6. Davis, L. S., and A. Rosenfeld. Applications of relaxation labelling, 2: Spring-loaded template matching, Technical Report TR-440, Computer Science Center, University of Maryland, February 1976.

7. Eastman, C. M. Automated space planning, Artificial Intelligence, Vol. 4, 1973, pp. 41-64.

8. Erman, L. D., and V. R. Lesser, A multi-level organization for problem solving using many, diverse, cooperating sources of knowledge, Adv. Papers Fourth International Joint Conf. Artificial Intelligence, Tbilisi, Sept. 1975, pp. 483-490.

9. Feldman, J. A., and Y. Yakimovsky. Decision theory and artificial intelligence: I. A semantics-based region analyzer, Artificial Intelligence, Vol. 5, 1974, pp. 349-371.

10. Fikes, R. E. REF-ARF: A system for solving problems stated as procedures, Artificial Intelligence, Vol. 1, 1970, pp. 27-120.

11. Fischler, M. A. and R. A. Elschlager. The representation and matching of pictorial structures, IEEE Trans. Computers, Vol. C-22, 1973, pp. 67-92.

12. Gaschnig, J. A constraint satisfaction method for inference making, Proc. 12th Annual Allerton Conf. on Circuits and System Theory, Univ. of Illinois, Urbana-Champaign, 1974.

13. Gregory, R. L. The Intelligent Eye, McGraw-Hill, New York, 1970.

14. Haar, R. Computational models of spatial relations, Technical Report (in preparation), Computer Science Center, University of Maryland.

15. Hartigan, J. A. Clustering Algorithms, Academic Press, New York, 1973.

16. Hinton, G. Using relaxation to find a puppet, Draft, Cognitive Studies Programme, University of Sussex, 1976.

17. Huffman, D. A. Impossible objects as

nonsense sentences, in B. Meltzer and D. Michie (eds.), Machine Intelligence 6, Edinburgh U. Press, Edinbrugh, 1971, pp. 295-323.

18. Koffka, K. Principles of Gestalt Psychology, Harcourt, Brace, and World, Inc., New York, 1935.

19. Kowalski, R. A proof procedure using connection graphs, Journal of the ACM, Vol. 22, No. 4, 1975, pp. 572-595.

20. Lev, A., Zucker, S. W., and A. Rosenfeld. Iterative enhancement of noisy images, Technical Report TR-445, Computer Science Center, University of Maryland, March 1976.

21. Mackworth, A. K. Consistency in networks of relations, Technical Report 75-3, Department of Computer Science, University of British Columbia, Vancouver, July 1975.

22. Marr, D. Early processing of visual information, A.I. Memo 340, Artificial Intelligence Lab., M.I.T., Dec. 1975.

23. Matelli, F. The perception of transparency, Scientific American, Vol. 230, No. 4, 1974, pp. 90-98.

24. Montalvo, F. S. Consensus versus competition in neural networks: A comparative analysis of three models, Int. J. Man-Machine Studies, Vol. 7, 1975, pp. 333-346.

25. Montanari, V. Networks of constraints: Fundamental properties and applications to picture processing, Information Sciences, Vol. 7, 1974, pp. 95-132.

26. Newell, A., and H. A. Simon. Human Problem Solving, Prentice Hall, Englewood Cliffs, N. J., 1972.

27. Paz, A. Introduction to Probabilistic Automata, Academic Press, New York, 1971.

28. Penrose, L. S., and R. Penrose. Impossible objects: A special type of illusion, Brit. J. Psychol., Vol. 49, 1958.

29. Rosenfeld, A. Networks of automata: Some applications, IEEE Trans. Systems, Man, and Cybernetics, Vol. SMC-5, 1975, pp. 380-383.

30. Rosenfeld, A., Hummel, R., and S. W. Zucker. Scene labelling by relaxation operations, IEEE Trans. Systems, Man, and Cybernetics, in press, 1976.

31. Rosenfeld, A., and A. C. Kak. Digital Picture Processing, Academic Press, New York, 1976.

32. Rosenfeld, A., and M. Thurston. Edge and curve detection for visual scene analysis, IEEE Trans. Computers, Vol. C-20, 1971, pp. 562-569.

33. Sussman, G. J. A computational model of skill acquisition, A. I. TR-297, Artificial Intelligence Lab., M.I.T., August 1973.

34. Tenenbaum, J. M., et. al. An interactive facility for scene analysis research, Technical Note 87, Stanford Research Institute, Menlo Park, 1974. See also, Barrow, H. G., and J. M. Tenenbaum. Experiments in interpretation guided segmentation, Technical Note 123, Stanford Research Institute, March 1976.

35. Ullmann, J. R. An algorithm for subgraph isomorphism, Journal of the ACM, Vol. 23, No. 1, 1976, pp. 31-42.

36. VanderBrug, G. J. Experiments in iterative enhancement of linear features, Technical Report TR-425, Computer Science Center, University of Maryland, 1975.

37. VanderBrug, G. J. and A. Rosenfeld. Two-stage template matching, Technical Report TR-364, Computer Science Center, University of Maryland, March 1975.

38. Waltz, D. Understanding line drawings of scenes with shadows, in P. H. Winston (ed.), The Psychology of Computer Vision, McGraw-Hill, New York, 1975.

39. Zahn, C. T. Graph-theoretical methods for detecting and describing Gestalt clusters, IEEE Trans. Computers, Vol. C-20, 1971, pp. 68-86.

40. Zucker, S. W. Region growing: Childhood and adolescence, Computer Graphics and Image Processing, in press, 1976.

41. Zucker, S. W., Hummel, R., and A. Rosenfeld. An application of relaxation labelling to line and curve enhancement, IEEE Trans. Computers, in press.

42. Zucker, S. W., Rosenfeld, A., and L. S. Davis. General purpose models: Expectations about the unexpected, Adv. Papers Fourth International Joint Conf. Artificial Intelligence, Tbilisi, Sept. 1975, pp. 716-721.

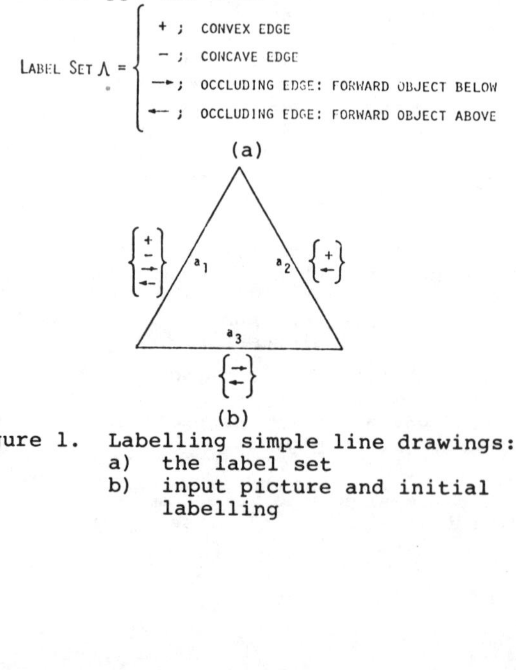

(b)

Figure 1. Labelling simple line drawings:
a) the label set
b) input picture and initial labelling

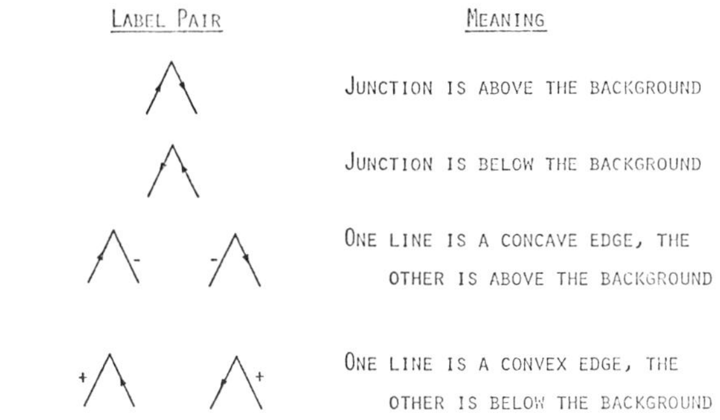

Figure 2. The constraint set of allowable label pairs $\Lambda_{ij}(\lambda,\lambda')$.

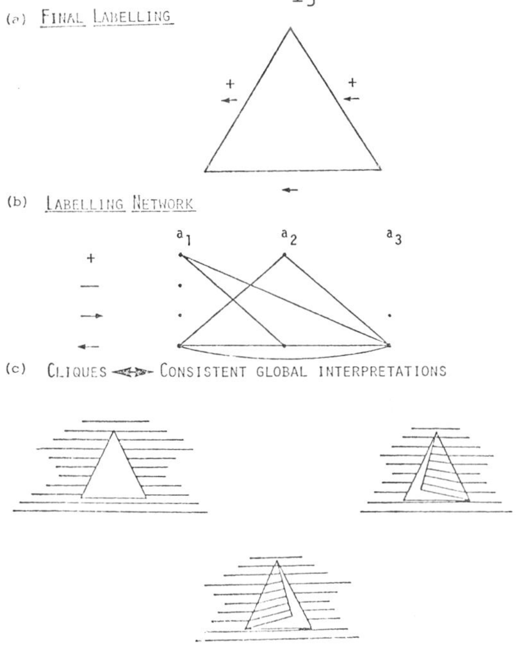

Figure 3. The line drawing final labelling and its interpretations.

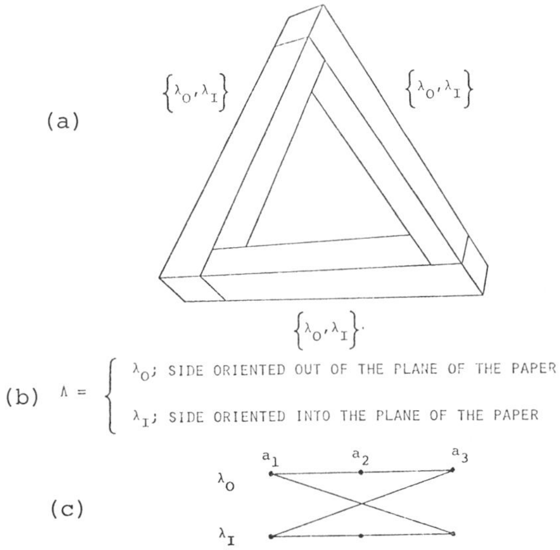

Figure 4. a) Penrose impossible triangle and initial labelling
b) label set
c) A spanning cycle in the labelling network.
Note that the object spanning cycles are of length six.

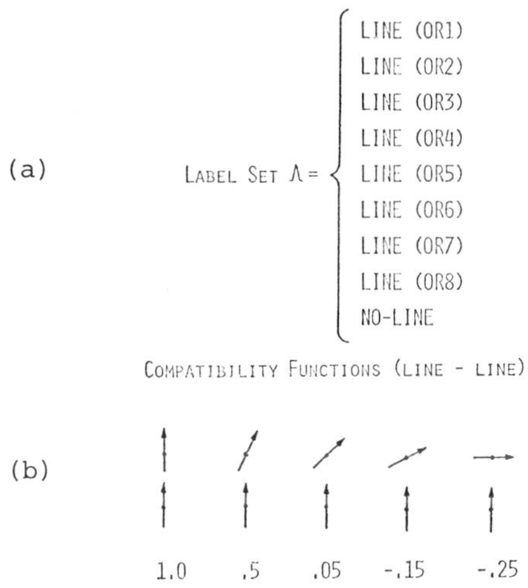

Figure 5. a) Label set for line and curve enhancement.
b) Line-to-line compatibilities

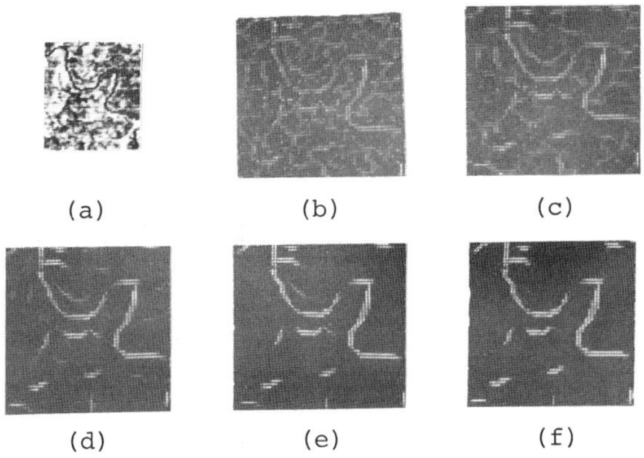

Fig. 6. Curve enhancement.
a) LANDSAT subimage containing river.
b) The initial probability assignments for (a). The intensity for each line segment is proportional to the maximum probability at that position.
c-f) Iterations 1, 3, 5, and 7 for relaxation process applied to (a).

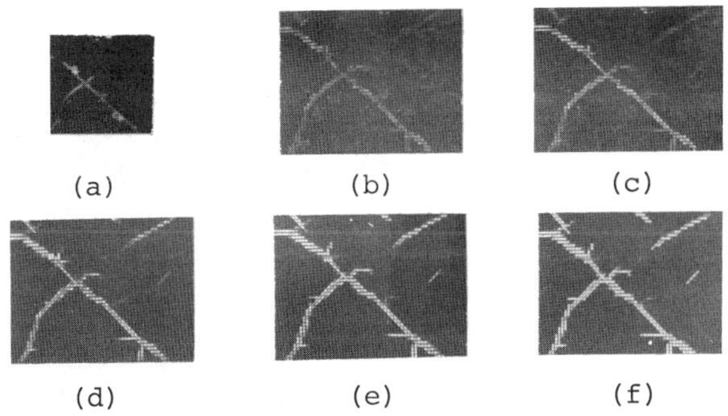

Fig. 7. a) SKYLAB subimage containing two crossing roads.
b) Initial probability assignments for (a).
c-f) Iterations 1, 3, 5, and 7 of the relaxation process applied to (a).

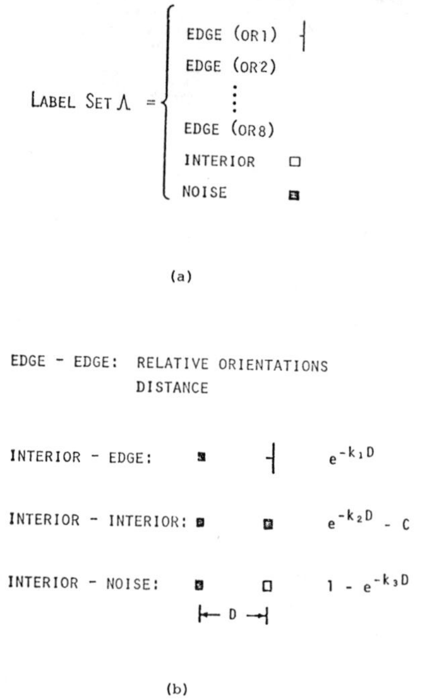

Figure 8. Labelling dot clusters:
a) label set and symbols
b) compatibility functions.

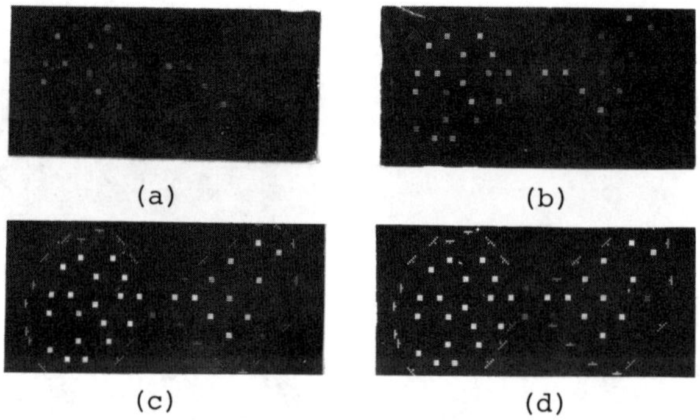

Figure 9. Labelling dot clusters. (Note that these images are negatives of those sketched in Fig. 8.)
a) Initial probability assignments
b-d) Iterations 5, 10, and 20 of the relaxation process applied to (a).

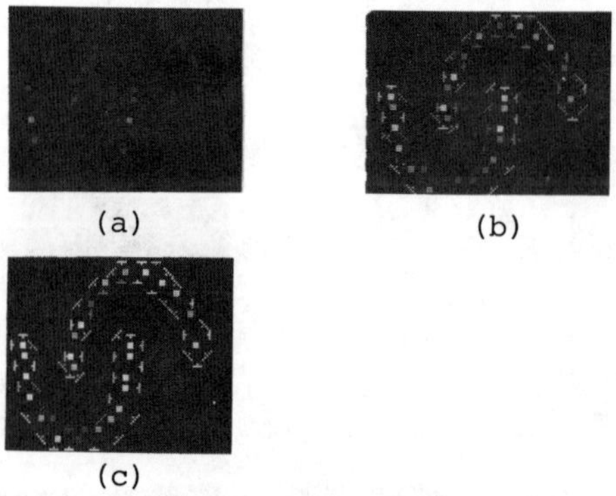

Figure 10. Intertwined clusters
a) Initial probability assignments
b-c) Iterations 10 and 20 of the relaxation process.

Bibliography

The references that follow have been broken up into eight parts, the last seven of which correspond to the seven parts of this book. This grouping of the references is designed to make it easier for the reader to locate the references dealing with a particular topic. To aid the reader further, a first section of general references on the subject of image processing and analysis has been added. This section consists of three parts:

A. References on image processing (coding and compression, digital filtering, enhancement, and restoration; these topics are not covered in this book).

B. Texts and bibliographies that cover image and scene analysis.

C. Journals, special journal issues, and hardcover paper collections that contain papers on the subject.

In each section, the references are arranged alphabetically by author.

General References

A. Image Processing

[1] J. K. Aggarwal and R. O. Duda, Guest Eds., Special Issue on Digital Filtering and Image Processing, *IEEE Trans. Circuits Syst.*, vol. CAS-22, pp. 161-304, Mar. 1975.

[2] H. C. Andrews (with contributions by W. K. Pratt and K. Caspari), *Computer Techniques in Image Processing.* New York: Academic, 1970.

[3] H. C. Andrews and L. H. Enloe, Guest Eds., Special Issue on Digital Picture Processing, *Proc. IEEE*, vol. 60, pp. 766-898, July 1972.

[4] E. L. Hall and C. F. George, Jr., Guest Eds., Special Issue on Two-Dimensional Digital Signal Processing, *IEEE Trans. Comput.*, vol. C-21, pp. 633-820, July 1972.

[5] T. S. Huang and O. J. Tretiak, Eds., *The MIT Picture Bandwidth Compression Symposium.* New York: Gordon and Breach, 1971.

[6] T. S. Huang, Ed., *Picture Processing and Digital Filtering.* New York: Springer, 1975.

B. Texts and Bibliographies

[1] R. O. Duda and P. E. Hart, *Pattern Classification and Scene Analysis.* New York: Wiley, 1973.

[2] A. Rosenfeld, *Picture Processing by Computer.* New York: Academic, 1969.

[3] A. Rosenfeld and A. C. Kak, *Digital Picture Processing.* New York: Academic, 1976.

[4] A. Rosenfeld, "Picture processing by computer," *Computing Surveys*, vol. 1, pp. 147-176, 1969.

[5] A. Rosenfeld, "Progress in picture processing: 1969-71," *ibid.*, vol. 5, pp. 81-108, 1973.

[6] A. Rosenfeld, "Picture processing: 1972," *Computer Graphics Image Processing*, vol. 1, pp. 394-416, 1972.

[7] A. Rosenfeld, "Picture processing: 1973," *ibid.*, vol. 3, pp. 178-194, 1974.

[8] A. Rosenfeld, "Picture processing: 1974," *ibid.*, vol. 4, pp. 133-155, 1975.

[9] A. Rosenfeld, "Picture processing: 1975," *ibid.*, vol. 5, pp. 215-237, 1976.

C. Serials and Collections

[1] *Proc. Int. Joint Conf. on Artificial Intelligence*, 1969ff.

[2] *Proc. Int. Joint Conf. Pattern Recognition*, IEEE, New York, NY, 1973ff.

[3] R. S. Ledley, Ed., *Pattern Recognition.* New York: Pergamon, 1968ff.

[4] A. Rosenfeld, H. Freeman, T. S. Huang, and A. van Dam, Eds., *Computer Graphics Image Processing.* New York: Academic, 1972ff.

[5] G. C. Cheng, R. S. Ledley, D. K. Pollock, and A. Rosenfeld, Eds., *Pictorial Pattern Recognition.* Washington, DC: Thompson, 1968.

[6] K. S. Fu, Guest Ed., Special Issue on Feature Extraction and Selection in Pattern Recognition, *IEEE Trans. Comput.*, vol. C-20, pp. 965-1120, Sept. 1971.

[7] A. Grasselli, Ed., *Automatic Interpretation and Classification of Images.* New York: Academic, 1969.

[8] L. D. Harmon, Guest Ed., Special Issue on Digital Pattern Recognition, *Proc. IEEE*, vol. 60, pp. 1117-1233, Oct. 1972.

[9] L. N. Kanal, Ed., *Pattern Recognition.* Washington, DC: Thompson, 1968.

[10] P. A. Kolers and M. Eden, Eds., *Recognizing Patterns: Studies in Living and Automatic Systems.* Cambridge, MA: MIT Press, 1968.

[11] B. S. Lipkin and A. Rosenfeld, Eds., *Picture Processing and Psychopictorics.* New York: Academic, 1970.

[12] A. Rosenfeld, Ed., *Digital Picture Analysis.* New York: Springer, 1976.

[13] J. Sklansky, Ed., *Pattern Recognition: Introduction

and Foundations, Stroudsburg, PA: Dowden, Hutchinson and Ross, 1973.

[14] L. Uhr, Ed.,*Pattern Recognition*. New York: Wiley, 1966.

[15] S. Watanabe, Ed., *Methodologies of Pattern Recognition*. New York: Academic, 1969.

[16] S. Watanabe, Ed., *Frontiers of Pattern Recognition.* New York: Academic, 1972.

[17] P. H. Winston, Ed., *The Psychology of Computer Vision*. New York: McGraw-Hill, 1975.

References on Part I Correlation and Matching

[1] P. A. Anuta, "Spatial registration of multispectral and multitemporal digital imagery using fast Fourier transform techniques," *IEEE Trans. Geosci. Electron.*, vol. GE-8, pp. 353-368, 1970.

[2] W. Doyle, "Operations useful for similarity-invariant pattern recognition," *J. ACM*, vol. 9, pp. 259-267, 1962.

[3] M. J. B. Duff, D. M. Watson, T. J. Fountain, and G. K. Shaw, "A cellular logic array for image processing," *Pattern Recognition*, vol. 5, pp. 229-247, 1973.

[4] G. L. Fischer, Jr., D. K. Pollock, B. Radack, and M. E. Stevens, Eds., *Optical Character Recognition*. Baltimore, MD: Spartan, 1962.

[5] S. B. Gray, "The binary image processor and its applications," Document 90365-SC, Information International Inc., Los Angeles, CA, 1972.

[6] U. Grenander, "A unified approach to pattern analysis," in *Advances in Computers*, vol. 10. New York: Academic, 1970, pp. 175-216.

[7] A. W. Holt, "Comparative religion in character recognition machines," *Computer Group News*, vol. 2, no. 6, pp. 3-11, 1968.

[8] R. Y. Kain, "Autocorrelation pattern recognition," *Proc. IRE*, vol. 49, pp. 1085-1086, 1961.

[9] V. A. Kovalevsky, *Character Readers and Pattern Recognition.* New York: Spartan, 1968.

[10] B. H. McCormick, "The Illinois pattern recognition computer—ILLIAC III," *IEEE Trans. Electron. Comput.*, vol. EC-12, pp. 791-813, 1963.

[11] J. A. McLaughlin and J. Raviv, "*N*th-order autocorrelation in pattern recognition," *Inform. Contr.*, vol. 12, pp. 121-142, 1968.

[12] R. N. Nagel and A. Rosenfeld, "Ordered search techniques in template matching," *Proc. IEEE*, vol. 60, pp. 242-244, 1972.

[13] K. Preston, Jr., *Coherent Optical Computers.* New York: McGraw-Hill, 1972.

[14] E. A. Smith and D. R. Phillips, "Automated cloud tracking using precisely aligned digital ATS pictures," *IEEE Trans. Comput.*, vol. C-21, pp. 715-729, 1972.

[15] J. R. Ullmann, *Pattern Recognition Techniques.* London: Butterworths, 1973.

[16] S. H. Unger, "A computer oriented toward spatial problems," *Proc. IRE*, vol. 46, pp. 1744-1750, 1958.

[17] A. Vander Lugt, "Coherent optical processing," *Proc. IEEE*, vol. 62, pp. 1300-1319, 1974.

[18] B. Widrow, "The 'rubber mask' technique," *Pattern Recognition*, vol. 5, pp. 175-211, 1973.

References on Part II Segmentation: Regions

[1] C. K. Chow and T. Kaneko, "Boundary detection of radiographic images by a threshold method," in S. Watanade, Ed., *Frontiers of Pattern Recognition.* New York: Academic, 1972, pp. 61-82.

[2] L. S. Davis, A. Rosenfeld, and J. S. Weszka, "Region extraction by averaging and thresholding," *IEEE Trans. Syst., Man, Cybern.*, vol. SMC-5, pp. 383-388, 1975.

[3] S. B. Gray, "Local properties of binary images in two dimensions," *IEEE Trans. Comput.*, vol. C-20, pp. 551-561, 1971.

[4] J. N. Gupta and P. A. Wintz, "A boundary finding algorithm and its applications," *IEEE Trans. Circuits Syst.*, vol. CAS-22, pp. 351-362, 1975.

[5] J. L. Muerle and D. C. Allen, "Experimental evaluation of techniques for automatic segmentation of objects in a complex scene," in G. C. Cheng *et al.*, Eds, *Pictorial Pattern Recognition.* Washington, DC: Thompson, 1968, pp. 3-13.

[6] R. B. Ohlander, "Analysis of natural scenes," Ph.D. dissertation, Computer Science Dep., Carnegie-Mellon Univ., Pittsburgh, PA, Apr. 1975.

[7] J. M. S. Prewitt and M. L. Mendelsohn, "The analysis of cell images," *Annals N.Y. Acad. Sci.*, vol. 128, pp. 1035-1053, 1966.

[8] A. Rosenfeld, "Connectivity in digital pictures," *J. ACM*, vol. 17, pp. 146-156, 1970.

[9] A. Rosenfeld and J. L. Pfaltz, "Sequential operations in digital picture processing," *J. ACM*, vol. 13, pp. 471-494, 1966.

[10] C. B. Shelman, "The application of list processing techniques to picture processing," *Pattern Recognition*, vol. 4, pp. 201-210, 1972.

[11] S. Tsuji and F. Tomita, "A structural analyzer for a class of textures," *Computer Graphics Image Processing*, vol. 2, pp. 216-231, 1973.

[12] S. W. Zucker, A. Rosenfeld, and L. S. Davis, "Picture segmentation by texture discrimination," *IEEE Trans. Comput.*, vol C-24, pp. 1228-1233, 1975.

References on Part III Region Representation and Description

[1] F. L. Alt, "Digital pattern recognition by moments," *J. ACM*, vol. 9, pp. 240-258, 1962. (Also in G. L. Fischer,

Jr. *et al.*, Eds., *Optical Character Recognition*. Baltimore, MD: Spartan, 1962, pp. 153-179.)

[2] R. Bajcsy, "Computer identification of visual surfaces," *Computer Graphics Image Processing*, vol. 2, pp. 118-130, 1973.

[3] H. Blum, "Biological shape and visual science (Part I)," *J. Theoretical Biology*, vol. 38, pp. 205-287, 1973.

[4] L. Calabi and W. E. Hartnett, "Shape recognition, prairie fires, convex deficiencies and skeletons," *Amer. Math. Monthly*, vol. 75, pp. 335-342, 1968.

[5] V. E. Giuliano, P. E. Jones, G. E. Kimball, R. F. Meyer, and B. A. Stein, "Automatic pattern recognition by a Gestalt method," *Inform. Contr.*, vol. 4, pp. 332-345, 1961. (See also R. F. Meyer, V. E. Giuliano, and P. E. Jones, "Analytic approximation and translation invariance in character recognition," in G. L. Fischer *et al.*, Eds., *Optical Character Recognition*. Baltimore, MD: Spartan, 1962, pp. 181-195.)

[6] J. K. Hawkins, "Textural properties for pattern recognition," in B. S. Lipkin and A. Rosenfeld, Eds., *Picture Processing and Psychopictorics*. New York: Academic, 1970, pp. 347-370.

[7] C. J. Hilditch, "Linear skeletons from square cupboards," in B. Meltzer and D. Michie, Eds., *Machine Intelligence*, vol. 4. Edinburgh, Scotland: Edinburgh Univ. Press, 1969, pp. 403-420.

[8] M. K. Hu, "Pattern recognition by moment invariants," *Proc. IRE*, vol. 49, p. 1428, 1961.

[9] G. Levi and U. Montanari, "A grey-weighted skeleton," *Inform. Contr.*, vol. 17, pp. 62-91, 1970.

[10] B. H. McCormick and S. N. Jayaramamurthy, "A decision theory method for the analysis of texture," *Int. J. Comput. Inform. Sci.*, vol. 4, pp. 1-38, 1975. (See also *ibid.*, vol. 3, pp. 329-343, 1974.)

[11] J. L. Pfaltz and A. Rosenfeld, "Computer representation of planar regions by their skeletons," *Commun. ACM*, vol. 10, pp. 119-122, 125, 1967.

[12] K. Preston, Jr., "Feature extraction by Golay hexagonal pattern transformations," *IEEE Trans. Comput.*, vol. C-20, pp. 1007-1014, 1971.

[13] K. Preston, Jr. and P. E. Norgren, "Interactive image processor speeds pattern recognition by computer," *Electronics*, vol. 45, pp. 89-98, Oct. 23, 1972.

[14] A. Rosenfeld, "A characterization of parallel thinning algorithms," *Inform. Contr.*, vol. 29, pp. 286-291, 1975.

[15] A. Rosenfeld and J. L. Pfaltz, "Distance functions on digital pictures," *Pattern Recognition*, vol. 1, pp. 33-61, 1968.

[16] F. W. Smith and M. H. Wright, "Automatic ship photo interpretation by the method of moments," *IEEE Trans. Comput.*, vol. C-20. pp. 1089-1095, 1971.

[17] R. Stefanelli and A. Rosenfeld, "Some parallel thinning algorithms for digital pictures," *J. ACM*, vol. 18, pp. 255-264, 1971.

[18] J. S. Weszka, C. R. Dyer, and A. Rosenfeld, "A comparative study of texture measures for terrain classification," *IEEE Trans. Syst., Man, Cybern.*, vol. SMC-6, pp. 269-285, 1976.

References on Part IV Segmentation: Boundaries

[1] L. S. Davis, "A survey of edge detection techniques," *Computer Graphics Image Processing*, vol. 4, pp. 248-270, 1975.

[2] A. K. Griffith, "Mathematical models for automatic line detection," *J. ACM*, vol. 20, pp. 62-80, 1973.

[3] M. H. Hueckel, "A local visual operator which recognizes edges and lines," *J. ACM*, vol. 20, pp. 634-647, 1973.

[4] C. Kimme, D. Ballard, and J. Sklansky, "Finding circles by an array of accumulators," *Commun. ACM*, vol. 18, pp. 120-122, 1975.

[5] L. S. G. Kovasznay and H. M. Joseph, "Image processing," *Proc. IRE*, vol. 43, pp. 560-570, 1955.

[6] R. S. Ledley, J. Jacobsen, and M. Belson, "BUGSYS: A programming system for picture processing—not for debugging," *Commun. ACM*, vol. 9, pp. 79-84, 1966.

[7] A. Martelli, "Edge detection using heuristic search methods," *Computer Graphics Image Processing*, vol. 1, pp. 169-182, 1972.

[8] J. W. McKee and J. K. Aggarwal, "Finding the edges of the surfaces of three-dimensional curved objects by computer," *Pattern Recognition*, vol. 7, pp. 25-52, 1975.

[9] U. Montanari, "On the optimal detection of curves in noisy pictures," *Commun. ACM*, vol. 14, pp. 335-345, 1971.

[10] F. O'Gorman and M. B. Clowes, "Finding picture edges through collinearity of feature points," *IEEE Trans. Comput.*, vol. C-25, pp. 449-456, 1976.

[11] K. K. Pingle and J. M. Tenenbaum, "An accommodating edge follower," in *Proc. 2nd Int. Joint Conf. on Artificial Intelligence*, London, England, 1971, pp. 1-7.

[12] A. Rosenfeld, M. Thurston, and Y. H. Lee, "Edge and curve detection: Further experiments," *IEEE Trans. Comput.*, vol. C-21, pp. 677-715, 1972.

[13] S. D. Shapiro, "Transformations for the computer detection of curves in noisy pictures," *Computer Graphics Image Processing*, vol. 4, pp. 328-338, 1975.

References on Part V Boundary Representation and Description

[1] H. Freeman, "Computer processing of line-drawing images," *Computing Surveys*, vol. 6, pp. 57-97, 1974.

[2] R. L. Lillestrand, "Techniques for change detection," *IEEE Trans. Comput.*, vol. C-21, pp. 654-659, 1972.

[3] R. D. Merrill, "Representations of contours and regions for efficient computer search," *Commun. ACM*, vol. 16, pp. 69-82, 1973.

[4] U. Montanari, "A note on minimal length polygonal approximation to a digitized contour," *Commun. ACM*, vol. 13, pp. 41-47, 1970.

[5] J. H. Munson, "Experiments in the recognition of hand-printed text: Part I—Character recognition," in *Proc.*

AFIPS 1968 Fall Joint Computer Conf. Washington, DC: Thompson, 1968, pp. 1125-1138.
[6] R. Nevatia, "Depth measurement by motion stereo," *Computer Graphics Image Processing*, vol. 5, pp. 203-214, 1976.
[7] T. Pavlidis, "Segmentation of pictures and maps through functional approximation," *Computer Graphics Image Processing*, vol. 1, pp. 360-372, 1972.
[8] T. Pavlidis, "Waveform segmentation through functional approximation," *IEEE Trans. Comput.*, vol. C-22, pp. 689-697, 1973.
[9] T. Pavlidis, "Optimal piecewise polynomial $L2$ approximation of functions of one and two variables," *IEEE Trans. Comput.*, vol. C-24, pp. 98-102, 1975.
[10] T. Pavlidis and G. S. Feng, "A segmentation technique for waveform classification," *IEEE Trans. Comput.*, vol. C-21, pp. 901-904, 1972.
[11] T. Pavlidis and S. L. Horowitz, "Segmentation of plane curves," *IEEE Trans. Comput.*, vol. C-23, pp. 860-870, 1974.
[12] A. Rosenfeld, "Digital straight line segments," *IEEE Trans. Comput.*, vol. C-23, pp. 1264-1269, 1974.
[13] A. Rosenfeld and E. G. Johnston, "Angle detection on digital curves," *IEEE Trans. Comput.*, vol. C-22, pp. 875-878, 1973.
[14] J. Sklansky, R. L. Chazin, and B. J. Hansen, "Minimum-perimeter polygons of digitized silhouettes," *IEEE Trans. Comput.*, vol. C-21, pp. 260-268, 1972.
[15] S. H. Unger, "Pattern detection and recognition," *Proc. IRE*, vol. 47, pp. 1737-1752, 1959.
[16] C. T. Zahn and R. Z. Roskies, "Fourier descriptors for plane closed curves," *IEEE Trans. Comput.*, vol. C-21, pp. 269-281, 1972.

References on Part VI Scene Analysis: The Blocks World

[1] M. B. Clowes, "On seeing things," *Artificial Intelligence*, vol. 2, pp. 79-116, 1971.
[2] G. Falk, "Interpretation of imperfect line data as a three-dimensional scene," *Artificial Intelligence*, vol. 3, pp. 101-144, 1972.
[3] G. R. Grape, "Model based (intermediate-level) computer vision," AIM 201, Stanford Artificial Intelligence Lab., Stanford, CA, May 1973.
[4] A. K. Griffith, "Edge detection in simple scenes using a priori information," *IEEE Trans. Comput.*, vol. C-22, pp. 371-381, 1973.
[5] A. K. Mackworth, "Interpreting pictures of polyhedral scenes," *Artificial Intelligence*, vol. 4, pp. 121-137, 1973.
[6] Y. Shirai, "A context sensitive line finder for recognition of polyhedra," *Artificial Intelligence*, vol. 4, pp. 95-119, 1973.
[7] Y. Shirai, "Analyzing intensity arrays using knowledge about scenes," in P. H. Winston, Ed., *The Psychology of Computer Vision.* New York: McGraw-Hill, 1975, pp. 93-113.
[8] S. A. Underwood and C. L. Coates, Jr., "Visual learning from multiple views," *IEEE Trans. Comput.*, vol. C-24, pp. 651-661, 1975.
[9] D. Waltz, "Understanding line drawings of scenes with shadows," in P. H. Winston, Ed., *The Psychology of Computer Vision.* New York: McGraw-Hill, 1975, pp. 19-91.
[10] P. H. Winston, "Learning structural descriptions from examples," in P. H. Winston, Ed., *The Psychology of Computer Vision.* New York: McGraw-Hill, 1975, pp. 157-209.

References on Part VII Scene Analysis: The Natural World

[1] G. J. Agin and T. O. Binford, "Computer description of curved objects," *IEEE Trans. Comput.*, vol. C-25, pp. 439-449, 1976.
[2] R. Bajcsy and L. Lieberman, "Texture gradient as a depth cue," *Computer Graphics Image Processing*, vol. 5, pp. 52-67, 1976.
[3] R. T. Chien and Y. H. Chang, "Recognition of curved objects and object assemblies," in *Proc. 2nd Int. Joint Conf. on Pattern Recognition*, Copenhagen, Denmark, Aug. 1974, pp. 496-510.
[4] M. Eden, "Handwriting and pattern recognition," *IRE Trans. Inform. Theory*, vol. IT-8, pp. 160-166, 1962.
[5] T. G. Evans, "Descriptive pattern analysis techniques," in A. Grasselli, Ed., *Automatic Interpretation and Classification of Images.* New York: Academic, 1969, pp. 79-95. (See also T. G. Evans, "Descriptive pattern analysis techniques: Potentialities and problems," in S. Watanabe, Ed., *Methodologies of Pattern Recognition.* New York: Academic, 1969, pp. 147-157.)
[6] J. A. Feldman and Y. Yakimovsky, "Decision theory and artificial intelligence: I. A semantics-based region analyzer," *Artificial Intelligence*, vol. 5, pp. 349-371, 1974.
[7] K. S. Fu, *Syntactic Methods in Pattern Recognition.* New York: Academic, 1974.
[8] C. A. Harlow, "Image analysis and graphs," *Computer Graphics Image Processing*, vol. 2, pp. 60-82, 1973.
[9] F. Hayes-Roth, "Representation of structured events and efficient procedures for their recognition," *Pattern Recognition*, vol. 8, pp. 141-150, 1976.
[10] B. Horn, "Obtaining shape from shading information," in P. H. Winston, Ed., *The Psychology of Computer Vision.* New York: McGraw-Hill, 1975, pp. 115-155.
[11] S. Kaneff, Ed., *Picture Language Machines.* New York: Academic, 1970.
[12] G. J. Kaufman, Jr., and K. J. Breeding, "The automatic recognition of human faces from profile silhouettes," *IEEE Trans. Syst., Man, Cybern.*, vol. SMC-6, pp. 113-121, 1976.
[13] R. A. Kirsch, "Computer interpretation of English text and picture patterns," *IEEE Trans. Electron. Comput.*, vol. EC-13, pp. 363-376, 1964.
[14] W. F. Miller and A. C. Shaw, "Linguistic methods in pic-

ture processing—A survey," in *Proc. AFIPS 1968 Fall Joint Computer Conf*. Washington, DC: Thompson, 1968, pp. 279-290.

[15] F. Nake and A. Rosenfeld, Eds., *Graphic Languages*. Amsterdam: North-Holland, 1972.

[16] R. Narasimhan, "Labeling schemata and syntactic description of pictures," *Information Control*, vol. 7, pp. 151-179, 1964.

[17] R. Narasimhan, "Syntax-directed interpretation of classes of pictures," *Commun. ACM*, vol. 9, pp. 166-173, 1966.

[18] R. Nevatia and T. O. Binford, "Structured descriptions of complex objects," in *Proc. 3rd Int. Joint Conf. on Artificial Intelligence*, Stanford, CA Aug. 1973, pp. 641-647.

[19] A. Rosenfeld, R. A. Hummel, and S. W. Zucker, "Scene labeling by relaxation operations," *IEEE Trans. Syst., Man, Cybern.*, vol. SMC-6, pp. 420-433, 1976.

[20] T. Sakai, M. Nagao, and T. Kanade, "Computer analysis and classification of photographs of human faces," in *Proc. 1st USA-Japan Computer Conf.*, Oct. 1972, pp. 55-62.

[21] J. M. Tenenbaum and H. G. Barrow, "MSYS: A system for reasoning about scenes," Annual Rep. SRI Project 1187, Stanford Res. Inst., Menlo Park, CA, Dec. 1975.

[22] K. J. Turner, "Computer perception of curved objects using a television camera," Ph.D. dissertation, Dep. Machine Intelligence, Univ. of Edinburgh, Edinburgh, Scotland, 1974.

Author Index

Numbers after the authors' names are either page numbers or reference numbers. A number without a decimal point is a page number, and a number with a decimal point is a reference number, e.g., "321" means page 321, "B.1" means Reference 1 in Section B of the General References, and "III.4" means Reference 4 in the References on Part III.

A

Aggarwal, J. K., A.1, IV.8, 271
Agin, G. J., VII.1
Allen, D. C., II.5
Alt, F. L., III.1
Ambler, A. P., 397
Andrews, H. C., A.2-3
Anuta, P. A., I.1

B

Bajcsy, R., III.2, VII.2
Ballard, D., IV.4
Barnea, D. I., 23
Barrow, H. G., 397, 435, VII.21
Bartz, M. R., 59
Belson, M., IV.6
Binford, T. O., VII.1, VII.18
Blum, H., 153, III.3
Breeding, K. J., VII.12
Brice, C. R., 79
Burstall, R. M., 397

C

Calabi, L., III.4
Caspari, K., A.2
Chang, Y. H., VII.3
Chazin, R. L., V.14
Cheng, G. C., C.5
Chien, R. T., VII.3
Chow, C. K., II.1
Clowes, M. B., IV.10, VI.1
Coates, C. L., Jr., VI.8

D

Davis, L. S., II.2, II.12, IV.1
Dinstein, I., 141
Doyle, W., I.2
Duda, R. O., A.1, B.1, 204, 271
Duff, M. J. B., I.3
Dyer, C. R., III.18

E

Eden, M., C.10, VII.4
Elschlager, R. A., 31
Enloe, L. H., A.3
Essinger, P., 209
Evans, T. G., VII.5

F

Falk, G., VI.2
Feldman, J. A., 426, VII.6
Feng, G. S., V.10
Fennema, C. L., 79
Fischer, G. L., Jr., I.4
Fischler, M. A., 31
Fountain, T. J., I.3
Freeman, H., C.4, 247, V.1
Fu, K. S., C.6, 369, VII.7

G

George, C. F., Jr., A.4
Giuliano, V. E., III.5
Golay, M. J. E., 172
Granlund, G. H., 256
Grape, G. R., VI.3
Grasselli, A., C.7
Gray, S. B., I.5, II.3
Greanias, E. C., 209
Grenander, U., I.6
Griffith, A. K., IV.2, VI.4
Gupta, J. N., II.4
Guzmán, A., 324

H

Hall, E. L., A.4
Hansen, B. J., V.14
Haralick, R. M., 141
Harlow, C. A., VII.8
Harmon, L. D., C.8
Hart, P. E., B.1, 204
Hartnett, W. E., III.4
Hawkins, J. K., III.6
Hayes-Roth, F., VII.9
Hilditch, C. J., III.7
Holt, A. W., I.7
Horn, B., VII.10
Horowitz, S. L., 101, V.11
Horwitz, L. P., 12
Hu, M. K., 113, III.8
Huang, T. S., A.5-6, C.4
Hueckel, M. H., 191, IV.3
Huffman, D. A., 338
Hummel, R. A., VII.19

J

Jacobsen, J., IV.6
Jayaramamurthy, S. N., III.10
Johnston, E. G., V.13
Jones, P. E., III.5
Joseph, H. M., IV.5

K

Kain, R. Y., I.8
Kak, A. C., B.3
Kanade, T., VII.20

Kanal, L. N., C.9
Kaneff, S., VII.11
Kaneko, T., II.1
Kaufman, G. J., Jr., VII.12
Kelly, M. D., 228
Kimball, G. E., III.5
Kimme, C., IV.4
Kirsch, R. A., VII.13
Kolers, P. A., C.10
Kovalevsky, V. A., I.9
Kovasznay, L. S. G., IV.5

L

Ledley, R. S., C.3, C.5, IV.6
Lee, Y. H., IV.12
Lendaris, G. G., 122
Levi, G., III.9
Lieberman, L., VII.2
Lillestrand, R. L., V.2
Lipkin, B. S., C.11

M

Mackworth, A. K., VI.5
Martelli, A., 217, IV.7
McCormick, B. H., I.10, III.10
McKee, J. W., IV.8
McLaughlin, J. A., I.11
Meagher, P. F., 209
Mendelsohn, M. L., II.7
Merrill, R. D., V.3
Meyer, R. F., III.5
Miller, W. F., VII.14
Montanari, U., III.9, IV.9, V.4
Muerle, J. L., II.5
Munson, J. H., V.5

N

Nagao, M., VII.20
Nagel, R. N., I.12
Nake, F., VII.15
Narasimhan, R., VII.16-17
Nevatia, R., V.6, VII.18
Norgren, P. E., III.13
Norman, R. J., 209

O

O'Gorman, F., IV.10
Ohlander, R. B., II.6

P

Pavlidis, T., 101, V.7-11
Pfaltz, J. L., II.9, III.11, III.15
Phillips, D. R., I.14
Pingle, K. K., IV.11
Pollock, D. K., C.5, I.4
Pratt, W. K., A.2
Preston, K., Jr., I.13, III.12-13
Prewitt, J. M. S., II.7

R

Radack, B., I.4
Raviv, J., I.11
Roberts, L. G., 285
Rosenfeld, A., B.2-9, C.4-5, C.11-12, I.12, 69, II.2, II.8-9, II.12, III.12, III.14-15, III.17-18, 183, IV.12, V.12-13, VII.15, VII.19
Roskies, R. Z., V.16

S

Sakai, T., VII.20
Shanmugam, K., 141
Shapiro, S. D., IV.13
Shaw, A. C., VII.14
Shaw, G. K., I.3
Shelman, C. B., II.10
Shelton, G. L., Jr., 12
Shirai, Y., VI.6-7
Silverman, H. F., 23
Sklansky, J., C.13, IV.4, V.14
Smith, E. A., I.14
Smith, F. W., III.16
Stanley, G. L., 122
Stefanelli, R., III.17
Stein, B. A., III.5
Stevens, M. E., I.4
Swain, P. H., 369

T

Tenenbaum, J. M., IV.11, 435, VII.21
Thurston, M., 183, IV.12
Tomita, F., II.11
Tretiak, O. J., A.5
Tsuji, S., II.11
Turner, K. J., VII.22

U

Uhr, L., C.14
Ullmann, J. R., I.15
Underwood, S. A., VI.8
Unger, S. H., I.16, V.15

V

van Dam, A., C.4
Vander Lugt, A., 5, I.17

W

Waltz, D., VI.9
Watanabe, S., C.15-16
Watson, D. M., I.3
Weszka, J. S., II.2, III.18
Widrow, B., I.18
Winston, P. H., C.17, VI.10
Wintz, P. A., II.4
Wright, M. H., III.16

Y

Yakimovsky, Y., 426, VII.6

Z

Zahn, C. T., 263, V. 16
Zucker, S. W., II.12, 445, VII.19

Subject Index

A

Adaptive thresholding, 59
Adjacency (tree), 69
Aerial photograph, 122
Annular-ring signature, 122
Aperture effects, 122
Autocorrelation, 12

B

Bicurve, 69
Binary pattern, 69, 153, 172, 263
Black-and-white pattern, 69, 153, 172, 263
Blocks world, 285, 324, 338, 445
Border (*see also* contour, curve, and edge), 69
Bottom-up merging, 79, 101, 426, 435

C

Central moments, 113
Chain coding, 247
Change detection, 271
Character recognition, 12, 59, 113, 209, 256
Characteristic function, 113
Closed (polygonal) curve, 263
Cloud motion, 271
Clustering by relaxation, 445
Coarseness, 141
Collinear points, 204
Concatenation, 369
Connected (components), 69
Connectivity, 69
Constraint satisfaction, 435, 445
Contour (*see also* border, curve, and edge), 153, 209, 217, 228, 256, 263
  description, 264
  detection, 209, 217, 228
  fitting, 79, 285
  following, 209, 217, 228
Contrast, 59
Crease, 338
Cross correlation, 5, 12, 23
Curvature point, 263
Curve (*see also* border, contour, and edge)
  description, 247
  detection, 183, 204
  enhancement, 445
  smoothing, 263
Cyclic set, 338

D

Decision theory, 426
Depth index, 338
Depth perception, 285
Descriptors of shape, 153
Diffraction pattern sampling system (DPSS), 122
Digital picture, 69
Dynamic programming, 31, 217

E

Edge detection, 183, 191, 217, 228, 285
  using least squares fitting, 191
  using multiple masks, 183
  using planning, 228
Edge following (*see also* contour following), 228
Edge operator, 183, 191
Efficient cross correlation, 23
8-adjacent, 69
8-connectivity, 69
8-neighbor, 69
8-path, 69
ERTS (LANDSAT), 141

F

Fast cross correlation, 23
Fast matching, 23
Filtering (for region interpretation), 435
4-adjacent, 69
4-connectivity, 69
4-neighbor, 69
4-path, 69
Fourier descriptor, 256
Fourier transform 5, 12, 122, 256
Fraunhofer diffraction pattern, 12, 122
Freeman coding, 247

G

Gestalt psychology, 183
Gradient, 285
Grammar, 369
Grass-fire propagation, 153
Grouping, 101
Growing and shrinking, 172

H

Heuristic optimization, 31, 426
Heuristic search, 217
Hexagonal grid, 69, 172
Hidden line, 285, 338
Hidden line elimination, 285
Hilbert space, 191
Homogeneous coordinates, 285
Homomorphism, 397
Hough transformation, 204
Human head, 228

I

Illusion, 338
Image segmentation, 79, 101, 426, 435
Impossible object, 338
Initial partition, 79
Interactively guided segmentation, 435
Interpretation guided segmentation, 435
Invariant (feature), 113, 122, 141, 256, 263
Isomorphism, 397

L

Labeling constraint, 338, 435, 445
Learning, 426
Least squares edge fitting, 191
Line
  detection, 204
  drawing, 285
  enhancement, 445
  fitting, 79, 285
  interpretation, 338
  labeling, 338
  width, 59
Linear embedding algorithm, 31
Linguistic pattern recognition, 369

M

Machine perception, 285
Machine printing, 59
Matching, 5, 12, 23, 31, 285, 397
Measure of texture, 141
Medial axis (function), 153
Merging, 79, 101, 426, 435
Model matching, 285
Moment generating function, 113
Moment invariant, 113
Moment, 113
Monomorphism, 397
Motion, 271
Moving image, 271

N

Nearest neighbor logic, 172
Nonsense sentence, 338

O

Object detection, 31
Object recognition, 79
Occlusion, 271, 324
Optical
  filtering, 5
  illusion, 153
  methods, 5, 12
  reader, 59
Optimization, 31, 217, 426
Overlapping figures, 271

P

Parallel processing, 172, 445
Parsing, 369
Partition, 79
Pattern
  grammar, 369
  primitive, 369
  transformation, 172
Phagocyte heuristic, 79
Photomicrograph, 141
Picture
  description, 31
  grammar, 369
  graph, 338
  primitive, 191, 369
  representation, 31
  segmentation, 79, 101, 426, 435
Plane curve, 247
Planning, 228
Point-line transformation, 204
Polygonal figure, 79, 263, 271
Polyhedra, 285, 324, 338
Polyhedral scene, 285, 324, 338
Primitive, 369
Principal axis, 113
Probabilistic syntactic analysis, 369

R

Registration, 12, 23
  efficiency, 23
  invariance, 12
Region
  analysis, 79, 101, 426, 435
  boundary, 79
  linking, 324
  merging, 79, 101, 426, 435
  of uniform gray, 79
  representation, 153
Relational structure, 397
Relaxation labeling, 445
Rubber template, 31

S

Sampling (geometry), 122
Satellite imagery, 141
Segmentation, 79, 101, 228, 263, 426, 435
Semantic interpretation, 426, 435
Sequential
  logic, 209
  registration, 23
  similarity detection, 23
Shape, 153
  analysis, 247
  descriptor, 153
  feature, 113, 153, 209, 256
Shift register, 12
Shortest path, 217
Sidewise merging, 101
Signal detection, 5
Signature analysis research facility (SARF), 122
Similarity detection algorithm, 23
Simply connected, 69
Skeletonizing, 153, 172
Smooth object, 338
Spatial
  cross correlation, 23
  dependence, 141
  dependence matrix, 141
  filtering, 5
Split-and-merge scheme, 101
Splitting, 101
Spot detection, 183
Statistical textural feature, 141
Streak detection, 183
Structural description, 397
Symmetric
  axis (distance), 153
  shape, 113, 256
Syntactic pattern recognition, 369
Syntax, 369

T

Template matching, 12
Textural feature, 122, 141
Textural tone, 141
Texture, 122, 141
  discrimination, 183
  edge, 183

Three-dimensional object, 285, 324, 338
Threshold operator, 59
Thresholding, 59
Top-down splitting, 101
Topological change, 271
Topology, 69
Translation invariance, 12
Tree, 69
Trihedral solid, 338
Two-dimensional
  filtering, 5
  Fourier analyzer, 5
  Fourier transform, 122
  impulse, 122
  moment, 113
Typewriter printing, 59

U

Utility function, 426

V

Vertex classification, 324, 338
Video thresholding, 59
Visual perception, 153, 183

W

Waltz algorithm, 445
Wave number (spectrum), 12
Weakness heuristic, 79
Wedge signature, 122

Editors' Biographies

J. K. Aggarwal (S'62–M'65–SM'74–F'76) was born in Amritsar, India, on November 19, 1936. He was educated in India, England, and the United States. He received the B.S. degree in mathematics and physics from the University of Bombay, Bombay, India, in 1956, the B. Eng. degree from the University of Liverpool, Liverpool, England, in 1960, and the M.S. and Ph.D. degrees from the University of Illinois, Urbana, in 1961 and 1964, respectively.

He joined the University of Texas, Austin, in 1964 as an Assistant Professor and has since held positions as Associate Professor (1968) and Professor (1972). Currently, Dr. Aggarwal is a Professor of Electrical Engineering and of Computer Science at the University of Texas at Austin. Further, he was a Visiting Assistant Professor to Brown University, Providence, RI, in 1968, and a Visiting Associate Professor at the University of California, Berkeley, during 1969–1970.

He has published numerous technical papers and a textbook, *Notes on Nonlinear Systems*, published by Van Nostrand Reinhold in 1972, in the series *Notes on System Sciences*. His current research interests are digital filters, computational methods, image processing, and computer pattern recognition.

Dr. Aggarwal is an active member and a Fellow of IEEE. He was an AdCom member of the Circuits and Systems Society (1972–1975), Chairman of the Technical Committee on Signal Processing (1974–1975), the Associate Editor of the Society TRANSACTIONS (1973–1975), Co-Editor of its Special Issue on Digital Filtering and Image Processing (March 1975), and the Editor of the *Circuits and Systems Society Newsletter* (1971–1972). He is also a member of Eta Kappa Nu.

Richard O. Duda (S'57–M'58) was born in Evanston, IL, on April 27, 1936. He received the B.S. and M.S. degrees in engineering from the University of California, Los Angeles, in 1968 and 1969, respectively, and the Ph.D. degree in electrical engineering from the Massachusetts Institute of Technology, Cambridge, in 1962.

He joined the staff of the Stanford Research Institute, Menlo Park, CA, in 1962, where he participated in research on character recognition, statistical pattern classification, adaptive and learning systems, visual scene analysis, and artificial intelligence, together with their applications to meteorology and industrial automation. He has been the author or coauthor of a number of papers in these fields, and the coauthor with Dr. P. E. Hart of the book *Pattern Classification and Scene Analysis* (New York: Wiley-Interscience, 1973). During the 1973–1974 academic year, he was a Visiting Associate Professor in the Department of Electrical Engineering and the Department of Computer Sciences at the University of Texas, Austin.

Dr. Duda is a member of Tau Beta Pi, Sigma Xi, Phi Beta Kappa, the Association for Computing Machinery, and the American Association for the Advancement of Science. He is a past Associate Editor of the IEEE TRANSACTIONS ON COMPUTERS.

Azriel Rosenfeld (M'60–F'72) was born in New York, NY, on February 19, 1931. He received the B.A. degree in physics from Yeshiva University, New York, NY and the Ph.D. degree in mathematics from Columbia University, New York, NY, in 1950 and 1957, respectively.

From 1954 to 1956 he was a Physicist with the Fairchild Controls Corporation, New York, NY; from 1956 to 1959 he was an Engineer with the Ford Instrument Company, Long Island City, NY; and from 1959 to 1964 he was Manager of Research with the Budd Company Electronics Division, Long Island City, NY, and Information Sciences Center, McLean, VA. Since 1964 he has been a Professor of Computer Science at the University of Maryland, College Park. He has published seven books and nearly 150 papers, most of them dealing with the computer analysis of pictorial information. He edits the journal *Computer Graphics and Image Processing*, and was Chairman of the Third International Joint Conference on Pattern Recognition.